Geotechnical Characteristics of Soils and Rocks of India

Geotechnical Characteristics of Soils and Rocks of India

Edited by

Sanjay Kumar Shukla

School of Engineering, Edith Cowan University, Joondalup, Perth, Australia

CRC Press
Taylor & Francis Group
Boca Raton   London   New York   Leiden

CRC Press is an imprint of the
Taylor & Francis Group, an **informa** business

A BALKEMA BOOK

Cover image: Sanjay Kumar Shukla

First published 2022
by CRC Press/Balkema
Schipholweg 107C, 2316 XC Leiden, The Netherlands
e-mail: enquiries@taylorandfrancis.com
www.routledge.com – www.taylorandfrancis.com

CRC Press/Balkema is an imprint of the Taylor & Francis Group, an informal business

Library of Congress Cataloging-in-Publication Data
A catalog record has been requested for this book

ISBN: 978-1-032-01098-4 (hbk)
ISBN: 978-1-032-01104-2 (pbk)
ISBN: 978-1-003-17715-9 (ebk)

DOI: 10.1201/9781003177159

Typeset in Times New Roman
by codeMantra

Contents

Preface

Every country needs to construct and maintain infrastructures, such as buildings, bridges, roads, railway tracks, runways, ports, tunnels, towers, dams and retaining structures. Transport and building projects are two everyday examples as we may experience their performance, even without being engineers and technical professionals. We also need similar infrastructures in industrial areas and mining, agriculture, aquaculture and energy sectors. The stability and performance of all these facilities greatly depend on the ground strength and compressibility characteristics of soils and rocks. For their analysis and design, the project professionals need the basic and technical details of the project ground sites.

This book presents the basic description of soils, rocks and other similar materials of India with a focus on the subsurface investigation in its 28 states and 8 union territories. An attempt has been made to cover more geological and geotechnical engineering aspects. It includes the ground subsurface details in terms of boring logs, project site photographs, engineering properties of soils/rocks and specific field challenges. Some details are also presented on how soils and rocks are used as construction materials. Challenges caused by natural hazards, case studies and effects of geoenvironmental impacts are also presented as observed in specific areas of India.

This book has adopted a simple and practical approach for easy learning of geotechnical characteristics and related aspects, even by self-reading, in a single volume. The engineering professionals (e.g., planners, managers, practising engineers, consultants and contractors) and university students, including researchers from several areas (civil, mining, agriculture and aquaculture), may find relevant details in this book as often required in analysis, design, construction and maintenance of the ground infrastructure/field projects, especially for project site selection and getting the design parameters during the initial planning and preliminary design stages of the projects. Those dealing with geology and geological engineering, geography and soil science may also find useful information in this book. This book may be recommended as a reference book in courses dealing with soils and rocks. High school students and common people may also get some useful information of their level.

I would like to extend special thanks and recognition to all the authors for their valuable contributions to this book. The authors have worked hard during the COVID-19 pandemic to present their contributions as per the need of the readers/users. I am truly grateful to them.

I would like to thank Alistair Bright, Acquisitions Editor and the staff of CRC Press/Balkema, Taylor and Francis, and Assunta Petrone, Codemantra for their support and cooperation at all the stages of preparation and production of this book.

I wish to extend sincere appreciation to my wife, Sharmila, for her encouragement and support throughout the preparation of the manuscript. Thanks to my daughter, Sakshi, and my son, Sarthak, for their patience during my work on this book at home.

Finally, I welcome suggestions from the readers and users of this book for improving its contents in future editions.

Sanjay Kumar Shukla
Perth, 2021

About the editor

Sanjay Kumar Shukla is Founding Editor-in-Chief of the *International Journal of Geosynthetics and Ground Engineering*, Springer Nature, Switzerland. He is Founding Research Group Leader (Geotechnical and Geoenvironmental Engineering) at the School of Engineering, Edith Cowan University, Joondalup, Perth, Australia. He holds the Distinguished Professorship in Civil Engineering at Delhi Technological University, Delhi, VIT University, Vellore, Chitkara University, Solan, Himachal Pradesh, VR Siddhartha Engineering College, Vijayawada, Amity University, Noida, and Amrita University, Coimbatore, India. He has over 25 years of experience in teaching, research and consultancy in the field of Civil (Geotechnical) Engineering. He collaborates with several world-class universities, research institutions, industries and individuals on academic and field projects. As a consulting geotechnical engineer, he has successfully provided solutions to the challenging field problems faced by many engineering organisations. He has authored more than 280 research papers and technical articles, including over 175 refereed journal publications. He is also author/ editor of 23 books, including seven textbooks and 22 book chapters. In 2020/2021, his ICE textbooks, namely *Core Principles of Soil Mechanics* and *Core Concepts of Geotechnical Engineering*, have been ranked number 1 by Amazon. His research and academic works have been cited well. Shukla's Generalized Expressions for Seismic Active Thrust (2015) and Seismic Passive Resistance (2013) are routinely used by practising engineers worldwide for designing the retaining structures. Shukla's wrap-around reinforcement technique, developed during 2007–2008, is a well-established ground improvement technique. He has been honoured with several awards, including 2021 ECU Aspire Award from the Business Events Perth, Australia, and the most prestigious IGS Award 2018 from the International Geosynthetics Society (IGS), USA, in recognition of his outstanding contribution to the development and use of geosynthetics. He serves on the editorial boards of several international journals. He is a fellow of American Society of Civil Engineers and Engineers Australia, a life fellow of the Institution of Engineers (India) and Indian Geotechnical Society, and a member of several other professional bodies.

Contributors

B. Arun WRD, TNPWD, Environmental Cell Division, Chennai, India

Ketan Bajaj Risk Management Solution, Noida, India

Kaushik Bandyopadhyay Department of Construction Engineering, Jadavpur University, Kolkata, India

Sukumar Baruah NEEPCO Limited, Itanagar, Arunachal Pradesh, India

Arun Bhave Marhsal Geo Test Laboratory, Raipur, India

Sanjoy Bhowmik Engineers India Limited, New Delhi, India

Lalit Borana Discipline of Civil Engineering, IIT Indore, Indore, India

Vinay Bhushan Chauhan Department of Civil Engineering, MMMUT, Gorakhpur, India

Harsh Chittora Landmark Material Testing and Research Laboratory Pvt. Ltd., Jaipur, India

Hemant S. Chore Department of Civil Engineering, Dr B.R. Ambedkar National Institute of Technology, Jalandhar, India

Anil K. Choudhary Department of Civil Engineering, NIT Jamshedpur, Jamshedpur, India

Awdhesh K. Choudhary Department of Civil Engineering, NIT Jamshedpur, Jamshedpur, India

Utpal Kumar Das Department of Civil Engineering, Tezpur University, Tezpur, India

Ramani Mohan Das Public Works Department, Chandmari, Guwahati, India

M. Dhanasekaran Public Works Department, Kottar, India

Anil Dixit Landmark Material Testing and Research Laboratory Pvt. Ltd., Jaipur, India

Avinash Dubey Department of Civil Engineering, Indian Institute of Technology Jammu, Ban, India

Ashish D. Gharpure Genstru Consultants Pvt. Ltd., Wakad, India

Chandan Ghosh National Institute of Disaster Management, Ministry of Home Affairs, Government of India, New Delhi, India

Sima Ghosh Civil Engineering Department, National Institute of Technology, Agartala, India

K.S. Gill Civil Engineering Department, Guru Nanak Dev Engineering College, Ludhiana, India

Radha J. Gonawala Department of Civil Engineering, Sardar Vallabhbhai National Institute of Technology, Surat, India

Diganta Goswami Department of Civil Engineering, Assam Engineering College, Guwahati, India

Manish Kumar Goyal Department of Civil Engineering, IIT Indore, Indore, India

Ashish Gupta Department of Civil Engineering, BIET Jhansi, Jhansi, India

Devashish Gupta Department of Civil Engineering, Delhi Technological University, New Delhi, India

S.S. Gupta Department of Civil Engineering, College of Technology, G.B. Pant University of Agriculture and Technology, Pantnagar, India

Sunny Deol Guzzarlapudi Department of Civil Engineering, NIT Raipur, Raipur, India

Abhipriya Halder Department of Construction Engineering Jadavpur University, Kolkata, India

Chennarapu Hariprasad Department of Civil Engineering, Ecole Centrale School of Engineering, Mahindra University, Hyderabad, India

Ashwani Jain Department of Civil Engineering, National Institute of Technology (NIT), Kurukshetra, India

Jayamohan J. LBS Institute of Technology for Women, Thiruvananthapuram, India

J.N. Jha Formerly Muzaffarpur Institute of Technology, Muzaffarpur, India,

Anil Joseph Geostructurals (P) Ltd., Cochin, India

Nitin Joshi Department of Civil Engineering, Indian Institute of Technology Jammu, Ban, India

Ajanta Kalita Department of Civil Engineering, North Eastern Regional Institute of Science and Technology, Nirjuli, India

Abhishek Kanoungo Chitkara School of Engineering and Technology, Chitkara University, Chandigarh, India

Manvi Kanwar Chitkara School of Engineering and Technology, Chitkara University, Chandigarh, India

Varinder S. Kanwar Chitkara School of Engineering and Technology, Chitkara University, Chandigarh, India

T. Kaviarasu Department of Civil and Structural Engineering, Annamalai University, Chidambaram, India

J.R. Kayal Civil Engineering Department, National Institute of Technology Agartala, Agartala, India, (Formerly Geological Survey of India), Kolkata, India

Sreevalsa Kolathayar Department of Civil Engineering, National Institute of Technology Karnataka, Mangalore, India

Bappaditya Koley Department of Construction Engineering, Jadavpur University, Kolkata, India

Rajashri Shashikant Kulkarni Department of Civil Engineering, Dr B. R. Ambedkar Institute of Technology, Port Blair, India

Ashish Kumar Department of Civil Engineering, Muzaffarpur Institute of Technology, Muzaffarpur, India

Gaurav Kumar SNF India, 1408 Laurel Building, Nahar Amrit Shakti, Chandivali, India

Naveen Kumar Genstru Consultants Pvt. Ltd., Wakad, India

Prabhat Kumar Department of Civil Engineering, Indian Institute of Technology Jammu, Ban, India

Rakesh Kumar Department of Civil Engineering, Sardar Vallabhbhai National Institute of Technology, Surat, India

Vijay Kumar Department of Civil Engineering, Muzaffarpur Institute of Technology, Muzaffarpur, India

Sunita Kumari Department of Civil Engineering, National Institute of Technology, Patna, India

Mahasakti Mahamaya O.P. Jindal University, Raigarh, India

Nirbhay Mathur Landmark Material Testing and Research Laboratory Pvt. Ltd., Jaipur, India

Bashir Ahmed Mir Department of Civil Engineering, National Institute of Technology Srinagar, Srinagar, India

Manas Chandan Mishra School of Infrastructure, IIT Bhubaneswar, Khordha, India

Mohit K. Mistry Department of Civil Engineering, Sardar Vallabhbhai National Institute of Technology, Surat, India

Sravanam Sasanka Mouli Department of Civil Engineering, VNR VJIET, Hyderabad, India

M. Muthukumar School of Civil Engineering, Vellore Institute of Technology (VIT), Vellore, India

Saptarshi Nandi Department of Construction Engineering, Jadavpur University, Kolkata, India

Prashant Navalakha Genstru Consultants Pvt. Ltd., Wakad, India

Sadanand Ojha Swati Structure Solutions Pvt. Ltd., New Delhi, India

Yachang Omo Department of Civil Engineering, Central Institute of Technology, Kokrajhar (CIT), India

Nizar P.K. Calicut Central Subdivision, CPWD, Calicut, India

V.K. Panwar Engineers India Limited, Gurugram, India

C.R. Parthasarathy Sarathy Geotech & Engineering Services Pvt. Ltd., Bangalore, India

Manali S. Patel Department of Civil Engineering, Sardar Vallabhbhai National Institute of Technology, Surat, India

Vikrant Patel Department of Civil Engineering, BIET Jhansi, Jhansi, India

Rajesh Pathak Thapar Institute of Engineering and Technology, Patiala, India

Aman Pawar Department of Civil Engineering, Delhi Technological University, New Delhi, India

Vikas Poonia Department of Civil Engineering, IIT Indore, Indore, India

K. Premalatha Department of Civil Engineering, Anna University, Chennai, India

R. Premkumar Department of Civil Engineering, Pondicherry University, Puducherry, India

Akash Priyadarshee Department of Civil Engineering, Muzaffarpur Institute of Technology, Muzaffarpur, India

Nitish Puri Environment Division, AECOM India Private Limited, Chennai, India

N.J.L. Ramesh Department of Civil and Structural Engineering, Annamalai University, Chidambaram, India

Bendadi Hanumantha Rao School of Infrastructure, IIT Bhubaneswar, Khordha, India

G.V. Rama Subba Rao Department of Civil Engineering, Velagapudi Ramakrishna Siddhartha Engineering College, Vijayawada, India

S. Rupali Department of Civil Engineering, Dr B.R. Ambedkar National Institute of Technology, Jalandhar, India

Sunil Saha Department of Geography, University of Gour Banga, Malda, India

Smrutirekha Sahoo Department of Civil Engineering, National Institute of Technology Meghalaya, Shillong, India

Subhajit Saraswati Department of Construction Engineering, Jadavpur University, Kolkata, India

J. Saravanan Department of Civil and Structural Engineering, Annamalai University, Chidambaram, India

Raju Sarkar Department of Civil Engineering, Delhi Technological University, New Delhi, India

V.A. Sawant Department of Civil Engineering, Indian Institute of Technology Roorkee, Roorkee, India

Purnanand P. Savoikar Department of Civil Engineering, Goa College of Engineering, Ponda, India

Swagatika Senapati Institute of Technical Education and Research (ITER), SOA University, Bhubaneswar, India

Sangeeta Shougrakpam Department of Civil Engineering, Delhi Technological University, New Delhi, India and, Department of Civil Engineering, Manipur Institute of Technology, Imphal, India

Harvinder Singh Civil Engineering Department, Guru Nanak Dev Engineering College, Ludhiana, India

Moirangthem Johnson Singh Discipline of Civil Engineering, IIT Indore, Indore, India

S. K. Singh Department of Civil Engineering, Punjab Engineering College (Deemed to be University), Chandigarh, India

Chandresh H. Solanki Department of Civil Engineering, Sardar Vallabhbhai National Institute of Technology, Surat, India

Chava Srinivas Department of Civil Engineering, Velagapudi Ramakrishna Siddhartha Engineering College, Vijayawada, India

Ashutosh Trivedi Department of Civil Engineering, Delhi Technological University, New Delhi, India

Balunaini Umashankar Department of Civil Engineering, Indian Institute of Technology Hyderabad, Hyderabad, India

Surinder Kumar Vashisht Himachal Pradesh Housing and Urban Development Authority (HIMUDA), Dharamshala, India

Neetu Yadav Department of Civil Engineering, Shree Starambhai Naranjibhai Patel Institute of Technology and Research Centre, Surat, India

Laxmikant Yadu Department of Civil Engineering, NIT Raipur, Raipur, India

Arunkumar Yendrembam Department of Civil Engineering, Manipur Institute of Technology, Imphal, India

Falak Zahoor Department of Civil Engineering, National Institute of Technology Srinagar, Srinagar, India

Chapter 1

Introduction

Sanjay Kumar Shukla
Edith Cowan University

CONTENTS

1.1   SOIL AND ROCK

The materials that constitute the Earth's crust are arbitrarily divided by civil engineers into soil and rock. Soil, a natural aggregate of mineral grains, comprises all the materials in the surface layer of the Earth's crust that are loose enough to be moved by a spade or shovel. In general, soil is a particulate and multiphase system consisting of three phases, namely solid, liquid and gas. The space in a soil mass occupied by liquid and/or gas is known as the void. Dry soil has only air in the void, while the void volume of a fully saturated soil is occupied by water only. There are several phase relationships and inter-relationships as detailed in the textbook of Shukla (2014). Based on the method of formation, soils are classified as residual soils, sedimentary soils, organic soils and fills (or man-made soils). Figure 1.1 shows a project site view of the soil mass. Permeability, compressibility and shear strength are the most important properties of soil, which are used in a suitable form by civil engineers for designing geotechnical structures and other ground-related infrastructures.

Rock is a natural aggregate of mineral grains connected by strong and permanent internal cohesive forces and occurs in large masses and fragments. Rocks generally require blasting for their excavation. Based on their formation by geological

DOI: 10.1201/9781003177159-1

Figure 1.1 A project site view of the soil mass.

Figure 1.2 A project site view of the rock mass.

processes, rocks are classified as igneous rocks, sedimentary rocks and metamorphic rocks. Granite, basalt, dolerite, gabbro and rhyolite are examples of igneous rocks. Sandstone, limestone, shale and conglomerate are examples of sedimentary rocks. Quartzite, marble, slate, mica schist, graphite and gneiss are examples of metamorphic rocks. A rock mass generally behaves as an inhomogeneous and anisotropic material because of the presence of discontinuities in different forms such as joints, fractures and bedding planes. Figure 1.2 shows a project site view of the rock mass. Permeability, strength/stiffness and durability are important properties of rocks.

The index properties of soils and rocks are total unit weight, void ratio, specific gravity of solids, water content, degree of saturation and water absorption. For soils, consistency limits, namely, liquid limit, plastic limit and shrinkage limit, and particle size classification are the index properties, which help in classifying the soil and predicting its behaviour during the initial planning/designing of a project site.

Soil and rock support the structural foundations, and they are used as construction materials in various civil/geotechnical/infrastructure engineering projects.

The engineering properties of soil and rock at and/or beneath the ground surface at a project site can vary dramatically. A thorough site investigation (also known as site exploration) is therefore a prerequisite for the design of civil engineering structures and mining excavations. The purpose of site investigation is to conduct a scientific examination of the site for collecting as much information as possible at a minimal cost about the existing topographical and geological features of the site, for example, exposed overburden, course of streams/rivers nearby, rock outcrop, hillock or valley, and vegetation, and mainly the subsurface conditions underlying the site (Shukla, 2015).

The subject dealing with soil, rock and similar materials (e.g., coal ashes, mine tailings) is known as geotechnical engineering. This subject involves analysis, design, construction, maintenance and renovation of geotechnical structures, which can be categorised into the following seven basic types: foundation, retaining wall, slope, embankment, earth dam, tunnel and pavement (Shukla, 2014, 2015). Geotechnical engineering presents cost-effective, sustainable and environmentally friendly solutions to soil and rock problems. Sound background knowledge of engineering geology, soil mechanics and rock mechanics is essential for becoming a geotechnical engineer.

1.2 INDIA AND ITS STATES AND UNION TERRITORIES

India, officially the Republic of India and also known as Bharat, Hindustan and Aryavarta, is a constitutional country in South Asia with New Delhi as its capital. It has 28 states and 8 union territories (UTs) as listed below (Figure 1.3):

States of India

1. Andhra Pradesh
2. Arunachal Pradesh
3. Assam
4. Bihar
5. Chhattisgarh
6. Goa
7. Gujarat
8. Haryana
9. Himachal Pradesh
10. Jharkhand
11. Karnataka

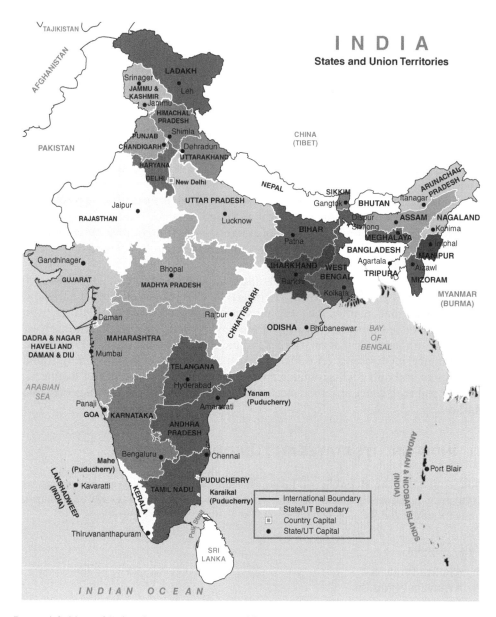

Figure 1.3 Map of India showing its capital, 28 states and 8 UTs with their capitals, state boundaries and international boundaries (Maps of India, 2021).

12. Kerala
13. Madhya Pradesh
14. Maharashtra
15. Manipur
16. Meghalaya
17. Mizoram

18. Nagaland
19. Odisha
20. Punjab
21. Rajasthan
22. Sikkim
23. Tamil Nadu
24. Telangana
25. Tripura
26. Uttarakhand
27. Uttar Pradesh
28. West Bengal

Union Territories of India

1. Andaman and Nicobar Islands
2. Chandigarh
3. Daman and Diu and Dadar and Nagar Haveli
4. Delhi
5. Jammu and Kashmir
6. Ladakh
7. Lakshadweep
8. Puducherry

With reference to land area, India ranks the seventh-largest country in the world and is the second-most populous country (Britannica, 2021). Though the country's population remains largely rural, Mumbai, Kolkata and Delhi are the three most populous cities of India. Bengaluru, Chennai and Hyderabad are amongst the world's fastest-growing high-technology centres. India has roughly one-third coastline, bounded by the Indian Ocean on the south, the Arabian Sea on the southwest and the Bay of Bengal on the southeast. India is bounded on the northwest by Pakistan, on the north by Nepal, China and Bhutan, and on the east by Myanmar and Bangladesh. In the Indian Ocean, Sri Lanka is in the vicinity of India, about 65 km off the southeast coast of India, and also the Maldives. Andaman and Nicobar Islands of India share a maritime border with Thailand and Indonesia (Britannica, 2021; Wikipedia, 2021). The Himalayas, which is the loftiest mountain system in the world and is geologically young, form the northern limit of India.

There is a wide range of soil and rock types in India. Igneous, sedimentary and metamorphic rocks are found in different states/UTs of India. As the products of natural environmental weathering processes, soils can be broadly divided into two groups: residual (or in situ) soils and transported soils. The type of soil is determined by numerous factors, including parent rock/material, topography, climate, organisms and time (Sivakugan et al., 2013; Shukla, 2014). Figure 1.4 shows the major soil map of India with soils classified as alluvial, desert, black, mixed red and black, red, grey and black, laterite and mountain soils. Chapters 2–37 present the details of soils and rocks and related aspects focussing on their geological/geotechnical characteristics as observed in the respective Indian state/UT.

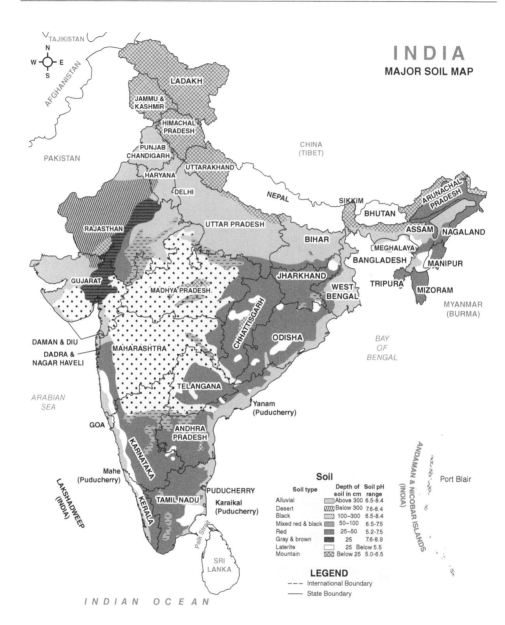

Figure 1.4 Major soil map of India (Maps of India, 2021).

1.3 STRUCTURE AND USE OF THIS BOOK

This book presents the geotechnical characteristics of soils and rocks of all the 28 states and 8 UTs of India. There are 37 chapters, including this chapter, in this book, and Chapters 2–37 have been presented according to the alphabetical order of the states and UTs. Each chapter, contributed by a team of authors, follows a common section format as applicable. The sections are described below.

1.3.1   Contents

This section provides the contents of the chapter.

1.3.2   Introduction

This section includes very basic information about the geology, soils and rocks of an Indian state/UT. A general map of the state/UT is also presented.

1.3.3   Major types of soils and rocks

This section describes major types of soils and rocks as observed during the geological and geotechnical site investigations.

1.3.4   Properties of soils and rocks

This section provides typical engineering geological information and geotechnical properties of soils/rocks as observed in open pits/boreholes. The properties have been included in tabular, graphical or other suitable forms with relevant discussion.

1.3.5   Use of soils and rocks as construction materials

In most of the chapters, an attempt is made to explain how soils and rocks are used in the particular state for the construction of geotechnical and other structures.

1.3.6   Foundations and other geotechnical structures

Soils and rocks are often used as foundations of structures (buildings, bridges, towers, retaining structures, embankments, slopes, dams, canals and so on), including foundations for electrical and mechanical infrastructures/poles/machines. This section highlights some related aspects.

1.3.7   Other geomaterials

In addition to soils and rocks, there are some waste materials, such as fly ash, bottom ash and mine tailings, which are often utilised in civil/geotechnical engineering applications in the way we use soils and rocks. This section covers some aspects in this regard, as applicable.

1.3.8   Natural hazards

Most of the Indian states/UTs face geotechnical problems caused by natural hazards (e.g., earthquakes/volcanoes, landslides/erosion, floods, avalanches). This section covers some details of the hazards focussing on how they affect foundations and other geotechnical structures.

1.3.9 Case studies and field tests

As possible, some case studies of success and/or failure of geotechnical structures as observed/reported in the state have been described with specific technical details, including a brief description of the field tests.

1.3.10 Geoenvironmental impact on soils and rocks

This section highlights some aspects of the geoenvironmental impact caused by temperature change, climate change, global warming, and interaction with wastes, chemicals and so forth on soil, rock and geotechnical structures.

1.3.11 Concluding remarks

Each chapter ends with some key points about the soils, rocks and other related aspects of that particular state or UT.

1.3.12 References

The relevant references are included.

All the chapters with the above sections cover highly practical information and technical data for application in ground infrastructure projects, including foundations of structures (buildings, towers, tanks, machines and so on), highway, railway and airport pavements, embankments, retaining structures/walls, dams, reservoirs, canals, ponds, landfills and tunnels. Engineering professionals (e.g., practising engineers, consultants, contractors) and university students, including researchers from several areas (civil, mining, agriculture, aquaculture), may get some useful basic technical details as often required in the analysis, design, construction and maintenance of the ground infrastructure/field projects, especially during the initial project planning and preliminary design stage. Though this book covers the Indian ground characteristics, the information provided in this book is quite helpful to the professionals of other countries having similar ground conditions.

It is important to note that no two project sites/locations have identical subsoil profiles, water table locations and other geological/topographical features. In most cases, the variation in geotechnical properties of soil and rock is extremely large. Additionally, the climate at the project site differs significantly. As the geographical area of states and UTs is large, it is difficult to present the variation in the subsoil profile from a geotechnical point of view in a comprehensive form. Certainly, the information presented in this book provides basic information only. It is mandatory/obligatory to conduct detailed site investigations and testing for any specific construction/infrastructure project. The geological and geotechnical information presented in this book may help practising engineers to plan a proper soil investigation and testing programme. Inadequate soil investigation or change of project location of a construction site without carrying out the studies at its new place may result in considerable damage to constructed facilities.

It is also important to note that the information provided in this book does not imply the expression of any opinion whatsoever on the part of the chapter's contributors, editor or publisher concerning the legal status of the state, territory, city or area. Moreover, the data presented do not necessarily represent the exact status of the soil/rock, and only give an idea about the subsurface profile while the detailed investigations are essentially required for any ground projects in the civil, mining, agriculture, aquaculture and other related areas.

1.4 CONCLUDING REMARKS

1. Soil, a natural aggregate of mineral grains, comprises all the materials in the surface layer of the Earth's crust that are loose enough to be moved by a spade or shovel.
2. Rock is a natural aggregate of mineral grains connected by strong and permanent internal cohesive forces and occurs in large masses and fragments. Rocks generally require blasting for their excavation.
3. India has 28 states and 8 UTs, and this book covers several aspects of soil, rock and other related details as observed in all the states and UTs of India.
4. Chapters 2–37 have been presented according to the alphabetical order of the states and UTs. This book provides only basic information about soil, rock and related aspects. The evaluation of soil and rock for any ground-related project shall be based on site-specific investigation and testing.
5. The information in this book has been presented only for learning the several geotechnical aspects of the soils and rocks of India. The contributors, editor or publisher of this book has no legal responsibility for the correctness and accuracy of the details as included, although the contributors have taken all the care.

REFERENCES

Britannica (2021). https://www.britannica.com/place/India, accessed 4 April 2021.
Maps of India (2021). https://www.mapsofindia.com/maps/, accessed 4 April 2021.
Shukla, S.K. (2015). *Core Concepts of Geotechnical Engineering*. ICE Publishing, London.
Shukla, S.K. (2014). *Core Principles of Soil Mechanics*. ICE Publishing, London.
Sivakugan, N., Shukla, S.K. and Das, B.M. (2013). *Rock Mechanics – An Introduction*. CRC Press, Taylor & Francis, Boca Raton, FL.
Wikipedia (2021). https://en.wikipedia.org/wiki/India, accessed 4 April 2021.

Chapter 2

Andaman and Nicobar Islands

Rajashri Shashikant Kulkarni
Dr B. R. Ambedkar Institute of Technology

CONTENTS

2.1   INTRODUCTION

The Andaman and Nicobar Islands are a pristine world of silver sands, clear blue sea, coral reefs, swaying palms, tropical forests, volcanic mountains, and gently undulating landscapes. The undulating islands are covered with dense forests and an endless variety of indigenous and exotic flora and fauna. The islands are described as "green islands in marigold sun" on which horizons are fixed. The islands are also called "KALAPANI" or the land of dark water because of the inhuman and degrading treatment of deported convicts and prisoners by the British government.

About 150 million years ago, there were continuous hill ranges stretching from Burma to Indonesia. Due to geological activity and violent movement of the Earth's crust, these hills got submerged into the surrounding seas with the peaks and ridges standing out. These are the present Andaman and Nicobar Islands (Bandopadhyay and Cater, 2017a).

The island is situated between 6°45″N and 13°41″N latitudes and between 92°12″E and 93°57″E longitudes extending over an area of 8,249 km. The island comprises 572 islands, out of which only 38 islands are inhabited. There are two groups of islands, namely, the Andaman and the Nicobar groups separated by the Ten Degree Channel (Bandopadhyay and Cater, 2017b). Figure 2.1 shows the map of the island.

The Andaman and Nicobar Islands have varied topographical features. The Andaman group of islands generally features a mountainous terrain with long ranges of hills and narrow valleys. The maximum altitude of these islands is at Saddle Peak, which is about 730 m above the mean sea level. The peak is formed of sandstone, limestone and clay. There are no great elevations, and the slopes are also moderate.

DOI: 10.1201/9781003177159-2

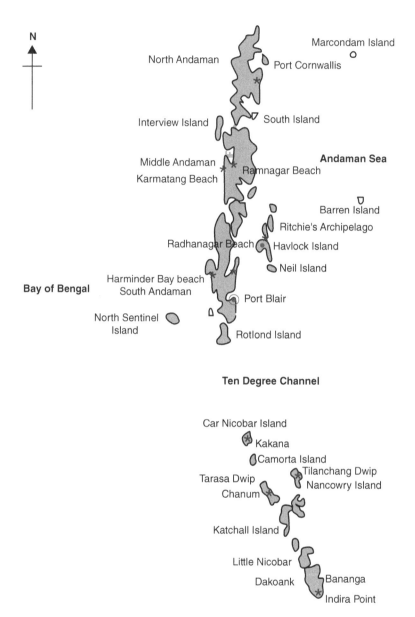

Figure 2.1 Map of Andaman and Nicobar Islands. (http://www.mapsofindia.com/, accessed 18 February 2021.)

The Nicobar Islands are surrounded by shallow seas and coral reefs. The topography of the Nicobar Islands features long, sandy beaches. Katchal and Car Nicobar have almost flat terrain. In Great Nicobar and Little Nicobar, the land is very irregular, having steep hills and valleys.

The main rock system consists of sedimentary rocks, made up of sandstones and conglomerates, as well as igneous rocks that resulted due to the volcanic process.

Besides these, quartzite, limestones, shale and grey sandy stones are found. The Andaman group has primarily siliceous sedimentation, while in the Nicobar group, calcareous soils are obtained. Basaltic and dolerite rocks are found in Little Andaman and Nancowrie group, but Kamorta and Terressa Islands, where the tree cover was removed in the past, have developed hard iron-rich shale with shallow soils and poor nutrients in the subsoil.

Soils are mostly of alluvial type in valleys and are highly porous with little water holding capacity. In the hills, hard clayey soils are obtained. Sandy loam and alluvial soils have accelerated water runoff due to porosity and poor water holding capacity. The soils are generally acidic with pH varying between 4.5 and 6.0. The soil in the hilly region is low in organic matter, strongly leached, rich in clay and more or less yellow in colour. The texture is clayey. The soil in the valley region is also clayey having more than 5% clay (Das, 1982). It is rich in silica and the soil of the coastal region is highly saline giving rise to halophytic vegetation. The soil of villages is of alluvial type, which becomes sandy near the seacoast. The soil varies greatly in texture and fertility. The soil of the hilly regions is less fertile than that of the valley where thick alluvial soils have been deposited.

The Andaman and Nicobar Islands have a tropical climate. The islands are exposed to both south-west and north-east monsoons and the average rainfall (Port Blair) is 292 cm with an average of 138 rainy days per year. The temperature variation is slight between 30° and 22° and the average relative humidity is 79% (ISRO, 2009).

The mineral resources in Andaman are very limited. Minerals were discovered near Port Blair. They are chromium, copper, iron and sulphur (Radicliffe Brown, 1992). But no important deposits of any mineral have been found to occur except for some traces of lignite deposits of pure limestone. Also, none are found economically exploitable. Coal is of very poor quality and no coal is economically exploitable. Off-shore drilling undertaken recently near Swarajya Dweep (Havelock Island), 19 km away from Port Blair, has showed rich deposits of gases.

This chapter presents some details of the types of soils and rocks and their properties, geological formation, disasters and so on as found in this union territory.

2.2 MAJOR TYPES OF SOILS AND ROCKS

The soils in the study area encounter three orders, viz., Entisols, Inceptisols and Alfisols, showing alluvial and colluvial origin. They vary in depth, texture and chemical composition and are acidic in nature. The soils have fine to medium texture on the surface and medium to heavy texture/gravelly nature in the subsoil. The soil formation process begins first with the breakdown of rock into regolith. The parent rocks mainly undergo mechanical and chemical weathering processes with different degrees of weathering, leading to the development of a soil profile. According to the morphological observations, the soils are of sedimentary formations with a carbonaceous grey clay material, while the volcanic formation has a silt-clay material with a coarse grainy texture. They possess very little internal friction between soil particles when wet. During rain runoff, water causes saturation of the soil material which in turn causes caving in the lands. A soil profile study reveals that unweathered and partly weathered fragments of parent materials are found throughout the profile. Humus is generally absent in soil at hilly uplands and owing to heavy rain, the soil gets abraded and broken down into finer particles.

The borehole data were collected from the site of an earthen dam over Kamsarat Nallah near Wimberlygunj in South Andaman. The reservoir area and the area around the dam site were devoid of the rock exposures except for the Nallah bed and a few exposures on the slopes. The general hill slopes range from 30° to 45° with a slightly higher angle on the right bank, whereas the underlying rocks show a high angle between 65° and 75° (Shahid et al., 2016). The hill slopes are generally covered with a slope wash material consisting of fragments of sandstone, siltstone and shale. In the upstream of dam, sandstone, siltstone and a shale sequence of the Mithakhari group of rocks are found on the Nallah bed. The shale is black to greenish-black and soft and splintery in nature. The siltstones are grey to brownish, fine-grained and usually fresh. The sandstones range from fine to coarse and are brownish, moderately hard and often stained on the surface.

Table 2.1 shows the index properties of soil and the percentage of soil particles of soil samples collected from different islands.

In the Andaman and Nicobar Islands, the rocks are highly folded due to frequent and violent tectonic movements in the past. The 69% of the Andaman Island underlined by the Andaman Flysch group constitutes the major lithology unit, the Mithakhari and Ophiolite groups cover areas of about 16% and 14%, respectively, and less than 1% area is covered by the Archipelago Group. The sedimentary Mithakhari and Nicobar Groups are most pervasive and occupy nearly 80% of the entire area of the Nicobar Islands while the igneous group covers nearly 5% and the rest 15% goes to the coralline and limestone formations.

Late Cretaceous igneous rocks – the "Ophiolite suite", marine sedimentary rocks of the Palaeocene to the Oligocene age and Recent to sub-Recent beach sands, mangrove clay, and alluvium and coral rags are exposed in the islands. The Ophiolite group forms the basement of the Andaman Island, a part of the Indo-Burma accretionary complex. It comprises several dismembered slices of Ophiolite and interleaves with Ophiolite derived clastic sediments bound by faults which generally trend in the N–S direction. The Ophiolite complex of the area mainly consists of Tholeitic Basalt (mainly occurs as pillowed lavas), layered ultramafic and mafic cumulates, felsic rocks (mainly Plagiogranites), tectonite, non-cumulates gabbro (in patches) and sedimentary mainly ribbon chert. The Ophiolite group is superposed by younger metasedimentary rocks, namely, quartzite, slate, phyllite and schist.

Table 2.1 Properties of soils from different parts of the island

Properties	Area			
	South Andaman	North Andaman	Campbell Bay	Middle Andaman
Liquid limit (%)	37.80	44.09	24.47	42.93
Plastic limit (%)	28.16	28.27	17.81	30.13
Plasticity index (PI)	9.64	15.82	6.66	12.80
Gravel (%)	23.17	5.52	2.83	7.32
Silt (%)	21.20	20.04	23.88	23.68
Clay (%)	54.98	73.09	73.28	69.00
Activity of clay	0.17	0.21	0.09	0.18

The Tertiary sediments classified as the Mithakhari and Andaman Flysch Group comprising thinly bedded alternations of sandstones and siltstones, grit, conglomerate, limestone, shale and so on are of the Upper Cretaceous to the Upper Eocene age. The Tertiary Group is overlain successively by the Archipelago Group, Nicobar Group and the Quaternary Holocene Group, intervening with unconformity.

2.3 PROPERTIES OF SOILS AND ROCKS

The properties of Andaman and Nicobar soils are discussed in this section based on laboratory tests on soil samples collected from four boreholes. These boreholes are taken from the construction site of the earthen dam over Kamsarat Nallah at Wimberlygunj (Shahid et al., 2016).

The grain size analysis of tested soil samples from borehole #1 indicates that the tested soil possesses predominantly silt-size particles followed by medium sand size. The soil has low to medium plasticity characteristics except for two soil samples that exhibited non-plasticity characteristics. From the value of activity of clay, the soil possesses normal activity. Based on the SPT 'N' values, it may be inferred that the foundation strata in borehole #1 possess medium to dense compactness. The value of in-situ permeability tests indicates that the foundation strata possess semi-pervious characteristics. The value of in-situ dry density tests indicates that the foundation strata possess medium compactness. The results of the triaxial shear tests conducted on the soil samples indicate that the soil samples are likely to exhibit good shear strength characteristics. The data of tested soil samples from the rest of the island show that the soil consists of predominantly clay-size particles followed by silt-size particles. The soil generally possesses low to medium plasticity characteristics. The value of activity of clay in soil indicates that the amount of kaolinite mineral is more, and the soil is inactive in nature (Shahid, et al., 2016).

2.4 USE OF SOILS AND ROCKS AS CONSTRUCTION MATERIALS

The town of Port Blair, from the beginning, had been built mainly with the locally available resources, which gave it a distinct character. The local building materials shaped the design and construction method to a large extent. The use of wood and associated materials formed the basic urban complex of Port Blair. Port Blair's urban complex was almost entirely based on the use of wood as the main construction material. Bricks do not form a standard building material in the Andaman and Nicobar Islands. The construction of RCC structures is not entirely devoid of local materials. Furthermore, stone as a building material is not available in adequate quantities in the islands. The standard of sand used in a cement mortar is inferior, and it is locally obtained from various parts of the islands (there are no rivers in the Andaman Islands). The use of local sand has been controversial, as there was a large-scale movement of this material from other parts of the islands to Port Blair; sand is being imported from the mainland also at an exorbitant cost. In recent times, the use of other techniques such as hollow block cement bricks in framed reinforced steel concrete structures with slab roofs is becoming common. Many public sector buildings have also been built with these materials. The standardised designs of the various central government departments are just replicated in the island environment without much consideration.

2.5   FOUNDATIONS AND OTHER GEOTECHNICAL STRUCTURES

The water table in the island is shallow. Also, the island falls under seismic zone V as per IS 1983–2002. The soil in the Andaman and Nicobar Islands faces the problem of liquefaction. The commonly used foundation type is pile foundation which mainly depends on the type of structure and subsoil. However, for some structures, mat foundation or a combination of shallow and pile foundations is also used.

2.6   OTHER GEOMATERIALS

The use of other geomaterials is not yet practised on the island. Landslides and soil erosion are common problems. In some areas, geojute is used for reducing erosion along the coastal side. The major production and cultivation of coconuts in the island are concentrated in South Andaman, Campbell Bay, Car Nicobar and Katchal Islands. Coconut plantation is 42.83% of the land out of 50,000 hectares of land available on the island for agriculture purposes. It is estimated that 280 million nuts are available annually for industrial exploitation. The fibres from the coconuts can be used as geomaterials for soil stabilisation, but only a negligible amount is being used presently for making coir products, and the rest of the coir fibres are used as domestic fuel. Coconut coir has about 48% of lignin which adds strength and elasticity to the cellulose-based fibre walls. Since lignin resists biodegradation, high lignin content also imparts longevity to outdoor applications. Coir fibre nearly takes more than 20 years to decompose. Coir fibres are available in different forms (e.g., ropes, grids, sheets, mats, boards, blocks) based on the necessity of the market, such as for household uses, for commercial uses involving lightweight boards for sound-proof walls in auditoriums and so on.

2.7   NATURAL HAZARDS

The Andaman and Nicobar Islands are surrounded by sea and contain a single live volcano at the Barren Island due to which the island occasionally experiences natural hazards such as cyclones, earthquakes, and Tsunamis. On 24 December 2004, the island experienced the most severe earthquake of a magnitude of 9.0 followed by a Tsunami that resulted in a great loss of life and property (Martin, 2005). The 2004 earthquake generated a variety of land level changes reflected on the coastal features, micro atoll and mangroves; it also caused ground cracking, liquefaction and sand blows. These features serve as indicators of co-seismic deformation and also provide telltale evidence of similar events in the past. Several buildings in Port Blair settled possibly due to foundation failure, erosion or settlement of the underlying soil due to inundation of the region by Tsunami waves and liquefaction. In most of the cases, buildings located near the sea settled down and signs of liquefaction, if any, were washed away by the Tsunami.

2.8   CASE STUDIES AND FIELD TESTS

In some areas, soil erosion, slope instability, mass movement of soil and damage were observed in Port Blair. Field and laboratory tests were carried out which reveal that the area is unstable and prone to landslides. Figure 2.2 shows the landslides at the VIP

Figure 2.2 A view of a landslide at the VIP road.

road. The main causes of instability are soil characteristics, poor subsoil drainage conditions and extra overburden pressure due to illegal construction in that area. Some preventive measures in the past were taken which have achieved limited success. Repair of a hill slope in the Port Blair area was carried out in 1994. The successive stone layers with bamboo piling in the soil were aimed at achieving better stability and subsoil drainage in the area. This repair was quite satisfactory for few years after which due to excessive pressure and heavy rainfall, landslides occurred at many places. There is a need for proper calculation of overburden pressure and analysis of failure planes while designing the slope stabilisation measures.

2.9 EFFECTS OF ENVIRONMENTAL CHANGES ON SOILS AND ROCKS

Due to the impact of the environment, there is a continuous rise in the water level, which is ultimately increasing the mean sea level (MSL) in many coastal areas. This increase in the MSL leads to an increase in the rate of erosion and consequently decreases the rate of accretion in the coastal areas. This is being predominantly observed after the disaster of the 2004 Tsunami. Figure 2.3 shows the exposure of rocks at the coastline due to erosion.

2.10 CONCLUDING REMARKS

Many areas of islands are untouchable, even those areas that are occupied. The main rock system of the island consists of sedimentary rocks, made up of sandstone and conglomerates. The soils in the island mainly are of alluvial type in valleys and hard clay type in hills. A detailed subsurface investigation is required for soil profiling, characterisation and slope stability assessment. Geosynthetics can be used in erosion control and stability of slopes. Research can be done on the use of local materials in construction.

Figure 2.3 Exposure of rocks due to erosion.

REFERENCES

Bandopadhyay, P.C. and Carter, A. (2017a). Chapter 2 Introduction to the geography and geomorphology of the Andaman and Nicobar Islands. Geological Society, London, Memoirs, 47, 75–93, 2 February 2017, doi:10.1144/M47.2.

Bandopadhyay, P.C. and Carter, A. (2017b). Chapter 6 Geological framework of Andaman – Nicobar Islands. Geological Society, London, Memoirs, 47, 75–93, 2 February 2017, doi:10.1144/M47.6

Das, S.T. (1982). *The Andaman and Nicobar Islands*, Sagar Publishers, New Delhi. http://www.mapsofindia.com/, accessed 18 February 2021.

IS 1893 (2002). Indian Standard Criteria for Earthquake Resistant Design of Structures: Part 1–General Provisions and Buildings, Bureau of Indian Standards, New Delhi.

ISRO (2009). National Wetland Atlas, *Andaman & Nicobar Island, SAC/RESA/AFEG/NWIA/ATLAS/04/2009*, Space Applications Centre (ISRO), Ahmedabad, India.

Martin, S.S. (2005). Intensity distribution from the 2004 M 9.0 Sumatra-Andaman earthquake. *Seismological Research Letters*, Vol. 76, No. (3), pp. 321–330.

Radicliffe Brown, A.R. (1992). *The Andaman Islanders*, The Free Press, Glencoe, IL.

Shahid, N., Singh, A., Chitra, R. and Gupta, M. (2016). Geotechnical investigations for foundation of earthen dam over Kamsarat Nallah for water supply scheme at Andaman and Nicobar–A case study. *International Journal of Innovation in Engineering and Research and Technology (IJERT)*, Vol. 3, No. 3, pp. 1–11.

Chapter 3

Andhra Pradesh

G.V. Rama Subba Rao and Chava Srinivas
Velagapudi Ramakrishna Siddhartha Engineering College

CONTENTS

3.1   INTRODUCTION

Andhra Pradesh is a state located in the south-eastern part of India. It is the eighth-largest state in India, covering an area of 1,60,205 km^2. Andhra Pradesh has the second-longest coastline of about 974 km among the states of India. This state is made up of two major regions, namely Rayalaseema and Coastal Andhra. Rayalaseema is in the inland south-western part of the state and Coastal Andhra is in the east and north-east parts, bordering the Bay of Bengal. Physiographically, the state has landforms as a plain land, including rugged plains, hills, isolated hillocks, dykes, uplands, rolling areas and flat land (Raju, 2012). A general map of the state indicating key cities/towns is presented in Figure 3.1.

Alluvial soils are found in the deltas of Godavari River, Krishna River and Penna River. Alluvial soils consist of varying proportions of silt, sand and clay. Black cotton soils are found in many parts of the state in the semi-arid regions. Black soils or soils containing montmorillonite clay minerals tend to undergo swelling during absorption and shrinkage during evaporation causing serious problems to civil engineering constructions. These soils are called expansive soils irrespective of the colour. Red soils are often found in some parts of the state. Red soils are composed of weathered crystalline and metamorphic rocks and get their colour from a high diffusion of iron. The red soil region is characterised by a highly undulating terrain with slopes (Vijayalakshmi et al., 1989). Marine soils are found along the coastline.

DOI: 10.1201/9781003177159-3

Figure 3.1 Map of Andhra Pradesh. (https://ap.meeseva.gov.in/DeptPortal/UserInterface/ LoginForm.aspx, accessed 24 January 2021.)

3.2   MAJOR TYPES OF SOILS AND ROCKS

During the geotechnical investigations, we used to encounter fine-grained soils at the surface frequently in the state of Andhra Pradesh. These fine-grained soils are usually silty clays of high compressibility (CH/MH-CH) or sometimes sandy clays of inter-mediate compressibility (CI/MI-CI). Coarse-grained soils especially poorly graded fine to medium sandy soils (SP) exist below the fine-grained soils, particularly in the delta regions. Expansive soils are found in several parts of the state as it is a semi-arid region. The red soils are present at the top and are clayey or silty gravel (GC/GM) in the dry lands of several districts of the state.

3.3   PROPERTIES OF SOILS AND ROCKS

Geotechnical properties of soils and rocks as the data obtained from geotechnical investigations carried out in the state are presented in this section. Soil classification is done as per the guidelines of IS: 1498-1970. Soils with a wide range of geotechnical properties are present in the state. The liquid limit of CH soils is in the range of 70%–86% and the plastic limit of CH soils is in the range of 26%–40%. The percentage of fines in CH soils is in the range of 90%–100%. The liquid limit of MH-CH soils is in the range of 50%–70% and the plastic limit of MH-CH soils is in the range of 27%–35%. The percentage of fines in MH-CH soils is in the range of 85%–100%. The liquid limit of CI soils is in the range of 42%–50% and the plastic limit of CI soils is in the range of 18%–24%. The percentage of fines in CI soils is in the range of 80%–90%. The liquid limit of MI-CI soils is in the range of 37%–45% and the plastic limit of MI-CI soils is in the range of 23%–27%. The percentage of fines in MI-CI soils is in the range of 70%–95%. Coarse-grained soils, especially poorly graded fine to medium sandy soils (SP), exist below fine-grained soils, particularly in the delta regions of the state. Expansive soils are found in several parts of the entire state as it is a semi-arid region. Expansive soils in the state have a degree of expansivity from a low degree to a high degree. The red soils are mainly found in the dry lands of several districts of the state and these soils have little or no presence of gravel. The subsurface explorations carried out in several locations within the state show rocks of varying quality. Rock fragments have a core recovery (CR) of 50%–65% and a rock quality designation (RQD) of 0%. The CR of poor-quality rock is in the range of 60%–75% and its RQD is in the range of 25%–40%. The CR of rock having fair quality is in the range of 90%–98% and its RQD is in the range of 50%–60%.

3.4 USE OF SOILS AND ROCKS AS CONSTRUCTION MATERIALS

Soil investigations are essential for the selection of soil from borrowed areas with the intended use as construction materials. The rural water supply and sanitation department of the Krishna district has planned to enhance the storage capacity of the summer storage tank at Allampuram. The area of the tank is about 10 acres and the storage capacity of the tank is 3 lakh cubic metres. A constant head permeability test was conducted on soil samples collected from the tank bed and the test results are presented in Table 3.1. The presence of permeable soil at the bed of the tank leads to a seepage loss. It is recommended to place a compacted clay liner of 50 cm thickness as the bed carpet. Liner soil should be impervious, and it should have a coefficient of permeability less than 10^{-7} cm/s. The coefficient of permeability of the liner soil sample is 4.86×10^{-8} cm/s.

Table 3.1 Permeability of soil

Observation	Trail-1	Trail-2	Trail-3	Trail-4	Trail-5
Constant head, H (cm)	128	128	128	128	128
Length of the soil specimen, L (cm)	12.5	12.5	12.5	12.5	12.5
Diameter of the soil specimen, D (cm)	10	10	10	10	10
Cross-sectional area of the soil specimen, A (cm^2)	78.55	78.55	78.55	78.55	78.55
Amount of water collected, V (ml or cc)	200	200	200	200	200
Elapsed time for collection of water, t (s)	22	21	23	22	21
Discharge, q (cc/s)	9.09	9.52	8.69	9.09	9.52
Coefficient of permeability, k (cm/s)	0.0113	0.0118	0.0108	0.0113	0.0118

Red earth is commonly used in the construction of unpaved roads. Red earth possesses all the unfavourable qualities such as high degree of water absorption, plasticity and volume changes (Bhavannarao, 2006). The properties of the red earth obtained from the Gollanapally quarry in the Krishna district are presented in Table 3.2. It is proposed to replace red earth with crushed stone dust (CSD) to reduce its plasticity. The geotechnical properties of the CSD are presented in Table 3.3. The plasticity index values of red earth replaced with various percentages of quarry dust are presented in Table 3.4. From Table 3.4, it can be clearly noticed that red soil replaced with 40% CSD has low plasticity.

Table 3.2 Properties of red earth

Property	Value
Specific gravity	2.69
Gravel (%)	5.60
Sand (%)	49.60
Fines (%)	44.80
Liquid limit (%)	47
Plastic limit (%)	19
Plasticity index (%)	28
Classification of the soil (ISSCS)	SC-CI
Differential free swell index (%)	10.00
Degree of expansivity	Low
Modified compaction characteristics	
Maximum dry density (Mg/m^3)	2.03
Optimum moisture content (%)	10.2

Table 3.3 Properties of crushed stone dust

Property	Value
Specific gravity	2.66
Gravel (%)	41.28
Coarse sand (%)	23.26
Medium sand (%)	21.04
Fine sand (%)	9.33
Fines (%)	5.09
Coefficient of uniformity, C_u	20.83
Coefficient of curvature, C_c	2.13
Classification of soil (ISSCS)	SW
Maximum dry density (g/cc)	1.962
Minimum dry density (g/cc)	1.600

Table 3.4 Plasticity index of red soil replaced with quarry dust

% of Quarry dust replaced	Liquid limit, LL (%)	Plastic limit, PL (%)	Plasticity index, PI = (LL − PL) (%)
100% Red soil	47.0	19.48	27.52
90% Red soil + 10% Quarry dust	44.3	21.10	23.2
80% Red soil + 20% Quarry dust	39.9	21.73	18.17
70% Red soil + 30% Quarry dust	33.4	22.98	10.42
60% Red soil + 40% Quarry dust	29.8	24.02	5.78

3.5 FOUNDATIONS

Foundation transmits loads due to various engineering activities on or into soil/rock mass. Allowable bearing capacity can be calculated based on the results of field and laboratory tests, and it is used for the structural design of foundations. The subsoil profile at a location in Tanuku, West Godavari district is shown in Figure 3.2.

Subsoil investigation is carried out for the construction of a G+3 residential building. A standard penetration test (SPT) is routinely performed as a part of the geo-technical investigation in the state. Boring is carried by using a combination of shell

Depth below ground level (meter)	Type of soil	Symbol	Test depth (meter)	SPT value (N-value)
1.00	Yellowish brown silty clay (MH-CH)		0.55-1.00	6
2.00			1.55-2.00	10
3.00	Yellowish brown fine sandy clay (CI)		2.55-3.00	12
4.00			3.55-4.00	13
5.00			4.55-5.00	15
6.00			5.55-6.00	20
7.00	Yellowish brown fine to medium sand (SP)		6.55-7.00	22
8.00			7.55-8.00	22
9.00			8.55-9.00	23
10.00			9.55-10.00	25

Figure 3.2 Subsoil profile at a location in Tanuku, West Godavari district.

and auger methods with casing pipe depending upon the type of strata met within the borehole location using a hand boring machine. Subsoil investigation is carried out as per the guidelines of IS: 1892-1979. Soil boring of 150 mm diameter up to 10 m depth is done. The SPT test is conducted at every 0.5 m regular intervals within the borehole. The subsoil mainly consists of silty clay (MH-CH) at the top followed by fine sandy clay (CI) up to 5.00 m from the existing ground level (EGL) and finally fine to medium sand which is poorly graded (SP). The ground water table (GWT) is observed at a depth of 4.50 m below the EGL. Considering the type of superstructure and the subsoil characteristics, isolated footings are selected.

Details of the subsoil profile explored for the construction of a cellar+G+5 commercial building at a location in Gannavaram, Krishna district are shown in Figure 3.3. The subsoil mainly consists of silty clay (MH-CH) at a depth 2.00 m below the EGL followed by fine sandy clay (CI) up to 3.00 m. The subsoil clayey gravel (GC) extends from 3.00 to 4.00 m and finally meets with soft disintegrated rock fragments (SDR) at a depth of 4.00 m from the EGL. Understanding that the cellar is proposed at a depth of 4.50 m below the EGL. Hydraulic rock breaking stone hammer tools are used for breaking the hard gravel which is present at 4.00 m below the EGL. Raft foundation is recommended for the proposed commercial building.

Depth below ground level (meter)	Type of soil	Symbol	Test depth (meter)	SPT value (N-value)
1.00	Yellowish red silty clay (MH-CH)		0.55-1.00	4
2.00			1.55-2.00	5
3.00	Yellowish red fine sandy clay (MI-CI)		2.55-3.00	13
4.00	Yellowish red clayey gravel (GC)		3.55-4.00	26
5.00	Soft disintegrated rock (SDR)		4.55-5.00	>100

Figure 3.3 Subsoil profile at a location in Gannavaram, Krishna district.

Depth below ground level (meter)	Type of soil	Symbol	Test depth (meter)	SPT value (N-value)
1.00			0.55-1.00	4
2.00			1.55-2.00	5
3.00	Yellowish brown silty clay (MH-CH)		2.55-3.00	6
4.00			3.55-4.00	9
5.00			4.55-5.00	11
6.00			5.55-6.00	13
7.00			6.55-7.00	14
8.00	Black silty clay (MH-CH)		7.55-8.00	15
9.00			8.55-9.00	17
10.00			9.55-10.00	21

Figure 3.4 Subsoil profile at a location in Bhavanipuram, Vijayawada city, Krishna district.

The subsoil profile (Figure 3.4) was obtained as a part of the geotechnical investigation carried out for the construction of a stilt+G+5 residential apartment building in the housing board colony at Bhavanipuram, Vijayawada. The subsoil mainly consists of two layers. The top layer is silty clay (MH-CH) and it is up to 6.00 m below the EGL. The bottom layer is black silty clay (MH-CH). GWT is placed at a depth of 6.50 m below the EGL. It is identified that in the top layer up to 1.50 m below the EGL, the soil is expansive in nature (the differential free swell index of the soil is 60%). The uplift of foundations caused by swelling soils leads to the development of tension. Hence, tension-resistant foundations like under-reamed pile foundations are recommended.

Subsoil investigation is carried out for the construction of a high-level bridge at the Valiveru–Vetapalem road, Tsunduru mandal in Guntur district. The subsoil profile obtained is shown in Figure 3.5. The subsoil profile consists of fine sandy clay (CI) up to

Depth below ground level (meter)	Type of soil	Symbol	Test depth (meter)	SPT value (N-value)
1.00	Yellowish brown fine sandy clay (MI-CI)		0.55-1.00	7
2.00			1.55-2.00	9
3.00			2.55-3.00	10
4.00	Yellowish brown silty sand (SM)		3.55-4.00	14
5.00			4.55-5.00	16
6.00			5.55-6.00	17
7.00			6.55-7.00	21
8.00			7.55-8.00	23
9.00			8.55-9.00	24
10.00			9.55-10.00	26
11.00			10.55-11.00	27
12.00			11.55-12.00	28
13.00			12.55-13.00	30
14.00			13.55-14.00	31
15.00	Yellowish brown fine to medium sand (SP)		14.55-15.00	31
16.00			15.55-16.00	33
17.00			16.55-17.00	35
18.00			17.55-18.00	36
19.00			18.55-19.00	36
20.00			19.55-20.00	38
21.00			20.55-21.00	39
22.00			21.55-22.00	40
23.00			22.55-23.00	40
24.00			23.55-24.00	42
25.00			24.55-25.00	44

Figure 3.5 Subsoil profile at Valiveru–Vetapalem road, Guntur district.

Figure 3.6 (a) Soft disintegrated rock fragments; (b) rock of poor quality; and (c) rock of fair quality.

3.00 m followed by silty sand (SM) up to 6.00 m. Fine to medium SP exists from 6.00 to 25.00 m within the borehole. Well foundation is recommended for abutments and piers. The depth of the well foundation is 15.00 m below the EGL. The diameter of the well foundation is 8.00 m and the grip length of the well is 5.00 m. The safe bearing capacity is evaluated based on the shear criterion as per the guidelines of IS: 6403-1981 and also the allowable bearing pressure is evaluated as per the guidelines of IS: 3955-1967. An allowable bearing capacity of 390 kPa is recommended for the well foundation.

Subsoil investigation is carried out across the Gundlakamma river at km 12/10 of Markapur–Tarlupadu road in Prakasam district for the recommendation of a foundation for a high-level bridge. Rotary core drilling is carried out for making advancing boreholes at the site. A borehole of 150 mm diameter up to 10.00 m is drilled within the location. The collected samples were brought to the laboratory in an airtight wooden box. A closer view of the collected samples is shown in Figure 3.6. GWT is not observed in the borehole. In this case, isolated footings are recommended.

The subsoil profile (Figure 3.7) mainly consists of SDR up to 5.00 m below the EGL followed by rocks of poor quality up to 8.00 m. Rocks of fair quality exist at 8.00–10.00 m in the borehole and the unconfined compressive strength of the rock core samples is evaluated as per IS: 9143-1979 and found to be 52 MPa. Isolated footing with an allowable bearing capacity of 300 kPa is recommended at a depth of 3.50 m below the EGL.

3.6 OTHER GEOMATERIALS

The utilisation of waste materials is a wise choice and is also one step towards accomplishing sustainable development. One of the abundantly available materials is fly ash, which is obtained from the nine thermal power stations situated in the state. The annual generation of fly ash from the thermal power plants in this state is presently around 6.8 million t/annum. The utilisation of fly ash obtained from the thermal power plants reaches about 95% in the state. This is possible by bulk utilisation of fly ash in various infrastructure activities. Fly ash is one of the hazardous industrial wastes generated in huge quantities in thermal power plants to meet the demand for electricity which is otherwise disposed of involving huge costs and environmental problems. In some cases, a by-product is inferior to traditional earthen material, but its cost makes it an attractive alternative if adequate performance can be obtained. The major areas

Depth below ground level (meter)	Type of soil	Symbol	Test core run (meter)	CR (%)	RQD (%)
1.00	Soft disintegrated rock fragments (SDR)		0.00-1.00		
				58	0
2.00			1.00-2.00		
				60	0
3.00			2.00-3.00		
				63	0
4.00			3.00-4.00		
				66	0
5.00			4.00-5.00		
				68	0
6.00	Rock of poor quality (SR)		5.00-6.00		
				80	25
7.00			6.00-7.00		
				83	27
8.00			7.00-8.00		
				85	30
9.00	Rock of fair quality (HR)		8.00-9.00		
				96	50
10.00			9.00-10.00		
				97	52

Figure 3.7 Subsoil profile across the Gundlakamma river at km 12/10 of the Markapur–Tarlupadu road in Prakasam district.

Figure 3.8 Fly ash used as a backfill for a back-to-back reinforced wall at the Vijayawada inner ring road, Gunadala.

of fly ash utilisation are making bricks/blocks, cellular concrete products and light-weight aggregates, manufacture of cement, construction of roads and embankments and backfill for retaining structures (Hazara and Patra, 2008). Fly ash can be used effectively in bulk quantities for several civil engineering applications such as fill material, lightweight aggregate and subbase material (Leelavathamma et al., 2005; Ghosh and Subbarao, 2001). About 2,500 numbers of fly ash brick manufacturing units operate in the state. Each unit has a production capacity of between 10 and 30 t/day. Fly ash is used as a backfill for the construction of a back-to-back reinforced earth wall at Vijayawada inner ring road, Gunadala (Figure 3.8). The geotechnical properties of the fly ash used are presented in Table 3.5.

Table 3.5 Geotechnical properties of fly ash

Property	Value
Specific gravity	2.10
Fine sand (%)	25.07
Fines (%)	74.93
Maximum dry unit weight (kN/m^3)	13.63
Optimum moisture content (%)	21.4
Unsoaked CBR value (%)	10.73
Soaked CBR value (%)	0.95

CSD is a waste product that is produced in huge quantities from crushing plants. The state has about 274 stone quarries of reasonable size and 2,500 other small crushing units. Crushing plants generate huge quantities of stone dust. CSD can be used as an alternative to natural river sand, which is used as a coarse pavement material. CSD can avoid detrimental effects on the environment caused by the excessive mining of river sand (Sanjay et al., 2016). The utilisation of CSD is possible through geotechnical applications such as embankments, backfill material and subbase material (Sridharan et al., 2006).

3.7 CASE STUDIES AND FIELD TESTS

This section throws light on case studies related to inadequate subsoil exploration, improper construction of under-reamed pile foundation and strengthening soft ground by using timber piles (casuarina ballies).

3.7.1 Case study #1: inadequate subsoil exploration

This case study deals with improper subsoil exploration carried out for the construction of an overhead balancing reservoir at Tarakaturu village, Gudur mandal, Krishna district. Initially, as a part of achieving the safe bearing capacity of the foundation soil, only one undisturbed soil sample was collected by the construction agency at a depth of 2.70 m below the EGL. The foundation selected for the proposed OHBR is the annular raft foundation and it was designed for a safe bearing capacity of 70 kPa. Excavation was carried out up to 2.70 m for the construction of the foundation. During the construction, authorities have observed an excessive settlement. Then, they looked for an investigation of soil underneath. The SPT was conducted at the bottom of the excavation (Figure 3.9) as per the guidelines of IS: 2131-1981.

Figure 3.10 shows the subsoil profile. After examination of the subsoil profile, it was realised by the authorities that soft clay exists up to a depth of 9.00 m from the EGL followed by silty sand (SP-SM) up to 15.00 m below the EGL. The liquid limit of

(a)   (b)   (c)

Figure 3.9 (a) Tripod setup; (b) hammer blows; and (c) the presence of soft clay.

Depth below ground level (meter)	Type of soil	Symbol	Test depth (meter)	SPT value (N-value)
0.00–3.00	Excavation			
4.00			3.55–4.00	1
5.00			4.55–5.00	1
6.00	Black plastic clay (CH)		5.55–6.00	1
7.00			6.55–7.00	1
8.00			7.55–8.00	1
9.00			8.55–9.00	2
10.00			9.55–10.00	9
11.00			10.55–11.00	11
12.00	Black silty sand (SP-SM))		11.55–12.00	12
13.00			12.55–13.00	14
14.00			13.55–14.00	16
15.00			14.55–15.00	17

Figure 3.10 Subsoil profile at a location in Tarakaturu village, Gudur mandal, Krishna district.

the soft clay deposit is 81% and the natural moisture content of the soil is 69%. Inadequate subsoil exploration has led to affecting the project schedule and a huge loss of money. There was a need to give more significance to geotechnical investigations. A detailed site investigation involving deep boreholes, in-situ tests like SPT, plate load test, etc. and further laboratory testing of soil/rock samples are necessary for any construction activity. Proper geotechnical investigations need to be done as per the guidelines of IS: 1892-1979 and proper investigations are useful to understand the behaviour of geomaterials like soil/rock at a site.

3.7.2 Case study #2: improper construction of under-reamed pile foundation

Subsoil investigation was carried out for the construction of a stilt+G+5 residential apartment building in Bhavanipuram, Vijayawada city, Krishna district and the subsoil profile is shown in Figure 3.4. The foundation selected for this structure is the under-reamed pile foundation of 380 mm stem diameter and 5.00 m length with double bulbs. A 16 mm rod bent at an angle of 45° at the bottom was used as a reaming tool for making bulbs (Figure 3.11). Figure 3.12 shows the right practices in the construction of an under-reamed pile foundation as per the guidelines of IS: 2911 (Part 3)-2015.

3.7.3 Case study #3: strengthening soft ground using casuarina ballies

A case study is presented which highlights the use of casuarina piles (ballies) for strengthening soft soil which exists up to a depth of 5.00 m below the canal bed. A single-lane bridge is proposed across the Lavala medium drain in Kalidindi mandal,

Figure 3.11 A 16 mm diameter rod bent at an angle of 45° at the bottom is used as a reaming tool.

(a)   (b)   (c)

Figure 3.12 (a) Augering with a tripod setup; (b) under reaming tool; and (c) tremie concreting.

Figure 3.13 Driving casuarina ballies.

Krishna district. The soil at the foundation level was filled with sand of 1.0 m thickness over soft clay and ballies of 10 cm diameter and 6 m length were driven at an interval of 0.5 m c/c in a staggered manner (Figure 3.13).

A plate load test was conducted on an improved ground where the foundation of the bridge pier was planned. The test was performed on the sand bed to a size of

Figure 3.14 Plate load test setup.

2.25 m × 2.25 m (five times the size of the test plate) as per IS: 1888-1982. The size of the plate was 450 mm × 450 mm, and the thickness was 25 mm. The reaction for the vertical compression load test on the plate was obtained by means of kentledge (Figure 3.14).

The sandbags were placed on the platform made with ISMB450 steel beams of length about 6.0 m to create the required dead load necessary for the reaction. The reaction available for the test plate was 50% more than the test load on the plate. The load was applied by means of a hydraulic jack and the reaction to the jack was provided by kentledge. The dead load placed on the platform was about 20 t. The settlements were recorded by means of four dial gauges with a sensitivity of 0.01 mm placed on immovable supports placed 1.5 m from the test plate. The test was carried out by applying a series of vertical downward incremental loads of 0.56 t. The displacements in each loading stage were measured at a regular time interval of up to 1 hour. The plate was loaded up to 12.32 t. A total settlement of 27.01 mm was observed for the test load of 12.32 t. Then, the plate was unloaded and the net settlement was found to be 25.45 mm. A plot drawn between the applied load and average settlement is shown in Figure 3.15, to find the safe bearing capacity of the soil.

The size of the foundation was 17.6 m × 5.2 m and the size of the plate was 450 mm × 450 mm. The plate settlement was determined for an allowable foundation settlement of 25 mm, and it is 10.0 mm. From the load-settlement curve (Figure 3.15), the load corresponding to a plate settlement of 10.0 mm is 5.0 t. This load induces a pressure intensity of 24.7 t/m^2. From the load-settlement curve, the ultimate load is found to be 7 t and the ultimate load-carrying capacity of the soil is 35 t/m^2 for a plate of size 450 mm × 450 mm. The safe bearing capacity of the foundation soil is obtained by dividing the ultimate bearing capacity with a factor of safety. The safe bearing capacity is equal to 11.7 t/m^2.

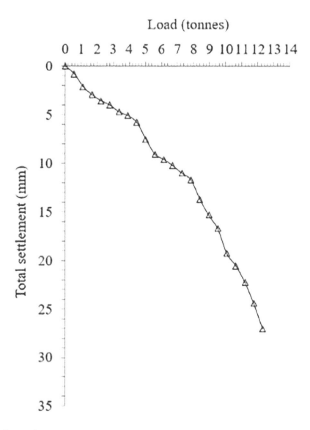

Figure 3.15 Load-settlement curve.

3.8   CONCLUDING REMARKS

As presented in the previous sections, the following key points may be mentioned:

1. The major soils found in Andhra Pradesh are alluvial soils, marine soils, black cotton soils and red soils. The state consists of igneous, sedimentary and metamorphic rock formations.
2. High-quality geotechnical investigations are the need of the hour. The inadequate geotechnical investigation leads to project delay which in turn raises the project cost, disputes and sometimes overdesign of the foundation, resulting in a highly uneconomical foundation.
3. The cost involved for geotechnical investigations is likely to be 0.01% of the project cost and sometimes less. There is a need to give more importance to geotechnical investigations.
4. The desired load-carrying capacity of foundations may not be achieved due to improper construction practices. Appropriate foundation construction practices must be adopted.
5. Using casuarina piles (ballies) for strengthening soft soil ground is a sustainable and cost-effective solution.

6. Accumulation of practical experience done in similar situations is highly useful for professional geotechnical engineers throughout the state. Good interaction among all the stakeholders, that is, owners, engineers and contractors, ensures good geotechnical practice.

REFERENCES

Bhavannarao, D. (2006). Use of red earth in the name of gravel in poor conditions of roads, *Proceedings of National Conference on Corrective Engineering Practices in Troublesome Soils (CONCEPTS)*, JNTU College of Engineering, Kakinada, India, pp. 27–29.

Ghosh, A. and Subbarao, C. (2001). Microstructural development in fly ash modified with lime and gypsum. *Journal of Materials in Civil Engineering*, Vol. 13, No. 1, pp. 65–70.

Hazara, S. and Patra, N.R. (2008). Performance of counterfort walls with reinforced granular fly ash backfills: Experimental investigation. *International Journal of Geotechnical & Geological Engineering*, Vol. 26, No. 23, pp. 259–267.

https://ap.meeseva.gov.in/DeptPortal/UserInterface/LoginForm.aspx, Meeseva Official Portal, Government of Andhra Pradesh, accessed 24 January 2021.

IS 1498 (1970). Classification and identification of soils for general engineering purposes. Bureau of Indian Standards, New Delhi, India.

IS 1888 (1982). Method of load test on soils. Bureau of Indian Standards, New Delhi, India.

IS 1892 (1979). Code of practice for subsurface investigation for foundations. Bureau of Indian Standards, New Delhi, India.

IS 2131 (1981). Method for standard penetration test for soils. Bureau of Indian Standards, New Delhi, India.

IS 2911 (2015). Code of practice for design and construction of pile foundations: Under-reamed piles, Part 3. Bureau of Indian Standards, New Delhi, India.

IS 3955 (1967). Code of practice for design and construction of well foundations. Bureau of Indian Standards, New Delhi, India.

IS 6403 (1981). Code of practice for determination of bearing capacity for shallow foundations. Bureau of Indian Standards, New Delhi, India.

IS 9143 (1979). Method for the determination of unconfined compressive strength of rock materials. Bureau of Indian Standards, New Delhi, India.

Leelavathamma, B., Mini, K.M. and Pandian, N.S. (2005). California bearing ratio behaviour of soil-stabilized class F fly ash systems. *Journal of Testing and Evaluation*, Vol. 33, No. 6, pp. 406–410.

Raju, A.S. (2012). Current status of soil management in Andhra Pradesh. *Bharathi Integrated Rural Development Society (BIRDS), Strategic Pilot on Adaptation to Climate Change (SPACC)*, pp. 15.

Sanjay, M., Sindhi, P.R., Vinay, C., Ravindra, N. and Vinay, A. (2016). Crushed rock sand: An economical and ecological alternative to natural sand to optimize concrete mix. *Perspectives in Science*, Vol. 8, pp. 345–347.

Sridharan, A., Soosan, T.G., Babu T.J. and Abraham, B.M. (2006). Shear strength studies on soil-quarry dust mixtures. *Geotechnical and Geological Engineering*, Vol. 24, No. 5, pp. 1163–1179.

Vijayalakshmi, K., Vittal, K.P.R. and Rao, U.M.B. (1989). Minimal irrigation on small agricultural watersheds with red soils in the semi-arid tropics of Andhra Pradesh, India. *Agricultural Water Management*, Vol. 16, No. 4, pp. 279–291.

Chapter 4

Arunachal Pradesh

Ajanta Kalita
North Eastern Regional Institute of Science and Technology (NERIST)

Sukumar Baruah
NEEPCO Limited

Yachang Omo
Central Institute of Technology (CIT) Kokrajhar

CONTENTS

4.1   INTRODUCTION

Arunachal Pradesh, also known as the 'Land of Rising Sun' with its breath-taking picturesque landscape, towering snow-clad mountains, steep gorges, lush green valleys and innumerable tributaries and rivulets is situated in the north-eastern tip with a longitude between 91°30'E and 97°30'E and latitude between 26°30'N and 29°31'N with an area of 83,743 km^2. It comprises 25 districts and touches the international boundaries with Bhutan (approx. 160 km) in the west, China (approx. 1,080 km) in the north and north-east, Myanmar (approx. 440 km) in the east and south-east and the plains of Assam to the south (APSG, 2020). Almost the entire state of Arunachal Pradesh is

DOI: 10.1201/9781003177159-4

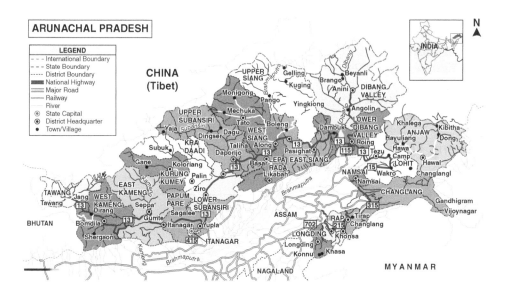

Figure 4.1 Map of Arunachal Pradesh. (https://www.mapsofindia.com/maps/arunachalpradesh, accessed 15 June 2020.)

a largely inaccessible terrain with dense forest and unpredictable climatic conditions and poor road communication thereby making the state a hotspot for various scientific studies. Figure 4.1 shows the map of Arunachal Pradesh.

Arunachal Pradesh has a forest cover of more than 90% out of the total utilisable land. The balance 10% utilisable land includes the grazing lands, culturable waste land, fallow lands, land under tree crops and the habitation land. The forest land is further classified into reserve forest (20.46%), protected forest (18.49%) and unclassified forest (61.05%), which is owned by the state, except for 0.3% owned by individuals where LPCs (Land Possession Certificates) are issued by the State Government as a proof of land ownership (ISFR, 2009). Geologically, the state of Arunachal Pradesh is also a lesser-known region. This state can be subdivided into four physiographic divisions, each division characterised by a distinctive geological and tectonic history. The four divisions are the following: (a) Himalayan range, (b) Trans-Himalayan range, (c) Naga-Patkai range, and (d) Brahmaputra plains. Figure 4.2 shows the geological map of Arunachal Pradesh.

The Himalayan range of Arunachal Pradesh exhibits a gradual increase in elevation from south to north where the elevation of the range in the south is about 100 m in comparison to a general elevation of over 1,000 m above the mean sea level (MSL) in the north. A few peaks are also observed to have attained heights of more than 7,000 m above the MSL towards the northern side of this range. In general, the Himalayan landforms are characterised by mega folds, faults and a thrusted terrain. The high degree of relief of this range can be attributed to the presence of tectonic lineament and heterogeneity in lithology with approaching immaturity of dissection. Simultaneous rejuvenation along with the operating erosional cycle has rendered mainly rivers of aggrading nature as a result of which macro relief of parallel ridges and valleys has been imprinted over the entire region. The Higher Himalayas are bounded by the Tibetan

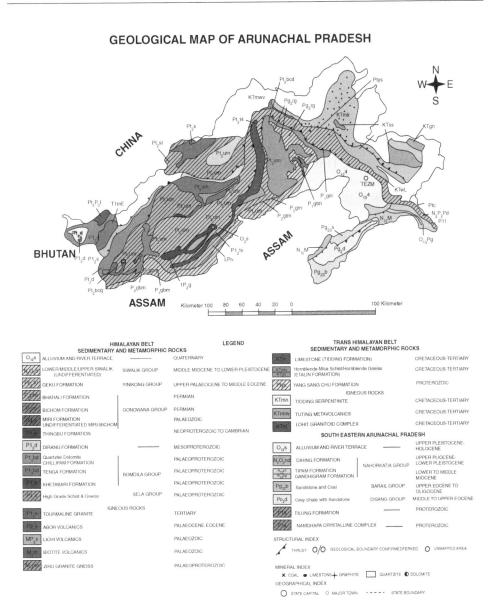

Figure 4.2 Geological map of Arunachal Pradesh. (After GSI, 2013.)

Himalayas in the north and the Lesser Himalayas in the south, the southern extent of which is marked by the main central thrust (MCT). In the western part near Bhutan, the trend of this zone is ENE-WSW which gradually alters to NE-SW towards the east. The distinguished characteristics of this zone are rugged topography with high ridges, average relief of about 6,000 m above the MSL, precipitous slopes and narrow deep gorges. This zone is primarily composed of the Paleoproterozoic high-grade gneiss and schists of the Se La Group along with some tertiary intrusive granites. The Tibetan

Himalayas characterised by high-altitude, low-relief gentle slopes and sparse alpine type vegetation occupies the north-western part of Arunachal Pradesh bordering Bhutan and Tibet. The Tibetan Himalayas is a NE-SW trending zone primarily composed of high-grade schists and gneisses of the Se La Group, Proterozoic metasediments of Lumla Formation and a probable part of the Tethyan sequence. This zone exhibits a varying altitude ranging from 3,000 to 6,000 m and has a thickness of about 30–40 km. Figure 4.3 shows the snow-cladded mountains of the Higher Himalayas in Arunachal Pradesh. Figure 4.4 shows the high-altitude barren valley of Arunachal Pradesh. The Naga-Patkai range defines the southern limit of the Upper Brahmaputra plains and has an ENE-WSW trend. This range is a part of the Arakan Youma Mountain chain, which assumes an arcuate pattern in the vicinity of the Mishmi Thrust. This range is primarily composed of the tertiary sequences of Assam and south-eastern Arunachal

Figure 4.3 Snow-cladded mountains of the Higher Himalayas in Arunachal Pradesh.

Figure 4.4 High-altitude barren valley of Arunachal Pradesh.

Figure 4.5 Brahmaputra Plains in Arunachal Pradesh.

Pradesh. The Brahmaputra Plains in Arunachal Pradesh is bounded by the Himalayan range in the north and east and by the Naga-Patkai range in the south and is underlain by post-Siwalik Quaternary sediments (Kesari, 2010). These plains are a part of the easternmost extremity of the vast Brahmaputra plains. Figure 4.5 shows the Brahmaputra Plains in Arunachal Pradesh.

4.2 MAJOR TYPES OF SOILS AND ROCKS

4.2.1 Soils

Based on various geotechnical investigations carried out at different places of Arunachal Pradesh, it is found that the properties of the soil layer vary significantly from place to place. Starting from Itanagar, the capital of Arunachal Pradesh, which is situated in the middle part of the state, the topsoil layer consists of mainly fine to medium sand (SW) of whitish-grey colour up to a depth of maximum 5 m beyond which boulder strata is encountered. Subsoil investigation reports of Daporijo, Chimpu and Kurung kumey, which are in the upper reaches of the middle portion of Arunachal Pradesh, reveal that the soil layer up to a depth of 3.0 m consists of mainly silty sand (SM) to gravel mixed with silty sand, whereas the soil layer up to a depth of 3.0 m of Ziro, which is also a place in the same geographic boundary, consists of clay with low compressibility (CL).

In the north-eastern part of Arunachal Pradesh, the subsoil investigation reports of three places Along, Pashighat and Kombu village reveal that the topsoil layer mainly consists of silty clay (CM) to clayey sand (SC) within a depth of 1.50 m. In various stretches, the soil layer is found to be mixed with gravels and pebbles.

In the south/south-eastern part of Arunachal Pradesh, the investigation report of the Namsai area shows that the pattern of soil encountered is SC up to a depth of 8.0 m, which is caused by the alluvial deposition in the entire area.

The soil types mainly encountered in the areas in the western reaches of Arunachal Pradesh, i.e., East Kameng, West Kameng and Tawang, vary between silty sand (SM) and SC mixed with gravel. The periodical freezing and thawing in the areas in the upper reaches makes the soils vulnerable due to which detail geotechnical investigation has to be carried before starting any construction activity.

4.2.2 Rocks

Arunachal Pradesh makes one experience varied lithology if one traverses across the state. Quaternary sediments are found to occur in the foothill regions adjoining the Brahmaputra plains, which are essentially composed of boulders, cobble, pebble, sand and sandy clay beds. These are fluvial sediments and occur as valley filled deposits on either side of the river valleys exposed at different levels.

The Sub-Himalayan range occurs as a linear belt along the foothills of Arunachal Pradesh and extends from Bhutan in the east to Pasighat in the west. This range is essentially a sedimentary terrain and is primarily composed of boulder conglomerate, pebble sandstones, felspathic sandstone, red arkosic sandstone and black carbonaceous shale with thin impersistent lenticular coal. Few intrusives in the form of tourmaline bearing leucogranite and dark green basic volcanics are found within this terrain.

The Lesser Himalayan range delineated by the main boundary thrust (MBT) and the MCT in the south and north, respectively, is comprised of a sedimentary sequence and a sequence of low- to medium-grade metamorphic rocks. The major rock type of this range includes black and carbonaceous shale, sandstones, quartzites, calc-silicate, garnetiferous phyllite and schist, carbonaceous mica schist, micaceous quartzite and biotite granite gneiss.

The Higher Himalayan range is a ~30 km thick sequence of medium- to high-grade metamorphic rocks with granitic intrusions in many places. This range is essentially comprised of biotite gneiss, garnetiferous biotite gneiss, kyanite-sillimanite schist/gneiss, hornblende bearing gneiss and migmatites.

The Trans-Himalayan range is an NNW-SSE sequence of high-grade metamorphic rocks that are found along the Tidding Suture in Upper Siang, Upper Dibang and Lohit valley areas. The major rock types of this zone are chlorite/graphitic schist, amphibolite gneiss, granodiorite, garnet-kyanite gneiss, meta-volcanic, two-mica leucogranites and anorthosite.

The Naga-Patkai range constitutes the south-eastern part of Arunachal Pradesh and is essentially a sedimentary sequence comprised of quaternary and tertiary sediments. The primary rock types of these regions are sandstones, shales, sandstones with intercalation of minor coal lenses, quartzites and quartz mica schists.

4.3   ENGINEERING PROPERTIES OF SOILS AND ROCKS

4.3.1   Soils

In this section, the engineering properties of soils such as cohesion (c), angle of internal friction (ϕ), specific gravity, and bulk density will be discussed. From the available data, it is found that in the Papumpare district, which is in the middle part of

Arunachal Pradesh, the ϕ value varies from 36° to 39° and the c value is zero because of the presence of sandy strata. The field density varies between 2.22 and 2.28 g/cc, and the specific gravity is around 2.65. However, in the Rono hills area within the district, the c value can be found varying from 0.30 to 0.08 kg/cm^2 and the ϕ value varying between 14° and 32°.

In the upper reaches of the middle part, in Chimpu and Ziro, the ϕ value is found to be varying from 34° to 40° depending on the depth. The c value is found to be zero because of the predominant sandy strata. The specific gravity is found to be around 2.65, and the field density varies between 1.95 and 2.22 g/cc depending upon the depth of the soil.

In the north-eastern part, in Pashighat, the ϕ value varies between 36° and 38°. The specific gravity is around 2.65 and the field density is found to be varying between 2.0 and 2.2. In the south/south-eastern part, in Namsai, the ϕ value varies with a depth from 27° at 2.0 m up to 40° at 40.0 m. Due to the alluvial deposition in the region, a minimal clay portion is found in the topsoil. In Namsai, the specific gravity is found to be around 2.65 and the field density is found to be varying from 1.7 to 2.3 g/cc.

At Sela Pass, located in the remote western part of Arunachal Pradesh, which is at an elevation of around 4,145 m from the MSL, the ϕ value is found near 43°. The c value is found to be 0.40 kg/cm^2 in Sela Pass.

Subsurface investigation reports of some areas of Arunachal Pradesh in the form of boring logs are presented in Figures 4.6–4.10 for a clear understanding of the different underlying soil strata, which may be helpful for planning new infrastructure projects.

4.3.2 Rocks

As mentioned, the primary rock types found in Arunachal Pradesh are shales, sandstones, quartzites, phyllites, schist, gneiss, leucogranites, meta-volcanic and carbonates.

Shales in Arunachal Pradesh can be generalised as grey and black coloured laminated fine-grained sedimentary rocks composed of silt and clay-sized particles. Due to the very small particle size, the interstitial spaces are very small, and hence the porosity is low. Permeability is also observed to be very low. Due to the mineralogy and the grain size of the shales, fissility planes are developed within the shales making them very prone to slides. These shales are found to exhibit a hardness of 3 on the Mohs scale and have a specific gravity between 2.4 and 2.6; their compressive strength varies in the range of 30–35 MPa.

Varied types of sandstones are found in Arunachal Pradesh. Those occurring in the lower structural levels are observed to be less matured as compared to those found in the higher structural level. Age and diagenetic conditions are the principal controlling factors determining the maturity of the sandstones. The sandstones found in the state are Arenite to Litharenite with silica/carbonate or iron oxide as their cementing materials. The primary constituents of the sandstones can be generalised as quartz, feldspar, mica, rock fragments with some minor components of clay, carbonaceous matter and other opaques. These are generally medium to coarse-grained sandstones making them highly porous and intermediately permeable. The common structures seen in these sandstones are stratification, current beddings and ripple marks

Boring method: Drilling

Depth (m)	Types of Sample	Observed N-Value	Group Symbol	Visual description of soil	% gravel > 4.75 mm	%Sand 4.75-0.075 mm	Silt 0.075-0.002	% Clay <0.002 mm	Field density, g/cc	Specific Gravity	cohesion "C" Kg/cm²	Angle of Shearing resistance (f°)
0	P	R			Core recovery = 20.00% RQD=Nil							
0.5	P	R										
1	P	R							2.2	2.65		36
1.5	P	R										
2	P	R		Bouldery strata 6m	Core recovery = 23.33% RQD=Nil							
2.5	P	R										
3	P	R										
3.5	P	R			Core recovery=12% RQD=Nil							37
4	P	R							2.02	2.65		
4.5	P	R										
5	P	R			Core recovery = 4.67% RQD=Nil							
5.5	P	R										
6	P	R										
6.5	D	R	GP	Whitish gray gravel mix with some fine SAND 7.5 m	55	45			2.07	2.66		38
7	P	R										
7.5	P	55										
8	P					100						
8.5	P		SP									
9	P	60										
9.5	D					100						
10	D											
10.5	P	R										
11	D	R			70	30				6.67		
11.5	P	R	GP									
12	P	R										
12.5	D	R			60	40						
13	D	R										
13.5	D	R										
14	P	R			65	35						

U : Undisturbed sample:: D: Disturbed sample :: P : Standard Penetration test
DS : Direct shear test :: R: Refusal, N-value>100

Figure 4.6 Subsurface investigation report of the RCC Bridge site at chainage 11.360 km on the PLTK road at Pasighat, Arunachal Pradesh, 2015. (Courtesy of Public Works Department, Govt. of Arunachal Pradesh.)

Boring method: Auger & Wash boring   **GROUND WATER LEVEL**: Not encountered

| Depth (m) | Type of Sample | SPT | | | N- Value | Visual Description of Soil. | LOG | "C" Kg/cm² | Angle of Shearing resistance (f°) |
		15 CM	15 CM	15 CM					
0-1.45	D					Medium to fine sand 1.45M		0	34
1.5	P	R	R	R	R	Sandy gravely strata with boulder 4.5 M			
3	P	R	R	R	R				
3.5	P	P	P	P	P				
4.5	P	R	R	R	R				
6-6.45	P	7	11	19	30	Whitish gray coarse SAND 6.5 M			
6.5	D								
7.5	P	R	R	R	R	Boulder strata with coarse sand 8 M			
9-9.45	P	8	19	25	44	Brownish gray coarse SAND 15.5 M			
9.5	D								
10.5-10.95	P	10	20	42	62				
11	D								
12-12.45	P	12	31	53	84				
12.5	D								
13.5-13.95	P	P	R	R	R				
14	D								
15-15.45	D	P	R	R	R				

U : Undisturbed sample:: D: Disturbed sample :: P : Standard Penetration test
DS : Direct shear test :: R: Refusal, N-value>100

Figure 4.7 Subsurface investigation report of steel girder composite bridge site over river Dollung at Paro (Span: 3 × 40.00 m), Ziro, Arunachal Pradesh, 2015. (Courtesy of Public Works Department, Govt. of Arunachal Pradesh.)

developed due to fluvial and marine dynamics. Joints are observed to have traversed the entire sedimentary sequence which had developed mainly due to the Himalayan Orogeny. The hardness of sandstones of Arunachal Pradesh has been noted to be between 5 and 7, the specific gravity varies from 2.0 to 2.8 and the compressive strength is between 180 and 200 MPa.

Quartzites constitute a major portion of the rock types of Arunachal Pradesh and are found to occur in both the Higher and Lesser Himalayan Sequences. These rocks are hard, dense, siliceous metamorphic rocks essentially composed of quartz with small amounts of mica, tourmaline, graphite and iron minerals. Quartzites are observed to be compact with interlocking quartz grains and exhibit a granular texture.

Boring method: Shell & Auger Wash boring   **Boring dia**: 150 mm
BH-1   **Depth of water table** =Not encountered

Depth (m)	Types of Sample	Observed N-Value	Group Symbol	Visual description of soil	% gravel > 4.75 mm	%Sand 4.75-0.075 mm	Silt 0.075-0.002	% Clay <0.002 mm	Field density, g/cc	cohesion "C" kg/cm²	Angle of Shearing resistance (f°)
1.5-1.95	P	R									
2	D		SW	Whitish gray fine SAND 3.5 M		100			1.99	2.65	38-DS
3-3.45	P	R									
3.5	D					100					39-DS
4.5-4.95	P	R		Whitish brownish gray Boulder							
5	D										
6-6.45	P	R									
6.5	D										
7.5-7.95	P	R		Soft rock 15.5 M							
8	D										
9.0-9.45	P	R									
9.5	D										
10.5-10.95	P	R									
11	D										
12-12.45	P	R									
12.5	D										
13.5-13.95	P	R									
14	D										
15-15.45	P	R									
15.5	D										
16.5-16.95	P	R		Soft rock 30.5 M							
17	D										
18-18.45	P	R									
18.5	D										
19.5-19.95	P	R									
20	D										
21-21.45	P	R									
22	D										
22.5-22.95	P	R									
23	D										
24-24.45	P	R									
24.5	D										
25.5-25.95	P	R									
26	D										
27.0-2745	P	R									
27.5	D										
28.5-28.95	P	R									
29	D										
30-30.45	P	R									
30.5	D										

D: Disturbed Sample:: U : Undisturbed Sample :: P: Standard Penetration test
DS: Direct shear :: EGL: Existing ground level:: R: Refusal, N>100

Figure 4.8 Subsurface investigation report of Bridge site over river Panchin in between Jullang and Chimpu, Arunachal Pradesh, 2009. (Courtesy of Public Works Department, Govt. of Arunachal Pradesh.)

Due to high compaction amongst its constituent minerals, porosity and permeability of quartzite are observed to be low to very low. The effect of the Himalayan Orogeny has also acted upon the quartzites of Arunachal Pradesh as joints are one of the most common structural properties of these rocks; however, as observed, these rocks generally break with a rough fracture surface. These quartzites have a hardness of

Boring method: Wash boring

Depth (m)	Moisture content (%)	Specific gravity	Bulk density (g/cc)	Dry Density (g/cc)	Submerged Density (g/cc)	Void Ratio	Clay (%)	Silt (%)	Fine Sand (%)	Medium Sand (%)	Coarse sand (%)	Fine Gravel (%)	Porosity (%)	Saturated density (g/cc)	"C" kg/cm²	φ°
2	25.73	2.63	1.92	1.53	0.95	0.722	21	66	13				42	1.95	0.35	14
3.5	24.21	2.64	1.93	1.55	0.96	0.695	18	65	17				41	196	0.39	11
4.5	2263	2.64	1.98	1.61	1	0.635	9	38	26	12		15	39	2	0.12	29
6	20.57	2.65	2.02	1.67	1.04	0.582		39	33	11		17	37	2.04	0.09	30.5
9	18.76	2.65	2.04	1.72	10.7	0.543		37	27	15		21	35	2.07	0.06	31
12	16.51	2.65	2.05	1.76	1.1	0.512		36	31	11		22	34	2.1	0.08	32.5

Figure 4.9 Subsurface investigation report of proposed 100-seated Women Hostel of Rajiv Gandhi University, Doimukh, Arunachal Pradesh, 2008. (Courtesy of Public Works Department, Govt. of Arunachal Pradesh.)

Boring method: Wash boring   Boring dia: 150 mm Date Commenced: 28-01-2014
Depth of water table =Not encountered   Date completed : 28-01-2014

Depth (m)	Types of Sample	Observed N-Value	Group Symbol	Visual description of soil	% gravel >4.75 mm	%Sand 4.75-0.075 mm	Silt 0.075-0.002	% Clay <0.002 mm	Field density g/cc	Specific Gravity	cohesion "C" kg/cm²	Angle of Shearing resistance (F°)
1.5-1.95	P	5	SC	Brownish gray fine SAND some clay trace silt 3.0M		75	10	15	1.72	2.65		27-DS
2	D											
3-3.45	P	0	SW	Whitish gray fine to coarse SAND 7.5 M								
3.5	D					100						
4.5-4.95	P	14										
5	D					100			1.93	2.66		30-DS
6-6.45	P	67										
6.5	D					100						
7.5-7.95	P	R	SP	Whitish gray gravelly coarse SAND 15.5 M								
8	D				35	65			2.02	2.65		36-DS
9.0-9.45	P	R										
9.5	D				45	65						
10.5-10.95	P	R										
11	D				35	65			2.12	2.63		37-DS
12-12.45	P	R										
12.5	D				45	65						
13.5-13.95	P	R										
14	D				35	65			2.18	2.62		38-DS
15-15.45	P	R										
15.5	D				35	65						
16.5-16.95	P	R										
17	D		GP	Whitish gray sandy GRAVEL	65	35			2.23	2.62		39-DS
18-18.45	P	R										
18.5	D				65	35						
19.5-19.95	P	R										
20	D				60	40			2.31	2.66		40-DS

Figure 4.10 Subsurface investigation report of the proposed RCC Bridge at Namsai, Arunachal Pradesh, 2014. (Courtesy of Public Works Department, Govt. of Arunachal Pradesh.)

about 7, a specific gravity between 2.6 and 2.8, and a compressive strength varying from 250 to 295 MPa and thus act as an excellent constructing material. Phyllites are generally found in the Lesser Himalayan Sequences of Arunachal Pradesh. These are low-grade metamorphic rocks, fine-grained, foliated lustrous rocks. These rocks are essentially composed of quartz, chlorite, muscovite and splits along foliation planes with an uneven surface. The porosity and permeability of these rocks vary from very low to medium, the hardness varies between 1 and 2 on the Mohs scale, the specific gravity varies between 2.5 and 2.7, and the compressive strength varies between 2 and 10 MPa, thereby making these rocks not suitable to be used as constructing materials.

Schistose rocks are found both in the Higher and Lesser Himalayan Sequences of Arunachal Pradesh and are the product of regional metamorphism. In general, the schistose rocks of Arunachal Pradesh are medium to high grade of metamorphism. These rocks are essentially composed of quartz and mica, usually muscovite and biotite and some accessory minerals such as garnet, staurolite, kyanite, sillimanite, epidote and hornblende. Schistose rocks of the state of Arunachal Pradesh are observed to be coarse-grained rocks having a prominent schistose structure and they tend to split along the plane of schistosity. The porosity of these rocks varies from very low to low and so does the permeability, the hardness varies between 4 and 5 on the Mohs scale, the specific gravity ranges between 2.5 and 3.0 and the compressive strength ranges between 70 and 200 MPa. Joints are common in these rocks. Schistose rocks can be used as construction materials.

Gneissic rocks are generally found in the Higher Himalayan Sequences of Arunachal Pradesh. These are medium to high-grade metamorphic rocks, coarse-grained and irregularly banded. Quartz and feldspar form the light-coloured bands with alternate dark bands of flaky ferromagnesian minerals, such as biotite and hornblende. Due to high compaction amongst the constituent grains, these rocks exhibit very low to low porosity and permeability and possess hardness ranging between 6 and 7, specific gravity between 2.6 and 3.0 and compressive strength varying between 250 and 300 MPa. The Himalayan Orogeny has developed prominent joints in these rocks. However, these rocks can be used as construction materials.

Leucogranites are essentially found in the Higher Himalayan Sequences. Leucogranites of Arunachal Pradesh are observed to have intruded the high-grade metamorphic rocks of the Higher Himalayan Sequences and are medium to coarse-grained occurring both as massive bodies and dykes. These rocks are composed of microcline perthite, albite, oligoclase and quartz with interstitial muscovite showing inclusions of biotite, tourmaline, zircon and opaques as accessories. These granitic rocks are good construction materials and are observed to be very less porous and permeable.

Carbonate rocks are found to occur as mainly crystalline limestones, dolomites and marbles. These rocks are found in almost all the stratigraphic horizons except for the tertiary sequence of Arunachal Pradesh. The hardness of these rocks ranges from 3 to 4, the specific gravity ranges between 2.3 and 2.7 and the compressive strength ranges between 100 and 200 MPa. The occurrence of these carbonates in some places has been already classified as deposits having the potentiality to develop the Cement Industry. However, these rocks act as good construction materials.

4.4   USE OF SOILS AND ROCKS AS CONSTRUCTION MATERIALS

Using construction materials that are available locally can significantly reduce the cost and time taken for a project. Arunachal Pradesh being rich in natural resources has various locally available construction materials such as rocks, sand, timbers, and bamboo . However, construction materials such as bricks, cement and reinforcement bars are imported from neighbouring states. There are very few small-scale manufacturing industries that cannot cater to the needs of this developing state. Thus, many modern construction materials are being imported from other states.

Construction materials such as boulders and sand are abundantly available on the banks of the rivers and rivulets which cater for the need of local construction activities. The small size boulders are normally broken by manual means to produce aggregates of required sizes. It is observed that good quality sands of Zones II and III are available at some depth below the surface of sand deposits in riverbanks. Timbers, bamboos and hard tree leaf for use in the roof are being utilised as major house building materials for ages by the people of Arunachal Pradesh. Bamboo is available abundantly in Arunachal Pradesh. It is reported that the north-east region of India has 89 different species of bamboo, out of which 40% are found in Arunachal Pradesh. Bamboo being natural, biodegradable and environmentally friendly with a high tensile strength can be used as a reinforcement to improve the mechanical properties of soil.

The special style of houses known as "Chang-ghar", that is, house built on timber piles with its floor above the ground level, is a part and parcel of their culture and traditional values. Houses with cement plastered split bamboo walls have been a low-cost alternative in the area compared to brick or RCC houses due to the availability of timber and bamboos.

Due to the non-availability of bricks, abundantly available boulders are frequently used in plinth works of buildings. It is observed that boulder masonry walls and gabion walls provide a low-cost solution to slope failure problems in the region. Nowadays, it is observed that locally made cement concrete bricks are also used in construction in place of clay bricks as the former gives a low-cost alternative due to the availability of sand and aggregate in most of the areas.

Some site views showing the use of locally available construction materials are presented in Figures 4.11 and 4.12.

There is a large scope of infrastructure developmental activities or projects which may be undertaken by the government or private parties which in turn will develop the living standard of the people of Arunachal Pradesh. Being a mountainous state with numerous rivers and rivulets, Arunachal Pradesh is a potential hub for the development

Figure 4.11 Boulder masonry work as a slope protection measure.

Figure 4.12 Gabion wall as an erosion protection measure.

of hydropower projects with an estimated capacity of more than 50,000 MW out of which only 815 MW of hydropower has been harnessed till now. There are numerous challenges that are being faced by the ongoing construction projects or the projects in the pipeline in the state. Some of the challenges are discussed below in detail.

4.5 NATURAL HAZARDS

4.5.1 Earthquake

Arunachal Pradesh has experienced many earthquakes out of which there are about 20 major earthquakes with a magnitude greater than 7 (Richter) since 1897 (Kumar, 1997). The earthquakes of 1897 and 1950 are the two major earthquakes with a magnitude of more than 8 on the Richter scale, which has caused many damages to the entire region. The epicentre of the 1950 earthquake was near the N-E part of Arunachal Pradesh which signifies the vulnerability of the region. As per the Seismic Zoning Map of Indian Standard, IS 1893 (Part 1) (2002), the entire Arunachal Pradesh has been placed in Zone V, which is susceptible to major earthquakes.

4.5.2 Landslides

Landslide is a common phenomenon in Arunachal Pradesh. Every year, reports are found in various areas regarding road blockages, mudslides in the dwelling area and damages to irrigation structures and other public assets. The main causes of landslides in Arunachal Pradesh can be narrowed down to two factors, one being heavy rainfall and the other being anthropogenic activities. Arunachal Pradesh is one of the highest rainfall recipient states in the country. In over a period of 8–9 months, the annual rainfall is more than 3,500 mm with most of the rainfall occurring between May and September. The runoff water saturates the soil on steep slopes and the water infiltration causes a rapid rise in the groundwater levels. Due to the rise in the water level, the slope becomes unstable and causes a landslide.

Some of the anthropogenic activities that lead to landslides are unplanned settlement, improper protection of cut slopes during construction, deforestation and lack of a well-planned drainage system. Many landslides have been reported along with the hillslope cutting for construction of highways which leaves the hillslope highly vulnerable to slope failure during rainy days. The encroachments along the hill sides for settlement and Jhum field by cutting the trees, levelling of hill tops and blockage of a natural drainage system also lead to environmental degradation such as soil erosion of the catchment area.

The methods that are being used normally to contain landslides in susceptible areas are the provision of subsurface drainage, removal of unstable deposits from slopes and provision of retaining walls wherever feasible. Ground modifications like mechanical, hydraulic, physical and chemical modifications of the soil can be carried out with the help of compaction. Different ground improvement techniques may be used at slopes to minimise the risk of landslides in slopes.

Apart from common engineering practices, appropriate planning norms and standards along with planning principles related to sustainable urban development practices have to be adopted. Restriction of developments along hill faces, which includes identification and declaration of ecologically sensitive areas, should also be done. The felling of trees and earth cutting should be completely forbidden on the hill slopes and awareness should be given to promote the plantation on the slopes through social forestry schemes and conservation of soils.

Photographs showing landslides in Arunachal Pradesh are given in Figures 4.13 and 4.14.

4.5.3 Flood inundation and soil erosion

In Arunachal Pradesh, the chance of occurrence of a flood is minimal due to its hilly topography except at the foothill areas and the flood plains on the south-eastern side. But the runoff through the steep terrain causes flash floods and erosion in many areas

Figure 4.13 Collapse of baily bridge due to excessive debris.

Figure 4.14 Shoulder and edge failure of a highway due to landslide triggered by heavy rainfall.

during the monsoon period. Many times, the construction activities in the affected areas suffer very badly due to scouring of foundation bases or filling up of the area with flood deposits. The flash floods also erode the agricultural fields and the bases of dwelling units, thereby causing losses to properties.

4.5.4   Siltation

The rivers of Arunachal Pradesh normally have a steep gradient in the upper reaches and become stable in the foothills due to which the silt carried by the rivers tends to get deposited in the foothills making the mighty floodplains. The energy dissipated by the rivers in the foothill region causes braiding and spreading overland. This problem is normally being faced by Seijosa in East Kameng, Likhabali in West Siang, Pasighat in East Siang, Roing in Lower Dibang Valley, Tezu and Namsai in Lohit District, Diyun in Changlang district and so on, due to which the submergence of agricultural land, towns and other public assets occurs during every monsoon.

4.6   CASE STUDIES AND FIELD TESTS

Two case studies of geotechnical problems encountered in Arunachal Pradesh along with field tests undertaken for a detailed study and the remedial measures are discussed here.

4.6.1   Slope stability

The powerhouse of the 600 MW Kameng HE Project is located on the right bank of Kameng River in the West Kameng district of Arunachal Pradesh, which is developed by NEEPCO Ltd., a public sector unit under the Ministry of Power, Government of India. The powerhouse is a semi-underground type with a size of 110 m by 30.0 m, housing four units of Francis turbines of 150 MW capacity. The excavation work of the powerhouse was started from El. 288.00 m to reach a foundation level at El. 212.00 m providing slopes and berms of varying dimensions depending on the rock encountered. The slopes varied from 1H:0.5 V up to 1H:4 V and the intermediate berms were up to 10.0 m width. The slopes were protected using 5.0 m long rock bolts with 100 mm thick shotcrete. In November 2006, a major slope failure occurred in the northern side slope when the excavation reached an elevation of 227.00 m. The failure area was of 80.0 m width and started from El. 285.0 m up to the toe of the excavated slope at El. 227.00 m, that is, for a height of 50.0 m. Settlement of berms, bulging of the slope toe and multiple vertical cracks with cracks on the shotcreted face were observed in the affected area.

A detailed subsoil exploration was done in the area with four 40.0-m deep boreholes along with the measurement of further movement of cracks. From these studies, the failure reason was interpreted to be the presence of a prominent folded sequence of rock on the northern slope, which is not suitable for deep excavation. Furthermore, stronger sandstone was found above El. 260.0 m whereas below this layer, a relatively weak shale layer prevailed. The slope failure was a combination of creeping, toppling and sliding.

Remedial measures undertaken to stabilise the slope included flattening of the excavated slope of 1H):4 V above El. 248.0 m and provision of RCC piles of 300 mm diameter over two berms. A total of 210 piles of 25.0 m depth and 201 piles of 22.0 m depth with the provision of pile cap were installed at El. 248.00 m and El. 237.00 m berms in three rows in each berm at an effective spacing of 0.50 m, respectively. For monitoring of the slope movement after installation of piles, 19 inclinometers and nine pore pressure metres were installed in the area. The intermediate excavated slopes were protected by the construction of a thin PCC cladding wall (Sarma, 2014).

After completion of the piling works in August 2009, excavation work resumed below El. 227.00 m and reached up to El. 223.00 m in December 2009 when inclinometer data started showing active displacement inside the slope. Further cracks were also observed in the PCC done over the berms. As a remedial measure, the cracks were filled with a cement slurry to stop the ingress of water inside the cracks and RCC blocks as toe protection measures were constructed at El. 222.00 m. These blocks were a part of the structural concrete of the powerhouse raft and subsequently, it was integrated with the main base raft. These measures worked and the excavation work could be completed for the entire powerhouse. Some units of this project with 300 MW capacity have been commissioned.

4.6.2 Squeezing of rock and cavity formation

The water conductor system of the Kameng Hydro Electric Project contains a 6.70 m diameter and 14.45 km long head race tunnel of modified horse shoe shape from its main dam at Bichom up to Surge shaft. In the maximum portion of the tunnel excavation, the rock type encountered was Class-IVA and IVB which is the weakest comprising shear zones constituted by fractured and crushed rock with clay gauge having low compressive strength. Depending on the encountered rock mass, either full face excavation or a heading and benching method was applied using conventional drilling and blasting methodology. Normally 1.5–2.0 m pulls were undertaken and the stretch was protected with the erection of the D-shaped steel rib made by ISMB 200 × 100 at a spacing of 0.50–1.00 m, fixing of laggings and backfilling behind the rib with M20 grade of concrete.

In several stretches, the squeezing phenomenon was observed over a period of 1 month after excavation and fixing of the rock support system. The squeezing effect of the rock mass can be visualised by evidence of the development of cracks in laggings and buckling of steel ribs, which reduces the tunnel section. The squeezing effect was mainly observed in stretches where the rock cover is above 800 m combined with the presence of very week rock mass. Furthermore, it was also observed from MPBX data that the squeezing phenomena showed a gradual downturn with time and stops after 6–7 months. Laboratory tests of different samples reveal that the compressive strength of rock mass in these zones is found to be below 7.00 MPa. It is also found that the presence of clay minerals such as kaolinite and montmorillonite in rock mass in these vulnerable zones contributes to squeezing (Alam and Sharma, 2014).

As a remedial measure, the tunnel section was reviewed and the tunnel diameter in the effected region was increased by 300 mm. Due to this, the tunnel sections after squeezing becomes stable at the required sectional diameter. Presently this tunnel is successfully conveying water to run the 600 MW Kameng powerhouse.

4.7   CONCLUDING REMARKS

The soil of Arunachal Pradesh along with the underlying rock strata significantly varies from place to place. The following conclusions can be made on the soils and rocks of Arunachal Pradesh:

1. Arunachal Pradesh is a geologically complex region constituting mega folds, faults, thrust and shear zones combined with abruptly changing orientation of bedding planes.
2. The overall soil of Arunachal Pradesh is found to be of sandy nature mixed with gravels and pebbles except for some pockets and the flood plain region on the south-eastern side.
3. The primary rock types found in Arunachal Pradesh are shales, sandstones, quartzites, phyllites, schist, gneiss, leucogranites, meta-volcanic and carbonates.
4. Natural construction materials, such as sand, boulders, wood, bamboo and so on, are abundantly available on the riverbanks; however, cement, steel and so on are to be brought from other states.
5. Landslides and soil erosion are the main problems in the hilly stretches whereas siltation occurs in the flood plains. Earthquake is also a major factor as the whole area is in the highly active Himalayan region.
6. The case studies reveal that geological surprises were encountered while undergoing construction work due to which the projects got delayed causing significant time and cost overrun of a project other than the contractual problems. As such, subsurface investigations should be carried out thoroughly to avoid geological surprises during construction.

ACKNOWLEDGEMENTS

The authors would like to thank Dr Atop Lego, Chief Engineer, S&I, Design and Planning (PWD), Arunachal Pradesh, for his kind cooperation and for providing the subsurface investigation reports of various places of Arunachal Pradesh. The authors also acknowledge the kind help and support of Dr Bashab Nandan Mahanta, Suptdg. Geologist, Geological Survey of India, NER, Shillong for his guidance in the geological incorporation in the chapter. The authors would also like to convey special thanks to Mr Hari Shankar Nath, Sr. Manager (C), NEEPCO Ltd., for his constant help and support. Also, special thanks go to Ms Nisha Kumari, Postgraduate student, NERIST, for her kind cooperation.

REFERENCES

Alam, H. and Sharma, M.K. (2014). Criticalities encountered in construction of head race tunnel (HRT) of Kameng Hydro Electric Project (600 MW), Arunachal Pradesh, India. *International Journal of Scientific & Engineering Research*, Vol. 5, No. 9, pp. 217–231.

APSG (2020). Arunachal Pradesh State Government, http://www.arunachalpradesh.gov.in, accessed on June 2020.

GSI (2013). *North Eastern Region*, Briefing Book, Geological Survey of India, New Delhi.

IS 1893 (Part 1) (2002). Criteria for earthquake resistant design of structures, Bureau of Indian Standards, New Delhi, India.

https://www.mapsofindia.com/maps/arunachalpradesh, Map of Arunachal Pradesh, accessed 15 June 2020.

ISFR (2009). Forest survey of India, Indian State Forest Report, Arunachal Pradesh, pp. 58–61.

Kesari, G.K. (2010). Geology and mineral resource of Arunachal Pradesh. *Geological Survey of India, Miscellaneous Publications*, No. 30, Part IV, Vol 1(i), Arunachal Pradesh.

Kumar, G. (1997). *Geology of Arunachal Pradesh*. Geological Society of India Publication, Bangalore.

Sarma, R. (2014). Powerhouse excavation slope failure and remedial works in Kameng hydroelectric project (600 MW), Arunachal Pradesh, *International Society for Rock Mechanics*, Vol. 3, No. 1, pp. 11–15.

Chapter 5

Assam

Utpal Kumar Das
Tezpur University

Diganta Goswami
Assam Engineering College

Ramani Mohan Das
Public Works Department

CONTENTS

5.1   INTRODUCTION

The state of Assam lies in the north-eastern (NE) part of India. It is the gateway to all other six states in NE India. The state is of immense importance in view of the act east policy of India. The mighty river Brahmaputra enters the Indian plains in Assam and flows through a length of over 700 km covering the entire length of the state constituting the Brahmaputra valley as shown in Figure 5.1. The Barak river flows through the southern districts of Assam constituting the Barak valley.

A major part of the state is covered by the recent alluvium of the Brahmaputra and Barak river plains. The hill districts, which are in the central Assam, comprise gneisses and schists with granite intrusions as reported by the Geological Survey of India. The Brahmaputra and most of its 13 major tributaries in its north bank have originated mostly from the Himalayan range. A sudden flattening of a gradient of these rivers as they enter the flood plains in Assam results in a large amount of sediment deposition and, consequently, the formation of a braiding pattern of the river takes place. The rivers

DOI: 10.1201/9781003177159-5

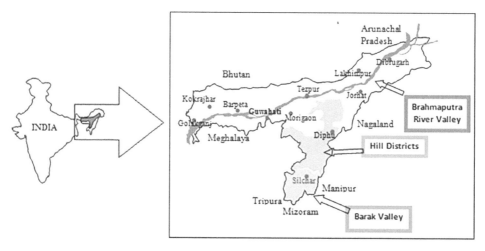

Figure 5.1 Physical map of Assam.

Figure 5.2 Infrastructure development in plains and hilly areas of Assam.

carrying rock and sand deposits are a good source of construction materials. The river system of Assam offers huge industrial potential in terms of hydroelectric power, irrigation, water supply and navigation, which is still much underutilised. The hill slopes, composed mainly of weathered residual soil, often pose stability problems due to cutting to make way for roads and railway lines. Figure 5.2 shows a view of infrastructure building in the plains and hills of Assam.

The climate of Assam is oppressive humid, tropical type in the plains and pleasant sub-alpine type in the hills. The relative humidity rarely goes below 75%. The average temperature ranges between 6°C and 38°C. The state is one of the highest rainfall receiving areas of India and receives average yearly rainfall of more than 2,000 mm. Although rainfall occurs almost throughout all the seasons of the year, the south-west monsoon mostly sets in from June to September when geotechnical construction activities become very difficult to execute. The state is in one of the most seismically active zones of India and the world as well. The state experienced two major earthquakes of magnitude more than 8 on the Richter scale in the years 1897 and 1950 (Singnar and Sil, 2017). Assam is famous for tea cultivation and one-horned rhinoceros, and is rich in natural resources. The state has five national reserved parks and 18 wildlife sanctuaries with a wide biodiversity. In addition to crude oil, the state has reserves of coal, limestone, iron ore and so on. Assam is politically divided into 33 districts with a total population of 32.17 million (as per the 2011 census). Although all the district headquarters have developed into towns with good infrastructures, Dibrugarh in upper Assam has developed as the centre of tea and oil industries, Silchar as the centre of Barak valley and Guwahati as the capital city of Assam. Guwahati city is located on the bank of the river Brahmaputra covering an area of approximately $600 \, km^2$. As per the 2011 census, the city holds a population of approximately 1 million residents.

5.2 MAJOR TYPES OF SOILS AND ROCKS

The soil deposits in the plains of Assam are mostly of alluvial type. The subsoil, as seen from boring logs, mostly consists of layers of coarse-grained soils as well as fine-grained soils. The boring logs up to a depth of 15–30 m were prepared by soil testing by different agencies. The thickness of the layers is found to vary considerably from one borehole site to another. The coarse soil layers are found to be mixed with varying fines content and are broadly classified as silty sand or clayey sand. The fines content in the sandy soil layers is often observed to decrease with the increasing depth of the borehole. The clayey soil layers are reported to have low to high plasticity. These fine-grained layers are also found to be silty clay. The silty soil layers are generally found to be inorganic silts or clayey silts with none to low plasticity. Although both coarse and fine layers are mostly observed to exist in the subsoil, boreholes with only coarse or fine soil throughout the full depth are also found. The SPT N values of up to 10 m borehole depth indicate that soil deposits up to 10 m depth are generally loose to medium dense.

In particular reference to the capital city Guwahati, the geology of the city comprises granitic rocks as outcrops in different locations. Alluvial deposits over the undulated and faulted basement of granitic rocks constitute the plain areas of the city (Raghu Kanth and Dash, 2010). The Geological Survey of India in the Seismotronic Map of India (Dasgupta et al., 2000) has noted four important faults running across Guwahati: (i) a NE-SW trending fault, about 15 km long; (ii) an N10°E–S10°W trending fault, about 10 km long; (iii) an N40°E–S40°W trending fault, about 20 km long; and (iv) an E-W trending fault.

The hill soils of Assam are generally made up of residual soil at different levels of decomposition. The hills surrounding Guwahati city are mostly composed of porphyritic granites and quartzo feldspathic gneiss transversely cut by amphibolite intrusive and quartz veins (Singar and Sil, 2017). Layers of silty clay and fragile silty sand are typically found in the hill soils. The stability problems of hill slopes are, in most cases, anthropogenic. In central Assam, Karbi Anglong and North Cachar (Dima Hasao) hills, clayey soil, brownish to yellowish in colour, is found mixed with sand and different sizes of pebbles. Rockfalls, particularly during rainy seasons, are evidenced in the hill areas of the state. Rock outcrop can be seen at locations like Kalapahad, Japorigog, Jalukbari and Fatasil of Guwahati.

Riverbank erosion is a major problem in Assam. Apart from the flow characteristics of the river, the bank soil is also responsible for its susceptibility to erosion. Sarma and Acharjee (2012) in a detailed study on bank erosion of Brahmaputra river identified the presence of a layer of loose medium to coarse sand of about 2 m thickness as a cause of bank erosion in Rohmoria, one of the most severely affected erosion sites of Brahmaputra, near Dibrugarh in upper Assam. This layer is overlain by a thick layer of cohesive soil. Undercutting of this loose sandy layer by flood water leaves the upper clayey layer overhanging, which eventually falls into the river due to its self-weight. This is a typical pattern of bank erosion in most of the rivers in the state.

The level of groundwater plays a crucial role in the response of foundation soil to static and dynamic loading. A study conducted on about 100 boreholes scattered throughout Guwahati city has found the existence of a shallow water table at a depth below 2.0 m in the central area mostly along the Guwahati–Shillong road and the south-western area surrounding Maligaon, covering more than half of the city area. The northern and southern areas of the city have water tables at a greater depth, in the range of 2–6 m (Singar and Sil, 2017). Assam, being located in a seismic zone, has a shallow ground water level which is duly taken into account in the discussion of liquefaction potential later in this chapter.

5.3 PROPERTIES OF SOILS AND ROCKS

The subsoil of Assam typically has alternate layers of fine- and coarse-grained soil. A study of borehole data obtained from different locations across the state of Assam gives an idea of this subsoil stratification up to the borehole exploration depth. The data were collected from geotechnical testing reports of various geotechnical investigation firms. The wash boring method was applied in these explorations carried out mostly up to a depth of 15 m. The boring logs along with field N values for test sites in six different towns across the state are reported in Tables 5.1a and b.

In the city of Guwahati, a great deal of variability in subsoil stratification is observed according to previous studies of SPT boreholes. The detailed soil profile and N values obtained from boreholes made in different locations of Guwahati are shown in Tables 5.2a and 5.2b. The N values reported in Tables 5.2a and b are uncorrected field values.

The borehole data show a low N value for the top 3 m of the soil in most of the test locations, whereas the topsoil displays higher resistance in few other locations. The silty/sandy clay layers and silt/sand layers mixed with clay have generally shown low N values in the top 3 m of the subsoil. This indicates loose deposits in the top layer of subsoil.

Laboratory test results of the top layers of some of the above borehole soils showed low shear strength with cohesion (c) values of less than 50 kPa and angle of internal friction ϕ values less than 20°. Average values of bulk density of 1.70 g/cc, specific gravity of 2.65 and field water content of about 25% were reported. The bearing capacity of the topsoil is found to be generally low. But for the deeper sandy strata, the internal friction angle is found to be as high as 42°. Deep foundations like cast-in-situ RCC piles are frequently recommended for high-rise buildings in the state instead of isolated footings. Foundations are discussed in detail later in this chapter. It is noted that the values mentioned here are site-specific only. Any geotechnical design is to be based on the geotechnical exploration of the site of construction.

Table 5.1a Soil profile and SPT N value obtained from boreholes in towns belonging to lower Assam

Location →	Golakganj		Barpeta		Morigaon	
Depth (m)	Soil description	Field N value	Soil description	Field N value	Soil description	Field N value
0.0–1.50	Silty clay (CL)	3	Sandy clay	4	Silt with traces	6
1.50–3.00		5	with traces	5	of clay (MI)	7
3.00–4.50	Fine sand with clay (SC)	9	of silt (SC)/Silty clay (CI)	7		8
4.50–6.00		12	Fine sand	12		9
6.00–7.50		14	with silt	20		10
7.50–9.00		23	traces (SP)	28		11
9.00–10.50	Poorly graded	32		31	Fine sand with	14
10.50–12.00	fine sand (SP)	40		36	silt (SM)	16
12.00–13.50		53	Dense med.	40		18
13.50–15.00		58	to fine sand	45		20
					Fine to medium sand (SW) at a depth of 20 m	30
					Poorly graded fine sand (SP) at a depth of 30 m	53

Table 5.1b Soil profile and SPT N value obtained from boreholes in towns belonging to central and upper Assam

Location =>	Tezpur		Lakhimpur		Dibrugarh	
Depth (m)	Soil description	Field N value	Soil description	Field N value	Soil description	Field N value
0.0–1.50	Silty sand (SM)	4	Clayey	4	Clay with int.	3
1.50–3.00		5	sand with	10	plasticity (CI)	5
3.00–4.50	Poorly graded sand (SP)	12	traces of silt (SC)	15	Silty sand (SM)	20
4.50–6.00		13	Silty sand	25		24
6.00–7.50		22	with traces of clay (SM)	26		26
7.50–9.00		21	Fine to	30		32
9.00–10.50		29	medium	34		-
10.50–12.00		30	sand with	38		-
12.00–13.50	Well graded	45	little silt	41		-
13.50–15.00	sand (SW)	50	(SP)	46		-

Table 5.2a Soil profile and SPT *N* value obtained from boreholes in different locations of Guwahati

Location →	Central Guwahati Sarabhati Chariali		West Guwahati Borjhar, near airport		East Guwahati Beltola (wireless locality)	
Depth (m)	Soil description	Field N value	Soil description	Field N value	Soil description	Field N value
0.0–1.50	Silty sand	3	Silty clay	2	Sandy clay	4
1.50–3.00	GWT 0.7 m	3	(water	3	GWT 1.0 m	5
3.00–4.50	Clayey silt with	3	table at	5	Silty clay	9
4.50–6.00	traces of	11	0.75m)	7		13
6.00–7.50	sand	13		7		12
7.50–9.00		10		17	Sandy/silty clay	25
9.00–10.50	Sand with	14	Silty sand	28	Sandy clay	23
10.50–12.00	traces of silt	21		37		18
12.00–13.50		21	Fine sand/silt	42		20
13.50–15.00		24		46		22

Table 5.2b Soil profile and SPT *N* value obtained from boreholes in different locations of Guwahati

Location →	Kahikuchi, Guwahati Borehole # 1		Kahikuchi, Guwahati Borehole # 2		Jalukbari, Guwahati	
Depth (m)	Soil description	Field N value	Soil description	Field N value	Soil description	Field N value
0.0–1.50	Reddish clay	13	Clay with	16	Silty clay	19
1.50–3.00		11	intermediate	17		13
3.00–4.50	Clay with sand	4	plasticity	13	Sandy clay	26
4.50–6.00		15		16		20
6.00–7.50	Fine sand	8	Fine sand	28	Silty clay	22
7.50–9.00		9	mixed with	37		33
9.00–10.50		12	clay	37		31
10.50–12.00		14		37		22
12.00–13.50		17		37	Clay with high	24
13.50–15.00		20		37	plasticity	23

In an attempt to give an idea of the spatial variation of the soil profile for a given site in Guwahati, the data obtained from 25 boreholes along a 500 m stretch of road are presented in Figure 5.3. The boreholes were made up to a depth of 25 m. The soil profile data are simplified into three major divisions according to density/stiffness.

A generalised soil profile drawn from lithology (Borah et al., 2016) of the Borjahar area of South Guwahati shows silty clay of 0–10 m depth from the ground surface, fine sand of 10–30 m depth, medium to coarse sand of 30–80 m depth, clayey sand of 80–100 m depth, clay mixed with gravel of 100–130 m depth, gravel mixed with clay of 130–150 m depth, fine to medium sand of 150–160 m depth, medium to coarse sand of 160–170 m depth, coarse to very coarse sand of 170–190 m depth, clay mixed with gravel of 190–195 m depth, gravel of 195–200 m and bedrock of 200 m depth.

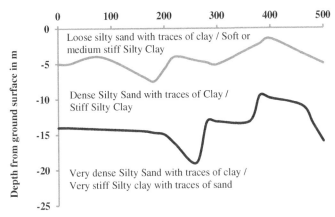

Figure 5.3 Variation of the general soil profile within 500 m distance in Gansehguri, Guwa-hati obtained from data of 25 boreholes.

Erosion of soil from hills in and around Guwahati city due to hill cutting and high rainfall perennially blocks the drainage lines and results in artificial floods during the rainy season. The hills of Guwahati are made up of two types of residual soils, reddish silty clay (in-situ dry density 1.49 g/cc, liquid limit 49%, plastic limit 27%) overlaying yellowish silty sand (in-situ dry density 1.63 g/cc, liquid limit 39%, non-plastic). The particle size distribution of the two soil types is presented in Figure 5.4.

There are several low-lying localities in Guwahati and they need to be filled with soil for building and other construction activities. Until hill cutting was banned in the city in the 1990s, land filling was mostly done using hill cut soil. The information of the hill

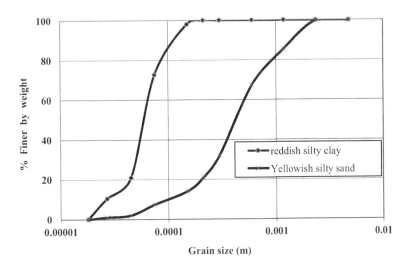

Figure 5.4 Grain size distribution curves of reddish silty clay and yellowish silty sand.

Table 5.3 Geotechnical properties of the Jia Baharali riverbank

Depth (m)	Bank soil classification	Dry density (g/cc)	Liquid limit	Plastic limit	Cohesion (kPa)	Friction angle (°)
0.3	Silty sand	1.33	22.8	19.8	01	13.6
0.6		1.26	36.1	30.6	17	8.08
0.8		1.45	29.5	23.3	22	3.02
1.1	Clayey sand	1.45	32.9	23.3	21	0.8
1.6		1.50	31.0	24.1	20	0.52

soil of Guwahati gives an idea about the expected top soil properties in such already earth-filled sites where construction/reconstruction is planned.

The state of Assam is crisscrossed by a network of rivers. A generalised profile of Brahmaputra and Barak river soil inferred from 50 m deep boring log data in locations near Guwahati and Silchar shows a 10 m thick top layer of silty clay with a fine sand/loose fine sand (average SPT field N value of less than 10) followed by 15 m thick medium dense fine sand with silt/clay (N value in the range of 10–25, increasing with depth), 10–15 m thick dense fine sand with silt (N value 25–36, increasing with depth) and very dense fine to medium sand with silt (N value 36–50, increasing with depth up to 50 m).

The geotechnical properties are found to vary significantly with depth in the study on bank soil of the Jiyabharali river, one of the major north bank tributaries of Brahmaputra (Das, 2016). The bank soil profile is shown in Table 5.3.

5.4   USE OF SOILS AND ROCKS AS CONSTRUCTION MATERIALS

Several tributaries of the river Brahmaputra flowing into its north bank have descended from the Himalayan range and carry a considerable amount of aggregates. These river-borne aggregates are used in construction activities in the state. The engineering properties of these river-borne natural coarse aggregates are studied and provided in Table 5.4 (Pegu and Gogoi, 2015).

The physical properties of the river-borne coarse aggregates are found to meet material specifications of granular subbase, water-bound macadam, bituminous macadam and bituminous concrete for road construction works in India (Specifications for road and bridge works, 5th. Rev. 2013, Ministry of Road Transport and Highways, MORTH, India). So, the river gravels of Assam, particularly those on the north bank of Brahmaputra, are suitable for construction works. Laboratory determination of the physical properties of commercially available crushed granite coarse aggregates obtained from a central Assam quarry has shown a flakiness index of 12.3%, a crushing value of 17.8%, an abrasion value of 28.5%, an impact value of 19.4% and a specific gravity of 2.64.

Table 5.4 Physical properties of Champabati river aggregate of Assam

Flakiness index	12.19	Elongation index	20.75	Specific gravity	2.67
Impact value	15.61	Crushing value	19.78	Abrasion value	24.34
Soundness value	1.2	Water absorption	0.6%		

Soil is used as the primary construction material in different construction projects in Assam. Earthen embankments are the most common type of structures constructed for flood protection in the state. Application of geotextile bags, geotextile tubes and so on is a recent addition to the construction practice adopted for these embankments. Soil from borrow pits in the vicinity of the construction site is preferably used due to its economy. Borrow pit soil is also used as a material for the construction of guide bunds of bridge approaches of national highways over major rivers of the state. With the rising popularity of construction of reinforced earth walls in approaches of national highway overbridges, the demand for soil as a construction material has also increased. In Table 5.5, some properties of borrow pit soils collected from different locations of Assam as determined by laboratory tests are presented. These properties in general indicate the availability of soil for construction across the state of Assam. The borrow pit soils are found suitable for use as retained fill in reinforced earth walls, the recommended properties for which are $\phi > 25°$ and plasticity index < 20 as per the Indian Roads Congress recommendation (IRC-SP-102-2014: Guidelines for design and construction of reinforced soil walls). IRC recommends that soils with %fines $< 15%$, plasticity index ≤ 6 and $C_u > 2$ are suitable for reinforced soil fill. These borrow soils are also found to be suitable for the construction of homogeneous earth embankments and guide bunds (IS 8826-1978) preferably with protection on a water-retaining face due to low erosion resistance. The borrow soil filled in geotextile bags can be effectively used for the water face erosion protection.

In Assam, river sand is the major source of fine aggregate used in cement concrete. Laboratory tests on riverbed sand obtained from three rivers, one each from lower, central and upper Assam, have revealed that all these samples are classified as poorly graded sand (SP). The river sand obtained from lower Assam was found to be coarser (Zone-II classification of IS 383-2016) than the other two river sands (Zone IV classification of IS 383-2016) and, therefore, they are more suitable for heavy construction.

Table 5.5 Physical properties of fine aggregate collected from different borrow pit sites of Assam

		Borrow Site-I, Assam	Borrow Site-II, Assam	Borrow Site-III, Assam	Borrow Site-IV, Assam	Borrow Site-V, Assam	Borrow Site-VI, Assam
Grain size	% Fines	4.7	11	14.7	55	83	57
	C_c	1.30	0.81	0.81	-	-	6.00
	C_u	2.80	2.67	3.00	-	-	11.40
	D_{30} (mm)	0.19	0.11	0.13	-	-	0.06
USCS classification		SP	SP-SM	SM	CL-ML	CI	ML
Plasticity index (%)		NP	NP	NP	6.8	17	4.5
Shear properties	c (kPa)	18	19	24	25	50	-
	ϕ	39.7°	39.2°	37.2°	31°	23.3°	-
Compaction properties	OMC%	-	-	-	11.4	20.2	11.7
	MDD (g/cc)	-	-	-	1.91	1.60	1.88
Coeff. of permeability		Between 1.0×10^{-4} cm/s and 1.0×10^{-5} cm/s					

Note: C_c is the coefficient of curvature; C_u is the coefficient of uniformity; NP is non-plastic; c is the cohesion value; ϕ is the angle of internal friction; OMC refers to the optimum moisture content and MDD is the maximum dry density.

5.5   FOUNDATIONS AND OTHER GEOTECHNICAL STRUCTURES

Assam has many fast-growing cities, where various infrastructures are rapidly coming up. The population has increased manifold in the last decade. The population of Guwahati city has increased from nearly 1 million (as per the 2011 census) to about 2.4 million in 2020. With the increase in the population, there has been a huge demand for the development of infrastructure and so the foundation engineers have been facing new challenges to provide a safe foundation for various structures.

The soils of Assam can be categorised under three distinctive sub-headings: alluvial soil, lateritic soil and hill soil. Wherever a structure is intended to be built, detailed information about the subsoil must be collected at least up to a significant depth to which the subsoil is likely to be affected by the load from the superstructure. IS: 1904-1986 should be strictly adhered to while the geotechnical investigation is carried out. The shallow foundation safe bearing capacity of the supporting soil may be calculated for various depths of foundation, as well as, for various footing plan dimensions following provisions of IS: 6403-1981. It is worth mentioning here that the entire state of Assam lies in seismic zone V (IS: 1893 – pt. 1–2016) and therefore a frame type structure consisting of beam and column is preferred to load-bearing structures. The vertical and horizontal loads on each column should be calculated by accurately modelling the structure and its elements. While designing a foundation for a given structure, the first endeavour should always be to provide a shallow foundation, as it is generally the most economical one, if not found feasible because of the following reasons:

- The topsoil is either very loose having a low safe bearing capacity or the column load is quite large and/or is likely to give rise to settlement (total and/or differential) beyond the permissible limit of settlement for the structure.
- To achieve the desired bearing capacity, if the depth of the foundation comes out to be large, heavy shoring to support the sidewalls of the foundation pit needs to be provided, which in turn makes the construction time consuming and expensive or practically impossible.
- The ground water table is near to the surface causing continuous seepage into the foundation pit and dewatering is practically impossible.
- The topsoil tends to have a high degree of swelling or shrinking when there is a rise or drop in its moisture content. Black cotton soil is such a soil that is very sensitive to change in moisture, with respect to shrink–swell characteristics and is encountered in certain isolated pockets of upper Assam.
- Organic soil is found to occur at a shallow depth in some locations which may give rise to a huge amount of settlement and may cause significant structural damage if footings are placed on it.

Although the subsoil may vary to a great extent within even a small distance for low-rise buildings (up to three storeys), or sometimes even for mid-rise buildings (three to five storeys), shallow isolated or combined RCC footings are generally found to be suitable when placed on hill soil and on lateritic soil (occurring in the foothills) of the state of Assam. In Assam, lateritic soil occurs almost all over the North Cachar Hills district, some parts of southern Karbi Anglong, the southern part of the Golaghat district, the northern part of the Barak valley and at the foothills of the many hillocks of Guwahati.

However, in the hilly areas and in the foothills, due to rainwater or due to anthropogenic activities, eroded soil coming down from the hill gets deposited in some depressed areas. In such locations, consolidation of the soil takes a very long time and footings should not be placed on these soils, because of very low strength and the tendency for high settlement. Unless appropriate measures are taken for ground improvement (by the installation of sand drains for accelerated consolidation, deep compaction, use of suitable admixture, chemical grouting, vibratory rolling and so on), these soils do not perform well under superstructure load. Figure 5.5 shows a typical boring log of Dokmoka, Karbi Anglong, where isolated or combined footings for a mid-rise building are expected to be feasible if placed at a depth of around 2.7 m from the existing ground level.

Figure 5.6 shows isolated footing by open excavation in the hill soil of Kharghuli at Guwahati for a proposed mid-rise RCC building.

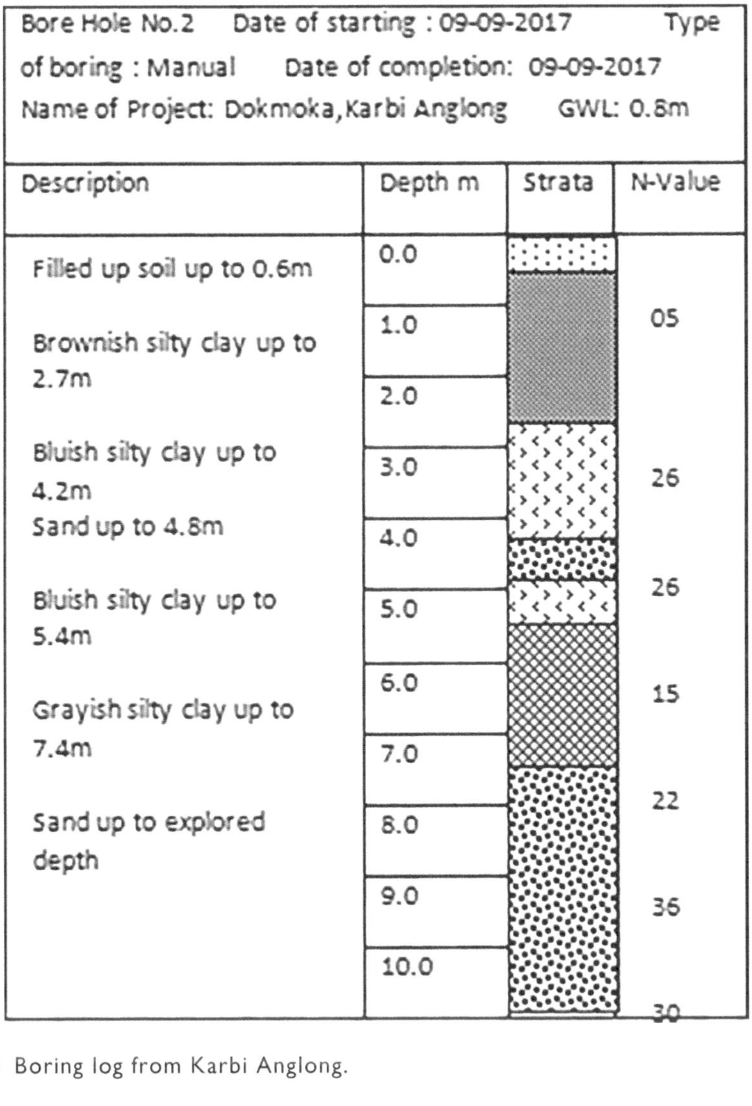

Bore Hole No.2 Date of starting : 09-09-2017 Type of boring : Manual Date of completion: 09-09-2017 Name of Project: Dokmoka, Karbi Anglong GWL: 0.8m			
Description	Depth m	Strata	N-Value
Filled up soil up to 0.6m	0.0		
Brownish silty clay up to 2.7m	1.0		05
	2.0		
Bluish silty clay up to 4.2m	3.0		26
Sand up to 4.8m	4.0		
Bluish silty clay up to 5.4m	5.0		26
Grayish silty clay up to 7.4m	6.0		15
	7.0		
Sand up to explored depth	8.0		22
	9.0		36
	10.0		
			30

Figure 5.5 Boring log from Karbi Anglong.

Figure 5.6 Open excavation for isolated footing in the hill soil of Guwahati.

Guwahati, which is said to be the gateway to NE India, has the peculiarity of a highly variable, treacherous soil type. At places, the thickness of the top quaternary sedimentary deposit of newer and the older alluvium is of the order of 100 m or more, and at many other places, pre-Cambrian rock outcrops also exist. Once rock outcrop is seen, or it is established that rock bed occurs at a shallow depth, the shallow isolated or combined footing is generally found to be suitable even for high-rise structures. If the quality of the rock mass is found to be good (which can be determined from parameters like low volumetric joint, high uniaxial compressive strength, favourable joint dip and strikes, less joint thickness, no movement of the joints and so on) and it is planned to place the foundation on this rock, sometimes even the spread of the footing may not be advisable. Under such conditions, column reinforcing bars may be anchored directly to the bed rock by drilling up to a suitable length for deriving the desired bond and inserting the rebars into it and grouting the annular space between the hole and rebar suitably. Assam being situated in an active seismic zone, all the structures may be subjected to high horizontal seismic load. When footings are placed directly on the rock bed at shallow depth or anchoring of column rebars is carried out into the bed rock, it is to be ascertained that the foundation mechanism is good enough to resist possible horizontal shear force at the superstructure–bed rock interface, during the highest possible earthquake event.

The paucity of land in city areas is compelling people to expand the structures vertically. For tall structures where the foundation load is high, and/or the columns are spaced closely, isolated footings may overlap with one another and raft foundation is found to be a better option. It also reduces the differential settlement. Many structures have basement floors or semi-basement floors, and raft foundation is quite advantageous in such cases, in the sense that the bottom floor needs to be concreted anyways and in many cases, the demand for a special waterproofing treatment at the basement/semi-basement floor arises. By provision of a raft foundation, both these factors are automatically taken care of. However, raft foundations are found to be difficult to be constructed in many locations due to the shallow ground water table.

Another type of foundation for heavy structures, which is very common in Assam, is pile foundation. Where shallow footing is not possible, the next option is pile foundation. Manually auger-bored or wash-bored cast-in-situ concrete piles are more common for up to mid-rise buildings and up to a diameter of 500 mm. For high-rise structures, road and railway bridge pile foundations and for piles for flyovers and so on, where large diameter and long piles are used, mechanically bored and cast-in-situ piles are found suitable. Wherever there is a possibility of side soil collapse, which is more common in cohesionless soil below the ground water table, direct mud-circulation technique with a bentonite slurry of appropriate viscosity must be used. In Assam, driven and precast piles are still not very common. Although precast piles are better compared to bored and cast-in-situ piles, in terms of both quality and installation time, driving the precast piles creates a lot of vibration and noise in the surroundings. Therefore, in thickly populated areas with heavy structures nearby, installation of driven precast piles is somewhat not always preferable. For important projects, both initial and routine tests are mandatory to assess the load capacity of piles as per IS: 2911 (Part 4)-2013.

In Assam, the structures are to be designed to withstand a very high lateral load arising out of earthquake shaking. The magnitude of this lateral load is the highest at the foundation level and piles are to be designed so that they can carry this load, especially at the level of pile cap and pile interface. The lateral load capacity of piles can be derived by knowing the horizontal modulus of the subgrade reaction of the soil. Once executed, the lateral pile load test needs to be carried out as per IS: 2911 (Part 4)-2013. Appropriate care also needs to be taken in detailing the reinforcement at the pile–pile cap interface. One other type of pile, which is also used in medium-stiff alluvium, is an under-reamed pile. The single or double bulbs created by the under-reaming tool enhance the end bearing capacity and the pullout resistance of the pile. However, one needs to ascertain the condition of the bulb/bulbs near the pile tip, which can be done by concrete volume measurement.

A combined piled raft foundation (CPRF) is also a very good and economical foundation alternative for heavy structures to be rested on loose to medium-stiff soil of the state. For a CPRF, a rigorous numerical analysis is to be carried out considering the superstructure–raft–pile–soil interaction. Due to load sharing by the raft and the piles, the settlement of a CPRF system is found to be much less than the foundations having raft only. The CPRF system is gaining popularity, especially where thick alluvial soil is encountered, in cities like Guwahati where a number of high-rise structures (G+12 and above) are coming up fast.

Assam has several zones susceptible to liquefaction. Many parts of the Sonitpur district, Cachar district, especially both the banks of river Barak, Dibrugarh district and Kamrup district, especially south Kamrup, need a study for evaluation of liquefaction potential before selecting the foundation type. Under normal conditions, even if a shallow foundation is found to be suitable, the presence of liquefiable zones underneath may demand bypassing that zone by long and large diameter piles. Raft foundation also may not behave satisfactorily if rested over liquefiable soil.

In certain locations of Assam, for example, most of the regions of Bongaigaon and Kokrajhar is underlain by a thick gravelly deposit. The topsoil comprises mostly loose newer alluvium with thickness in the range of 5–8 m. Here, pile foundation is difficult to be executed as boring through the gravelly deposit is time consuming and expensive as well. The sandy gravelly deposit is highly permeable and as such, not susceptible to liquefaction. Raft foundation may be a suitable alternative in such a type of subsoil strata.

Wherever medium-stiff alluvial soil is encountered up to a large depth, for heavy structures, many very long and large diameter piles may be required. In such cases, ground improvement techniques, especially, stone columns or granular piles may be a good alternative. Stone columns, when installed properly with an appropriate compactive effort, in layers of around 2.5 times the bore diameter, can increase the allowable bearing pressure of the subsoil significantly. In Assam, improving the bearing capacity of subsoil by the installation of a stone column is quite common. However, appropriate quality control measures and subsequent field plate load test is the key to the good performance of this ground improvement technique.

5.6   OTHER GEOMATERIALS

The application of fly ash in the construction industry has gained popularity in India in recent times. Due to the availability of good quality limestone in Assam and its neighbouring state of Meghalaya, several cement manufacturing plants have been established in the two states. Fly ash is used in these plants as an admixture in the manufacture of cement. But the application of fly ash in geotechnical construction works in Assam is still in an experimental phase. Near non-production of fly ash in Assam can be stated as a reason for this scenario. For bulk consumption, fly ash has to be carried from the states of West Bengal, Bihar or Orissa, which adds to the cost of construction. Research results on the behaviour of Assam soil on mixing with fly ash are encouraging. The compressive strength of the red silty clay hill soil and Brahmaputra sand was increased by four times at 35% soil replacement by Class F Fly ash after a curing period of 28 days (Sumesh et al., 2010). Further research is needed to extend these initial findings to the economical application of fly ash in road embankment and subgrade construction in the state.

As Brahmaputra sand is the most abundantly available soil for construction projects in Assam, the improvement of the load-carrying capacity of this soil by adopting soil improvement techniques is of significant interest to geotechnical engineers and researchers. In research in this direction, an effort was made to improve the bearing capacity and settlement of granular foundation bases made of Brahmaputra river sand by confining it in a geocell made from a locally available waste material (Doley et al., 2020). The geocell was made by stitching waste curtain strips having 0.8 mm thickness and a tensile strength of 24 kN/m. The study found that under a square footing load, the load-carrying capacity and resistance to settlement were improved after providing the geocell reinforced bed. For the reported optimum cell height and cell size, the bearing capacity got improved by 2.5 times over unreinforced conditions and settlement of the footing was reduced by more than 50%. There was a notable increase in the bearing capacity with an increase in the relative density of the geocell infill Brahmaputra sand.

5.7 NATURAL HAZARDS

In Assam, being located in one of the highest seismically active zones of India, there is always a possibility of liquefaction of its subsoil. The state has experienced several major earthquakes in the last century, the most devastating ones being the Shillong Earthquake (in the year 1897 with 8.1 Richter magnitude) and the Assam Earthquake (in the year 1950 with 8.7 Richter magnitude). The predominance of silty soil of the state increases the vulnerability of liquefaction in the state. In a post-earthquake report of 1950, the earthquake caused the occurrence of extensive liquefaction in the alluvial plains of the river Brahmaputra and its tributaries (Poddar, 1952). Sand blows, lateral spreads and sinking and collapse of heavy structures like buildings and bridges were observed and recorded in the report. However, scientific studies on the possibility of liquefaction in Assam have mostly centred around its capital city Guwahati due to its fast-growing infrastructure and importance as the gateway of NE India (Raghu Kanth and Dash, 2010; Ayothiraman et al., 2012; Sarma and Hazarika, 2013; Singnar and Sil, 2017). These studies are based on borehole data distributed over the city. A factor of safety (FoS) against liquefaction, defined as the ratio of the cyclic resistance ratio and cyclic stress ratio, was determined from the SPT borehole data for estimation of the liquefaction potential of the city. These studies have tried to assess the vulnerability of the city soil to liquefaction with due consideration to the thickness and depth of soil layers and the depth of the ground water table.

In a study conducted on 82 SPT borehole data of the city, Singnar and Sil (2017) demarcated that the areas along the GS Road, Panbazar, Bharalumukh and Pandu are in a high liquefaction risk zone. With an assumed peak ground acceleration (PGA) of 0.36 g, the FoS was found to be in the range of 0.13–1.35. The depth of groundwater in high-risk areas was found to be less than 2.0 m.

Raghu Kanth and Dash (2010) in their simulation study based on the SPT profile of Guwahati subsoil concluded that the central part of the city along the GS Road and Dispur has a high vulnerability to liquefaction. This was attributed to the large thickness of soft soil deposits and shallow depth of the ground water table (0.3–0.6 m). A PGA of 0.15–0.25 g was taken in this prediction based on past three major earthquakes in 1869, 1897 and 1950.

Although all the studies have agreed that the subsoil of Guwahati city is prone to liquefaction, there is a diverse opinion on the extent of the areas vulnerable to liquefaction within the city. In another study (Ayothiraman et al., 2012), based on 100 SPT boreholes, it was concluded that the whole of Guwahati metropolitan area is susceptible to liquefaction for an earthquake with a PGA of 0.36 g. On the other hand, in a study undertaken by Sarma and Hazarika (2013) based on 200 soil boring data of 30 m depth distributed over the city and its downtown areas, it was justified that nearly three-fourths of the city area is not susceptible to liquefaction under an earthquake event of 0.36 g PGA. The areas found susceptible to liquefaction in this study include Jalukbari, Pandu, Bharalumukh, Panbazar, Chandmari, parts of GS Road and Gorchuk within the city and Azara, Dharapur and parts of the northern bank of the Brahmaputra in the outskirts of the city. This diversity of opinion on the liquefaction potential of Guwahati is due to the predominance and variation of the silt content in the subsoil and difference in field identification of the predominant layers with a high silt content and difference in the approach adopted towards the evaluation of liquefaction susceptibility of silt/silty clay layers.

Another important natural hazard related to soil in Assam is landslide. The state every year faces landslide related damage to human life and property. Slopes in the hilly areas of Assam and particularly in Guwahati, which are mostly made up of residual soils, often exist in a state of partial saturation. Due to the deep ground water table, these slopes have low saturation and high matric suction in the near-surface region. The soil properties that determine the suction evolution are water permeability and the shape of the soil–water characteristic curve (SWCC), which gives the relationship between the degree of saturation and suction. During the rainy season, many of these slopes undergo slope failure. Conventional slope stability analysis is usually based on the effective stress parameters of the slope. For the residual soil hill slopes, which are unsaturated to a great extent, it is seen that the conventional slope stability analysis is not sufficient to explain the existence of steep geometrical configurations of the slopes. The cut soil slope in these hills is found to have remained stable at a slope as steep as 75°.

Field investigation of the hill slopes around Guwahati revealed that they are made of two types of residual soils. A top layer of reddish silty clay residual soil with the thickness varying from a few centimetres to about 30 m is underlain by yellowish silty sand residual soil. The index properties and the particle size distribution of the two soil types are already presented in Section 5.3.

There is a positive contribution of soil suction to the shear strength of unsaturated soil and this can be estimated from the SWCC of the soil. The SWCCs, established in the laboratory by using modified triaxial testing equipment, are shown in Figure 5.7a and b for the two types of residual soils of Guwahati hill soils, i.e., reddish silty clay and yellowish silty sand respectively (Das and Saikia, 2012).

The SWCC for the silty clay in Figure 5.7a shows that the water content of the soil decreases very slowly up to a suction of 35 kPa of the soil, after which it shows a more pronounced decrease in the water content of the soil with the increase of suction. This indicates a desaturation of the sample when suction exceeds 35 kPa. This

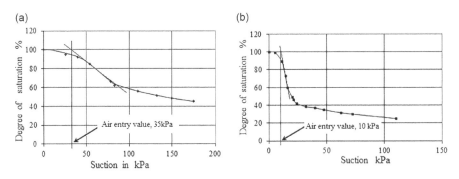

Figure 5.7 Soil–water characteristic curves: (a) reddish silty clay soil and (b) yellowish silty sand soil.

value of suction after which the soil starts to desaturate is referred to as the air entry value of the soil. The air entry value of the reddish silty clay soil under study is 35 kPa. Figure 5.7a also shows that the water content falls at a considerably slower rate after the suction exceeds 80 kPa, which indicates that the rate of desaturation of the silty clay considerably decreases beyond a suction of 80 kPa. The SWCC for the yellowish silty sand in Figure 5.7b shows that the air entry value for this residual soil is 9 kPa and the rate of desaturation decreases considerably beyond a suction of 30 kPa. The effective cohesion and angle of internal friction of the silty clay soil were determined to be 10 kPa and 31°, respectively, and those of the silty sand soil were determined to be 0 kPa and 38.5°, respectively (Das and Saikia, 2012). This information will help in a more realistic stability analysis of man-made slopes in these hills.

As already stated, every year Assam faces the immense problems of floods and riverbank erosion. The causes of riverbank erosion in Assam, particularly the Brahmaputra river, are rather complex. Although there are some studies carried out on this problem, a convergence of opinion on the causes of this river erosion and sustainable solution is yet to be arrived at. However, the very steep bed slopes of the rivers as they enter the state from Arunachal Pradesh and Bhutan along with the sediment load are a major factor. Kotoky et al. (2005) observed that in some locations in the bank line of Brahmaputra, older alluvium projecting towards the river channel offers high resistance to the flow regime causing constant changes in the flow direction producing bank caving along the reach. Due to this bank caving and meandering, intense undercutting takes place resulting in overhanging of the bank materials and slumping of the bank soil into the river. Slumping is more visible in locations where the bank soil is composed of silty clay or clayey silt.

5.8 CASE STUDIES AND FIELD TESTS

The lateral, as well as vertical, variation of the soils of Assam can be significant within short distances, and thus an accurate soil map of Assam is quite difficult to be prepared. The selection of the most suitable foundation and its satisfactory performance, for a given structure, depend on the adequacy of geotechnical investigation, appropriate design, good workmanship and adequate testing. Many structures in Assam have been performing quite satisfactorily, even under challenging soil conditions, wherever all the above-mentioned exercises were carried out sincerely. The absence or slight negligence of any one of the above may lead to failure or distress in the structures. For the same reason, structure failure in Assam is not very uncommon. The failure of retaining structures, man-made slope failures, the sinking of roads in hilly areas, tilted buildings, cracks in buildings due to excessive differential settlements and so on are found quite often. In this section, a brief overview of a few failed structures is provided and a few case studies, where appropriate measures were taken to prevent any possible failures, are also presented.

5.8.1 Building settlement

Guwahati has witnessed several cases where structures have failed, mostly from a functionality point of view, the reason for which may primarily be attributed to geotechnical failure. Newly built but tilted structures can be seen in several places in Guwahati.

Figure 5.8 Tilting of building in Guwahati.

In the Rajgarh locality of the city, many structures are found to have undergone differential settlement up to 600 mm, as shown in Figure 5.8. A case study has shown that the effected buildings are supported on isolated footings. The study revealed the existence of an about 50.0 m wide and 1 km long strip of organic soil up to a depth of about 10.0 m from the existing ground level. Ignorance of this organic layer during the design and construction of buildings has resulted in this settlement problem.

5.8.2   Slope stabilisation by soil nailing

In another case study of slope instability, a 12.0 m high hill slope face was found to exist in a near-vertical position in the Sarania hills of Guwahati with a residential house very close to the foot of the slope. The occupants of the residential house noticed signs of instability of the slope. In a case study, it was noticed that a community water reservoir also existed at the top, very near to the slope. Quite often water spilled over the water tank and fell on the soil slope. Frequent wetting due to water spillage created instability in the soil slope and the house down the slope was in danger. Geotechnical investigation of the slope soil and slope stability analysis, using finite element software Geo-Slope, were carried out. Three layers of soil nails gave a FoS greater than 1.5 and the slope has been performing satisfactorily for the last 15 years. Figure 5.9 presents the details of the slope and the installation of soil nails.

5.8.3   Prevention of subsidence of structure by cement grouting

During excavation for the basement floor of a proposed multi-storeyed shopping mall complex at R.G. Baruah Road, Guwahati, the shoring piles designed and installed for supporting the side soil collapsed. As a result of the failure of the shoring piles, side

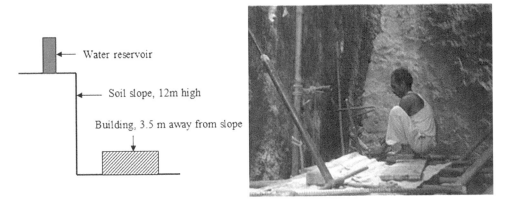

Figure 5.9 Stabilisation of slope by soil nailing in Guwahati.

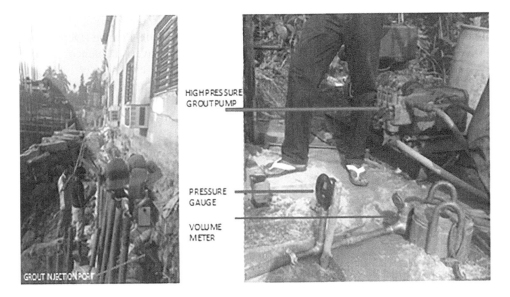

Figure 5.10 Disturbance of the foundation soil of existing building due to adjacent new excavation (left) and stabilisation using cement grout (right)

soil also started to fall off. A two-storey office building stood very close to the excavation line, and because of the gradual falling of the soil from below the structure into the excavation pit, the structure was in danger of possible tilt and subsequent toppling. The possible movement of the office building was arrested by injecting cement grout under controlled pressure and filling all voids underneath, created by the slope movement. Figure 5.10 shows the dangerously standing two-storeyed office building and the arrangement for cement pressure grout with which the foundation of the structure was stabilised.

5.9 GEOENVIRONMENTAL IMPACT ON SOILS AND ROCKS

Climate change has touched all aspects of human life worldwide. A study on the climate change modelling by Ravindranath et al. (2011) for India shows that the Indian sub-continent is likely to experience warming of over 3°C–5°C and significant changes (increase and decrease) in flood and drought frequency and intensity. Although the study predicts a 25% increase in flood intensity in the NE Indian states, some parts of the region may see a 5%–10% decrease in flood during 2021–2050. Drought weeks may also rise by about 25% during monsoon months. An Assam government study report (Department of Environment, Government of Assam, 2015) indicates an increase in extreme rainfall events by 5%–38% by the year 2050. This study also reports a prediction of up to 25% increase in the flood in the southern parts of Assam with the heavy precipitation days likely to dominate by the mid of the current century. Projections also indicate that the extreme precipitation days are likely to increase further by the end of the century. Himalayan glacier melt due to global warming will add to the flood intensity in the state.

With the projected increase in the flood in the plains of Assam as a consequence of climate change, the focus of the government will be on the measures of containing it. In a draft report of Assam State Action Plan on Climate Change (Department of Environment, Government of Assam, 2015), strategies to avert incidences of flood escalation have been outlined. The strategies include the development of integrated flood, erosion and sediment management action plan drawing upon knowledge on improved technical standards, global best practices and technologies, the extension of erosion control through riverbank stabilisation work using geotextile materials and application of the modern technology like using geotextile and so on in the riverbank stabilisation work and embankment construction and maintenance.

5.10 CONCLUDING REMARKS

The large floodplains of Assam are composed of younger alluvium of loose fine to medium sand and/or clayey or silty soil underlain by thick older alluvium consisting of horizontally bedded lenses of semi-consolidated sandy, silty or clayey soil and occasionally mixed with pebbles. Lateritic soils occurring in the hilly areas are clayey, yellowish to reddish in colour. Hill soils occurring in the hilly terrain of the state of Assam are highly permeable, dark greyish brown to yellowish red and fine to coarse loamy soil.

Raft and pile foundations are now taking over the isolated footings in buildings due to low bearing capacity at shallow depths in the state. Precast driven piles are yet to gain popularity in the state. Case studies on the installation and performance of precast driven piles in Assam may be carried out for the benefit of geotechnical designers.

Past earthquakes and recent research have indicated that Assam soil is prone to liquefaction in the event of a major earthquake. But there is a difference of opinion on the extent of the vulnerable areas. There is much scope of research on the liquefaction vulnerability assessment of the entire state.

Shallow landslides in the hilly areas of Assam occur every year due to human activities. These slides are mostly triggered by heavy rainfall. The slopes constituted of silty clay and silty sand residual soil layers are often under unsaturated conditions. There is a scope for further study towards prediction and early warning of failures and recommendation of appropriate mitigation measures.

Riverbank erosion due to undercutting of loose non-cohesive layers of bank soil and slumping of the overlaying topsoil has been a perennial problem of the state. There is also a scope of research in studying the performance of existing bank protection measures and finding sustainable solutions to this problem.

The tributaries of Brahmaputra are good sources of fine and coarse aggregates. Initial research shows the suitability of these aggregates for various construction projects, but there is a great deal of locational variability of the material properties and further study is required to have a wider database of the river aggregates of Assam.

REFERENCES

Ayothiraman, R., Raghu Kanth, S.T.G. and Sreelatha, S. (2012). Evaluation of liquefaction potential of Guwahati: gateway city to North-eastern India. *Natural Hazards*, Vol. 63, pp. 449–460.

Borah, S., Pathak, J. and Goswami, D. (2016). Site response analysis: Guwahati city and CMP2025. *6th International Conference on Recent Advances in Geotechnical Earthquake Engineering and Soil Dynamics*, 1–6 August 2016, Roorkee, India, pp. 1–8.

Das, U.K. and Saikia, B.D. (2012). A modified method for stability analysis of unsaturated residual soil hill slopes. *ICETCESD*, 10–12 March 2012, Sylhet, Bangladesh, pp. 109–122.

Das, U.K. (2016). A case study on performance of Jiabharali riverbank protection measure using geotextile bags. *International.Journal of Geosynthetics and Ground Engineering*, Vol. 2, No. 12, pp. 1–9.

Dasgupta, S., Pande, P., Ganguly, D., Iqbal, Z., Sanyal, K., Venkataraman, N.V., Sural, B., Harendranath, L., Mazumdar, K., Sanyal, Roy, A., Dasgupta, S., Das, L.K., Mishra, P.S. and Gupta, H. (2000). *Seismotronic Atlas of India and Its Environs*. Special Publication Geological Survey of India, 86.

Department of Environment, Government of Assam. (2015). Assam State Action Plan on Climate Change (2015–2020). Draft Report, http://moef.gov.in/wp-content/uploads/2017/08/ASSAM-SAPCC.pdf, accessed 14 November 2021.

Doley, C., Das, U.K. and Shukla, S.K. (2020). Effect of cell height and infill density on the performance of geocell-reinforced beds of Brahmaputra river sand. *Sustainable Civil Engineering Practices. Lecture Notes in Civil Engineering*, Vol. 72. Springer, Singapore. doi:10.1007/978-981-15-3677-9_17.

IRC SP 102 (2014). Guidelines for design and construction of reinforced soil walls. Indian Roads Congress, New Delhi, India.

IS 383 (2016). Coarse and fine aggregate for concrete – specification. Bureau of Indian Standards, New Delhi, India.

IS 1893 (2016). Criteria for earthquake resistant design of structures: general provisions and buildings. Part I. Bureau of Indian Standards, New Delhi, India.

IS 1904 (1986). Design and construction of foundations in soils: general requirements. Bureau of Indian Standards, New Delhi, India.

IS 2911 (2013). Design and construction of pile foundations-code of practice: load test on piles, Part IV, Bureau of Indian Standards, New Delhi, India.

IS 6403 (1981). Determination of bearing capacity of shallow foundations. Bureau of Indian Standards, New Delhi, India.

IS 8826 (1978). Guidelines for design of large earth and rockfill dams. Bureau of Indian Standards, New Delhi.

Kotoky, P., Bezbaruah, D., Baruah, J. and Sarma, J.N. (2005). Nature of bank erosion along the Brahmaputra river channel, Assam, India. *Current Science*, Vol. 88, No. 4, pp. 634–640.

Pegu, A. and Gogoi, I.B. (2015). A comparative geotechnical study of natural, crushed and mixture of natural and crushed river borne aggregates of the river Champabati. *International Journal of Innovative Research in Advanced Engineering*, Vol. 2, No. 7, pp. 25–29.

Poddar, M.C. (1952). Preliminary report of the Assam earthquake. *Issue 2, Bulletins of the Geological Survey of India (Series B) Engineering Geology and Ground Water*, 15 August 1950.

Raghu Kanth, S.T.G. and Dash, S.K. (2010). Evaluation of seismic soil-liquefaction at Guwahati city. *Environmental Earth Science*, Vol. 61, pp. 355–368.

Ravindranath, N.H., Rao, S., Sharma, N., Nair, M., Gopalakrishnan, R., Rao, A.S., Malaviya, S., Tiwari, R., Sagadevan, A., Minsi, M., Krishna, N. and Bala, G. (2011). Climate change vulnerability profiles for north east India. *Current Science*, Vol. 101, No. 3, pp. 384–394.

Sarma, J.N. and Acharjee, S. (2012). Bank erosion of the Brahmaputra river and neotectonic activity around Rohmoria, Assam, India. *Comunicações Geológicas*, Vol. 99, No. 1, pp. 33–38.

Sharma, B. and Hazarika, P. J. (2013) Assessment of liquefaction potential of Guwahati city: a case study. *Geotechnical & Geological Engineering*, Vol. 31, pp. 1437–1452.

Singnar, L. and Sil, A. (2017). Assessment of liquefaction potential of Guwahati city based on geotechnical standard penetration test (SPT) data. *Disaster Advances*, Vol. 10, No. 12, pp. 10–21.

Sumseh, M., Kalita, A. and Singh, B. (2010) An experimental investigation on strength properties of fly ash blended soils treated with cement. *Journal of Environmental Research and Development*, Vol. 5, No. 2, pp. 322–329.

Chapter 6

Bihar

J.N. Jha
Muzaffarpur Institute of Technology

S. K. Singh
Punjab Engineering College

Sunita Kumari
National Institute of Technology Patna

Akash Priyadarshee, Vijay Kumar, and Ashish Kumar
Muzaffarpur Institute of Technology

CONTENTS

6.1 INTRODUCTION

Bihar is the third largest Indian state population-wise and the twelfth largest state area-wise. It was formed on 22 March 1912. It is situated in the north-eastern part of the country located between 24°-20'-10" N and 27°-31'-15" N latitudes and between 83°-19'-50"E and 88°-17'-40"E longitudes. Nepal is situated in the north, Jharkhand is in the south, Uttar Pradesh is in the east and West Bengal is situated in the west (Figure 6.1). The total area of Bihar is 94,163 km². The urban area is 1,095.49 km² and the rural area is 92,257.51 km². Patna is the capital of Bihar. Muzaffarpur, Bhagalpur, Gaya and Katihar are some of the major cities of the state. The climate of the state is diverse in nature. The average temperature of the state is 27°C and the average precipitation is 1,200 mm. The climate of the state is sub-tropical. The average elevation of Bihar above the mean sea level is 53 m. The Ganga river is the major river in the state. It flows from the west to east and divides the state into two parts. All other rivers join the river Ganga from the north and south. Gandak, Bagmati, Kosi and Kali join the river

DOI: 10.1201/9781003177159-6

Figure 6.1 Map of Bihar (MSMEDI, 2016).

Ganga from the north, while Sone, Phalgu and Punpun join from the south (https://state.bihar.gov.in).

Most of the land of Bihar falls in the plain of the Ganga basin. Due to this reason, the major land contains alluvium soil and is suitable for agriculture. Based on the physical and structural conditions, Bihar can be divided into three parts: the Shivalik range region, the Bihar plain region and the southern plateau region. The Shivalik region is situated in the northern part of the West Champaran district. The Bihar plain is spread between the Himalayas and the Chhota Nagpur plateau. The area of the Bihar plain is 90,650 km². It is a part of the bed of Tethys Sea, which was filled by the deposition of the rivers that originated from the Himalayas and the Chhota Nagpur plateau. The northern region of Bihar is mostly a plain area, whereas in the plain of the south region Gaya, Rajgir, Kaimur and Kharagpur, hills are found. The North Bihar plain is formed by the deposits of the rivers, namely Gandak, Kosi, Bagmati, Mahananda, Kamla-Balan and their tributaries. The districts of East Champaran and West Champaran lie in the northern part of the North Bihar plain that has a higher elevation and these regions are known as the Terai area. The soils for these reasons are porous in nature and have a higher percolation rate. Due to this reason, groundwater tables in these regions are higher. The south of Terai consists of a marshy land and in these regions, oxbows popularly known as 'Chaurs' are found. The northern plain is greater in area than the southern plain and the southern plain is triangular i.e., wider in the west and narrower in the east.

Soils in the northern plain are formed by deposition of the rivers like Gandak and Burhi Gandak flowing on the north side. Piedmont swamp soil, Terai soil and Gangetic alluvium soil are the types of soil found in the northern plain of Bihar (Ranjan and Kumar, 2013). Piedmont swamp soil is brown to grey in colour and found in the north-west region of West Champaran. This type of soil is mostly clayey and contains organic matter. Terai soil is found in the northern part of Bihar near the Nepal border and it is grey to yellow in colour. The Gangetic alluvial soil is loamy in nature with varying thickness. It is thicker in the north and thinner in the south. In the upland area, this soil is known as the 'Bhangar soil', while in the low land area, it is known as the 'Khadar soil'. The Bhangar soil is rich in carbonaceous compounds and it has a high clay content. The Khadar soil has lesser carbonaceous compounds and a lesser clay content. The soil in the southern plain is mainly formed by the deposition of rivers (Falgu, Sone and Punpun). Karail-kewal soil, Tal soil and Balthar soil are the major soil types found in the southern plain of Bihar (Ranjan and Kumar, 2013). Karail-kewal soil is found in some different districts (Gaya, Munger and Bhagalpur) of Bihar. These types of soil have a higher clay content and the colour varies from brown to yellow. Tal soil is found in the region from the Buxar to the Banka district. This soil is light grey to dark grey in colour. Balthar soil is found from the south of the Ganga plain to the Chhota Nagpur plain. This type of soil has less water absorption capacity with its colour being red. The soils in the southern plateau of Bihar, formed after the disintegration of the igneous and metamorphic rocks, respectively, are red and yellow soil. These are found in the districts of Nawada, Gaya, Banka and Aurangabad while in the Kaimur and Rohtas districts, the soil is found to have a higher percentage of sand known as red sandy soil.

There are five power plants in Bihar, namely, Barauni Thermal Power Station, Kanti Thermal Power Station, NTPC Kahalgaon Super Thermal Power Station, NTPC Barh Super Thermal Power Plant and Nabinagar Super Thermal Power Project. Fly ash production from these power plants during 2016–2017 was 7.4 million tonnes and utilisation was 32.4% (CEA, 2017).

This chapter is divided into different sections that include the major types of soils, properties of the soil and rocks, construction materials used in Bihar, foundation and geotechnical structures, other geomaterials, natural hazards, case study and effect of change in environment on the soils and rocks.

6.2 MAJOR TYPES OF SOILS

Based on physical and structural conditions, Bihar can be classified into three major regions: the Shivalik range region, the Bihar plain region and the southern plateau region. Considering the borehole data of ten different places covering all the regions, Tables 6.1–6.10 are presented. Tables 6.1–6.4 show the borehole data up to 10–11 m depth in Bettiah, Madhubani, Samastipur and Kishanganj districts, respectively. Data show that in Madhubani, Samstipur and Kishanganj, soils are mainly silty in the upper and lower strata. North Bihar is a region where flood is a common phenomenon in rivers every year and this turns out to be the reason why soils in the upper layers are silty. Table 6.5 shows the soil data obtained from a borehole near a rail bridge over river Ganga between Hathidah and Barauni stations. In this region, in the upper

Table 6.1 Soil properties near Raidhurva, Bettiah

Depth below GL (m)	Sample no.	N value observation	Visual description of soil with IS classification	Depth (m) from	Depth (m) to	Thickness (m)	Liquid limit	Plastic limit	Bulk density (g/cc)	Natural moisture content (%)	Specific gravity	Cohesion, c (kg/cm²)	Angle of friction, φ (°)
1.0				0.0									
1.5	S1	7	Greyish sandy clayey silt ML			2.5	33.0	27.3	1.94	28.6	2.65	0.07	14.7
2.5					3.0								
3.0	S2	10		3.0		8.0			1.89	31.4	2.62	0.00	28.9
4.0													
4.5	S3	12	Greyish silty sand SM-SP						1.88	32.1	2.62	0.00	29.2
5.5													
6.0	S4	15							1.87	32.9	2.62	0.00	29.8
7.0													
7.5	S5	18							1.86	33.8	2.62	0.00	30.4
8.5													
9.0	S6	22							1.85	34.6	2.62	0.00	31.2
10.0													
10.5	S7	25			10.5				1.85	34.2	2.61	0.00	31.6

Table 6.2 Soil properties near Madhubani

Depth below GL (m)	Sample no.	N value observation	Visual description of soil with IS Classification Observation	Depth (m) from	to	Thickness (m)	Liquid limit	Plastic limit	Bulk density (g/cc)	Natural moisture content (%)	Specific gravity	Type of test	Cohesion, c (kg/cm²)	Angle of friction, ϕ (°)	Compression index (C_c)
1.0			Greyish	0.0	–							–	–	–	–
1.5	S1	2	sandy	–	–	–	26.8	22.0	1.92	29.4	2.63	–	–	–	–
2.5			clayey silt	–	–							–	–	–	–
3.0	S2	4	ML	–	–				1.94	28.2	2.64	–	0.05	13.2	–
4.0			Greyish	3.0	–							–	–	–	–
4.5	S3	7	sandy	–	–	–	26.8	23.3	1.94	28.8	2.64	–	0.07	14.5	–
5.5			clayey silt	–	–							–	–	–	–
6.0	S4	9	ML	–	–				1.96	28.0	2.68	–	0.09	14.9	–
7.0			Greyish	6.0	–							–	–	–	–
7.5	S5	12	clayey silt ML	–	7.5	–			1.98	26.4	2.67	–	0.12	16.2	–
8.5				7.5	–							–	–	–	–
9.0	S6	16	Greyish silty	–	–	3.0			1.98	26.8	2.67	–	0.15	17.2	–
10.0			clay ML	–	–				1.99	26.2	2.68		0.20	18.1	–
10.5	S7	20		–	10.5							–	–	–	–

Table 6.3 Soil properties near ITI, Samastipur

Depth below GL (m)	Sample no.	N value observation	Visual description of soil with IS Classification (Observation / Classification)	Depth (m) from	to	Thickness (m)	Liquid limit (%)	Plastic limit (%)	Plasticity index (%)	Bulk density (g/cm³)	Natural moisture content (%)	Specific gravity	Type of Test	Cohesion, c (kg/cm²)	Angle of friction, ϕ (°)	Compression index (C_c)
1.0	–	–	Greyish	0.0	–	–	–	–	–	–	–	–	–	–	–	–
1.5	S1	10	sandy	–	–	3.0	–	–	–	1.94	28.4	2.65	–	0.09	15.2	–
2.5	–	–	clayey silt,	–	–	–	–	–	–	–	–	–	–	–	–	–
3.0	S2	12	ML	–	3.0	–	29.2	25.6	3.6	1.94	28.2	2.64	–	0.11	15.4	–
4.0	–	–	–	3.0	–	–	–	–	–	–	–	–	–	–	–	–
4.5	S3	15	Greyish	–	–	4.5	–	–	–	1.89	31.3	2.62	–	0.00	28.5	–
5.5	–	–	with grits	–	–	–	–	–	–	–	–	–	–	–	–	–
6.0	S4	17	silty sand,	–	–	–	–	–	–	1.88	31.3	2.62	–	0.00	28.7	–
7.0	–	–	SP-SM	–	–	–	–	–	–	–	–	–	–	–	–	–
7.5	S5	20	–	–	7.5	–	–	–	–	1.88	32.2	2.62	–	0.00	29.0	–
8.5	–	–	–	7.5	–	–	–	–	–	–	–	–	–	–	–	–
9.0	S6	23	Greyish	–	–	3.0	–	–	–	1.87	32.9	2.62	–	0.00	29.3	–
10.0	–	–	silty sand,	–	–	–	–	–	–	–	–	–	–	–	–	–
10.5	S7	27	SP-SM	–	10.5	–	–	–	–	–	–	–	–	–	–	–

Table 6.4 Soil properties near Terhagachh, Kishanganj

Depth below GL (m)	Sample no.	N value observation / Observation	Visual description of soil with IS Classification	Depth (m) from to	Thickness (m)	Liquid limit	Plastic limit	Plasticity index	Bulk density (g/cc)	Natural moisture content (%)	Specific gravity	Type of Test	Cohesion, c (kg/cm²)	Angle of friction, φ (°)	Compression index (Cc)
1.0				0.0					–	–	–	–	–	–	–
1.5	S1	6	Clayey silt, ML	–	3.0	39.1	29.3	9.8	1.94	28.5	2.65	–	0.06	14.2	–
2.5				– 3.0											
3.0	S2	5		3.0		–	–	–	1.94	28.8	2.66	–	0.06	14.0	–
4.0			Sandy silt, ML	–	3.0										
4.5	S3	19		–		37.7	28.5	9.2	1.93	27.3	2.65	–	0.22	12.2	–
5.5				– 6.0											
6.0	S4	8		6.0		–	–	–	1.94	28.2	2.64	–	0.08	15.0	–
7.0			Greyish sandy clayey silt, ML	–											
7.5	S5	10		–	4.5	30.8	23.8	7.0	1.94	28.1	2.64	–	0.10	16.2	–
8.5				–											
9.0	S6	10		–		–	–	–	1.94	28.1	2.64	–	–	–	–
10.0				–											
10.5	S7	11		10.5		–	–	–	1.94	28.1	2.64	–	0.12	16.0	–

Table 6.5 Soil properties near rail bridge over river Ganga between Hathidah and Barauni stations

Depth (m)	N value	Soil description	Gravel (%)	Sand (%)	Silt (%)	Clay (%)	Liquid limit(%)	Plastic limit (%)	Plasticity index (%)	Specific gravity	Bulk density (gm/cc)	Moisture content (%)	Dry density (g/cc)	Compression index (cc)	Cohesion intercept (kg/cm²)	Angle of friction, φ (°)
GL																
1.0		Greyish sandy	0.0	44.1	45.6	10.3	29.5	24.3	5.2	2.65	1.94	28.5	1.51	–	0.10	14.9
1.5	12	clayey silt	0.0	6.7	57.6	35.7	29.3	24.2	5.1	2.64	–	28.2	–	–	–	–
3.0	14	ML	0.0	9.3	60.3	30.3	29.1	24.4	4.7	2.64	–	28.2	–	–	–	–
4.0		Greyish clayey	0.0	43.5	44.5	12.0	30.8	26.7	4.1	2.63	1.92	2.9	1.87	–	0.03	26.1
4.5	21	sandy silt	0.0	17.6	50.1	32.4	30.7	26.6	4.1	2.63	–	29.3	–	–	–	–
6.0	23	SC-ML	0.0	15.4	49.5	35.1	30.5	26.5	4.0	2.63	–	30.1	–	–	–	–
7.0		Greyish sandy	0.0	5.3	59.1	35.6	30.8	25.0	5.8	2.65	1.94	28.5	1.51	–	0.10	14.7
7.5	11	clayey silt	0.0	10.2	57.5	32.3	30.6	24.9	5.7	2.65	–	24.4	–	–	–	–
9.0	13	ML	0.0	14.4	55.3	30.4	30.5	24.9	5.6	2.64	–	28.3	–	–	–	–
10.0			2.7	3.2	58.5	35.6	46.8	25.0	21.8	2.71	2.06	22.6	1.68	0.110	1.15	5.3
10.5	36	Brownish grey	5.7	6.9	55.1	32.3	46.5	25.2	21.3	2.71	–	23.1	–	–	–	–
12.0	31	silty clay CI	5.7	11.7	52.3	30.4	46.6	25.3	21.3	2.71	–	23.3	–	–	–	–
13.0			3.8	5.6	58.0	32.7	46.8	27.2	19.6	2.71	2.01	24.9	1.61	0.130	0.68	5.0
13.5	17		2.6	5.1	56.2	36.2	46.6	26.1	20.5	2.70	–	24.7	–	–	–	–
15.0	21		2.6	8.8	58.4	30.3	46.5	25.9	20.6	2.70	–	24.2	–	–	–	–
16.0		Greyish yellow	1.4	9.9	56.4	32.3	35.2	21.9	13.3	–	–	–	–	–	–	–
16.5	13	sandy silty					35.6	22.2	13.4	2.70	–	25.3	–	–	–	–
18.0	17	clay CI					–	–	–	2.70	–	24.7	–	–	–	–
19.0							–	–	–	–	–	–	–	–	–	–
19.5	20	Yellowish grey	4.7	8.1	56.9	30.3	34.1	20.4	13.7	2.70	–	24.0	–	–	–	–
21.0		silty clay CL	8.3	18.5	60.6	12.7	34.2	20.4	13.8	2.70	2.04	24.1	1.65	0.119	0.89	5.2
22.0	24		8.3	21.1	60.0	10.7	29.8	17.9	11.9	2.70	–	23.5	–	–	–	–
22.5	26		5.8	46.4	42.2	5.7	31.2	19.1	12.1	2.70	–	23.4	–	–	–	–
24.0	30		–	–	–	–	30.1	25.6	4.5	2.68	–	28.0	–	–	–	–
25.0		Greyish sandy	5.8	46.4	39.2	8.7	30.2	25.5	4.7	2.68	1.96	27.8	1.53	–	0.18	20.2
25.5	39	clayey silt	0.0	59.1	33.2	7.7	30.0	25.4	4.6	2.68	–	27.9	–	–	–	–
27.0	41	ML					–	–	–	–	–	–	–	–	–	–
28.0		Greyish silty	0.0	64.4	30.0	5.7	–	–	–	2.63	1.90	30.8	1.45	–	0.00	30.8
28.5	46	clayey sand	0.0	59.1	30.6	10.3	–	–	–	2.63	–	30.7	–	–	–	–
30.0	48	SC	–	–	–	–	–	–	–	2.63	–	30.9	–	–	–	–

Table 6.6 Soil properties near Mithapur, Patna

Depth (m)	N value	Sample no.	Soil description	Gravel %	Sand %	Silt (%)	Clay (%)	Liquid limit (%)	Plastic limit (%)	Plasticity index (%)	Specific gravity	Bulk density (g/cc)	Moisture content (%)	Dry density (g/cc)	Compression index (C_c)	Cohesion intercept (kg/cm²)	Angle of friction, ϕ (°)
GL																	
1.5	0	D1		0.0	4.7	59.7	35.6	-	-	-	-	-	-	-	-	-	-
3.0	10	D2		-	-	-	-	43.2	23.5	19.7	2.71	2.00	26.2	1.58	0.140	0.49	5.0
4.5	12	D3		-	-	-	-	42.4	23.2	19.2	2.71	2.01	25.5	1.60	0.135	0.62	5.0
6.0	14	D4		-	-	-	-	42.1	22.5	19.6	-	-	-	-	-	-	-
7.5	4	D5	Greyish silty clay, CI	0.0	2.3	61.0	36.7	36.7	21.0	15.7	2.70	1.95	29.2	1.51	0.166	0.20	2.6
9.0	6	D6		-	-	-	-	-	-	-	-	-	-	-	-	-	-
10.5	0	D7		0.0	1.6	61.8	36.6	36.0	20.4	15.6	2.70	1.94	29.9	1.49	0.171	0.10	1.5
12.0	11	D8		-	-	-	-	-	-	-	-	-	-	-	-	-	-
13.5	10	D9		5.3	6.6	57.6	30.5	36.5	20.7	15.8	2.71	2.00	26.1	1.59	0.141	0.50	5.0
15.0	22	D10		0.0	1.6	61.8	36.7	-	-	-	-	-	-	-	-	-	-
16.5	20	D11		5.6	3.9	59.9	30.6	37.2	20.9	16.3	2.70	2.02	24.7	1.62	0.128	0.90	5.2
18.0	15	D12		1.4	1.6	60.9	36.2	-	-	-	-	-	-	-	-	-	-
19.5	18	D13		2.8	1.9	59.7	35.6	30.1	20.2	9.9	2.70	2.02	24.7	1.62	0.129	0.84	5.1
21.0	37	D14		-	-	-	-	-	-	-	-	-	-	-	-	-	-
22.5	16	D15	Yellowish grey silty clay, CL	2.8	1.9	59.6	35.8	44.5	25.7	18.8	2.71	2.02	24.8	1.62	0.130	0.71	5.1
24.0	22	D16		-	-	-	-	-	-	-	-	-	-	-	-	-	-
25.5	16	D17		0.0	2.7	61.2	36.1	35.5	22.4	13.1	2.71	2.02	24.9	1.62	0.13	0.72	5.1
27.0	15	D18		-	-	-	-	-	-	-	-	-	-	-	-	-	-
28.5	18	D19		-	-	-	-	-	-	-	-	-	-	-	-	0.84	5.1
30.0	31	D20		1.8	5.2	62.8	30.3	31.20	22.20	9.0	2.70	2.02	24.7	1.62	0.13	0.84	5.1

Table 6.7 Soil properties near Kaimur, Bhabhua

Depth below GL (m)	Sample no.	N value observation	Visual description of soil with IS Classification Observation	Depth (m) from to	Thickness (m)	Liquid limit (%)	Plastic limit (%)	Plasticity index (%)	Bulk density (g/cc)	Natural moisture content (%)	Specific gravity	Type of test	Cohesion, c (kg/cm²)	Angle of friction, ϕ (°)	Compression index (C_c)
1.0				0.0											
1.5	S1	6				32.2	18.5	13.7	1.96	28.6	2.70		0.30	3.8	–
2.5															
3.0	S2	8							1.98	27.0	2.70		0.40	4.8	–
4.0															
4.5	S3	14	Light grey			30.6	18.2	12.4	2.01	25.3	2.70		0.63	5.0	0.132
5.5			silty clay,		10.5											
6.0	S4	21	CL						2.03	24.2	2.70		0.80	5.1	–
7.0															
7.5	S5	25							2.03	24.0	2.69		0.88	5.2	–
8.5															
9.0	S6	30							2.05	22.6	2.70		0.98	5.3	–
10.0				10.5											
10.5	S7	29							2.05	22.7	2.70		0.96	5.3	–

Table 6.8 Soil properties near Bodh Gaya, Gaya

Depth blow GL (m)	Sample no.	N value observation / Observation	Visual description of soil with IS Classification	Depth (m) from to	Thickness (m)	Liquid limit (%)	Plastic limit (%)	Plasticity index (%)	Bulk density (gm/cc)	Natural moisture Content (%)	Specific gravity	Type of test	Cohesion, c (kg/cm²)	Angle of friction, φ (°)	Compression index (Cc)
1.0				0.0											
1.5	S1	12				37.5	22.0	15.5	2.01	25.4	2.70		0.57	5.0	
2.5	S2	15							2.01	25.1	2.69		0.64	5.1	
3.0															
4.0															
4.5	S3	20	Greyish silty		10.5	37.4	20.6	16.8	2.02	24.7	2.70		0.75	5.2	0.127
5.5			clay, CI												
6.0	S4	18							2.02	24.8	2.70		0.70	5.2	
7.0															
7.5	S5	22							2.03	24.2	2.70		0.79	5.2	
8.5															
9.0	S6	27							2.04	23.5	2.70		0.89	5.2	
10.0															
10.5	S7	33		10.5					2.05	22.6	2.70		1.02	5.3	

Table 6.9 Soil properties near Nathnagar, Bhagalpur

Depth below GL (m)	Sample no.	N value observation / Observation	Visual description of soil with IS Classification	Depth (m) from – to	Thickness (m)	Liquid limit	Plastic limit	Bulk density (g/cc)	Natural moisture content (%)	Specific gravity	Type of tst	Cohesion, c (kg/cm²)	Angle of friction, φ (°)	Compression index (Cc)
1.0 1.5	S1	11	Greyish silty clay CL	0.0 – 1.5	1.5	32.4	22.0	2.01	25.3	2.70	–	0.54	5.0	–
2.5 3.0	S2	14	Reddish greyish silty clay CL	1.5 – 3.0	1.5	–	–	2.01	25.1	2.69	–	0.62	5.0	–
4.0 4.5	S3	17	Reddish greyish gritty silty clay CL	3.0 – 4.5	1.5	30.4	20.5	2.02	24.8	2.70	–	0.68	5.1	–
5.5 6.0	S4	21	Greyish yellowish gritty silty clay CL	4.5 – 6.0	1.5	–	–	2.02	24.5	2.69	–	0.76	5.1	–
7.0 7.5 8.5	S5	24	Reddish greyish gritty silty clay CL	6.0 –	3.0	–	–	2.03	24.2	2.70	–	0.82	5.2	–
9.0	S6	28	Reddish clay CL	– 9.0	1.5	–	–	2.04	23.5	2.70	–	0.90	5.3	–
10.0 10.5	S7	34	Reddish yellowish gritty silty clay CL	9.0 – 10.5		–	–	2.05	22.7	2.69	–	1.02	5.3	–

Table 6.10 Soil properties near Dhorai, Banka

Depth below GL (m)	Sample no. Observation	N value observation Observation	Visual description of soil with IS Classification	Depth (m) from to	Thickness (m)	Liquid limit (%)	Plastic limit (%)	Bulk density (g/cc)	Natural moisture content (%)	Specific gravity	Shear test Type of tst	Cohesion, c (kg/cm²)	Angle of friction, φ (°)	Compression index (Cc)
1.0			Reddish	0.0		–	–	–	–	–	–	–	–	–
1.5	S1	12	yellowish	–	3.0	28.0	17.6	2.01	25.2	2.70	–	0.55	5.0	–
2.5			sandy silty	–		–	–	–	–	–	–	–	–	–
3.0	S2	17	clay CL	3.0	3.0	–	–	2.02	24.6	2.70	–	0.68	5.0	0.128
4.0				3.0		–	–	–	–	–	–	–	–	–
4.5	S3	20		–		–	–	1.86	33.5	2.62	–	0.00	29.9	–
5.5				–		–	–	–	–	–	–	–	–	–
6.0	S4	23	Reddish	–		–	–	1.86	33.6	2.62	–	0.00	30.2	–
7.0			silty sand	–	7.5	–	–	–	–	–	–	–	–	–
7.5	S5	25	SM-SP	–		–	–	1.86	33.7	2.62	–	0.00	30.5	–
8.5				–		–	–	–	–	–	–	–	–	–
9.0	S6	28		–		–	–	1.85	34.5	2.62	–	0.00	30.8	–
10.0	S7	33		10.5		–	–	1.85	34.7	2.62	–	0.00	31.2	–

Table 6.11 Summary of regional variation in the type of soils

Region of Bihar	Area/city	Type of soil up to 10 m depth	Reference borehole
Shiwalik range region	West Champaran	Fine to medium sand overlain by gravels with bounders	Table 6.1 (soil stratigraphy in the West Champaran district
Northern Bihar Plain region	Bettiah, Madhubani, Sitamarhi and Kishanganj	Silty soil/silty sand	Tables 6.1–6.4
Middle and southern Plateau region	Patna, Bhabhua, Gaya, Bhagalpur and Banka region	Clay with low to medium plasticity clayey silt	Tables 6.5–6.10

strata of soil, mainly silty soil is present with a significant amount of clay particles. In the lower strata, the soil has mainly sand mixed with silt sized particles. Borehole data of different regions of South Bihar, namely Patna, Bhabhua, Gaya, Bhagalpur and Banka, are presented in Tables 6.6–6.10, respectively. In Patna (Table 6.6), the soil is mainly clay with intermediate and low plasticity. In Bhabhua (Table 6.7), the upper strata soil is silty clay, while in the lower strata, the soil is clayey with intermediate plasticity. The soil in the Gaya region (Table 6.8) is mainly clayey with the silt content having intermediate plasticity. In Bhagalpur (Table 6.9), the soil is mainly clay with intermediate plasticity. In Banka (Table 6.10), the soil is clayey up to a depth of 3 m with low plasticity while below this level, the soil is mainly silty sand. Since the geographical area of Bihar is quite large, it is difficult to present the variation in spatial soil stratigraphy from a geotechnical point of view in a comprehensive form Ground Water Yearbook Bihar, 2013–2014. However, a summary of regional variation of soil characteristics is summarised in Table 6.11, which may be referred to for preliminary study only. A detailed soil investigation should be carried out for the actual design of the structure at a particular site.

6.3 PROPERTIES OF SOILS

The properties of soils of different regions are presented in Tables 6.1–6.10. Soil properties of the Bettiah region can be seen in Table 6.1. The field density of the soil is found to vary from 1.85 to 1.94 g/cc. Soil is non-cohesive in nature with an angle of internal friction in the range of 14.7°–33.6°. Standard penetration test data show that the N value starts from 7 and increases with depth and is found to be 25 at a depth of 10.5 m. The depth of water table is found at 1 m below the ground level. In Madhubani (Table 6.2), the density varies from 1.92 to 1.99 g/cc. The cohesion of soil varies from 0.05 to 0.20 kg/cm^2 and the angle of internal friction varies from 13.2° to 18.1°. The N value is found in the range of 2–20 from the top to a depth of up to 10.5 m. The water table level is found to be 1.3 m below the ground level. The variation in density, angle of internal friction, the N value and depth of water table for other different regions of the state can be found in Tables 6.3–6.11.

6.4   USE OF SOILS AND ROCKS AS CONSTRUCTION MATERIALS

For the construction of the geotechnical structures in the state, different types of earth materials are used. The sand used in Bihar is mainly river sand obtained from different sand ghats approved by the Government of Bihar. There are currently 105 approved sand ghats. The maximum number of sand ghats in the state is 37 in the Patna district. The sand is mostly collected from the Ganga river. The sands are of mainly three categories, namely plastering grade, Zone-II/III/IV sands and white micro-silica sands. The white micro-silica sands mainly come from the tributaries of Ganga and they are mostly used for mortar making and embankment construction or as fill materials. The graded zoned sands and plastering grade sands are mostly being brought from the banks of the Sone river. Sone sands are being used for reinforced cement concrete, foundation and substructure works.

In Bihar, only Thuri has been a single quarry site for the production of aggregates, but it has been currently closed due to environmental obligations. This site has weak stones with a lower specific gravity. Currently, the aggregates are being brought from the neighbouring states, mainly from Jharkhand and West Bengal. These aggregates are being used in the construction of buildings, pavements and bridges and other construction works. Due to the shortage of these aggregates, many construction activities are being delayed or sometimes being closed for intermittent periods. The utilisation of local sands, soils and industrial byproducts is also highly necessary to use in lieu of these aggregates.

Locally available soils are mainly alluvial in nature and spread along the Gangetic plains of North and South Bihar. The southern districts have moorum type of soils. These soils are used as fill materials and in the construction of embankments, pavement subgrades and unpaved shoulders with and without modifications or stabilisations. As these soils are fine-grained, mostly silty and clayey, the modifications for these soils are necessary for utilisation. Bihar has a huge deficiency of aggregates, not having a single stone quarry. Modifications with different stabilisers for pavement and other construction purposes are highly beneficial.

Bihar has fly ash/pond ash as major byproducts from all five thermal power plants. These waste products are being utilised in reinforced earth walls and for making fly ash bricks. They are also being used for various purposes such as in the construction of granular subbase, fine aggregates for concreting and some other substructure purposes.

6.5   FOUNDATIONS AND OTHER GEOTECHNICAL STRUCTURES

Depending upon the requirement of the structure and strength of the foundation soil, the types of foundation used in the state are strip footing, isolated footings, raft foundation and pile foundation. In the North Bihar region, the soil in the upper strata is silty in nature, while in South Bihar, the soil is clayey with silt content. Isolated types of shallow foundations are often used in the state for light structures, such as residential buildings. For the construction of relatively heavy structures, where the isolated foundation is not sufficient, mat/raft foundation is commonly used. At present, different academic buildings of the newly established engineering colleges of Bihar are under construction. Most of the academic buildings, the engineering colleges, are being constructed over the raft foundation.

Pile foundation is used for the region where the water table is at a shallower depth or the bearing capacity of the soil is not sufficient to take the load. In cities like Patna, Muzaffarpur, Bhagalpur and Gaya, there are many infrastructural developments. In all these cities, pile foundation is popular among geotechnical engineers as the upper strata are made up of soft to medium clay and the water table is high. To support the bridge structure, a well foundation is commonly used. The depth of the well foundation varies from 20 to 30 m depending upon the scour depth. Mahatma Gandhi Setu, Koilvar Bridge, Rajendra Setu, Nehru Setu, JP Setu and so on are some remarkable bridges of Bihar. Mahatma Gandhi Setu is one of the longest bridges in India. At present, many bridges are under construction over the Ganga and other rivers. In the current construction practice, large-diameter pile foundations are being preferred in place of well foundations on the scourable riverbeds due to difficulties in the sinking operation of the well foundation.

6.6 OTHER GEOMATERIALS

Bihar is very rich in mineral resources. The important minerals found in the state are limestone in Kaimur (Bhabhua) and Monghyr; Mica in the Nawada district; quartz/silica sand in the Bhagalpur, Jamui, Monghyr and Nalanda districts. Iron ores, hematite and magnetite are found in the Bhagalpur and Gaya districts, respectively. Lead, zinc and pyrites are also found in the Rohtas district (Indian Minerals Yearbook, 2013).

From the thermal power plants, a huge amount of waste ash is produced in the form of fly ash and pond ash. Bulk production of these wastes creates environmental problems. These wastes can be used in the construction sector in the production of brick and cement and as fill materials. In Table 6.12, the properties of fly ash obtained from the Kanti thermal power plant, Muzaffarpur are presented. Rai et al. (2018) have shown that the specific gravity of the fly ash obtained is lighter than the local soil of Muzaffarpur. Hence, this fly ash can be used in the construction of light-weight geotechnical structures. It is also shown in the study that through the mixing of fly ash in soil, the load-carrying capacity of the local soil can be improved.

Other than the utilisation for geotechnical purposes, fly ash is also popular in the preparation of fly ash bricks. Malty and Chandran (2016) have reported a detailed

Table 6.12 Properties of the fly ash obtained from the Kanti thermal power plant (Rai et al., 2018)

Properties	Value
Specific gravity	2.32
Clay (%)	3
Silt (%)	70
Sand (%)	27
Plasticity characteristics	NP
Cohesion (kPa)	27
Angle of friction	29°
CBR (Unsoaked)	18
CBR (Soaked)	16

Figure 6.2 Utilisation of fly ash as a fill material near the guest house of the MIT Muzaffarpur, Bihar.

investigation on the fly ash brick industry in Bihar. As per their studies, major fly ash brick industries are growing around the thermal power plants located in the state. Policies and programmes prepared by the Ash Utilisation Division have a major focus on the maximum utilisation of fly ash. As per the study by Malty and Chandran (2016), fly ash can be utilised in cement, bricks, reclamation of low-lying areas, mine filling, roads and so on. The cement industry is the largest consumer of fly ash compared to others. Figure 6.2 shows the utilisation of fly ash as a fill material near the guest house of the Muzaffarpur Institute of Technology, Muzaffarpur.

6.7 NATURAL HAZARDS

Bihar is prone to many disasters, such as floods, drought, earthquakes, fire, cyclones, heat waves, and cold waves. Every year, many lives are lost by these disasters. Geographical and climatic conditions make the state vulnerable to these hazards. Some disasters like the earthquake that occurred on 15 January 1934 or the flood that occurred in 2008 in North Bihar are examples of worst disasters in the country. Figure 6.3 shows the multi-hazard map of Bihar. This map includes the hazards due to cyclones, earthquakes and floods. From an earthquake point of view, most of the region lies in Zone IV or V. Zone V is considered as the region of the most severe earthquake.

Figure 6.4 shows the earthquake map of Bihar. It can be observed that the districts of the northern part of Bihar like Sitamarhi, Madhubani, Supaul, Araria, Madhepura and some parts of Darbhanga lie in Zone V. This region covers about 15.2% of the total area of Bihar. The remaining part of North Bihar lies in Zone IV, while major parts of South Bihar are also in Zone IV. About 63.7% of the total geographical area of Bihar lies in this zone. The region of Bihar is earthquake-prone because of the tectonic

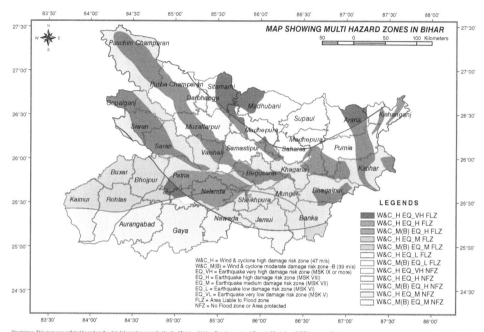

Figure 6.3 Multi-hazard map of Bihar (BSDMA, 2020).

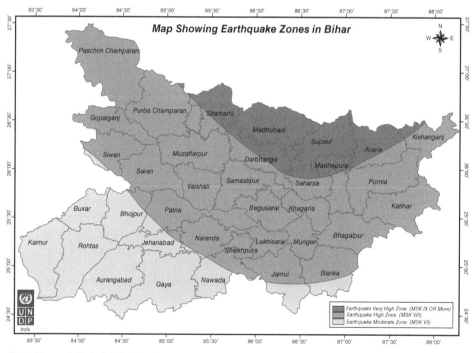

Figure 6.4 Earthquake zones of Bihar (BSDMA, 2020).

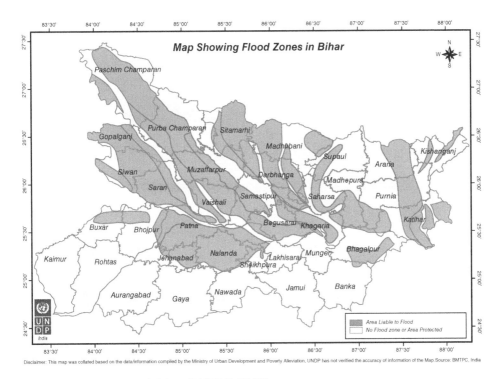

Figure 6.5 Flood zones of Bihar (BMTPC, 2020).

activities that occur in the Himalayan region. A total of six districts of Bihar out of 38 districts fall in Zone III.

Many perennial and non-perennial rivers flow through the state. The majority of the rainfall occurs during 3 months of the monsoon. Around 73% of the geographical area of Bihar is vulnerable to floods. This includes 28 districts of Bihar. Annually 30%–40% of food damages take place due to floods. There are four categories: Class I (flash floods), Class II (river floods), Class III (drainage congestion in river confluence) and Class IV (permanent waterlogged area). Figure 6.5 shows the flood map of Bihar. Most of the Northern districts of Bihar lie in the flood zone. The districts of the southernmost region of Bihar are free from floods. The 'Kosi' river is also known as the sorrow of Bihar. A recent collapse of a newly constructed bridge in the Gopalganj district took place due to heavy rainfall during floods (Figure 6.6). Figures 6.7 and 6.8 show the flood scenario where water is crossing over the road and the protection of the embankment and bank from the floods in the East Champaran district, respectively.

6.8   GEOENVIRONMENTAL IMPACT ON SOILS AND ROCKS

With the growing population, different human activities, including agriculture and construction, are increasing day by day, resulting in adverse impacts on the environment. Different pollutants get produced in the form of liquids, gases and solids. Solid wastes are becoming a great threat. Due to these wastes, pollution of the soil, groundwater, surface water and so on is taking place. Industrial wastes, urban wastes,

Figure 6.6 New constructed bridge collapse in Gopalganj after heavy rainfall in 2020.

Figure 6.7 Bihar floods in the East Champaran district.

Figure 6.8 Embankment and bank protection works against flood.

agricultural practices, pesticides, insecticides and so on are major sources of soil pollution. Solid wastes generated from industries include toxic pollutants, heavy metals, organic compounds, inorganic complexes and so on. Tannery industry is one of the major industries of the state. Muzaffarpur, Bettiah, Patna, Purnia, Munger and Aurangabad are the centres of tannery industry. From the tannery industry, chemicals like ammonia, calcium hydroxide, hydrochloric acid, sulphides and so on are generated (Dixit et al., 2015). Due to these wastes, groundwater and soil get polluted. Figure 6.9 shows the effluent pond of the leather industry of the Muzaffarpur district. In Muzaffarpur city alone, 478 tonnes of solid wastes are generated, out of which 258 tonnes of domestic waste (54%) get produced (Anand, 2009). Other than industrial wastes or domestic wastes, the excessive use of different chemicals in the form of fertilisers, pesticides and so on is also the cause for the pollution of soil and water. Since Bihar is one of the agricultural states, utilisation of such chemicals in agriculture is very frequent. Excessive utilisation of these chemicals is becoming a reason for the accumulation of toxic chemicals in soil and water.

Water quality assessment of Gaya district, Bihar was done between 2012 and 2014 and three villages were found to be fluoride endemic. A health survey was also conducted and the residents of these villages were found to suffer from dental, skeletal and non-skeletal fluorosis (Ranjan and Yasmin, 2012). Anomalies were also found in the thyroid function (Yasmin et al., 2013) and haematological parameters (Yasmin et al.,

Figure 6.9 Effluent pond near the leather industry of Muzaffarpur, Bihar.

2014). A detailed investigation found that around 50% of fluoride intake was through drinking water, while the rest 50% came through food crops grown locally and irrigated with the same fluoride-rich groundwater (Ranjan and Yasmin, 2015). Analysis of the composition of rock and soil samples can reveal the source of fluoride in the groundwater of the region. Therefore, XRD analysis of rock and soil was carried out to confirm the presence of fluoride contributing minerals. The fluoride contamination in groundwater in the Gaya district is not due to anthropogenic origin as there are no industries in and around. The Public Health and Engineering Department (PHED) of the Bihar Government has installed filtration units in the fluoride endemic areas for safe drinking water, but regular maintenance of such units is required. Provision for alternative safe water for irrigation and household purposes should also be made through supply from neighbouring fluoride non-endemic regions.

6.9 CONCLUDING REMARKS

Since Bihar is a large state, it is difficult to characterise the subsoil throughout the state with the help of a few boreholes. However, it can be broadly classified into three regions, viz. Gangetic alluvium of the Indo-Gangetic plain region, Piedmont Swamp soil which is found in the north-western part of the West Champaran district and Terai soil which is found in the eastern part of Bihar along the border of Nepal. In addition to this, the following facts and conditions regarding subsoil-related issues are presented:

- The state of Bihar has 96% of plains and is divided by the river Ganga into north and south regions.
- In the North Bihar plains, the soil has been observed to be silty in the upper layers and the areas near the flood plains are having thicker layers of silty soil with a higher silt content.
- In the South Bihar plains, the soil has been observed as clayey in the upper layers with low to intermediate plasticity.
- Sand is available as a major construction material in Bihar. The river sands are being utilised as mortar and being mined from the Ganga river and its Himalayan tributaries. The single source of graded sands is the river Sone.
- Isolated footings have been preferred for light structures and raft foundations have been preferred for heavy structures. Areas having shallow water tables and low bearing capacity viz. Patna, Muzaffarpur, Bhagalpur and Gaya pile foundation have been preferred.
- Bihar is prone to earthquakes and most of the regions are in high to medium risk zones. The silty layers with shallow water tables increase the vulnerability levels. Also, the north part of Bihar gets flooded by the Himalayan tributaries of Ganga causing deterioration of soils and infrastructure facilities.
- Due to the increase in urbanisation and inflow of industrial pollutants, the soil gets polluted. The Gaya district has been found to be affected by the fluoride endemic. However, other districts have not shown any serious environmental effects on soil.

REFERENCES

Anand, A. (2009). Planning for solid waste management of Muzaffarpur City Bihar. *Thesis for Master of Urban and Rural Planning*, IIT Roorkee.

BMTPC (2020). Building Materials & Technology Promotion Council, India. www.bmtpc.org, accessed 4 January 2020.

BSDMA (2020). Bihar state disaster management authority, Patna. http://www.bsdma.org/Natural%20Hazards.aspx, accessed 4 January 2020.

MSMEDI (2016). Annual progress report 2015–16, *Micro, Small and Medium Enterprises Development Institute, Patna*. http://msmedipatna.gov.in, accessed 01 January 2021.

CEA (2017). A report on fly ash generation at coal based thermal power stations and its utilization in the country for the year 2016–17, Central Electricity Authority, New Delhi.

Dasgupta, S. and Mukhopadhyay, B. (2015). Historical notes: Historiography and Commentary on the Nepal–India Earthquake of 26 August 1833. *Indian Journal of History of Science*, 50 (3): 491–513. doi: 10.16943/ijhs/2015/v50i4/48319

Dixit, S., Yadav, A., Premendra, D., Diwedi, D. and Das, M. (2015). Toxic hazards of leather industry and technologies to combat threat: A review. *Journal of Cleaner Production*, 87: 39–49.

Ground Water Yearbook Bihar (2013–2014). Central Ground Water Board, Ministry of Water Resources, Govt. of India, Mid-Eastern Region, Patna, India.

Indian Minerals Yearbook (2013). *Indian Bureau of Mines, Government of India*, 52nd edition, State reviews, Bihar.

Malty, S. and Chandran K. (2016). *The Fly Ash Brick Industry in Bihar, Development Alternatives*, New Delhi.

Rai, A.K., Priyadarshee, A., Kumar, V. and Kumar, A. (2018). Comparison of geotechnical behaviour of Muzaffarpur soil with locally available fly ash. *International Conference on Sustainable Waste Management through Design*, 125–132.

Ranjan, S. and Yasmin, S. (2012). Assessment of groundwater quality of Gaya region with respect to Fluoride. *Journal of Ecophysiology and Occupational Health*, 12: 21–25.

Ranjan, S. and Yasmin, S. (2015). Assessment of Fluoride intake through food chain and mapping of endemic areas of Gaya district of Bihar, India. *Bulletin of Environmental Contamination and Toxicology*, 94 (2): 220–224.

Ranjan, N.K. and Kumar, S. (2013). *Know Your State Bihar*. Arihant Publishers, New Delhi.

Yasmin, S., Ranjan, S. and Hilaluddinand D'Souza, D. (2013). Effect of excess fluoride ingestion on human thyroid function in Gaya region, Bihar, India. *Toxicological and Environmental Chemistry*, 95 (7): 1235–1243.

Yasmin, S., Ranjan, S. and D'Souza, D. (2014). Haematological changes in fluorotic adults and children in fluoride endemic regions of Gaya district, Bihar, India. *Environmental Geochemistry and Health*, 36 (3): 421–425.

Chapter 7

Chandigarh

S. K. Singh
Punjab Engineering College

CONTENTS

7.1   INTRODUCTION

Chandigarh, known as the City Beautiful, was planned by the famous French architect Le Corbusier. It is a union territory (UT) located at the foothills of the Shiwalik, about 250 km north of New Delhi. It lies between north latitudes 30° 40' and 30° 46' and east longitudes 76° 42' and 76° 51'. Chandigarh has an area of 114 km^2, out of which 36 km^2 is rural and the remaining 78 km^2 is urban (CDPR, 2010). The city has the specialty of being the joint capital of Punjab and Haryana states even though it is a UT itself. It is considered to be a part of the Chandigarh capital region or tricity, which includes Chandigarh, the city of Panchkula (in Haryana) and Mohali (in Punjab) as the satellite towns. The general slope of the land is towards the south. The elevation ranges from about 400 m above the mean sea level (MSL) in the foot hills to about 200 m above the MSL in the plains.

There are two major streams, Sukhna Choe and Patiali ki Rao. They originate from the Shiwalik Hill ranges and form the natural drainage of the city. The Sukhna Choe flows from the north to the south, drains the eastern part and joins the Ghaggar river. The other important stream is Patiala-ki Rao, which flows from the northeast to the southwest and drains the northern parts of the city. There is another N-Choe that flows through the leisure valley and drains the major parts of the city. It flows from the northeast to the southwest direction and traverses the north central part of the city. The Sukhna Choe has been dammed on the northeast side of the city, which has given rise to an artificial lake covering an area of about 1.62 km^2. A general map of Chandigarh along with the satellite towns Mohali and Panchkula has been shown in Figure 7.1.

DOI: 10.1201/9781003177159-7

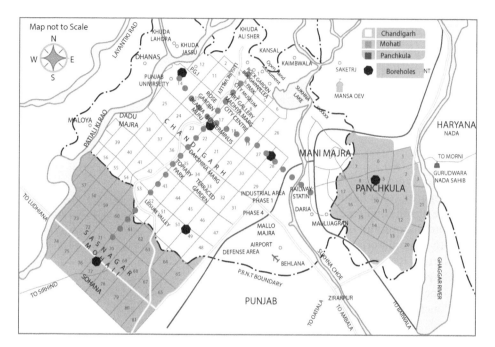

Figure 7.1 A general map of Chandigarh. (Source: https://chandigarhofficial.com, accessed, 5 Oct. 2020.)

Near Chandigarh within a 50 km radius, there are two thermal power plants, namely the Guru Gobind Singh Super Thermal Power Plant (GGSSTP) Ropar (6×210 MW) and the Rajpura super thermal power plant (2×700 MW). Coal ash generated from the super power plant is hydraulically deposited in the ash pond. Therefore, there is an abundance of fly ash and pond ash available within the 50 km distance from Chandigarh and that can also be used as a construction material in the subgrade, embankment, filling of low lying areas, cement industries and concrete production.

The subsoil stratigraphy of the tricity is depicted with data from representative boreholes located in different parts of the city. The location of six boreholes, four in the Chandigarh area and one each in the Mohali and Panchkula areas, has been marked as black dots on the map of Chandigarh. The borehole data have been taken from the geotechnical investigation programme for preparing a detailed project report for the proposed Chandigarh metro rail submitted to the Chandigarh Administration by the Delhi Metro Rail Corporation (DMRC, 2011) in Feb 2012.

The soil strata in Chandigarh mostly comprises clayey silt soils in the top layers followed by sandy silt to silty sands up to 30 m depth in most of the sectors. Soils are light yellowish-brown to pale brown in colour. In the northern parts of Chandigarh, the soil is sandy to sandy loam whereas it is loamy to silty loam in the southern parts. Ground water occurs under unconfined to semi-confined conditions at shallow depth (15–20 m) and under confined conditions in deeper aquifers (150–200 m) (CGWB, 2013).

In the Panchkula region, the sediments comprise boulders, pebbles, gravel and sand with clays mixed in varying proportions. Gravel and pebbles also occur occasionally.

In the hills and foothills, the soil is a mixture of sand and clay with coarse ingredients of pebbles and gravels. On the other hand, in Mohali, most of the area is plain having loam to silty clay and clayey silt except along the Ghaggar river and choes where some sandy patches are found. The quaternary sediments in the area, chiefly, are composed of clay, silt, sand or mixtures of the above with or without pebbles. Subsurface geological formations comprise fine to coarse-grained sand, silt and clay.

The Chandigarh city is situated in the foot hills of the Shivalik range. The Siwalik range forms the north-eastern boundary of Chandigarh covering the Kishangarh village, Manimajra (UT) and Panchkula (Haryana). It is exposed in a small patch on the north-eastern side constituting conglomerates, sandstone and claystone. Conglomerates have a sandy base with cobbles and pebbles of quartzite. Stray pebbles of granite, limestone, sandstone and lumps of clay stones are also present. An excavated vertical cut showing the stratigraphy of the soil in Sector 24 Panchkula is shown in Figure 7.2.

The south-western slopes of the foothills are covered with a loose talus material deposited by hill torrents forming alluvial fans. These alluvial fans coalesce to form piedmont Kandi formation running parallel to the hill ranges.

The ground water level has been observed to be nearer to the surface in the range of 3 to 5 m towards the western side (Mohali) and varied to significant depths of 6 to 20 m towards the eastern side. The water table elevation difference between the northern and south-western parts is 20 m. Due to this hydraulic difference, the groundwater moves from the north to the south-western direction. In the western area, groundwater flow is towards Patiala-ki-Rao and it flows parallel to Sukhna Choe.

The subsoil characteristics up to a depth of 30 m are presented in this chapter with the help of the borehole data for all the three regions, that is, Chandigarh (UT), Mohali (Punjab) and Panchkula (Haryana) collectively known as the tricity due to their adjacent locations. Their soil type and typical engineering properties are also compiled to facilitate the understanding of the subsoil stratigraphy of Chandigarh and its adjacent cities. In order to design seismic resistant structures, the liquefaction potential of the area has been identified based on the previous studies carried out by Dharmaraju et al. (2008) and Kandpal et al. (2009, 2018).

Figure 7.2 Open cut face for construction in Sector 24, Panchkula.

7.2   MAJOR TYPES OF SOILS AND ROCKS

The subsoil stratigraphy up to a depth of 30 m has been depicted based on the perusal of various borehole data obtained from the soil investigation programme. However, a set of six boring log data are presented in Figures 7.3 to 7.8 (Borehole # 1–6) to get an

BH-01 Near Gurudwara, Sector 77, Mohali

Depth (m)	Penetration resistance (N)	Samples	Graphic log	Material description	Grain size distribution % wt. retained				Atterberg limits			Bulk density (g/cc)	Moisture content (%)	Dry density (g/cc)	Specific gravity	Cohesion (kg/cm^2)	Angle of internal friction (degree)
					Clay	Silt	Sand	Gravel	L.L.	P.L.	P.I.						
	-			Filled up Strata	-	-	-	-	-	-	-						
2	-			Clayey Silt	14.38	84.14	1.48	0.00	32	19	13						
4	10.5				13.48	85.18	1.34	0.00	31	19	12						
6	-			Silty Sand with Clay	8.59	19.57	71.84	0.00	30	21	9						
	9.40				7.15	20.82	72.03	0.00	29	22	7						
8	-			Clayey Silt with Sand	12.25	51.49	35.94	0.32	32	22	10						
10	8.91				13.28	63.01	23.33	0.37	32	20	12						
	-				14.37	71.96	12.94	0.73	33	20	13						
12	9.94				17.38	72.96	9.00	0.66	33	18	15						
14	-				20.38	74.03	5.18	0.41	37	19	18						
	9.45				8.39	57.11	33.13	1.37	24	15	9						
16	-	UDS		Silty Sand with Clay	15.39	19.63	64.98	0.00	28	15	13	2.15	28.88	1.67	2.65	0.19	19.5
18	11.40			Clayey Silt with Sand	15.93	77.22	6.03	0.82	34	20	14						
20	-				17.38	74.55	7.40	0.67	35	20	15						
	10.40				16.03	77.58	5.54	0.85	33	19	14						
22	-	UDS		Silty Sand	0.00	12.10	87.90	0.00	25	NIL	NP	1.86	27.61	1.46	2.66	0	27.5
24	9.00			Clayey Silt with Sand	16.38	77.69	5.27	0.66	35	20	15						
26	-				13.82	78.78	6.84	0.58	32	20	12						
	9.66			Clayey Silt	19.28	77.43	1.96	0.00	36	19	17						
28					20.73	77.96	1.84	0.00	36	18	18						

Figure 7.3 Boring log (Borehole #1) in Sector 77, Mohali.

BH-02 Junction 49, Chandigarh

Depth (m)	Penetration resistance (N)	Samples	Graphic log	Material description	Grain size distribution % wt. retained				Atterberg limits			Bulk density (g/cc)	Moisture content (%)	Dry density (g/cc)	Specific gravity	Cohesion (kg/cm^2)	Angle of internal friction (degree)
					Clay	Silt	Sand	Gravel	L.L.	P.L.	P.I.						
	-			Clayey silt with sand	9.82	52.90	37.28	0.00	29	20	9						
2	-				13.23	49.80	36.97	0.00	32	21	11						
	8.19				11.21	49.66	39.13	0.00	33	23	10						
4	-				7.54	53.56	39.00	0.00	30	22	8						
6	9.40				8.43	56.44	35.33	0.12	30	21	9						
8	-				8.05	58.56	33.27	0.32	31	23	8						
	12.96				9.24	55.93	34.51	0.41	30	21	9						
10	-	UDS			9.24	58.05	32.30	0.36	30	21	9	2.20	23.82	1.78	2.67	0.11	22
12	14.91			clayey silt	10.95	84.84	3.85	0.00	31	21	10						
14	-				9.61	86.33	4.06	0.00	30	21	9						
	15.06				14.67	80.38	4.95	0.00	32	19	13						
16	-	UDS			18.47	77.24	4.29	0.00	37	21	16	2.12	20.71	1.76	2.64	0.41	5
18	15.20				18.26	77.24	4.20	0.00	36	20	16						
20	-				19.47	75.92	4.94	0.40	38	21	17						
22	15.56			Clayey silt with sand	18.36	77.29	4.29	0.64	37	21	16						
	-	UDS			16.74	51.97	4.20	0.63	35	20	15	2.25	27.70	1.76	2.68	0.37	8
24	16.62				16.78	50.63	4.61	1.63	36	21	15						
26	-				15.70	52.34	3.95	2.03	35	21	14						
	18.06				15.46	52.69	30.65	1.63	36	22	14						
28	-				14.70	53.61	31.96	1.32	33	20	13						

Figure 7.4 Boring log (Borehole #2) in Sector 49, Chandigarh.

overview of the subsoil stratigraphy of the Chandigarh region, including its satellite towns as Mohali (Punjab) and Panchkula (Haryana).

Three distinct types of subsoil stratigraphy exist in the tricity of Chandigarh (UT), Mohali (Punjab) and Panchkula (Haryana). Chandigarh, which is divided into differ-ent sectors from 1 to 56, has predominantly subsoil characteristics as clayey silt with

BH-03 Lekha bhavan Sector 17

Depth (m)	Penetration resistance (N)	Samples	Graphic log	Material description	Grain size distribution % wt. retained				Atterberg limit			Bulk density (g/cc)	Moisture content (%)	Dry density (g/cc)	Specific gravity	Cohesion (kg/cm^2)	Angle of internal friction (degree)
					Clay	Silt	Sand	Gravel	L.L.	P.L.	P.I.						
	-			Clayey Silt with Sand	13.5	75.83	9.14	1.78	33	21	12						
2	-				11.42	76.06	10.35	2.17	30	20	10						
4	20.4			clayey silt with sand and gravels	12.67	68.88	10.11	7.44	30	19	11						
	-			Clayey Silt with Sand	12.17	75.87	9.82	2.14	32	21	11						
6	21.3				12.83	77.17	10	0.00	33	21	12						
8	-	UDS			18.34	72.98	8.68	0.00	36	20	16	1.85	11.90	1.65	2.67	0.33	7
10	19.9				15.83	72.85	10.26	1.06	33	19	14						
12	-				13.72	70.16	13.68	2.42	32	20	12						
	27.3			Silty sand with gravel	0.00	10.37	80.53	9.10	25	NIL	NP						
14	-				0.00	7.13	83.91	8.96	24	NIL	NP						
16	18.0			Silty Sand	0.00	13.37	85.80	0.83	24	NIL	NP						
	-	UDS			0.00	7.16	92.33	0.51	25	NIL	NP	2.03	17.00	1.74	2.68	0	28.5
18	15.4				0.00	7.78	91.49	0.73	24	NIL	NP						
20	-				0.00	8.77	90.21	1.02	24	NIL	NP						
22	18.6				0.00	6.46	92.81	0.73	25	NIL	NP						
	-			Clayey Silt with Sand	23.16	62.50	12.5	1.84	43	22	21						
24	18.0				23.54	64.64	9.66	2.16	44	20	24						
26	-	UDS			29.42	61.63	5.81	2.51	45	18	27	1.99	24.39	1.60	2.64	0.51	6
	12.3				22.54	67.30	7.65	1.2	46	26	20						
28	-				21.43	68.96	8.41	0.00	43	24	19						

Figure 7.5 Boring log (Borehole #3) in Sector 17, Chandigarh.

sand of low plasticity/sandy silt of low plasticity underlain by non-plastic silty sand (Borehole # 2–5). The ground water table ranges from 5 to 25 m below the normal surface level (NSL) approximately as we move from the north to the south direction. The typical subsoil characteristics in the Mohali region (Borehole # 1) consists of clayey silt/silty sand in alternate layers up to a depth of 30 m. The water table in the Mohali region is at a shallow depth ranging from 3 to 5 m from the NSL.

BH-04 Near Punjab University, Sector 14

Depth (m)	Penetration resistance (N)	Samples	Graphic log	Material description	Grain size distribution % wt. retained				Atterberg limits			Bulk density (g/cc)	Moisture content (%)	Dry density (g/cc)	Specific gravity	Cohesion (kg/cm^2)	Angle of internal friction (degree)
					Clay	Silt	Sand	Gravel	L.L.	P.L.	P.I.						
	-			Filled up strata	-	-	-	-	-	-	-						
2	-			Clayey Silt with Sand	18.89	70.82	10.29	0.00	37	20	17						
	30.25				18.27	67.76	13.97	0.00	37	21	16						
4	-				19.42	61.12	18.81	0.65	37	20	17						
6	28.42				20.18	65.10	13.82	0.90	26	18	18						
8	-			Silty Sand	0.00	22.31	76.82	0.87	24	NIL	NP						
	32.76			sand	0.00	3.23	96.77	0.00	25	NIL	NP						
10	-	UDS			0.00	1.27	98.73	0.00	25	NIL	NP	1.79	10.26	1.62	2.66	0	28
12	36				0.00	3.50	96.50	0.00	24	NIL	NP						
14	-				0.00	3.45	96.55	0.00	25	NIL	NP						
	26.80			Silty Sand	0.00	8.42	74.95	0.00	25	NIL	NP						
16	-	UDS		Gravely sand with clay and silt	0.00	8.48	74.63	16.57	24	NIL	NP	1.99	6.34	1.87	2.66	0	28
18	16.24			Silty sand with gravel	0.00	18.31	77.82	7.06	24	NIL	NP						
20	-				0.00	14.17	87.77	8.01	25	NIL	NP						
22	15.56			Silty Sand	0.00	11.78	82.95	0.45	24	NIL	NP						
	-	UDS			0.00	12.05	8.06	0.00	25	NIL	NP	1.84	10.86	1.66	2.67	0	28
24	11.52			Clayey Silt with Sand	19.28	72.66	8.83	0.00	37	20	17						
26	-				17.93	73.24	23.04	0.00	36	20	16						
28	15.49			clayey silt with sand and gravels	19.23	50.34	19.87	9.95	37	20	17						
	-				20.13	51.35	19.80	8.65	37	19	18						

Figure 7.6 Boring log (Borehole #4) in Sector 14, Chandigarh.

The Panchkula (Haryana) region mostly falls in the Ghaggar river basin. Ghaggar is a perennial river, descending from the Himalayas in Himachal Pradesh. The Panchkula region is characterised by clayey silt with pebbles and gravels or silty sand ranging from a depth of 8 to 12 m underlain by gravel, kankar and boulders extending up to a depth of 30 m (Borehole # 6). The water table in this region ranges from 10 to 20 m from the NSL.

BH-05 Near Haryana panchayat bhavan, Sector 28

Depth (m)	Penetration resistance (N)	Samples	Graphic log	Material description	Grain size distribution % wt. retained				Atterberg limit			Bulk density (g/cc)	Moisture content (%)	Dry density (g/cc)	Specific gravity	Cohesion (kg/cm^2)	Angle of internal friction (degree)
					Clay	Silt	Sand	Gravel	L.L.	P.L.	P.I.						
	-			clayey silt with sand and gravels	16.93	50.76	19.05	13.26	35	21	14						
2	-			Clayey Silt with Sand	14.26	61.04	23.90	0.80	32	20	12						
4	39.0			Silty Sand	0.00	17.46	79.82	2.72	26	NIL	NP						
	-				0.00	18.82	79.92	1.26	25	NIL	NP						
6	26.7			clayey silt with sand and gravels	20.15	44.33	21.11	14.41	37	20	17						
8	-	UDS			27.35	48.00	8.93	15.72	47	23	24	1.74	12.78	1.54	2.62	0.43	4
10	27.2			Clayey silt	25.36	68.33	2.68	3.60	43	21	22						
	-				28.29	66.68	3.80	1.23	46	21	25						
12	25.8			Clayey Silt with Sand	23.06	56.66	18.94	1.34	43	23	20						
14	-				21.35	59.53	17.07	2.03	39	21	18						
16	25.8				19.26	67.39	9.09	4.26	38	22	16						
10	-	UDS			23.65	44.74	29.74	1.87	42	21	21	1.92	18.37	1.62	2.59	0.41	4.2
	32.4			Silty sand with gravel	0.00	18.01	69.64	12.35	26	NIL	NP						
20	-				0.00	17.70	71.28	11.02	24	NIL	NP						
22	23.4				0.00	15.85	73.95	10.20	26	NIL	NP						
	-				0.00	16.09	70.65	13.26	25	NIL	NP						
24	22.6				0.00	10.03	82.23	7.74	26	NIL	NP						
26	-			Silty Sand	0.00	17.92	77.87	4.21	25	NIL	NP						
28	17.1			Clayey silt	22.35	73.29	3.62	0.74	38	19	19						
	-			Clayey Silt with Sand	16.42	75.12	7.81	0.65	36	22	14						

Figure 7.7 Boring log (Borehole #5) in Sector 28, Chandigarh.

In Chandigarh and Mohali regions, boreholes for the soil exploration can be made easily with shell and auger boring in the clayey silt/sandy silt/silty sand strata and the standard penetration test (SPT) can be performed. In the Panchkula region, the bore-hole is to be advanced with rotary drilling with an NX drill bit in the gravelly/ boulder strata. Conducting SPT in these strata is difficult, but the dynamic cone penetration test can be performed; however, at many places, refusal is encountered.

| BH-06 Near Panchkula bus stand Sector 5 | | | | | | | | | | |

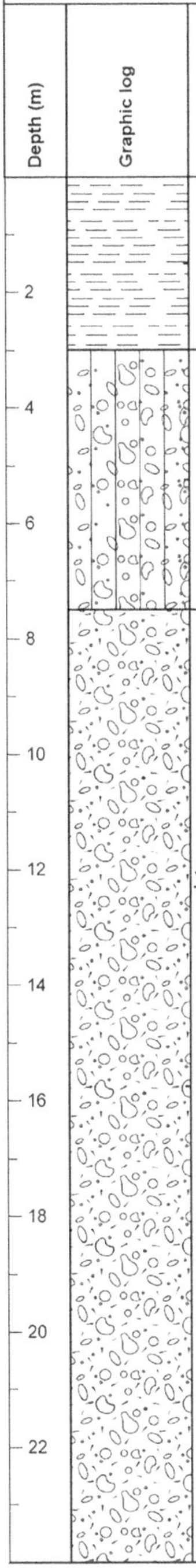

Depth (m)	Graphic log	Material description	Grain size distribution % wt. retained				Atterberg limits		
			Clay	Silt	Sand	Gravel	L.L.	P.L.	P.I.
		Clayey Silt with Sand	15.26	45.20	37.15	2.39	33	20	19
2			10.23	52.99	36.78	0.00	24	15	9
4		Silty sand with gravel	9.97	45.44	39.13	5.46	24	NIL	NP
			0.00	10.19	80.16	9.65	26	NIL	NP
6			0.00	9.15	80.59	10.26	25	NIL	NP
8		Boulders	-	-	-	-	-	-	-
10			-	-	-	-	-	-	-
			-	-	-	-	-	-	-
12			-	-	-	-	-	-	-
14			-	-	-	-	-	-	-
16			-	-	-	-	-	-	-
18			-	-	-	-	-	-	-
20			-	-	-	-	-	-	-
22			-	-	-	-	-	-	-
			-	-	-	-	-	-	-

Figure 7.8 Boring log (Borehole #6) in Sector 5, Panchkula.

7.3   PROPERTIES OF SOILS AND ROCKS

As observed from the borehole data in Chandigarh presented in Section 7.2, the predominant soil type is clayey silt with low plasticity (ML)/ non-plastic silty sand (SM) up to 30 m depth. The water table ranges from 5 to 20 m as the terrain slope is downgrading from the north to the south. As we move from the north to the south (Mohali side) in Chandigarh, the depth of the water table from the NSL decreases. Based on the number of borehole data, the soil profile is delineated from the north to the south direction covering sectors 3, 16, 17, 34, 43, 51 and 62 as depicted in Figure 7.9.

The locations of the boreholes are also shown in grey circles on the map of Chandigarh as shown in Figure 7.1. Similarly, the cross-section from the east to the west direction covering sectors 14, 17, 18, 19 and 28 and the Chandigarh railway station has been depicted in Figure 7.10.

A range of geotechnical properties has been drawn from the site investigation and laboratory testing at different borehole locations, as presented in Table 7.1.

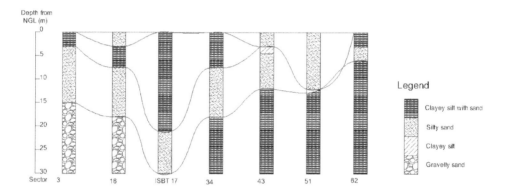

Figure 7.9 Subsoil profile from the north to the south direction in Chandigarh.

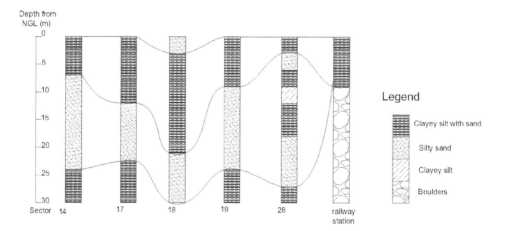

Figure 7.10 Subsoil profile from the east to the west direction in Chandigarh.

Table 7.1 Geotechnical properties of major soils in Chandigarh

Depth	Soil types	Field density (g/cc)	Specific gravity	Cohesion (kg/cm^2)	Angle of internal friction (°)	Water table level from the NSL
0–12 m	Clayey silt with sand/ silty sand	1.75–2.20	2.63–2.69	0.05–0.60/0–0.05	3–22/25–30	Southern sectors: 5–10 m Northern sectors: 10–25 m
12–30 m	Clayey silt/silty sand	1.80–2.01	2.64–2.90	0.10–0.70/0–0.05	4–18/25–30	

Table 7.2 Geotechnical properties of major soils in Mohali

Depth	Soil types	Field density (g/cc)	Specific gravity	Cohesion (kg/cm^2)	Angle of internal friction (°)	Water table level from the NSL
0–30 m	Clayey silt with sand/ clayey silt/ sandy silt	1.75–2.20	2.63–2.69	0.01–0.60/0. 10–0.70/ –	3–22/4– 18/25–30	3–5 m

Table 7.3 Geotechnical properties of major soils in Panchkula

Depth	Soil types	Field density (g/cc)	Specific gravity	Cohesion (kg/cm^2)	Angle of internal friction (°)	Water table level from the NSL
0–5 m	Silty sand with gravel	1.54–2.04	2.61–2.69	0	26–29	10–20 m
5–30 m	Boulders	–	–	–	–	

Similarly, the geotechnical properties for the Panchkula and Mohali regions are also given in Tables 7.2 and 7.3, respectively.

It can be noted that as we move from the east to the west direction, the layer of boulder mixed with gravel starts from the railway station Chandigarh to the Panchkula side. After perusal from the geotechnical properties (in-situ density, specific gravity and strength parameters) of different boreholes and subsoil profiles, a range of in-situ geotechnical properties are suggested and may be used for reference purposes.

The chemical analysis of the soil reveals that it is alkaline in nature and the pH ranges from 8 to 9.5. Chloride and sulphates are less than 0.005%. The chemical analysis of water used for construction purposes reveals that the pH value ranges from 7 to 8 with the ranges of chloride content and sulphate content ranging as 80–130 and 60–120 mg/l, respectively. The requirement of water as per IS 456 (2000) is that the pH value should not be less than 6, the maximum chloride content is 2,000 mg/l for

PCC and 500 mg/l for RCC and the maximum sulphate content is 400 mg/l. Therefore, the chemical contents for soil and water are within safe limits for construction purposes.

7.4   USE OF SOILS AND ROCKS AS CONSTRUCTION MATERIALS

Many geotechnical structures have been built in the tricity such as earthen dams, foundations of many multistorey buildings, basements and so on. The local soil has been used as a construction material for many earthen dams such as Kaushalya dam, Siswan dam and Jayanti dam. Out of these, the Kaushalya dam (Figure 7.11) is the largest earth-fill embankment dam having a length of 700 m and a height of 34 m. It is built on the Kaushalya river, which is a tributary of the Ghaggar river. Since the soils are of general nature, any specific problem is rarely observed in the construction projects.

7.5   FOUNDATIONS AND OTHER GEOTECHNICAL STRUCTURES

It is necessary to have sufficient information about the arrangement and behaviour of the underlying materials and their physical properties for adopting and designing the structural foundation. Soil/rock exploration through field investigation and relevant laboratory testing of the substrata material are essential to arrive at the required parameters for the design of the foundation. After perusal of subsoil data through a number of the exploratory boreholes covering the tricity region, it is observed that there is not any typical specific issue regarding the design and construction of foundation and geotechnical structures. Since the water table in the Mohali region is at a shallow depth ranging from 3 to 5 m, it requires dewatering for the construction of deep foundations. Also, since the type of soil is sandy silt/clayey silt and the water table is high, a detailed site-specific study for liquefaction potential should be carried out. In the Panchkula region, weathered rock mass and boulders are met at the depth of 5 to 7 m onwards. A photograph for the construction of the mat foundation to be laid in the Panchkula region is shown in Figure 7.12.

Figure 7.11 A view of Kaushalya Dam in Panchkula (Haryana).

Figure 7.12 Construction of mat foundation in Panchkula (Haryana).

7.6   OTHER GEOMATERIALS

Fly ash and pond ash generated from the GGSSTP, situated near village Ghanauli on the Chandigarh–Manali highway, are being used in various construction activities. Coal ash is being discharged in the form of slurry in three ash ponds spread in an area of 974 acres. A view of pond ash generated from GGSSTP is shown in Figure 7.13.

Fly ash is being used in cement industries and pond ash is being used for road construction, in landfills and also in cement industries. According to a report on fly ash generation and utilisation of the GGSSTP for the financial year 2018–19, fly ash and

Figure 7.13 Pond ash generated from the Guru Gobind Singh Super Thermal Power Plant (GGSSTP) Ropar.

Table 7.4 Properties of fly ash and pond ash

Properties	Fly ash	Pond ash
Specific gravity	2.12	2.26
% sand	2.90	83.20
% silt	90.40	16.50
% clay	6.70	0.40
Maximum dry unit weight (kN/m^3)	10.67	11.64
OMC (%)	35.5	26.8
c (kPa)	0	0
ϕ ($^\circ$)	30.6	34.2

pond ash that were utilised for different construction activities were 3.70 lakh metric tonnes and 9 lakh metric tonnes, respectively. Out of these, a maximum of 5.2 lakh metric tonnes of pond ash was used in road construction. Typical geotechnical properties of fly ash and pond ash are given in Table 7.4.

Geotechnical properties of pond ash are quite comparable with those of natural sand except for that it is a lighter material. However, once pond ash is compacted to the maximum dry density at OMC, the strength parameters are similar to those of natural sand. Therefore, the compacted pond ash could be a substitute for the natural sand material for all construction purposes.

7.7 NATURAL HAZARDS

Chandigarh, located just south of the Himalayan Frontal Belt, has been included in seismic Zone IV of the Seismic Zonation Map (IS: 1893, 2002) and has experienced earthquakes of magnitude 6 and above since historical times. The seismo-tectonic status of the area reveals that Chandigarh/Mohali is broadly associated with a seismic intensity of VIII on the MSK scale and has been categorised in the high hazard zone. The seismic effects in this zone vary from site to site depending on the geological, geomorphological and geotechnical conditions.

Liquefaction during earthquakes has always been a major cause of concern, particularly with respect to the destruction of constructed facilities. This is quite common in the case of saturated loosely packed sand deposits and silts while subjected to ground vibration. In most of the sectors in Chandigarh, it is noted that the soils in the top few metres are mostly clayey to clayey silt followed by sands/silty sands. However, the boreholes available from a few sectors indicate poorly graded and fine sands with low N-values based on the SPT, showing signs of potential to liquefaction in the case of major earthquakes. The ground water level has been observed to be near the surface, with the depth ranging from 1.8 to 8.0 m towards the western side (Mohali) and varying from 2.0 to 14.5 m towards the eastern side. A few sectors (6, 12, 25, 27, 30, 36, 37, 42, 43, 52, 54 and 63) have significant amounts of non-plastic silts and fine sands with low N-values and high water table conditions, which are susceptible to liquefaction in case of a major earthquake (Kandpal et al., 2009). Dharmaraju et al. (2008) conducted the study of the liquefaction potential assessment of the Chandigarh city based on local geology, seismo-tectonics and geotechnical variances. The susceptibility level of liquefaction hazard in sectors 6, 10, 12, 51 and 51A of Chandigarh for a peak ground acceleration of 0.08g was rated very high while it was low in sectors 1, 7, 8, 14, 17, 17A, 20C, 21A, 33, 35A, 38 West, 40B, 42, 44B, 45, 48, 52 and 63. The soil in sectors 9, 11, 15B, 16, 19B, 25, 30, 31, 37A and 54 was not found to be prone to liquefaction on account of the deeper ground water table and/or soil conditions.

7.8 CASE STUDIES

This is the case study (Datta and Singh, 2010) regarding the frequent breach of a micro hydel channel that takes of the water from the boiler of the Ropar thermal power plant and runs adjacent to the northern peripheral dyke of the ash pond. The canal is lined and the average temperature of the water at the beginning is 40°C. At the downstream of the ash dyke, the canal water energy is used to produce small electricity and after that, the tail water merges to the Sutlej River. This microhydel channel has witnessed breaches in 2008 and 2011. A photograph of such a breach that occurred in 2008 is shown in Figure 7.14.

The microhydel channels run above the ground (i.e. in filling) and are very close to the dyke of the adjacent ash pond as shown in Figure 7.15.

Figure 7.14 Canal breach occurred in 2008.

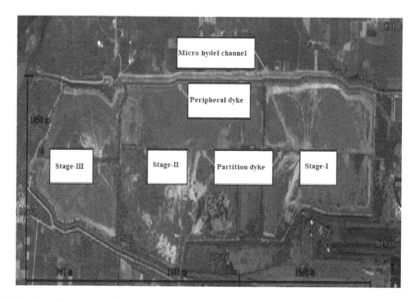

Figure 7.15 Satellite view of Ropar thermal power pond ash (2018).

The slope of the channel is observed to be steep (1.5H:1.0V). Stability analysis for channel and ash pond embankment is taken in conjunction. Stability analyses of the outer slope of the hydel channel section under low, medium and high phreatic conditions were carried out. The minimum FOS (Factor of Safety) for all the three cases comes out to be 1.5, 1.5 and 0.44, respectively. It implies that the outer slope of the hydel channel remains stable as long as the phreatic line remains low to medium. The downstream slope of the hydel channel becomes highly unstable (FOS = 0.44) with a high phreatic line as depicted in Figure 7.16.

Due to the adjacent ash dyke, whenever the ponding of water in the ash dyke raised the phreatic line in the hydel channel embankment, a breach in the hydel channel had occurred. Therefore, strengthening measures are required to make the hydel channel stable under high phreatic line conditions.

As a part of the strengthening measure, a 3 m wide berm has been provided in the downstream slope of the microhydel channel. Now, the stability analysis with berm under high phreatic conditions has been carried out and the FOS increased to 2.2 (Figure 7.17), which is a stable condition.

The following remedial measures have been recommended for additional safety of the breached portion of the embankment:

- provide an inclined chimney drain in the channel embankment (surrounded by transition filters) at the interface of impervious fill and random fill to intercept the seepage water before it reaches the toe,
- provide horizontal transition filter layers above and below the horizontal blanket drain at the base of the embankment below the random fill,
- provide an inspection road all along the toe of the embankment,

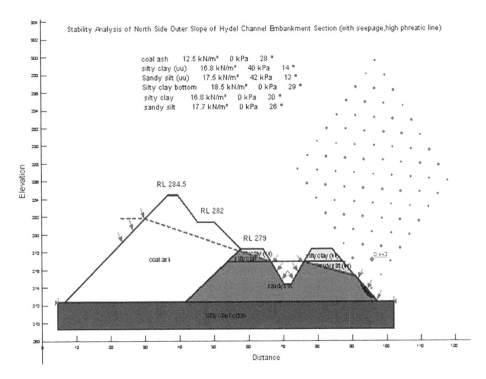

Figure 7.16 Stability analysis before strengthening.

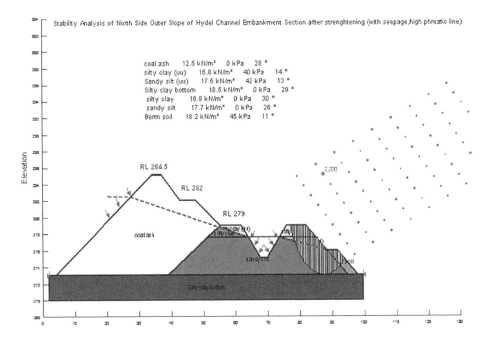

Figure 7.17 Stability after strengthening.

- increase the width of the rock toe by 0.6 m to provide a walking path along the embankment at the toe level and
- keep the rock toe free of vegetation.

7.9 GEOENVIRONMENTAL IMPACTS ON SOILS AND ROCKS

Improper storage, handling, transportation, treatment and disposal of waste result in an adverse impact on the ecosystem, including the human environment. When discharged on land, heavy metals and certain phytotoxic organic compounds can adversely affect soil and aquifers. In Chandigarh, the municipal solid waste (MSW) is disposed of in a landfill located in Daddu Majra (Sector 38-West). Chandigarh produces 479 metric tonnes of MSW per day (MoHUA, 2020). The available area for the landfill is 45 acres, out of which 25 acres have already been capped. Out of the 20 acres of land available for waste disposal, there is an engineered landfill of 8.3 acres.

The air and water pollutions are of great concern due to the impact of the landfill as the landfill site has come in the middle of Chandigarh due to development over the years. At present, in order to reduce the amount of waste to be disposed of directly in the landfill, a refuse-derived fuel (RDF) plant has been installed which processes the segregated inorganic waste to convert into RDF, thus, only a small amount of the residue from the RDF plant goes to landfill now (Johal and Kaur, 2020).

In Chandigarh, there are two motor markets located in sectors 28 and 38 in which repair, maintenance and servicing of motor vehicles are carried out. The area in the motor market gets contaminated due to the disposal/spill of engine oil and lubricants. The geotechnical property of the soil gets altered in the presence of these contaminants. It has been observed through field investigation that these contaminants are found up to a depth of about 1.5 m and they may further go down as time elapses.

7.10 CONCLUDING REMARKS

The subsoil characteristics in the city of Chandigarh (UT) and its neighbouring satellite cities Mohali (Punjab) and Panchkula (Haryana) vary significantly. A general observation regarding the soil profile, position of water table and construction practices has been documented for preliminary assessment of the site only. The following subsoil stratigraphy exists in the tricity:

1. In Chandigarh, the subsoil predominantly consists of clayey silt with sand of low plasticity or sandy silt of low plasticity underlain by non-plastic silty sand. In the northern parts of Chandigarh, the soil is sandy to sandy loam whereas it is loamy to silty loam in the southern parts. The ground water table ranges from 25 to 5 m below the NSL as we move from the north to the south direction.
2. The typical subsoil characteristics in the Mohali region consists of clayey silt or silty sand in alternate layers up to a depth of 30 m. The water table in the Mohali region is at a shallow depth ranging from 3 to 5 m from the NSL.
3. The Panchkula (Haryana) region mostly falls in the Ghaggar river basin. The Panchkula region is characterised by clayey silt with pebbles and gravels or silty sand ranging up to a depth of 8 to 12 m underlain by gravel, kankar and boulders

extending up to a depth of 30 m. The water table in this region ranges from 10 to 20 m from the NSL.

4. As such, there is no problematic soil in the tricity. Construction practices vary with the subsoil conditions. The details as presented in this chapter could not be a substitute for the detailed subsoil investigation, and hence, as required, site-specific investigation must be undertaken before construction of any structure in the Chandigarh tricity region.

REFERENCES

CDPR. (2010). *City Development Plan Report of Chandigarh*, Chandigarh administration. Jawahar Lal Nehru Urban Renewable Mission, Ministry of Urban Development, GOI, pp. 50–110.

CGWB. (2013). Hydrogeology of Chandigarh, *Report of Central Ground Water Board*, New Delhi, India.

Datta, M. and Singh, S.K. (2010). Raising of Dykes for Ash Pond at Guru Govind Singh Super Thermal Plant, Ropar, *Technical consultancy report*, PEC Chandigarh (India), submitted to GGSSTP, Ropar.

DMRC. (2011). *Geotechnical investigation for detailed project report for proposed phase 1 of Chandigarh metro rail approx. 33 km*, Carried out by CEG test house and research centre private Ltd. Jaipur and submitted to Delhi Metro Rail Cooperation Ltd., New Delhi.

Dharmaraju, R., Ramakrishna, V.V.G.S.T. and Devi, G. (2008). Liquefaction potential of Chandigarh city – a conventional approach, *Proceedings of the 12th International Conference of International Association for Computer Methods and Advances in Geomechanics*, Goa, India.

https://chandigarhofficial.com/, accessed on 5th Oct., 2020.

IS 456. (2000). *Indian Standard code of practice for plain and reinforced concrete*. BIS, New Delhi.

IS: 1893. (2002). *Indian standard code of practice on criteria for earthquake resistant design of structure*. BIS, New Delhi.

Johal, R.K. and Kaur, A. (2020). Local governance for sustainable solid waste management: a case study of Chandigarh (Union Territory), India. In: Ghosh, S. (eds) *Sustainable Waste Management: Policies and Case Studies*. Springer, Singapore. https://doi.org/10.1007/978-981-13-7071-7_8

Kandpal, G.C., John, B. and Joshi K.C. (2009). Geotechnical studies in relation to seismic microzonation of union territory of Chandigarh. *Journal of Indian Geophysics Union*, Vol. 13, No. 2, pp. 75–83.

Kandpal, G.C. and Agarwal, K.K. (2018). Assessment of liquefaction potential of the sediments of Chandigarh area. *Journal Geological Society of India*, Vol. 91, pp. 323–328.

MoHUA. (2020). Ministry of Housing and Urban Affairs, "*Swachhata Sandesh Newsletter*," January 2020.

Chapter 8

Chhattisgarh

Laxmikant Yadu and Sunny Deol Guzzarlapudi
National Institute of Technology Raipur

Mahasakti Mahamaya
O.P. Jindal University

Arun Bhave
Marhsal Geo Test Laboratory

CONTENTS

8.1 INTRODUCTION

Chhattisgarh is formed on November 1, 2000, by a partition from Madhya Pradesh. Soils of Chhattisgarh are mainly developed by the action and interactions of relief, parent material and climate. Biotic features, mainly the natural vegetation, follow climatic patterns. As per the Chhattisgarh Environment Conservation Board (CECB, 2016), this state has been broadly divided into three distinct plains according to agro-climatic zones. It comprises Chhattisgarh plains covering an area of 51%, 28% Bastar plateau and 21% northern hills zone (CECB, 2016). Primarily, Chhattisgarh plains comprise

DOI: 10.1201/9781003177159-8

red-yellow soil, silty sand, silt soils, sandy clay soils and clayey type soils. The Bastar plateau consists of red-yellow soil, silty sand, silt soils, laterite soils, red sandy soils and sandy clay soils. Northern hills consist of silty clay type soils, red-yellow soil, silty sand, silt soils, laterite soils, red sandy soils and sandy clay soils (CECB, 2016). These zones mainly comprise Entisols (alluvial type mostly: Bhata (laterite), Tikara (red sandy soil)), covering 19.5% area, Inceptisols (Matasi: red-yellow soil with mostly silty sand and silt) covering 14.8% area, Alfisols (Dorsa, Mal, Chawar: sandy clay with medium plasticity) covering 39% area, Vertisols (Kanhar, Gabhar, Bahra: clayey) covering 26.4% area and Mollisols (silty clay) covering 0.3% area (Singh et al., 2006).

Occurrences of garnet, amethyst, beryl, andalusite, kyanite, sillimanite and rare precious mineral alexandrite are also reported from different parts of the state. Deposits of grey, pink, red, and black (dolerite, amphibolite, and gabbro), granites, and flagstone of grey, black and purple shades are widely available. These are suitable for dimension stone and decorative purposes. The detailed map indicating the location of various minerals in Chhattisgarh is available in the CG Mineral Map (2021).

Chhattisgarh is a mineral-rich state having large economic deposits and other non-economic minerals. As per the Mineral Resources Department, the Government of Chhattisgarh, large deposits of coal, iron ore, limestone, dolomite and bauxite are available in various parts of the state. A substantial quantity of diamondiferous kimberlites has been identified in the Raipur district to yield diamonds (Deputy Director (Geology), 2018, 2020). A moderate quantity of deposits of tin (cassiterite) bearing pegmatites have been identified. Medium to small deposits of gold, base metals, quartzite, soapstone/steatite, fluorite, corundum, graphite, lepidolite and amblygonite of workable size are also noticed.

Chhattisgarh has a wide variety of rocks such as igneous, sedimentary and metamorphic rocks. Mainly limestone and shale rock types are available in various parts (Deputy Director (Geology), 2018, 2020). As per the district mineral survey report, this state is also a major hub for the cement-manufacturing industries. The availability of cement grade limestone at a large scale is found in various locations of the Baloda Bazar district. Mining leases have been given to various cement-manufacturing companies such as Ambuja cement, Shree cement, Emami cement, Ultratech, Grasim and Lafarge as shown in the CG Mineral Map (2021).

Chhattisgarh is a power hub since various thermal power stations are situated in the Korba region. This region is undergoing mining activities at a large scale to extract coal as shown in the CG Mineral Map (2021).

This chapter presents the overall geology, major soil formations and industrial by-products of Chhattisgarh. The rocks and soils of the state are investigated using boring data. A series of tests have been conducted to classify the soils and rocks in terms of their geotechnical properties. Based on the available data on the properties of industrial wastes, fly ash, ground granulated blast-furnace slag (GGBS), basic oxygen furnace slag (BOFS) and mine overburden are used as alternate construction materials.

8.2   MAJOR TYPES OF SOILS AND ROCKS

The geological structure of Chhattisgarh mainly consists of Archean and Cuddapah rocks, but dharwar, gondwana, deccan trap and old alluvial rock systems are also found in some parts of the state. Mainly five major types of soils are found in

Chhattisgarh, namely red-yellow soil, red sandy soil, red loam soil, black soil and laterite soil. These soils are named locally; for example, reddish-yellow soil as Matasi, red sandy soil as Tikra (Plateau of Bastar), black soil as Kanhar, and laterite soil as Bhata.

Table 8.1 presents the boring log data of various locations of different districts. The soil type available at different depths was collected from various districts of Chhattisgarh such as Surajpur, Sarguja, Raigarh, Korba, Mungeli, Balodabazar, Raipur, Durg, Rajnandgaon and Bastar. Predominantly, clay and sandy clay type soils are present at shallow depths in various locations. Similarly, Table 8.2 depicts boring log information of rock strata at the mine locations.

Table 8.1 Boring data for soil at various locations of Chhattisgarh

District	Location	Depth (m)	Soil description
Durg	Incubation centre	0.00–2.50	Medium sandy clay
	Bhilai, Nehru	2.50–3.00	Gravelly clay
	nagar, District:	3.00–3.75	Sandy clay
	Durg	3.75–10.00	Soft shale
	180 Men Barrack for	0.00–1.50	Medium coarses
	CISF RTC Utai,	1.50–3.75	Inorganic clay of low to intermediate compressibility (CL-CI)
	District: Durg		
		3.75–4.50	Strata change
		4.50–6.00	Gravelly sandy clay
		6.00–7.50	Weathered rock
		7.50–10.00	Disintegrated rock
Raipur	Kamal vihar, V. Y.	0.00–1.50	Inorganic clay of low to intermediate compressibility (CL-CI)
	Hospital, Raipur		
		1.50–3.00	Strata change
		3.00–4.05	Moderately strong
		4.05–5.26	Strong rock
		5.26–5.83	Moderately strong
		5.83–15.05	Strong rock
	Siltara area, Raipur	0.00–1.50	Inorganic clay of low to intermediate compressibility (CL-CI)
		1.50–3.00	Clayey sandy gravel
		3.00–4.50	Clayey gravelly sand
		4.50–5.80	Strata change
		5.80–15.10	Hard rock without lamination
Baloda bazar	Crusher building,	0.00–1.50	Clayey soil
	Bhatapara	1.50–20.00	Greyish limestone
Mungeli	Jawahar Navodaya	0.00–2.30	Soft clay (black)
	Vidyalaya, Mungeli	2.30–3.60	Medium stiff clay (yellowish)
		3.60–4.50	Gravelly clay with boulders
		4.50–6.00	Sandy gravelly clay
		6.00–7.50	Gravelly clay
		7.50–9.00	Sandy clay
		9.00–10.00	Sandy clayey gravel
Raigarh	Rambhata, Raigarh	0.00–1.50	Clayey sand
		1.50–2.90	Sandy gravel
		2.90–5.40	Moderately strong rock
Korba	Bhandarkar MCCPL,	0.00–1.50	Brown sandy soil
	Korba	1.50–10.10	Compacted sand

(Continued)

Table 8.1 (Continued) Boring data for soil at various locations of Chhattisgarh

District	Location	Depth (m)	Soil description
Surguja	Ambikapur	0.00–3.50	Clayey sand
		3.50–5.50	Inorganic clay of intermediate compressibility (CI)
		5.50–6.00	Sandy gravelly clay
		6.00–6.40	Strata change
		6.40–10.30	Moderately weak rock
Surjapur	Vishrampur	0.00–2.20	Yellowish colour clay
		2.20–3.70	Light reddish colour fine sand with clay
		3.70–10.00	Yellowish colour fine sand with clay
Rajnandgaon	Senior boy's dormitory,	0.00–3.00	Inorganic clay of low to intermediate compressibility (CL-CI)
	Dongargarh	3.00–3.75	Clayey sand
		3.75–5.00	Inorganic clay of low to intermediate compressibility (CL-CI)
		5.00–9.00	Disintegrated rock
		9.00–10.50	Clayey sand
Bastar	Super speciality hospital, Jagdalpur	0.00–1.50	Inorganic clay of low to intermediate compressibility (CL-CI) (reddish)
		1.50–2.00	Strata change
		2.00–7.50	Inorganic clay of low to intermediate compressibility with shale (CL-CI) (yellowish)
		7.50–10.50	High compressible clay with shale (CH) (yellowish)
		10.50–12.00	Inorganic clay of low to intermediate compressibility with shale (CL-CI) (yellowish)
		12.00–15.00	High compressible clay with shale (CH) (yellowish)

Table 8.2 Boring log data for rock of Raigarh coal field in Raigarh district, Chhattisgarh

Location	Depth (m)	Soil description
Gare Palma IV/8 sub-block, Gare Palma block, Mand Raigarh coalfield in Gharghor/Tamnar Tehsil	0.00–10.50	Soil sandy
	10.50–12.00	Sandstone (fine grained)
	12.00–14.00	Carb shale
	14.00–24.62	Sandstone
	24.62–26.00	Shaly coal
	26.00–29.00	Shaly shale
	29.00–41.40	Sandstone (coarse grained)
	41.40–44.62	Coal
	44.62–64.02	Sandstone
	64.02–64.72	Shaly coal
	64.72–68.00	Sandstone
	68.00–70.10	Shale dark grey
	70.10–76.85	Sandstone
	76.85–80.00	Shaly coal
	80.00–105.70	Sandstone (coarse grained)

(Continued)

Table 8.2 (Continued) Boring log data for rock of Raigarh coal field in Raigarh district, Chhattisgarh

Location	Depth (m)	Soil description
	105.70–107.18	Coal
	107.18–110.00	Shale, dark grey
	110.00–119.00	Sandstone (coarse grained)
	119.00–122.00	Shale
	122.00–155.92	Sandstone
	155.92–157.71	Sandstone
	157.71–158.32	Carb shale
	158.32–160.32	Coal
	160.32–179.74	Sandstone
	179.74–181.28	Coal
	181.28–196.35	Sandstone (coarse grained)
	196.35–197.25	Coal
	197.25–198.38	Sandy shale
	198.38–213.86	Sandstone (coarse grained)
	213.86–215.05	Shaly coal
	215.05–221.00	Sandstone
Location-II	0.00–10.50	Loose soil (alluvial soil)
	10.50–15.00	Shale
	15.00–25.50	Fine to coarse-grained sandstone
	25.50–27.00	Coal and shale
	27.00–43.50	Medium coarse-grained sandstone
	43.50–46.50	Coal
	46.50–52.50	Fine grained sandstone
	52.50–58.00	Coarse-grained sandstone
Ambuja cement, Raigarh	15.00–26.54	Fine to medium grained sandstone
	26.54–27.74	Grey shale
	33.0–36.56	Coal and carb. Shale
	42.00–46.28	Coal and shaly coal
	46.28–57.00	Sandstone shale intercalation
	63.00–66.00	Fine grained sandstone
	69.00–75.00	Coal and shaley coal
	75.00–90.61	Fine grain sandstone
	90.61–94.91	Grey shale
	94.91–104.49	Fine grain sandstone
	104.49–119.14	Coal and shaley coal
	119.14–135.00	Fine grain to coarse grain sandstone
	135.00–148.50	Medium grain sandstone
	148.50–152.01	Coal
	152.01–188.60	Medium grain sandstone
	188.60–190.61	Coal
	190.61–216.00	Fine to coarse grain sandstone
	216.00–225.40	Carb Shale and shaily coal
	225.40–266.36	Medium to fine grain sandstone
	266.36–270.95	Coal
	270.95–295.43	Fine to medium grain sandstone
	297.00–324.00	Medium grain sandstone

8.3 PROPERTIES OF SOILS AND ROCKS

Typical geotechnical properties of diverse types of soils at various locations in the state of Chhattisgarh have been reported here. These geotechnical properties include physical, volumetric, chemical and strength properties to understand the significant behaviour of soils for a wide variety of engineering applications. Tables 8.3–8.6 depict the descriptive statistics of typical soil properties identified in various districts of Chhattisgarh. Figure 8.1 shows typical photographs of soils collected from different locations of Chhattisgarh.

Table 8.3 Summary of geotechnical properties of problematic soil of various locations of Dhamadha and Gunderdehi block of the Durg district

Property	Mean	Standard deviation	Coefficient of variation	Minimum	Maximum	No. of soil samples
Liquid limit (W_L), %	45	5.00	0.11	35	56	14
Plastic limit (W_P), %	25	3.79	0.16	19	33	14
Shrinkage limit (W_s), %	15.98	3.44	0.21	5.9	21.3	14
Plasticity index (I_P)	20	3.98	0.19	16	27	14
Specific gravity (G)	2.39	0.12	0.05	2.2	2.6	14
Maximum dry density (γ_{dm}), kN/m^3	17.90	0.05	0.03	17.0	18.8	14
Optimum moisture content (W_O), %	14.01	1.00	0.07	12.6	16.1	14
Unconfined compressive strength (UCS), kPa	122.94	89.96	0.73	26.9	354.6	14
Cohesion I, kPa	39.07	41.64	1.07	5.9	157.0	14
Angle of internal friction (φ)	25.21	9.17	0.36	3.6	37.4	14
California bearing ratio (CBR) soaked	2.51	0.71	0.28	1.9	4.8	14

Table 8.4 Geotechnical properties of soils at various locations of the Narayanpur district

Property	Mean	Standard deviation	Coefficient of variation	Minimum	Maximum	No. of soil samples
Gravel fraction, %	10.21	5.55	0.54342	0.70	20.78	29
Coarse sand fraction, %	6.57	2.69	0.40925	2.20	13.20	29
Medium sand fraction, %	39.64	11.21	0.28277	19.20	57.60	29
Fine sand fraction, %	19.28	6.43	0.33337	9.40	34.78	29
Silt + clay, %	24.27	18.77	0.77352	0.50	59.00	29
W_L, %	31	3.60	0.11506	25	39	29
W_P, %	21	3.47	0.15483	15	27	29
I_P, %	10	4.61	0.48222	5	29	29
W_s, %	12.52	2.44	0.19462	7.00	18.00	29
Free swell index, %	10.75	5.15	0.47855	0.00	20.00	29
Specific gravity	2.60	0.01	0.00431	2.57	2.61	29
Organic content (%)	1.77	0.15	0.08469	1.30	1.97	29
Sulphate content (%)	0.08	0.07	0.81498	0.01	0.20	29
Carbonate content, (%)	0.17	0.02	0.09889	0.12	0.20	29
γ_{dm}, kN/m^3	18.46	0.86	0.04641	17.00	20.00	29
W_O, %	11.14	1.40	0.12612	8.20	14.00	29
CBR, %	5.18	0.96	0.1863	3.00	7.20	29
UCS, MPa	0.35	0.20	0.57989	0.12	0.95	29
Plasticity product	2.17	1.74	0.80282	0.03	5.60	29
Uniformity coefficient, C_u	24.26	17.18	0.70805	3.90	63.00	29

Table 8.5 Geotechnical properties of soils at various locations of the Rajnandgaon
district

Property	Mean	Standard deviation	Coefficient of variation	Minimum	Maximum	No. of soil samples
Gravel fraction, %	4.46	5.89	1.32	0.00	25.60	18
Coarse sand fraction, %	8.29	9.64	1.16	0.00	28.60	18
Medium sand fraction, %	20.56	16.59	0.81	1.60	43.80	18
Fine sand fraction, %	7.33	5.75	0.78	1.80	25.80	18
Silt + clay, %	59.35	27.95	0.47	24.40	95.60	18
W_L, %	36	5.96	0.17	18	46	18
W_P, %	24	3.50	0.15	17	30	18
I_P, %	12	3.92	0.31	1	19	18
W_s, %	15.01	5.09	0.34	5.70	25.30	18
Free swell index, %	11.28	6.64	0.59	5.00	29.00	18
Specific gravity	2.60	0.01	0.00	2.59	2.62	18
Organic content (%)	1.74	0.16	0.09	1.31	1.95	18
Sulphate content (%)	0.12	0.03	0.25	0.06	0.18	18
Carbonate content, (%)	0.18	0.02	0.13	0.12	0.20	18
γ_{dm}, kN/m^3	18.18	1.72	0.09	16.00	20.70	18
W_O, %	11.68	1.79	0.15	8.60	14.40	18
CBR, %	6.37	3.98	0.62	2.40	18.80	18
UCS, MPa	0.36	0.27	0.77	0.04	0.91	18
Plasticity product	7.98	5.21	0.65	0.19	18.00	18
C_u	6.42	4.18	0.65	2.38	15.55	18

Table 8.6 Geotechnical properties of soils at various locations of the Kanker district

Property	Mean	Standard deviation	Coefficient of variation	Minimum	Maximum	No. of soil samples
Gravel fraction, %	4.46	4.85	1.09	0.39	12.60	12
Coarse sand fraction, %	2.93	3.71	1.27	0.30	10.00	12
Medium sand fraction, %	22.48	6.53	0.29	12.80	34.00	12
Fine sand fraction, %	19.56	8.42	0.43	9.00	40.37	12
Silt + clay, %	50.56	12.15	0.24	24.76	63.00	12
W_L, %	30	6.00	0.20	20	43	12
W_P, %	20	4.58	0.24	14	32	12
I_P, %	11	3.28	0.32	4	15	12
W_s, %	11.78	4.69	0.40	4.20	20.70	12
Free swell index, %	4.05	1.62	0.40	1.50	7.20	12
Specific gravity	2.63	0.04	0.02	2.57	2.69	12
Organic content, %	1.71	0.32	0.19	1.18	1.99	12
Sulphate content, %	0.17	0.02	0.11	0.14	0.19	12
Carbonate content, %	0.18	0.02	0.10	0.14	0.19	12
γ_{dm}, kN/m^3	19.55	0.55	0.03	18.60	20.60	12
W_O, %	9.05	1.77	0.20	6.00	12.20	12
CBR, %	6.54	6.07	0.93	2.50	24.50	12
UCS, MPa	0.52	0.32	0.62	0.05	1.26	12
Plasticity product	5.11	1.97	0.39	1.73	8.32	12
C_u	9.55	7.29	0.76	5.10	30.00	12

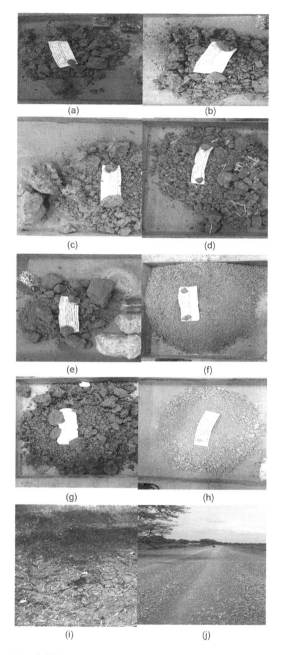

Figure 8.1 Typical soils of Chhattisgarh. (a) Clay type soil: T-07 to Udri Navagaon road, Khairagarh. (b) Silty clay type soil: Dumariya to Benni, Chhuikhadan. (c) Clay type soil: Malud to Gondra road, Khairagarh. (d) Silty clay type soil: Dumardih to Rengakatera, Rajnandgaon district. (e) Silty clay type soil: Birkha to Khera Navapara, Chuyidakhadan. (f) Red sandy soil: Choutna to Bharritola, Dongargarh. (g) Black clay soil: T05 to Mudukusara, Dongagarh. (h) Silty gravel soil: Nadiya to Basavar; Chhuikhadan. (i) Soil type in Tatibandh Raipur district. (j) Soil type in Baloda Bazar district.

The samples are collected from several locations of the district to investigate the different properties of soil. Some selected descriptive parameters such as mean, standard deviation and coefficient of variation of each property of soil samples are reported in Tables 8.3–8.6.

Similarly, lithological information has been collected from several locations of the state and the same has been reported in Tables 8.7 and 8.8. Typical photographs of the rock strata are given in Figure 8.4. The lithological information includes both the physical and mechanical properties of the identified rocks.

Table 8.7 Lithological information of Raigarh coalfield in the Raigarh district, Chhattisgarh

Depth (m)	Uniaxial compressive strength (MPa)	Density (kN/m³)	Slake durability index (%)
0.00–10.50	–	–	–
10.50–15.00	18.86	22.65	98.25
15.00–25.50	9.73	21.68	76.80
25.50–27.00	9.70	12.75	98.00
27.00–43.50	12.50	21.78	68.81
43.50–46.50	28.30	13.24	98.64
46.50–52.50	19.03	21.68	96.73
52.50–58.00	12.07	20.89	91.84

Table 8.8 Lithological information of Ambuja cement in the Raigarh district, Chhattisgarh

Depth (m)	U (MPa)	T (MPa) average		ts (MPa)	W (kN/m³)	S (MPa)	Y (MPa)	P (MPa)
		Cohesion	φ					
15.0–26.54	16.33	2.58	48.10	2.30	27.57	1.13	2,110	1.79
26.54–27.74	12.91	2.31	46.66	2.08	23.05	1.65	842	0.68
33.00–36.56	12.98	–	–	1.93	19.81	1.08	–	0.19
42.00–46.28	6.22	–	–	1.02	16.18	0.52	–	0.11
46.28–57.00	13.23	3.54	33.44	1.95	24.53	1.25	1,500	0.86
63.00–66.00	8.87	–	–	1.78	24.91	1.11	–	0.71
69.00–75.00	–	–	–	–	17.17	–	–	0.32
75.00–90.61	12.15	4.42	31.73	1.70	22.95	1.27	1,430	1.25
90.61–94.91	8.17	–	–	1.77	22.17	1.02	–	0.70
94.91–104.49	10.31	–	–	1.37	24.32	1.66	–	1.05
104.49–119.14	8.73	–	–	1.20	20.61	1.53	–	0.85
119.14–135.00	5.05	1.64	32.83	0.81	23.44	1.23	750	0.74
135.00–148.50	8.40	1.48	50.67	1.60	21.68	1.55	1,000	1.05
148.50–152.01	15.92	–	–	1.75	13.53	1.15	–	0.26
152.01–188.60	6.26	2.17	16.38	0.97	23.25	1.30	670	0.70
188.60–190.61	–	–	–	–	19.72	–	–	0.28
190.61–216.00	9.72	2.93	26.87	1.27	24.03	1.39	2,000	0.52
216.00–225.40	18.71	–	–	2.66	21.09	2.30	–	0.98
225.40–266.36	7.14	–	–	1.11	22.17	1.45	–	1.07
266.36–270.95	6.39	–	–	1.33	12.26	0.59	–	0.57
270.95–295.43	11.78	1.56	43.70	1.50	24.32	2.04	1,210	1.06
297.00–324.00	6.95	1.21	41.85	1.35	24.32	1.29	1,600	1.24

Note: U = average uniaxial compressive strength, T = average triaxial strength, ts = average tensile strength, W = unit weight, S = shear strength, Y = Young's modulus of elasticity, P = point load strength.

8.4   USE OF SOILS AND ROCKS AS CONSTRUCTION MATERIALS

Soil and rock reserves available in Chhattisgarh are identified to be good construction materials. Various types of naturally available soils and rocks are being used in diverse civil engineering works as backfill materials, pavement materials, embankment materials, building materials and so on. The most common type of soils and rocks used as construction materials are soil moorum, sand (coarse and fine), crushed stone aggregates and stone dust. Some of the significant applications of these materials in geotechnical and other civil engineering structures are discussed below.

8.4.1   Soil moorum

The most common type of soil being used as a good foundation material for various civil engineering structures in the state of Chhattisgarh is lateritic soil (soil moorum). This soil is being used as (i) a good backfill material in the construction of buildings, (ii) foundation soil, (iii) subgrade and subbase in flexible and rigid pavements (roadways, railways, and air fields), (iv) embankment for approach roads and (v) earthen dams. This type of soil has been collected from various locations of the state for the investigation and some details are presented in Table 8.9 as per their applications. The plasticity index of this type of soil varies from 12% to 20%, cohesion (UU) from 7.5 to 94.6 kPa and the corresponding internal friction angle value from 33° to 37°. Similarly, the maximum dry unit weight varies from 20 to 23 kN/m^3 and California bearing ratio (CBR) from 15% to 24%. Moorum soil is primarily used as the best substitute for clayey soils. It significantly enhances the structural integrity of foundations (footing/subgrade) and thereby reduces the cost of the superstructure by minimising the quantity of other building materials to be used. Moorum soil is also used as a subbase material (unbound and bound) in low volume flexible pavements by partial or full replacement of conventional crushed stone aggregates. Thus, its use minimises the consumption of natural aggregates.

8.4.2   Natural stone aggregates

Crushed stone aggregates (coarse and fine) obtained from naturally available limestone is identified to be the predominant construction material in Chhattisgarh. This type of rock is being used as a (i) railway track ballast, (ii) granular material (subbase and base) in flexible and rigid pavements (roads and airfields), (iii) key ingredient in cement and bituminous concrete, (iv) retaining wall (Gabion) and (v) insulating material in electric substations. The typical properties of these rocks from different locations are investigated and some details are presented in Table 8.10 according to their applications. Some of the key properties of rocks, namely, abrasion value, impact value, water absorption, specific gravity and soundness, are presented in Table 8.10. Partial and full replacement of sand with crushed stone dust has become a common practice as a fine aggregate in preparing cement and bituminous concrete mixes. The crushed stone dust material has become the best substitute for conventional sand that satisfies the physical and mechanical requirements, thereby reducing the overall cost of the material to be used.

Table 8.9 Typical geotechnical properties of soil moorum at various locations of Chhattisgarh

Type of geotechnical structure	Property/test	Location-I		Location-II		Location-III		Location-IV	
Embankment for approach road	W_L, %	46		47		18		46	
	W_P, %	30		28		Non-plastic		28	
	W_s, %	21.60		17.90		19.20		22.30	
	Grain size analysis	Size (mm)	% finer	Size (mm)	% finer	Size (mm)	% finer	Size (mm)	% finer
		4.75	76.35	4.75	76.62	4.75	83.53	4.75	36.8
		2.36	61.26	2.36	63.02	2.36	76.35	2.36	27.64
		1.18	34.08	1.18	43.35	1.18	55.27	1.18	22.91
		0.600	31.86	0.600	42.37	0.600	49.96	0.600	22.45
		0.425	25.75	0.425	39.48	0.425	26.81	0.425	22.00
		0.300	24.89	0.300	39.14	0.300	23.05	0.300	21.79
		0.150	22.29	0.150	37.8	0.150	15.87	0.150	21.24
		0.075	21.07	0.075	36.55	0.075	14.59	0.075	21.00
		0.064	20.97	0.069	36.44	0.065	14.47	0.061	20.88
		0.055	20.59	0.054	36.06	0.056	14.3	0.052	20.73
		0.041	20.56	0.049	36.03	0.044	14.27	0.04	20.68
		0.031	20.47	0.032	35.89	0.03	14.2	0.03	20.64
	Permeability (cm/sec)	1.08×10^{-6}		1.04×10^{-5}		3.60×10^{-4}		5.05×10^{-6}	
	Compaction test	$\gamma_{dm} = 22.27$ kN/m³ $W_O = 8.50\%$		$\gamma_{dm} = 20.30$ kN/m³ $W_O = 11.50\%$		$\gamma_{dm} = 21.38$ kN/m³ $W_O = 8.0\%$		$\gamma_{dm} = 22.46$ kN/m³ $W_O = 6.50\%$	
	Triaxial test (UU)	$c = 45$ kN/m² Friction angle $(\Phi) = 37°$		$c = 76.8$ kN/m² Friction angle $(\Phi) = 35.90°$		$c = 7.5$ kN/m² Friction angle $(\Phi) = 33°$		$c = 94.8$ kN/m² Friction angle $(\Phi) = 35.10°$	
	Triaxial test (CD)	Cohesion I = 19.5 kN/m² Friction angle $(\Phi) = 22.5°$		Cohesion I = 20.90 kN/m² Friction angle $(\Phi) = 19.60°$		Cohesion I = 15.0 kN/m² Friction angle $(\Phi) = 20.0°$		Cohesion I = 17.8 kN/m² Friction angle $(\Phi) = 23.5°$	
	CBR (soaked), %	17.80		15.30		19.70		18.40	
Subgrade for airfield pavement	W_L, %	33		34		35			
	W_P, %	15		15		14			
	I_P, %	18		19		21			
	W_s, %	16		17		18			
	Gravel fraction, %	24.38		27.54		23.72			
	Coarse sand fraction, %	46.18		43.79		42.76			
	Medium sand fraction, %	5.26		4.70		4.81			
	Fine sand fraction, %	4.97		3.95		6.17			
	Silt + clay, %	19.21		20.01		22.54			
	γ_{dm}, kN/m³	21.79		21.61		21.58			
	W_O, %	10.43		10.51		10.65			
	CBR, %	23.40		22.10		21.19			

Table 8.10 Typical properties of limestone at various locations of Chhattisgarh

Type of geotechnical structure	Property/test	Location-I	Location-II	Location-III
Railway track ballast	Abrasion value (hardness)	16.80%	15.89%	–
	Impact value(toughness)	8.50%	10%	–
	Water absorption value (porosity)	0.39%	0.17%	–
Railway track ballast	Abrasion value (hardness)	10.3%	14.7%	13.5%
	Impact value(toughness)	5.4%	7.5%	3.3%
	Water absorption value (porosity)	0.23%	0.34%	0.37%
Pavement materials (subbase, base, bituminous concrete and cement concrete)	Aggregate impact value	11.14%	13%	8.1%
	Water absorption	0.220%	0.269%	0.4%
	Soundness (sodium sulphate)	1.45%	1.28%	1.32%
	Specific gravity	2.653	2.675	2.669

8.5   FOUNDATIONS AND OTHER GEOTECHNICAL STRUCTURES

Problematic soils and rocks often cause a challenge for civil engineers in ensuring the structural health of foundations and other geotechnical structures. Some of the key challenges identified along with the conventional and non-conventional solutions are discussed in this section.

8.5.1   Foundation

Foundations built over expansive soil adversely affect the structural integrity of the overall building. Figure 8.3 shows a strip footing founded on expansive soil. The expansive behaviour of this foundation soil leads to uneven settlement. Furthermore, progresses to cause cracks and damages in various locations of the existing building are shown in Figure 8.2.

8.5.2   Pavements

Some of the common failure theories of flexible and rigid pavements that often challenge pavement engineers are rutting failure (deformation along wheel path) and differential settlements. One of the common causes of this form of failure is poor subgrade soil, that is, expansive soil. Few engineering practices are being implemented in the state of Chhattisgarh such as (i) replacement of expansive soil up to shallow depth with selected soil preferably of low to medium plasticity and high bearing capacity, (ii) mechanical or chemical stabilisation of expansive soil to reduce the plasticity, swelling and shrinkage characteristics, and (iii) laying of the geotextile layer.

(a)   (b)

Figure 8.2 Failures in the foundation on expansive soil. (a) Strip foundation rests on expansive soil. (b) Defects in the structure due to uneven settlement of foundation rests on expansive soil.

8.5.3   Embankment of the flyover approach

Construction of the flyover approach with reinforced soil walls is an involved process requiring due diligence and quality control. Moreover, repairs and remedial are often laborious, difficult, time-consuming, expensive, often ineffective in the long run and in most cases impossible to implement. Careful consideration shall be given to construction procedures by the construction agency. One specific case study in Chhattisgarh has been reported to be reinforced earth wall failure due to improper drainage of the embankment that further progressed to the failure of the embankment as shown in Figure 8.3a and b. The major issues diagnosed after a thorough forensic investigation are mainly

(a)   (b)

Figure 8.3 Heavy settlements in the approach road. (a) High severity settlements and depression up to 550 mm depth. (b) Dislocation of RS wall panels.

(i) poor drainage system in the embankment, (ii) poor compaction of the embankment material and (iii) catastrophic bulging of abutment fascia and dislocation of abutment and edge panels. Among the major issues reported, one case relates to the foundation (embankment) failure due to poor compaction, that is, inadequate dry density.

8.5.4   Stability of rock mass

The stability of rock masses located adjacent to the transport facilities such as roads and railway tracks often causes a challenge in ensuring the safety criteria in terms of stability to overcome sliding. One such case has been reported in the state of Chhattisgarh as a low to medium severity sliding of rock masses. It results in blocking the railway track section partially during the monsoon season. Typical photographs of the rock mass are shown in Figure 8.4a–d. Preliminary investigations revealed that the rock mass undergoing a continuous weathering process develops fractures, fissures and dislocation, and finally it collapses and blocks the railway track, especially during the monsoon seasons.

Figure 8.4 Typical rock strata from the Manendragarh station to the Udalkachar station railway line in Chhattisgarh.

8.6 OTHER GEOMATERIALS

8.6.1 Industrial wastes/by-products

Chhattisgarh is a mineral-rich state that primarily comprises large deposits of commercial minerals such as coal, iron ore, limestone, dolomite, bauxite and other mineral ores. These deposits encouraged the development of various industries such as thermal power plants, steel manufacturing plants, sponge iron industries, cement-manufacturing industries and so on. These industries generate a huge volume of wastes which are further converted to industrial by-products by utilising in various sectors of the construction industry. Some of the common industrial by-products being used in the construction industry are fly ash, bottom ash, pond ash, steel slag (GGBS, twin hearth furnace slag, BOFS, air-cooled blast-furnace slag, blast-furnace flue dust and banded haematite quartz), mine tailings and rice husk ash. Figure 8.5 shows the dumpsite of steel slag at the Bhilai steel plant.

8.6.2 Applications of industrial wastes/by-products

Depletion of natural resources such as sand, aggregates and good quality soil required for the construction industry forced the civil engineers to look after alternate/substitute materials that partially or fully replace the natural resources (Yadu and Tripathi, 2018). The following are some of the major areas of utilisation of these industrial by-products directly in the construction industry of Chhattisgarh:

1. Use of fly ash and bottom ash as retained fill in flyover approaches, pavement material (subgrade).
2. Use of fly ash in the manufacturing of building materials such as bricks, cement, paver blocks, and stabilisation of weak foundation soil (footings/subgrade layer).

Figure 8.5 Views of steel slag dumpsites.

3. Use of pond ash and bottom ash used as a filter medium in dams.
4. Use of GGBS in the manufacturing of cement.

Figure 8.6 shows the use of fly ash/pond ash as reinforced and retained fill in reinforced soil wall, constructed in the flyover approach of the national highway, Durg district. Apart from the cement industries, fly ash and GGBS can be used for soil stabilisation in the area of the Korba district. Fly ash/pond ash has been used as a subgrade in rural roads as shown in Figure 8.7a and b.

Figure 8.6 Use of fly ash as retained and reinforced fill in the reinforced soil wall flyover approach of the national highway, Durg district.

(a)   (b)

Figure 8.7 Utilisation of Industrial wastes in road construction. (a) Madwarani to Amlipara rural road construction using fly ash as subgrade, block: Kartla, district: Korba. (b) T08 to Indalbhata rural road construction using fly ash as subgrade block: Kartla, district: Korba.

8.6.3 Properties of industrial waste/by-products

The physical and mechanical properties of some industrial by-products such as pond ash, bottom ash, fly ash and steel slag (GGBS and BOFS) determined for utilisation in various civil engineering applications are summarised in Tables 8.11–8.15 (Yadu and Tripathi, 2018; Tripathi et al., 2018). These properties of the industrial by-products assist in understanding the basic characteristics and utilisation of these materials in various civil engineering applications for sustainable development by preserving natural resources.

Table 8.11 Geotechnical properties of pond ash and bottom ash of ash dyke at lagoon-3 NTPC, Sipat

Test/ property	Pond ash (location-I)		Pond ash (location-II)		Bottom ash (location-I)		Bottom ash (location-II)	
Specific gravity	1.95		1.89		1.89		1.88	
W_L, %	31		32		36		35	
W_P, %	NP		NP		NP		NP	
Gradation test	Size (mm)	% Finer	Size (mm)	% Finer	Size (mm)	% Finer	Size (mm)	% Finer
	4.75	100.0	4.75	100.0	4.75	100	4.75	100
	2.36	100.0	2.36	100.0	2.36	100	2.36	100
	1.18	100.0	1.18	100.0	1.18	97.5	1.18	97.2
	0.600	100.0	0.600	100.0	0.600	92.6	0.600	92.7
	0.425	99.9	0.150	99.9	0.425	83.2	0.425	83
	0.300	99.7	0.300	99.8	0.300	59.2	0.300	58.4
	0.150	93.7	0.150	94.3	0.150	23.7	0.150	22.8
	0.075	73.7	0.075	73.8	0.075	3.9	0.075	3.7
	0.064	65.6	0.063	66.8	0.064	3.1	0.067	3.0
	0.055	61.9	0.054	63.8	0.054	1.2	0.055	1.6
	0.047	60.6	0.046	63.0	0.048	0.5	0.047	0.4
	0.038	57.5	0.037	60.1				
	0.024	47.3	0.026	49.3				
	0.019	38.4	0.018	39.4				
	0.011	24.6	0.012	25.6				
	0.009	18.5	0.009	19.2				
	0.006	11.3	0.007	12.7				
	0.002	8.4	0.003	8.9				

Table 8.12 Properties of pond ash for the national highway in the Bilaspur district

Test	Location-I		Location-II	
Specific gravity	1.93		1.97	
W_L, %	51		42	
W_P, %	NP		NP	
Compaction test	$\gamma_{dm} = 12.38$ kN/m^3	$W_O = 23.80\%$	$\gamma_{dm} = 11.99$ kN/m^3	$W_O = 23.90\%$
Direct shear test	$c = 0.0$ kN/m^2	$\Phi = 30°$	$c = 0.0$ kN/m^2	$\Phi = 30.4°$
Permeability test	2.03×10^{-4} cm/sec		1.24×10^{-4} cm/sec	
	4×10^{-4} cm/sec		3.21×10^{-5} cm/sec	
Consolidation test	Coefficient of consolidation $(c_v) = 3.10$ m^2/year		Coefficient of consolidation $(c_v) = 3.20$ m^2/year	

Table 8.13 Geotechnical properties of flyash, rice husk ash and GGBS

Properties	Fly ash	Rice husk ash	Granulated blast-furnace slag
Specific gravity	2.09	2.04	2.57
W_L, %	84	94	35
W_P, %	Non-plastic	Non-plastic	Non-plastic
pH value	–	9.2	8.4
γ_{dm} (kN/m^3)	13.2	8.73	19.8
W_O (%)	27	45.5	9.30

Table 8.14 Geotechnical properties of basic oxygen furnace slag (coarse) from Bhilai steel plant, Chhattisgarh

Name of the test	Test property	Sample batch	Results obtained	
Water absorption	Porosity, %	I. Current arisal	0.60	
		II. Approx. 5 years old	0.40	
		III. Approx. 10 years old	0.80	
Aggregate impact test	Toughness, %	I. Current arisal	4.60	
		II. Approx. 5 years old	5.70	
		III. Approx. 10 years old	4.00	
Combined flakiness and elongation index test	Particle shape, %	I. Current arisal	27	
		II. Approx. 5 years old	28	
		III. Approx. 10 years old	25	
Los Angeles abrasion test	Hardness, %	I. Current arisal	10.60	
		II. Approx. 5 years old	10.40	
		III. Approx. 10 years old	11.00	
Soundness test (sodium sulphate)	Durability, %	I. Current arisal	0.37	
		II. Approx. 5 years old	0.35	
		III. Approx. 10 years old	0.40	
Specific gravity test	Specific gravity	I. Current arisal	3.23	
		II. Approx. 5 years old	3.22	
		III. Approx. 10 years old	3.04	
Bulk density	Bulk density, kN/m^3	I. Current arisal	Loose	15.85
		II. Approx. 5 years old	Compacted	17.66
		III. Approx. 10 years old	Loose	16.23
		I. Current arisal	Compacted	18.29
		II. Approx. 5 years old	Loose	15.82
			Compacted	17.52
Crushing test	Crushing strength, %	I. Current arisal	17.80	
		II. Approx. 5 years old	18.30	
		III. Approx. 10 years old	18.60	

8.6.4   Problems with industrial waste/by-products

Currently, safe disposal of industrial waste/by-products is a challenging task being faced by the industries, such as thermal power plants, steel plants, sponge iron and other manufacturing industries in Chhattisgarh. Some of the major problems are listed below:

1. Scarcity of space for disposal of industrial waste/by-products.
2. Failure of high raised ash dyke towards the downstream side in the fly ash dumpsites of thermal power plants.

Table 8.15(a) Geotechnical properties of basic oxygen furnace slag (fine) from Bhilai
steel plant, Chhattisgarh

Name of the test	Test property	Sample batch	Test result	
Deleterious content	Cleanliness	I. Current arisal	15.8%	
		II. Approx. 5 years old	19.4%	
		III. Approx. 10 years old	21.2%	
Particle size analysis (%)	Gravel fraction	I. Current arisal	11.60%	
	Coarse sand fraction		11.20%	
	Medium sand fraction		47.20%	
	Fine sand fraction		16.80%	
	Silt + clay (75 micron passing)		13.20%	
	Gravel fraction	II. Approx. 5 years old	11.00%	
	Coarse sand fraction		12.00%	
	Medium sand fraction		42.80%	
	Fine sand fraction		17.40%	
	Silt + clay (75 micron passing)		16.80%	
	Gravel fraction	III. Approx. 10 years old	7.00%	
	Coarse sand fraction		12.40%	
	Medium sand fraction		47.20%	
	Fine sand fraction		14.60%	
	Silt + clay (75 micron passing)		18.80%	
Atterberg's limits	W_L, %	I. Current arisal	35	
	W_P, %		*NP	
	I_P, %		–	
	W_s, %		–	
	W_L, %	II. Approx. 5 years old	36	
	W_P, %		*NP	
	I_P, %		–	
	W_s, %		–	
	W_L, %	III. Approx. 10 years old	34	
	W_P, %		*NP	
	I_P, %		–	
	W_s, %		–	
Free swell index	Swell potential, %	I. Current arisal	0	
		II. Approx. 5 years old	0	
		III. Approx. 10 years old	0	
Specific gravity (G)	Specific gravity	I. Current arisal	3.36	
		II. Approx. 5 years old	3.37	
		III. Approx. 10 years old	3.21	
Modified proctor test	γ_{dm} (kN/m^3)	I. Current arisal	23.50	
	W_O (%)		4.16	
	γ_{dm} (kN/m^3)	II. Approx. 5 years old	22.80	
	W_O (%)		5.90	
	γ_{dm} (kN/m^3)	III. Approx. 10 years old	21.81	
	W_O (%)		4.90	
CBR test (soaked)	Bearing capacity, %	I. Current arisal	38.50	
		II. Approx. 5 years old	36.70	
		III. Approx. 10 years old	35.80	
Bulk density	Bulk density, kN/m^3	I. Current arisal	*L	20.09
			*C	22.60
		II. Approx. 5 years old	L	19.50
			C	21.30
		III. Approx. 10 years old	L	18.20
			IC	20.70

Table 8.15(b) Particle size analysis of basic oxygen furnace slag (coarse) from Bhilai steel plant, Chhattisgarh

Sample batch	Cumulative percentage passing different sieve sizes in mm, %									
	63	50	40	31.5	25	20	16	12.5	10	6.3
I. Current	87.3	9.7	7.8	3.8	0.9	0.9	0.1	0.0	0.0	0.0
II. Approx. 5 years old	100.0	100.0	100.0	62.6	6.6	2.8	1.2	0.3	0.0	0.0
III. Approx. 10 years old	100.0	100.0	100.0	66.9	27.7	26.2	13.4	4.1	1.3	0.0

3. Deposition of mine overburden dump and dump stabilisation with ash.
4. Occasional slope failure and environmental pollution of the nearby areas.
5. Pollution of water bodies through leaching in the vicinity of dumpsites.

8.7   NATURAL HAZARDS

Chhattisgarh is placed at a location where the risk of natural calamities like earthquakes, volcanoes, floods, avalanches and so on is limited. Moreover, hilly terrain is situated in some locations of some districts (Sarguja, Bastar and so on) where the risk of landslides/erosion is possible. However, to date, no such major landslides or erosion is reported.

8.8   GEO-ENVIRONMENTAL IMPACT ON SOILS AND ROCKS

The effect of weathering on the structural stability of rock strata over a considerable period is a significant parameter to be considered in the design of civil engineering structures. A case study has been discussed briefly in Section 8.5.4. The impact of climate changes over a considerable duration of time on the steel slag is substantial and the same has been identified. The variations in the geotechnical properties of the steel slag of different ages exposed to the atmosphere over a specific period are reported in Tables 8.14 and 8.15. Preservation of the environment (air and ground) within the vicinity of fly ash dump sites is a daunting task. Changes in climatic conditions such as during monsoon season, the flow of fly ash with rainwater and thereby pollution of water bodies are major concerns. Similarly, during dry summer seasons, fly ash dumps also pollute the air and nearby land.

8.9   CONCLUDING REMARKS

1. Chhattisgarh has been divided into three natural zones: the plain zone, the Bastar Plateau and the north hills zone.
2. The soils of the state are broadly classified under five formations: Entisols, Inceptisols, Alfisols, Mollisols and Vertisols.
3. As per the visual classification, the five major types of soils are red-yellow soil, red sandy soil, red loam soil, black soil and laterite soil.

4. The most common types of soils and rocks used as construction materials are soil moorum, sand, crushed stone aggregates and stone dust.
5. Coal, iron ore, limestone, dolomite and bauxite are the major mineral reserves.
6. The availability of natural resources encourages industrial growth.
7. The industrial wastes and mine overburden can be used as alternatives to the construction materials.
8. Boring logs at salient points and their description can help professional engineers.

REFERENCES

CECB. (2016). Chhattisgarh Environment Conservation Board, http://chtenvis.nic.in/soil.html, accessed 26 January 2021.

CG Mineral Map. (2021). https://industries.cg.gov.in/cgmap/CG_MINERAL.pdf, accessed 26 January 2021.

Deputy Director (Geology). (2018). District mineral survey report of Baloda Bazar district, Directorate of Geology and Mining, Mineral Resources Department, Government of Chhattisgarh, Raipur, India.

Deputy Director (Geology). (2020). District mineral survey report of Baloda Bazar district, Directorate of Geology and Mining, Mineral Resources Department, Government of Chhattisgarh, Raipur, India.

Singh, R., Chaudhari, S.K., Kundu. D.K., Sengar, S.S. and Kumar, A. (2006). Soils of Chhattisgarh: characteristics and water management options. Research Bulletin 34, Water Technology Centre for Eastern Region, Indian Council of Agricultural Research, Bhubaneshwar, India.

Tripathi, R.K., Guzzarlapudi, S.D. and Yadu, L.K. (2018). Performance evaluation of some selected pavement sections in the state of Chhattisgarh. Final report of research project submitted to National Rural Road Development Agency, Ministry of Rural Development, New Delhi, India.

Yadu, L.K. and Tripathi, R.K. (2018). Construction of low volume roads on soft subgrade soil improved by locally available marginal materials. Final report of research project submitted to HSMI-HUDCO, New Delhi, India.

Chapter 9

Dadra and Nagar Haveli and Daman and Diu

Rakesh Kumar and Radha J. Gonawala
Sardar Vallabhbhai National Institute of Technology

Ketan Bajaj
Risk Management Solution Noida

Neetu Yadav
Shree Starambhai Naranjibhai Patel Institute of Technology and Research Centre

Moirangthem Johnson Singh and Lalit Borana
Indian Institute of Technology Indore

Chandresh H. Solanki
Sardar Vallabhbhai National Institute of Technology

CONTENTS

9.1   INTRODUCTION

Daman is the capital city of the union territory (U.T.) Dadra and Nagar Haveli and Daman and Diu, located in the western part of India. Daman is a small port on the Arabian Sea and is situated between 20° 22′ 58″ and 20° 29′ 58″N latitudes and 72° 49′ 42″ and 72° 54′ 43″E longitudes as shown in Figure 9.1. The altitude is 12 m above the mean sea level (MSL). Daman covers 11 km from the north to the south and a width of 8 km from the east to the west (Gupte, 2013).

After Liberation on 19th December 1961 from the Portuguese Rule of more than four centuries, Daman and Diu became a part of the U.T. of Goa, Daman and Diu under the Government of India. After the delinking of Goa, which attained statehood,

DOI: 10.1201/9781003177159-9

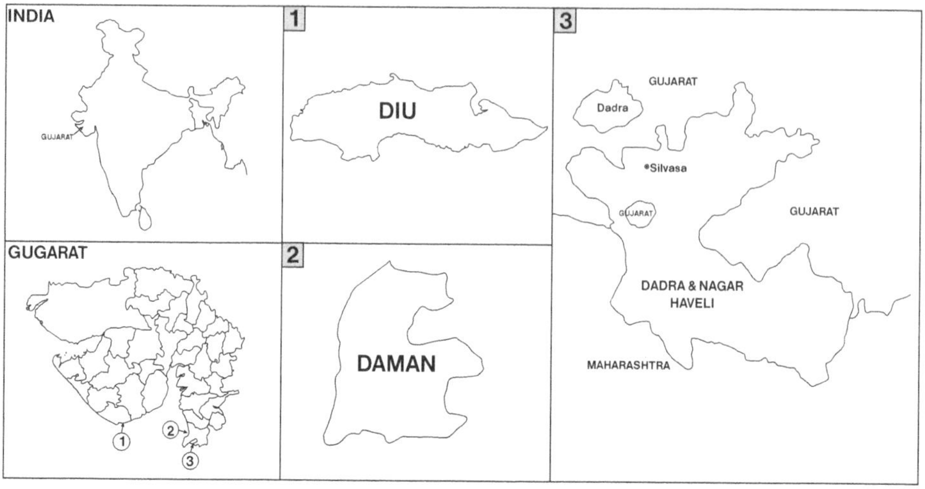

Figure 9.1 Location map of Diu, Daman and D&NH districts. (https://www.mapsofindia. com/daman-diu, accessed 15 December 2020.)

the U.T. of Daman and Diu came into existence on 30th May 1987 (GOI, 2020). Daman is located 170 km north of Mumbai and lay 10 km west of the Vapi Railway station, on the Bombay–Delhi Railway line. Daman is a bounded Valsad district of Gujarat from the north, east and south side and west by the Arabian Sea. The river Damanganga flowing through the middle of Daman divides the small land territory (72 km^2) into two parts, namely Moti Daman and Nani Daman.

Furthermore, the Kolak and Kalu rivers flow along the northern boundary and the southern boundary of Daman. Three rivers flow almost parallel to each other and enter Daman from the southeast boundary and follow the westerly course. Out of 7,200 ha, nearly 4% of the area is barren and uncultivated with mostly marshy land usually inundated during high tides. Daman geography is categorised by a plain terrain with a few mounts and knolls dotting the landscape. The Damanganga River drains the central part of the area. The geological U.T. area was formed by Deccan traps's basaltic flows, recent alluvial deposits in the southwestern segment and beach sands all along the coastline (Gupte, 2013).

The distance between Daman to Diu by road is 657 km and by aerial transport is 195 km. Diu is an island isolated from mainland Gujarat by an east-west spreading wet low land enclosed by the Arabian Sea's tidal waters. Diu's geographical extent varies from 70° 52′ 16″ to 71° 27′ 47″ E longitude and 20° 44′ 33″ to 20° 41′ 55″ N latitude. Diu is a tiny island with 43.8 km^2 on the Saurashtra Peninsula's southern peripheral by a 250 m width swampy creek. Diu is extended nearly 19.2 km from the east to the west, and its width varies from 1 to 2.5 km from the north to the south. The eastern part of the island forms an undulatory topography composed of extended equivalent elevations of miliolite, whereas the western part is a flat terrain with crescent-shaped calcareous dunes, which characterise the central region. The wavecut raised area is noticeable on the coast, and at creek zones, clay beds cover tidal flats. The primary rock forms exposed are miliolite, calcarenite, calcrudite and micrite. These rocks spread from Diu in the east to Molola in the west and continue further below dunes with a total thickness of 28 m, as established by drilling (GSI, 2020).

Dadra is a small enclave within the state of Gujarat. Nagar Haveli is a C-shaped enclave located between Gujarat and Maharashtra which contains a counter enclave of Gujarat around the village of Maghval. The D&NH is situated on the western side of Western Ghats' foothills and has an undulating terrain with 41.63% of the total geographical area covered with forests offering a look of woodland (Jain, 2013). The major river Damanganga and its tributaries crisscross the Dadra Nagar Haveli and drain into the Arabian Sea at Daman. The D&NH has spread over 491 km² area landlocked between Gujarat in the north and Maharashtra in the south and is close to India's western coast. It is located in the middle of the Damanganga River's undulating watershed, which flows through Nagar Haveli and later forms the short southern border of Dadra. The towns of Dadra and Silvassa are situated on the north bank of the river. The Western Ghats range rises to the east, and the range's foothills occupy the eastern portion of the district. The main southern area's stretch is hilly terrain, especially towards the northeast and east, surrounded by ranges of the Sahyadri mountains (Western Ghats). The central alluvial region of the land is almost level, and the soil is fertile and rich. The river Damanganga rises in the Ghat 64 km from the western coast and discharges itself in the Arabian Sea at the port of Daman after crossing D&NH. Its three tributaries, Varna, Pipri and Sakartond, join Damanganga within the territory (Jain, 2013).

9.2 MAJOR TYPES OF SOILS AND ROCKS

Diu has high saline and alkaline soil with a high percentage of silt up to a kilometre from the coast. The formation took place due to the degeneration of coastal soil by the entrance of salinity. Slowly soils turn into yellowish-brown calcareous soils having a mixture of miliolite shell pieces including medium to coarse-grained materials. The layer has a thickness in the range of 0.3 to 1 m. At some low-lying areas, a gathering of organic materials combined with extreme weathering conditions gave rise to black cotton soil covers which vary from few centimetres to a metre in thickness. On the middle, elevated land, the blown sand sediment is weathered from the crumbly miliolite limestone and is highly calcareous. The area containing miliolite limestone is about 50 m thick and of Pleistocene to Modern period. It is extremely porous limestone and could levigate excluding the one or two strata near the top surface. The calcification process of limestones gave rise to a calcium carbonate solution that offers a challenging and compact crust.

In Diu, the miliolite limestone of high-grade has minimal contents of magnesium and is insoluble. Solution activity has resulted in the development of caverns of different sizes. The karstic activity is observed predominantly in the zone where the water level fluctuates and near the lower contact with the underlying formations of clay. The limestones exposition to strike is approximately parallel to the sea and the dips are undulant like usual sand-dune deposits. Gaj formations of Miocene age underlie miliolite limestones. The Gaj formations comprise upper yellowish-white clays caused by interlayered marls, grits, impure limestones, calcareous sandstones and clays.

Table 9.1 The geological formation in Diu

Age	Formation	Lithology	Max. thickness/ remarks
Recent to Pleistocene	Coastal alluvium and miliolite limestone	Sand, clays, miliolite limestone	40–50 m
Miocene (Tertiary)	Gaj beds	Clay, marl, calc. sandstone, limestone and so on	+200 m, not exposed
Upper Cretaceous to Eocene	Deccan trap	Basaltic lava	Not exposed

The formation of Gaj found to be extending down to the explored depths of 200 m and on the base of Gaj formation rests over the Deccan trap basalt. The generalised geological succession in the area is mentioned in Table 9.1.

In Daman, soils can be classified into two types based on their origin. The soil layers along the coast and at the creek banks are of alluvial type, while in the rest, the soil layers resulted from weathered basaltic rocks. The shore soil, which is settled in saline water, showed salinity and alkalinity with a nearly identical texture, that is, clayey loam to silty loam. The soil strata found are in dark grey to black colour with too high pH and high electrical conductivity values. These soils' reclamation was too tricky due to higher content, low permeability and a high-water table and salinity level.

With the degree of weathering, the basaltic soils show extensive dissimilarity in the texture of the earth. The initial stage of basalt's weathering contributes to growth in light soil involving weathered basalt pieces (locally known as murom). The depth of such weathered soils differs from a small number of centimetres to 50 cm. In the plateau and nearby the hillocks, the soil's colour is dark yellowish-brown, whereas in the flat valleys, the colour is brownish-black to black and the soil is found to have medium to fine texture. Grounds are found to be non-calcareous and have moderate water holding capacity.

In Moti Daman, basalt layers are found as primary basement rock at variable depths and are exposed at the surface, namely in Marwad, Devka Kadiya in the northwest part of Daman. Basaltic ridges have an elevation of about 111 m above the MSL exposed around Kunta and Wankad villages. Basalt sheet rocks are visible in Damanganga, Kalu and Kolak's rivers bordering the U.T. of Daman. The deposition of alluvial soil is found covering the basalts, in Moti Daman, Dabhel and Kachigam areas with 12 to 40 m. On the banks of the Damanganga River, deposits of alluvium can be observed. Basalt takes place in the manner arrangement of movements containing enormous and dense basalt at the bottom and step by step goes into vesicular basalt at the top. The basalts colour varies from dark green to pink and different sets of joints could be found.

The joint structures are limited to separate flow, rarely cutting through other flows. With the weathered spheroidal surface, weathering is characterised. The Great Trap region of the Deccan covers a substantial part of the Basin with volcanic formation. The volcanic portion consists of compact, stratified basalts, and an earthy trap. The basalts are the most conspicuous geological features.

To the west, they lie in flat-topped ranges, separated by valleys, trending from the west to the east. In some flows, the basalt is columnar, and then it weathers into fantastic shapes. The traps base is the formation of amygdaloidal, containing quartz in vertical veins with crystals and zeolitic minerals and apophyllite weathering into a grey soil. The absence of laterite, which caps the hill's summits to the south, is a curious feature in the area's geology. The basalt is fine textured, or it is coarse and nodular.

In D&NH, the three types of soils found are vertisols, inceptisols and ultisols. The vertisols are deep black soils. They usually form from basaltic materials with a high clay content of the north-to-south axis, including the central and northwestern parts of D&NH. The inceptisols are shallow black soils in the western part of D&NH. The ultisols are lateritic soils said to have low clay components and are found predominantly in the northeastern part, the eastern part abutting the Damanganga Reservoir and the southwestern part of D&NH.

Physiographically the part of the D&NH formed from Deccan Plateau. It divides into three physiographic units, i.e., residual plateaus, denudational slopes and valley plains. The residual plateaus are flat-topped crests surrounded by steep scarps. These residual plateaus occur in the western fringe, the southeastern part and the isolated patches in the northeastern part of D&NH. The elevation of the residual plateau's category ranges between 100 and 300 m above the MSL and a peak of more than 200 m above the MSL is observed in the northeastern and southeastern part of the region. The denudational slopes form the scarp zone between plateaus and plains and comprise moderate to steep slopes (Jain, 2013). The primary area under the denudational slopes category lies almost in the eastern part in the north–south alignment in dissected form ranging roughly between 50 and 100 m above the MSL. The valley plains form the flat topography with a gentle slope of SE-NW forming the Damanganga River and its tributaries. It ranges in elevation below 50 m above the MSL. It comprises weathered rock fragments and soils.

9.3   PROPERTIES OF SOILS AND ROCKS

Typical boring logs of the Diu and Daman soil strata show that mostly sandy clay deposits on the top and sandy silt in the further depth are found. The boring logs data of Diu are provided in Table 9.2. Two boring log trial pits were bored in Daman till 10 m depth with the standard penetration test (SPT) (IS:2131, 1981), and the same boring logs data are provided in Tables 9.3 and 9.4.

From Diu's boring logs data, as mentioned in Table 9.5, the upper soil strata were found to be sandy clay with a plasticity index of 20.2%. After one metre depth from the OGL, mostly non-plastic silty sand is found. The field density was observed to be 1.61 g/cc with 11% moisture for the depth below 1 m in the SPT. The specific gravity of soil layers was found to be 2.64, and the angle of friction is higher than 25°. These properties of soil make it suitable as a filling material in construction work. The top clayey soil is classified as CI and is almost prevalent till 1 m thickness (PMAY, 2020).

Daman's boring logs data as mentioned in Tables 9.6 and 9.7 show a deposit of alluvial soil, and in further depth, moorum (sand mix with clay), hard moorum and weathered basalt were also found in the borehole investigation. Moorum is also a type of soil, mostly used for construction purposes. Generally, it is deep brown or red in colour. Moorum is used in plinth filling, road pavements, backfilling in trenches, footing pits and so on. It is a suitable type of soil in the construction field, since it does not contain any organic matter and can be compacted quickly, forming hard surfaces.

Table 9.2 Boring log of Diu

Depth (m)	Thickness (m)	Soil classification	Description	Sample type	Depth (m)	SPT 'N' value
1	1	CI	Blackish yellow stiff clayey soil of intermediate plasticity with kankars	DS	0	
2	9	SM	Yellowish silty fine sand	DS	1.5	
				STP	1.8	24
3				DS	3	
				STP	3.3	25
4				DS	4.5	
				SPT	4.8	25
5				DS	6	
6				SPT	6.5	24
7				DS	7.5	
8				SPT	7.8	24
9				DS	9	
10				DS	10	

Table 9.3 Boring log #1 of Daman

Depth (m)	Thickness (m)	Soil classification	Description	Sample type	Depth (m)	STP 'N' value
0	1.5	CI	Filled-up soil	DS	1.5	–
1				SPT		–
2	3.5	SM	Sandy soil with silt and gravel	UDS	3.5	–
2.45				SPT		26
3				UDS		–
4				UDS		–
4.45				SPT		33
5.55	1.0	SM	Black colour silty clay	SPT	1.0	43
6	1.0	SP	Weathered basalt recovered as moorum	DS	1.0	–
7	3.0	GM	Yellowish colour amygdaloidal basalt rock	SPT	3.0	>100
8				DS		
9				SPT		>100
10				DS		

The sandy soil and soft to hard moorum are present till 6 metres depth, and after 6 m depth, weathered basalt and subsequent hard basalt strata were observed. After 1 to 1.5 m depth, the earth can be classified as SM and SP type with a specific gravity of 2.62 and an angle of internal friction of 24°. Alluvial soil deposits were observed on the top, sandy silt and clay were found below and basalt rock was found in a greater depth.

The recovered undisturbed samples were taken from the borehole for laboratory testing. The rock strata of Diu belong to the Palaeozoic Era, as prevalent in the Saurashtra region of Gujarat. The rocks are sedimentary consisting of sandstones and limestones. The soils of the area are loamy, clays and sandy which are very adhesive. In the northern part of the region, the river chassis and the grounds are marshy due to the influence of tides. In the southwestern part, the rock strata contain sandstone and

Table 9.4 Bore log #2 of Daman

Depth (m)	Thickness (m)	Soil classification	Description	Sample type	Depth (m)	STP 'N' value
0	1.5	CI	Filled-up soil	DS	1.5	–
1						–
2	3		Sandy soil with silt and gravel	UDS	3	–
2.55				SPT		28
3		SM		UDS		–
4				UDS		–
4.2				SPT		34
5	2		Black colour silty clay	DS	2	–
6		SP		SPT		48
6.55				SPT		–
7	3.5		Yellowish colour amygdaloidal basalt rock	SPT	3.5	>100
8				DS		–
8.5		GM		SPT		>100
9				DS		–
10				DS		–

limestone. At present, it is devoid of vegetation cover. The central portion of the region covers a layer of productive soils and is the only area available for agriculture. Sea beaches have golden sands and sheet rocks. Sand for construction purposes is available along the sea beaches of Nagoa, Jallandhar and Malala. For many constructions, sand was taken from the Nagoa beach as it was available in good quality and quantity. All infrastructure construction works in the region are done with sand, sandstone and cement. No clay or bricks are available for construction works. In some pockets of marshy areas, salt is extracted from that area. The seawater is rich in saline contents, and due to favourable climatic conditions in Diu, the salt is of good quality than in many other places.

The soil of D&NH is mainly drained by the Damanganga River and its tributaries (382.32 km). Kolak river and its tributaries drain the northern part (69.32 km). The Kalunadi and its tributaries drain a small portion (37.26 km) of the west territory. The boring log data show the presence of clayey strata up to 12 m in depth, as mentioned in Table 9.8. After that presence of the gravel soil with silty loams, it is evident from borehole soil investigation, as mentioned in Table 9.9.

The soils occurring on the Deccan Plateau are generally black cotton soils. Three significant soil types occur in D&NH, i.e., lateritic soils, shallow black cotton soils and deep black soils. The lateritic soils are well stabilised and have good porosity leading to good permeability, aeration and non-calcareous soils with low pH. The shallow black cotton soil is medium-textured soil with gravels and pebbles, and hard rock is met early below the soil restricting free water and air movement. The soils are permeable enough, have a low moisture-holding capacity and thus call for specific moisture management for sustainable land use under cultivation. The black cotton soil is available in very deep layers with very high fertility and is capable of supporting a variety of crops.

The primary water-bearing formation in D&NH is Deccan trap basalt. The basaltic lava flows are massive and fine-grained with negligible primary porosity and

Table 9.5 Boring log of Diu with soil properties

Sr. no.	Type of sample	Depth	Properties														
			Bulk density (g/cc)	FDD (g/cc)	FMC (%)	Grain size analysis				Atterberg limits			I. S. classification	Specific gravity	Shear parameters			FSI (%)
						Gravel (%)	Sand (%)	Silt (%)	Clay (%)	L.L. (W₁) (%)	P.L. (Wp) (%)	P.I. (%)			Type of test	C (kg/cm²)	ø deg. (°)	
1	DS	0	—	—	—	6	32	34	28	38.7	18.4	20.2	CI	—	—	—	—	—
2	DS	1.5	—	—	—	10	69	21		20.7	NP	NP	SM	—	—	—	—	—
3	DS	3	1.843	1.622	11.98	9	71	20		23	NP	NP	SM	2.641	BS	0	25	—
4	DS	4.5	—	—	—	12	63	25		22.1	NP	NP	SM	—	—	—	—	—
5	DS	6	1.833	1.631	11.02	14	65	21		19.5	NP	NP	SM	2.643	BS	0	35	—
6	DS	7.5	—	—	—	11	69	20		21.2	NP	NP	SM	—	—	—	—	—
7	DS	9	—	—	—	15	62	23		21.9	NP	NP	SM	—	—	—	—	—
8	DS	10	—	—	—	11	67	22		22.2	NP	NP	SM	—	—	—	—	—

BS: box shear or direct shear test, DS: disturbed sample.

Table 9.6 Boring log #1 of Daman with soil properties

Lab no.	Type of sample	Depth	Properties														
			Bulk density (g/cc)	FDD (g/cc)	FMC (%)	Grain size analysis				Atterberg limits			I.S. classification	Specific gravity	Shear parameters			FSI (%)
						Gravel (%)	Sand (%)	Silt (%)	Clay (%)	L.L. (W_L) (%)	P.L. (W_p) (%)	P.I. (%)			Type of test	C (kg/cm²)	ø deg. (°)	
234.0	DS	1.5	—	—	—	—	—	—	—	—	—	—	—	—	—	—	—	—
234.1	UDS	2	1.92	1.67	14.70	24	41	23	12	25	20	5	SM	2.62	BS	0.05	22	—
234.2	SPT	2.45	—	—	15.32	21	46	23	10	23	17	6	SM	—	—	—	—	—
234.3	UDS	3	1.87	1.64	13.80	25	40	22	13	25	18	7	SM	2.62	BS	0.05	24	—
234.4	UDS	4	1.85	1.62	14.20	25	44	20	11	26	20	6	SM	2.63	BS	0.04	24	—
234.5	SPT	4.45	—	—	16.52	27	51	20	2	24	NP	NP	SM	—	—	—	—	—
234.6	SPT	5.55	—	—	14.62	23	57	15	5	26	NP	NP	SM	—	—	—	—	—
234.7	DS	6	—	—	18.47	25	53	15	7	23	NP	NP	SM	—	—	—	—	—
234.8	SPT	7	—	—	17.51	39	51	10		—	—	—	SP.	—	—	—	—	—
234.9	DS	8	—	—	11.32	34	61	5		—	—	—	SP.	—	—	—	—	—
235.0	SPT	9	—	—	10.67	49	37	14		—	—	—	G.M.	—	—	—	—	—
235.1	DS	10	—	—	12.92	45	39	16		—	—	—	G.M.	—	—	—	—	—

BS: box shear or direct shear test, DS: disturbed sample, UDS: undisturbed sample, SPT: standard penetration test.

Table 9.7 Boring log #2 of Daman with soil properties

Lab no.	Type of sample	Depth (m)	Properties													Shear parameters			FSI (%)
			Bulk density (g/cc)	FDD (g/cc)	FMC (%)	Grain size analysis				Atterberg limits			I.S. classification	Specific gravity	Type of test	C (kg/cm²)	ø deg. (°)	
						Gravel (%)	Sand (%)	Silt (%)	Clay (%)	L.L. (WL) (%)	P.L. (Wp) (%)	P.I. (%)						
	DS	1.5																
234.0	UDS	2	1.91	1.67	14.50	25	41	22	12	27	21	6	SM	2.62	BS	0.04	23	—
234.1	SPT	2.6	—	—	11.36	21	43	28	8	25	19	6	SM	—	—	—	—	—
234.2	UDS	3	1.91	1.68	13.90	24	41	25	10	21	16	5	SM	2.62	BS	0.04	23	
234.3	UDS	4	1.84	1.61	14.00	25	43	18	14	25	19	6	SM	2.63	BS	0.05	24	
234.4	SPT	4.2	—	—	13.56	28	49	14	9	25	NP	NP	SM	—				
234.5	DS	5	—	—	15.80	26	51	12	11	25	NP	NP	SM	—				
234.6	SPT	6	—	—	16.41	23	55	16	6	21	NP	NP	SM	—				
234.7	SPT	6.6	—	—	19.41	25	47	28		—	—	—	SP.	—				
234.8	SPT	7	—	—	16.14	31	59	10		—	—	—	SP.	—				
234.9	DS	8	—	—	13.62	29	47	24		—	—	—	SM.	—				
235.0	SPT	8.5	—	—	14.78	41	39	20		—	—	—	G.M.	—				
235.1	DS	9	—	—	14.01	46	39	15		—	—	—	GM					
235.2	DS	10	—	—	10.25	42	37	21		—	—	—	GM					

BS: box shear or direct shear test, DS: disturbed sample, UDS: undisturbed sample, SPT: standard penetration test.

Table 9.8 Boring log of D&NH

Depth (m)	Thickness (m)	Soil classification	Description	Type	Depth (m)	SPT 'N' value
1	1	–	Filled-up soil	DS	1.0	6
2	11		Yellowish silty clay	SPT	2	16
3				DS/UDS	3	–
4				SPT	4	23
5				DS/UDS	5	–
6				SPT	6	24
7		CI/SC		DS/UDS	7	–
8				SPT	8	28
9				SPT	9	31
10				SPT	10	34
11				DS/UDS	11	–
12				SPT	12	39
13			Gravelly soil with silt	DS/UDS	13	–
14				SPT	14	>100
15				DS/UDS	15	–
16				SPT	16	>100
17		GM		DS/UDS	17	–
18				DS/UDS	18	–
19				DS/UDS	19	–
20				DS/UDS	20	–

transmissivity. The area occurs in the vicinity of the western coast, which has witnessed many tectonic disturbances. These have caused the development of joints and fractures in the basaltic strata. Also weathered zones of about 10–20 m thickness have developed in plains and depressions. Thus, the weathered, jointed and fractured zones of vesicular and massive flow units constitute the main water-bearing horizons. However, these zones are not continuous and uniformly developed laterally or vertically, and this factor plays an essential role in the success and failure of wells in the area.

9.4   USE OF SOILS AND ROCKS AS CONSTRUCTION MATERIALS

Most civil engineering tasks comprise more or less earthwork, altering the ground surface structure. For construction activities, if the soil or rock layer is removed, it is called cut or excavation and, it is named fill or embankment when soil or rock layers are added. The site is made suitable for construction by cutting and filling for the projected development. All the structure works require improvement of the bearing capacity of the soil. For example, road construction work requires an effective 8% California bearing ratio (CBR). It is necessary to match the requirements, and a contractor has to choose soils from several borrowers. Based on the engineering properties required, soil selection is made. However, the cost of borrowing a lead distance of more than 12 km becomes uneconomical.

In Diu and Daman, sandy silt is the most commonly found soil and having a right angle of friction with less cohesion makes it suitable for filling work. Gravels are good materials for filling because of their high strength and low compressibility even under

Table 9.9 Boring log of D&NH with soil properties

Sr no.	Type of sample	Depth (m)	Bulk density (g/cc)	FDD (g/cc)	FMC (%)	Grain size analysis				Atterberg limits			I.S. classification	Specific gravity	Shear parameters			FSI (%)
						Gravel (%)	Sand (%)	Silt (%)	Clay (%)	L.L. (W_L) (%)	P.L. (W_p) (%)	P.I. (%)			Type of test	C (kg/cm^2)	ø deg. (°)	
1	UDS	2	1.84	1.6	14.8	8	41	51		39	18	21	CI	2.64	BS	0.26	23	22
2	UDS	3	1.84	1.61	14.3	12	40	48		38	18	20	SC	2.64	BS	0.11	21	18
3	UDS	4	1.84	1.61	14.1	10	44	46		39	17	22	SC	2.62	BS	0.19	20	17
4	UDS	6	1.84	1.62	13.8	8	50	42		40	20	20	SC	2.63	BS	0.09	23	19
5	UDS	7.5	1.85	1.63	13.6	9	46	45		39	18	21	SC	2.63	BS	0.21	17	16
6	UDS	9	1.86	1.64	13.4	11	41	48		39	19	20	SC	2.61	BS	0.13	22	17
7	UDS	10.5	1.87	1.65	13.5	10	42	48		39	18	21	SC	2.63	BS	0.06	27	18
8	UDS	12	1.95	1.78	9.7	65	24	11		—	N.P.	—	GP-GM	2.61	BS	0	29	—
9	UDS	13.5	1.97	1.8	9.6	55	31	14		—	N.P.	—	G.M.	2.62	BS	0	31	—
10	UDS	15	1.98	1.81	9.5	64	21	15		—	N.P.	—	G.M.	2.62	BS	0	31	—
11	UDS	16.5	1.99	1.83	8.9	64	22	14		—	N.P.	—	G.M.	2.62	BS	0	31	—
12	UDS	18	2.01	1.85	8.6	59	27	14		—	N.P.	—	G.M.	2.63	BS	0	32	—
13	UDS	19.5	2.01	1.86	8.5	64	23	13		—	N.P.	—	G.M.	2.63	BS	0	32	—

BS: box shear or direct shear test, UDS: undisturbed sample.

wet conditions. GW and GP have a high hydraulic conductivity that permits water to drain speedily, mainly on highways. Even sands make good fills because of high strength, low compressibility and a right angle of friction.

The SW and SP soils keep all of their strength even under wet conditions. However, SP and SW soils became problematic when exposed to the ground surface, because vehicles can become stuck and grades are difficult to maintain. The sandy soils are vulnerable to surface erosion except when protected, such as with vegetation. The SM and SC soils found are approximately identical for many applications, as their hydraulic conductivity is adequate. ML and CL soils are less desired than SM or SC due to their loss of strength under wet conditions and the requirement of careful moisture control. Even they are difficult to dry if the initial moisture content is high.

The MH, MI, CI and CH soils are choices when low hydraulic conductivity is mandatory, like serving as in landfill caps. If not, they are tough to handle and compact, specifically when the preliminary moisture content is beyond the optimum. These soils expand under wet conditions because pavements and other foundation work include lightly loaded foundations, flatwork concrete and similar projects. These are not the correct choice for backfill in retaining walls of their expansive nature, weak wet strength and low hydraulic conductivity.

9.5 FOUNDATIONS AND OTHER GEOTECHNICAL STRUCTURES

When it comes to building a structure, it is crucial to construct a firm base that holds the superstructure under all climatic conditions without collapsing or decaying on a given soil stratum. Different types of foundation bases utilise different structures, and each one has a unique design and specific configuration, which makes a particular built structure more durable and firmer. In the U.T. region, the combined footing is used for residential buildings as Daman has low soil bearing capacity. A deep foundation is a type of foundation that transfers structure loads to the earth. The depth of the ground in the deep foundation is above 3 m. In the construction of high-rise structures/bridges, it is necessary to go deep into the ground to provide the support for the superstructures. Deep foundations transfer loads to the soil through both ends bearing and side friction.

9.6 NATURAL HAZARDS

Daman is geographically part of Gujarat on the Arabian Sea Coast. During the last 200 years, Gujarat recorded nine earthquakes of moderate to severe intensity in 1819, 1845, 1847, 1848, 1864, 1903, 1938, 1956 and 2001. The last one of the worst earthquakes in history was in 2001 with a death toll of 26. On 4th October 1851, Daman suffered a moderate earthquake. According to one recorded version, it sounded like underground explosions and heavy rumblings which continued for some seconds. According to India's earthquake hazard map, Daman is located in the moderate damage risk zone with a probable earthquake of 5.0–6.0 magnitudes on the Richter scale. Natural hazards peculiar to developed sites such as coastal storms, flooding, landslides and earthquakes may damage infrastructure and reduce long-term benefits.

The significant damage to the infrastructure of Diu and Daman is because of Hurricane type cyclone. In 1982, a hurricane cyclone majorly impacted the embankment structure at Diu port. Furthermore, hurricane damaged the houses in the coastal creek area and the Government-owned "Vanakbara" lighthouse was entirely washed away. Also, hurricane in June and October of 1996 cyclones affected the Diu district, causing damages to salt pans, industries, roads, the foundation of electricity poles and so on. According to the Cyclone Hazard Map of India, Daman is located in the moderate damage risk zone, with a probable maximum wind speed of 44 m/s. The historic cyclone had footprints of the damage to building foundations.

9.7   CASE STUDY AND FIELD TESTS

Field investigations were very significant for assessing the in-situ pavement strength characteristics and the pavement layer material properties. The purpose of the analysis is to measure the structural condition of roads, in terms of the remaining year's life, and to indicate the strengthening required to achieve the capacity to withstand the anticipated traffic loading over a design life of 10 years.

A field examination was performed for a pavement design of road at km 0/0 to 13/200 in Fansa village on the Vapi–Daman road. The main objective was to study the design of the road for 12 MSA traffic. To fulfil this objective, field investigation was to be done. Also, by using the falling weight deflectometer (FWD), the analysis was carried out to evaluate the structural capacity of existing pavement and the over thickness of the investigated road. A FWD survey is required to carry out the workout requirement of overlay and residual life of the pavement by evaluating the pavement component's modulus of the resilient (M_R) (IRC:37, 2018). The soil investigation was carried out at different trial pits to find the present soil properties on-site. The detailed soil investigation and filed data are presented in Table 9.10, and the site investigation picture is given in Figure 9.2.

The details of exciting crust thickness are shown in Table 9.11. The soil investigation was done by digging the different trial pits on the side of the road. Figure 9.3 illustrates the trial pit location on the road, and the crust thickness varies from 295 to 555 mm at the different trial pits.

The soil investigation shows that the soil classification is CI with the CBR (IS 2720-16, 1987) of soil varying from 2.1% to 2.9%. The entire stretch's laboratory test indicates that the liquid limit ranged from 38% to 47%, the plastic limit from 20% to 25% and the plasticity index from 16% to 23% with FSI of 28% to 42%. The maximum dry density (MDD) (IS 2720-7, 1980) does not show much difference having values of 1.8 to 1.9 g/cc with the optimum moisture content (OMC) varying from 14.9% to 17.8%. The crust thickness of the existing pavement varies from 295 to 555 mm from the trial pit analysis.

The structural evaluation of road layers was carried out for every 500 m interval along the considered road using a FWD as per IRC:115 (2014). FWD uses a nondestructive method for evaluating the structural capacity for flexible pavements as the project corridor consists of flexible pavements. The data were analysed to estimate overlay layer thickness to strengthen the exiting flexible pavement.

The design traffic has been calculated using the fourth power law and a standard axle load of 82 kN. A compound traffic growth of 6% for all vehicle types has been

Table 9.10 Soil data analysis of the Fansa site

Sr no.	Chainage no.	Grain size analysis				Atterberg limits			Free swell index (%)	I.S. classifications	Heavy		CBR (soaked) (%)
		Gravel (%)	Sand (%)	Silt (%)	Clay (%)	L.L. (%)	P.L. (%)	P.I. (%)			Maximum dry density (g/cc)	Optimum moisture content (%)	
1	0/000–0/500	0	15	12	73	46	25	21	35.0	CI	1.80	16.9	2.3
2	0/500–1/100	1	19	15	65	41	22	19	31.0	CI	1.83	16.3	2.5
3	1/000–1/500	3	17	13	67	44	24	20	33.0	CI	1.84	17.4	2.4
4	1/500–2/000	2	21	18	59	39	22	17	28.0	CI	1.86	15.3	2.9
5	2/000–2/500	1	18	12	69	40	22	18	30.0	CI	1.85	15.8	2.7
6	2/500–3/000	0	15	11	74	43	24	19	33.0	CI	1.83	17.2	2.6
7	3/000–3/500	0	19	15	66	42	23	19	31.0	CI	1.81	16.4	2.5
8	3/500–4/000	2	22	18	58	41	23	18	28.0	CI	1.82	16.0	2.8
9	4/000–4/500	1	17	12	70	40	21	19	31.0	CI	1.84	16.7	2.6
10	4/500–5/000	0	14	10	76	47	24	23	38.0	CI	1.79	17.5	2.2
11	5/000–5/500	1	18	12	69	43	23	20	34.0	CI	1.80	16.8	2.5
12	5/500–6/000	2	22	17	59	39	22	17	29.0	CI	1.85	15.3	2.8
13	6/000–6/500	0	16	11	73	45	23	22	38.0	CI	1.79	17.1	2.2
14	6/500–7/000	0	19	18	63	42	24	18	31.0	CI	1.82	16.0	2.4
15	7/000–7/500	1	21	19	59	40	21	19	32.0	CI	1.84	15.8	2.6
16	7/500–8/000	3	23	20	54	38	20	18	30.0	CI	1.86	14.9	2.9
17	8/000–8/500	2	20	17	61	43	22	21	38.0	CI	1.83	16.2	2.7
18	8/500–9/000	0	15	11	74	45	23	22	37.0	CI	1.78	17.5	2.1
19	9/000–9/500	2	18	14	66	42	20	22	34.0	CI	1.82	15.9	2.5
20	9/500–10/000	0	16	12	72	44	23	21	36.0	CI	1.80	16.7	2.2
21	10/000–10/500	0	17	13	70	46	23	23	42.0	CI	1.78	17.8	2.1
22	10/500–11/000	1	19	15	65	43	23	20	35.0	CI	1.83	16.1	2.4
23	11/000–11/500	0	16	14	70	45	24	21	38.0	CI	1.79	17.2	2.3
24	11/500–12/000	2	22	18	58	39	21	18	32.0	CI	1.84	15.7	2.8
25	12/000–12/500	1	20	17	62	42	23	19	35.0	CI	1.82	16.4	2.5
26	12/500–13/000	3	23	19	55	38	22	16	30.0	CI	1.86	15.3	2.9
27	13/000–13/200	0	18	14	68	44	23	21	40.0	CI	1.80	16.9	2.4

Figure 9.2 Side soil in a trial pit of the Fansa site.

Table 9.11 Existing road crust composition and CBR

| Chainage | Crust thickness, mm | | | Subgrade CBR* (%) |
	Bituminous layer	Granular layer	Total	
1/600	220	210	430	2.9
4/250	175	380	555	2.6
5/500	140	240	380	2.5
7/000	225	70	295	2.5
10/400	260	150	410	2.1

* Note: CBR values are taken from laboratory soaked CBR tests.

Figure 9.3 Trial pit for finding crust thickness at chainage no. 1/600.

applied over a design life of 10 years. The FWD deflection analysis described in the preceding sections and the traffic loading of 12 MSA requiring overlay have been identified. Subgrade modulus has been calculated by a four-day soaked CBR test. The input range has been considered from 11 to 46 MPa. The pavement's residual life with the current thickness of the pavement layer is given in Table 9.12. As the pavement has a low granular layer thickness and low subgrade modulus value, strengthening the top bituminous layer by providing a bituminous overlay to meet the design traffic volume may not sustain the pavement for a long duration. Hence, to enhance the pavement for a long duration, a strong base layer was recommended. For upgrading the base and subgrade layers, it is recommended to have reconstruction for the given stretch with a material having an effective 8% CBR. The design for the reconstruction of the stretch was done after soil investigation of the stretch and nearby borrow areas; the details are given in Table 9.13.

9.8   GEOENVIRONMENTAL IMPACT ON SOILS AND ROCKS

Soil formation depends on climate, relief, parent material, organisms and time. Thousands of years were taken for the earth to form, and most soils are still emerging into the changes. Climate could be considered one of the supreme significant factors that affect the formation of soil. Soil erosion is observed as one of the critical implications for soil strata changes. Therefore, the ecological management of the upper strata of soil is necessary regarding the soil structure, topsoil water holding capacity stability and erosion affecting the climatic conditions.

Due to an increase in the construction and tourism activities in Diu and Daman, soil erosion is one of the significant issues in these places. Sand mining and depletion of coastal resources for infrastructure development increased the degradation of land resources. Initial environmental examinations of the sites include the assessment of impacts due to the infrastructure development projects. Sand is mined on a large scale from beaches for use in the construction of hotels and resorts. It is another cause of soil erosion in the Diu and Daman coasts. Proper care needs to be taken to prevent this. It is necessary to prevent the coastline from getting eroded, and infrastructure cum construction activities should conform to government regulations except where relaxations are specified. Another corrective measure to avoid soil erosion is to provide suitable slope stabilisation measures. Runoff erosion during rains from unprotected excavated areas resulting in excessive soil erosion can be very damaging in Diu. It can be lessened by careful planning of cut-and-fill to minimise erosion, including resurfacing/revegetation of exposed areas. Furthermore, dikes' provision to hold runoff to settle out soil particles can be another solution for the coastal region of Diu and Daman.

The threat to the D&NH region's environment is urbanisation, deforestation, land degradation, groundwater dependence, inefficient wastewater systems and growing industrialisation. Over the years, many industrial units have been set up in the D&NH region due to its accessibility to industrial hubs like Vapi, Surat, Mumbai and more so due to the tax concessions. An unmanaged and scattered growth of industries is exerting pressure on the geo-environment of U.T. Several brick-making kilns are observed in terraced fertile agricultural land in the region, reducing the soil's top surface fertility, which would disturb the crop productivity in the area.

Table 9.12 Remaining life of existing pavement as per the homogeneous section

HS.	Start chainage	End chainage	15th percentile modulus, MPa			15th percentile layer thickness, mm		Calculated strain value		Residual life	
			Bituminous layer	Granular layer	Subgrade	Bituminous layer	Granular layer	Tensile strain at the bottom of the bituminous layer	Vertical strain at the top of the subgrade	Fatigue failure	Rutting failure
HS-1	0.00	5.28	553.35	73.66	34.71	175.0	210	6.42E−04	1.18E−03	0.85	0.27
HS-2	5.28	6.70	1,372.04	76.19	34.71	140.0	240	4.69E−04	1.02E−03	1.33	0.51
HS-3	6.70	10.30	566.10	77.33	34.71	225.0	70	5.22E−04	1.22E−03	1.85	0.23
HS-4	10.30	12.01	558.90	95.80	34.71	260.0	150	3.88E−04	8.54E−04	5.95	1.15

Table 9.13 Overlay requirement as per the homogeneous section (H.S.)

H.S.	Start chainage	End chainage	Design life (MSA)	15th percentile modulus, Mpa			15th percentile layer thickness, mm		Required overlay of B.C. (mm)	Calculated strain value		Residual life	
				Bituminous layer	Granular layer	Subgrade	Bituminous layer	Granular layer		Tensile strain at the bottom of the bituminous layer	Vertical strain at the top of the subgrade	Fatigue failure	Rutting failure
HS-1	0.00	5.28	12.000	553.35	73.66	34.71	175.0	210	140	2.63E−04	5.02E−04	27.23	12.80
HS-2	5.28	6.70	12.000	1,372.04	76.19	34.71	140.0	240	110	2.33E−04	4.99E−04	19.98	13.13
HS-3	6.70	10.30	12.000	566.10	77.33	34.71	225.0	70	150	2.13E−04	5.05E−04	60.52	12.44
HS-4	10.30	12.01	12.000	558.90	95.80	34.71	260.0	150	100	2.30E−04	5.03E−04	45.79	12.72

The solid waste and industrial waste collected at present are not disposed of into a properly designated sanitary landfill site, resulting in soil degradation and water contamination (Aravind and Das, 2007). The water quality in D&NH is tested with samples taken by the Pollution Control Committee. The dissolved oxygen is found as low as 1.5 to 2.5 mg/l in industrial areas like Pipariya, Dapada and Kharadpada. As an impact of this, brownish, reddish-orange water, low pH, low dissolved oxygen and high biological oxygen demand are observed (Gupte, 2013). The solid waste and industrial waste are collected and dumped beyond urban boundaries with inadequate facilities of a modern sanitary landfill in the region. It would result in the contamination of groundwater (Tamminana, 2014). Therefore, there is a need to balance natural soil strata and development in the U.T. of Daman, Diu and D&NH.

9.9   CONCLUDING REMARKS

Diu and Daman's soil strata vastly influence the inhabitants, rivers and sea due to their topography and geology. D&NH is landlocked and its topography varies from plain to hilly. The following are the significant observations listed from the detailed study of the terrain:

*   The Daman and Diu geographical area falls into five major physiographic subdivisions, viz., coastal alluvial plain (25.69%), undifferentiated hillside slopes (1.37%), pediments (6.96%), upper pediplains (6.79%) and lower pediplains (26.61%) of the total geographical area.
*   The major geology covered an area of 3,514.87 ha (33.71%) in the survey, the area is basalt, followed by coastal alluvium 2,678.74 ha (25.69%) and sandstone 836.05 ha (8.02%).
*   Diu is an isolated island having miliolite limestone and dunes with weathered soil on its terrain.
*   In Daman, the alluvium deposition is found in rivers and coastal areas, and the clay and silt dunes formation shows a seawater effect on the parent soil.
*   The geographical strata of D&NH belong to the Deccan trap made up of basic volcanic rocks of basaltic composition. The soils are a mixture of clay and moorum. The texture, composition and depth vary from place to place depending upon the erosion rate. Generally, on hill slopes and plateaus, the soil is shallow and rocky, while in plains, it is bouldery and deep and common moorum shallow type.
*   The D&NH region has alluvium deposits as well as weathered basalt rock at a greater depth.
*   Rapid industrialisation and urbanisation are evident with environmental change and soil degradation in Diu and Daman and D&NH.

Diu is an isolated island, Daman is a minor city and D&NH is situated between major states namely Gujarat and Maharashtra. Still, the soil variation and effect of different soil claimant and human activity on the soil properties are evident. A geotechnical engineer plays a vital role in protecting structures on the soils and further protecting soil from any adverse effects by different environmental and human activities.

REFERENCES

Aravind, K. and Das, A. (2007). Industrial waste in highway construction, Department of Civil Engineering, IIT Kanpur, India.

GOI. (2020). U.T. Administration of Dadra and Nagar Haveli and Daman and Diu, Govrnment of India, https://daman.nic.in/about-daman.aspx, accessed 15 December 2020.

GSI. (2020). Geological Survey of India (GSI), https://www.gsi.gov.in, accessed 15 December 2020.

Gupte, P.R. (2013). Ground Water Brochure Daman, UT of Daman and Diu, Central Ground Water Board, West Central Region, Ahmedabad, Ministry of Water Resources, Government of India.

https://www.mapsofindia.com/daman-diu/, accessed 15 December 2020.

IRC:115. (2014). Guidelines for Structural Evaluation and Strengthening of Flexible Road Pavement using Falling Weight Deflectometer (FWD) Technique, Indian Roads Congress, New Delhi.

IRC:37. (2018). Guidelines for the Design of Flexible Pavements. Indian Roads Congress, New Delhi.

IS:2131. (1981). Method for standard penetration test for soils. Bureau of Indian Standards, New Delhi.

IS 2720-16. (1987). Methods of Test for Soils, Part 16: Laboratory Determination of CBR. Bureau of Indian Standards, New Delhi.

IS 2720-7. (1980). Methods of Test for Soils, Part 7: Determination of Water Content-Dry Density Relation Using Light Compaction. Bureau of Indian Standards, New Delhi.

Jain, P.K. (2013). Ground water information U.T. D&NH, Central Ground water Board, Ministry of water resources, Government of India, New Delhi.

PMAY. (2020). Soil Investigation of PMAY at Ghoghla & Bucharwada in Diu/Report No./MPPL/SBC/BRH/36/220318/1/1-16, 2020.

Tamminana, V. (2014). Ground Water Brochure Diu, UT of Daman and Diu, Central Ground Water Board, West Central Region, Ahmedabad, Ministry of Water Resources, Government of India, New Delhi.

Chapter 10

Delhi

Ashutosh Trivedi
Delhi Technological University

Sadanand Ojha
Swati Structure Solutions Pvt Ltd.

CONTENTS

10.1   INTRODUCTION

In India, there have been ancient traditions to study all the natural geomaterials as life on the planet is at the bequest of earth. In the texts of antiquity, some references indicate that people respected soil, water and air in the proximity of forests and river valleys with sacredness for being a natural benefactor of human life and its support systems. There has been a history of natural geomaterials being used as construction materials ever since the human settlements began in India, whereas modern Delhi has been central to activities in many details. The frequently referred ancient civilisations are coincidental with Mahabharat (a reference to Indraprastha, which was in proximity to the geographical location of modern Delhi) and Ramayan periods when there had been uses of soil, mud and rocks extensively for the construction of royal palaces, castles and monasteries (Dutt, 1988). There were courtyard structures of monasteries and temples that have been used for public assemblies for educational and recreational purposes. Many cottages to be built for the saintly livelihood of the people involved in higher orders of attainments across India had been constructed with locally available well-identified soils mixed with certain fibre reinforcement conditioners and plasticisers found in cow dung. Several genera of bacteria identified (*Bacillus* and *Pseudomonas*) are known for their antagonistic properties against viruses, bacteria and fungi, which

DOI: 10.1201/9781003177159-10

supported the use of cow dung as a purifier in practices and for disease suppression by organic conditioning (Girija et al., 2013) relevant even to modern times.

The historical account of hundreds of Chinese monks who travelled to India in a millennium, namely Faxian (AD 399–412) in his notable and authentic work Foguoji (a record of Buddhist kingdom, project Gutenberg, 2006), referred to the centrality of ancient cities in the proximity of modern Delhi assigning richest possible livelihood of the people in India and thereby the use of construction materials and civil structures in the contemporary Mathura (near Delhi) and Magadha empire (Sen, 2006). There is a well-maintained monastery of Faxian at Nanjing China (2019), which has the original storage, relics and reflections of the ancient imprint.

Delhi is the national capital territory (NCT) of India. It is located in the northern part of the country. The geographical area of Delhi is estimated to be about $1482\,km^2$, of which nearly 70% is rural while the remaining 30% is urban. It has just $18\,km^2$ of the area occupied by surface water. It has the second-highest population density in India next to Mumbai. It has a population density of 11,312 persons per km^2 as per a census of the year 2011 and therefore experiences higher pressure on its land, air and water uses for residential and engineering activities compared to the national and global average of 500 and 25 persons per square kilometre. Delhi has higher population pressure per unit area compared to New York, Tokyo and Beijing; therefore, it poses complex problems related to underground and surface habitation, road communication, metro movement, safe navigation, intelligent transportation, water supply, air quality, drainage disposal, bio-safety, livelihood, healthcare, life safety, hazard mitigation and pandemics and so on, which need special understanding of Delhi soils. The NCT of Delhi situated along the sacred river, the Yamuna, also houses the capital of the world's largest democracy and has a population of almost 1.38 billion in 2020. According to the Yamuna river project, it faces unprecedented pressure on its eco-system, whereas Delhi soil bears the overburden. The sacred river, the Yamuna, is currently facing one of the highest pollutions in the world, thereby contaminating nearby soil systems. The entire quantum of freshwater flowing into New Delhi is redirected to fulfil the freshwater requirements of the city. From the Wazirabad barrage where it enters the capital city to the Okhla barrage and where it exists in the south, the Yamuna consists of only treated and untreated sewage and other toxic effluents forming new toxicity in the soil crust. The water to the north is rendered dead, with 0% dissolved oxygen, posing serious bio-hazards to the biota of Delhi soils (http://www.yamunariverproject.org/, accessed 12 May 2020).

10.2   MAJOR TYPES OF SOILS AND ROCKS

The surface soil formation in Delhi mostly has Yamuna sand with percentages of silt varying from location to location and significantly from a point in the ground to another depending upon its typical geological deposits (Ojha, 2015), land uses, biological changes, manmade depositions and disposal of wastes (Goel, 2018). Delhi has interesting geology on account of its location at the edge of the exposed ancient Aravalli mountain ranges extending northeast in this area. Delhi and its adjoining regions are surrounded in the northeast by Indo-Gangetic Yamuna plains, while the west has a plane extension of the Thar desert and the south has the Aravalli mountain ranges.

Table 10.1 Tentative depth of bedrocks
in NCT of Delhi

Area	Depth (m)
Vikas Bhawan	100–110
Patel Nagar (W)	60–65
Embassy Area	20–50
Pusa	10–15
Palam	15–20
Shahadra	175–200
Kichripur	100–115
Madanpur	130–140
Alipur	25–30

The rock masses around Delhi have undergone multiple jointing, folding, faulting and gauge deposition amid themselves in different phases of metamorphism along with some transverse features. The depth of bedrock in the Delhi region is shown in Table 10.1. The depth of bedrock in Delhi ranges from 10–20 m (Pusa and Palam) to 175–200 m (Shahadara). Geologically, the alluvial deposits belong to the Pleistocene period, that is, older alluvial deposits and of recent age, that is, newer alluvium. Older alluvium deposits consist of mostly fragmented deposits of clay, silt and sand along with Kankar. Broadly, Delhi can be considered as a geographical zone with an intricate surface network of roads, public utilities and civil structures. A map of the NCT of Delhi with the boundary of districts is shown in Figure 10.1.

Several field tests for geotechnical investigation as described in the textbook (Shukla, 2015) and relevant test standards have been conducted covering all the nine districts of Delhi further scattered at the varied locations, as shown in Figures 10.1 and 10.2 to assess the type and various index properties of soils for varied engineering uses.

The knowledge of index properties is very important (Shukla, 2014) for designing facilities constructed on various types of soil and rock masses. Some of the areas in the regions of northeast, east and north districts near the Yamuna flood plain having high water tables are prone to liquefaction and hence special care has to be taken while designing the foundation over liquefiable soil. The depth of the liquefaction prone zone ranges from 1.5 to 9 m using a soil investigation and microzonation report (https://moes.gov.in/, accessed 18 May 2020). The photographs of two prominent municipal landfill sites which seek serious attention of city administrators, political leaders, planners and habitants in the light of the recent pandemic, namely Bhalaswa and Gazipur, are shown in Figure 10.3. They show a low value of SPT often less than 4 and frequently 1–2.

Though the soil data for more than 70 locations are available, only a few have been presented here (Figure 10.4a–g) due to space constraints. Delhi being a metro city and a densely populated area needs special attention to create basic infrastructures like housing for all, sewage disposal, water supply scheme and construction of sewage treatment plants and water treatment plants. The soil profile for coral reef island Naiafru, Maldives (Figure 10.4h) mainly consists of marine soils, coral reefs and other biological calcareous deposits compared to the alluvial deposits of Delhi.

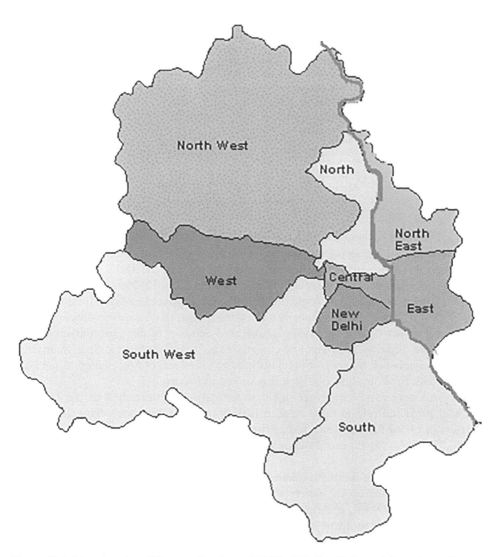

Figure 10.1 Map showing different districts of NCT of Delhi. (Adapted from https://moes.gov.in/, accessed 18 May 2020.)

Due to the peculiar nature of the soil, a specific site exploration is needed for all important works to have safe and economic structures. Landfill sites, waterlogged areas and the problem of uplift due to high water tables need a specific solution. The existence of liquefiable soil in some areas is also to be taken care of while designing a suitable foundation. A general map of the Delhi state, indicating the names of key segments and borehole locations as included with boring logs in this chapter, has been presented. Only selected boring logs from areas representing the main types of soils/rocks of Delhi have been included. Figure 10.4a shows the soil profile at the Ramjas College of Delhi University in North Delhi where lower SPT values can be noticed at shallow depths while a higher value of 60 and more can be observed at depths greater than 20 m.

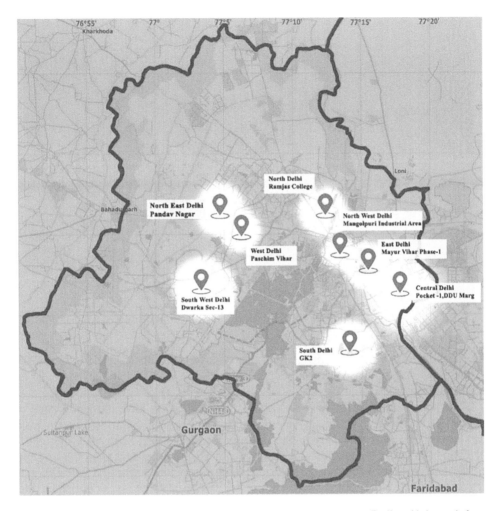

Figure 10.2 Map showing representative field test locations in Delhi. (Adapted from https://moes.gov.in/, accessed 18 May 2020.)

Based on the geological formation and geotechnical site investigations, Delhi soil can be classified into four broad divisions such as Khadar (new alluvium), Banger (old alluvium), Dabar soil low-lying areas and Kohi soil (rocky areas). The tentative depth of bedrocks in different parts of the NCT of Delhi is shown in Table 10.1. Based on the collected borehole data, soil profiles are made covering the almost entire region to study the subsoil heterogeneity. In the trans-Yamuna area, silt is very predominant. In the eastern block, the soils are sandy silts/silty sands with a high percentage of medium to fine sand of Yamuna river alluvium. The areas like Noida, Mayur Vihar, Yamuna Vihar, Abdul Fazal Enclave and Geeta Colony, which fall in the eastern block, have very soft soil deposits and a high water table. Northern and western blocks have silty sands with a reasonable percentage of clay. Areas such as Rohini, Punjabi bagh, Paschim Vihar, Janakpuri and Dwarka fall in these blocks. The south and central blocks

(a)   (b)

Figure 10.3 Photographs showing MSW landfills: (a) Bhalswa and (b) Gazipur landfill site.

have gravelly sands with varying percentages of gravels (Ghitorni, Maidan Garhi and Satbari). As per soil classification systems, these sands and silts are coarse and fine-grained granular materials.

The soils in the vicinity of the Yamuna flood plain do have nascent formation while elsewhere it often witnesses water logging in low-lying areas amid poor drainage conditions. Delhi forms a part of the Indo-Gangetic plain. It is an extension of the Aravalli hills from Rajasthan which enters Delhi through Gurugram on the southern border and expands into an elongated ridge of 5–6 km width forming ridges along the north to northeast and south to southwest separated by flatlands and depressions filled with transported materials. The Delhi alluvium can further be divided into two subdivisions, one for recent deposits and the other for old alluvium deposits. The alluvium soils of recent flood plains exhibit a distinct stratification being finer on the surface and a coarser layer below a certain depth. It is often silty sand in the upper layer and sandy silt in the lower strata. This type of soil is concentrated mostly in the eastern part of Delhi along the flood plain of Yamuna.

The old alluvial deposits are mainly concentrated in the northwestern part of Delhi in areas such as Rohini, Pitampura, Narela, and Paschim Vihar. In a large part of this region, soil deposits are affected by water logging during the rainy season due to poor drainage conditions. These soils are coarse loamy in texture and are generally fertile soils with high moisture contents. However, at some places, saline and alkaline patches are also present. In the southwest part of Delhi, most of the area is water-logged and the texture of the soil in these areas is normally sandy loam. Large areas are covered by saline and alkaline soils due to poor drainage. The residual strength of soil grains varies due to the wide dissimilarity in the topography of Delhi and the level of the underground water table in the particular area. In some areas, the water level varies from 1 to 1.5 m depths mainly in alluvial soil, whereas in some areas, the water table is as below as more than 50 m below the ground surface. The localised filled-up area is another matter of concern. It is found that in some cases, a small chunk of land-filling with Malba ranges from 1 to 3.5 m depth.

The shear strength of the soil is the principal criterion for selecting a suitable type of substructure for buildings, bridges, roads and related infrastructure works such as water treatment plants, sewage treatment plants (STP) and so on. Hence, it is always advisable to go for a detailed soil investigation before the design of the substructure since it is difficult to generalise the soil type in any particular area due to the buildup area and loading conditions. The varied intrinsic and engineering properties of soils frequently observed in Delhi soils are shown in Tables 10.2–10.4.

(a)

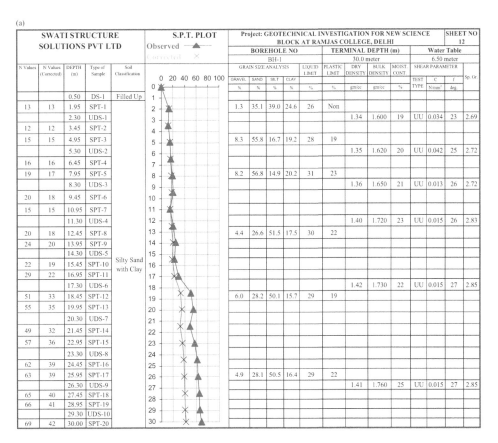

Figure 10.4 (a) Soil profile at Ramjas College, North Delhi. (b) Soil profile for plot no.131–133 Block C, at Mangolpuri Industrial Area Phase 1, Delhi. (c) Soil profile for Police Housing at Mayur Vihar, Phase-1, Delhi. (d) Soil profile for 120 GPRA Type-VII Flats Pocket-1 at DD Upadhyay Marg, New Delhi. (e) Soil profile for SS School Sec-13 Dwarka, New Delhi. (f) Soil profile for Sarvodaya Vidyalaya Paschim Vihar, New Delhi. (g) Soil profile for Police Post Pandav Nagar, New Delhi. (h) Soil profile for coral reef island Naiafru, Maldives. (i) Collective representation of penetration resistance observed for central Delhi.

(*Continued*)

(b)

SWATI STRUCTURE SOLUTIONS PVT. LTD.

S.P.T. PLOT — Observed ▲ / Corrected ✕

PROJECT: GEOTECHNICAL INVESTIGATION FOR PLOT NO. 131,132,133 BLOCK C, AT MANGOLPURI INDUSTRIAL AREA PHASE 1, DELHI   SHEET NO. 13

BOREHOLE NO	TERMINAL DEPTH (m)	Water Table
BH-1	15.0 meter	1.00 meter

N Values	N Corrected Values	DEPTH (m) from E.G.L.	Type of Sample	Soil Classification	GRAVEL %	SAND %	SILT %	CLAY %	LIQUID LIMIT %	PLASTIC LIMIT %	DRY DENSITY gm/cc	BULK DENSITY gm/cc	MOIST CONT. %	TEST TYPE	C N/mm²	f deg	Sp. Gr.
		0.50	DS-1														
12	12	1.95	SPT-1	SILTY SAND WITH GRAVEL	6.9	13.2	59.4	20.5	21	NP							
		2.30	UDS-1								1.48	1.80	22	UU	0.004	25	2.65
17	16	3.45	SPT-2		9.5	18.0	46.6	25.9	27	NP							
23	19	4.95	SPT-3														
		5.30	UDS-2								1.39	1.74	25	UU	0.010	25	2.68
28	22	6.45	SPT-4	SILTY SAND	0.0	36.6	44.3	19.1	30	NP							
41	28	7.95	SPT-5														
		8.30	UDS-3								1.40	1.78	27	UU	0.011	24	2.67
48	32	9.45	SPT-6		5.4	15.8	21.8	57.0	26	20							
51	33	10.95	SPT-7	CLAYEY SILT													
		11.30	UDS-4								1.35	1.74	29	UU	0.011	25	2.68
41	28	12.45	SPT-8														
42	29	13.95	SPT-9		8.8	54.4	24.4	12.4	22	NP							
		14.30	UDS-5	SANDY SILT							1.42	1.76	24	UU	0.010	26	2.65
51	33	15.00	SPT-10														

(c)

SWATI STRUCTURE SOLUTIONS PVT. LTD.

S.P.T. PLOT — Observed ▲ / Corrected ✕

PROJECT: GEOTECHNICAL INVESTIGATION FOR C/O POLICE HOUSING, MAYUR VIHAR, PHASE -1, DELHI.   SHEET NO. 13

BOREHOLE NO	TERMINAL DEPTH (m)	Water Table
BH-1	15.0 meter	1.00 meter

N Values	N Corrected Values	DEPTH (m) from E.G.L.	Type of Sample	Soil Classification	GRAVEL %	SAND %	SILT %	CLAY %	LIQUID LIMIT %	PLASTIC LIMIT %	DRY DENSITY gm/cc	BULK DENSITY gm/cc	MOIST CONT. %	TEST TYPE	C N/mm²	Ø deg	Sp. Gr.
		0.50	DS-1														
8	8	1.95	SPT-1	SILTY SAND WITH GRAVEL													
		2.30	UDS-1								1.65	1.7	3	UU	0	24	2.65
12	12	3.45	SPT-2		2	79	16	3	-	NP							
15	15	4.95	SPT-3														
		5.30	UDS-2														
13	13	6.45	SPT-4	SILTY SAND													
16	16	7.95	SPT-5														
		8.30	UDS-3								1.34	1.72	28	UU	0	26	2.63
18	17	9.45	SPT-6		3	76	18	3	-	NP							
20	18	10.95	SPT-7	CLAYEY SILT													
		11.30	UDS-4														
24	20	12.45	SPT-8														
28	22	13.95	SPT-9	SANDY SILT													
		14.30	UDS-5														
34	25	15.00	SPT-10														

(Continued)

(d)

SWATI STRUCTURE SOLUTIONS PVT. LTD.

S.P.T. PLOT — Observed ▲ — Corrected ✕

PROJECT: CONSTRUCTION OF 100 NOS. (MODIFIED TO 120 NOS.) GPRA TYPE-VII FLATS IN POCKET-1 AT DEEN DYAL UPADHYAY MARG, NEW DELHI   SHEET NO 22

EOREHOLE NO	TERMINAL DEPTH (m)	Water TaEle
BH-4	25.0 meter	5.00 meter

N Values	N Values (Corrected)	DEPTH (m)	Type of Sample	Soil Classification	GRAVEL %	SAND %	SILT %	CLAY %	LIQUID LIMIT %	PLASTI C LIMIT %	DRY DENSITY gm/cc	EULK DENSITY gm/cc	MOIST CONT %	TEST TYPE	C N/mm²	Ø deg	Sp. Gr.
		0.50	DS-1	Filled Up													
0	0	1.95	SPT-1														
		2.30	UDS-1														
12	12	3.45	SPT-2														
15	15	4.95	SPT-3		3.1	67.5	27.3	2.0	23	NP							
		5.30	UDS-2								15.4	18.2	18	UU	0.019	27	2.64
15	15	6.45	SPT-4														
20	18	7.95	SPT-5		3.5	58.7	32.8	4.9	21	NP							
		8.30	UDS-3								14.7	17.8	21	UU	0.019	28	2.65
22	19	9.45	SPT-6														
23	19	10.95	SPT-7														
		11.30	UDS-4								15.2	18.1	19	UU	0.020	27	2.66
27	21	12.45	SPT-8	Silty Sand	1.4	60.9	33.6	4.1	23	NP							
30	23	13.95	SPT-9														
		14.30	UDS-5														
32	24	15.45	SPT-10														
36	26	16.95	SPT-11														
		17.30	UDS-6								16.3	19.5	20	UU	0.020	27	2.66
37	26	18.45	SPT-12		0.3	64.1	31.8	3.8	24	NP							
41	28	19.95	SPT-13														
		20.30	UDS-7														
43	29	21.45	SPT-14														
45	30	22.95	SPT-15		0.5	63.8	31.4	4.3	24	NP							
		23.30	UDS-8								15.7	19.2	22	UU	0.019	27	2.67
48	32	25.00	SPT-16														

(e)

SWATI STRUCTURE SOLUTIONS PVT. LTD.

S.P.T. PLOT — Observed ▲ — Corrected ✕

PROJECT: GEOTECHNICAL INVESTIGATION REPORT FOR SENIOR SECONDARY SCHOOL BUILDING, SECTOR -13, DWARKA, DELHI   SHEET NO 13

BOREHOLE NO	TERMINAL DEPTH (m)	Water Table
BH-1	20.0 meter	5.50 meter

N Values	N Values Corrected	DEPTH (m)	Type of Sample	Soil Classification	GRAVEL %	SAND %	SILT %	CLAY %	LIQUID LIMIT %	PLASTIC LIMIT %	DRY DENSITY gm/cc	BULK DENSITY gm/cc	MOIST CONT %	TEST TYPE	C N/mm²	f deg	SPECIFIC GRAVITY	VOID RATIO 'oo'
		0.50	DS-1	Silty Sand with Clay														
13	13	1.95	SPT-1															
		2.30	UDS-1		0	10	72	17	28	15	1.39	1.7	22	CU	0	22	2.88	22
16	16	3.45	SPT-2															
21	18	4.95	SPT-3															
		5.30	UDS-2								1.42	1.72	21	CU	0	23	2.72	23
23	19	6.45	SPT-4		13	47	28	12	-	NP								
27	21	7.95	SPT-5															
		8.30	UDS-3															
39	27	9.45	SPT-6															
46	31	10.95	SPT-7															
		11.30	UDS-4	Silty Sand							1.51	1.74	15	CU	0	24	2.68	24
50	33	12.45	SPT-8															
55	35	13.95	SPT-9		0	51	32	17	-	NP								
		14.30	UDS-5															
60	38	15.45	SPT-10															
0	0	16.95	SPT-11															
		17.30	UDS-6															
0	0	18.45	SPT-12															
0	0	20.00	SPT-13															

(Continued)

(f)

SWATI STRUCTURE SOLUTIONS PVT. LTD.

S.P.T. PLOT — Observed ▲ / Corrected ✕

PROJECT: C/O GOVT SARVODAYA CO-ED VIDYALAYA B-4 PASCHIM VIHAR SCHOOL ID1617008 — SHEET NO 13

BOREHOLE NO BH-1 | TERMINAL DEPTH 20.0 meter | Water Table 5.50 meter

N Values	N Values Corrected	DEPTH (m)	Type of Sample	Soil Classification	Gravel %	Sand %	Silt %	Clay %	Liquid Limit %	Plastic Limit %	Dry Density gm/cc	Bulk Density gm/cc	Moist Cont %	Shear Test Type	C N/mm²	f deg	Specific Gravity	Void Ratio
		0.50	DS-1	Silty Sand with Clay														
10	10	1.95	SPT-1		0.4	75.2	18.4	6.0	29	17								
		2.30	UDS-1								1.52	1.78	17	UU	0.026	27	2.65	0.74
11	11	3.45	SPT-2		1.8	39.5	39.9	18.8	20	NP								
12	12	4.95	SPT-3															
		5.30	UDS-2								1.51	1.77	17	UU	0.025	27	2.65	0.75
17	16	6.45	SPT-4		0.0	49.0	36.5	14.4	23	NP								
18	17	7.95	SPT-5															
		8.30	UDS-3								1.45	1.75	21	UU	0.026	26	2.66	0.84
18	17	9.45	SPT-6		0.0	46.3	37.6	16.1	23	NP								
23	19	10.95	SPT-7															
		11.30	UDS-4	Silty Sand							1.50	1.80	20	UU	0.026	27	2.66	0.77
27	21	12.45	SPT-8															
31	23	13.95	SPT-9		2.1	41.5	36.4	20.0	24	NP								
		14.30	UDS-5								1.48	1.78	20	UU	0.025	26	2.67	0.80
34	25	15.45	SPT-10															
38	27	16.95	SPT-11															
		17.30	UDS-6								1.51	1.80	19	UU	0.026	26	2.65	0.75
41	28	18.45	SPT-12		2.8	49.5	36.7	11.0	23	NP								
39	27	20.00	SPT-13															

(g)

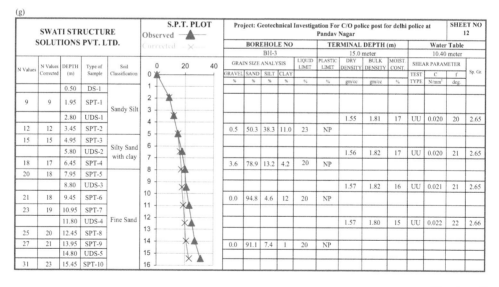

SWATI STRUCTURE SOLUTIONS PVT. LTD.

S.P.T. PLOT — Observed ▲ / Corrected ✕

Project: Geotechnical Investigation For C/O police post for delhi police at Pandav Nagar — SHEET NO 12

BOREHOLE NO BH-3 | TERMINAL DEPTH 15.0 meter | Water Table 10.40 meter

N Values	N Values Corrected	DEPTH (m)	Type of Sample	Soil Classification	Gravel %	Sand %	Silt %	Clay %	Liquid Limit %	Plastic Limit %	Dry Density gm/cc	Bulk Density gm/cc	Moist Cont %	Shear Test Type	C N/mm²	f deg	Sp. Gr.
		0.50	DS-1														
9	9	1.95	SPT-1	Sandy Silt													
		2.80	UDS-1								1.55	1.81	17	UU	0.020	20	2.65
12	12	3.45	SPT-2		0.5	50.3	38.3	11.0	23	NP							
15	15	4.95	SPT-3														
		5.80	UDS-2	Silty Sand with clay							1.56	1.82	17	UU	0.020	21	2.65
18	17	6.45	SPT-4		3.6	78.9	13.2	4.2	20	NP							
20	18	7.95	SPT-5														
		8.80	UDS-3								1.57	1.82	16	UU	0.021	21	2.65
21	18	9.45	SPT-6		0.0	94.8	4.6	1	20	NP							
23	19	10.95	SPT-7														
		11.80	UDS-4	Fine Sand							1.57	1.80	15	UU	0.022	22	2.66
25	20	12.45	SPT-8														
27	21	13.95	SPT-9		0.0	91.1	7.4	1	20	NP							
		14.80	UDS-5														
31	23	15.45	SPT-10														

(Continued)

(h)

N Values	N Corrected Values	DEPTH (m)	Type of Sample	Soil Classification	S.P.T. PLOT (Observed / Corrected) 0 20 40 60 80		GRAIN SIZE ANALYSIS GRAVEL %	SAND %	SILT %	CLAY %	LIQUID LIMIT %	PLASTIC LIMIT %	DRY DENSITY gm/cc	BULK DENSITY gm/cc	MOIST. CONT. %	SHEAR PARAMETER TEST TYPE	C N/mm²	f deg.	Sp. Gr.
		0.50	DS-1	MARINE SAND WITH CORAL STONE		0													
23	19	1.95	SPT-1				21.9	66.9	9.6	1.6	21	NP	1.26	1.50	19	UU	0.130	30	2.66
		2.30	UDS-1																
28	22	3.45	SPT-2				57.7	38.0	3.2	1.2	25	NP	1.34	1.57	17	UU	0.015	30	2.65
35	25	4.95	SPT-3				48.1	39.0	10.6	2.3	27	NP	1.36	1.60	18	UU	0.015	31	2.65
		5.30	UDS-2																
		6.45	SPT-4																
		7.95	SPT-5																
REFUSAL		8.30	UDS-3	REFUSAL	REFUSAL														
		9.45	SPT-6																
		10.00	SPT-7																

PROJECT: GEOTECHNICAL INVESTIGATION REPORT FOR CONSTRUCTION OF SEWERAGE FACILITIES AND WATER SUPPLY SYSTEM AT I.H. NAIAFRU, MALDIVES

BOREHOLE NO: BH-5 (PS-3) TERMINAL DEPTH (m): 10.0 meter Water Table: 0.80 meter

(i)

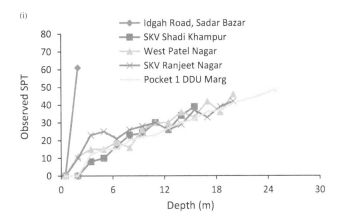

Figure 10.4 (a) Soil profile at Ramjas College, North Delhi. (b) Soil profile for plot no.131–133 Block C, at Mangolpuri Industrial Area Phase I, Delhi. (c) Soil profile for Police Housing at Mayur Vihar, Phase-I, Delhi. (d) Soil profile for 120 GPRA Type-VII Flats Pocket-I at DD Upadhyay Marg, New Delhi. (e) Soil profile for SS School Sec-13 Dwarka, New Delhi. (f) Soil profile for Sarvodaya Vidyalaya Paschim Vihar, New Delhi. (g) Soil profile for Police Post Pandav Nagar, New Delhi. (h) Soil profile for coral reef island Naiafru, Maldives. (i) Collective representation of penetration resistance observed for central Delhi.

The authors recommend the uses of strength and dilatancy concepts, which suggest that the difference between triaxial and plane strain friction angles is related to the dilatational characteristics of the Yamuna sand (Figure 10.5). The rate at which a material dilates in shear is primarily dependent on the observed peak friction angle ϕ_{peak}. Bolton's (1986) stress–dilatancy equation in terms of peak and constant volume friction angles (ϕ_c) and an angle of dilatancy (ϕ_d) at peak strength can be expressed by a common term I_{af}, called the relative dilatancy, as

$$\phi_{peak} - \phi_c = AI_{af} \tag{10.1}$$

Table 10.2 Frequently observed properties in the geotechnical investigation of selected areas in Delhi

S. no.	Name of soil property	Values (range)	Reference
1.	Water content	6%–25%	Test results
2.	Specific gravity	2.60–2.68	obtained from
3.	Plasticity index	NP/6	field tests on
4.	In situ unit weight of soil	17–18 k/Nm^3	Delhi soil
5.	Relative density	30%–80%	
6.	Cohesion (c)	0–25 kPa	
7.	Friction angle (ϕ)	22°–38°	
8.	SPT	3–60	
9.	Rock classification	Heavily jointed rock masses	

Table 10.3 Grain size characteristics of sandy soils in selected areas of Delhi compared to standard sandy soils based upon laboratory analysis

Sand type	(%) Fines	D_{10} (mm)	D_{30} (mm)	D_m (mm)	D_{60} (mm)	C_c	C_u	References
Delhi soils (represented by Yamuna sand and silt)	0–25	0.12 to 0.08	0.20 to 0.15	0.25 to 0.20	0.25 to 0.28	1.0 to 2.0	1.0 to 3.0	Ojha and Trivedi (2013a, 2013b)
Ghaggar sand	0–25	0.1 to 0.2	0.33 to 0.18	0.50 to 0.37	0.56 to 0.48	1.00 to 2.00	2.9 to 4.7	Gupta and Trivedi (2009)
Ottawa sand	0	0.18	–	0.39	0.27	1.09	1.48	Salgado et al. (2000)
Ticino sand	0	–	–	0.55	–	–	1.6	Lo Presti (1987)
Ham river	0	–	–	0.22	–	–	–	Bishop and Gordon (1965)

Table 10.4 Intrinsic variables of sandy soils in selected areas of Delhi compared to standard sandy soils based upon laboratory analysis

Sand type	% Fines	e_{min}	e_{max}	ϕ_c	D_m	References
Delhi soils (represented by Yamuna sand and silt)	0–25	0.50 to 0.31	0.78 to 0.62	24.2 to 30.7	0.225 to 0.208	Ojha and Trivedi (2013a, 2013b),
Ottawa sand	0–25	0.48 to 0.29	0.78 to 0.62	29.0 to 33.0	–	Salgado et al. (2000)
Ham river sand	0	0.92	0.59	33	0.22	Bishop and Gordon (1965)
Monterey sand	0	0.57	0.86	37	–	Houlsby (1993)
Sacramento river sand	0	0.53	0.87	33.2	0.3	Lade and Yamamuro (1997)

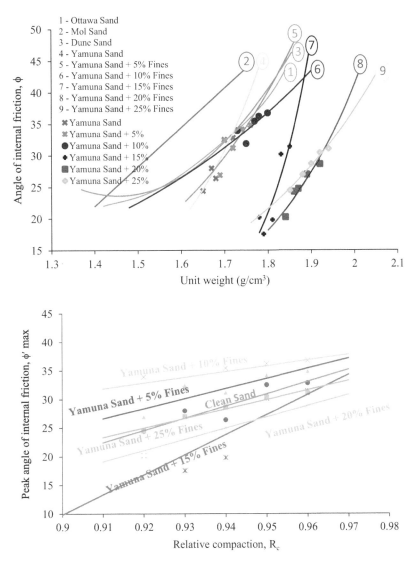

Figure 10.5 (a) Variation of unit weight with the angle of internal friction (Ojha, 2015).
(b) Variation of the angle of internal friction with relative compaction (Ojha, 2015).

with

$$I_{af} = R_c(Q_{af} - \ln 100\, p' / P_A) - R_{af} \qquad (10.2)$$

In Eqs. (10.1) and (10.2), A is an empirical constant having a value of 3 for triaxial strain conditions and takes care of the scaling effect, inclination factor and shape factor, R_c is the relative compaction defined as the ratio of natural dry unit weight to maximum unit weight, expressed as a number between zero and 1, p' is the mean effective stress at peak strength in kPa, P_A is the reference stress (100 kPa) in the same units as p' and Q_{af} and R_{af} are non-dimensional and non-linear shear strength parameters, respectively. It has been observed that the dilatancy angle at peak strength was a function of the relative compaction R_c and mean normal effective stress confinement p' and strain condition (Ojha, 2015). In Eq. (10.1), Q_{af} and R_{af} are non-linear shear strength parameters for Yamuna sand with silt evaluated in the present work for a varied proportion of fines. The maximum dilatancy angle (10°) was useful to relate the strength of silty sand to the liquefaction potential.

The Delhi ridge area and most of south Delhi extending from Mayapuri towards Dhaula Kuan, Saket, Malviya Nagar, Greater Kailash, Nehru Place, Ghitorni and so on consist of rocky soil. It is normally composed of quartzite or sandstone of the Delhi ridge. The texture of such soils varies from sandy loam to clay loam. Due to the uneven topography, these soils are subjected to a severe degree of erosion. The Delhi ridge which is the northernmost extension of the Aravalli mountain ranges consists of jointed rocks (Figure 10.6a and b) and extends from the southern parts of the territory

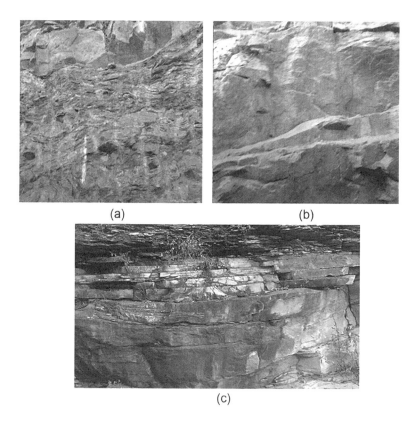

(a)   (b)

(c)

Figure 10.6 (a) Exposed depth of vertical cut in bedrock in south Delhi (30 cm white marker along with descriptive waviness and entrapped gauge classified with low GSI and RQD values). (b) Exposed depth of 25 m deep vertical cut in rock mass in a south Delhi location (the selected section has been classified for high GSI and RQD values). (c) Exposed depth of more than 50 m deep vertical cut in a rock mass with horizontal joints and fragmented infills at a location in Kota Rajasthan (selected section classified for low GSI and RQD values).

Table 10.5 Results of site investigation in the south Delhi area

Sr. No.	Depth (m)	Core recovery (%)	Rock quality designation (%)	Rock quality (IS:11315 (1985), Part 11)
BH-1	1.0–1.3	76	12	Very poor
BH-2	1.0–1.2	54	10	Very poor

Table 10.6 Sample evaluation of rock masses based on RMR

Description	Rock mass rating increment
Point load index (R3)	4
RQD is less than 25% (R1)	3
Joint spacing is more than 5 mm (R2)	10
Surface of joints take slightly rough (R3)	6
Ground water condition (R4)	7
Total	30
Unfavourable orientation for foundation (R6)	−15
RMR	15

to the western bank of Yamuna. Several cases of jointed rock formations are found at other locations in India (Figure 10.6c). The alluvial formations overlying the quartzite bedrock have different natures on either side of the ridge.

The subsurface investigations to interpret the engineering properties and RQD of some jointed rocks observed in Delhi are shown in Tables 10.5–10.8, which provide the scope of the evaluation of deformation modulus (E_{mr}) and compressive strength of rock masses (σ_{mr}) based upon the interpretation of Rock Mass Rating (RMR), GSI and in situ RQD values. The strength and modulus ratio of rock masses has been presented as

$$s_{mr} = E_{mr} \exp[\alpha\, C_{hs} \exp\,(\beta\, RQD)];$$ (10.3)

where C_{hs} is the plasticity parameter for the rock masses, while α and β depend upon the relationship of J_{fg} (joint factor of rock masses) and RQD (Trivedi 2013a, 2015, 2017).

India is divided into four zones as per the seismic map. Delhi comes under zone IV of the seismic zone. Based on soil dynamics to assess the correct spectrum to be used in the evaluation of design acceleration coefficient, the type of soil on which structure is placed is classified into four categories as per IS:1893 (2016):

• Soil type A – Rock or hard soil
• Soil type B – Medium or stiff soil
• Soil type C – Soft soil
• Soil type D – Liquefiable soil

The above soil classification is based on the N (standard penetration test) values obtained from the SPT field test and grain size distribution curve as given in IS:1893 (2016). A summary of N values for central Delhi has been plotted along with depth in Figure

Table 10.7 Strength assessment from rock mass characterisation systems (RMR, GSI and RQD)

Rock mass characterisation system (scale)	RQD	Joints spacing	UCS (σ_r)	Joint water pressure ($J_w = u/\sigma_1$), inflow (J_k)	Gouge (t/t_a) (J_a), roughness parameter (J_r)	Orientation ($n\beta$)	SRF	Current classification
RMR = R1 + R2 + R3 + R4 + R5 + R6; a $\sigma_{mr}= [exp\{(RMR-100)/9\}]^{0.5}$ b $\sigma_{mr}= exp[(RMR-100)/24]$								
GSI (0–100)	R1	0.01–10	300–0	0–0.6	–	–	–	Tables 10.5 and 10.6
GSI = 10 + R1 + R2 + R3 + R4 (from RMR); a $\sigma_{mr}= [exp\{(GSI-100)/9\}]^{0.5}$ for GSI > 25								
RQD[d,e] (0–100)	R1	f_2(R2)	f_3(R3)	f_4(R4)	$c_g=f_5$(R5) $r=f(\sigma_m)$	$n\beta = f_6$(R6)	$a_p = \lambda/C$	Tables 10.5 and 10.6 Insitu application
R4[c] $J_{fg} = \alpha\ [exp\ (\beta\ RQD)]$; $\sigma_{mr} = E_{mr}\ exp\ [\ \alpha\ C_{hs}\ exp\ (\beta\ RQD)]$; α and β depend upon the relationship of J_{fg} and RQD; C_{hs} is a pressure and damage-sensitive plastic parameter								

[a] Hoek and Brown (1980); [b] Kalamaras and Bieniawski (1995); [c] Barton (2002); [d] Trivedi (2013b); [e] Trivedi (2015, 2017).

Table 10.8 Strength ratio directly from in situ characterisations and RQD

Rock mass characterisation system (scale)	RQD	Application	Current classification
RQD (0–100)	$\sigma_{mr} = E_{mr}\ exp\ [\alpha\ C_{hs}\ exp\ (\beta\ RQD)]$; α and β depend upon the relationship of J_{fg} and RQD^a	In situ strength and deformations	RQD > 10 Tables 10.5–10.7

[a] Trivedi (2013b, 2015).

10.4i as a collective representation of penetration resistance observed for central Delhi. Figure 10.4a–g shows some of the typical boring log profiles based on field and laboratory tests carried out by Swati Structural Solutions, Delhi. This chapter provides a guideline about the basic nature of Delhi soil showing the soil profile of various random locations selected to represent the particular region. Figure 10.5a shows a variation of the angle of internal friction with unit weight as per the experimental programme by Ojha (2015) on Yamuna sand with varying percentages of fines in comparison to the other standard sand. The range of variation of engineering parameters is also shown in Tables 10.3 and 10.4. Table 10.3 shows grain size characteristics of sandy soils in selected areas of Delhi compared to standard sandy soils based upon laboratory analysis. Table 10.4 shows intrinsic variables of sandy soils in selected areas of Delhi compared to standard sandy soils based upon laboratory analysis by Ojha (2015). The strength properties of silty sand from the Yamuna basin for varied packing densities (relative compaction R_c between 0.92 and 0.98) show a friction angle in the range of 28°–38°.

From the soil profiles as presented here, it is clear that Delhi soil is mostly sandy with varying percentages of silt and clay depending upon its location away from River Yamuna and depth from the ground level. It has been observed that soil is non-plastic except in the old Delhi region (Ramjas College) where it is plastic in nature below 5 m. However, it has also been observed that a thin layer of plastic soil is embedded between silty sand at a certain depth in various parts of the Delhi region.

In the east Delhi area where the water table is as high as 1.5 m and strata are sandy, the soil is liquefiable up to a depth of 6.5 m below which no liquefaction is normally encountered. A detailed method for evaluation of the liquefaction potential is available in the literature and is mostly based on Seeds and Idriss's method (1971). The authors have also proposed to evaluate the liquefaction potential of soil based on relative compaction which is much simple and can be adopted. Details are given in a separate section in this chapter. The rock formation is mostly concentrated in the south and southwest Delhi area and the rock quality is soft disintegrated rock (SDR), evaluated based on RMR and RQD methods of analysis. The detailed procedure is given in IS:11315 (1985) and IS:1892 (1979).

10.3 PROPERTIES OF SOILS AND ROCKS

Based on various soil profiles derived from field tests for silty sand and silty sand with clay, the physical and engineering properties of Delhi soil can be obtained from the fractional presence of silt in Yamuna sand. The behaviour of Yamuna sand with varying percentages of silt has been investigated extensively, compared with standard sandy soils, and summarised in Tables 10.2–10.4.

Figure 10.2 shows the map of Delhi with various boring log locations where detailed soil investigations have been carried out to determine the nature and strength of soil (IS:1948 (1970), IS:2131 (1981), IS:2720 (1980)) for evaluation of the safe bearing capacity of soil for the design of foundation for various types of civil structure. The moisture content of soil plays a very important role in the evaluation of the dry density of soil. It is important to know the optimum moisture content of the soil to achieve maximum density compaction and therefore the maximum bearing capacity of the soil. One of the most important uses is in the construction of an embankment, where the density of the compacted embankment is dependent upon the moisture content. It reflects the history of weathering. It gives an idea about the suitability of the soil as a construction material; the higher value of specific gravity gives more strength for roads and foundations. It is also used in the calculation of void ratio, porosity, degree of saturation and other soil parameters. The specific gravity obtained for Delhi soil is in the range of 2.65–2.68, which is normal for sandy soil or silty sand. The higher specific gravity of soil results in the higher shearing strength of the soil. Also, the CBR value of soil increases with an increase in the specific gravity, which is an important criterion for pavement design. Since most of the soil sample tested has a PI value less than 12 or none-plastic, it can be concluded that mostly Delhi soil exhibits a low degree of serviceability and is not critical. Relative density (RD) is a measure of the state of compaction of soil between the two extreme states, that is, its densest and loosest states. RD values between 30 and 80 show that Delhi soil lies between loose and very dense states.

The grain size characteristics of sandy soil depend upon the presence of varied proportions of fines in the form of silt and clay. The mean size (D_{50}) of Yamuna sand with varied proportions of fines is in the range of 0.208 to 0.225 mm. Average particle sizes finer than 60% (D_{60}) vary from 0.24 to 0.25 mm. Grain sizes finer than 30% lie between 0.15 and 0.19 mm, while the effective size (D_{10}) is in the range of 0.08 to 0.13 mm. The coefficient of uniformity ranges from 1.852 to 3.307 for Yamuna sand with a silt percentage of up to 10% which shows that the sample characteristics change very significantly for a higher silt content.

The engineering behaviour of soil is greatly affected by its strength parameters known as cohesion and angle of internal friction. For evaluating the angle of friction and cohesion, a direct shear test was conducted on undisturbed samples collected at various depths, and the average values are shown in Tables 10.2 and 10.4. The critical state is defined as the state at which the sand is sheared without changes in either shear strength or volume. The critical friction angle depends upon the minimum and maximum void ratios, which are dependent on the varied proportion of fines and mean particle size. It has been established that the maximum and minimum void ratio decrease as the proportion of fines increases and the critical state friction angle increases with the increase in the proportion of fines in the soil. Table 10.5 shows the results of site investigation for the south Delhi area. Table 10.6 shows sample evaluation of rock masses based on RMR. Table 10.7 gives the strength assessment from rock mass characterisation systems (RMR, GSI and RQD). Table 10.8 gives the strength ratio from laboratory and in situ characterisation systems and RQD. The strength and modulus ratio of the rock masses are defined as the ratio of compressive strength and deformation modulus of the rock mass to that of intact rock under similar conditions of testing.

The rock formation in Delhi is mostly concentrated in the south and southwest Delhi area and the rock quality is SDR evaluated based on the RMR, GSI and RQD methods of analysis. Analysis of allowable bearing capacity on rock can be done by the following three methods:

- Presumptive values as published in IS:12070, IRC:78,
- Based on RMR and
- Based on compressive strength of intact rock specimen using the procedure given in IS:12070 (1987).

If rock is encountered at a shallow depth, the foundation shall be embedded minimum 0.5 m in rock. If the foundation is laid over the rock it should be checked for structural stability for sliding. The shear key shall be provided as per the design given by the structural engineer. The method of evaluation of safe bearing pressure on the rock as per IS:12070 (1987) is given below:

- Good rock with wide (1 to 3 m) or very wide (>3 m) spacing of discontinuities is analysed based on rock mass classification.
- Rock mass with closed discontinuities at moderately close (0.3 to 1 m) spacing shall be analysed by evaluating the core strength.
- A pressure metre is generally used for evaluation of strength characteristics of rock of low to a very low strength (<500 kg/cm^2): rock mass with discontinuities at close (5 to 30 cm) or very close (<5 cm) spacing, fragmented or weathered rock.
- A plate load test is generally recommended for rocks of very low strength (<250 kg/cm^2): rock mass with discontinuities at very close spacing, fragmented or weathered rock.

10.4 USE OF SOILS AND ROCKS AS CONSTRUCTION MATERIALS

The use of geomaterials is referred to in the introduction; even in the recent past, about 600–700 mm thick walls were constructed using mud mortar for the construction of homes in villages and even in urban villages using wood and tiles as covering materials.

Soil properties and behaviour are of great importance in the various fields of civil engineering. Many earthen dams are constructed for storage of water and production of electricity in India. Banasura Sagar Dam located at Kalpetta, in the Wayanad district of Kerala in the Western Ghats, is the largest earthen dam in India and the second largest in Asia. The dam is built up with a massive stack of stones and boulders. Approaches of most of the flyover are built with reinforced earth retaining walls. In hilly areas, structural walls built with stone masonry are an excellent example of the use of soils and rocks as construction materials.

Central Public Works Department, Delhi has published a detailed guideline for the construction industry to use locally available soil as a construction material. Various options/considerations in this area are adobe bricks, adobe pouring construction, compressed earth blocks, soil-based building blocks and cob walls. These materials used in the construction reduce carbon emission and create an energy-efficient structure. Also, since these materials are easily available, there is cost saving in the transportation of materials to the construction site. By using sustainable materials in construction, we are saving the conventional materials, which are depleting and are available in scarce.

10.5   FOUNDATIONS AND OTHER GEOTECHNICAL STRUCTURES

Foundation is an important element of any structure that transmits the load of superstructure to the soil. Soils and rocks are often used as foundations of structures (buildings, bridges, towers, retaining structures, embankments, slopes, dams, canals and so on), including the foundations for electrical and mechanical infrastructure/poles/machines. Foundations cannot be constructed on the loosely filled-up ground with non-uniform density or consistency unless adequate strengthening of the soil is made by applying ground improvement techniques. In the weak subsoil, the design of shallow foundation for structures and equipment may present problems for sizing of foundation as well as control of foundation settlements. Under such situations, if the shifting of the structure is not possible, then a viable alternative is to improve the subsoil to an extent that the subsoil would develop an adequate bearing capacity and the foundations constructed after the subsoil improvement would have resultant settlements within the acceptable limits (IS:1904 (1986), IS:6403 (1981), IS:12070 (1987), IS:1080 (1986) and IS:1892 (1979)). The selection of ground improvement techniques may be done under good practice. At some places, there is the existence of loose sand with a high water table and the soil is liquefiable. A specific measure is required under such situations. Deep excavation near an existing structure is a common problem faced by foundation engineers. Construction of tunnels required for water and sewerage pipelines passing through or near an old non-engineered structure poses a serious threat to the safety of the existing structure. Delhi Metro Rail Corporation has encountered such problems at many places in Delhi. Some of the specific foundation problems normally encountered in the practical application and their tentative solutions are explained here.

A multi-storey structure consisting of three basements, ground and ten storeys above was proposed to be constructed near an existing significant multi-storey building which has a single basement and is approximately 9 m away. Excavation up to 15 m below the ground was made for laying of raft foundation for the proposed structure. The water table was encountered at about 7 m depth below the ground level. It was a challenge for the executing agency to cast the foundation with due precaution and to avoid any damage due to the settlement of the existing structure. The aim was to prevent the foundation from uplift till the casting of the basement as long as the total upward thrust is being balanced by the total downward force with the required factor of safety. A diaphragm wall with soil anchors has been proposed about 3 m away from the line of excavation up to a depth of 20 m below the ground level. Continuous dewatering was done under the supervision of the second author to keep the water well below the foundation level up to the construction of all three basements. The raft has additionally been provided with soil anchors as an additional factor of safety against uplift.

A ground-supported STP was constructed with its base at 5.3 m above the existing ground level. The safe bearing capacity of soil at the site is 110 kPa at 2 m depth. The slope of the inlet pipeline warranted the tank to be constructed at more than 5.3 m above the natural ground level. The biggest challenge was to attain the required safe bearing capacity at the level of construction of the base of the tank. There were two options for this problem. The first was to provide a pile foundation below the existing ground level to be capable of taking the 5.3 m surcharge load and the load of the sewage and its dead weight. Though the option was technically feasible, it was too expensive and was out of the budgetary provision. The second option was to build the tank on compacted filled-up stabilised soil using suitable soil improvement techniques. The second option was selected after due deliberation with various stakeholders. The details of the project are shown in Figure 10.7.

Murrum was placed and compacted in horizontal layers of 20 to 50 cm of loose thickness with a standard Proctor maximum dry unit weight of more than 18 kN/m³ as recommended. The Murrum having not more than 20% fines was having a plasticity index of 7–17 (medium plastic). The particle size grading was to be 20 mm and downsize to fill and to be compacted to at least 90% maximum dry unit weight. The fill was required to be compacted by vibratory rollers. The vibratory roller for compacting fill was of smooth steel-drum, either self-propelled or towed type. The total static weight of the roller was 8 to 10 tonnes. The backfill material was observed to have a minimum CBR (soaked) of 8 (looking at the site conditions).

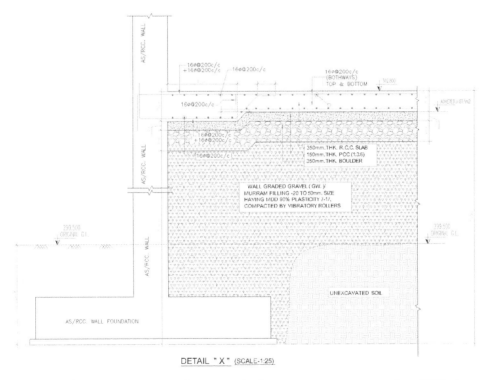

DETAIL " X " (SCALE-1:25)

Figure 10.7 Soil improvement technique applied at a sewage treatment plant (STP) site.

10.6   OTHER GEOMATERIALS

This section provides information about the geomaterials similar to soils and rocks, including three types of wastes, namely coal ash, construction and demolition waste (CDW) and municipal solid waste (MSW), as found in the state.

Coal has been used for power generation in Delhi since 1974 at Badarpur (767 acres of ash pit beside the plant) and 1990 at Rajghat (24 acres of ash pit beside the plant) which had nameplate power generation capacity of 705 and 135 MW, respectively but owing to environmental concerns, they were closed for power generation in 2018 and 2015, respectively. The mass production of power by thermo-electrical plants was brought to a halt but the problem of ash disposal in the past few decades remains. The ash produced by the coal-fired power plants consists of fly ash composed of particle sizes normally less than 300 µm and bottom ash made up of significantly coarser particles. The mixture of fly ash and bottom ash is eventually disposed of into a slurry containment facility known as an ash pond. Worldwide, more than 65% of fly ash produced from coal power stations is disposed of in landfills and ash ponds. Therefore, typical characterisation and behaviour of ash fill remained a matter of great interest among the engineers, scientists, planners and developers, contractors and owners in the past three decades.

According to the literature (Trivedi, 2013b), the composite ash collected from electrostatic precipitators and the bottom of the hopper of a thermal power plant may be classified as coal ash. The coarse ash collected from the furnace bottom is known as bottom ash. It is around 20% to 25% of the total ash produced. The ash is disposed into a pond by mixing it with water to form a slurry. The slurry usually contains 20% solids by weight. This method of ash disposal is called the wet method. The landfill of ash may be used as a construction fill if the suitable ashes are properly characterised. The fine ashes may collapse upon wetting. To avoid excessive settlement upon wetting, the suitability of coal ash should be examined as per the criteria of collapse and liquefaction. The chemical and physical characteristics of the ash produced depend upon the quality of coal used, the performance of wash-units, the efficiency of the furnace and several other factors (Trivedi, 2013b).

The grain size distribution of varied ash samples indicates particle size in the range of coarse sand, silt to clay. However, the maximum frequency of particles is in the range of fine sand to silt. Figure 10.8a shows the grain size distribution of typical samples from ash hopper (F1–7), mix hopper (MH), ash pond (PA1–2) and uniform sand (Trivedi and Sud, 2004). The effective size and mean sizes may control the shear strength and compaction of ashes. The pond ash which is examined for mass behaviour as a fill may contain 5%–10% of the particle in coarse and medium sand size, 35% to 50% in fine sand size and 40%–60% of particles in the range of silt. The presence of superfine (size ~0.01 mm) transforms inter-particle friction, agglomeration and formation of pendular bonds in the presence of moisture which results in the collapse of the structure of ash fill when there is sudden inundation of infills. Figure 10.8b shows a criterion of collapse based on the mean size for typical samples from ash hopper (F1–7), MH and ash pond (PA1–2) (Trivedi and Sud, 2004).

Trivedi and Singh (2004) reported standard cone resistance of compacted ash fill. The settlement of ash fills at varying relative densities is obtained by CPT using a method modified by Trivedi and Singh (2004). The following relation gives the

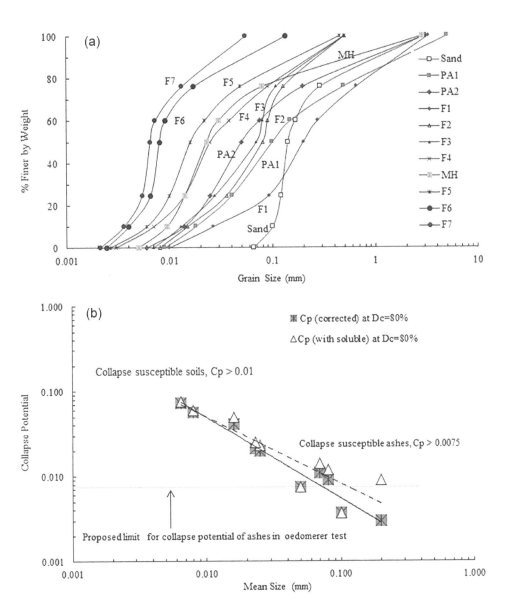

Figure 10.8 (a) Ash from hopper (FI–7), mix hopper (MH), ash pond (PAI–2) and uniform sand. (Adapted from Trivedi and Sud, 2004.) (b) Criteria of collapse based on mean size from ash hopper (FI–7), mix hopper (MH) and ash pond (PAI–2). (Adapted from Trivedi, 2013b.)

settlements of ash fill which is less than 50% to 10% of the settlement obtained by the Meyerhof (1956) method as

$$S_c = \Delta pB / [n \times RD + m)]q_c \tag{10.4}$$

where Δp is the net foundation pressure. The cone resistance (q_c) is taken as the average over a depth equal to the width of the footing (B). RD is the relative density while n and m are constants that take approximate values of 10 to 3.5, respectively. Although no significant study has been reported on the in situ behaviour of Rajghat and Badarpur ash ponds, the results of variation of SPT for compacted ash deposits are presented in Tables 10.9 and 10.10 for understanding the penetration resistance of ash fills. Since there is an anticipated similarity in the engineering behaviour of compacted ash fills, the settlement of selected footings on compacted ash is shown in Table 10.10.

Delhi also has three sewage dumping grounds located at three landfill sites – Bhalswa, Ghazipur and Okhla. Delhi generates about 3,800 million litres of sewage per day as per the estimates of the Central Pollution Control Board (CPCB). These landfill sites pose a risk hazard to the people living in the adjoining areas. Due to the presence of inflammable gases in the sewage, regular fire is reported throughout the year and is a major cause of pollution in Delhi. Photographs showing sewage landfill sites in Delhi at Bhalswa and Ghazipur are shown in Figure 10.3.

Another major problem in Delhi is the production of more than 4,000 tonnes of CDW due to multiple types of construction activity and infrastructure work being carried out throughout the city. Due to improper disposal, this waste is choking storm-water drains and polluting the Yamuna river bed. To address the above issues and to understand the need for CDW, a recycling plant has been set up in Burari, Delhi. This plant can cater to approximately 500 tonnes per day of CDW and converts

Table 10.9 Variation of the SPT for ash deposits

Ash type	D_c (%)	N value	Ratio N_a/N_s
Bottom ash	–	75	1.25
Compacted ash dyke	95	4–27	0.06–0.45
Hydraulically deposited ash	Loose state	0	0
Well compacted fly ash in a valley	–	75	1.25
Hydraulically deposited ash	Loose state	0–1	0
Hydraulically compacted ash	Dense state	1–10	0–0.16

Source: Adopted from Trivedi and Sud (2007).

Table 10.10 Settlement of footings on compacted ash

Plate size (mm) and shape	Degree of compaction (%)	Moisture content	$(S/B)_n$ of an ash fill and $(S/B)s$ (N =50) at 100 kPa
900, square	85.24	Wet of critical	1.55
600, square	85.24	Wet of critical	1.75
300, square	85.24	Wet of critical	1.25
600, square	90.29	Wet of critical	1.11
900, square	90.29	Wet of critical	0.94
600, square	<95	Wet of critical	0.61
300, square	<95	Wet of critical	0.64
300, square	Sand, N = 50	–	1

Source: Adopted from Trivedi and Sud (2007).

Table 10.11 CBR of construction and demolition waste

Description of materials	CBR at OMC	CBR soaked
Concrete aggregates (CG)	85	128
CG + bricks	87	83
CG + 10% sand	93.5	93
CG + 10% cement	139	>150
Demolition waste aggregates	72.5	68.6
Concrete waste aggregate	87.4	79.8
Mortar and brickwork waste aggregate	40.4	34.8

Source: Adopted from Goel (2018).

it to recycled finish products like floor tiles, stone aggregates, paver tiles and other useful materials. A similar plant has also been set up in Okhla, New Delhi. The values of CBR of CD waste are presented in Table 10.11, which shows that CD waste has a very high potential for its use as a subgrade of highway sections.

Figure 10.9a shows a gradation of various recycled aggregates from a Delhi plant (Goel, 2018). Figure 10.9b shows the dry density–moisture content plot for various recycled materials to check their utility compared to other engineering materials. Its behaviour in compaction has been observed similar to natural aggregates. A pavement layer thickness using CDW aggregates constructed adjoining the Crossing Republic Delhi–Meerut Expressway was proposed and constructed in 2015. The performance of a pitched road constructed using CDW aggregates adjoining the Crossing Republic Delhi–Meerut Expressway was found satisfactory in the past 5 years (Goel, 2018).

10.7 NATURAL HAZARDS

The hazard potential of living in an area is associated with the potential of risk to human lives and property and the commercial viability of economic growth fuelled by engineering activities in that area. However, the traditional lifestyle of settlements in India prompted complete convergence of biological, sociological, physiological and engineering spheres of human life as a part of biota. This concept supports the division of hazard potential into various natural and manmade hazard potentials of living in a congested area of population density as of Delhi. According to the WHO, inadequate shelter and overcrowding are the major factors in the transmission of diseases with an epidemic potential of acute respiratory infections, meningitis, typhus, cholera, scabies, COVID-19 and so on. Outbreaks of these diseases are more frequent and more severe when the population density puts higher pressure on land uses. Decreasing overcrowding by providing extra facilities and a proper organisation of the sites or services in healthcare facilities need more land. Under such manmade circumstances, the occurrence of natural hazards becomes critical.

There have been multi-levels of assessments for the microzonation of NCT of Delhi, namely source characterisation, ground characterisation and engineering seismology of built environment for hazard evaluation. The geological, geomorphological and geotechnical characterisation together with basin configuration of soft sediments have been assessed for first and second level microzonations. The site characterisation,

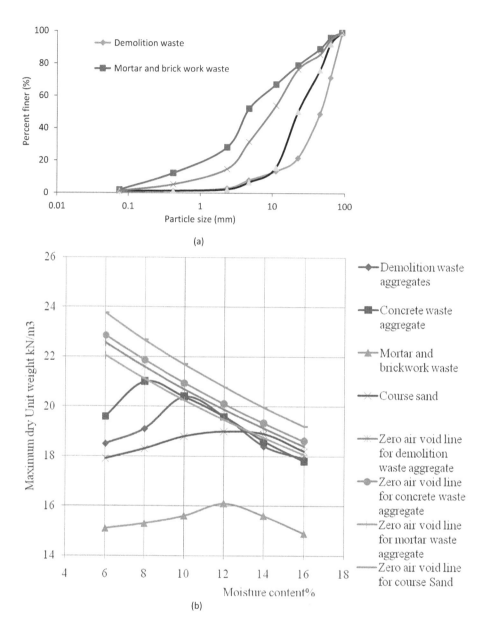

Figure 10.9 (a) Gradation of various recycled aggregates. (Adapted from Goel, 2018.) (b) Dry density vs. moisture content for recycled materials. (Adapted from Goel, 2018.)

site response and details of the built environment have been captured for third- and fourth-level microzonations (Shukla et al., 2001). The scientific community made attempts to connect the dynamic shear modulus of soil for site characterisation.

The NCT of Delhi is located in the seismic zone IV of the seismic zoning map of India (IS:1893 (2016), Part 1). Geographically, the region is located on folded crustal ramp represented by basement rocks of the Delhi supergroup, bounded by two regional faults, namely the Mahendragarh–Dehradun subsurface fault in the west and the great boundary fault in the east Delhi. An important structural element of the belt is the NW-SE trending Delhi–Sargodha ridge passing through Delhi skirted by basins on either side, namely the Sahaspur basin in the north and the Bikaner basin in the southwest. Delhi being the national capital of India has given great importance to safety due to cultural and administrative reasons. It houses the embassy of various countries of the world including the Indian parliament, president house and PM house.

Because of its high importance, the Ministry of Earth Sciences, Government of India prepared the seismic microzonation map for Delhi (https://moes.gov.in/, accessed on 18 May, 2020). The committee has done a detailed investigation covering about 500 locations of Delhi and has proposed a seismic microzonation map of Delhi (2020) at a scale of 1:10,000 incorporating the effects of local soil conditions. Based upon the observation of SPT, Figure 10.10a shows the values adopted for microzonation of Delhi soils for a critical depth of 3–6 m, which displays a spectrum for SPT > 10, SPT 10–20, SPT 20–30 and SPT > 30, to classify relative liquefaction resistance of the Delhi zone. Similarly, Figure 10.10b displays factor of safety (FoS) of the Delhi region for 0–9 m, depth against liquefaction: for FoS < 1 (liquefaction certain), for FoS 1–1.2 (liquefaction probable) and for FoS > 1 (no liquefaction).

Liquefaction potential is defined as the capacity of soil to resist liquefaction. Hence, the evaluation of the liquefaction potential of soils at any site requires the determination of two sets of parameters, namely the cyclic stress ratio (CSR) due to seismic action and soil properties which describe the soil resistance under these loads termed as the cyclic resistance ratio (CRR). Liquefaction generally takes place in loose fine-grained sands (fines < 10%, D_{60}, 0.20 to 1.0 mm and C_u between 2 and 5) with an N value less than 15. In the case of soil strata having $N > 15$, liquefaction of soil will not take place normally. Delhi falls in the seismic zone IV. Considering the history of past earthquakes and available seismic data, an earthquake of magnitude 7.5 having peak ground acceleration $a_{max} = 0.24$ g is considered in the present analysis as provided in IS:1893 (2016).

The modified Chinese criteria by Wang (1979) for silty and clayey soils for sandy soils represent the most widely used criteria for defining potentially liquefiable soils over the last two decades. The Chinese criteria consider that silty sands are potentially of liquefiable type if:

- The fraction is finer than 0.005 mm ≤ 15%,
- Liquidity index ≤ 0.75,
- Liquid limit, LL ≤ 35% and
- Natural water content ≥ 0.9 LL.

The most widely used criterion for the assessment of liquefaction potential of foundation strata is made by the simplified approach proposed by Seed and Idriss (1971) from the SPT data and peak ground acceleration likely to occur at the site for a horizontal ground surface. In this method, cyclic shear stress likely to be induced in the foundation strata by design basis earthquake is first evaluated. Next, threshold cyclic shear stress, which is good enough to cause liquefaction, is determined and the empirical

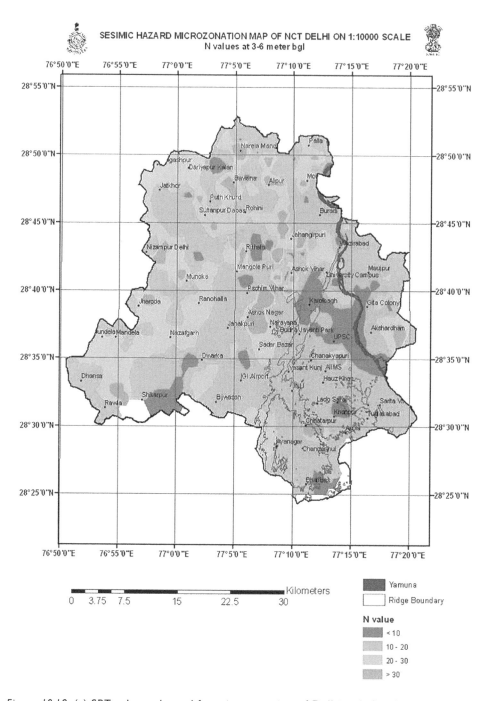

Figure 10.10 (a) SPT values adopted for microzonation of Delhi soils for the critical depth
of 3–6 m, colour codes used: saffron for SPT > 10, yellow for SPT 10–20, grey
for SPT 20–30 and light purple for SPT > 30. (Adapted from https://moes.gov.
in/, accessed 18 May 2020.)

(*Continued*)

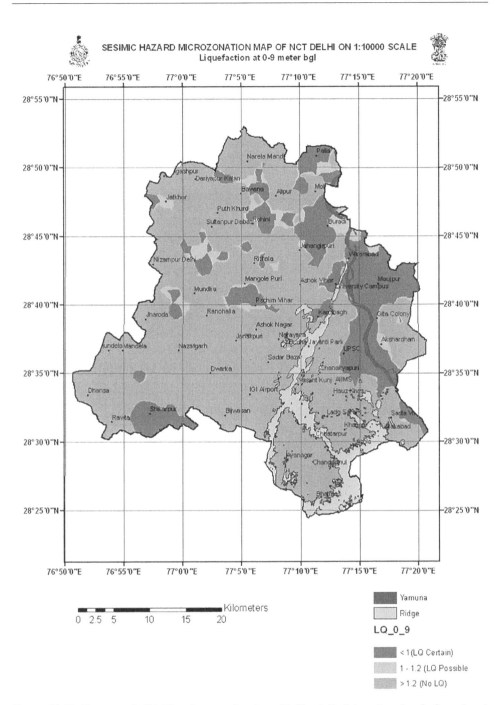

Figure 10.10 (Continued) (b) The factor of safety (FoS) of Delhi region for 0–9 m depth against liquefaction; colour codes used: red for FoS < 1 (liquefaction certain), yellow for FoS 1–1.2 (liquefaction probable) and green for FoS > 1 (no liquefaction). (Adapted from https://moes. gov.in/, accessed 18 May 2020.)

relations of CSR and CRR are estimated from field tests (corrected SPT, CPT or shear wave velocity). Finally, a comparison of CSR and CRR is used in the estimation of liquefaction susceptibility of the relevant strata.

10.8 CASE STUDIES

In this section, a case study of the success of geotechnical structures as observed and reported in the state has been described, including technical details. Several examples and case studies on deep foundation practices are listed in the proceedings of Testing and Technology for Load Carrying Capacity of Deep Foundations by Trivedi and Shukla (2019).

Figure 10.11 shows the quick installation of double-wall corrugated pipes of underground sewerage line and manhole in soft soil at the Arocon Rainbow housing project (NH-24), Delhi–Meerut expressway. These pipes are manufactured using HDPE polymer, which is resistant to various types of gases and chemicals that are generated due to the putrefaction of various ingredients flowing in the system. These pipes have a smooth internal surface and corrugated external surface and are manufactured as per the relevant standard. The corrugated external surface provides greater stiffness, withstands soil movement and takes higher loads (static and dynamic), whereas the internal surface helps in the smooth flow of sewerage. Double-wall corrugated pipes are available in SN 4 and SN 8 stiffness classes with sizes from 100 to 1,000 mm internal diameter. These pipes have a standard length of 6 m and are provided with rubber ring joints. The nominal stiffness for these pipes is evaluated as

(a)   (b)

Figure 10.11 Installation of double-wall corrugated pipes for quick installation of underground sewerage line: (a) angular and (b) linear alignment and manhole in soft soil at the Arocon Rainbow housing project, Delhi–Meerut expressway (NH-24).

$$S_N = EI / (Dn)^3 \qquad\qquad (10.5)$$

where S_N is the nominal stiffness (kN/mm^2), E is the modulus of elasticity (kN/mm^2), I is the moment of inertia (mm^4/mm) and Dn is the pipe diameter (mm).

For installation, a trench is excavated up to the line of the soil such that the uniformity of the base is maintained. In case the excavation encounters stones, rocks and brickbats, it is recommended to excavate below to perform the following filling. The soil from the excavation should be mounted enough from the edge of the trenches to avoid their collapse or that the landslide of it could not endanger the workers in case of deep trenches. The filling of the excavations performed below the base level is normally uniform soils. The filling material is normally loose sand, gravel or crushed stone, provided that the maximum size is not greater than 2 cm. It is avoided to use unsuitable lands for this purpose.

10.9 GEOENVIRONMENTAL CONCERNS ON SOILS AND ROCKS

As discussed in previous sections, the NCT of Delhi has undergone exponential pressure due to its demography and human settlements; therefore, civil engineering engagements have changed significantly over the past few decades. There are huge power demands in peak summer to keep the livelihood of the city in order. The city's temperature is recording a new high every year. There are current concerns for developments of energy geo-structures that need fresh planning, research and long-term policy push. The energy geo-structures (Laloui and Loria, 2019) are one of the most promising areas where the NCT of Delhi should take strong initiatives. There are several geoenvironmental concerns related to waste disposal, namely MSW, electronic wastes and medical wastes. Some effects of such problems on air, water and soil quality are visible and are reflected. Fresh concerns related to an outbreak of biological hazards and its implications on engineering evolution need more research.

10.10 CONCLUDING REMARKS

The NCT of Delhi is structurally expanding explosively all around. The geotechnical investigation of the soils and rocks found in the national capital region of Delhi has localised the availability of non-traditional geomaterials in addition to soils and rocks, mostly in the southwestern region, coal ash and CDW as construction materials. The soil investigation of the central Delhi region for 20 m deep boring logs indicated near-gradually increasing SPT values (0–40). The geotechnical investigation of Delhi presented several interesting engineering problems. There have been several significant deposits of silty sand obtained from selected depths of investigations at varied densities (1.65 to 1.95 g/cm^3). The strength properties of silty sand from the Yamuna basin for varied packing densities (relative compaction R_c between 0.92 and 0.98) show a friction angle in the range of 28°–38°. The maximum dilatancy angle was found to be 10° to relate the strength of silty sand to the liquefaction potential. The natural hazard potential of the Delhi region has shown the vulnerability to liquefaction in the Yamuna basin and for several lenses (FoS < 1; liquefaction certain and FoS in the range of 1–1.2; liquefaction probable) with reference to the geotechnical investigation and microzonation. The useful avenues for potential wastes have been captured for CDW as a highway material and performance of coal ash as a structural fill material. A couple of problems of working on difficult grounds for a deep foundation, a rock slope and an STP on loose soil have been captured to show the examples of engineering solutions.

ACKNOWLEDGEMENTS

The authors place on record their appreciation of Ms Swati Structure Solutions Pvt. Ltd for providing data of test locations from Delhi and outside Delhi. The authors are thankful to Dr Rajiv Goel, CEO, Arocon Rainbow Housing Project for providing photographs and comments on the use of double-wall corrugated pipes for quick installation sewerage line in soft soil and uses of CDW.

REFERENCES

Barton, N. (2002). Some new Q-value correlations to assist in site characterisation and tunnel design. *International Journal of Rock Mechanics and Mining Sciences*, 39(2), 185–216.

Bishop, A. W., and Gordon, E. G. (1965). The influence of end restraint on the compression strength of a cohesionless soil. *Geotechnique*, 15(3), 243–266.

Bolton, M. D. (1986). The strength and dilatancy of sands. *Geotechnique*, 36(1), 65–78.

Dutt, S. (1988). *Buddhist monks and monasteries of India: their history and their contribution to Indian culture*. Motilal Banarsidass Publisher, India.

Girija, D., et al. (2013). Analysis of cow dung microbiota – a metagenomic approach. *Indian Journal of Biotechnology*, 12: 372–378.

Goel, R. (2018). C&D waste as a geomaterial for highways, Department of Civil Engineering, Faculty of Technology, University of Delhi, Delhi.

Gupta, R., and Trivedi, A. (2009). Effects of non-plastic fines on the behavior of loose sand – an experimental study. *Electronic Journal of Geotechnical Engineering*, 14(B), 1–14.

Hoek, E., and Brown, E. T. (1980). Empirical strength criterion for rock masses. *Journal of Geotechnical and Geoenvironmental Engineering*, 106 (ASCE 15715).

Houlsby, G. T. (1993). Interpretation of dilation as a kinematic constraint. *Modern Approaches to Plasticity*, 119–138.

http://www.yamunariverproject.org/ (2020). Accessed 12 May, 2020.

https://moes.gov.in/writereaddata/files/Delhi_microzonation_report.pdf/ (2020). Accessed 18 May, 2020.

IS:1080 (1986). Code of practice for design and construction of shallow foundation on soils (other than raft, ring and shall) (second revision), Bureau of Indian Standards, New Delhi, India.

IS:11315 (1985). Indian Standard method for the quantitative descriptions of discontinuities in rock masses: Part 11 Core recovery and RQD, Bureau of Indian Standards, New Delhi, India.

IS:12070 (1987). Indian standard (reaffirmed 2010) code of practice for design and construction of shallow foundations on rocks, Bureau of Indian Standards, New Delhi, India.

IS:1892 (1979). Code of practice for sub surface investigations for foundations, v.

IS:1893 (2016). Criteria for earth quake resistant design of structure: (Part-1) general provisions and buildings.

IS:1904 (1986). Code of practice for design and construction of foundation in soils: (first revision), Amendment 1, Bureau of Indian Standards, New Delhi, India.

IS:1948 (1970). Classification and identification of soils for general: Engineering purposes (first revision), Amendment 2, Bureau of Indian Standards, New Delhi, India.

IS:2131 (1981). Method of standard penetration tests for soils, Bureau of Indian Standards, New Delhi, India.

IS:2720 (1980). Methods of test for soils. Bureau of Indian Standards, New Delhi, India.

IS:6403 (1981). Code of practice for determination of bearing capacity of shallow foundation: (first revision), Amendment 1, Bureau of Indian Standards, New Delhi, India.

Kalamaras, G. S., and Bieniawski, Z. T. (1995). A rock mass strength concept for coal seams incorporating the effect of time. In 8th ISRM Congress. International Society for Rock Mechanics and Rock Engineering.

Lade, P. V., and Yamamuro, J. A. (1997). Effects of nonplastic fines on static liquefaction of sands. *Canadian Geotechnical Journal*, 34(6), 918–928.

Laloui, L., and Loria, A. F. R. (2019). *Analysis and design of energy geostructures: theoretical essentials and practical application*. Academic Press.

Lo Presti, D. (1987). Mechanical behaviour of Ticino sand from resonant column tests. PhD diss., Ph. D. thesis, Politecnico di Torino, Torino, Italy.

Meyerhof, G. G. (1956). Penetration tests and bearing capacity of cohesionless soils. *Journal of the Soil Mechanics and Foundation Division*, 82(1), 1–19.

Ojha, S., and Trivedi, A. (2013a). Comparison of shear strength of silty sand from Ottawa and Yamuna river basin using relative compaction. *Electronic Journal of Geotechnical Engineering*, 18(K), 2005–2019.

Ojha, S., and Trivedi, A. (2013b). Shear strength parameters for silty-sand using relative compaction. *Electronic Journal of Geotechnical Engineering*, 18(1), 81–99.

Ojha, S. (2015). Non-linear behaviour of silty soil and its engineering implications, Department of Civil Engineering, Faculty of Technology, University of Delhi, Delhi.

Salgado, R., Bandini, P., and Karim, A. (2000). Shear strength and stiffness of silty sand. *Journal of Geotechnical and Geoenvironmental Engineering*, 126(5), 451–462.

Sen, T. (2006). The travel records of Chinese pilgrims Faxian, Xuanzang, and Yijing. *Education about Asia*, 11(3), 24–33.

Seed, H. B., and Idriss, I. M. (1971). Simplified procedure for evaluating soil liquefaction potential. *Journal of the Soil Mechanics and Foundations Division*, 97(9), 1249–1273.

Shukla, A. K., Prakash, R., Singh, D., Singh, R. K., Pandey, A. P., Mandal, H. S., and Nayal, B. M. S. (2001). Seismic microzonation of NCT–Delhi. In Proceedings of the Workshop on Microzonation, 39–43.

Shukla, S. K. (2014). *Core principles of soil mechanics*, ICE Publishing, London.

Shukla, S. K. (2015). *Core concepts of geotechnical engineering*, ICE Publishing, London.

Trivedi, A., and Shukla, S. K. (2019). *Testing and technology for load carrying capacity of deep foundations*. ISBN: 978-93-5391-519-3.

Trivedi, A., and Singh, S. (2004). Cone resistance of compacted ash fill. *Journal of Testing and Evaluation*, 32(6), 429–437.

Trivedi, A. (2013a). Estimating in situ deformation of rock masses using a hardening parameter and RQD. *International Journal of Geomechanics*, 13(4), 348–364.

Trivedi, A. (2013b). *Engineering behaviour of ash fills. Advanced carbon materials and technology*. Wiley-Scrivener, 419–474.

Trivedi, A. (2015). Computing in-situ strength of rock masses based upon RQD and modified joint factor: using pressure and damage sensitive constitutive relationship. *Journal of Rock Mechanics and Geotechnical Engineering*, 7(5), 540–565.

Trivedi, A. (2017). Comments on modulus ratio and joint factor concepts to predict rock mass response. *Rock Mechanics and Rock Engineering*, 50(5), 1357–1362.

Trivedi, A., and Sud, V. K. (2004). Collapse behavior of coal ash. *Journal of Geotechnical and Geoenvironmental Engineering*, 130(4), 403–415.

Trivedi, A., and Sud, V. K. (2007). Settlement of compacted ash fills. *Geotechnical and Geological Engineering*, 25(2), 163–176.

Wang, W. (1979). *Some findings in soil liquefaction*, Water Conservancy and Hydroelectric Power Scientific Research Institute, Beijing, China.

Chapter 11

Goa

Purnanand P. Savoikar
Goa College of Engineering

CONTENTS

11.1   INTRODUCTION

Goa is the 28th state in India, liberated from the Portuguese rule on 19th December 1961 and merged into India as the Union territory. On 30th May 1987, Goa was granted statehood. The population of Goa is about 14.6 lakhs. Goa is flanked by Sahyadris (Western Ghats) on the eastern side and the Arabian Sea on the western side and is bounded on the north by Maharashtra and on the south and east by the Karnataka state. Goa lies between longitudes 73°40′33″E and 74°20′13″E and latitudes 14°53′54″ N

DOI: 10.1201/9781003177159-11

and 15°40′00″N. Goa has six important estuarine rivers, namely Mandovi, Zuari, Sal, Terekhol, Chapora and Talpona, and receives about 2,500 mm annual rain spread across June to September. Goa has a sprawling coastal line of 105 km. Goa is spread across two districts, North Goa and South Goa consisting of 12 talukas. The capital city is Panaji. Figure 11.1 shows the geographical map of Goa.

Figure 11.1 Map of Goa. (https://indiascheme.com/goa-land-records/, accessed on 20 January 2021.)

Physiographically, the coastal state of Goa is divided into three terrain types:

- the western coastal estuarine plain consisting of sandy beaches, estuarine mud-flats, khazan lands, salt pans, mangroves, fields and settlement areas,
- the central undulating region or midlands featuring hills ranging from 100 to 600 m and
- the Western Ghats or Sahyadris ranging from 600 to 1,000 m high on the eastern and southern parts.

The geology of Goa indicates that Goa is part of the Indian Precambrian shield consisting of green schist supracrustal rocks lying over the trondhjemitic gneiss. The north-eastern periphery of Goa consists of late Cretaceous Deccan Traps. Laterite cover is observed on the residual hills. The alluvium and sand cover is observed on the coastal estuarine plains in most of the geological formations of Goa.

Goa is covered by rocks of Dharwar Super Group of the Archaen Protozoic age, which extend in the NW-SE direction, while along the NE corner, it is covered by Deccan trap of the upper Cretaceous–lower Eocene age. The Western Ghats, which are clad by evergreen forests, have a general altitude ranging from 600 to 1,000 m. In the western margin of the territory along the Arabian Sea, there are long and narrow strips of sandy plains and low flat-topped laterite hills. Most of Goa's soil is made up of laterites rich in ferric-aluminium oxides and is reddish in colour. In inland and along the riverbanks, the soil is alluvial and loamy. The oldest rocks in the Indian subcontinent are found in Goa between Molem and Anmod on Goa's border with Karnataka. The rocks are estimated to be 3,600 million years old. Two-thirds of the area of Goa is enclosed by laterite having a thickness of 2–25 m. Almost all rock types of Goa have been lateralised to some extent. In general, iron manganese phyllites show a laterite profile of about 100 m depth; laterite cover is seen over a depth of 10 m in ultramafites and 5 m over the Deccan trap. Lateritic soils are softer when wet and harder when dried. These are formed as a result of leaching out of lime, silica and organic matter under the action of high rainfall and high temperature. They are rich in iron and aluminium content and hence, mostly red in colour.

Goa is known for its mineral reserves. Iron ore is found in the northern part of Goa and in some locations in South Goa. Manganese ore is found in the southern part of Goa. Bauxite reserves are found in the western coastal belt of Goa while limestone reserves are in the north-east corner of Goa.

The various types of rock found in Goa are detailed in the geological map of Goa (Dessai, 2018). As per this map, the geological classification of rocks in Goa comprises Barcem formation, Ponda (Bicholim and Sanvordem formation) and Vageri formation. Vesicular metabasalts of the Barcem formation are found in Canacona taluka while non-vesicular metabasalt extend from Pollem to Usgao-Dharbandora in central Goa. These rocks rest on tonalite–trondhjemite–granodiorite gneiss. The Sanvordem formation overlies the Barcem formation and comprises metagreywackes (metamorphosed sandstones) observed in Sanvordem, Calem and also Ribandar Goa and argillites observed in Aguada and Chapora. The Bicholim formation overlies the Sanvordem formation and comprises ferruginous and manganiferous phyllites, banded quartzites and limestones found in the NW-SE belt of Goa from Naibag in North Goa to Salgini in South Goa. The Vageri formation rocks overlie

Figure 11.2 The 6th-century laterite rock-cut Pandava Caves at Aravalem Goa. (Photo courtesy: Ms Roshani Majik.)

the Bicholim formation and comprise metagreywackes and metabasalts found in the Valpoi region. Granite and gneiss are observed in the Canacona and Chandranath regions.

The trondhjemite gneiss is the oldest rock in India, about 3.4 billion years old, and is located at Palolem Canacona Goa (Fernandes, 2009). Another monument, the ancient rock-cut Pandava Caves, constructed with laterite rock at Aravalem Goa, which dates back to the 6th Century, is shown in Figure 11.2. These caves are five in number and are believed to be cut during Mahabharata times by the Pandavas during their exile period.

In this chapter, the details of the main soil types occurring in Goa are discussed. Boring logs at typical locations are presented to understand the types of strata occurring at some of the prominent places in Goa. Few shallow test samples at different locations have been collected and analysed for the geotechnical properties. Figure 11.3 shows the map of Goa showing the borehole and open-pit soil sampling carried out at various locations in Goa.

A detailed table showing the geotechnical properties at various locations marked in Figure 11.3 is presented in Table 11.1. Though sampling does not cover the entire area of Goa and that soil/rock type may vary from site to site, the information presented herein will be helpful for readers to understand the geological strata and their properties.

11.2   MAJOR TYPES OF SOILS AND ROCKS

As mentioned in the preceding section, the lateritic highlands are observed on the southern and eastern parts of Goa, and lateritic lowlands and lateritic plateaus are observed in the western and southern parts of Goa. Most of the part of Goa is covered with laterites and lateritic soils. These laterite plateaus are: (i) Pernem, (ii) Mapusa, (iii) Porvorim, (iv) Panjim, (v) Ponda, (vi) Vasco da Gama, (vii) Dabolim–Madgaon, (viii) Quepem, (ix) Cabo de Rama and (x) Canacona.

Lateritic rock found in Goa is rich in iron and aluminium and has formed in tropical warm and humid regions. Laterite is developed by intensive and prolonged chemical weathering of the underlying parent rock which is termed as lateritisation. Laterites

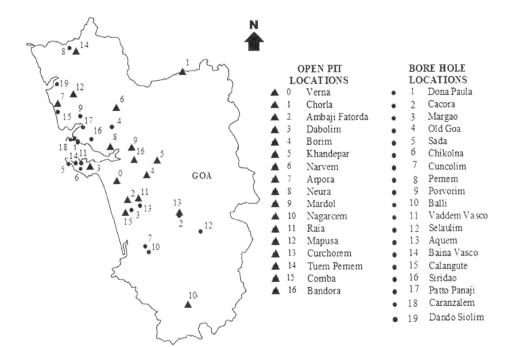

	OPEN PIT LOCATIONS			BORE HOLE LOCATIONS
▲ 0	Verna		● 1	Dona Paula
▲ 1	Chorla		● 2	Cacora
▲ 2	Ambaji Fatorda		● 3	Margao
▲ 3	Dabolim		● 4	Old Goa
▲ 4	Borim		● 5	Sada
▲ 5	Khandepar		● 6	Chikolna
▲ 6	Narvem		● 7	Cuncolim
▲ 7	Arpora		● 8	Pemem
▲ 8	Neura		● 9	Porvorim
▲ 9	Mardol		● 10	Balli
▲ 10	Nagarcem		● 11	Vaddem Vasco
▲ 11	Raia		● 12	Selaulim
▲ 12	Mapusa		● 13	Aquem
▲ 13	Curchorem		● 14	Baina Vasco
▲ 14	Tuem Pemem		● 15	Calangute
▲ 15	Comba		● 16	Siridao
▲ 16	Bandora		● 17	Patto Panaji
			● 18	Caranzalem
			● 19	Dando Siolim

Figure 11.3 Boring log locations in Goa.

are of rusty-red colouration due to the high content of iron oxide. Laterite covers the major portion of Goa and typically occurs as plateau landforms (Widdowson and Gunnel, 1999). Laterite, alluvium and sand cover most of the geological formations in Goa, as one expects in the wet tropical climate. A laterite profile is a typical field section of laterite from the surface to the parent rock. The term 'profile' refers to an 'outline' and not a 'cross-section' of an outcrop. The upper part of the section is generally red, hard and massive, which is referred to as 'lateritic crust'. It consists of a fretwork of oxides and hydroxides of iron and aluminium. The lower part of it is vermiform in nature and grades into tubular, semi-indurated laterite. Below this, there is a prominent thick horizon of variegated colours and is dominated by clay minerals called as 'lithomarge'. It is generally nodular concretionary at the top and mottled below. The mottled portion consists of kaolinite along with patches and accumulations of iron oxyhydroxides. The saprolite beneath the mottled zone is relatively impermeable and corresponds to the 'plasmic zone' of the profile. A significant portion of the saprolite shows bleaching and constitutes the 'pallid zone'.

Some of the boring logs at the various places as mentioned in Figure 11.3 are presented here. Though boring logs and the soil layers may vary from site to site and within a site also, the logs attached herewith will give an approximate idea of the strata in a particular area. However, for more detailed investigations on the soil profile, a fresh borehole should be driven and the layers and geotechnical properties are ascertained afresh. Figures 11.4–11.12 show the typical boring logs in Goa at the locations marked in Figure 11.3.

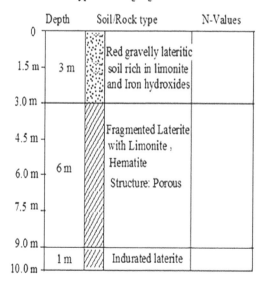

Figure *II.4* Typical boring logs in: (a) the Balli area and (b) the Cuncolim area.

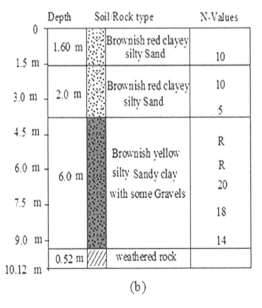

Figure 11.5 Typical boring logs in: (a) the Aquem Margao area and (b) the Gogol Margao area.

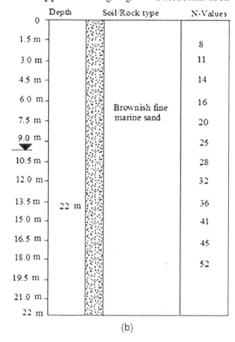

Figure 11.6 Typical boring logs in: (a) the Siolim area and (b) the Candolim area.

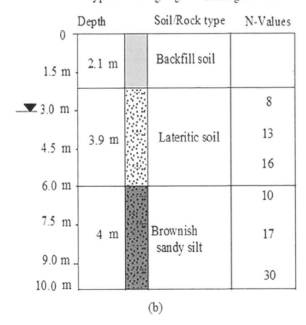

Figure 11.7 Typical boring logs in: (a) the Siridao area and (b) the Calangute area.

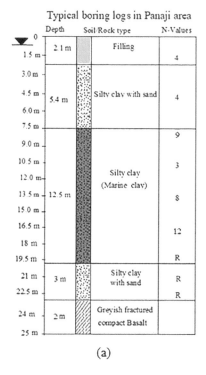

(a)

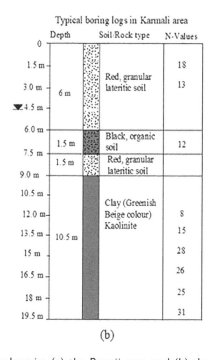

(b)

Figure 11.8 Typical boring logs in: (a) the Panaji area and (b) the Karmali area.

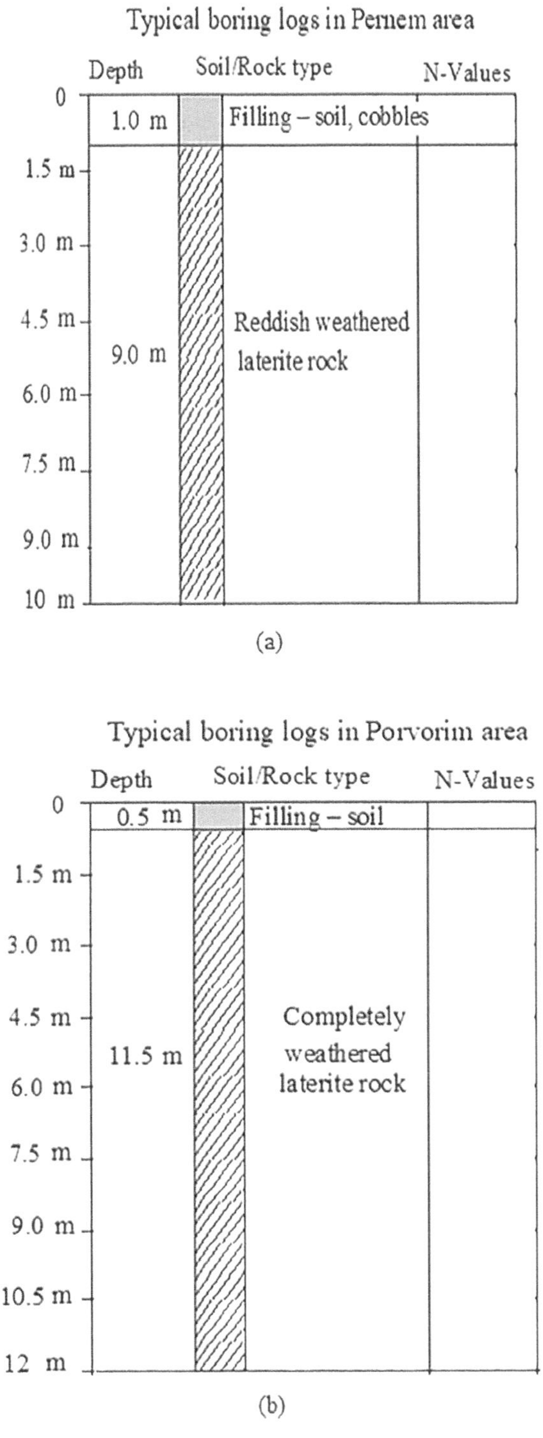

Typical boring logs in Pernem area

Depth	Soil/Rock type	N-Values
0		
1.0 m	Filling – soil, cobbles	
1.5 m		
3.0 m		
4.5 m	Reddish weathered	
9.0 m	laterite rock	
6.0 m		
7.5 m		
9.0 m		
10 m		

(a)

Typical boring logs in Porvorim area

Depth	Soil/Rock type	N-Values
0		
0.5 m	Filling – soil	
1.5 m		
3.0 m		
4.5 m	Completely	
11.5 m	weathered	
6.0 m	laterite rock	
7.5 m		
9.0 m		
10.5 m		
12 m		

(b)

Figure 11.9 Typical boring logs in: (a) the Pernem area and (b) the Porvorim area.

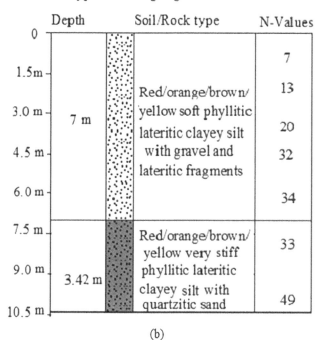

Figure 11.10 Typical boring logs in: (a) the Cacora area and (b) the Salaulim area.

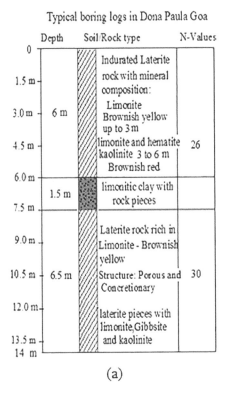

Typical boring logs in Dona Paula Goa

(a)

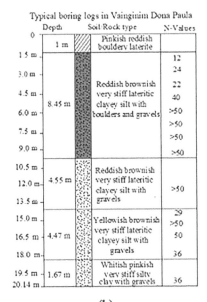

Typical boring logs in Vainginim Dona Paula

(b)

Figure 11.11 Typical boring logs in the Dona Paula Goa area.

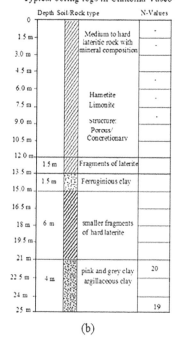

Figure 11.12 Typical boring logs in the Vasco da Gama area.

11.3   PROPERTIES OF SOILS AND ROCKS

Some typical soil/rock profiles in Goa are shown in Figure 11.13 depicting the various soils layers, lithomarge, boulders and so on interspersed in the soil. These are at various sites, which are cut either to make new roads or for making plots for construction purposes. These cuts in many cases are the potential landslide sites as the clay/silt portion in these open cuts gets washed away due to heavy rains. Also, the alternate spells of high temperature in day and cool night spells disintegrate these lateritic soils, resulting in loosening of gravels and boulders leading to landslides.

The geotechnical properties of soils and rocks from the different borehole sites shown in Figures 11.4–11.12 are presented in Table 11.1. Table 11.2 shows the geotechnical properties of soils and rock at the shallow open pits and at some landslide sites. In most of the sites, lateritic soils and silts were observed. At many locations, the silty clays are observed with layers extending to as much as 15–20 m with no hard rock observed at depths of more than 20–25 m. In addition to this, the water table is observed at shallow depths. Such sites are problematic sites requiring the strengthening of subsoils by adopting ground improvement techniques. For low rise structures, various ground improvement techniques are available, such as the replacement of soft soil with well-compacted murrum and/or boulders. For high rise structures, pile foundations are suitable, provided good bearing strata are available at reasonable depths. But, since rock is not available at shallow/reasonable depths and also friction piles cannot be adopted due to weak soils with shallow water tables, adopting rock-socketed piles, to be socketed at greater depths of more than 25 m, is uneconomical. In such cases, a combination of deep and shallow ground improvement techniques can be thought of.

In the case of lateritic rocks, it is observed that laterite is not a very hard and intact rock. At many places in Goa, the laterite rock observed has a porous and concretionary structure. In some cases, a layer of 1.5–3 m thick lateritic soil or lithomargic clay was observed within the lateritic rock strata. A typical boring log shown in Figure 11.12 at the Chikolna Vasco da Gama site in Goa may be seen where a 1.5 m thick layer of ferruginous clay was observed at the depth of 13.5 m and a soft argillaceous clay layer at the depth of 21 m. It was observed that for 10–12 storey structures proposed on such sites with raft foundations, the pressure bulb may extend beyond this soft layer. In such cases, individual footing may be economical and safe.

11.4   USE OF SOILS AND ROCKS AS CONSTRUCTION MATERIALS

11.4.1   Lateritic soil

Lateritic soil is found in abundance in Goa. It has been used in Goa as a construction material. Its compressibility varies from moderate to low and being mostly impervious, it finds use in building construction for walls, flooring and mud blocks. The local laterite soil is an excellent building material as it has the right amount of clay for adhesion, sand and murrum to prevent cracks. Blocks can also be cast with lateritic soil and water mix. Since ancient times, it has been used for the construction of rammed earth walls that are constructed using a formwork where the soil is placed and rammed with metal or wooden rammers. After testing, the strength of rammed earth walls is

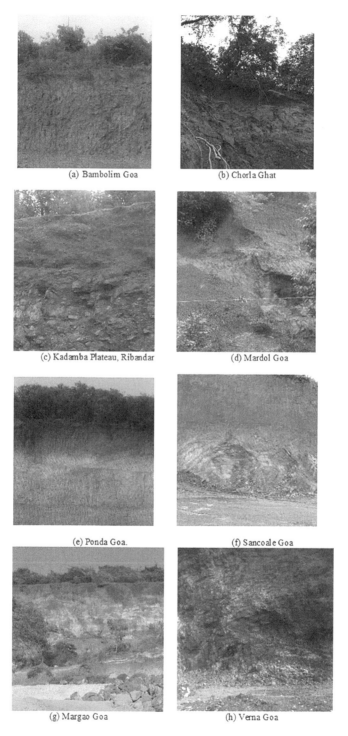

Figure 11.13 Typical soil/rock profiles in Goa. (Photo courtesy: Mr Leonardo Souza.) Note the different colours of lateritic soils in the profile.

Table 11.1 Geotechnical properties of soils/rocks – borehole data

Site	Soil/rock	Depth (m)	Specific gravity and total unit weight		NMC (%)	Free swells (%)	Atterberg limits		Sieve analysis				Strength properties			Remarks
			G	γ (kN/m³)			LL (%)	PL (%)	Gravel (%)	Sand (%)	Silt (%)	Clay (%)	Soil c (kPa)	ϕ (°)	Rock UCS (kPa)	
Dona Paula	Laterite rock	2–10	-	18.96–24.08	2.70–6.31	-	-	-	-	-	-	-	-	-	845.18–6,374.29	
Cacora	Lateritic soil	0–9	1.46–1.94	14.1–21.4	16.92–38	1.2–7	25–37	22–33	0	31.6–36	63.2–68.4	-	1–35	10–32	-	
Margao (Gogol)	Soil – silty sandy clay	3–7	2.632–2.662	-	-	-	-	-	5.88–13.40	31.2–57.8	36.99–55.35	-	35–47	0	-	
Margao (Aquem)	Soil – silty sandy clay	2.5–5	2.53–2.55	13.10	-	-	70–75	55–63	0–10	0–10	18–23	62–67	5	7.63	3	
Sada Vasco	Laterite rock	2–10	-	16.36–23.64	2.63–7.41	-	-	-	-	-	-	-	-	-	746.2–3,120	
Chikolna	Laterite rock	1.5–10	-	21.96–24.29	2.9–6.3	-	-	-	-	-	-	-	-	-	5,589–6,885	
Cuncolim	Laterite rock	2–10	-	12.88–31.61	-	-	-	-	-	-	-	-	-	-	882–7,342	
	Lateritic soil	1.5	-	16.22–17.88	-	-	31–37	27–32	-	-	-	-	7.1–9	25–28	-	
Pernem	Laterite rock	1.5–10	-	21.84–34.63	2.15–11.63	-	-	-	-	-	-	-	-	-	753–8,095	
Porvorim	Laterite rock	1.5–10	-	17.4–21.3	2.93–11.6	-	-	-	-	-	-	-	-	-	883–5,481	
Balli, Goa	Soil – sand	0–12	2.06–2.1	19.45–20.87	20.86–27.98	-	37–49	25–31	28–36.8	25.87–35.86	7.62–11.82	26.31–31.18	140–220	11–17	45–26.5	
	Rock – gneiss	0.15–14	-	29.4–44.7	2.02–2.17	-	-	-	-	-	-	-	-	-	796–104,159	
Vaddem, Vasco	Lateritic clay	5–17	2.53–2.7	17.5–19	21.10–70.10	-	33–60	26–44	1–13.5	5–20	66.5–92	-	5–21.5	6–21	*3,938.8–4,035.1	*bed rock GWT at 0.6 m
Selaulim	Lateritic soil	5.5–10.95	-	17.2–18.2	37–43.3	-	-	-	0–33.5	5.25–29.25	32.25–91.25	-	10–15	15–20.5	-	
Baina	Silty sand	2–4.6	2.63	-	-	-	-	NP	18–20	69–75	5–13	0	3.33–30	0	-	GWT at 3 m

(Continued)

Table 11.1 (Continued) Geotechnical properties of soils/rocks – borehole data

Site	Soil/rock	Depth (m)	Specific gravity and total unit weight G	Specific gravity and total unit weight γ (kN/m³)	NMC (%)	Free swells (%)	Atterberg limits LL (%)	Atterberg limits PL (%)	Sieve analysis Gravel (%)	Sieve analysis Sand (%)	Sieve analysis Silt (%)	Sieve analysis Clay (%)	Strength properties Soil c (kPa)	Strength properties Soil φ (°)	Strength properties Rock UCS (kPa)	Remarks
Calangute	Lateritic soil/sandy silty clay	3–10	-	-	-	-	20–55	12–37	-	15–30	15–62	16–70	35–47	0	-	-
Candolim	Fine sand	0–20	-	18	-	-	-	-	-	-	80–87	11–20	0	28	-	Non-plastic
Siridao	Silty sand	6–7	2.55–2.66	-	-	-	28–34	NP	5–8	60–75	12–22	8–10	50–67	0	-	-
Panaji (Patto)	Rock – basalt	23–27	2.66–2.79	21.3–27.7	-	-	-	-	-	-	-	-	-	-	1,952.4–9,470.5	GWT at 0.5 m
	Silty clay/sand	3–9	2.569–2.671	16.54–16.99	36.14–39.04	-	20–48	11–23	0–1	38–75	25–60	-	40–90	18–25	-	
Panaji (Miramar)	Sand	0–33	2.671	14.1–16.8	21–41	-	-	-	0–8	73–100	1–27	0	0	33	2,000*	*Bedrock at 33 m
Caranzalem	Silty sand	26	-	19.5	20.4	-	-	NP	1	89.5	9.5	0	0	13.6–22	*6,028.6	*Metagreywacke rock at 18–22 m
Siolim	Clayey silt	1–10	2.57–2.59	-	-	-	11–15	10–12	0	10–14	71–80	10–15	30–40	0	-	GWT at 1.1 m
DonaPaula Vainguinim	Lateritic clayey silt	4.5–10	2.09–2.91	18.7–21.3	14.24–24.52	-	-	NP	30.5–53	7–59.5	2–41	-	0–27	25–30	*3,506–4,918	*Lateritic rock at 0–7.5 m

Table 11.2 Geotechnical properties of soils/rocks – shallow open-pit test data

Site	Soil/rock	Depth (m)	Specific gravity and total unit weight		NMC/MC (%)	Free swells (%)	Atterberg limits		Sieve analysis				Strength properties			Remarks
			G	γ (kN/m³)			LL (%)	PL (%)	Gravel (%)	Sand (%)	Silt (%)	Clay (%)	Soil c (kPa)	φ (°)	Rock UCS (kPa)	
Verna	Soil	-	2.29	14.2–18	21.15	-	55	42	43	37	20		12–15	15–6	-	Landslide site
	Rock	1.5	-	27.96	-	-	-	-	-	-	-		-	-	2637	Landslide site
Chorla	Soil	-	2.31–2.82	18.9–19.6	27.53–30.91	-	34–37	24–30	4.2–28.6	48–59.6	23.4–36.2		13–14	18–23	-	Landslide site
Ambaji	Soil	1.5	2.66	15.3	-	-	41	29	28.06	65.61	6.33		8	28	-	Landslide site
Dabolim	Soil	1.5	-	17	-	-	-	-	74	17.6	8.4		9.8	29	-	Non-plastic
	Rock	2	-	28.75	-	-	-	-	-	-	-		-	-	2.297	
Borim	Soil	-	2.57	16.87	23.47	-	38	33	15.3	66.4	18.3		18	16.7	-	Landslide site
Khandepar	Soil	-	2.61	15.48	28.41	-	34	30	18.12	54.78	17.10		9	15	-	Landslide site
Narvem	Soil	-	-	18.8	10.58	-	-	-	35.44	64.56			13	17.0	-	Landslide site
Arpora	Soil	-	-	19.28	13	-	-	-	28.2	71.8			0	24	-	Landslide site
Neura	Soil	-	-	20	-	-	-	-					15	11	-	Landslide site
Mardol	Soil	-	2.47	18.3	22.9	-	27	22	32	64	4		5	15	-	Landslide site
Nagorcem	Soil	-	2.62	16.5	-	-	56	36	43	52.7	4.23		3	28	-	Landslide site
Raia	Soil	-	2.57	15.3–17.4	19.6	-	46	32	33.95	55.73	10.33		4–20	24–32	-	Landslide site
Mapusa	Soil	2.0	-	17.97	6.97	-	-	-	3.60	74.40	22		1	20	-	Non-plastic
Tuem, Pernem	Rock	1.5	-	19.50	5.98	-	-	-	-	-	-		-	-	3.130	-
Comba	Soil	1.5	-	17.7–18.30	31.19–37.24	-	-	-	-	-	-		14.5	19	-	-
Bandora	Soil	2	-	25.85–26.82	18.72–19.30	-	34–37	29–32	34.8	31.9	33.3		17	25	-	-
Quela-Ponda	Rock	2.5	-	28.2	-	-	-	-	-	-	-		-	-	6.823	-
	Soil	2.5	-	24.4–29.8	15.66–19.8	-	33–50	29–43	23.2–45.6	30–40.7	24.4–38.3		16–20	22–23	-	-
KRCL Margao	Soil	2–2.5	2.33	19.4–19.85	19.5–21.1	-	19–31	17–23	6.15–8.2	50.9–61.6	30.2–42.9		8.5–12	2–24	-	-
Piligao	Soil	2.5	2.55–2.79	16.85–18.72	12.57–32.4	-	25–27	21–24	3.4–6.4	61–43.9	35.6–49.7		-	-	-	-

found to be as strong as a lateritic rock. In addition to this, lateritic soil has been used as a building material by stabilising it with lime or jaggery since ancient times. The other stabilisers used in lateritic soils include cement and eggshell powder. Lateritic soil has also been used successfully as a base and sub-base material in road construction. Lateritic soil is also suitable for use as a fill material in embankment and dam construction.

11.4.2   Sand

Sand is found in Goa in large quantities, but its main sources are rivers and streams. There are no sources of land base reserves of sand. Sand has been extracted and used in large quantities in concrete and plastering works. In the past few years, restrictions have been imposed on sand extraction from rivers and streams due to environmental concerns.

11.4.3   Laterite stones

Laterite stone is a soft rock, and it can be easily cut when fresh. However, after drying in the air, it becomes a strong building material. It has been used abundantly as a construction material for building houses, monuments, temples, churches and mosques in Goa for centuries and it has proved to be the best construction material available in the state. Load-bearing houses have been constructed up to ground + three storeys using lateritic stones. It is used for carving and also for constructing exposed masonry structures in Goa.

11.4.4   Basalt

The basalt rock in Goa is composed of dolerite and gabbro and acid intrusives like matagreywacke, metabasalt, granite and granite gneiss. The north-eastern corner of the state is covered by Deccan Traps and represented by basalt rock. Wide outcrops of basalt are found in Pernem, Sattari, Salcete, Sanguem and Quepem talukas in Goa. Basalt is used in concrete works as coarse aggregates and as fine aggregates in the form of crushed sand, as a road metal and railway ballast. Apart from application in concrete and as a road metal, basalt was used in ancient times in the construction of temples. Figure 11.14 shows the famous Mahadeva Temple, Tambdi Surla which was built in the 12th century in the Kadamba style from the basalt rocks from the Deccan plateau.

  Laterite stone is the main building material in Goa for centuries. This is evident from the monuments, churches and temples which were constructed many years ago. Figure 11.15 shows some of the ancient monumental structures built using lateritic stone. The Viceroy's Arch in Old Goa was built using red laterite stone in 1597 by Francisco da Gama, the great-grandson of Vasco da Gama, after he became the Viceroy of Goa. During those days, every Governor who took charge of Goa had to pass through the arch. The UNESCO World Heritage site, the Basilica of Bom Jesus, a Roman Catholic Basilica located in Old Goa, was constructed using exposed lateritic

Figure 11.14 Ancient Mahadev temple at Tambdi Surla in basalt. (Photo Courtesy: Ms Roshani Majik.)

masonry during the period from 1594 to 1605. Safa Shahpuri Masjid and its precincts in Ponda are examples of ancient monuments in exposed lateritic masonry. Almost all the temples in Goa have been built with lateritic stones since ancient times. Several ancient cave temples in Goa are located in Khandepar, Paroda hill, Netravalli, Narve, Kundaim, Surla-tar and several other places. These temples are simple and carved out of red laterite stones. Famous temples dedicated to Lord Shiva at Mangeshi, Nageshi, Ramnathi, Kapileshwari and Dhavali and Goddess Mahalaxmi temple at Bandora, Navadurga temples at Madkai, Kundaim and Borim, Shantadurga in Kavlem, Mahalsa in Mardol Goa and many others have been built in red lateritic stones.

11.5 FOUNDATIONS AND OTHER GEOTECHNICAL STRUCTURES

11.5.1 Tunnels

The Konkan Railway Project was a prestigious rail project of the Government of India, commissioned in Goa–Maharashtra–Karnataka through the treacherous, loose-soiled and hilly Western Ghats terrain, stretching over 76 km with 91 tunnels and over 2,000 bridges along it from the Roha terminus in Maharashtra to Thokur in Karnataka. During its construction, several technologies were introduced and successfully completed like the longest tunnel, tallest viaduct with slipforming technology, incremental launching of girders at Ratnagiri, indigenously developed and patented anti-collision devices, slope monitoring techniques using the "Raksha Dhaga" technology, use of vetiver grass for slope protection, rock bolting, steel–concrete composite bridges and so on.

The three tunnels in Goa at Barcem in Canacona, Old Goa and Pernem Goa posed several geotechnical challenges due to soft and collapsible lateritic soils and dripping water issues and loss of life due to collapse of tunnels during the construction of the Konkan Railway.

(a) Viceroy's Arch, Old Goa

(b) Basilica of Bom Jesus, Old Goa

(c) Temple at Kapileshwari Ponda Goa

(d) Patto Bridge, Panaji – Goa

Figure 11.15 Some ancient monuments in laterite. (Photo Courtesy: Prof. Smita Aldonkar and Ms Roshani Majik.)

11.5.2   Dams

There are few earthen and masonry dams constructed across the rivers and major streams in Goa. The major dams are the Salaulim dam in South Goa and the Anjunem dam in North Goa, which are explained here. The other important dams are:

- the Tillari irrigation project which is a joint venture of Goa and Maharashtra (395 m masonry dam and 79 m earthen dam of height 70.5 m across the Tillari river catering to water supply and irrigation),
- 760 m long and 41 m high earthen dam across Chapoli nallah in Canacona catering to water supply and irrigation,
- 450 m long and 24.92 m high earthen dam across Amthane nallah in Bicholim catering to water supply and irrigation and
- 230 m long and 20 m high homogeneous earthen dam in Panchwadi, Ponda catering to water supply and irrigation.

11.5.2.1   Salaulim dam

This dam is located in Salaulim in Sanguem taluka in South Goa and caters to irrigation, domestic and industrial uses. The dam is a composite structure of earth-cum-masonry type of 42.70 m height above the deepest foundation level. The length of the dam at the crest is about 1 km and the reservoir water spread area within Goa is 24 km². The dam structure has a volume content of 2.714 (MCM). The gross storage capacity of the reservoir is 234.361 MCM with the live or effective storage capacity fixed at 227.157 MCM. The spillway which is of the unique Duckbill type (Morning Glory type) is an ungated structure located in the gorge section with a length of 44 m which is designed to pass an estimated design flood discharge of 1,450 m³. Figure 11.16a shows the panoramic view of the Salaulim dam.

11.5.2.2   Anjunem dam

Anjunem dam is located on the Costi river in Sattari taluka in Goa. The dam is a straight gravity type masonry dam of 176 m length and 42.8 m height. The spillway length is 9.48 m and is provided with four gates of radial type. Flip bucket type energy dissipation devices are provided at the downstream side. The dam has a gross command area of 2,624 hectares and a drainage area of 17.18 km². The maximum design flood is 485.71 cumecs. The full reservoir level capacity is 448 hectares. The dam caters mainly irrigation purposes. Figure 11.16b shows the panoramic view of the Anjunem dam.

11.6   OTHER GEOMATERIALS

In Goa, the backbone of the state's economy is the mining industry which has flourished over the past many years. The main ore available in large quantities is iron ore found in the northern part of Goa and in some locations in South Goa, which has been

(a)

(b)

Figure 11.16 (a) Salaulim dam and (b) Anjunem dam.

extracted to the extent possible. Other ores like manganese are found in the southern part of Goa, limestone reserves are located in the north-east corner of Goa and bauxite reserves are found in the western coastal belt of Goa. Because of the mining industry, huge dumps of mine tailing are available which are generally used for filling excavated pits. A lot of research work has been done in utilising mine wastes for making bricks and blocks. A similar plant for making bricks out of mining waste based on the research work done at the Goa Engineering College has been set up by Prof. K. G. Guptha, Professor and Head of the Civil Engineering Department of

Goa Engineering College, Farmagudi. This project was funded by M/s. Fomento Resources, Panaji and the eco-friendly material developed was considered as a sustainable material for construction by the Holcim Group in the year 2017.

There are no thermal power plants in Goa and hence the production of fly ash waste is not there. Due to mine ore processing in the industry, there is large-scale production of ground granulated blast furnace slag. Slag has been utilised in the production of blended cements popularly known as the slag cements. In addition, one of the biggest business houses in Goa M/s. Alcon Group under the able leadership of Mr Anil Counto has invented microfine products from slag under the brand name Alccofine at the R&D and production unit, M/s. Counto Microfine Products Pvt. Ltd. at Goa. This pioneering invention of a concrete additive product has found a market in India and also across the world for various applications like metro rail projects, dam projects, roads, flyovers, bridges and iconic structures. It has also been found useful as cementitious microfine injection grouts for soil stabilisation, tunnel grouting, permeation grouting and so on.

11.7 NATURAL HAZARDS

11.7.1 Seismic events

The state of Goa is located in seismic zone III. So far, no major seismic activity was reported in Goa. The nearest major earthquake events reported were at Koynanagar (1967) in Maharashtra ($M_w = 6.6$, Intensity VIII, shallow intra-plate earthquake), which is about 350 km away from Goa, and at Latur (1993) in Maharashtra ($M_w = 6.2$, Intensity VIII, Intra-plate earthquake with the epicentre at Killari), which was about 450 km from Goa. Minor tremors have been felt in Goa due to the above earthquakes with no damages reported in Goa.

11.7.2 Landslides

Landslides have been occurring in Goa regularly, especially during the rainy seasons. Several landslides have occurred in the Western Ghats causing inconvenience to locals. Slopes have been cut for many recent developments in the state of Goa, which can pose a great threat to inhabitants living there due to lack of checking the stability of the slopes hence causing landslides. As per the Geological Survey of India, 12.6% of the total landmass of India comes under the landslide prone hazardous zone, which constitutes about 0.42 million km². About 0.09 million km² out of the 0.42 million km² includes the Western Ghats and Konkan hills (Tamil Nadu, Kerala, Karnataka, Goa and Maharashtra). The probable causes of landslides are basically twofold. At first instance, there is an increase in shear stress, which may occur due to:

•   removal of lateral and underlying support caused during erosion, road cuts and quarries,
•   an increase of load due to weight of rain or snow, fills and vegetation,

- an increase of lateral pressures caused by hydraulic pressures, roots, crystallisation and swelling of clay,
- transitory stresses due to earthquakes, vibrations of trucks, machinery and blasting,
- regional tilting, which is due to geological movements, and
- toe excavation.

Second, the material strength reduction takes place due to:

- a decrease of material strength due to weathering,
- a change in the state of consistency,
- changes in inter-granular forces,
- an increase in the void ratio due to swelling, hence increasing the water content along the shear failure plane, and
- leaching of salts.

The other causes can be geological causes attributed to weak sensitive materials, jointed or fissured rock materials, sheared and weathered materials having discontinuities are disposed to landslides, morphological causes, physical and human causes. Table 11.3 shows the classification of landslides based on depth.

Most of the landslides in Goa are shallow landslides, except for those occurring in Chorla Ghat areas. Major landslides occur almost every year in the Chorla Ghat area (near the Goa–Karnataka border) during the rainy season and in some of the parts of Goa. The landslide inventory map of Goa is available on the Geological Survey of India website (GSI, 2021). The recent landslides that have occurred in Goa in the recent past are shown in Figure 11.17a–i. The details of some past major landslides are given below.

11.7.2.1   Landslide at Ambaji, Fatorda

This site is near the locality of the Collectorate Office of South Goa. Cutting of hill slopes was carried out on both sides of the hillock which resulted in a huge landslide in the rainy season in 2014. The debris travelled for about 200 m downstream and the retaining wall which was constructed was washed away. Figure 11.17a shows the landslide that had occurred at Ambaji Fatorda Goa.

Table 11.3 Landslide classification with respect to depth

Maximum depth (m)	Type
<1.5	Surface slides
1.5–20	Shallow slides
5.0–20	Deep slides
>20.0	Very deep slides

Figure 11.17 Typical landslides in Goa: (a) landslide at Ambaji – Fatorda in 2014; (b) landslide at Porvorim in 2014; (c) 120 m long and 10 m deep crack occurred before a landslide in Porvorim; (d)–(f) landslides in Chorla Ghat in 2017, 2019 and June 2020, respectively; (g) landslide in Mardol in 2016; (h) landslide in Khandepar in 2018 and (i) landslide in Dona Paula in 2019.

11.7.2.2 Landslide at Porvorim Goa

This site is near the Secretariat Complex of the Government of Goa. At this site, in 2014 in the rainy season, on the top of the hillock, a 120 m long and a 10 m deep crack was observed. This crack widened resulting in detaching of a huge mass of soil and boulders, which can run down on the highway (Figures 11.17b and c).

11.7.2.3 Landslide at Chorla Ghat

This site is a hill located at the boundary between Goa and Karnataka and is prone to landslides during monsoons every year. Figure 11.17d–f shows the typical landslides that occurred at this site in 2017, 2019 and 2020.

The other typical landslides that occurred in Goa were Nagarcem (2014), Raia (2014), Narvem (2015), Arpora (2015), Neura (2016), Ponda (2016, 2019), Mardol (2016), Aquem (2017), Borim (2018) and Khandepar (2018). Geotechnical properties of soils at some of these landslide sites have been presented in Table 11.2. Table 11.4 shows the important geotechnical parameters of these three major landslides in Goa.

11.8 CASE STUDIES AND FIELD TESTS

In this section, few interesting case studies of lateritic soils are presented.

Table 11.4 Characteristics of typical landslides in Goa

District	South Goa	North Goa	South Goa
Location	Ambaji, Margao	Chorla ghat, SH31 between Goa and Karnataka	Khandepar
Latitude	15.2954828	15.6325631	15.4201855
Longitude	73.9579137	74.1182475	74.0406736
Height	13.8 m	18 m	20 m
Inclination of slope	70°	65°	85°
Slide type	Debris type rapid slide	Debris type rapid slide	Debris type rapid slide
Failure mechanism	Shallow translational	Shallow translational	Shallow translational
Geomorphology	Moderately dissected hills	Highly dissected hills	Highly dissected hills
Land use/land cover	Sparse vegetation	Good vegetation	Good vegetation
Triggering	Rainfall	Rainfall and anthropogenic	Rainfall and anthropogenic
Geoscientific cause	Cutting of slope	Road cut leading to destabilisation and modification of slope	Road cut leading to destabilisation and modification of slope

11.8.1 Geotechnical properties of different coloured lateritic soils

Lateritic soil is rich in iron oxide and is a result of weathering of a wide variety of rocks that undergo leaching and oxidation due to natural agencies like heavy rains, high temperature and so on with silica leaching out of it. Hence, lateritic soil may contain various clay minerals in different quantities such as the iron oxide minerals goethite ($HFeO_2$), hematite (Fe_2O_3), limonite ($Fe_2O_3.nH_2O$), lepidocrocite ($FeO(OH)$), kaolinite ($Al_2Si_2O_5(OH)_4$), gibbsite ($Al_2O_3 \cdot 3H_2O$) and so on, which impart different colours to lateritic soil. In addition, lateritic soils are explored for bauxite, titanium oxides and hydrated oxides of aluminium and so on. Laterites in Goa are also tried for rare earth elements like lanthanides, scandium, yttrium and so on.

In the state of Goa, lateritic soils of different colours are observed in most of the parts of Goa. It was observed that reddish and brownish lateritic soils possess better strength properties. They have higher unit weight, higher specific gravity, lower water content, higher maximum dry density and higher gravel content. Yellowish, whitish and blackish lateritic soils on the other hand are weaker soils as they contain higher moisture content, lower unit weight and specific gravity, higher clay content and high plasticity index, which make them less suitable as construction materials.

11.8.2 Typical lateritic soft clay site in Goa

This case study refers to a typical lateritic soft clay site in Cacora in the Quepem taluka in South Goa where a G + 2 reinforced concrete structure was proposed. A schematic layout of the building and the reddish-yellowish soil observed at the site are shown in Figure 11.18. During excavations for foundations, hard lateritic rock was observed in Phases I and II, while in Phase III – A and B, hard lateritic soil was observed. In the case of excavations for Phase III-C, a very soft reddish-yellowish lateritic soil was observed. Hence, it was decided to drill boreholes of at least 10 m deep at three locations to ascertain the extent of soft soil and design the foundations accordingly. A typical borehole profile is shown in Figure 11.10 and the corresponding geotechnical properties of the soil are presented in Table 11.1 for the Cacora site. The soft soil encountered contained kaolinite, gibbsite and limonite up to 1.5 m depth with higher contents of gibbsite and kaolinite for a depth of up to 6 m. For the remaining portion of 6–10 m, limonitic argillaceous clay/silt was observed. The soil was observed to be reddish to yellowish in colour with a very fine texture.

A shallow water table was observed. Sieve analysis results indicated the absence of gravel, but the sand content was about 63.2%–68.4% while the silt + clay content was 31.6%–36.8%. The lowest value of safe bearing capacity at the site was about 79 kPa at the depth of 1.5 m while the plate load tests conducted indicated SBC values of 76.8 kPa. A typical load-settlement curve based on the plate load test is shown in Figure 11.19. Settlement calculations resulted in a value of 52.79 mm for isolated footings. Hence, raft foundations were recommended after ground improvement. It is also recommended to replace the existing soft soil with better quality soil to improve the load-settlement characteristics. It was proposed to construct the raft foundation at a depth of 1.5 m below the ground level at the site. This raft shall be placed on a 1 m thick compacted layer consisting of a well-graded and well-compacted murrum. For this,

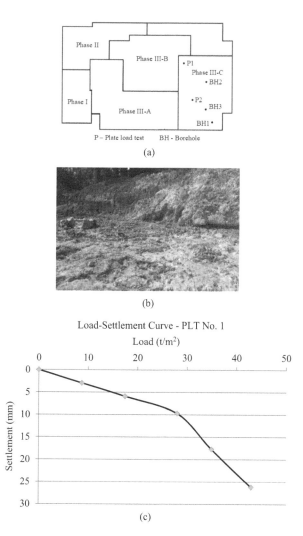

(a)

(b)

Load-Settlement Curve - PLT No. 1

(c)

Figure 11.18 Typical very soft lateritic soil site: (a) G + 3 Building layout; (b) soft lateritic soil and (c) plate load test results.

the soil may be excavated for a depth of 2.5 m below the existing ground level and may be replaced with well-graded murrum compacted in three layers (with a total thickness of 1 m) and the raft was proposed to be constructed on this compacted murrum layer, so that raft will be placed at a depth of 1.5 m below the existing ground level.

11.9 GEOENVIRONMENTAL IMPACT ON SOILS AND ROCKS

Lateritic rocks are sedimentary type and are formed due to the weathering action on the parent rock like basalt. A very soft pinkish yellowish lithomarge layer also known as saprolite is often seen in typical lateritic profiles. Many times due to a higher rate of

Figure 11.19 Cavities in lateritic rock at Bambolim. (Photo Courtesy: Mr Rajesh Raikar.)

internal drainage and erosion, the lithomarge layer may get washed away giving rise to caves or cavities as shown in Figure 11.19. Such cavities are often found in lateritic rocks in Bambolim, Dona Paula, Taliegao and Headland Sada.

11.10   CONCLUDING REMARKS

The state of Goa comprises the Western coastal estuarine plain consisting of sandy beaches, estuarine mudflats, khazan lands, salt pans, mangroves, fields and settlement areas, the central undulating region or midlands featuring hills ranging from 100 to 600 m and the Western Ghats or Sahyadris ranging from 600 to 1,000 m high on the eastern and southern parts. Laterite cover is observed on the residual hills while the alluvium and sand cover is observed on the coastal estuarine plains in most of the geological formations of Goa. The soils and rock cover mostly consist of lateritic soils and lateritic rocks. In South Goa, the subsoil consists of mainly silty clay. Laterites and lateritic soils have been used in Goa as construction materials since ancient times. Their compressibility varies from moderate to low and being mostly impervious, they find use in building construction for walls, flooring and mud blocks. Due to the presence of various minerals like limonite, hematite, kaolinite and so on, the colour of laterite varies. Reddish and brownish soils and rock have better strength properties as compared to yellowish and whitish soils. The major concern with laterites and lateritic soils is the presence of a soft lithomarge layer, which makes them susceptible to landslides especially in monsoons after heavy rains. Lateritic soils many times prove to be a poor foundation material when the water table is available at shallow depths. The safe bearing capacity of these soils was observed to be as low as 70 kPa to sometimes 220 kPa. Lateritic rocks are classified as soft rocks due to their porous and concretionary nature. The safe bearing pressures are often assumed to be up to 400 kPa.

ACKNOWLEDGEMENTS

The author is thankful to Prof. Ashoka Dessai, Prof. Smita S Aldonkar, Prof. Mandira Faldesai, Prof. Leonardo Souza, Er. Neil Agshiker (Water Resources Dept.), Ms Roshani Majik and Mr Rajesh Raikar and Soil Investigation Agencies for their contributions.

REFERENCES

Dessai, A.G. (2018). *Geology and Mineral Resources of Goa.* First Edition, New Delhi Publishers, New Delhi.

Fernandes, O.A. (2009). From geological to historical time: The Goan scenario. *Natural Resources of Goa: A Geological Perspective.* Mascarenhas, A. and Kalavampara, G. (Eds.), Geological Society of Goa, Panji, Goa, pp. 5–10.

GSI. (2021). Geological Survey of India, www.gsi.gov.in, accessed 10 February 2021.

Widdowson, M. and Gunnell, Y. (1999). Lateritization, geomorphology and geodynamics of a passive continental margin: The Konkan and Kanara coastal lowlands of Western Peninsular India. *Paleoweathering, Paleosurfaces and Related Continental Deposits.* Thity, M. and Simon-Coincon, R. (Eds.), International Association of Sedimentologists Special Publication 27, Blackwell, Oxford, pp. 245–274.

Chapter 12

Gujarat

Chandresh H. Solanki, Mohit K. Mistry, and Manali S. Patel
Sardar Vallabhbhai National Institute of Technology

CONTENTS

12.1 INTRODUCTION

The state of Gujarat is located on the western coast of India enclosed within the GPS coordinates of 22° 18′ 33.9300″ N and 72° 8′ 10.4280″ E. The state covers an approximate area of 2,00,000 km^2 along with a coastline of 1,600 km. It is the fifth-largest state of India which comprises extensive diversities in geology, climates, physiography and vegetation. This state offers a variety of soil types: black soil, mixed red and black soil, laterite, alluvial soil, hill soils and desert soils.

A significant part of the land in Gujarat is covered with black soil, which is also known as clayey soil. Based on its different characteristics further, it is classified as low plasticity black soil, medium plasticity black soil and high plasticity black soil. The low plasticity black soils are usually light brown-grey in colour. They mainly fall under sandy clay loam as per the textural classification. The medium plasticity black soils are usually categorised as silt loam to clay. The high plasticity black soils are dark brown to greyish in colour which possesses very poor drainage. These black soils are developed from gneiss, metamorphic rock and granite, an igneous rock.

The mixed red and black soils are usually clay loam to clay in texture. They are usually found at shallow depths in brown colour with the red shade at a higher elevation

DOI: 10.1201/9781003177159-12

and with grey shade at lower elevations. Due to the presence of stony materials, such types of soils offer ideal drainage conditions.

Laterite is a soil that possesses high iron and aluminium contents and is commonly developed in hot and wet tropical regions where the annual rainfall is greater than 200 cm. Texturally, such soils are clayey in nature and reddish in colour due to the high iron oxide. Such soils are generated due to the chemical weathering of rock. Usually, after a few hours of precipitation or water supply, the hard crust formed on its surface, which restricts the infiltration, and eventually it becomes hard. Usually, such types of soils are not found in the Gujarat state.

Alluvial soils are generally found alongside rivers and are transported by their streams during weathering of rocks. Such soils are generally covered by vegetation. Furthermore, they are classified as alluvial sandy to sandy loam soils, alluvial sandy to sandy clay loam soils and coastal alluvial soils.

Hill soils are usually found in hilly regions. Due to the sloppy terrain, such soils are frequently experiencing erosion, which reduces their fertility.

The desert soils usually do not possess any specific definite structure. They are deep and light grey in colour. Texturally, they fall under the category of sandy loam with silt clay loam in structure. The presence of a high amount of salt makes such types of soils saline in nature. Figure 12.1 shows the district map of Gujarat (Maps of India, 2019). This chapter provides the details of soils and rocks and other related aspects.

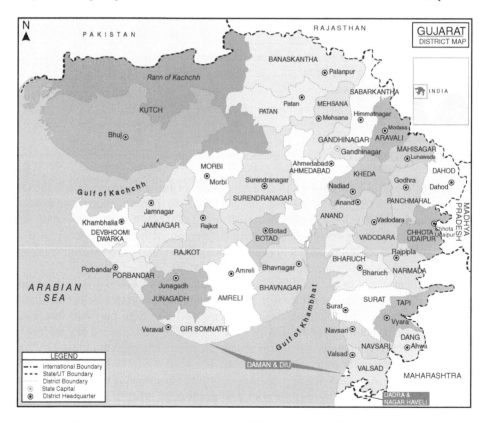

Figure 12.1 District map of the Gujarat state (Maps of India, 2019).

12.2   MAJOR TYPES OF SOILS AND ROCKS

The Gujarat state is divided into four regions as follows:

1. South Gujarat region,
2. Central Gujarat region,
3. Saurashtra and Kutch region and
4. North Gujarat region.

Figure 12.2 shows the soil map of Gujarat prepared in ArcGIS based on the collected data from various geotechnical projects.

12.2.1   South Gujarat region

The south region of Gujarat mainly covers the Dang, Valsad, Navsari, Surat, Tapi, Narmada and Bharuch districts. Texturally, the soils in the South Gujarat region generally fall under clayey to clayey loam. Generally, this type of soil is deep black in colour. At some places, they are also found to be in dark brown to very dark greyish brown colour. The clay mineral found in this type of soil is from the montmorillonite group, which is responsible for the high swell-shrink nature of clayey soil. The significant land along the riverside of Narmada is made up of alluvial soil. Figure 12.3 shows the soil profile of the Vesu area of the Surat district.

12.2.2   Central Gujarat region

The central region of Gujarat mainly covers Vadodara, Kheda, Anand, Dahod and Panchmahal districts. The significant parts of the southern area of Vadodara, Bharuch, Surat and Valsad are covered with high plasticity black soil. The soils in the core area of Panchmahal, Dahod, Kheda and Anand districts are covered with clayey soils

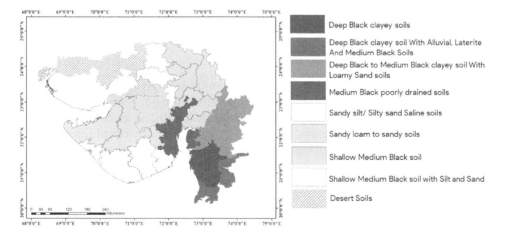

Figure 12.2 Soil map of the Gujarat state.

Figure 12.3 Soil profile of the Vesu area, Surat.

with some coarser sand. The soils of the southern part of Ahmedabad, as well as some parts of Central Gujarat and the coastal areas, are formed from the weathering of granite, gneissic and basaltic rocks (Merh, 1995). They are silty-loam to clay and sandy clay loam to clay with a smectite group of clay minerals. In general, the soils in Central Gujarat are dark to light grey, yellowish-red, reddish-brown to dark reddish-brown in colour.

12.2.3   Saurashtra and Kutch region

The Saurashtra and Kutch region mainly covers the Kutch, Jamnagar, Porbandar, Junagadh, Amreli, Bhavnagar and Rajkot districts. A significant part of the Saurashtra region is covered with clayey soils where the majority of the Kutch region is covered with sandy soils. They are originated from the weathering of granite, basalt and gneissic rocks under tropical semi-arid to humid climates (Merh, 1995). They are usually light grey, greyish brown and reddish-brown in colour. In general, the soils in Saurashtra and Kutch are sandy clay to clay and clay loam to clay, which are subangular and blocky.

12.2.4   North Gujarat region

The North Gujarat region mainly covers the Surendranagar, Ahmedabad, Gandhinagar, Sabarkantha, Mehsana, Patan and Banaskantha. A major part of the North Gujarat region is loamy. In Ahmedabad and Surendranagar districts, the significant area is covered with clayey soil whereas the Banaskantha district consists of sandy soil as a major type. Similarly, Gandhinagar, Mehsana and Sabarkantha districts also contain sand and sandy clay as their major portion. They are originated from the weathering of granite, basalt and gneissic rocks under tropical semi-arid climates (Merh, 1995).

Geologically, Gujarat comprises a Precambrian basement over which younger rocks commencing with Jurassic, continuing through Cretaceous, Tertiary and Quaternary have given rise to a varying sequence in different parts. Thus, the rocks in Gujarat belong to formations ranging in age from the oldest Precambrian to Recent

(Thaker, 2012). The igneous rocks encountered in the South Gujarat region are pierite basalt, rhyolite, obsidian, granophyre, felsite, aplite and diorite (Dave, 1971). The igneous complexes are also found in the Saurashtra region as Aiech and Barda Hills, Girnar and Osham Hills and Chogat Chamardi Hills. The sedimentary rocks like limestone are well found in the Jamnagar, Junagadh and Bhavnagar districts of Saurashtra and Kutch region, Banaskantha district of the North Gujarat region and Bharuch and Surat districts of the South Gujarat region. The soft brown sedimentary rock lignite, which is associated with Tertiary rock, is found in the Kutch and Bharuch region. Bauxite is also mined in Kheda, Kutch, Junagadh and Jamnagar districts (Prakash et al., 2012).

12.3 PROPERTIES OF SOILS AND ROCKS

This section comprises the description of the subsurface profile of different regions of Gujarat. To summarise the geological profile of entire Gujarat, nearly 4,000 borehole data were collected from various geotechnical projects. The entire Surat district comprises clayey soil and silty clay as its upper 10–15 m strata followed by the silty sand soil at a greater depth. Figures 12.4 and 12.5 show the soil profile of two sections of the Surat district, Khajod village to Kumbhariya and Dumas village to Paldikande, respectively. The ground water table varies between 2 and 4 m near the Dumas beach. In other parts of the Surat district, the water table is found at 10–12 m moderate depth.

The significant part of the Vadodara district in the Central Gujarat region comprises SM, SC and CI soils as its upper layer followed by SM and CI soil at the greater depth. The water table near the Mahisager river and Vishwamitri riverfront area are found at a shallow depth of 2–5 m. In other areas of the Vadodara district, the water table varies from 10 to 15 m. The subsurface profile of the Vadodara district is illustrated in Figures 12.6 and 12.7.

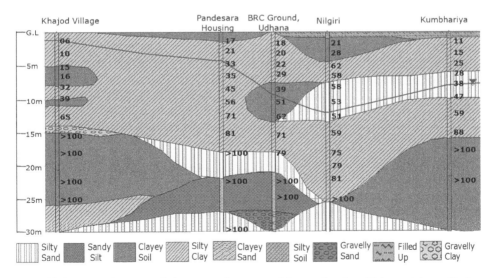

Figure 12.4 Fence diagram of the central region of Surat district. (Adapted from Thaker, 2012.)

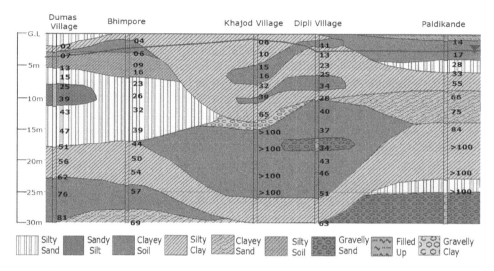

Figure 12.5 Fence diagram of the southwest region of Surat district. (Adapted from Thaker, 2012.)

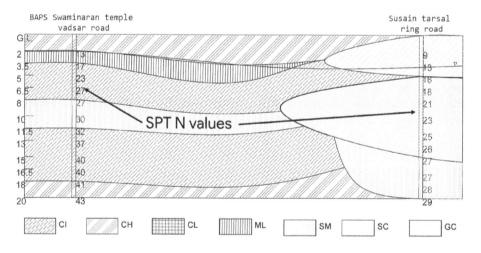

Figure 12.6 Fence diagram of the southwest region of Vadodara district. (Adapted from Mehta, 2021.)

The subsurface profile of the Ahmedabad district of the North Gujarat region is illustrated in Figure 12.8. The upper 3–5 m layer in a major part of the Ahmedabad district is covered with the CL and CI types of soil. Filled up soil at a shallow depth of 1–1.5 m was also encountered in few areas of the central Ahmedabad district (Figure 12.9). The majority of the upper land (up to 15–20 m) in Bharuch district is found as CH and CI soils having SPT-N values ranging from 6 to 30. The areas nearby the Narmada River comprise poorly graded sand with a maximum SPT-N value of 25. The Dahej taluka of the South Gujarat region consists of CH soil as its upper 10 m layer and CI soil at medium to a greater depth with SPT-N values ranging from 4 to 30.

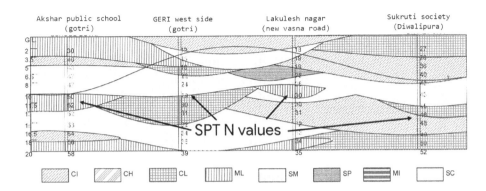

Figure 12.7 Fence diagram of the northwest region of Vadodara district. (Adapted from Mehta, 2021.)

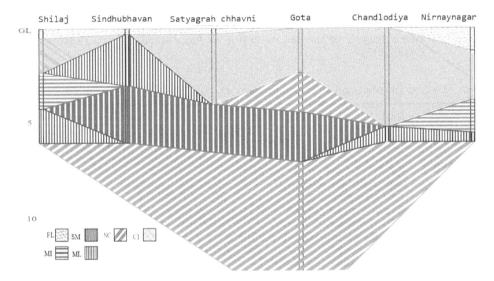

Figure 12.8 Fence diagram of the northwest region of Ahmedabad district. (Adapted from Dave, 2017.)

A certain portion of Dahej taluka comprises SM, ML and SP soils at a medium depth. Due to the Gulf of Khambhat and Narmada River, the water table in the Bharuch, Ankleshwar and Dahej area is found at a shallow depth of 4–8 m.

The southeast portions of the Saurashtra region like Bhavnagar, Amreli, Junagadh and Porbandar districts are covered with soft to very soft clay soils. Some regions of Junagadh and Amreli consist of dark brown silty sands at shallow depths sandwiched between black clayey soils. A significant part of the Kutch region is covered with silty sands. The upper portion of the entire Kutch region (10–12 m depth) consists of medium to dense brown sandy silt followed by dense to very dense dark brown sandy silt up to 17 m depth and stiff to very stiff clayey soils. The boring logs of Bhavnagar district, Jamnagar district and Kutch region are shown in Tables 12.1, 12.2 and 12.3, respectively.

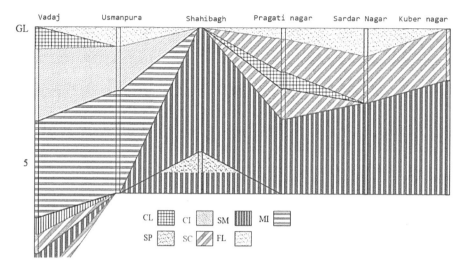

Figure 12.9 Fence diagram of the central region of Ahmedabad district. (Adapted from Dave, 2017.)

Table 12.1 Soil stratification of Bhavnagar district

Nari chokdi location (m)	Description	IS classification	Moti Talav location (m)	Devgana location (m)	Palitana location (m)
Upto 0.60	Blackish brown, very fine-grained, clays of high plasticity	CH	0–0.8	0–0.4	0–0.32
0.60–2.80	Greyish brown, very fine-grained, silty clay to high plasticity	CH	0.8–3.2	0.4–6.5	0.32–5.2
2.80–5.40	Brownish fine to medium-grained sandy clays of high plasticity with occasional gravels	CH	3.2–12	6.5–10	5.2–16
5.40–8.10	Brownish fine to medium-grained clayey sand with occasional gravels	SC	12–15	10–16.5	16–20
8.00–12.00	Brownish fine to medium-grained intermediate plasticity with occasional gravels	CI	15–18	16.5–22.5	20–28.5
12.00 downwards	Dark brownish, fine to medium-grained, clayey sand with moderate to high gravel content	SC	18–30	22.5–30	-

Source: After A. Pandya, Personal communication, 2020.

Table 12.2 Soil stratification of Jamnagar district

Morkanda location (m)	Description	IS classification	Lavadia location (m)
0–1	Filled up soil – Black sand with clay and fine gravels	-	1.2
1–3	Very soft to soft black clay	CH	1.2–5
3–4.5	Dark brown to black silty sand with clay and fine gravels	SM-SC	5–8
4.5–9	Hard yellowish to radish brown clay silt with sand and fine gravels	CH	8–15
9–20	Hard yellowish to radish brown clay silt with sand and fine gravels	CH	15–18

Source: After G. Goel, Personal communication, 2019.

Table 12.3 Soil stratification of Hajipir village, Kutch

Depth (m)	Description	IS classification
0–1.2	Filled up soil	-
1.2–12	Medium to dense brown sandy silt	ML
12–16.5	Dense to very dense dark brown to black sandy silt/silty sand	ML-SM
16.5–22.5	Very stiff dark brown to black clayey silt	CL
9–20	Stiff to very stiff greyish black clayey silt	CI

Source: After A. Pandya, Personal communication, 2020.

12.4   USE OF SOILS AND ROCKS AS CONSTRUCTION MATERIALS

Geomaterials are the materials that arise from the geological systems, originated from the billion years' long history of the earth. However, due to industrialisation, these materials are artificially processed to the outputs as construction materials or by-product wastes. The geomaterials may include rock, clay, granular materials, treated soils and industrial wastes like plastics, tyre products, fly ash, natural and synthetic materials, which are widely used in geotechnical practice. In the state of Gujarat, significant land is covered with expansive soils. Depending upon the site condition, to improve weak soil strata as well as to support the excavated weak soil strata, various techniques are being used. Few practical applications with site images are described in the following section.

Due to the large number of power plants (23 plants + 9 thermal power plants), fly ash is easily available throughout the state and hence its use in pavement subgrade material is very popular. Figure 12.10 shows the application of fly ash in pavement subgrade material at the Olpad site. In Gujarat, gabion wall is one of the popular techniques used as soil retaining structures. The underpass of the Sarkhej–Gandhinagar highway was built using gabion wall (Figure 12.11).

The foundation soils of bridge abutments of the Ahmedabad–Vadodara expressway are retained with the use of gabion wall (Figure 12.12). Figure 12.13 shows the sheet pile wall construction, which was used to support the adjacent structure at the

Figure 12.10 The fly ash application in the pavement at Olpad, Sahol State Highway. (After Bakare, 2020.)

(a)   (b)

Figure 12.11 Gabion wall construction at the Sarkhej–Gandhinagar (SG) highway.

Figure 12.12 Gabion wall construction at the Ahmedabad–Vadodara expressway (National Expressway-I).

Figure 12.13 Sheet wall pile construction at Navsari.

Figure 12.14 Geogrid encased stone column for SMC multi-level parking building, Surat.
(After J. Patel, Personal communication, 2020).

Navsari district. Another popular technique for ground improvement in Gujarat is the use of granular aggregates as a stone column. Figure 12.14 shows the encased column used in the construction of a multi-storeyed parking building in Surat.

12.5 FOUNDATIONS AND OTHER GEOTECHNICAL STRUCTURES

In the development of Gujarat, state dams have played an important role. There are nearly 21 large dams in Gujarat. Details of the top five largest dams of Gujarat are provided in Table 12.4. The Sardar Sarovar dam is constructed on river Narmada, which flows along the east-north-east to west-south-west trending Narmada-Son Fault which is longer than 1,000 km.

The Narmada canal is the main canal that carries water from the Sardar Sarovar dam north through 460 km in the Gujarat state only and then 74 km into the Rajasthan state. It is the world's largest lined irrigation canal with a length of 750 km consisting of

Table 12.4 Dams of Gujarat

Sr. No.	Name	Type	Height (m)	Location
1	Sardar Sarovar Dam	Gravity Dam	163	Narmada River
2	Ukai Dam	Earth-Masonry Dam	105	Tapti River
3	Kanada Dam	Masonry with Embankment Dam	66	Mahi River
4	Dantiwada Dam	Mud-Masonry Dam	61	Banas River
5	Kamleshwar Dam	Rockfill Embankment Dam	25	Hiran River

42 branch canals. Due to the extensive length, the foundation soil for the construction of embankment affected the design of canal embankments. The entire canal line was built on the high plasticity clay soil.

The Statue of Unity, the world's tallest statue of Late Sardar Vallabhaiji Patel, a great national leader, is constructed at around 3.50 km downstream of the Sardar Sarovar Dam (Ghan, 2021). The statue is located on a rocky promontory known as Sadhu hill; one hill islanded in the Narmada River waterway. It consists of sedimentary rocks and quartzitic sandstone. The Sadhu hill is around 250 m inside the Narmada waterway and the top of the hill is at RL 72.0 m. The statue is constructed on the right bank of river Narmada which flows along the east-north-east to west-south-west trending Narmada-Son Fault, which is now inactive. As per IS 1893 (2016), the site of the Statue of Unity falls under seismic zone III where earthquakes of 6.0 magnitude can be expected. The effects of lithology, rock joints, stratigraphy and other engineering parameters on foundation were studied. Furthermore, contact between the overburden and in-situ foundation rock condition and other geotechnical inputs like rock permeability, porosity, slope stability analysis and so on were incorporated to support the design mechanism of the support system. The following tests like borehole logging, packer permeability test, high-pressure dilatometer, cross-hole shear test and electrical resistivity test were carried out for detailed rock properties. The Statue of Unity is built on the raft foundation having a thickness of 3 m with a large base of 44 m × 34 m. The bearing capacity of the founding sandstone strata was computed to be 1,500 kPa. The complex geology and the associated issues like the stability of rock mass, erosion and scouring protection of shale from flood water, seismicity of area and usual wind speed effect have a strong influence on the foundation design.

Ahmedabad being the most populated city of Gujarat, the traffic need of the city cannot be met only by the road-based system. The Ahmedabad metro rail project has been started by the Gujarat Metro Rail Corporation (GMRC) Limited in the year 2016 (GMRC, 2016). The total length of Phase-I is about 40.03 km, out of which 18.87 km will connect the north and south regions and the remaining 21.16 km will connect the east and west regions of the Ahmedabad district. It also consists of approximately 6.5 km underground construction in the old Ahmedabad region. A significant portion of Ahmedabad contains CL and CI types of soil at shallow depths followed by the SM-SC type of soil. After necessary treatments, the above-ground sections of the metro project are constructed on a single pier with pile foundations. The underground sections are being tunnelled using the North Austrian tunnelling method whereas the underground stations are being constructed using cut and cover with a top-down technique. Currently, the construction of Phase-I is in progress.

In the South Gujarat region, the considerable land is covered with expansive clayey soils and the groundwater table is also found at moderate depths. Hence in South Gujarat, usually, diaphragm walls are a much popular construction technique used for supporting the stability of excavated soils and slopes. Figure 12.15 shows the construction of the diaphragm wall with soil nails in the Surat district. In the Ahmedabad district, nowadays the majority of the high-rise buildings and malls are being constructed with 2–3 basements. At the majority of the construction site, the available soils are dry, poorly graded and cohesionless and the groundwater table is also found at shallow depths. So, in Ahmedabad and Central Gujarat districts as a retaining structure, in place of diaphragm walls, the soil nailing technique is much popular. In this technique, the excavated soil slope is being stitched with steel bars along with shotcrete. Figure 12.16 shows a view of soil nailing at a construction site where the G + 12 corporate office with two basements was constructed in the Makarba area of Ahmedabad city.

Figure 12.15 Diaphragm wall construction with soil nail at Surat. (After J. Patel, Personal communication, 2020)

Figure 12.16 Soil nailing at the Makarba area, Ahmedabad district.

12.6   NATURAL HAZARDS

Gujarat has faced the most devastating natural disasters in the past. In the recent past, the whole of Gujarat and/or a significant part of its land has witnessed extreme climate conditions and/or drastic natural events. Owing to its topographical position, climatic and geological background, Gujarat has had a fair share of disasters of varying magnitudes from ancient times. Out of the 25 districts of Gujarat, the Ahmedabad, Bharuch, Valsad and Surat districts are highly susceptible to all four natural hazards: geophysical, hydrological, climatological and metrological ones whereas other districts are prone to three different hazards.

The state is located in the Himalayan collision zone where the Indo-Australian tectonic plate slides under the more northern Eurasian plate in a predominantly northern direction at a rate of 2 cm/year. As per IS 1893, part XVI, entire Gujarat is having earthquake hazards of different levels from moderate to high as zone III to V except for the Dahod district which falls under zone II. The Kutch region falls under zone V where earthquakes of 5–8 or more magnitude on moment magnitude scale and intensity ≥ X can be expected. The nearby area, like the eastern part of Kutch and north Saurashtra, falls under zone IV where ground motion with the intensity of VIII on the Mercalli intensity scale can be expected due to the earthquakes of Kutch and local earthquakes due to the Kathiawar fault in north Saurashtra. The rest of Gujarat falls under zone III where ground motion with the intensity of VII on the Mercalli intensity scale can be expected due to the moderate to strong earthquakes of Kutch. The details of the three major earthquakes are mentioned here.

12.6.1   1819 Rann of Kutch earthquake

The earthquake of Rann of Kutch (1819) was one of the disastrous earthquakes in the history of Gujarat. The estimated range of magnitude recorded was 7.7 to 8.2 M_w. The convergent boundary between the Indian plate and the Eurasian plate was reported as a prime reason for the earthquake. The maximum intensity was reported as XI (extreme) on the Mercalli intensity scale.

12.6.2   1956 Anjar earthquake

The earthquake of Anjar (1956) was due to reverse faulting, similar in type to that which is thought to have caused the 1819 Rann of Kutch earthquake. It had an estimated magnitude of 6.1 M_w and the maximum intensity was reported as IX on the Mercalli intensity scale.

12.6.3   2001 Bhuj earthquake

The Bhuj earthquake was the intraplate earthquake having a magnitude of 7.7 M_w and the intensity as X on the Mercalli intensity scale. The variations in stress caused by the Coulomb stress transfer due to the 1819 Rann of Kutch earthquake was reported to trigger the 2001 Bhuj earthquake. The epicentre was at 16 km depth.

Table 12.5 Liquefaction failure in earthen dams during the Bhuj earthquake

Sr. No.	Dam	Height (m)	Epicentral distance, (R)/crest length (L)
1	Chang	15.5	35.15
2	Fatehgadh	11.6	19.75
3	Kaswati	8.8	75.60
4	Rudramata	27.4	89.14
5	Shivlakha	18.0	93.33
6	Suvi	15.0	17.64
7	Tapear	13.5	10.60

Source: Singh et al. (2005).

Earthquake liquefaction is one of the prime contributors to urban seismic risk. Many failures of earth retaining structures, slopes and foundations have been attributed in the literature to the liquefaction of sand. Liquefaction often occurs in the form of sand fountains. The 2001 Bhuj earthquake is an illustration of the liquefaction phenomenon causing catastrophic damages to structures and resulting in loss of life and properties. The significant damage in the Ahmedabad district was reported during the Bhuj earthquake. The investigation from the borehole studies reported that the soil profile of the major part at a moderate depth comprises sandy soil, which may trigger liquefaction induced earthquake. The studies conducted by Vipin et al. (2013), Mehta and Thaker (2020), Patel and Thaker (2020) and Patel et al. (2021a, 2021b) reflect the liquefaction and seismic severity of different parts of Gujarat state.

Due to the Bhuj earthquake, several dams within 150 km of epicentre got damaged and some of them completely got collapsed. The liquefaction failure in earthen dams during the Bhuj earthquake is illustrated in Table 12.5.

In each dam, the foundation soil was found as loose to medium dense, alluvial and silt–sand mixture which is highly susceptible to liquefaction. Moreover, it was reported by EERI (2001) that liquefaction susceptibility of the foundation soils was not considered in the design of these dams. The distress factors in each aforementioned dam are as follows:

- possible liquefaction in the foundation near the upstream toe,
- shallow failure in upstream slope and
- cracking.

Additionally, the failure of upstream and downstream slopes was found in Chang and Shivlakha dams. The leakage in the dam was reported in Kaswati dam, whereas slumping occurred in the Fatehgadh dam.

12.7 GEOENVIRONMENTAL IMPACT ON SOILS AND ROCKS

Due to the various anthropogenic activities, soils are subjected to various pollutants. In Gujarat, the major industrial anthropogenic sources are situated in seven districts; Jamnagar, Ahmedabad, Vadodara, Bharuch, Dahej, Surat (Hazira) and Valsad (Vapi) where significant land is contaminated. The 2001 Bhuj earthquake has resulted in cracks at the bottom of storage tanks and contaminated $10,000\ m^2$ area with spillage of acrylonitrile (AN). A spill simulation was carried out by the Council of Scientific & Industrial Research (CSIR). In field, under natural environmental conditions, acrylonitrile can be degraded slowly. Hence, field remediation was conducted using acrylonitrile degrading bacterial cultures and the soil was completely remediated in few days (Chakrabarti, 2020). The excessive contamination of metals like Cu, Cr, Co and Zn are found in Hazira, the industrial area of Surat (Krishna and Govil, 2007).

Geologically, the Gujarat state has a wide range of rock types from old ages. The geomorphic diversity of Gujarat reflects the influence of various geologic factors like mechanical and chemical weathering as well as various climatic factors (rainfall, wind and temperature). The long coastline from the Valsad district in the south to the Jamnagar district in southwest and Kutch, other alluvial plains starting from Aravalli hills to other regions of central and western Gujarat, large saline wasteland in Bhavnagar as well as saline deserts of Kutch and hilly regions of mainland Gujarat, all possess their own distinctive geomorphic characteristics.

Gujarat, located on the Tropic of Cancer, falls in the sub-tropical climatic zone and a large part of the state lies between 35°C and 50°C isotherms. The overall rainfall in the state is moderate to high. The state is located in between the heavy monsoon areas of Maharashtra and the very dry area of Rajasthan. Due to that, the climatic conditions are greatly varying in the state.

In accordance with various geological and environmental factors, the soils of Gujarat are divided into five categories, namely (i) Entisols, (ii) Inceptisols, (iii) Vertisols, (iv) Aridisols and (v) Alfisols (Merh, 1995).

Entisols have developed over the traps, granite, gneiss, quartzite and alluvium and are well distributed in Saurashtra, North Gujarat and in parts of Kutch and mainland Gujarat. Under the tropical semi-arid climate and low to moderate precipitation, the rocks get weathered and formed the aforementioned area with sandy clay, loam or clay loam to clay, which are structurally weak and mainly found in subangular particles. These soils are calcareous and alkaline.

Inceptisols have formed over basaltic, granitic, gneissic and alluvial parents. They occur on gentle to moderate and steep pediments, valley bottoms and moderately sloping interfluves in the parts of Sabarkantha and Panchmahal districts, Rajkot, Surendranagar, Jamnagar and Bhavnagar districts of Saurashtra, north Mehsana, southeastern Banaskantha as well as the southern part of Ahmedabad and Central Gujarat. Under the tropical semi-arid to humid climates with moderate to high precipitation, the rocks get weathered and formed silty-loam to clay soils which are neutral to alkaline in nature.

Vertisols have formed over granitic, basaltic and gneissic parents. They occur in the districts of Bharuch, Surat and Valsad in South Gujarat, Mehsana and eastern parts of Ahmedabad districts, northern Kheda and Vadodara districts in Central Gujarat as well as Bhal and Ghed tracts of Saurashtra. Under the semi-arid, sub-humid and humid climate with moderate to high rainfall, the rocks get weathered into black cotton soil. Morphologically, they are confined to uplands, piedmont plains, flood plains and intervening valleys.

Aridisols have mainly developed over the Aeolian silts and dune sands and are distributed on residual hummocky dunes and ridges, pediment surfaces, mudflats and dissected flood plains of Kutch, North Gujarat and in plains of Central Gujarat. They develop under an arid climate with low to moderate precipitation and get weathered into sandy loam with silty clay loam.

Alfisols are usually found in the central Banaskantha and northeastern parts of Surendranagar districts. They have developed mainly over the sandstone and at places over alluvial deposits. These soils mainly occur on gently sloping pediments in the warm semi-arid to sub-humid regions with moderate rainfall.

12.8   CASE STUDIES AND FIELD TESTS

The type and characteristics of soils vary within limited distances. The most common practice to identify the soil stratigraphy is by collecting the disturbed and undisturbed soil samples from borings, test pits, trenches, open cuts and so on and testing them in a laboratory. The standard penetration test (SPT), cone penetration test, dynamic cone penetration test are the field tests used to get the site soil profile. Specifically, at the dam sites, sometimes for the identification of variation in soil strata in the horizontal as well as the vertical direction, the electrical resistivity method is being used. For the construction of foundation at a shallow depth at some particular site and to identify the mud lines, the sonic method is being used. Before the construction of important heavy structures like the Sardar Sarovar dam and the Statue of Unity, non-invasive techniques like the multichannel analysis of surface waves (MASW) test is used for geotechnical characterisation of near-surface materials. Figure 12.17 shows the MASW testing conducted by authors at the Dahej site.

12.9   CONCLUDING REMARKS

The details about soils and rocks, including various soil improvement techniques, used in Gujarat are discussed in this chapter, which would be helpful to the geotechnical engineers. Based on these details, the key points about soils and rocks of Gujarat are mentioned as follows:

Figure 12.17 MASW testing at Dahej.

- The significant land of Gujarat comprises clayey soils with montmorillonite, kaolinite and illite clay minerals. Nearby regions of the Narmada River consist of alluvial soils whereas a major part of Kutch is covered with silty sand. The significant region of Saurashtra is also covered with clayey soil with some percentage of silt and sand.
- A significant part of Gujarat is formed from the weathering of granite, gneissic and basaltic rocks. The igneous rocks like pierite basalt, rhyolite, obsidian, granophyre, felsite, aplite and diorite are found in the South Gujarat region. Saurashtra, Kutch and some parts of the central and north regions comprise sedimentary rocks like limestone.
- The North Gujarat region comprises loamy sand and sandy clay as its major portion; the sandy clay to clay and clay loam to clay originated from the weathering of granite, basalt and gneissic rocks are found in the Saurashtra and Kutch region.
- The soils in Central Gujarat are dark to light grey, yellowish-red, reddish-brown to dark reddish-brown in colour and classified as high to medium plasticity clay soils with some coarser sand.
- The South Gujarat region consists of clayey to clayey loam soils. Generally, this type of soil is deep black in colour and comprises montmorillonite minerals. At some places, they are found in dark brown colour with a mixture of kaolinite and montmorillonite minerals.
- Although the Gujarat state is under seismic zones III, IV and V, as well as significant lands of Gujarat have problematic clayey soil as its upper strata, with the help of great geotechnical engineering knowledge and practice, the large structures like the Statue of Unity, the world's tallest statue, Sardar Sarovar Dam, which serves water and electricity to the four states of India, Narmada canal, the world's longest irrigation line canal and so on have been successfully constructed and survived all-natural disasters.

ACKNOWLEDGEMENTS

The generous help of Mr Gaurav Goel, Senior Engineer at the Grimtech Projects India Pvt. Ltd., Surat; KCT consultancy, Ahmedabad; M. K. Soil testing laboratory, Ahmedabad; Ahmedabad Municipal corporation; Ahmedabad Urban Development Authority; Geoengineering Services, Vadodara; Dr Jignesh B. Patel, Assistant Professor, Geotechnical Divison, Civil Engineering Department, SVNIT, Surat; Dr Tejas Thaker, Head of Civil Engineering Department, PDEU, Gandhinagar; Dr Alpesh Pandya, Associate Professor, Government Engineering College, Bhavnagar, Mayuresh Bakare, Former MTech Student, SVNIT Surat and Mrs Payal Mehta, PhD research scholar, PDEU, Gandhinagar was very much appreciated.

REFERENCES

Bakare, M. (2020). Former MTech student, SVNIT Surat, India. Personal communication.
Chakrbarti, T. (2020). Remediation of contaminated sites: Two case studies. *Proceedings of Geoenvironment-2020, An International Seminar on Contaminated Sites, Geo-2020*, 17–19 February 2020, Geotechnical and Geoenvironmental Group, Civil Engineering Department, IIT Delhi, p. 22.

Dave, M. (2017). Analysis of Existing Subsurface Exploration Data for Assessment of Lique-faction Potential for Ahmedabad City, M.Tech. dissertation, Dharamsingh Desai University, Nadiad, Gujarat.

Dave, S. S. (1971). The geology of the igneous complex of the Barda hills, Saurashtra, Gujarat state (India). *Bulletin of Volcanology*, Vol. 35, pp. 619–632, doi:10.1007/BF02596832.

EERI (2011). Earthquake Reconnaissance Report, Bhuj, India Republic Day, January 26, 2001, Earthquake Engineering Research Institute.

Ghan, S. (2021). Geological and Geotechnical Investigations and Interpretations Thereof for Statue of Unity Foundation. In: Patel, S., Solanki, C.H., Reddy, K.R., and Shukla, S.K. (eds). *Indian Geotechnical Conference 2019*. Lecture Notes in Civil Engineering, vol. 140. Springer, Singapore. https://doi.org/10.1007/978-981-33-6590-2_10

GMRC (2016). Gujarat Metro Rail Corportaion Ltd, https://www.gujaratmetrorail.com/, accessed 28 January 2021.

IS 1893 (2016). Criteria for Earthquake Resistant Design of Structures, Part I: General Provisions for Buildings. Bureau of Indian Standards, New Delhi.

Krishna, A. K. and Govil, P. K. (2006). Soil contamination due to heavy metals from an industrial Area of Surat, Gujarat, Western India. *Environmental Monitoring and Assessment*, vol. 124, No. 1–3, pp. 263–275. doi: 10.1007/s10661-006-9224-7.

Maps of India (2019). District Map of Gujarat State, https://www.mapsofindia.com/maps/gujarat/gujarat.htm, accessed 25 January 2021.

Mehta, P. and Thaker, T.P. (2020). Seismic hazard analysis of Vadodara Region, Gujarat, India: Probabilistic & deterministic approach. *Journal of Earthquake Engineering*. doi: 10.1080/13632469.2020.1724212.

Mehta, P. (2021). Seismic Hazard Assessment of Vadodara Region, PhD. thesis, Pandit Deendayal Petroleum University, Gandhinagar.

Merh, S. (1995). Geology of Gujarat. Geological Society of India. ISBN: 81-85867-14-3.

Patel, M.S. and Thaker, T.P. (2020). Examination of present subsurface investigation data for valuation of liquefaction potential for Ahmadabad city by means of SPT-N value. In: Prashant, A., Sachan, A., and Desai, C. (eds) *Advances in Computer Methods and Geomechanics*. Lecture Notes in Civil Engineering, Vol. 56. Springer, Singapore. doi: 10.1007/978-981-15-0890-5_7

Patel, M., Thaker, T. and Solanki, C. (2021a). Examination and appraisal of liquefaction vulnerability between Idriss-Bolulanger method and Andrus-Stroke method. In: Sitharam T.G., Dinesh S.V., and Jakka R. (eds) *Soil Dynamics*. Lecture Notes in Civil Engineering, Vol. 119. Springer, Springer. doi: 10.1007/978-981-33-4001-5_21

Patel, M., Solanki, C. and Thaker, T. (2021b). Deterministic seismic hazard analysis of Ankleshwar city, Gujarat. In: Hazarika, H., Madabhushi, G.S.P., Yasuhara, K., and Bergado, D.T. (eds) *Advances in Sustainable Construction and Resource Management*. Lecture Notes in Civil Engineering, Vol. 144. Springer, Singapore. doi: 10.1007/978-981-16-0077-7_57

Prakash, I., Gupta, K., Dey, A., Singh, P., Dhote, P., Bahseer, H., Dalia A. and Modi, V. (2012). *Geology and Mineral Resources of Gujarat, Daman and DiU*. Geological Survey of India, ISBN: ISSN:0579-4706.

Singh, R., Roy, D. and Jain, S. (2005). Analysis of earth dams affected by the 2001 Bhuj Earthquake. *Engineering Geology*, Vol. 80, No. 3–4, pp. 282–291, doi: 10.1016/j.enggeo.2005.06.002.

Thaker, T. (2012). Seismic Hazard Analysis and Microzonation Studies for Surat City and Surrounding Region, PhD. thesis, Indian Institute of Technology Delhi, New Delhi, India.

Vipin, K.S., Sitharam, T.G. and Kolathayar, S. (2013). Assessment of seismic hazard and liquefaction potential of Gujarat based on probabilistic approaches. *Natural Hazards*, Vol. 65, pp. 1179–1195. doi: 10.1007/s11069-012-0140-6

Chapter 13

Haryana

Ashwani Jain
National Institute of Technology Kurukshetra

Nitish Puri
AECOM India Private Limited

CONTENTS

DOI: 10.1201/9781003177159-13

13.1   INTRODUCTION

Haryana is an agrarian state located between 27°39′ and 30°35′ N latitude and 74°28′ and 77°36′ E longitude. It was created in 1966 and is surrounded by the states of Himachal Pradesh, Uttarakhand, Uttar Pradesh, Delhi, Rajasthan and Punjab. It is a moderate-size state having an area of 44,212 km^2 and ranks 22nd in terms of area in the country. It has six administrative divisions, 22 districts, 72 sub-divisions, 93 revenue tehsils, 50 sub-tehsils, 140 community development blocks, 154 cities and towns, 6,848 villages and 6,222 village panchayats. Faridabad is the most populous city of the state and is covered under the National Capital Region (NCR). Gurugram is a leading financial hub of the NCR with major Fortune 500 companies located in it. The state experiences rapid infrastructural development. Haryana has a large network of roadways, railways and irrigation canals with many bridges and flyovers. A nuclear power plant is planned to be constructed in the Fatehabad district. The chapter details the main types of soils, rocks and waste materials found in the state, and assesses water table conditions and seismic hazards. In the end, a few case studies focussing on the geotechnical aspects have also been discussed.

13.2   GEOLOGY OF THE STATE

The soils and rocks of Haryana can be divided into the following three geological systems (Sachdev et al., 1995):

i.   Aravalli system – These are the oldest formations present in the southwestern part of the state covering districts of Bhiwani, Charkhi Dadri, Mahendragarh and Gurugram and are composed of quartzites, quartzitic sandstone, mica schists, phyllites and crystalline limestone.
ii.   Siwalik system – This is in the northern parts of the Ambala district and is composed of sedimentary rocks, mainly sandstones and shales, with clays and boulders.
iii.   Indo-Gangetic alluvial plain – This system is formed by deposition of alluvial sediments between the Siwalik and Aravalli system and consists of sands, silts, clays and occasional gravel beds.

The major area of the state is a part of the Indo-Gangetic alluvial plain. Except for the river Yamuna, flowing along the eastern boundary of the state, the only other stream is Ghagghar. Some important tributaries of these rivers are Markanda, Saraswati, Chautang and Tangri. Other seasonal streams, Sahibi, Dohan and Krishnawati, originate from Aravalli ridges and flow from south to north.

The wind-blown sand deposits are found in the form of sandy plains and sand dunes over alluvial deposits in parts of Bhiwani, Charkhi Dadri, Hisar and Sirsa districts. Figure 13.1 shows desert sand at the outskirts of Charkhi Dadri.

The age of the soil deposit is an important geomorphological unit. The Geological Survey of India (GSI, 1973) has developed a map describing the age of soil deposits in Haryana. Most soil deposits in Haryana belong to the Holocene and Pleistocene age groups.

Figure 13.1 A farmer ploughing desert sand at the outskirts of Charkhi Dadri with *Kapoori* Hillock (Aravalli System) on the background.

An assessment of groundwater conditions in the state has been done using groundwater data collected by the Central Ground Water Board, Chandigarh (CGWB, 2013). A groundwater table map has been developed using the nearest neighbour interpolation (reported elsewhere) (Puri and Jain, 2014). There are areas with shallow water tables, i.e. between 0 and 10 m, in the districts of Sonipat, Rohtak and Jhajjar and several other rural areas scattered across the state. Some areas in the districts of Ambala, Hisar and Mewat are facing the problem of water logging. The depth of groundwater table varies between 30 and 40 m in Gurugarm, Faridabad, Kurukshetra and in some areas of Bhiwani and Sirsa districts bordering the state of Rajasthan.

13.3 LOCAL SOIL

13.3.1 Development of geotechnical database

Geotechnical data have been collected from several government and private organisations to assess the groundwater table and soil conditions in the state. The developed database has information for distinct locations for boreholes up to 50 m depth covering various districts of Haryana. A map of the state, showing various districts and 133 representative borehole locations, has been presented in Figure 13.2. The type of soil encountered (IS:1498, 1970), SPT N-values and level of ground water table of a few boreholes in various districts have been reported in Table 13.1.

While designing and assessing geotechnical problems using the SPT data collected from different agencies, the knowledge of the equipment used and the method followed for conducting SPT is essential to accurately calculate the correction factors to be applied to the observed SPT N-values. Specifically, special care should be given to select an appropriate correction factor for borehole diameter, as different agencies often use boring equipment and casings of different dimensions. Moreover, correction for liners is also crucial, as many geotechnical investigation agencies avoid the use of liners, as it is a consumable item with a recurring cost.

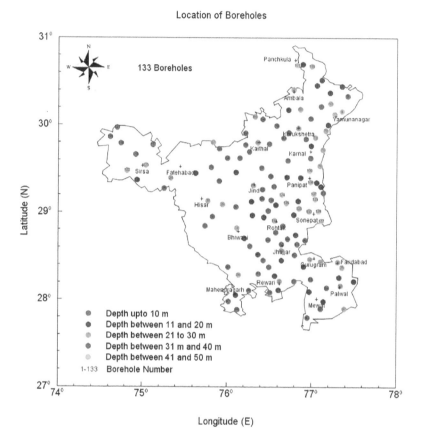

Figure 13.2 Map of the state of Haryana showing various districts and borehole locations.

13.3.2   Local subsoil conditions

In Ambala, Yamunanagar and Panchkula districts, most of the soil in substrata is sandy, mainly silty sand and poorly graded sand. Poorly graded gravel is also found along the foothills. Limestone of grey colour is often encountered in the Yamunanagar district in areas close to Siwalik formations at the boundary with the state of Himachal Pradesh.

Fine-grained soils are found in Kurukshetra, Kaithal and Jind districts. Low to medium plasticity clays are the predominant soils in Kaithal and Jind districts with silts as sub-dominant soil types. In the Kurukshetra district, largely, low plasticity and clayey silts are found in the subsurface.

In Sirsa district, the major soil type is silty sand, with low plasticity and clayey silt also encountered in some pockets.

In Fatehabad and Hisar districts, fine-grained soils are encountered. In Hisar, clay is the most common soil type. However, clayey sands are also found. In Fatehabad, the most common types of soil are low plasticity and clayey silts. Silty clay and low to medium plasticity clays are also found in some areas.

Table 13.1 Description of soil properties in various districts based on some selected boreholes[a]

Borehole no. as in Figure 13.2	Site name	District	Profile IS soil classification (IS:1498, 1970) (with depth in metres)	Profile depth (m)	SPT N-value	Water table (m)
1	Brada	Ambala	SM (0–4.5; 16.75–18.35; 23.5–26.5), SP-SM (4.5–10.5; 12–13; 15.25–16.75; 21.25–23.5), SP (10.5–12), CL (13–15.25; 18.35–21.35; 26.5–32)	32	7–46	25.9
2	Patti Shekhan	Ambala	SM (0–7.5; 17.5–21; 26.5–30), ML (7.5–13.5), ML-CL (21–26.5), CL (13.5–17.5)	30	11–51	0.9
5	Bilaspur	Yamunanagar	SM (0–3; 9–12), ML (3–9)	12	12–16	5.5
8	Buria	Yamunanagar	CL (0–2.4, 20.5–31.5), SP (2.4–8.5, 11.8–15.7; 34–39.8; 42.5–45), SP-SM (8.5–11.8; 39.8–42.5), ML (15.7–17.2), SM (17.2–20.5, 32–34)	45	27–92	2.9
12	Tosham	Bhiwani and Charkhi Dadri	ML (0–4.8), ML-CL (4.8–9.45)	9.5	6–14	1.8
14	Lajpat Nagar		CL (0–2.8; 5.2–6.6), ML (2.8–5.2; 11.5–12.6; 17.8–18.6), SM (6.6–11.5; 12.6–17.8; 18.6–25)	25	5–44	2.65
15	Charkhi Dadri		SC (0–35)	35	12–91	8.0
17	Daruhera	Rewari	ML (0–1.5; 3.75–4.2; 5.1–9), SM (1.5–3.75; 4.2–5.1)	10	7–12	NE
21	Bawal		ML-CL (0–0.7; 6–9.8), SM (0.7–2.5; 3.6–4.9), ML (2.5–3.6; 4.9–6; 9.8–20)	20	4–30	1.6
23	Asaoti	Faridabad	ML (0–3; 4.5–7.5), SM (3–4.5), SP-SM (7.5–20)	20	8–35	3.6
25	Sector 76		CL (0–1.2), M (1.2–3.3), SM (3.3–12.8), SP-SM (12.8–20.6), SP (20.6–30)	30	13–50	21.5
26	Daiyar Village	Fatehabad	ML (0–1.5; 6–7.5), CL-ML (1.5–6)	7.5	10–13	6.0
28	Bhattu Mandi		ML-CL (0–2.0), CL (2–5.9; 17.1–21.8), SM (5.9–17.1; 21.8–25)	25	9–42	3.8
31	Manesar	Gurugram	SM (0–2.5), ML (2.5–15)	15	8–53	NE
32	Basai Village		SM (0–4; 8–10; 15–25), ML-CL (4–8; 10–15)	25	4–23	2.4
34	Chamar Khera		ML (0–6)	6	6–12	3.75
39	Hansi	Hisar	CL (0–1.9; 12.6–16.4), ML (1.9–4; 16.4–21.6; 23–25), SM (4–12.6), ML-CL (21.6–23)	25	4–32	3.55
41	Bupnia, Shahpur		ML (0–6)	6	9–14	4.5
52	Site on Rohtak to Bawal Road (NH-71)	Jhajjar	ML (0–2.5; 12–18; 33–36), CI (2.5–4.5), SM (4.5–12; 18–33; 36–39)	40	9–64	3.25

(Continued)

Table 13.1(Continued) Description of soil properties in various districts based on some selected boreholes[a]

Borehole no. as in Figure 13.2	Site name	District	Profile IS soil classification (IS:1498, 1970) (with depth in metres)	depth (m)	SPT N-value	Water table (m)
53	Village Bhag Khera		ML (0–6)	6	8–12	3.25
60	Site on Kalayat-Sajuma-Narwana Road	Jind	SM (0–2; 7–8; 29–31), CL-ML (2–7), CL (8–10; 22–29), ML (10–22)	31	5–42	3.0
62	Village Kharkan Mundri Village	Kaithal	ML (0–4.5)	6	6–10	NE
67		Kaithal	CL-ML (0–4.5; 24–27), SM (4.5–8; 27–30.5), SP (8–20; 33–34), CL (20–24; 30.5–33)	34	6–46	21.0
70	Kurali Village		ML (0–12)	12	9–40	3.5
77	Site on NH-1	Karnal	SM (0–8.5; 14.5–16; 17–20; 22–23; 25–26.5; 29.5–31; 35–40), ML (8.5–14.5; 16–17; 20–22; 23–25; 32.5–33.5), SP (26–29.5; 31–32.5; 33.5–35)	40	6–27	NE
78	Village Babain		ML (0–1.5), SM (1.5–3), SP-SM (3–6)	6	2–13	NE
82	Site on Thol Road	Kurukshetra	ML (0–2.5), SP (2.5–16), CI (16–19), ML (19–30)	30	8–37	NE
84	Nangal Chowdhury	Mahendragarh HQ -	ML-CL (0–1.2; 2.5–6), ML (1.2–2.5)	6	8–10	NE
85	Site on Rewari Road	Narnaul	CL (0–2.8), SM (2.8–4.8; 6.6–15), ML (4.8–6.6)	15	11–38	NE
90	Ferozepur Zhirka	Mewat	ML (0–1.8), CL (1.8–10)	10	7–21	NE
92	Nuh Village Mankaula		CL (0–1.4; 4.9–7.2; 9.6–13.5), ML (1.4–2.5), SM (2.5–4.9; 7.2–9.6)	13.5	11–29	NE
94			ML (0–1), CL-ML (1–6)	6	2–6	1.5
95	Alwalpur	Palwal	CL (0–3; 4.5–7.5), ML-CL (3–4.5), ML (7.5–9; 9–10.5;10.5–12; 13.5–15), CI (12–13.5)	15	8–50	NE
99	Sector 31	Panchkula	ML (0–2.8), SP-SM (2.8–5.6), CI (5.6–11.6; 17–20), CL (11.6–17)	20	8–50	NE

(Continued)

Table 13.1 (Continued) Description of soil properties in various districts based on some selected boreholes[a]

Borehole no. as in Figure 13.2	Site name	District	Profile IS soil classification (IS:1498, 1970) (with depth in metres)	Profile depth (m)	SPT N-value	Water table (m)
102	Site on Panipat - Assandh Road	Panipat	ML (0–3), SM (3–12)	12	12–20	NE
107	PIET		SM (1–1.5; 10.5–18; 22.5–28.5; 37.5–45), SP (1.5–10.5), ML-CL (18–22.5; 28.5–37.5)	45	9–95	35.0
110	Meham	Rohtak	SM (0–3; 4.5–6; 12–15), ML (3–4.5; 6–12)	15	10–16	NE
115	Site on SH-9		ML-CL (0–3.0; 9–12), SM (3–9; 12–18; 21–40), CI (18–21)	40	11–67	0.9
116	Abubshehar	Sirsa	ML (0–6)	6	3–10	1.5
123	Court Colony		SM (0–4.45; 9.90–16.95; 18.90–30), ML-CL (4.45–9.90; 16.95–18.90)	30	10–75	NE
124	Site near Sunder Branch Canal	Sonipat	SM (0–2.8), CI (2.8–16.95), SP-SM (16.95–19.95; 23.8–40.45), CL (19.95–23.80)	41	8–82	3.0
133	Barwasni village		ML (0–6; 11–19; 23.6–25), ML-CL (6–11), CL (19–23.6)	24	3–42	7.65

[a] Geotechnical data obtained from Public Works Department (PWD) Haryana; Northern Railways (NR); Haryana Urban Development Authority (HUDA); Delhi Metro Rail Corporation (DMRC); Prof. R K Bansal, NIT Kurukshetra; Dr. Ghumman & Gupta Geotech Consultants, Chandigarh; Jindal Consortium, Ambala; Mananda Test House, Dera Bassi, Mohali and Magma Infrastructures Private Limited, Delhi.

Abbreviations:
GP: Poorly graded gravel
GM: Gravel with silty fines
GC: Gravel with clayey fines
SM: Sand with silty fines
SP: Poorly graded sand
SP-SM: Poorly graded sand with silty fines
SW-SM: Well-graded sand with silty fines
SW: Well-graded sand
SC: Sand with clayey fines
ML: Silt of low plasticity or non-plastic silt
ML-CL: Clayey silt
CL: Clay of low plasticity
CL-ML: Silty clay
CI: Clay of medium plasticity
MI: Silt of medium plasticity
NE: Not encountered
HQ: Headquarter

In Mahendragarh, Rewari, Gurgaon and Faridabad districts, silty sands are found predominantly, while non-plastic, low plastic and clayey silts are also present in sub-layers. In Faridabad and Gurgaon districts, weathered grey colour fractured quartzite is often met near the depth of refusal, which is normally between 15 and 30 m.

In Mewat and Palwal districts, the substrata consist of silty and low to medium plastic clays, while non-plastic, low plastic and clayey silts are the other common types of soils.

In Rohtak and Bhiwani districts, fine-grained soils are found. In Bhiwani, low plasticity clays are predominantly found, while other soil types are non-plastic, low plastic and clayey silts, and clayey sands. In Rohtak, low plastic and clayey silts are predominantly found in substrata, while at some places, medium plastic clays and silty sands are also present. In the Charkhi Dadri district, clayey sand is the most common type of soil.

In Karnal, Panipat and Sonipat districts, the subsurface mainly consists of poorly graded, well-graded and silty sands. However, in Panipat, low plastic and clayey silt, and in Sonipat, silty and low plastic clays are also encountered.

13.3.3   Properties of Yamuna sands

Yamuna sand is a locally available fluvial deposit, grey in colour, commonly found along the banks of river Yamuna. It is used extensively as an aggregate in construction as well as a fill material. Its morphological characteristics (shape and surface) indicate that the particles are solid without any intraparticle voids (Jakka et al., 2010) and are angular (De and Basudhar, 2008). It comprises quartz (40%), feldspar (40%), mica (1%–2%) and carbonate (18%) (Rahim, 1989).

Five typical sandy soils from the region have been procured from the river Yamuna and its branches, Markanda, Tangri and Indri, flowing through various parts of Haryana: Yamuna fine sand from Kamalpur Gadria (Karnal), the Yamuna coarse from Khizrabad (Yamunanagar), Markanda sand from Shahabad Markanda (Kurukshetra), Tangri sand from Shahpur (Ambala) and Indri sand from Indri (Karnal). The grain size distribution curves of the procured sands have been plotted in Figure 13.3. It has been observed that the Yamuna fine sand, Markanda sand, Tangri sand and Indri sand have almost similar grain size distribution curves. This can be attributed to the fact that all the rivers originate from the Siwalik system. All the soils have been classified as poorly graded sands (*SP*), as per the Indian Standard Soil Classification System (IS:1498, 1970). The fines content is less than 5% and therefore, these can be classified as clean sands. The limiting gradation curves for most liquefiable zone and potentially liquefiable zone as given by Tsuchida (1970) have also been shown in Figure 13.3. It has been observed that except for Yamuna coarse sand, the grain size distribution curves of the soils fall in the potentially liquefiable zone.

13.4   MATERIALS/WASTE MATERIALS FOR GEOTECHNICAL REUSE

Badarpur sand is quarry-based sand, manufactured by crushing weathered quartzite rock, from which fines are removed by continuous washing. It is widely used as a construction material in Delhi-NCR. Its chief constituent mineral is quartz (98%). It is red

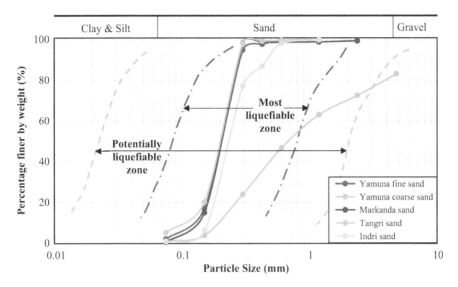

Figure 13.3 Plot of gradation curves of the procured sands with limiting gradation curves by Tsuchida (1970).

in colour due to the coating of ferric oxide and ferrosoferric oxide on quartz particles. It is uniformly graded medium sand with sub-angular particles. Its angle of internal friction (ϕ) varies from 27.6° to 42.5° as the dry unit weight increases from 1.23 to 1.67 g/cc (Rao and Dutta, 2006; Pothal and Rao, 2008).

Some of the other materials/waste materials available in Haryana with a potential for geotechnical reuse are rice husk ash, fly ash, bottom coal ash, stone dust, burnt press mud, bagasse ash, marble dust, mild steel sludge, granulated tyre rubber and so on.

The rice husk ash, fly ash, bottom coal ash, stone dust, press mud and bagasse ash have been used as raw materials in bricks, as subgrade materials or as additives in the stabilisation of problematic soils (e.g. rice husk ash (Rahman, 1987; Ali et al., 1992; Muntohar and Hashim, 2002; Basha et al., 2005; Jha and Gill, 2006), fly ash (Wright and Ray, 1957; Chu et al., 1975; Cokca, 2001), bottom coal ash (Rogbeck and Knutz, 1996; Schreurs et al., 2000; Singh and Siddique, 2013), press mud (Dass and Malhotra, 1990; Azme et al., 2019) and bagasse ash (Faria et al., 1990; Osinubi et al., 2009)). Stone dust and bottom coal ash are being widely used as a replacement for coarse sand due to their comparable grain size distribution curves (Figure 13.4). A good amount of fly ash is now being also utilised in the cement industry.

Materials like marble dust, mild steel sludge and granulated tyre rubber have vast applications in civil engineering (e.g. marble dust (Okagbue and Onyeobi, 1999; Aliabdo et al., 2014; Hamed et al., 2014), mild steel sludge (Das et al., 2007) and granulated tyre rubber (Patil et al., 2011; Anvari et al., 2017)). Granulated rubber is also being used for seismic isolation of structures in earthquake-prone areas (Tsang et al., 2012; Bandyopadhyay et al., 2015).

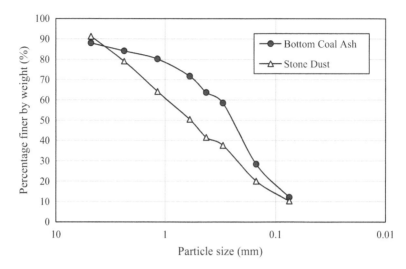

Figure 13.4 Particle size distribution of stone dust procured from stone crushers in Yamu-nanagar and bottom coal ash from Panipat Thermal Power Plant, Haryana.

The physical properties of common waste materials found in Haryana having potential for geotechnical reuse have been reported in Table 13.2.

13.5 EARTHQUAKE HAZARD

13.5.1 Earthquake damage report of Haryana

The state of Haryana is situated at the foothills of Himalayas to the south of Himalayan Frontal Thrust (HFT), and its geographical position makes it susceptible to huge damage due to earthquakes occurring in the Himalayan Thrust System. Moreover, the alluvial land cover also makes it prone to hazards due to wave amplification and soil liquefaction. The gradual increase in population density has also increased the seismic vulnerability of the state. Apart from this, there are a number of local lineaments, faults and tectonic features in and around Haryana, e.g. the Sargodha–Lahore–Delhi ridge on the northeastern side and the Aravalli–Delhi fold belt on the south. However, no major earthquake has been attributed to the Sargodha–Lahore–Delhi ridge.

The earthquake of maximum observed magnitude of 6.5 occurred on July 15, 1720, near Mathura, and the event was associated with the Mathura fault. Other notable earthquake events that have occurred in the region are M 6.0 near Faridabad on August 27, 1960, M_L 4.2 near Charkhi Dadri on February 28, 2001 and M 4.9 Bahadurgarh (Haryana-Delhi Border) on March 5, 2012. The 2012 M 4.9 Bahadurgarh earthquake was a non-Himalayan earthquake event that caused maximum intensity in Jhajjar and Rohtak districts (Gupta et al., 2013) and resulted in causalities, injuries and structural damage in these areas. The event has been assigned MMI values of V and VI, corresponding to strongly felt, and development of crack and wall collapse respectively, for the affected areas.

Table 13.2 Physical properties of common waste materials found in Haryana having potential for geotechnical reuse

Waste materials	Source location	Grain size distribution			Indian standard classification	Specific gravity (G_s)	Plasticity index (PI)	Colour
		Gravel (%)	Sand (%)	Silt + clay (%)				
Rice husk ash	Food Processing Unit, Grand Trunk Road, Murthal	0	23.22	76.78	ML	1.95	Non-plastic	-
Bottom coal ash	Thermal Power Plant, Panipat	-	-	-	-	-	Non-plastic	-
Fly ash	Thermal Power Plant, Panipat	0	7.5	92.5	ML	2.09	Non-plastic	-
Stone dust	Stone crushers, Yamunanagar	8.67	79.81	11.49	SM	2.64	Non-plastic	-
Burnt press mud	Sugar Mill, Yamunanagar	-	-	-	-	-	-	Blackish Grey
Bagasse ash	Sugar Mill, Shahabad Markanda, Kurukshetra	-	-	-	-	1.24	Non-plastic	Dark Black
Marble dust	Local Marble Cutting and Polishing Industry, Kurukshetra	-	-	-	-	2.69	Non-plastic	-
Mild steel sludge	-	-	-	-	-	7.86	Non-plastic	-
Granulated tyre rubber	Locally procured	Particle size between 1 and 10 mm			-	0.9–1.10	Non-plastic	-

13.5.2 Earthquake hazard assessment

Analysis of seismic hazard has been carried out for the state of Haryana using a probabilistic approach (Puri and Jain, 2019). The tectonics of the seismic study region has been comprehensively examined, and an earthquake catalogue has been developed based on the data from both historic and instrumental periods. Based on the available tectonic information, a tectonic map for the seismic study region has been developed (Figure 13.5), and active tectonic features have been identified by superimposing epicentres of earthquakes from the earthquake catalogue on the tectonic map. The maximum magnitude potential of the seismogenic sources has been calculated based on the total fault length, rupture length (RL) and maximum observed magnitude (M_{obs}) (Table 13.3).

Seismicity parameters have also been obtained in order to calculate the return period corresponding to the expected earthquake magnitude (Table 13.4). The seismicity parameters show that the Himalayan Thrust System can generate bigger earthquakes at lower return periods as compared to the Aravalli–Delhi Fold Belt and Sargodha–Lahore–Delhi ridge. The threat of a major earthquake in this region is imminent.

13.5.2.1 Peak ground acceleration (PGA)

The PGA(g) values assuming rock outcrop have been calculated for return periods of 475, 2,475 and 4,975 years with 10%, 2% and 1% probability of exceedance, respectively, for a time frame of 50 years. The PGAs for the state have been calculated at the

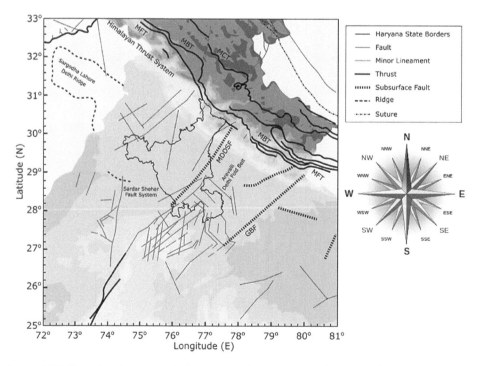

Figure 13.5 Map showing tectonic features in the seismic study region. (Adapted from Puri and Jain, 2019.)

Table 13.3 Maximum magnitude potential (M_{max}) values for potential seismogenic sources

Fault name	Fault code	M_{obs}	RL (km)	Maximum M_{max}
Delhi Fold Belt	DFB	6.7	-	7.2
Great Boundary Fault	GBF	5.5	105.33	7.55
Jawala Mukhi Thrust	JMT	5.5	96.67	7.48
Mahendragarh–Dehradun Subsurface Fault	MDSSF	5.4	99.00	7.5
Main Boundary Thrust	MBT	8	275.00	8.5
Main Crustral Thrust	MCT	7.3	256.33	7.8
Mathura Fault	MF	6.8	39.00	7.3
Main Frontal Thrust (near Saharanpur)	MFT2	5.5	15.33	6.15
Moradabad Fault	MDF	5.6	54.00	7.28
Ropar Fault	RF	4.1	12.67	6.83
Rohtak–Dehradun Lineament	RDL	5	67.33	5.5
Sardar Shehar Fault	SSF	7.1	90.33	7.6
Sargodha–Lahore–Delhi Ridge	SLDR	6.5	-	7

Source: Adapted from Puri and Jain (2019).

Table 13.4 Seismicity parameters for different area sources

Area source	b	a	Range of magnitude (M_w)
Himalayan Thrust System	0.75	3.8	4.0–8.0
Aravalli–Delhi Fold Belt	0.69	2.61	4.0–7.0
Sargodha–Lahore–Delhi Ridge	0.85	3.27	4.0–6.5

Source: Adapted from Puri and Jain (2019).

grid points, and contour maps have been drawn corresponding to each return period. Figure 13.6 shows a typical PGA map of Haryana for a 10% probability of exceedance in 50 years (return period of 475 years). The PGA_{rock} ranging from 0.05 to 0.35 g has been observed for a return period of 475 years with a 10% probability of exceedance in 50 years. For a return period of 2,475 years, it ranges from 0.1 to 0.6 g with a 2% probability of exceedance in 50 years, while for a return period of 4,975 years, it ranges from 0.1 to 0.7 g with a 1% probability of exceedance in 50 years.

It has been observed that at 10%, 2% and 1% probability of exceedance in 50 years, the north and northeastern parts of Haryana, which include districts of Panchkula, Ambala, Yamunanagar and some parts of Kurukshetra, are prone to very high ground motions during earthquakes with severe-most hazard reaching up to 0.7 g in terms of PGA. This is attributed to the proximity of this region to the Himalayan Thrust System.

For a 10% probability of exceedance in 50 years, the rest of the area in Haryana is prone to low seismic hazard. For a 2% probability of exceedance in 50 years, for the Jind district, the lowest PGA values have been observed. The rest of the area of the state except north and northeastern parts has been observed to be susceptible to moderate earthquake ground motions. It has been observed that at a 1% probability of exceedance in 50 years, the northwestern part is prone to low to moderate earthquakes and the southern part of the state is prone to severe hazard due to its proximity to the Aravalli–Delhi fold belt.

For further analysis, based on the PGA observed at 10% probability of exceedance in 50 years for a return period of 475 years, the study area has been geographically delineated into 7 zones each having different earthquake hazard potentials (Figure 13.6). Even though the PGA values in some areas range from low to moderate, the motions could get amplified due to local site effects.

13.5.2.2   Response spectrum

Response spectra have also been developed for the state to assist structural engineers and designers in better designing earthquake-resistant structures. The response spectra at rock outcrop corresponding to severe-most hazard at 10% probability of exceedance in 50 years, for various zones of the study area, have been plotted along with response spectra for rock sites for seismic zones II, III and IV as per IS:1893-Part 1 (2016). Figure 13.7 shows the typical response spectrum at rock outcrop for geographical Zone G. It has been observed that in Zones A, B and C, short-period spectral accelerations

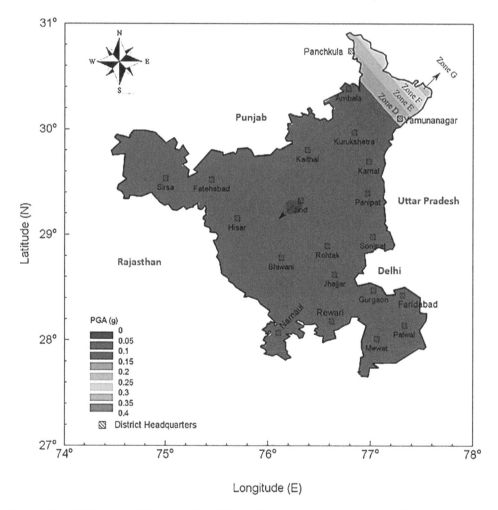

Figure 13.6 PGA map of Haryana for 10% probability of exceedance in 50 years (return period of 475 years). (Adapted from Puri and Jain, 2019.)

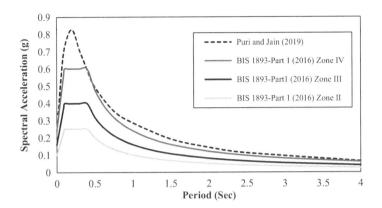

Figure 13.7 Response spectrum at rock outcrop for Zone G vs. code response spectra at rock site. (Adapted from Puri and Jain, 2019.)

are low, reaching up to 0.35 g, suggesting that the structures having short periods would be safe during earthquakes. However, in the rest of the zones, Zone D to Zone G, the short period spectral accelerations are moderate to high, reaching up to 0.83 g, implying that the short period structures would be severely affected. Furthermore, it has been observed that spectral acceleration at longer periods ($\geq$1 s), for Zones A, B, C and D, is very low. For Zones E–G, it is low to moderate, reaching up to 0.29 g, and structures with longer periods of low frequency would be quite safe. However, the old structures designed without any seismic considerations may sustain damage even in low-risk zones.

It has also been observed that IS:1893-Part 1 (2016) underestimates the hazard in earthquake Zones F and G. However, the response spectrum obtained for Zone E of the study is comparable to the response spectrum of Zone IV given by the Indian seismic code. The shape of response spectra obtained for the study is quite like that of the Indian seismic code.

13.6 CYCLIC BEHAVIOUR OF YAMUNA SANDS

13.6.1 Shear modulus degradation (G/G_{max}-γ) and damping ratio (D-γ) curves

The region-specific shear modulus degradation and damping ratio curves for geological deposits from the riverbed of Yamuna, originating from the Himalayan seismic zone of North India, have been developed (Puri et al., 2020) by conducting resonant column tests (ASTM D4015, 2015) and cyclic simple shear tests (Nikitas et al., 2017; Cui et al., 2017). The G/G_{max}-γ and D-γ curves proposed by Seed and Idriss (1970) and Darendeli (2001) are widely accepted for use in ground response studies. The comparison of experimental data of the study with the standard curves proposed by Seed and Idriss (1970) and Darendeli (2001) has revealed a significant deviation in trend for both the soils. Hence, site-specific parameters for the hyperbolic model have been evaluated for the soils used in the study, and for that, regression analysis on the experimental data has been carried out to identify the best-fit curves.

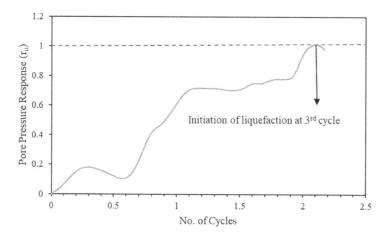

Figure 13.8 Pore pressure response (r_u) of Yamuna fine sand at 35% relative density for a CSR value of 0.2.

A general increase in the number of cycles to cause liquefaction has been observed with an increase in the relative density of the sample. This can be attributed to that for dense sands, the state of initial liquefaction does not produce large deformations due to increased stiffness and the dilative tendency of sand upon reversal of the shear stress. The chances of liquefaction and excessive settlement are, therefore, reduced with increased relative density.

For the range of parameters of the experimental programme, it has been observed that at a particular relative density, with the increase in CSR, cycles required to initiate liquefaction decrease. It has also been observed that for higher CSR values of 0.2 and 0.3, Yamuna fine sand is liquefiable at all the relative densities considered in the testing programme. Hence, Yamuna fine sand can undergo liquefaction during moderate to high earthquakes.

13.7   CASE STUDIES

13.7.1   Deformation behaviour of limestone under triaxial compression

With the rapid advancement in the construction of large structures, it has become necessary to make a detailed study of the shear strength of rocks under various conditions of loading. Investigations have been conducted to observe the behaviour of rocks under high confining pressures (Soni and Jain, 2006).

13.7.1.1   Test procedure

The rock selected for the present study is limestone of light grey colour of Siwalik formations collected from Kala Amb, a town at the border of the states of Haryana and Himachal Pradesh. A series of triaxial tests have been conducted on identical dry samples of limestone under confining pressures ranging up to 35.154 MPa at a strain rate of 0.0105 cm/min. All the specimens were obtained from a single slab to avoid variation in strength from specimen to specimen. Rock specimens were prepared finally for dimensions of 7.62 cm length and 3.81 cm diameter. The procedure included drilling of rock cores and their cutting and grinding to the final shape.

13.7.1.2   Test results

The unconfined compressive strength (σ_{ci}) of the sample was determined in the laboratory to be 47.716 MPa. The axial strain at failure ranges between 4.5% and 6.7% at low confining pressures, and between 9.5% and 11.8% at higher confining pressures. Secant modulus of elasticity under different confining pressures has been calculated for the linear portion of deviator stress versus axial strain curve. Secant modulus of elasticity and ultimate load increase with the increase in confining pressure. The values of m_i, as per the Hoek and Brown criterion (1980), have been calculated from the experimental observations. The values of m_i vary widely from 4.63 to 21.90, whereas the suggested value of m_i for such rocks is 7. Figure 13.9 shows a curve between deviator stress ($\sigma_1 - \sigma_3$) and confining pressure (σ_3). The plot is parabolic as suggested by Singh and Singh (2003), though the brittle–ductile transition is not reached, as in the present investigation, confining pressure has not been increased to a level so that it becomes equal to unconfined compressive strength. The values of parameter 'A' as suggested by Singh and Singh (2003) have also been calculated at various confining pressures. On the log-log plot between $(\sigma_1 - \sigma_3)/\sigma_3$ and σ_{ci}/σ_3 showing the proposed criterion for

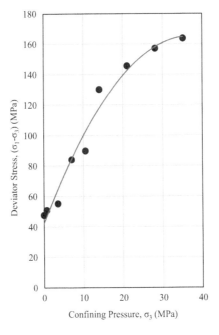

Figure 13.9 Deviator stress vs. confining pressure curve for limestone. (Adapted from Soni and Jain, 2006.)

chemical rocks (Ramamurthy, 1985), corresponding values obtained from the study have been plotted (reported elsewhere) (Soni and Jain, 2006). The criteria suggested by Ramamurthy (1985) and Singh and Singh (2003) for the strength of intact rocks find better agreement with the test data as compared to Hoek and Brown (1980) criterion.

13.7.2 Development of geotechnical correlations

In most of the developed countries, geotechnical explorations are done with caution, and data are kept well organised and readily available for research purposes. As geotechnical data were being collected from different sources for the study, it was noticed that there was a lot of missing data in some geotechnical reports, though the number of good reports was high. To overcome this difficulty of missing data, various machine learning techniques have been reviewed to develop region-specific correlations for different geotechnical parameters. Statistical correlations for in-place density using SPT N-value, compression index (C_c) using the liquid limit (LL) and void ratio (e), and cohesion (c) and angle of internal friction (ϕ) using SPT N-value developed by Puri et al. (2018) have been reported in Table 13.5.

The developed correlations for the determination of C_c using LL have been compared with the correlations developed by Skempton (1944) and Terzaghi and Peck (1967) (Puri et al., 2018). It has been observed that the modelled values of C_c are on the lower side of the Terzaghi and Peck (1967) and in close agreement with Skempton (1944). The correlations developed for the determination of C_c using void ratio have been compared with the correlations developed by Cozzolino (1961), Azzouz et al. (1976) and Kalantary and Kordnaeij (2012). It has been observed that the calculated values of C_c obtained using the correlations developed in the study are on the higher side of the values obtained by Azzouz et al. (1976), but closer to the values obtained by Cozzolino (1961). The values calculated using Kalantary and Kordnaeij (2012) are close to the values obtained using the developed correlations when the void ratio is less than 0.8. For the void ratio greater than 0.8, the values modelled by the present study are on the higher side (Puri et al., 2018).

A comparison of the developed correlation predicting angle of internal friction has been done with studies carried out by Shioi and Fukui (1982) and Wolff (1989). It has been observed that the correlation for the angle of internal friction developed in the study gives values very near to both the earlier studies (Puri et al., 2018). Also, the comparison has been done for the equation predicting cohesion with the studies carried out by Bowles (1982) and Kumar et al. (2016). It has been observed that the values of cohesion obtained in the study are on the lower side than those obtained from these studies. The proposed correlations are region-specific, and hence should only be used for sites in the state of Haryana and nearby areas.

13.7.3 Geotechnical failure – A case study of the forensic geotechnical investigation of building distress

The case study focuses on the forensic geotechnical investigation carried out for the SAC (Student Activity Centre) building constructed 33 years ago and located in the NIT Kurukshetra campus (Kumar and Jain, 2019). The possible causes of the distress in the building were explored with emphasis on geotechnical aspects. However, when engaging in forensic geotechnical investigations, the possibility of failure due to structural aspects should also be considered.

Table 13.5 Empirical relationships for various geotechnical parameters of soils of Haryana

S. No.	Geotechnical parameter	Soil type	Equation	Units	R^2	Remarks
1.	Bulk density (ρ_b)	Coarse grained	$\rho_b = 0.0096 \times N + 1.50$	g/cm^3	0.95	SPT N-value ranging from 1 to 39
			$\rho_b = 0.0141 \times N + 1.37$	g/cm^3	0.95	SPT N-value ranging from 40 to 50
2.	Dry density (ρ_d)	Coarse grained	$\rho_d = 0.0068 \times N + 1.56$	g/cm^3	0.96	SPT N-value ranging from 1 to 50
3.	Bulk density (ρ_b)	Fine grained	$\rho_b = 0.0080 \times N + 1.72$	g/cm^3	0.97	SPT N-value ranging from 1 to 50
4.	Dry density (ρ_d)	Fine grained	$\rho_d = 0.0114 \times N + 1.25$	g/cm^3	0.90	SPT N-value ranging from 1 to 50
5.	Compression index (C_c)	Clays	$C_c = (0.0092 \times LL) - 0.1091$	-	0.92	Liquid limit (LL) $\leq$ 29.25
			$C_c = (0.0017 \times LL) + 0.1235$	-	0.92	29.25 < LL < 37.35
			$C_c = (0.0064 \times LL) - 0.05237$	-	0.92	LL $\geq$ 37.35
			$C_c = (0.2945 \times e) - 0.0774$	-	0.95	Void ratio (e) $\leq$ 0.495
			$C_c = (0.2534 \times e) - 0.052$	-	0.95	0.495 < Void ratio (e) < 0.615
			$C_c = (0.7071 \times e) - 0.3471$	-	0.95	Void ratio (e) $\geq$ 0.615
6.	Cohesion (c)	Clays	$c = 0.0464 \times N + 0.0075$	kg/cm^2	0.93	SPT N-value ranging from 1 to 25
			$c = 0.0702 \times N - 0.5453$	kg/cm^2	0.93	SPT N-value ranging from 26 to 52
7.	Angle of internal friction (ϕ)	Sands	$\phi = 0.3125 \times N + 26.1261$	$^{\circ}$	0.99	SPT N-value ranging from 1 to 52

Source: Adapted from Puri et al. (2018).

13.7.3.1 Building inspection

In the preliminary survey, it was found that the building walls have large crack openings. Most of the columns were out of alignment from their position. On the backside of the building, a column in the left portion showed major cracks and some relative settlement. The doors and windows of the building were not functioning properly. The switchboards of electric fixtures were seen to be convergence points for cracks throughout the building. The sewerage system and water supply lines were also non-functional. There was a lot of unevenness on the building floor. The building showed dampness in the walls.

The all-around photography of the building was done to classify the type of cracks and to understand the causes of cracks in the building (Figure 13.10).

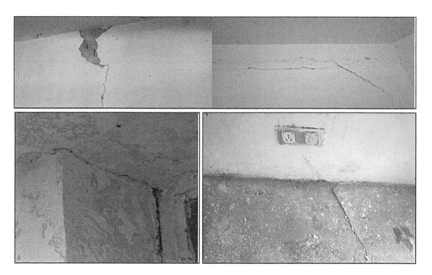

Figure 13.10 Major cracks in the walls/columns and differential settlement of floor.

13.7.3.2   Distress analysis

The distress analysis of the building was carried out based on preliminary investigation, visual inspection of the cracks and the field and laboratory tests, which have been discussed as follows.

13.7.3.3   Preliminary investigation

The preliminary investigation indicated that the building is constructed in a low-lying area with poor drainage conditions. There are mature trees near the building, which can influence the initiation of cracks. The cracks in the building could be caused by the entry of tree roots in the building elements and water penetrating into the ground due to the lack of a proper drainage system. Leakage from damaged supply lines could be responsible for moisture intrusion leading to dampness in the building. Leakage could also be taking place through damaged sewer lines, and subsequent seepage of sewage into the foundation soil increasing the settlement of the building. The building is resting on isolated footings, which now are showing differential settlement due to penetration of moisture into the soil below one or more footings. This could also be due to the change in properties of foundation soil due to nearby construction excavations. The provision of raft foundation would have been more suitable for the building as the same has been done for the adjoining buildings in the area to avoid differential settlement.

13.7.3.4   Crack analysis

The building inspection was carried out by measuring the width, length and orientation of cracks. The length and width of the cracks were marked, and monitoring was performed from time to time to see whether these increase with time. The monitoring of cracks indicated that cracks in the building are active cracks as seen by the appearance of new crack edges during the building inspection. Most of these cracks in walls are classified as very severe with an average crack width greater than 25 mm (Burland et al., 1977). The cracks in other structural elements are greater than the maximum specified allowable crack width of 0.3 mm (BS 8110, 1997). These cracks are more severe on the backside of the building in the left portion, which may be due to leakage in the supply line and damaged sewerage system of washrooms. These cracks are rapidly increasing in depth and width caused by the settlement of the foundation soil, and the possible sliding of some portion of the foundation. The horizontal cracks are wide and are of constant width throughout their lengths. These seem to have appeared at the same time, possibly caused by excessive settlement, because of poorly constructed foundation and poor workmanship. Diagonal cracks emerging from the corners of door and window edges and vertical cracks were identified. These may have been caused by the gradual vertical movement of the foundation and shrinkage of concrete. Due to settlement, there is a gap between the roof slab and top of the wall, and some of the columns have suffered severe cracks extending through the walls, possibly caused by uneven settlement of the foundation soil, putting the building under tension.

13.7.3.5 Field and laboratory tests

For soil investigation, boreholes were extended up to a depth of 6.0 m at four critical locations around the building. The samples were recovered from the desired levels. The subsoil water table was found to be below the depth of boreholes made at the time of field investigations. Standard Penetration Tests were conducted at various elevations in each borehole. Laboratory tests conducted included classification, strength and compressibility tests. The results show that the substrata comprise soil formation capped with silt and clay. Water content values at shallow depths are high indicating leakage from the water supply line and sewerage system, which was confirmed by visual inspection of the site also. The swell index values obtained indicate that the subsoil has a moderate swell potential. The development of centre doming (centre heave situation) of the floor slab may be due to intrusion of moisture in the foundation soil. The wet soil layer just below the foundation is found to show low strength and high compressibility causing settlement of the building. Since the subsoil has a high percentage of fine materials with high moisture content and plasticity index, all the cracks are active cracks with their width increasing with time.

The building is under investigation, and all other nearby buildings show visible distress. It could be due to the ageing of the infrastructure that has deteriorated at an exponential rate and lack of maintenance. The remedial measures to rectify distress in the building could include installation of prudent moisture control measures around the building with proper disposal of rainwater, installation of the new water supply line, repair of the sewerage system, filling of cracks with concrete to minimise the further expansion of cracks and jet grouting for correction in settlement and alignment of columns.

13.8 CONCLUDING REMARKS

In general, sands and silts are predominant soils in the state of Haryana and these serve as good foundation bearing materials. However, some areas of the state face the problems of soil salinity and sodicity, which can reduce the design life and serviceability of foundations, public utilities and other buried structures. The situation is quite alarming in some areas in the districts of Karnal, Kurukshetra, Rohtak, Panipat and Sonipat, which constitute about 9.7% of the total area of the state. The problem of differential settlement is also observed in some areas having a clay layer below the foundation level. A saucer type zone is formed comprising parts of Sonipat, Karnal, Kurukshetra, Yamunanagar and Faridabad districts due to the peculiar geomorphic setting of the state. This area accounts for 3.31% of the total area of the state. In these areas, due to the shallow groundwater table, problems of low bearing capacity, seepage and difficulties during construction are encountered. The state falls into three seismic zones as per the Indian Standard, IS:1893-Part 1 (2016), viz. Zone II, Zone III and Zone IV with expected PGA values of 0.10, 0.16 and 0.24 g, respectively, making it prone to risk from low to moderate earthquakes. In the north and northeastern parts of Haryana, the areas near the bank of river Yamuna could expect liquefaction of soils during earthquakes due to their proximity to the Himalayan Thrust System.

REFERENCES

Ali, F. H., Adnan, A. and Choy, C. K. (1992). Geotechnical properties of a chemically stabilized soil from Malaysia with rice husk ash as an additive. *Geotechnical and Geological Engineering Journal*, Vol. 10, No. 2, pp. 117–134.

Aliabdo, A. A., Elmoaty, A. M., Elmoaty, A. and Auda, E. M. (2014). Re-use of waste marble dust in the production of cement and concrete. *Construction and Building Materials*, Vol. 50, pp. 28–41.

American Society of Testing and Materials (2013). ASTM D5311/D5311M-13: Standard test method for load controlled cyclic triaxial strength of soil. American Society of Testing and Materials, West Conshohocken, PA.

American Society of Testing and Materials (2015). ASTM D-4015: Standard test methods for modulus and damping of soils by the resonant-column method. American Society of Testing and Materials, West Conshohocken, PA.

Anvari, S. M., Shooshpasha, I. and Kutanaei, S. S. (2017). Effect of granulated rubber on shear strength of fine-grained sand, *Journal of Rock Mechanics and Geotechnical Engineering*, Vol. 9, No. 5, pp. 936–944.

Azme, N. N. M., Murshed, M. F., Ishak, S. A., Azam, M. and Adnan, M. (2019). Utilization of sugarcane press mud as a natural absorbent for heavy metal removal in leachate treatment. *Proceedings of AWAM International Conference on Civil Engineering*, (AICCE 2019), pp. 1297–1307.

Azzouz, A. S., Krizek, R. J. and Corotis, R. B. (1976). Regression analysis of soil compressibility. *Soils and Foundations*, Vol. 16, No. 2, pp. 19–29.

Bandyopadhyay, S., Sengupta, A. and Reddy, G. R. (2015). Performance of sand and shredded rubber tire mixture as a natural base isolator for earthquake protection. *Earthquake Engineering and Engineering Vibration*, Vol. 14, No. 4, pp. 683–693.

Basha, E. A., Hashim, R., Mahmud, H. B. and Muntohar, A. S. (2005). Stabilization of residual soil with rice husk ash and cement. *Journal of Construction and Building Materials*, Vol. 19, No. 6, pp. 448–453.

Bowles, J. E. (1982). *Foundation Analysis and Design*. 3rd Edition, McGraw Hill, Inc., New York.

British Standard Institute (1997). BS 8110-1: Structural use of concrete. Code of practice for design and construction, British Standard Institute, UK.

Bureau of Indian Standards (1970). IS 1498: Indian standard methods of test for soils: Classification and identification of soil for general engineering purposes. Bureau of Indian Standards, New Delhi.

Bureau of Indian Standards (2016). IS 1893 (Part 1): Indian standard criteria for earthquake resistant design of structures, Part 1: General provisions and buildings. Bureau of Indian Standards, New Delhi.

Burland, J. B., Broms, B. and De Mello, V. F. B. (1977). Behaviour of foundations and structures. State of Art. Report. Session 2. *Proceedings of the 9th International Conference on Soil Mechanics and Foundation Engineering*, Vol. 2, pp. 495–545.

CGWB (2013). Ground water level information. Central Ground Water Board, *Ministry of Water Resources*, Government of India.

Chu, T. Y., Davidson, D.T, Goecker, W. L. and Moh, Z. C. (1975). Soil stabilization with lime-fly ash mixtures: Preliminary studies with silty and clayey soils. *Highway Research Board Bulletin*, Vol. 108, pp. 102–112.

Cokca, E. (2001). Use of class C fly ashes for the stabilization of an expansive soil. *Journal of Geotechnical and Geoenvironmental Engineering*, Vol. 127, No. 7, pp. 568–573.

Cozzolino, V. M. (1961). Statistical forecasting of compression index. *Proceedings of 5th International Conference on Soil Mechanics and Foundation Engineering*, Paris, pp. 51–53.

Cui, L., Bhattacharya, S. and Nikitas, G. (2017). Micromechanics of soil responses in cyclic simple shear tests. *Powders and Grains 2017 – 8th International Conference on Micromechanics on Granular Media*, Montpellier, France, EPJ Web of Conferences, Volume 140, Article id. 02008, pp. 1–4.

Darendeli, M. B. (2001). Development of a new family of normalized modulus reduction and material damping curves. PhD dissertation, University of Texas at Austin, USA.

Das, B., Prakash, S., Reddy, P. S. R. and Misra, V. N. (2007). An overview of utilization of slag and sludge from steel industries. *Resources, Conservation and Recycling*, Vol. 50, No. 1, pp. 40–57.

Dass, A. and Malhotra, S. K. (1990). Lime-stabilized red mud bricks. *Materials and Structures*, Vol. 23, No. 4, pp. 252–255.

De, S. and Basudhar, P. K. (2008). Steady state strength behavior of Yamuna sand. *Geotechnical and Geological Engineering*, Vol. 26, No. 3, pp. 237–250.

Faria, K. C. P., Gurgel, R. F. and Holanda, J. N. F. (1990). Recycling of sugarcane bagasse ash waste in the production of clay bricks. *Journal of Environmental Management*, Vol. 23, No. 4, pp. 252–255.

GSI (1973). Geological and mineral map of Haryana. Map and Cartography Division, *Geological Survey of India*, Kolkata.

Gupta, A. K., Chopra, S., Prajapati, S. K., Sutar, A.K. and Bansal B. K. (2013). Intensity distribution of M 4.9 Haryana-Delhi border earthquake. *Natural Hazards*, Vol. 68, No. 2, pp. 405–417.

Hamed, M. M., Ahmed, I. M. and Metwally, S. S. (2014). Adsorptive removal of methylene blue as organic pollutant by marble dust as eco-friendly sorbent. *Journal of Industrial and Engineering Chemistry*, Vol. 20, No. 4, pp. 2370–2377.

Hoek, E. and Brown, E. T. (1980). Empirical strength criterion for rock masses. *Journal of Geotechnical Engineering Division, ASCE*, Vol. 106, No. GT9, pp. 1013–1035.

Jakka, R. S., Ramana, G. V. and Datta, M. (2010). Shear behavior of loose and compacted pond ash. *Geotechnical and Geological Engineering*, Vol. 28, No. 6, pp. 763–778.

Jha, J. N. and Gill, K. G. (2006). Effect of rice husk ash on lime stabilization. *Journal of the Institution of Engineers (India),* Vol. 87, pp. 33–39.

Kalantary, F. and Kordnaeij, A. (2012). Prediction of compression index using artificial neural network. *Scientific Research and Essays,* Vol. 7, No. 31, pp. 2835–2848.

Kumar, K. and Jain, A. (2019). Forensic geotechnical investigation of building distress – A review. *Indian Conference on Geotechnical and Geoenvironmental Engineering (ICGGE-2019),* 1–2 March 2019, MNNIT Allahabad, India.

Kumar, R., Bhargava, K. and Choudhury, D. (2016). Estimation of engineering properties of soils from field SPT using random number generation. *INAE Letters,* Vol. 1, No. 3–4, pp. 77–84.

Muntohar, A. S. and Hashim, R. (2002). Silica waste utilisation in ground improvement: A study of expansive clay treated with LRHA. *The 4th International Conference on Environmental Geotechnics,* Rio de Janeiro, Vol. 1, pp. 515–519.

Nikitas, G., Arany, L., Aingarana, S., Vimalan, J. and Bhattacharya, S. (2017). Predicting long term performance of offshore wind turbines using cyclic simple shear apparatus. *Soil Dynamics and Earthquake Engineering,* Vol. 92, pp. 678–683.

Okagbue, C. O. and Onyeobi, T. U. S. (1999). Potential of marble dust to stabilise red tropical soils for road construction. *Engineering Geology,* Vol. 53, No. 3–4, pp. 371–380.

Osinubi, K. J., Bafyau, V. A. and Eberemu, O. (2009). Bagasse ash stabilization of lateritic soil, *Appropriate Technologies for Environmental Protection in the Developing World,* pp. 271–280.

Patil, U., Valdes, J. R. and Evans, T. M. (2011) Swell mitigation with granulated tire rubber. *Journal of Materials in Civil Engineering,* Vol. 23, No. 5, pp. 721–727.

Pothal, G. K. and Rao, G. V. (2008). Model studies on geosynthetic reinforced double layer system with pond ash overlain by sand. *Electronic Journal of Geotechnical Engineering,* Vol. 13, pp. 1–12.

Puri, N. and Jain, A. (2014). Preliminary investigation for screening of liquefiable areas in the state of Haryana, India. *ISET Journal of Earthquake Technology,* Vol. 51, No. 1–4, pp. 19–34.

Puri, N. and Jain, A. (2019). Microzonation of seismic hazard for the state of Haryana, India. *Journal of the Geological Society of India,* Vol. 94, No. 3, pp. 297–308.

Puri, N., Jain, A., Nikitas, G., Dammala, P. K. and Bhattacharya, S. (2020). Dynamic soil properties and seismic ground response analysis for North Indian seismic belt subjected to the great Himalayan Earthquakes. *Natural Hazards,* Vol. 103, No. 1, pp. 447–478.

Puri, N., Prasad, H. D. and Jain, A. (2018). Prediction of geotechnical parameters using machine learning techniques. *Procedia Computer Science,* Vol. 125, pp. 509–517.

Rahim, A. (1989). Effect of morphology and mineralogy on compressibility of sands, Ph.D. thesis, Department of Civil Engineering, Indian Institute of Technology, Kanpur, India.

Rahman, M. A. (1987). Effect of cement-rice husk ash mixtures on geotechnical properties of lateritic soils. *Journal of Soils and Foundations,* Vol. 27, No. 2, pp. 61–65.

Ramamurthy, T. (1985). Stability of rock mass, 8th Annual Lecture. *Indian Geotechnical Conference,* Roorkee, India, pp. 221–274.

Rao, G. V. and Dutta, R. K. (2006). Compressibility and strength behaviour of sand–tyre chip mixtures. *Geotechnical and Geological Engineering,* Vol. 24, No. 3, pp. 711–724.

Rogbeck, J. and Knutz, A. (1996). Coal bottom ash as light fill material in construction. *Waste Management,* Vol. 16, No. 1–3, pp. 125–128.

Sachdev, C. B., Lal, T., Rana, K. P. C. and Sehgal, J. (1995). Soils of Haryana: Their kinds, distribution, characterization and interpretations for optimising land use. *NBSS Publications,* 44.

Schreurs, J. P. G. M., Sloot, H. A. V. D. and Hendriks C. H. (2000). Verification of laboratory-field leaching behavior of coal fly ash and MSWI bottom ash as a road base material. *Waste Management,* Vol. 20, No. 2–3, pp. 193–201.

Seed, H. B. and Idriss I. M. (1970). Soil moduli and damping factors for dynamic response analyses. Earthquake Engineering Research Center, University of California, Berkeley, CA, Rep. No. EERC-70/10.

Shioi, Y. and Fukui, J. (1982). Application of N-value to design of foundation in Japan. *Proceedings of the Second European Symposium on Penetration Testing*, Amsterdam, Vol. 1, pp. 40–93.

Singh, M. and Siddique, R. (2013). Effect of coal bottom ash as partial replacement of sand on properties of concrete. *Resources, Conservation and Recycling*, Vol. 72, pp. 20–32.

Singh, M. and Singh, B. (2003). A simple parabolic strength criterion for intact rocks. *Proceedings of Indian Geotechnical Conference*, Roorkee, Vol. 1, pp. 555–558.

Skempton, A. W. (1944). Notes on the compressibility of clays. *Quarterly Journal of Geological Society*, Vol. 100, No. 1–4, pp. 119–135.

Soni, D. K. and Jain, A. (2006). Strength characteristics of limestone under high pressure. *Journal of Rock Mechanics and Tunnelling Technology*, Vol. 12, No. 1, pp. 53–63.

Terzaghi, K. and Peck, R. B. (1967). *Soil Mechanics in Engineering Practice*. 2nd Edition, John Wiley & Sons, New York.

Tsang, H., Lo, S. H., Xu, X. and Sheikh, M. N. (2012). Seismic isolation for low-to-medium-rise buildings using granulated rubber–soil mixtures: Numerical study. *Earthquake Engineering & Structural Dynamics*, Vol. 41, pp. 2009–2024.

Tsuchida, H. 1970. Prediction and countermeasure against liquefaction in sand deposits. Abstract of the Seminar of the Port and Harbour Research Institute, *Ministry of Transport*, Yokosuka, Japan, 3.1–3.33 (in Japanese).

Wolff, T. F. (1989). Pile capacity prediction using parameter function. *ASCE Geotechnical Special Publication*, No. 23.

Wright, W. and Ray, P. N. (1957). The use of fly ash in soil stabilization. *Magazine of Concrete Research*, Vol. 9, No. 25, pp. 27–31.

Chapter 14

Himachal Pradesh

Varinder S. Kanwar
Chitkara University

Surinder Kumar Vashisht
Himachal Pradesh Housing and Urban Development Authority (HIMUDA)

Abhishek Kanoungo
Chitkara University

Rajesh Pathak
Thapar Institute of Engineering and Technology

Manvi Kanwar
Chitkara University

CONTENTS

14.1 INTRODUCTION

The Indian state of Himachal Pradesh has an area of 55,673 km² and is arranged between 30° 22′ 40″–33° 12′ 20″ north and 75° 45′ 55″–79° 04′ 20″ east longitudes (Sharma and Dogra, 2011). The height in the state, an entirely precipitous state in the lap of the Himalayas, ranges from 350 to 6,975 m above the mean sea level (MSL). It is encompassed by Jammu and Kashmir in the north, Tibet in the northeast, Uttaranchal in the east-southeast, Haryana in the south and Punjab in the southwest. Himachal Pradesh is situated in the lower Himalayan region with numerous mountain ranges and rich natural resources (Attri, 2000). The geographical map of Himachal Pradesh is shown in Figure 14.1. There is a development in elevation as we go from north to south and from east to west. Shimla is the capital of Himachal Pradesh. To facilitate the administrative process, the state has been divided into 12 zones and has around 49 urban regions and towns (Sharma and Sharma, 2013). The state is picturesque and witnesses snowfall because of its propinquity to the Himalayas. The typical climate of different locations in the state fluctuates according to the elevation levels.

DOI: 10.1201/9781003177159-14

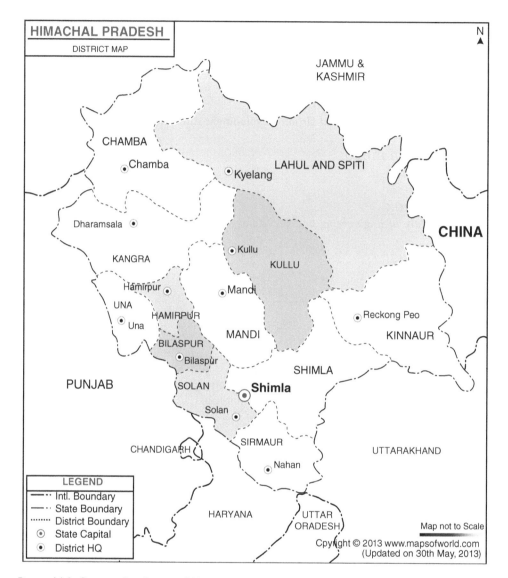

Figure 14.1 Geographical map of Himachal Pradesh (www.mapsofindia.com).

The state is separated into three physiographic partitions from the south to the north (Sharma and Dogra, 2011).

- The Great Himalayas – situated in the northern zone
- The lesser Himalayas – situated in the central zone
- The external Himalayas – referred to as Shivaliks

The Kangra valley is one of the well-known valleys in the state. It is situated on the longitudinal trough along the foot of the Dhauladhar range. Dhauladhar means "the white peak" in Hindi language. It has a height of around 4,550 m (Atri, 2000). The biggest mountain in the Himalayan range is the Pir Panjal and this range is different from the Greater Himalayan range that falls on the banks of the Sutlej. Various glaciers and passes exist crosswise over Pir Panjal. The Rohtang Pass, which is at an altitude of 3,978 m, is one of the most famous passes. The Shivalik range is at a lower height in the state and is at 600 m above the MSL. The slopes of these local areas contain rocks and soil which are prone to disintegration and deforestation.

The character and structures are unique in relation to and from one climatic and geographic zone to another. It can be categorised into three separate zones depending on their height (Sharma and Sharma, 2013).

Himachal is a rich wellspring of vegetation. The state has 38% of the region under the woods. The smallest town of the state is Naina Devi and the biggest city is the capital Shimla. The most conspicuous rivers of the state are Chenab, Beas, Ravi, Yamuna, Sutlej and Spiti. These rivers aid in broad horticulture in the state without the issue of an abundant water system (Randev and Parmar, 2009). They significantly support the natural beauty and prosperity of the state. The state, which is situated on the western corner of the Himalayas, works as an ideal combination of perfect streams that augment the state's water supply. The Himalayas, which are covered with thick snow, melt to shape significant rivers of the state. These rivers satisfy the water-related needs of the population in the summer season. The picturesque scenic beauty gives an impetus to the travel industry in the state. The fundamental land structures are mountainous territories that are encompassed by valleys everywhere. The most unmistakable part of Himalayan extents is the distinctive landforms like incline, height, microclimatic situation and vegetation. The perfect time to explore the state is from September to March.

There is an extraordinary variety in the climatic state of Himachal Pradesh because of variations in height. The climate varies from hot and sub-moist tropical in the southern tracts to cool, snow-capped and freezing in the northern and eastern mountain ranges with more prominent heights. The normal precipitation in Himachal Pradesh is 1,111 mm, differing from 450 mm in Lahaul and Spiti to more than 3,400 mm in Dharamshala. Precipitation decays from the west toward the east, and from the south toward the north. Winter precipitation happens as snow falls at heights over 1,800 m. A normal of 3 m of snow is experienced between December and March (Singh, 2015). The land use pattern of the state is presented in Table 14.1.

Table 14.1 Land use pattern of Himachal Pradesh

Land use	Area in '000 ha	Percentage
Total geographic area	5,567	--
Reporting area for land utilisation	4,549	100
Forests	1,101	24.2
Not available for cultivation	1,129	24.8
Permanent pastures and other grazing lands	1,496	32.8
Land under miscellaneous tree crops and groves	65	1.43
Culturable wasteland	138	3
Fallow lands other than current fallows	15	0.33
Current fallows	64	1.4
Net area sown	541	11.9

Source: http://mospi.nic.in/sites/default/files/cocsso/2_HimachalPradesh.pdf.

14.2   MAJOR TYPES OF SOILS, ROCKS AND THEIR PROPERTIES

The soils of the state can extensively be separated into nine categories. These are (i) alluvial, (ii) brown forests, (iii) brown earth, (iv) planosolic, (v) grey wooded, (vi) grey-brown podzolic, (vii) hill soil, (viii) iron podzols and (ix) alpine humus mountain. The physical substance that the soil gets is from the rocks, but the mineral and natural substances are affected by climatic conditions.

Himachal Pradesh enjoys different topography surrounding various landforms having different origins because of the tectonic variation and weathering processes. The soil properties of Himachal Pradesh were reported by the Indian Institute of Soil Science, Bhopal. The soil parameters taken were pH, electrical conductivity (EC), organic carbon (OC), nitrogen (N), phosphorous (P), potassium (K), sulphur (S), zinc (Z), iron (Fe), copper (Cu) and magnesium (Mn). The soil in the districts of Himachal Pradesh was mostly alkaline and acidic. Una, Kangra, Shimla and Hamirpur mostly have acidic soil while very few regions have neutral pH (Jreat, 2006). Most places of the state were found to be nitrogen deficient.

The state is the sole custodian of the country's stone salt assets. Barytes, limestone, salt (rock) and shale are the noteworthy rocks found in the state. Barytes is found in the Sirmaur region, limestone in Bilaspur, Chamba, Kangra, Kullu, Mandi, Shimla, Sirmaur and Solan areas and rock salt in Mandi. The fundamental stones found in the state are quartzite, schists, phyllites, dolomites, limestones, shales, records, gneisses and rocks. Quartzite produces sandy soil after breaking down, while stones, schists, shales and gneisses produce loamy and sandy-topsoil types (IBM, 2014). The stones of Himachal Pradesh have been exposed to exceptional deformation which at various spots has upset the first stratigraphic position. Different types of soil and rocks present in the state can be seen in Figures 14.2 and 14.3.

Himachal Pradesh has the potential for the occurrence of economic minerals, but it has not brought to light any worthwhile metallic mineral deposits so far. Though there are old workings of metallic minerals, there is no major metal mine in the whole state. However, the scenario is different for non-metallic minerals which have abundant reserves from cement to chemical grade limestone, dolomite and small reserves of barytes and gypsum. Slate and building materials are also important minerals found in the state (ENVIS, 2019). Limestone, either in its natural state or after calcination as lime, has numerous industrial uses, principally in building, chemical and agricultural industries, in the manufacturing of cement and limes, in flux in the metallurgical process, in glass, ceramic, paper, textile and tanning industries, in the manufacture of calcium carbide, alkali and bleaching powder, in sugar refining; as a fertiliser; as a filler in rubber, oil, cloth, cosmetics, toothpaste, shoe polish and so on. The total national consumption of limestone and other calcareous materials by different industries in 2002–2003 was around 146 million tonnes. Out of this, cement was the major consuming industry accounting for 95% of the total consumption followed by iron and steel (3%) and chemical and other grades (2%). As seen from Table 14.2, Himachal Pradesh possesses huge reserves of limestone of cement grade in the lower Himalayan zone (Bhandari and Verma, 1977).

The standard penetration test is very difficult to conduct due to the presence of gravel and boulders in most areas of Himachal Pradesh while in some areas, the dynamic cone penetration test is useful only for shallow depths. Strata involve

Figure 14.2 Types of soil present in different districts of the state.

Figure 14.3 Types of rocks present in the state.

Table 14.2 District-wise details of limestone reserves (in million tonnes)

District	Proved	Probable	Possible	Total
Bilaspur	370	150	500	1,020
Chamba	400	850	100	1,350
Kangra	10	20	10	40
Kullu			120	120
Mandi	500	20	600	1,120
Sirmour	150	200	1,200	1,550
Shimla		50	1,600	1,650
Solan	550	100	1,000	1,650
Lahaul and Spiti			100	100
Kinnaur			100	100
Total	1,980	1,390	6,230	9,600

Source: http://mospi.nic.in.

boulders, cobbles, pebbles and gravels, friable rocky mass, conglomerates of rock fragments and so on. By treating the supporting strata as a granular mass and using the correlation between the SPT (standard penetration test) and the DCPT (dynamic cone penetration test), a proportionate/equivalent value of 'N' is used to perform the bearing capacity analysis. Such an approach, besides being workable and economical, leads to reliably conservative values of allowable bearing capacity, which stand the test of time.

The nature of supporting strata met at the site is a boulder deposit. The allowable bearing capacity values of foundations resting in such deposits can conservatively be evaluated using the DCPT data. The compressibility and strength characteristics of such deposits vary from location to location and as such the bearing capacity analysis should be based on the average value of 'N_{cd}' observed within the zone of shear failure of the proposed footings. At refusal levels, the dynamic cone penetration resistance, $N_{cd} = 50$, is adopted.

Field test results of different sites located in various parts of the state are presented in the Tables 14.3–14.8.

14.3   USE OF SOILS AND ROCKS AS CONSTRUCTION MATERIALS

Mud construction is prevalent in Himachal and the two methods used are smashed earth development and sundried mudbrick development. Sundried mud blocks are utilised in the Kangra district where quality mud is available from riverbeds. The walls are made of sundried blocks about 0.60–0.90 ms thick and use mud phuska (Sarkar and Sharma, 2011). These walls are susceptible to disintegration because of rain and hence the structures are raised over stones to evade it. The floors are made of wood and mud is used to fortify the structure as shown in Figure 14.4.

Dry-stone construction is prevalent in the Kangra region where slate is in abundance (Figure 14.5). This type of construction is also found in the Kinnaur area, because of the availability of stone quarries. Distinctive stones are set over one another and compacted without the mortar. A solid bond is accomplished by interlocking the stone by using smaller stones to fill holes. The inside surface might be mud plastered.

Table 14.3 Geotechnical investigation for the construction of a CHC building site at Bathri, Chamba, Himachal Pradesh

Depth from the ESL (m)	Penetration resistance (N_{CD})	Soil composition (−80 mm fraction)			Atterberg limits (%)		Water content (%)	Bulk density (g/cc)	Specific gravity	Soil profile
		Gravel (%)	Sand (%)	Fines (%)	Liquid limit	Plastic limit				
0.9	38	0.0	41.6	58.4	Non-plastic		12.0	2.0	2.67	EGL to 1.8 m hard but friable rock (siltstone)
1.2	29									1.8–2.4 c compact to dense strata, beyond 2.4 m incompressible strata/refusal to DCPT
1.5	50 blows for 23 cm penetration (refusal)	0.0	44.0	56.0	Non-plastic		9.0	2.0	2.66	Subsoil water was not met up to the explored depth
1.8	42									
2.1	46									
1.5	50 blows for 18 cm penetration (refusal)									

Table 14.4 Geotechnical investigation for the construction site at Neri, Kullu, Himachal Pradesh

Depth from the ESL (m)	Penetration resistance (N_{CD})	Soil composition (−80 mm fraction)			Atterberg limits (%)		Water content (%)	Bulk density (g/cc)	Specific gravity	Soil profile
		Gravel (%)	Sand (%)	Fines (%)	Liquid limit	Plastic limit				
0.9	18	32.0	17.2	50.8	Non-plastic		15.0	2.0	2.68	EGL to 1.4 m compact matrix of boulders, cobbles and gravels
1.2	19									Compact matrix of cobbles, gravels, sand and silt
1.5	20									
1.8	21	15.0	58.7	26.3	Non-plastic		9.8	2.0	2.67	2.1–3.9 m compact to dense strata
2.1	24									
2.4	29									Beyond 39 m incompressible strata/refusal to DCPT
2.7	30									Subsoil water was not met up to the explored depth
3.0	30									
3.3	38									
3.6	40									
3.9	50 Blow for 20 cm Penetration (refusal)									

Table 14.5 Geotechnical investigation for the construction of RTO office building, Hamirpur, Himachal Pradesh

Depth from the ESL (m)	Penetration resistance (N_{CD})	Soil composition (−80 mm fraction)			Atterberg limits (%)		Water content (%)	Bulk density (g/cc)	Specific gravity	Soil profile
		Gravel (%)	Sand (%)	Fines (%)	Liquid limit	Plastic limit				
9.0 1.2	38 50 blows for 21 cm penetration (refusal)	0.0	72.5	27.5	Non-plastic		3.3	2.0	2.67	EGL to 2.1 m hard rock, mainly sandstone
1.5	50 blows for 17 cm penetration (refusal)	0.0	71.8	28.2	Non-plastic		4.2	2.0	2.67	Beyond 2.1 m incompressible strata/ refusal to DCPT
1.8 2.1	45 50 blows for 19 cm penetration (refusal)									Subsoil water was not met up to the explained depth

Table 14.6 Geotechnical investigations for the construction site at Leh, Himachal Pradesh

Depth from the ESL (m)	Penetration resistance (N_{CD})	Soil composition (−80 mm fraction)			Atterberg limits (%)		Water content (%)	Bulk density (g/cc)	Specific gravity	Soil profile
		Gravel (%)	Sand (%)	Fines (%)	Liquid limit	Plastic limit				
0.9	27	20.7	78.0	1.3	Non-plastic		3.9	2.0	2.67	ESL to 0.3 m fill
1.2	31									
1.5	26									0.3–1.7 m SP (dense sand)
1.8	29	6.3	39.0	0.7	Non-plastic		4.1	2.0	2.66	1.7–2.5 m dense matrix of cobbles and GP
2.1	31									Beyond 3.0 m incompressible strata
2.7	31									
3.0	50 blows for 21cm penetration (refusal)	62.7	37.0	0.3	Non-plastic		5.2	2.0	2.66	No subsoil water level was met up to the explored depth

Table 14.7 Geotechnical investigation for the construction of a water treatment plant at village Seu (near Seer Khad), Tehsil Ghumarwin, Bilaspur, Himachal Pradesh

Depth from the ESL (m)	Penetration resistance (N_{CD})	Soil composition (−80 mm fraction)			Atterberg limits (%)		Water content (%)	Bulk density (g/cc)	Specific gravity	Soil profile
		Gravel (%)	Sand (%)	Fines (%)	Liquid limit	Plastic limit				
0.25	48									EGL to 1.5 m hard but friable rock, mainly sandstone
0.55	50 blows for 15 cm penetration	73.6	26.4	Non-plastic		5.4	2.0	2.67	1.05–1.15 m dense strata
0.85	50 blows for 10 cm penetration	9.9	90.1	39.0	23.8				Beyond 1.15 m incompressible strata/refusal to DCPT
1.15	50 blows for 15 cm penetration	3.3	96.7	40.0	24.1	14.6	2	2.71	Sub-soil water level was not met up to the explored depth

Table 14.8 Geotechnical investigations for the construction of a hospital building in jail at Nahan, Sirmour, Himachal Pradesh

Depth from the ESL (m)	Penetration resistance (N_{CD})	Soil composition (−80 mm fraction)			Atterberg limits (%)		Water content (%)	Bulk density (g/cc)	Specific gravity	Soil profile
		Gravel (%)	Sand (%)	Fines (%)	Liquid limit	Plastic limit				
0.9	6	2.5	8.4	89.1	34.2	23.0	17.2	1.91		ESL to 1.2 m CL (firm clay)
1.8	6	5.0	4.8	90.2	38.4	23.9				1.2–5 m CL (firm to very stiff clay)
2.7	7	0	5.0	95	38.8	24				5–5.3 m compact to dense strata
3.6	12	0	6.6	93.4	37.8	23.5	4.5	2.0		Beyond 5.3 m incompressible strata/refusal to DCPT
4.5	25	0	2.2	97.8	37.5	23			2.67	Subsoil water level was not met up to the explored depth
5.0	37									
5.3	50 blows for 25 cm penetration									

Figure 14.4 View of mud construction.

Figure 14.5 View of dry-stone construction.

Figure 14.6 View of the wooden cantilevered upper floor with the ground floor of stone masonry.

Wooden construction is used in many parts of the state because of its easy availability and the role played by it in resisting the cool winters. Houses are fabricated 2–3 storeys high (Figure 14.6). Vertical wooden posts are erected to balance the load. Horizontal members are set at various levels with an infill of wooden battens. Mostly the ground floor is constructed in stone masonry with an upper floor made up of wood.

14.4 NATURAL HAZARDS

The province of Himachal Pradesh is susceptible to different hazards both common and artificial. Principal hazards comprise seismic tremors, avalanches, floods, blizzards and torrential slides, drafts, dam disappointments, fires – local and wild, mishaps – street, rail, air, charges, vessel inverting, natural, mechanical and perilous synthetics and so on (Randev and Parmar, 2009).

From a seismicity point of view, the state of Himachal Pradesh, which forms a part of the northwest Himalayas, is very sensitive. During the last century, the state has been shaken by several microearthquakes as well as macroearthquakes. Several damaging earthquakes have struck the state. Some of the prominent earthquakes that rocked the state are the Kangra earthquake in 1905 ($M=8.0$) in which 18,815 people were killed and the Kinnaur earthquake in 1975 ($M=6.7$) in which 60 people lost their lives (Randev and Parmar, 2009). Besides these major earthquakes, the state has been rocked by about 250 earthquakes with magnitude 4.0 and 62 earthquakes with magnitude more than 5.0. As per the earthquake hazard map of the state, the areas falling in districts Chamba, Kangra, Mandi, Kullu, Hamirpur and Bilaspur are very sensitive as they fall in the very high damage risk zone (Intensity IX or more), i.e. Zone IV, whereas the rest of the areas fall in the high damage risk zone (Intensity VIII) (ENVIS, 2019).

Soil breaking down is a marvel that is across the board and the most ordinary biological threat. Its topography, poor physical character, climatic conditions and anthropogenic interventions are the key drivers for soil break down. Extreme enduring/scree-bone risk is majorly found in the tribal areas of Kinnaur, Lahaul and Spiti and Chamba.

Landslide is a key biological danger faced in major parts of the state. The important slides in Himachal Pradesh that caused huge damage are as follows:

- Landslide on NH-22, District Solan, Himachal Pradesh (13th and 14th August 2018). Heavy precipitation set off the commencement of new landslides and reactivated existing ones, blocking and disturbing the traffic along NH-22 (ENVIS, 2019).
- Heavy rainstorm precipitation set off an enormous stone slide on NH-5 near Dhalli territory, Shimla area, Himachal Pradesh on 2nd September 2017 that led to parked vehicles getting damaged, blocking of the national highway and fractional harm of a sanctuary building (Attri, 2000).
- On 13 August 2017, a massive landslide took place at Kotrupi on NH-154, Padhar Tehsil, Mandi district, Himachal Pradesh. The landslide resulted in several casualties, which included passengers of Chamba–Manali and Manali–Katra HRTC buses.
- A landslide was reported at the well-known Sikh hallowed place Gurudwara Manikaran Sahib in Himachal Pradesh on 18 August 2015.
- Landslide in Himachal Pradesh on 25 July 2015. According to media reports, relentless downpour had caused huge landslides on NH-21 and NH-22 in Shimla, Kinnaur, Mandi and Kullu, hampering vehicular traffic.

In Himachal, winter begins in October and lasts till February. December is the coldest month of the year. Places such as Manali, Narkanda, Kufri and Shimla enjoy good snowfall. Places such as Rohtang Pass get quite heavy snowfall and are good places to enjoy winter sports such as heli-skating and skiing.

Snow avalanches are a common phenomenon in high mountain landscapes like the Himalayas where the shear force of the down sliding snow uproots rocks, trees or other materials and leaves behind a devastating picture of death and misery. Such phenomena occur in areas where the rapid accumulation of snow takes place and the most prone locations are the high slopes just after a snowstorm or a heavy snowfall (Chandel, 2015).

14.5 CONCLUDING REMARKS

Himachal Pradesh is one of the most picturesque hill states in India. The soil in the major part of the state is primarily alluvial soil except for Lahaul, Spiti and Chamba districts, which have humus mountainous soils. The plain regions of the state such as Hamirpur, Una and Solan have a high load-bearing capacity of foundation soils, indicating the occurrence of dense sand and gravel. The hilly regions of the state have comparatively low bearing capacity due to the presence of clay and silt. For the construction of massive structures, soil strengthening may be required depending upon the load of the structure. The hilly areas are prone to natural calamities like earthquakes, heavy rainfalls, cloud bursts, flash floods and frequent landslides. These natural disasters cause huge losses to life and property. Therefore, it is important to study the behaviour of soil critically and improve the quality of soil located in such areas, wherever required.

REFERENCES

Attri, R. (2000). *Introduction to Himachal Pradesh*, Sarla Publications, India.

Bhandari, S.M. and Varma, A.K. (1977). Report on preliminary investigation of limestone and mapping of Shali Formation Shimla and Mandi Districts, H.P, India.

Chandel, V.B. (2015). Snow Avalanche as disaster in mountain environment: A case of Himachal Pradesh. *International Journal of Geomatics and Geosciences*, Vol. 6, No 2, pp. 1578–1584.

ENVIS (2019). Annual Report on Disasters, ENVIS Centre, Disaster Management Authority, Ministry of Environment & Forest. Government of India, www.moef.gov.in., accessed 21 February 2020.

https://www.mapsofindia.com/, Maps of Indian States and Union Territories, accessed 30 January 2020.

http://mospi.nic.in/sites/default/files/cocsso/2_HimachalPradesh.pdf, Success Story & Policy Issues of Farmer's Welfare (Horticulture sector) in Himachal Pradesh, accessed 2 January 2021.

http://mospi.nic.in, District-wise details of limestone reserves, accessed 30 January 30, 2021.

IBM (2014). *Indian Minerals Yearbook*, 53rd edition, Government of India, Ministry of Mines, Indian Bureau of Mines, Nagpur.

Jreat, M. (2006). *Geography of Himachal Pradesh*, Indus Publishing Co, New Delhi.

Randev, A.K. and Parmar, Y.S. (2009). Impact of climate change on apple productivity in Himachal Pradesh-India. In *Proceedings of 60th International Executive Council Meeting & 5th Asian Regional Conference*, New Delhi, India.

Sarkar, A. and Sharma, S. (2011). Vernacular Architecture – Climate responsive features in the traditional shelter design in high altitude – An approach towards sustainable design in hilly region. Architecture Time Space & People: Council of Architecture, India, Vol. 11, No. 8, pp. 16–25.

Singh, K. (2015). *Green Growth and Transport in Himachal Pradesh*. www.teriin.org, accessed 10 January 2021.

Sharma, R.C. and Dogra, S. (2011). Characterization of the soils of lower Himalayas of Himachal Pradesh, India. Nature Environment and Pollution Technology. *An International Quarterly Scientific Journal*, Vol. 10, No. 3, pp. 439–446.

Sharma, S. and Sharma, P. (2013). Traditional and vernacular buildings are ecological sensitive, climate responsive designs-study of Himachal Pradesh. *International Journal of Chemical, Environmental & Biological Sciences*, Vol. 1, No. 4, pp. 605–609.

Chapter 15

Jammu and Kashmir

Falak Zahoor and Bashir Ahmed Mir
National Institute of Technology Srinagar

CONTENTS

15.1 INTRODUCTION

Jammu and Kashmir (J&K) is divided into a total of 20 districts and consists of two divisions – the Jammu division and the Kashmir division, with an area of 26,293 and 15,948 km^2, respectively. A map showing the boundaries of the region and of all the districts superimposed over the topographic relief prepared using an ArcGIS software is presented in Figure 15.1. J&K lies in the Himalayan region, which originated from the collision of the Indian plate with the Eurasian plate about 50 Ma ago (Searle et al., 1987; Thakur and Rawat, 1992). Himalayas extend in an NW-SE arc for over 2,400 km length along which the Indian plate is still moving northward and thrusting under the Eurasian plate. The Himalayas are composed of the following three major ranges from the north to the south: the crystalline High Himalayas, the Lesser Himalayas, composed of Precambrian to Tertiary metasedimentary and sedimentary rocks, and the Sub-Himalayas, which is predominantly the murree formation overlying the Siwalik Group (Hussain et al., 2009). The Zanskar range is a part of the Greater Himalayas, Pir Panjal is in the Lesser Himalayas and Siwalik hills represent the Sub-Himalayan region. Major rivers flowing through this region are Indus, Chenab, Jhelum and Tawi, while the major lakes are Manasbal, Dal, Nigeen, Anchar and Wular.

DOI: 10.1201/9781003177159-15

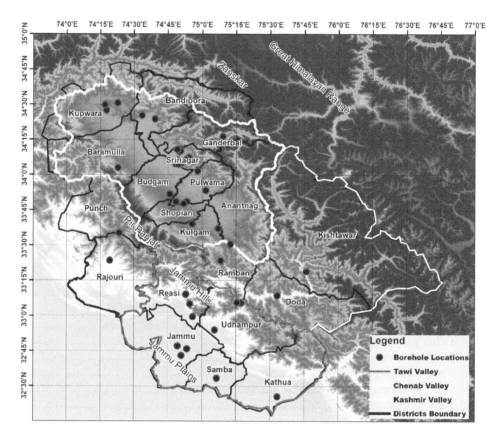

Figure 15.1 Map created using ArcGIS software showing the boundaries of J&K and of all its districts (as per Survey of India, www.surveyofindia.gov.in) superimposed over topographic relief.

J&K comprises complex landforms including mountains, hills, valleys as well as plains, which are products of the tectonic activity in the region (Yeats et al., 1992). In this chapter, J&K has been divided into four major geographical regions (Table 15.1). The stratigraphic succession for each district is provided in the Ground Water Information Booklets prepared by the Central Ground Water Board (CGWB, 2013). The topographical view and soil types in the key regions of J&K are shown in Figure 15.2, and the typical structures in this region are shown in Figure 15.3.

Kashmir is an NW-SE oriented valley about 140 km long and 50 km wide, bounded by the Zanskar range on the north and the Pir Panjal range on the south. Pir Panjal and the Great Himalayan range show differential tectonic activities, which have resulted in greater uplift of the Pir Panjal side (Madden et al., 2011; Shah, 2013; Dar et al., 2014). This is evident in the form of a topographic/drainage division that separates the Kashmir basin into two geomorphic divisions – a topographic elevation in the SE and a topographic depression in the NW (Dar et al., 2014; Shah and Malik, 2017). Sedimentation in the valley has been primarily controlled by fluvial

Table 15.1 Major physiographic units in the geographical divisions of Jammu and Kashmir

S. no.	Administrative division	Major physiographic units	Geographical division	Districts
I.	Kashmir division	• Mountains (Zanskar, Pir Panjal) • Karewas • Alluvial Plains	Kashmir/ Jhelum Valley	1. Anantnag 2. Bandipora 3. Baramulla 4. Budgam 5. Ganderbal 6. Kulgam 7. Kupwara 8. Pulwama 9. Shopian 10. Srinagar
II.	Jammu division	• High Hills/Mountains (Pir Panjal) • Dun belt • Siwaliks • Murrees • Outer plains – Sirowal and Kandi	Tawi Valley Pir Panjal Chenab Valley	1. Jammu 2. Kathua 3. Reasi 4. Samba 5. Udhampur 1. Poonch 2. Rajouri 1. Doda 2. Ramban 3. Kishtawar

Figure 15.2 (a) Excavation showing layers of silt and clay in Batpora, Srinagar; (b) coarse alluvial deposits at Harwan, Srinagar with the Zabarwan range in the backdrop; (c) fractured, cherty dolomite at the Anji Khad bridge site in Reasi (Courtesy: KRCL) and (d) Marshy land with organic soil at CUK, Ganderbal, Kashmir.

Figure 15.3 (a) Typical unreinforced brick masonry house in Kashmir; (b) traditional construction (Dhajji Dewari) in Kashmir; (c) well foundation provided for a foot-
bridge over Jhelum, Srinagar and (d) high rocky overburden at the portal of a
tunnel in Reasi, Jammu.

activity, climatic variations and the rise of Pir Panjal (Bhatt, 1989). Millions of years
ago, Kashmir had been a landform under the Tethys sea. The collision of the Indian
plate with the Eurasian plate caused compression and thrusting in this region, as a
result of which the primaeval fluvial system got impounded as a huge lake known as
the Karewa Lake. This lake received sediments from the surrounding uplifted mountain ranges. The rise of the Pir Panjal led to the steepening of river courses, causing
deposition of these sediments as intermontane basin-fill deposits. Furthermore, the
Karewa Lake started shrinking and shifting towards the NE portion of the valley
(Dar et al., 2014). The water in the lake got drained through the Baramulla gorge in
event of a large earthquake, leaving behind the remnants of the Karewa Lake – the
present-day Jhelum, Dal, Anchar and Wular, lying within the topographic depression. Sediments of the Karewa Lake are thus fluvial, glacio-fluvial and lacustrine
in nature and are now known as Karewa sediments (Burbank and Johnson, 1983;
Basavaiah et al., 2010). Basin-wide sedimentation due to Panjal uplift has terminated;
however, fluvio-lacustrine sedimentation in lakes and around various rivers is continuing. Aeolian (loessic) deposition occurred as a result of arid conditions created by
climatic changes due to the Panjal uplift, forming a capping layer over the Karewa
(Pant et al., 2005; Dar et al., 2014). The development of the Kashmir basin has thus

been governed mainly by interrelated tectonic, climatic and erosional (river incision, glaciation) processes (Dar et al., 2015).

Rivers on the SW flank of the valley originate from steep terrain of hard rocks, cut through gentler slopes of softer Karewa sediments and then reach soft alluvium in the plains. The hard rocks include Panjal traps, quartzite, agglomeratic slate, shale, gneissose, granite and limestone. The Karewa deposits can be considered as soft rocks. Rocks of all ages (from Archean to Recent) are found in J&K, forming a complete stratigraphic record (Dar et al., 2014). The Salkhala series and Dogra Slates are Precambrian and represent the oldest stratigraphic basement (Wadia, 1975). Bedrocks in the Kashmir Valley are formed by the Panjal volcanics (Panjal trap and agglomeratic slate), gneiss/gneissose, granite, schists and Triassic limestone which are of different geological times.

Jammu is situated south of Pir Panjal and is mainly divided into northern hilly areas known as Jammu hills (Siwaliks and Murree) and the southern plain area known as Jammu plains (outer plains) merging with the great Indian alluvial plain in the south. Murree formations are characterised by purple mudstone, claystone and shale interbedded with sandstone; on the other hand, Siwaliks are composed of shale, mudstone and sandstone with clayey layers. The outer plains are further divided into the Kandi belt and the Sirowal belt (CGWB, 2013). The plains are filled with sediments which are debris of the Siwaliks brought down by the action of rivers and gravity.

15.2   MAJOR TYPES OF SOILS AND ROCKS

J&K consists of several valleys surrounded by mountains and hilly terrain. The topography of this region has played a major role in modifying rocks and soils, whereas the formation of the Himalayas has directly controlled its physiography. Soils in this region are of alluvial, lacustrine and glacial origin. J&K consists of four main geographical divisions (Table 15.1):

a. Jhelum/Kashmir Valley – The Kashmir Valley is elongated and bowl-shaped, drained by Jhelum and surrounded by mountain ranges along the margins. Jhelum flows in a north-westerly direction and drains most of the Kashmir Valley along with its tributaries. It merges with Sindh near Srinagar and then forms a delta in the Baramulla district in the form of Wular lake. It suddenly changes its direction at Baramulla, flowing through a deep gorge at Kadanyar, and out into the Chenab in Pakistan. Ten districts of J&K are a part of the Kashmir Valley (Table 15.1). Karewas forms the most prominent geological formation. Recent or Holocene alluvium (silt, clay and sand) has been deposited in the flood plains of Jhelum along its course. The belt along the periphery comprises mountainous soils and hard rocks. Heterogeneous glacial and colluvial deposits comprising boulders, cobbles, pebbles and gravels are found in these margins, which gradually merge into the valley fill consisting of fine-grained sediments (CGWB, 2013). Terraces, alluvial fans, small hills of conglomerates and debris and smaller valleys formed by geomorphic processes are common features in the region. Peaty soils, locally known as Nambal soils, are found near rivers, lakes and wetlands.

b. Tawi Valley (Jammu hills and plains) – Tawi collects water from northeast of Jammu, flows around the Jammu city and drains the interior mountains. The river forms the Tawi Valley, which is surrounded by hills comprising ridges and steep escarpments. The valley is filled with alluvial sediments deposited by the river and its tributaries, brought down from the Siwaliks. This belt constitutes Jammu hills and Jammu plains. Jammu hills comprise Siwaliks and Murrees, whereas, Jammu plains are unconsolidated, loose alluvial deposits, merging with the Indian plains.

c. Pir Panjal – Pir Panjal is a mountain range running along the NW-SE direction and is a part of the Lesser Himalayas. It forms a barrier between the Kashmir Valley and the Jammu division. Pir Panjal passes through the districts of Poonch, Rajouri and Kishtawar, and some portions of Udhampur, Reasi and Ramban.

d. Chenab Valley – Chenab is another major river flowing through Jammu. The river originates from Pir Panjal and deposits high-velocity fluvial sediments into the Chenab Valley. It follows the regional trend of NW-SE, except at three locations – Kishtawar, Salal and Ramban – where it changes the direction southwards. The region is mostly hilly and lies between Pir Panjal (Lesser Himalayas) and Siwalik (Outer Himalayas) in the Jammu division. The basin is elongated and narrow and covers the districts of Doda, Ramban and some portions of Udhampur, Kishtawar, Jammu and Rajouri (Romshoo, 2016). The Chenab basin has hilly regions of Murrees and Siwaliks, mountain meadows, sub-mountain soils of debris of Siwaliks and alluvial soils which are sediments of Chenab in the valley portion (CGWB, 2013).

15.2.1 Rocks

The geology and physiography of this region is a product of complex tectonic as well as environmental processes that have occurred over millions of years. Accordingly, soils and rocks found in this region are of varied types. The major types of rocks found here are:

a. Igneous rocks – The collision of the Indian plate with the Eurasian plate created a high temperature and pressure in the crust, leading to volcanic eruptions in the region. This brought out magma, which spread over a large portion of the Kashmir Valley and cooled to form igneous rocks. In J&K, Panjal volcanic series is a manifestation of this igneous activity. It is a thick volcanic formation constituting the agglomeratic slate and the overlying Panjal traps. Agglomeratic slate is the lower pyroclastic portion and Panjal traps are the upper bedded flows (Middlemiss, 1910). Granites and basic intrusives are also found in the region.

b. Sedimentary rocks – Limestone and dolomite are spread over a wide area in the region. These are remnants of the uplifted floor of the Tethys sea. Murrees and Siwaliks comprise mostly sedimentary rocks including shale, claystone, siltstone, sandstone and mudstone.

c. Metamorphic rocks – Metamorphic rocks are also found including quartzite, slate, schist and phyllite formed from sedimentary rocks and gneiss formed from igneous rocks.

15.2.2 Soils

The valley portion is filled predominantly with fine-grained soils (silt and clay) whereas higher elevations/hilly terrains have coarse-grained soils and solid rock. Surficial soils are locally modified by agricultural practices, presence of forests and water bodies. The following types of soils are found in J&K based on origin:

a. Glacial Soils – Most of the region, especially the Kashmir Valley, is filled with thick granular sediments which were deposited by melting glaciers during the inter-glaciation period in the valley, subsequently overlaid by recent alluvial deposits. Places like Sonmarg, Gulmarg, Pahalgam, Yusmarg, Tangmarg and so on have cemented conglomerates resting on glacially moulded hills below Karewa deposits (Bhat, 2017).

b. Karewa (Older Alluvium) – The valley floor is overlaid by unconsolidated and semi-consolidated older alluvium and recent alluvium. Karewa or Karewa Group is composed of ~1,300 m thick Plio-Pleistocene glacio-fluvial sediments, comprising clays, fine to coarse-grained sands and conglomerates with lignite beds (Singh, 1982; Burbank and Johnson, 1983; Kotlia, 1985). The coarser deposits are restricted to the peripheral parts of the valley, whereas the finer variety prevails in the central portion. These were deposited in two stages – Upper Karewa and Lower Karewa, separated by an unconformity. The lower Karewa comprises silt and sand intermixed with gravel and occasional lignite seams. The upper Karewa is in the form of mounds, table lands and terraces, comprising marls and silts with medium to coarse-grained gravels and boulders. Singh (1982) subdivided the Karewa Group into Hirpur and Nagum formations, whereas Bhatt (1989) classified them further into Hirpur, Nagum and Dilpur formations. Horizons of volcanic ash are seen within these deposits (Burbank and Johnson, 1982). Karewa deposits are mostly present on the long gentle slopes of the southwestern flank of the Kashmir Valley (Dar et al., 2015). These are found as low flat mounds or elevated plateaus in the Kashmir Valley in areas like Kulgam, Shopian, Budgam, Qazigund, Tangmarg, Gulmarg, Baramulla, Pampore, Awantipora, Anantnag, Mattan, Tral and Ganderbal (Bhat, 2017). In the Jammu division, these are present in Kishtwar and Bhadarwah tracts.

c. Recent Alluvium – The top layer overlying the Karewas is formed by the recent (Holocene) to sub-recent alluvial sediments brought down by rivers in the valley, especially in the floodplains of Jhelum and its tributaries. These soils mainly consist of alternating layers of sand, silt and clay, with occasional gravels and boulders. The outer plains of Jammu are filled with sediments brought down from Siwaliks and alluvial deposits of rivers like Ravi and Tawi.

d. Colluvial Soils – Soil erosion and landslides due to gravity bring down sediments from mountains and hills.

e. Lacustrine Deposits – Several lakes like Dal, Manasbal, Wular, and Anchar present in the region contribute to the lacustrine deposits.

f. Aeolian Soils – Thin layers of Aeolian soils are also found in patches in Kashmir, which were deposited due to the arid conditions developed during the formation of the valley. These mostly cover the Karewa deposits on the SW flank (Pant et al., 1978). Horizons of volcanic ash deposited by the erupting volcanoes are seen within the Karewa deposits (Burbank and Johnson, 1982).

15.3   PROPERTIES OF SOILS AND ROCKS

The main sources of borehole data for this chapter are various consultancies, including Space Engineers Consortium Limited (SECL) Srinagar, Road Research and Materials Testing Laboratory (RRMTL) Srinagar, Konkan Railway Corporation Limited (KRCL) and Rail India Technical and Economic Service (RITES) Limited. In addition to these, information was collected from some published reports published by governmental agencies like Central Ground Water Board (CGWB), a few research papers and site investigation reports of NIT Srinagar. Boring logs have been created at significant locations in each district in order to give a representative picture of the main types of soils and rocks present throughout the region (Figures 15.4 and 15.5). Important geotechnical parameters and the actual depths of the boreholes have been presented in a tabular form in Table 15.2. The boring logs are not made up to the scale, instead, they are simply a pictorial representation of the strata present at various locations.

15.3.1   Kashmir Valley

The Srinagar district comprises mountain ranges; few hills like Sulaiman/Shankaracharya hill, Hariparbat and Zabarwan hills and Karewa deposits. A surficial layer of soft unconsolidated alluvium covers formations like Karewa, volcanics and Palaeozoic sedimentaries (CGWB, 2013). These have been deposited over the bedrock which comprises limestone, shale, andesites, slates, quartzites, siltstones and dolomites.

Rock exposures consisting of slates, phyllites and quartzites juxtaposed with Panjal volcanics (traps and agglomeratic slates) and granites are found near Faqir Gujri to Darawain, and near Alusteng (Mukhtar, 2008). Sulaiman hill, base of Zabarwan

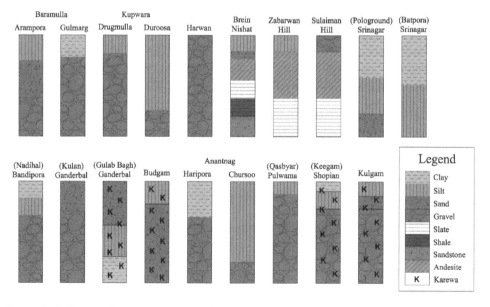

Figure 15.4 Pictorial representation of boring logs at various locations in the Kashmir Valley division created using CAD software (not to scale).

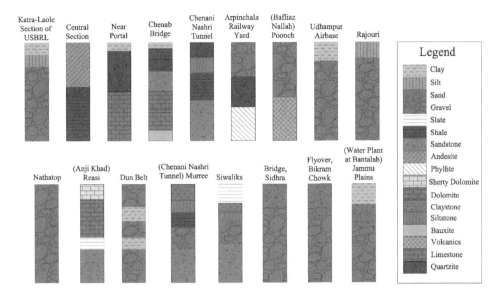

Figure 15.5 Pictorial representation of boring logs at various locations in the Jammu division created using CAD software (not to scale).

hill, Koh-Maran and Darawain in Srinagar are composed of Panjal traps overlying agglomeratic slates. The Brein hill is constituted of Panjal volcanics overlying shales and sandstones, while the Mahadev peak is composed of Panjal volcanics underlying Triassic limestones and shale.

Kashmir is a bowl-shaped basin with a depth of bedrock maximum at the centre. Boring logs indicate the maximum depth of soft unconsolidated soils at the centre, with decreasing thickness towards the basin edges. Flood plains of Jhelum and its tributaries are filled with recent flood outwash comprising interspersed soft silt, clay and sand layers. At the edges of the basin, the soft soil merges with sub-mountainous soil comprising coarse-grained sand, gravel, and stiff silt-clay layers.

Boring logs in Srinagar indicate the presence of soft, unconsolidated silt, clay and sand layers. Soils are very soft in the vicinity of water bodies like Dal, Anchar, Nigeen and Jhelum and have seams of organic soils and peat, especially in areas near wetlands. Waterlogging and decomposition leading to settlements are common problems in these soils. Lacustrine soils are present at the banks of lakes. Organic and peaty soils are present near rivers, lakes and wetlands in areas like Bemina, Qamarwari, Babademb, Dooldemb, Nigeen and so on. The Standard Penetration Test (SPT N) value varies significantly over the whole region. Very soft to soft soils with SPT N values ranging between 4 and 15 are found in most places especially in the vicinity of water bodies. Places like Anchar, Nawakadal, Bemina, Qamarwari, Rajbagh and Lalchowk and areas surrounding Dal have soft soils. Stiffer soils with SPT N values greater than 15 and even reaching refusal are found at the foot of hills and mountains, in areas like Hawal, Rainawari, Saidakadal, Harwan, Cheshmashi, Shalimar, Pandrethan and so on. Areas located on the mountains and hills like Brein, Harwan, Zabarwan and Cheshmashahi have stiff silty and clayey soils overlying gravels and hard rock.

Table 15.2 Variation of properties of soils and rocks based on SPT data from consultancies and available published reports

Division	District	Location	Source of data	Classification	SPT N	Consistency limits (%)	c (kg/cm²)	Φ (°)
								Description for rocks
Kashmir/ Jhelum Valley	Anantnag	Haripora	SPT data	Silty clay (CI-MI) over gravels	35–48 at surface, refusal beyond 6 m	$w_L = 24$–33, $w_P = 19$–21, $I_P = 04$–12	0.28– 0.38 0	17–19 32
		Chursoo	SPT data	Silt (ML)	6 at surface, 32 at 20 m depth	$w_L = 36$–39, $w_P = 25$–27, $I_P = 10$–13	0.23	9.5
	Bandipora	Gundpora Nadihal	SPT data	Silty Gravel (GM)	Refusal	NP	0	32
		Paribal	SPT data	Clay and silt (CL, ML)	22 at surface up to 61 at 15 m depth	$w_L = 34$–39, $w_P = 11$–16, $I_P = 11$–16	0.25– 0.29	17–18
	Baramulla	Gulmarg	SPT data	Well graded silty gravel (GW-GM)	Refusal	NP	0	32
		Arampora	SPT data	Layers of silt (ML), clay (CI-MI) and sand (SM, SC)	4–6 at surface, 20 at 20 m depth	NP silt, sand; Clay $w_L = 35$–39 $w_P = 10$–13 $I_P = 24$–28	0.05– 0.17	09–17
	Budgam	Chrari-i-Sharief	SPT data	Clayey and silty sand (CL, SM)	15 at the surface, 22 at 12 m depth	$w_L = 25$–28 $w_P = 18$–20 $I_P = 6$–8	0.16– 0.20	07–15
	Ganderbal	Gulabbagh	SPT data	Silt (M)	19–25	NP	0.12– 0.17	09–18
		Kulan	SPT data	Well graded sand (SW) at shallow depth overlying gravels (GP)	34 at surface, refusal at 6m depth	NP	0.08– 0.09 0	21–29 32

(Continued)

Table 15.2 (Continued) Variation of properties of soils and rocks based on SPT data from consultancies and available published reports

Division	District	Location	Source of data	Classification	SPT N	Consistency limits (%)	c (kg/cm²)	Φ (°)
								Description for rocks
	Kulgam	Gopalpora	SPT data	Silt (M) over gravels in sand (GP)	15 at surface, refusal at 2m depth	NP	0	31.5
		Chowgam	SPT data	Layers of silt and clay (MI, CI)	17 at surface, refusal at 8 m depth	$w_L = 35–38$ $w_P = 23–27$ $I_P = 10–12$	0.21–0.25	13–16
	Kupwara	Drugmulla	SPT data	Silt (ML) overlying gravels intermixed with sand (GW)	11–37 at surface, refusal at 6 m depth	NP	0	31
		Duroosa	SPT data	Layers of silt and clay (ML, CL)	2–6 at surface, 14 at 15 m depth, 34 at 20 m	$w_L = 30–41$ $w_P = 7–26$ $I_P = 07–27$	0.17–0.25	09–11
	Pulwama	Qasbyar	SPT data	Silty clay (CL) overlying gravels intermixed with sand (GW-SM)	16 at surface, refusal at 3 m depth	NP	0	32
		Wuyan, Pampore	SPT data	Silt (ML)	11–8 at surface, 15–20 at 20 m depth	$w_L = 30–41$ $w_P = 13–27$ $I_P = 7–27$	0.19–0.21	10

(Continued)

Table 15.2 (Continued) Variation of properties of soils and rocks based on SPT data from consultancies and available published reports

Division	District	Location	Source of data	Classification	SPT N	Consistency limits (%)	c (kg/cm²)	φ (°)	Description for rocks
	Shopian	Keegam	SPT data	Silt (ML) over gravel intermixed with sand (GW-SM)	16 at surface, refusal at 3 m depth	NP	0	32	
	Srinagar	Pologround, Lalchowk	SPT data	Silt (M) overlying silty sand (SM)	2–6 at the surface 10–14 at greater depths	Silt $w_L = 21–39$ $w_P = 25–28$ $I_P = 10–14$; NP Sand	0.17–0.27	10–15	
		Batpora	SPT data	Clay over silt (C,M)	15–24 at shallow depth, reaching refusal at 25 m	$w_L = 33–38$ $w_P = 25–29$ $I_P = 8–13$	0.16–0.27	08–16	
		Harwan	Mukhtar (2008)	Gravel intermixed with sand (GP)	Refusal at shallow depth	NP	0		18–21 for soil at the surface, ~32 for gravels
		Nishat, Brein Hill	Mukhtar (2008)	Gravel intermixed with sand, silt, clay (GP)	Refusal at shallow depth	NP	0		18–21 for soil at the surface, ~32 for gravels
		Zabarwan, Hill	Mukhtar (2008)	Gravel intermixed with sand, silt, clay (GP)	Refusal at shallow depth	NP	0		18–21for soil at the surface, ~32 for gravels
		Sulaiman Hill	Mukhtar (2008)	Gravel intermixed with sand, silt, clay (GP)	Refusal at shallow depth	NP	0		18–21for soil at the surface, ~32 for gravels

(Continued)

Table 15.2 (Continued) Variation of properties of soils and rocks based on SPT data from consultancies and available published reports

Division	District	Location	Source of data	Classification	SPT N	Consistency limits (%)	c (kg/cm²)	Φ (°)	Description for rocks
Tawi Valley	Jammu	Bridge over Tawi, Sidhra	SPT data	Silty sand (SM) up to 23 m, overlying cobbles, boulders and gravels in sandy matrix	15–35 up to 23 m, refusal beyond	NP	0	32–40	
		Flyover, Bikram Chowk	SPT data	Silty sand (SM) overlying cobbles, boulders and gravels in sandy matrix	20 at surface, refusal at 3 m depth	NP	0	32–40	
		Water Plant, Bantalab	SPT data	Silty sand (SM) overlying cobbles, boulders and gravels in sandy matrix	20 at surface, refusal at 1.5 m depth	NP	0	32–40	
	Kathua	Siwaliks	CGWB (2013)	Slate, siltstone, claystone, gravels, sandstone	–	–	–	–	
	Reasi	Anji Khad Bridge Site	SPT data	Cherty Dolomite, dolomite, slates and sandstones with criss-cross shear seams	Core drilling up to 50–60 m depth	–			Highly weathered at surface, highly fractured and jointed, with criss-cross shear seams filled with gouge material, poor RQD, moderately strong to strong

(Continued)

Table 15.2 (Continued) Variation of properties of soils and rocks based on SPT data from consultancies and available published reports

Division	District	Location	Source of data	Classification	SPT N	Consistency limits (%)	c (kg/cm²)	φ (°)	Description for rocks
		Chenab Bridge Site	Gupta and Singh (2015)	Clay, quartzite, shale, sandstone, limestone, dolomitic limestone, bauxite	Core drilling up to 150 m depth	—	56°		Rock strength (kg/cm²) 250–500; UCS (kg/cm²) 445–1,581; V_p (m/s) 4,658–6,527; V_s (m/s) 3,403–3,714; Young's Modulus (GPa) 61; Highly weathered, fractured, RQD as less as 13% at shallow depth and 98% at 150 m depth
		Dun Belt	CGWB (2013)	Silty gravels, and clay layers	Refusal, DCPT and plate load test is feasible	NP	0	30–38	
	Samba	Murree	CGWB (2013)	Sandstone, siltstone, claystone and clay-shale, shaly limestone, thin bands of sandstone	—	—			Highly weathered at surface, jointed, fractured, exhibit swelling due to absorption of water
	Udhampur	Udhampur Airbase	SPT data	Clay and Silt (C, M), overlying gravels (GP)	10–17 at surface, 34 at 4 m depth	NP	0.44– 0.51, 0	8–10, 32	
		Chenani-Nashri Tunnel	Palomba et al. (2013)	Sandstone, siltstone and claystone	—	—	—	—	
		Nathatop	SPT data	Silty gravel	47 at surface, refusal at shallow depth	NP	0.06– 0.11	26–30	
		—	Kapoor et al. (2017)	Silt up to 10–12 m, quartzites, schists	—	—	—	32–38	

(Continued)

Table 15.2 (Continued) Variation of properties of soils and rocks based on SPT data from consultancies and available published reports

Division	District	Location	Source of data	Classification	SPT N	Consistency limits (%)	c (kg/cm²)	Φ (°)	Description for rocks
Pir Panjal		Pir Panjal Tunnel from Qazigund to Banihal	Höfer-Öllinger and Millen (2014)	Andesite, basalt, quartzite, silicified limestone, agglomerates, shale, agglomeratic shale, limestone and tuff, quaternary deposits at portals	Core drilling	—	—	—	
	Poonch	Bridge over Bafliaz Nallah, Mughal Road	SPT data	Rock debris, gravel over rocks like quartzite breccia, cherty dolomite, dolomite; occasional thin bands of shale and slate, chert beds	SPT up to 10 m depth, core drilling up to 25 m depth	—			Fresh, hard and competent dolomite (90% RQD); soft, weathered to fresh breccia and chert
	Rajouri	GDC, Rajouri	SPT data	Silt (ML) over gravel (GP)	Refusal at 2m depth	NP	0.2 / 0	17 / 32	—
Chenab Valley	Doda	—	CGWB (2013)	Lacustrine and alluvial deposits, silt, quartzites, schists	—	—	—	—	
	Ramban	Katra-Laole Section of USBRL	SPT data	Silt, clay up to 17 m overlying gravels, boulders of quartzites, andesites, limestones	Refusal beyond 17 m	NP	0		
		Arpinchala Railway Station	SPT data	Silt, clay over gravels, boulders of quartzite, phyllite, gneiss	9–45 up to 8 m, refusal thereafter	—			Highly fractured, jointed, RQD is 0%–96%

The belt lying between Dal and Anchar lakes including areas like Habak, Soura, Buchpora, Batpora and so on have deposits of Karewa which are stiff in nature (Mukhtar, 2008; Bhat, 2017). Moreover, the water table is comparatively lower and hence, soils in this belt show high stiffness and high SPT N values.

Ganderbal is a neighbouring district lying east of Srinagar. It consists of mountains, hills and valley portions. Consequently, mountainous and glacial soils are present near hills and mountains and loose unconsolidated soils are present in the plains (e.g. in Sonmarg). Geotechnical data collected from this district show the predominance of sands and gravels in areas lying towards the mountains. The Dynamic Cone Penetration Test (DCPT) and plate load tests have been carried out instead of SPT in such areas. Silts and clays with N values ranging between 5 and 50 have been found in Alusteng, Buseerbagh, Gulabbagh, Gund Rehman, some areas of Kangan, Sehpora, Zazuna and so on. These fine-grained deposits are mostly non-plastic or of low plasticity. In many of the places like Laar and CUK Ganderbal, very soft organic soils with a high water content are present as the top layer, underlain by non-plastic silts and sands and gravels, which act as aquifers of water. Water under artesian pressure is present in the coarse-grained deposits fed by rivers and nallahs nearby. The top layer of organic soils is predominantly a product of rice and wheat cultivation in the region.

Borehole data from Bandipora show medium-stiff to stiff ($N > 15$) deposits of clays, silts and sands in areas like Mangnipora, Pethkote and Paribal. The fine-grained soils have low to medium plasticity. Most of the areas, however, have gravels and boulders showing high penetration resistance. Wular in the Bandipora district is a lake in a depression surrounded by mountain ranges that are constituted by several geological formations. These include hard rocks including Panjal volcanics, granite, gneiss, schists, limestone and slates with layers of shale and overlying recent alluvium comprising gravels, sand, silt and clay (Thakur and Rawat, 1992; Bhat, 2017).

Kupwara is characterised by Karewa tablelands, hard rock ridges, valleys and river terraces (CGWB, 2013). It is mostly hilly and mountainous with a valley portion in between. Igneous and metamorphic rocks of Panjal traps and Zewan limestone beds are found in this region. Geotechnical data reveal that the majority of the locations require DCPT, as the refusal is obtained at shallow depths due to the presence of gravels and boulders, except for a few areas like Chowkibal, Braripora, Durosa, Wadoora, Gotingoo and Wuyan, which have layers of soft to medium-stiff and stiff silt, clay and sand with N values ranging from 4 to 50. Karewa soils are found as tablelands in the southern areas like Handwara and Langate. Karewa soils are generally stiff with excellent bearing capacity under dry conditions. As these are mostly present as highlands in which the water table is low, the resulting N values are high suggesting high stiffness. However, in the presence of moisture, these soils become soft and hence problematic. Significant slope and foundation failures are reported from susceptible areas in Handwara and other areas.

Baramulla is a district located northwest of Srinagar. Most of the regions are mountainous and therefore, boring logs reveal deposits of conglomerates and sands. Data from a few areas like Daulatpora, Gutiyar, Khwajabagh, Shrakwara, Seer Jageer bridge, Mirpora Rohama and Pandithpora Bala suggest the presence of medium-stiff to stiff deposits of clays, silts and sands.

The slope of the valley increases gradually from the centre towards the south. The Budgam district lies south of Srinagar on this gentle slope formed mainly of Karewa deposits. Data indicate the predominance of stiff and dense deposits of silt, clay and sand showing N values greater than 15 and reaching refusal within 20 m depth. Lower N values (<10) have been observed in some regions like Bathar, Bunpora, Sebden Railway Station and Mazhama, which are nearer to the central plains of the valley. Places like Bugroo, Chrar-i-Sharief and Humhama have deposits of gravels and show refusal during the SPT.

The Shopian district is a fan-shaped region lying mostly on Karewa tablelands. The Karewa deposits are finer (silt, clay) in areas towards the central part of the valley (e.g. Alamganj, Herman and DK Pora), whereas they are coarser (boulders, gravels and sand) towards the Himalayan side (e.g. Dubjan, Aharbal, Sedow, Hirpur and Daygam). Areas like Largam, Sangarwani and Gund have deposits of medium-stiff to stiff $(N > 10)$ silt and clay with low to medium plasticity. Most of the region, however, has gravel intermixed with sand, which makes it imperative to perform DCPT or plate load tests instead of the SPT.

Pulwama and Anantnag lie on the southern flank of the valley consisting of hills and mountains of fractured and jointed Panjal trap formations. Karewa highlands, hard rock ridges, river terraces and intervening valleys are the major physiographic units in these districts (CGWB, 2013). Karewa plateaus are seen in regions like Pampore, Tral, Rohmoo, Newa and Malangpora in Pulwama and Mattan, Bijbehara, Qazigund and Shahabad. Rivers have deposited recent alluvium over Karewas and Panjal traps in the plains. Medium-stiff to stiff $(N > 15)$ Karewa deposits consisting of silt, clay and sands have been observed in boring logs obtained from the Pampore region of Pulwama. Soils with low N values (<10) have also been observed in a few places like Shariefabad and Wuyan in Pampore, which represent recent alluvial deposits. Many areas in Pulwama have deposits of gravel and few boulders of andesite and volcanic rocks towards the mountainous tracts. Boring logs in Anantnag reveal soft to medium-stiff alluvial deposits of silt, clay and sand in a few areas like Chursoo Awantipora, Donipawa, Khandaypora, Magraypora, Reshipora, Khandi and Noopora in the plains; most of the areas towards the higher reaches have gravels, pebbles and boulders intermixed with sands, silt and clay showing refusal at shallow depths.

Places like Yaripora, Chowgam, Tengsarg, Brazloo bridge site, Guffan and Naidgund in the Kulgam district have fine-grained soils with sand layers. These soils are medium-stiff to stiff showing SPTN values ranging from 11 to refusal. Soils are of low to medium plasticity with a plasticity index usually less than 14%.

15.3.2 Pir Panjal

Jammu is separated from Kashmir by the Pir Panjal. Mountain ranges and hills separated by deep narrow valleys, terraces, valleys with gentle slopes are found in Rajouri and Poonch. Bedrock is formed of Subathu, Salkhala and Sirban formations including limestone, shale, sandstone, phyllites and gneiss (CGWB, 2013). These are overlain by Murrees, Siwalik and the recent alluvial deposits in the form of gravels, boulders, sand, silt and clay.

Pir Panjal passes through the southern flank of Kashmir Valley, and mainly through the districts of Poonch and Rajouri of the Jammu division. Pir Panjal railway tunnel lies between Banihal and Qazigund, about 100 km south of Srinagar. Geological studies along the tunnel alignment show the presence of hard rocks like andesite and basalt (Panjal traps), quartzite, silicified limestone and agglomerates while the moderately hard rocks include shale, agglomeratic shale, limestone and tuff. Quaternary deposits are observed at portal sections (Höfer-Öllinger and Millen, 2014). Jawahar tunnel has been driven through rocks of flagstones, slates, shale and limestone at the portal sections and Panjal traps exposed at the central section (Verma, 1967).

Geotechnical report of the Katra/Reasi Katra-Laole section of Udhampur-Srinagar-Baramula Railway Line reveals fine-grained soil up to a maximum depth of 17 m overlying the fluvio-glacial deposit (soil mixed with boulders), including cobbles, boulders, rock fragments of quartzites, andesites and limestones.

Rajouri is a town in Pir Panjal belt. Boring logs from the Government Women's College, Rajouri indicate the presence of gravelly substrata with a thin layer of silty clay as overburden. Borehole data are available at a bridge site across Bafliaz Nallah on the Mughal road, about 3 km from District Headquarters of Poonch. Poonch town is located about 246 km north-west of Jammu city in the Lesser Himalayas (Pir Panjal). One of the most famous roads, the newly constructed Mughal road passes through a narrow belt in this region. The data at the nallah indicate the presence of quartzite breccia, cherty dolomite, massive to blocky dolomite from top to bottom. Occasional thin bands of shale and slate, chert beds and irregular quartz and hematite veins are seen. Dolomite is unweathered, hard and competent. On the other hand, breccia and chert are soft and mostly weathered.

Geotechnical investigation along the alignment of Mughal road reveals different series of rock groups – sandstone, siltstone, claystone and clay-shale in the area (Murree). Sandstones in this region are fine to coarse-grained, calcareous and ferruginous in some places, showing high to medium strength. Siltstones are fine-grained, mostly micaceous with veins of calcite, showing fair to high strength. Sandstones and siltstones exhibit prominent bedding laminations. Claystones are fine-grained showing soft to hard texture, which become friable on the loss of moisture or release of overburdened stresses. Clay shales are formed from the weathering of shale stones and are highly friable, showing swelling characteristics on the absorption of moisture. Clay shales are usually found in the form of thin bands/shear seams leading to instability in the foundation. Although sandstones exhibit high strength, they are highly jointed, fractured, sheared and thinly bedded at some places, which can again lead to instability.

15.3.3 Tawi Valley (Jammu hills and plains)

Jammu is a hilly and mountainous terrain made of rocks of Siwalik and Murree groups overlying rocks of Subathu, Salkhala and Sirban groups. Hard rock formations include granite and agglomeratic slates of the Panjal group. Near the base of the hills, the plains are underlain by coarse Kandi belt consisting of boulders, cobbles, sand and hard clay; and Sirowal belt which is a finer outwash comprising fine-grained soils overlying sands and gravels.

Kathua and Udhampur are characterised by mountain ranges, gorges, valleys and deep canyons. The northern portion has valleys known as the Dun belt separating Siwaliks from the Pir Panjal and comprising recent valleyfill deposits such as boulders, cobbles and clay layers (CGWB, 2013). The rest of the area is a part of the outer plains.

At Nathatop, Udhampur, gravels and pebbles in a mixture of hard clay and silt are found. N value is high, reaching refusal at shallow depths. Residual/transported foothill or stream deposits are observed in Udhampur. The area near the Udhampur airbase has silty gravel or poorly graded gravel-sand-silt mixtures (GM). The site is stiff and has high penetration resistance. N value reaches 30 at a depth of 4 m, beyond which refusal is encountered.

Borehole data are available for a bridge site on the banks of Tawi near Sidhra in Jammu. Quaternary sediments in the form of cobbles, gravels, pebbles in a sandy matrix and occasional boulders of mainly quartzite, sandstone and dolomite are found.

Layers of silt and traces of low plasticity clay are also present. SPT N values are high (>15 at the surface) with refusal obtained within 20 m depth.

Boring log obtained at the construction site of Bikram Chowk Flyover built over Tawi, near Jewel Chowk in Jammu, reveals the presence of gravels, pebble and boulders intermixed with coarse sand. The resistance to penetration is high and $N > 50$ is obtained at 3 m depth.

Information is also available for the Chenab Water supply plant, located near Jammu Cantonment, near Bantalab in Jammu plains. This site shows the presence of coarse sand with an N value of about 20, overlying gravels, pebbles and boulders in a matrix of sandy soil showing refusal during SPT.

15.3.4 Chenab Valley

Chenab Valley is surrounded by hills that mainly belong to Salkhalas, Dogra slates and Murrees. The basement rock is mostly quartzite and gneiss with overlying fine to coarse sediments of the Chenab. Several erosional hills formed by depositions from the mountains are found in the region. Soil is loose and sandy with very less moisture. The region is mostly hilly, and hence, landslides are frequent especially during rainy seasons. Kishtawar is a steep-sloped plateau divided into two segments by an EW trending ridge with alluvial covering (CGWB, 2013). It had once been a tectonic depression that was subsequently filled by lacustrine sediments. It has mountain ranges on the east and detached mounds of bedrock (schists) rising above the plateau up to around 100 m at the western boundary. Chenab and its tributaries drain this region, depositing alluvial sediments along with their courses. The plateau has silt in the top 10–12 m, underlain by 80 m silt and angular rock fragment of quartzites and schists.

Around 40 boreholes were drilled at the Chenab bridge site up to about 150 m depth, as a part of geotechnical and geological investigation for the construction of the bridge which is part of the USBRL (Gupta and Singh, 2015). Information collected from this source reveals that the bridge site is located on the mountains composed of soft to hard rocks. Layers of quartzite, shale, sandstone, limestone to shaly limestone, dolomitic limestone and bauxite have been observed in the drilled boreholes. The site is rocky and SPT shows refusal, however, the rocks are jointed and highly weathered at the surface, forming a layer of hard silty clay and clayey soil with rock fragments at the top.

Chenani-Nashri tunnel connects Chenani town in Udhampur district and Nashri town in Ramban district, passing mostly through the Chenab Valley. The tunnel alignment cuts through lower Murree formation which comprises alternate layers of sandstone, siltstone and claystone (Palomba et al., 2013). Siltsones and claystones are soft rocks. Sandstone has prominent joints containing silt-clay infills in some places.

Another major bridge crosses the river Chenab and connects the Katra–Reasi section of the USBRL. This is the Anji Khad bridge, and its location is geologically relatively unstable than the Chenab bridge site. Site investigation data from KRCL reveal that the Anji Khad nallah cuts through the competent cherty dolomite rock mass forming a narrow gorge. The banks expose an interbedded sequence of dolomite, cherty dolomite, siliceous dolomite, quartzite, slates and so on belonging to the Sirban Limestone group of rocks. These rocks are fine-grained showing a weak to strong nature. The rocks are fresh to highly weathered and fractured with occasional parallel as well as criss-cross shear seams.

The subsoil strata at Banihal to Qazigund is natural alluvium/flood outwash deposits, which generally include intermixed layers of silty clay or clayey-silt with potential seams of fine to medium sand. Soils are generally very soft to medium soft ($N < 15$) up to 12 m depth, beyond which medium-stiff to stiff/denser soils (N up to 55) exist.

Arpinchala railway yard in Ramban is a part of the USBRL. It is a stiff site characterised by silty and clayey soil, with pebbles, gravels, and boulders, overlying fractured and jointed, moderately strong to strong layers of quartzite, phyllite, gneiss and pegmatite.

15.4   USE OF SOILS AND ROCKS AS CONSTRUCTION MATERIALS

The majority of the residential houses in Kashmir are built of bricks made of clay (Figure 15.3a). The mortar used in between the bricks is either mud as found in older construction practice or cement as is prevalent in new constructions. In important structures like government buildings and schools, reinforced concrete or hollow concrete blocks with cement mortar are used. Stone masonry with mud or cement mortar is used in rural areas mostly in the hilly areas of North Kashmir. These structures are mostly laced by timber framework to provide more stability. Wooden truss supporting galvanised corrugated iron sheets form the roof. Old structures in Kashmir represent the traditional timber-brick masonry construction especially in old Srinagar city (Figure 15.3b). Timber forms the main framework with stone or burnt bricks forming the infill. This construction practice is of two types – Dhajji Dewari and Taq systems – in which the timber adds to the lateral load resistance of the structure. For the construction of the foundations, stone masonry is used for light residential buildings, whereas reinforced concrete is used for heavier structures. Silt and clay from borrow areas are used as fill materials and for the construction of embankments. Stone masonry is also used for the construction of retaining walls, most commonly in hilly terrains especially along roads leading to Gulmarg, Pahalgam, Sonmarg and so on.

Soils and rocks available locally are utilised for construction purposes. Bricks are manufactured from clay available locally in several brick kilns functioning in areas like Budgam, Pulwama, Anantnag, Kulgam and Baramulla. Aggregates for concrete and blocks of stones for masonry work are extracted in stone quarries from mountains, for example near Pandrethan in Srinagar. Gravels and sand are extracted from areas near riverbeds in regions like Ganderbal.

15.5 FOUNDATIONS AND OTHER GEOTECHNICAL STRUCTURES

The majority of the residential structures in J&K have strip foundations made of stone masonry. This type of foundation is mostly sufficient for such lightweight structures. In areas where soft soils and organic soils are present, there are problems of excessive settlement and tilting of structures. Raft foundation is mostly provided to overcome these types of failures. Organic soils undergo decomposition, waterlogging, and settlement, and have less bearing capacity. An alternative can be to excavate the top layer of organic matter and replace it with a compacted soil of better geotechnical parameters. For multistoreyed buildings and heavy structures like bridges and flyovers, pile foundations have been used in the whole region. Bridges over rivers with soft foundation soil like in Jhelum require well foundations, as has been provided for some bridges in Srinagar (Figure 15.3c).

Retaining walls are the most common geotechnical structures in hilly regions to prevent landslides. Wire meshes and berms are provided to contain rockfall. Soil nailing, use of anchor bolts and dowel bars are used to strengthen slopes by intercepting failure planes and containing rock joints. Shotcrete and geosynthetics are alternative options. Retaining walls are provided with drainage holes and pipes to remove rainwater in the backfill. In some cases, toe erosion of the retaining wall is prevented by using a wire mesh and providing deep foundations to the wall. Stone masonry is used for river protection works to prevent erosion of banks.

At rocky sites, even though the strength of rocks is supposedly quite high, factors like weathering, fractures, joints, presence of bedding planes, presence of shear zones filled with gouge material and soft rock beds cause problems like instability and reduction of strength. The construction of dams at such sites needs to cater for all the problems. Bridges need deeper foundations in the rock mass due to weathered and jointed planes. Necessary protection measures need be taken to protect rock slopes to account for the dip of rock formations, shear zones, folding and so on. Rock bolting, shotcreting and other remedial measures may be required.

15.6 OTHER GEOMATERIALS

The issue of siltation in Dal and Jhelum has been addressed by dredging the bed and flood channels of Jhelum and the shoreline of Dal. Dredging generates solid waste, which is normally disposed of at various sites, creating serious concerns for environmental concerns. The dredged soils from both Dal and Jhelum have thus been characterised to be put to various engineering applications.

Dredged material from Shalimar and Nishat basins has around 80%–90% sand, and some amount of gravel, silt and clay (Mir et al., 2015). Dredged soil obtained from lokut Dal and Bod Dal has poor to good characteristics (Mir et al., 2013) and can thus be used as fill material for agricultural use or land improvement in low-lying areas. Low compacted density and good drainage properties result in lower earth pressure, making it beneficial in the construction of embankments. The addition of stabilisers can improve the properties further.

Soil dredged from the Tailbal basin of Dal is classified as clayey/silty of low to medium compressibility (CL, CI, ML, MI) with a very high rate of loss of shear strength

(Mir, 2015). Similarly, dredged soil from Jhelum is observed to be composed of low to medium compressibility clay and silt, sand and organic matter (Mir et al., 2016). In the in-situ condition, dredged soil is not fit to be used for engineering purposes, which makes stabilisation of the material imperative before use as engineering construction material.

Geosynthetics have a great scope of use being lightweight and economically viable. These can be used for stabilising the slopes along the Srinagar-Jammu National Highway, for construction of roads, rockfall prevention, drainage, erosion control, prevention of mud-pumping in railway embankments along the Qazigund-Baramulla railway line which are built on cohesive soils, riverbank protection, surface protection, and containment of municipal wastes in landfills (Mir, 2014).

15.7   NATURAL HAZARDS

J&K is prone to several natural hazards, the most significant of which are the following:

a.   Earthquakes – Being in the Himalayan region and traversed by several faults, J&K has been struck by several major and minor earthquakes in the historic as well as recent past. Major earthquakes like 1555 (M_w=7.6) earthquake, 1885 (M_w=6.3) Pattan earthquake, 2005 (M_w=7.6) Muzaffarabad earthquake, 2013 (M_w=5–5.5) Kishtwar series of earthquakes, 2015 (M_w=7.5) Hindukush earthquake and 2019 (M_w=5.6) Mirpur earthquake have caused huge destruction to life and property. Major hazards like ground rupture, landslides, slope instability, permanent ground deformations, collapse of structures, liquefaction, subsidence of roads and collapse of tunnel portals are associated with these earthquakes.

b.   Ground motion amplification – Local sites effects due to soils can create a significant hazard for structures if not considered in the design. During the 1555 earthquake in Kashmir, more severe shaking was observed in villages surrounding the Martan Sun Temple in Mattan, Anantnag. The temple is built on a highland as compared to the surrounding valley. Higher amplification of seismic waves in the softer soils in the valley portion may be the reason for greater damage and stronger shaking experienced, as compared to the stiffer deposits in the temple region (Hough et al., 2009).

c.   Liquefaction – Liquefaction occurs when saturated soil loses shear strength as a result of shaking caused by earthquakes. During the 1555 earthquake, ground failure due to liquefaction displaced two towns Hasanpura and Hosainpora across Jhelum (Hough et al., 2009). Liquefaction was reported from Baramulla, Uri, Kupwara, along the banks of Jhelum in Kashmir Valley and Simbal village in Jammu after the 2005 Muzaffarabad earthquake. The types of soils, environmental conditions and high seismic activity make the region susceptible to significant liquefaction hazards in event of an earthquake (Zahoor et al.,2021a).

d.   Landslides and Rockfalls – Frequent instances of landslides and rockfall occur in areas of high relief in J&K. This is typically a major problem during rainfall and snowfall, especially in the weaker and softer rocks of Jammu hills. This has created a huge loss of life and property along the Srinagar-Jammu National Highway (NH-44), which forms the only road connection between Jammu and Srinagar.

Like the highway, the railway line connecting J&K has been constructed by cutting through the hills of Siwaliks, Murree and Pir Panjal. This section particularly has varied geological and strata conditions, presence of folds and faults, making it complex and creating major hurdles in construction projects Unplanned execution of excavation in these mountain ranges for road, railway and tunnel construction has increased these hazard manifolds. Geological formations of Murree and Siwaliks are prone to swelling, weathering and erosion on the absorption of water. Some of the critical slope failures along the Srinagar-Jammu highway are Samrauli landside, Nashri slide, Khoni Nallah slide, Shaitan Nallah slide (Mir, 2014), and Panthial landslide (Singh et al., 2010). The 2005 Kashmir earthquake caused widespread and destructive landslides in Punch, Uri, Baramulla (Kumar et al., 2005). Lakes were formed by landslide damming in Pir Panjal, leading to impounding of water. Such dams can fail due to overtopping of water causing loss of life and property on the downstream end.

e. Floods – Floods are common in catchment areas of rivers especially in the upper reaches. A widespread major flood affected the whole of Kashmir Valley including the low-lying Srinagar city in September 2014. Floodwaters from rivers breached several embankments and led to the collapse of several buildings (Gulzar et al., 2020) due to dynamic and static water pressure exerted on walls and foundations. The rise in the water table resulted in uplift pressure on floors and lateral hydrostatic pressure on walls of basements leading to their collapse and failure.

15.8 CASE STUDIES AND FIELD TESTS

a. Rakh-i-Arth (wetland) – Rakh-i-Arth is a marshy land located in the backwaters of Hokarsar wetland, Srinagar. It has been reclaimed for the construction of houses for relocating the dwellers of Dal and Nigeen. Several failures have been reported in the structures, like differential settlement leading to tilting of houses, cracks in the walls and floors and cracks in the strip foundations. The drainage system has also failed due to the submergence of pipelines below the water table. Due to insufficient compaction, during rainfall and snowfall, the filled land becomes soft creating problems for the residents.

b. Srinagar-Jammu National Highway – The highway passes through soft rocks prone to swelling and erosion on the absorption of water. Slope failures and erosion are thus common along this highway, even though several slope stability techniques have been implemented. Geosynthetics may work better and may prove to be a sustainable solution in such a scenario (Mir, 2014).

c. Multichannel analysis of surface waves (MASW) in Srinagar – MASW testing at 33 sites along the Srinagar metro alignment gave values of average shear wave velocity V_{S30} as low as 140 m/s in areas in the vicinity of Jhelum and Dal, and as high as 450 m/s in areas near the base of hills and mountains (Zahoor et al., 2021b). The Karewa belt between Dal and Anchar showed high V_{S30} values. High V_{S30} values signify stiff site conditions meaning better performance during earthquakes and low values represent soft site conditions with a high risk of amplification of seismic waves.

d. Geotechnical evaluation of soils in Srinagar – A detailed study was carried out by Zahoor et al. (2019) to obtain the geotechnical parameters of soils at 30 representative sites in Srinagar city. It was observed that the soils have the better bearing capacity and geotechnical parameters along the base of hills, mountains and in the northern part of the city between Dal and Anchar.

e. Bemina – Bemina is a low-lying marshy area in Srinagar, which was part of a wetland and flood absorption basin earlier. It has now been converted into a residential area; therefore, the water table is shallow. Soils are fine-grained, very soft, having high moisture content and a considerable amount of organic matter. It has been found that modification of the ground is important before the construction of any important building. Pile foundations are imperative for multistoreyed buildings. Raft foundations or strip foundations suffice for residential buildings after filling the land with suitable soil from a borrow area.

f. Tunnels – Tunnels (Figure 15.2d) excavated in the Himalayas as a part of the US-BRL project are deep with overburden as high as 1 km, e.g. in Chenani-Nashri tunnel (Palomba et al., 2013). The design and excavation works of deep tunnels in the Himalayas are influenced by high-stress conditions due to overburden, high water pressures, vulnerability to fault zones, presence of folds, poor and heterogeneous geology, high jointing and fracturing of rocks, exposure to squeezing, and swelling due to expansive soil and rock conditions. Moreover, tunnel portals are most susceptible to slope failures due to the presence of quaternary deposits. Heavy water inflow can cause a collapse of soluble rocks like limestone. Geologists and engineers faced great challenges in the excavation of tunnels through the Pir Panjal, Murrees and Siwaliks because of the aforementioned factors. Therefore, even though the tunnel supposedly passes through rocky terrain, it requires elaborate support systems and proper excavation techniques. Modified methods of tunnelling during excavation like NATM; rock bolting to prevent the collapse of fractured and jointed rocks from the ceiling; use of tunnel lining to stabilise the tunnel face and overcome water pressure are some of the solutions adopted. An example of the failure is the landslide at the portal (P2) of tunnel-47 of the Katra-Qazigund railway link in 2010, which caused widespread destruction in Sangaldan, Ramban (Singh et al., 2014).

g. Srinagar metro – A rapid transit plan is set to be implemented in Srinagar city. RITES Ltd. submitted a detailed project report, in which geotechnical investigations were carried out along the length of three proposed corridors. Owing to the soft soil conditions, high water table and low bearing capacity of soils in the region, it was found that underground corridors are not feasible and hence only elevated corridors have been finally included in the plan (Suryavanshi, 2020).

h. Chenab Bridge – Chenab bridge is the highest railway bridge in the world (359 m high), forming a crucial link in the Katra-Banihal rail link in J&K, connecting the Bakkal village on Jammu end to Kauri village on Kashmir end. It is an arch bridge with abutments in the rock slopes at the two ends, consisting of jointed and fractured dolomites, limestone, conglomerates, silt and clay. The weathered, jointed and fractured rocks and the presence of steep slopes made the task of construction challenges. Work for this bridge came to a halt when certain concerns were raised about the safety parameters, especially slope stability. However, 3D stability analysis of the bridge abutments and the results obtained in the form of displacements, stress and velocity values confirmed the stability of the rock slope (Varughese et al., 2010). Subsequently, the work was resumed, of which, around 80% has been completed at present.

i.   Central University of Kashmir, Ganderbal – Geotechnical investigations on the acquired land for the campus of Central University, Kashmir, Laar, Ganderbal reveal the presence of organic soil up to 3 m depth (Figure 15.2d), followed by silt and sand up to 6–10 m depth, overlying gravels with water under artesian pressure. To reduce the cost of construction, the engineers suggested the replacement of the top layer of organic soil with the soil of greater bearing capacity. The pile foundations could thus be replaced by shallow foundations ultimately reducing the cost of construction by manifolds.

15.9   GEOENVIRONMENTAL IMPACT ON SOILS AND ROCKS

Anthropogenic activities in construction projects in the region have affected the overall behaviour of soils and rocks in the region. Climate change has caused warming of the climate, flooding, melting of glaciers, and subsequent erosion, weathering and landslides. Flooding of rivers in 2014 lead to deposition of silt and clay in the flood plains of Jhelum in Srinagar, and sand and gravel by rivers in higher reaches. Glaciers melting at a rapid rate in places like Sonmarg and Gulmarg due to global warming bring down weathered rocks and debris leading to quicker deposition of glacial soils in the form of moraines and tills. The rocks which were once under glaciers are now exposed to direct agents of weathering like wind, snow, rain and heat, thus leading to greater breakdown and erosion. Road, railway and tunnel construction through the fragile ecology of the mountain between Kashmir and Jammu has greatly destabilised the region. This has increased the frequency and intensity of slope failures owing to the large-scale excavation carried out along the alignment. Landslides and rockfalls even block Chenab at some points and lead to the closure of the highway, especially during wet weather. Unplanned excavation and construction have a significant impact on the hydrogeology of rivers. The construction of dams on major rivers in J&K is another cause of concern for river basins. The creation of dams holds back silt which should have been normally flowing into the downstream lakes, increasing the scouring capacity of the rivers (Romshoo, 2016). This leads to excess scouring of the riverbanks.

15.10   CONCLUDING REMARKS

J&K has diverse landforms and varied geological formations. The unique environmental, geographical and tectonic setting of the region has led to the formation of different types of soils and rocks. The J&K divisions are geographically separated and have entirely different geological successions. Therefore, site-specific geotechnical tests are required before any important construction. The central region of the Kashmir basin is characterised by deep sedimentary deposits of soft alluvial soils having relatively less bearing capacity. This makes it essential to provide pile foundations for flyovers, bridges and multistoreyed buildings. The presence of clay and silt leads to the problem of large settlements for which provision of raft foundations becomes imperative. Organic soil around wetlands, near water bodies and in former rice/wheat fields creates problems since it is weak and decays over time. The mountainous and hilly regions have good soils in terms of strength and bearing capacity. However, due to frequent slope cutting for construction work, the strata become susceptible to major slope failures. A large area of Kashmir, especially the southern flank of the basin, is covered by lacustrine sediments known as Karewa deposits which are remnants of the ancient Karewa Lake. When dry, these are stiff soils with good bearing capacity and high penetration resistance. However, in presence of high moisture content, these become weak and can become problematic. Pir Panjal separating J&K divisions is composed of competent rocks. However, factors like jointing and fracturing of rocks under the existing stress conditions, high overburden, folding and faulting create a serious challenge during the execution of important construction projects in these formations. Jammu hills (Siwaliks) on the other hand have weaker rock strata locally known as Murree. High swelling and shrinkage in these soft rocks pose a major problem that is manifested in landslides along the Srinagar-Jammu highway. Jammu plains have stiff alluvial deposits, mostly gravel and sand, having good bearing capacity.

REFERENCES

Basavaiah, N., Appel, E., Lakshmi, B.V., Deenadayalan, K., Satyanarayana, V.V., Misra, S., Juyal, N. and Malik, M. A. (2010). Revised magnetostratigraphy and characteristics of the fluviolacustrine sedimentation of the Kashmir basin, India, during Pliocene-Pleistocene. *Journal of Geophysical Research*, Vol. 115, p. B08105.

Bhatt, D.K. (1989). Lithostratigraphy of the Karewa Group, Kashmir valley, India and a critical review of its fossil record. *Memoirs of Geological Survey of India*, Vol. 122, pp. 1–85.

Bhat, M.S. (2017). *Geomorphological field guidebook on Kashmir Himalaya* (Edited by M.N. Koul). Indian Institute of Geomorphologists, Allahabad, 28 p.

Burbank, D.W. and Johnson, G.D. (1982). Intermontane-basin development in the past 4 Myr in the north-west Himalaya. *Nature*, Vol. 298, pp. 432–436.

Burbank, D.W. and Johnson, G.D. (1983). The Late Cenozoic chronologic and stratigraphic development of the Kashmir intermontane basin, northwestern Himalaya. *Palaeogeography, Palaeoclimatology, Palaeoecology*, Vol. 43, pp. 205–235.

CGWB (2013). Ground water information booklets (for all districts of Jammu and Kashmir). Central Ground Water Board, Ministry of Water Resources, Government of India, North Western Himalayan Region, Jammu.

Dar, R.A., Romshoo, S.A., Chandra, R. and Ahmad, I. (2014). Tectono-geomorphic study of the Karewa Basin of Kashmir Valley. *Journal of Asian Earth Sciences*, Vol. 92, pp. 143–156.

Dar, R.A., Chandra, R., Romshoo, S.A., Lone, M.A. and Ahmad, M.A. (2015) Isotopic and micromorphological studies of Late Quaternary loess-paleosol sequences of the Karewa Group: Inferences for palaeoclimate of Kashmir Valley. *Quaternary International*, Vol. 371, pp. 122–134.

Gulzar, S.M., Mir, F.U.H., Rafiqui, M. and Tantray, M.A. (2020). Damage assessment of residential constructions in post-flood scenarios: A case of 2014 Kashmir floods. *Environment, Development and Sustainability*, doi:10.1007/s10668-020-00766-2

Gupta, P.S. and Singh, R.K. (2015). Geotechnical investigation and parameters for foundation of Chenab Bridge. *Him Parbat: USBRL Technical News Magazine*, Vol. 2, Issue 5, pp. 8–28.

Höfer-Öllinger, G. and Millen, B. (2014). Development of a geological model for a base tunnel. *Proceedings of the World Tunnel Congress 2014: Tunnels for a better Life*, 9–14 May 2014, Foz do Iguacu, Brazil, pp. 1–6.

Hough, S., Bilham, R. and Bhat, I. (2009). Kashmir valley megaearthquakes: Estimates of the magnitudes of past seismic events foretell a very shaky future for this pastoral valley. *American Scientist*, Vol. 97, pp. 42–49.

Hussain, A., Yeats, R.S. and MonaLisa (2009). Geological setting of the 8th October, 2005 Kashmir 518 earthquake. *Journal of Seismology*, Vol. 13, pp. 315–325.

Kapoor, U., Sharma, V., Singh, K.P. and Kanwar, P. (2017). Report on aquifer mapping and management plan of Kishtwar plateau, Kishtwar district, J&K (20 sq. km) *(AAP 2016–17)*. Ministry of Water Resources, Central Ground Water Board, North Western Himalayan Region, Jammu, 44 p.

Kotlia, B.S. (1985). Vertebrate fossils and paleoenvironment of the Karewa intermontane basin, Kashmir Northwestern India. *Current Science*, Vol. 54, No. 24, pp. 1275–1277.

Kumar, V.K., Martha, T.R. and Roy, P.S. (2005). Mapping damage in the Jammu and Kashmir caused by 8 October 2005 Mw 7.3 earthquake from the Cartosat-1 and Resourcesat-1 imagery. *International Journal of Remote Sensing*, Vol. 27, No. 20, pp. 4449–4459.

Madden, C., Ahmad, S. and Meigs, A. (2011). Geomorphic and paleoseismic evidence for late quaternary deformation in the southwest Kashmir Valley, India: Out-of-sequence thrusting, or deformation above a structural ramp. *American Geophysical Union Abstracts*, T54B-07.

Middlemiss, C.S. (1910). A revision of the Silurian-Trias sequence in Kashmir. *Records of Geological Survey of India*, Vol. 40, pp. 206–260.

Mir, B.A., Wani, B.A., Ahmad, N., Ayoub, R., Dar, L.A., Rashid, S.U. and Aziz, J. (2013). Physical and compaction behavior of dredged material from Dal lake Srinagar. *International Journal of Civil Engineering and Applications*, Vol. 3, No. 7, pp. 4–8.

Mir, B.A. (2014). Geosynthetics applications in highway construction in J&K. *Sustainable Infrastructure Development*, Vol. 3, No. 3, pp. 1–9.

Mir, B.A. (2015). Some studies on geotechnical characterization of dredged soil for sustainable development of Dal lake and environmental restoration. *International Journal of Technical Research and Applications*, Vol. 1, Special Issue 12, pp. 4–9.

Mir, B.A., Kumar, P., Kumar, A., Kumar, D., Kumar, M. and Mazumder, M. (2015). Geotechnical evaluation of dredged material from Dal lake as a construction material. *Proceedings of the Indian Geotechnical Conference 2011*, 15–17 Dec. 2011, Kochi, India, pp. 725–728.

Mir, B.A., Amin, F. and Majid, B. (2016). Some studies on physical and mechanical behavior of dredged soil from flood spill channel of Jhelum river, Srinagar. *Acta Ingenieria Civil*, Vol. 1, No. 1, pp. 1–7.

Mukhtar, G.A. (2008). Hydrogeology and hydrogeochemistry of Dachigam drainage basin. Doctoral Thesis, Department of Geology and Geophysics, University of Kashmir. *Shodhganga: A Reservoir of Indian Theses*, http://hdl.handle.net/10603/91872, Accessed 3 March 2020.

Palomba, M., Russo, G., Amadini, F., Carrieri, G. and Jain, A.R. (2013). Chenani-Nashri tunnel, the longest road tunnel in India: A challenging case for design-optimization during construction. *Proceedings of the World Tunnel Congress 2013: Geneva Underground – The Way to the Future!* 31 May–5 June 2013, Geneva, Switzerland, pp. 964–971.

Pant, R.K., Agrawal, D.P. and Krishnamurthy, R.V. (1978). Scanning electron microscopy and other studies on the Karewa beds of Kashmir, India. Norwich, Geological Processes, pp. 275–282.

Pant, R.K., Basavaiah, N., Juyal, N., Saini, N.K., Yadava, M.G., Appel, E. and Singhvi, A.K. (2005). A 20-ka climate record from Central Himalayan loess deposits. *Journal of Quaternary Science*, Vol. 20, No. 5, pp. 485–492.

Romshoo, S.A. (2016). *State of India's River Report: Jammu and Kashmir.* 3rd Edition of India River Week (IRW), 2016, New Delhi, India, 109 p.

Searle, M.P., Windley, B. F., CoWard, M.P., Cooper, D.J. W., Rex, A.J., Rex, D., Tingdong, Li, Xuchang, Xiao, Jan, M.Q., Thakur, V.C., and Kumar, S. (1987). The closing of Tethys and the tectonics of the Himalaya. *Geological Society of America Bulletin*, Vol. 98, No. 6, pp. 678–701.

Shah, A.A. (2013). Earthquake geology of Kashmir Basin and its implications for future large earthquakes. *International Journal of Earth Sciences*. Geologische Rundschau, Vol. 102, pp. 1957–1966.

Shah, A.A. and Malik, J.N. (2017). Four major unknown active faults identified, using satellite data, in India and Pakistan portions of NW Himalaya. *Natural Hazards*, Vol. 88, pp. 1845–1865.

Singh, I.B. (1982). Sedimentation pattern in the Karewa basin, Kashmir Valley, India, and its geological significance. *Journal of Palaeontological Society of India*, Vol. 27, pp. 71–110.

Singh, Y., Bhat, G.M., Pandita, S.K., Sharma, V., Kotwal, S.S., Kumar, K. and Koul, S. (2010). Engineering and geotechnical evaluation of Panthial landslide along national highway (NH), Jammu and Kashmir, India. *Journal of Nepal Geological Society 2010*, Vol. 41, p. 96.

Singh, Y., Sharma, V., Pandita, S.K., Bhat, G.M., Thakur, K.K. and Kotwal, S.S. (2014). Investigation of landslide at Sangaldan near Tunnel-47, on Katra- Qazigund railway track, Jammu and Kashmir. *Journal of Geological Society of India*, Vol. 84, pp. 686–692.

Suryavanshi, M. (2020). RITES submits revised DPR of Jammu and Srinagar Metro Rail projects. *Urban Transport News*. Retrieved from https://urbantransportnews.com/interview/rites-submits-revised-dpr-of-jammu-and-srinagar-metro-rail-projects, Accessed 30 Aug 2020.

Thakur, V.C. and Rawat, B.S. (1992). *Geologic Map of Western Himalaya*. Wadia Institute of Himalayan Geology, Dehra Dun, India.

Varughese, A., Rathod, G.W. and Rao, K.S. (2010). 3D stability analysis of Chenab bridge abutments. *Proceedings of the Indian Geotechnical Conference 2010*, 16–18 Dec. 2010, IIT Bombay, India, pp. 775–778.

Verma, R.S. (1967) Geological report on the landslides along the Jammu-Srinagar National Highway (1A), Jammu and Kashmir state. Geological Survey of India, Northern region, Lucknow. Field Season report 1966–67, https://www.gsi.gov.in/, Accession No. NRO-12925, Accessed 20 Mar. 2020, 22 p.

Wadia, D.N. (1975). *Geology of India*. 4th edition, Tata McGraw-Hill Publishing Co., New Delhi.

Yeats, R.S., Nakata, T., Farah, A., Fort, M., Mirza, M.A., Pandey, M.R. and Stein, R.S. (1992). The Himalayan frontal fault system. *Annals Tectonica*, Special Issue 6, pp. 85–98.

Zahoor, F., Rao, K.S., Banday, Z.Z., Bhat, M.A. and Farooq, M.F. (2019). Liquefaction potential analysis of Srinagar city, Jammu and Kashmir. *Proceedings of the 7th International Conference on Earthquake Geotechnical Engineering 2019*, 17–20 June 2019, Rome, Italy, pp. 5807–5814.

Zahoor, F., Rao, K.S., Malla, S.A., Tariq, B. and Bhat, W.A. (2021a). Seismic Site Characterization Using MASW of Sites Along Srinagar Metro Rail Alignment, Jammu and Kashmir. In: Patel S., Solanki C.H., Reddy K.R., Shukla S.K. (eds) *Proceedings of the Indian Geotechnical Conference 2019*. Lecture Notes in Civil Engineering, Vol. 138, pp. 581–593.

Zahoor, F., Rao, K.S., Hajam, M.Y., Kumar, I.A. and Najar, H.A. (2021b). Geotechnical Characterization and Mineralogical Evaluation of Soils in Srinagar City, Jammu and Kashmir. In: Patel S., Solanki C.H., Reddy K.R., Shukla S.K. (eds) *Proceedings of the Indian Geotechnical Conference 2019*. Lecture Notes in Civil Engineering. Vol. 133, pp. 59–72.

Chapter 16

Jharkhand

Anil K. Choudhary and Awdhesh K. Choudhary
National Institute of Technology Jamshedpur

CONTENTS

16.1 INTRODUCTION

Reliable information, associated with the location, quality and type of soil present, plays a major role in defining the agricultural development of a state. This also helps in favouring major construction activities taking place in that area. This chapter mainly focuses on the type of soil present in several districts of Jharkhand. A general map of the Jharkhand state of India is presented in Figure 16.1.

16.2 MAJOR TYPES OF SOILS AND ROCKS

The soil found here can be classified into five types. They are red soil, sandy soil, black soil, laterite soil and red micaceous soil. Red soil is mostly found in the Damodar valley and Rajmahal areas. They are formed by the decomposition of crystalline metamorphic rocks such as granite and gneiss. The colour of the soil is mostly chocolate dark except for a few places where it is red. They are porous and possess enough lime and potash but lack nitrogen, phosphorous and humus. Sandy soil is found in east Hazaribagh and Dhanbad. It is formed by the gradual decomposition of sandy loam and sandy soil. It is yellowish and has a poor humus content. Black soil is found in the Rajmahal areas and contains basalt, kaolin, potash and iron oxide. It contains a very higher proportion of clay and thus has a very high water-retaining capacity. The areas with higher elevation have a thin layer of black soil having light grey colour whereas those with lower elevation have a thick layer of soil having darker shades. Laterite soil is found in the western part of Ranchi, Palamu and parts of Santhal Parganas and Singhbhum. It mainly consists of aluminium, iron oxide and manganese oxide but lacks

DOI: 10.1201/9781003177159-16

Figure 16.1 General map of Jharkhand. (www.mapsofindia.com, accessed 13 May 2020.)

phosphorus, nitrogen, and potash. It has a low yield as it is prone to acid reactions. Red micaceous soil (containing parts of mica) is mainly found in Koderma, Mandu, Jhumri Telaiya and Barkagaon.

To have a greater idea about the type of soil found in the state, seven districts located in different parts of Jharkhand have been chosen. They are Ranchi in the central region, West Singhbhum in the southern region, East Singhbhum in the extreme corner of the southeastern region, Simdega in the extreme corner of the southwestern region, Dhanbad in the mid-eastern region, Dumka in the northeastern region and Palamu in the northwestern region. The detailed soil descriptions of the above-mentioned districts are discussed in the following section.

Ranchi mostly has the residual type of soil. High temperature and rainfall have helped in the formation of lateritic type of soils from rocks of Archean metamorphic complex exposed in most of the district. Texturally the soils of this district can be classified into the following four classes as per CGWB (2013a): (i) stony and gravelly soils found at the base of the hills, which are low-grade soil having poor fertility; (ii) red and yellow soils formed by the weathering of crystalline metamorphic rocks, which are rich in minerals like biotite, iron and so on; (iii) lateritic soil formed by lateralisation of weathered materials and found in Ratu, Bero and parts of Mandar Blocks, which is dark red or brown in colour and have a high iron content and (iv) alluvial soil found mostly along the river channels of the district, consisting mainly of coarse sand, gravel, silt and clay.

Mainly three types of soil are found in the West Singhbhum district as per the CGWB (2013b). They are (i) rocky soil found mainly in the southern, western and northwestern regions of this district. Being less fertile, it remains practically uncultivated in most of the seasons; (ii) red soil is spread throughout the district, which is texturally sandy and has poor fertility and (iii) black soil is found mostly in the lowlands of Kolhan. The soil texture is loamy and clayey. They are very fertile and support the cultivation of rice on it.

East Singhbhum mainly comprises five types of soil as per CGWB (2013c). They are (i) red gravelly soil found in Chakulia and parts of Bahragora blocks; (ii) red sandy soil found in Musabani, parts of Jamshedpur and Dumaria blocks; (iii) red loamy soil found in parts of Bahragora, Dhalbhumgarh and Jamshedpur Sadar blocks; (iv) red and yellow soil found in Patamda and Potka blocks and (v) lateritic soil found in small patches of the Bahragora block.

Soils in the Simdega district have been formed as a result of in situ decompositions of crystalline rocks as per CGWB (2013d). The various types of soil found in the district are (i) alluvial soil found mostly along the river channels deposited over consolidated rocks. They are of recent origin comprising coarse sand and gravel mixed with silt and clay; (ii) grey eroded scarp soil covers almost the entire district and is found as a thin covering over granitic rocks; (iii) red calcareous soil found mainly in the valley region. The soil texture is mostly sandy mixed with kankar and (iv) forest soil found mainly in the reserved forest area, present as a layer of organic matter over the surface.

Dhanbad mostly has the residual type of soils formed by the weathering of metamorphic rocks exposed in the greater part of the district. Texturally, the soils of this district have been classified into four categories as per CGWB (2013e). They are: (i) stony and gravelly soils found at the base of the hills. They are low grade in nature and contain a large admixture of cobbles, pebbles and gravels; (ii) sandy soil found generally near the river and streambeds. They are less fertile and contain more than 60% sand. They require heavy manuring and hence called hungry soils; (iii) loamy soils containing 30%–60% sand and mainly comprising decomposed rocks and vegetable matter and (iv) clayey soils which are very fertile and their yield can be significantly improved by adding sand, lime, manures and so on.

Soils in the Dumka district have been formed as a result of in situ weathering of crystalline rocks. They are non-calcareous and have good ion exchange capacity. The depth of soil is generally shallow on the ridges and plateaus and deep in the valleys. Extensive erosion, acidic character and low water-retaining capacity have adversely affected the fertility of the soil in this district as per CGWB (2013f).

In Palamu, three soil orders, namely Entisols, Inceptisols and Alfisols, can be found in the entire district. Alfisols are the dominant soils covering (53.9%) of the total gross area followed by Entisols (21.5%) and Inceptisols (20.0%) as per CGWB (2013g).

Till now, the types of soil found all over Jharkhand in general have been discussed. But in the case of detailed geotechnical design of foundation, soil data up to a deeper depth need to be investigated based on the footing size. To represent the soil profile with a significant amount of depth (10–20 m), some typical bore log data are presented in Figures 16.2–16.5 from different parts of Jharkhand.

Typical borehole data of Jamshedpur (East Singhbhum) are shown in Figure 16.2. Silty clay and fully weathered granite are the main soil strata found in Jamshedpur.

Depth (m)	Penetration resistance (N)	Sample	Graphic log	Material Description	Grain size distribution				Atterberg limits			Bulk density (g/cc)	Moisture content (%)	Dry density (g/cc)	Specific gravity	Cohesion (kg/cm²)	Angle of internal friction
					Clay	Silt	Sand	Gravel	LL	PL	PI						
2.5	25			Moorum mixed with quartzite (red colour)	16.5	27	46.5	10	30.80	20.85	9.95	2.24	10	2.03	2.69	0.8	16°
4.0	15																
5.5	18			Silty clay (yellowish grey)	26	41	29	4	54.15	24.84	29.11	2.17	15.09	1.88	2.65	0.5	14°
7.0	19																
8.5	20			Fully weathered granite (red colour)	29	34	34	3	28.15	22.47	5.68	2.31	9.29	2.11	2.59	1.1	14°

Figure 16.2 Bore log (Jamshedpur).

Depth (m)	Penetration resistance (N)	Sample	Graphic log	Material Description	Grain size distribution				Atterberg limits			Bulk density (g/cc)	Moisture content (%)	Dry density (g/cc)	Specific gravity	Cohesion (kg/cm²)	Angle of internal friction
					Clay	Silt	Sand	Gravel	LL	PL	PI						
1.5	15			Filed up soil consisting of slag, iron ore etc.	20	24	30	26	49	25.76	23.34	2.40	17.43	2.04	2.87		
3.0	16																
4.0	18			Weathered laterite with varying degree of weathering	18	32	46	4.0	33	15.27	17.73	2.58	14.08	2.26	2.51		
5.0	21																
6.0	24																
7.5	23			Yellowish colour mica schist mixed with laterites	28	27	41	4.0	33	16.01	16.99	2.72	11.13	2.44	2.52		
9.0	23																
10.5	26																

Figure 16.3 Bore log (Chaibasa).

Depth (m)	Penetration resistance (N)	Sample	Graphic log	Material description	Grain size distribution				Atterberg limits			Bulk density (g/cc)	Moisture content (%)	Dry density (g/cc)	Specific gravity	Cohesion (kg/cm²)	Angle of internal friction
					Clay	Silt	Sand	Gravel	LL	PL	PI						
1.5	5			Loose to medium dense sand continued and no strata encountered till the termination depth of borehole.	Nil	3	97	Nil	28.40	NP	NP	1.65	10.40	1.49	2.65	0.015	23°
3.0	6																
4.5	8																
6.0	9																
7.5	10																
9.0	12																
10.5	15																
12.0	16																
13.5	18																
15.0	20																

Figure 16.4 Bore log (Koel River, Daltonganj).

Depth	Penetration resistance	Sample	Graphic log	Material description	Grain size distribution				Atterberg limits			Bulk density (g/cc)	Moisture content (%)	Dry density (g/cc)	Specific gravity	Cohesion (kg/cm²)	Angle of internal friction
					Clay	Silt	Sand	Gravel	LL	PL	PI						
1.5 m	7			Loose to medium dense sand continued and no strata encountered till the termination depth of bore hole	Nil	2	96	2	NP	N.P	NP	1.65	11.43	1.48	2.65	0.065	22°
3.0 m	7																
4.5 m	9																
6.0 m	13																
7.5 m	14																
9.0 m	17																
10.5 m	16																
13.5 m	16			Medium dense to dense sand continued from depth of 12.5 m	Nil	3	97	Nil	NP	NP	NP	1.84	22.34	1.48	2.66	0.030	27°
15 m	16																
16.5 m	18																
18 m	20																
19.5 m	21																
21 m	23																

Figure 16.5 Bore log (Daltonganj).

The topsoil consists of moorum mixed with quartzite and a significant amount of fly ash. Typical borehole data of Chaibasa (West Singhbhum) are presented in Figure 16.3 indicating laterite soil as a major soil type. Figures 16.4 and 16.5 present the soil type of Daltonganj. The presence of loose to medium sand is shown in Figure 16.4. Loose to medium and medium to hard sand can be noticed in Figure 16.5.

16.3   PROPERTIES OF SOILS AND ROCKS

The amount of moisture present in the soil varies a lot in different parts of Jharkhand. As per the borehole data of Jamshedpur, the water table depth from the EGL ranges from 4.4 to 1.1 m. Water table variation with different borehole locations is shown in Figures 16.6 and 16.7. In Chaibasa, the depth of the water table varies from 1 to 3 m.

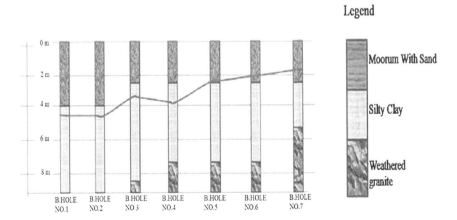

Figure 16.6 Water table depth variation from the borehole data of Jamshedpur (Borehole #1–#7).

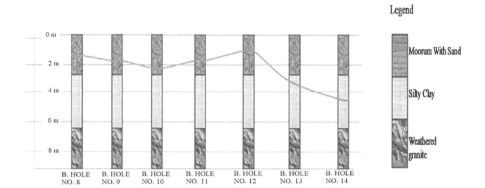

Figure 16.7 Water table depth variation from the borehole data of Jamshedpur (Borehole #8–#14).

This variation is shown in Figure 16.8. The position of the water table with respect to the EGL is less when Daltonganj is considered. Two borehole data of Daltonganj indicate water table depths of 1.4 and 0.9 m.

16.4   USE OF SOILS AND ROCKS AS CONSTRUCTION MATERIALS

The idea of using locally available earth materials in various civil engineering struc-tures is the best one from an economical point of view. Earth materials are frequently used in the construction of roads, building foundations, dams and so on. In Jharkhand, the Maithon dam is the largest and has an underground power station. This dam is made of rocks and soils. For this reason, this dam is called the earthen-rocky dam.

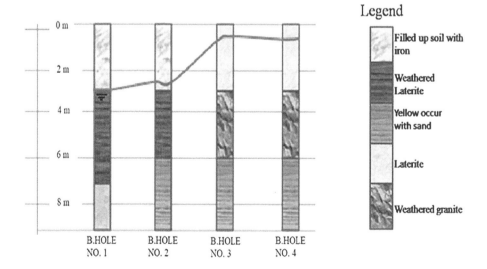

Figure 16.8 Water table depth variation from the borehole data of Chaibasa.

It is 4,789 m long and 50.3 m high and generates 60,000 kW of electric power. The Panchet dam across the Damodar river is another important dam in Jharkhand. It is an earthen dam and is famous for the fishery industry. This dam is situated at the border of Jharkhand and West Bengal. It is 6,777 m long and 148 m high.

16.5 FOUNDATIONS AND OTHER GEOTECHNICAL STRUCTURES

It is necessary to have sufficient information about the arrangement and behaviour of the underlying materials and their physical properties, for adopting and designing the structural foundation. Soil/rock exploration through field investigation and relevant laboratory testing of the substrata material are essential to arrive at the required parameters for the design of foundations. Due to the presence of steel plants in the Jamshedpur city, fly ash is dumped in some low-lying areas of the city. If the construction operation has to take place in those regions, engineers need to be very careful at the time of the design of the foundation and structures. A photograph showing the construction of NH-33 between Ghatsila and Jamshedpur is shown in Figure 16.9, which clearly shows the presence of moorum mixed with fly ash on the topsoil.

16.6 OTHER GEOMATERIALS

Nearly 73% of the country's total installed power generation capacity is thermal of which coal-based generation is 90%. They use bituminous and sub-bituminous coal and produce large quantities of fly ash. Indian coal has a high ash content of about 30%–40%, which contributes to the large volumes of fly ash (Dwivedi and Jain, 2013). Patratu Thermal Power Plant, Koderma Thermal Power Plant and Chandrapura Thermal Power Plant located in Jharkhand produce large quantities of fly ash every year. Over the years, fly ash has proven its worth as a raw material rather than just a waste. These days, after being treated, they are used as a soil ameliorating agent in agriculture and construction works, such as highway embankments, manufacturing of bricks and cement, and filler in low-lying areas.

Figure 16.9 Construction of NH-33 between Jamshedpur and Ghatsila.

16.7   NATURAL HAZARDS

According to the Ranchi Meteorological Centre, Jharkhand was earlier in seismic zone III which is one of the safest zones in the country, but this state is now shifting towards seismic zone IV with five districts added to the list. The last notable earthquake of slightly high intensity of 5 magnitudes was reported in the state with Hazaribagh being the epicentre back in 1952.

16.8   CASE STUDIES AND FIELD TESTS

Thermal power plants produce huge quantities of fly ash every year during their operation. The fly ash needs to be disposed of in ash ponds lying within its vicinity. But construction of ash ponds requires a huge quantity of soil which may not be available locally. This case study (Choudhary et al., 2009) mainly emphasised the advantageous use of fly ash and waste recycled product (WRP) as a fill material for the construction of ash pond dyke. Here, the WRP is basically recycled blast furnace slag generated from Tata Steel. The ash pond dyke was constructed for a local thermal power plant (Tata Power). Numerous trial mixes were formed by using both the wastes in different proportions. The geotechnical properties of these trial mixes were determined to get an optimum mix proportion. Furthermore, this optimum mix is fluxed with several admixtures such as cement or clayey soil in varying proportions which proved to be very effective for the construction of ash pond dyke. Based on the geotechnical characterisation of this optimum mix, initially, two cross-sections of the proposed ash pond dyke were suggested. Out of them, the one possessing maximum shear strength was adopted for construction. The use of clay stabilised mix proved to be better than cement stabilised mix as it exhibited the brittle behaviour of the mix leading to increased chances of granulation, making it more vulnerable to internal erosion throughout the dyke section.

Initially, grain size distributions were carried out for the WRP, fly ash and clayey soil and accordingly classified as SP, ML and CI, respectively, as per the Indian standard soil classification system.

From the above soil classification, it is evident that WRP and fly ash are poorly graded. Fly ash containing mostly silt size particles and when mixed with WRP containing mostly sand to gravel size particles in right proportion results in a mix having better gradation, stability, and strength. Keeping this in view, four trial mixes were prepared, namely M_1, M_2, M_3 and M_4 in which fly ash was mixed with varying proportions (20%–50% of the dry weight of WRP) and standard Proctor compaction tests were carried out for all the four mixes. The maximum dry density (MDD) and optimum moisture content (OMC) values are presented in Table 16.1.

Observations indicated a higher content of fly ash in M_1 and a lower content in M_4, whereas M_2 and M_3 were quite stable. Furthermore, unconsolidated undrained triaxial compression tests and falling head permeability tests were carried out on M_2 and M_3 with specimens prepared at MDD and the corresponding OMC. The shear strength parameters and permeability values are shown in Table 16.2. It can be seen that the value of permeability for the specimen lies in the range of 10^{-5} cm/s whereas it was desirable to be in the range of 10^{-7} cm/s. Therefore, admixtures in the form of cement and clayey soil were added to improve its permeability characteristics.

Table 16.1 Compaction Parameters of different mixes of fly ash and WRP

Sr. No.	Mix designation	Compaction parameters	
		MDU (kN/m³)	OMC (%)
1.	M_1	18.64	14.00
2.	M_2	17.84	14.40
3.	M_3	16.66	15.60
4.	M_4	16.09	17.20

Table 16.2 Shear strength and permeability characteristics of different mixes

Sr. No.	Mix designation	Shear strength parameters		Permeability (cm/s)
		Cohesion (kPa)	Angle of internal friction (°)	
1.	M_2	60	30	1.78×10^{-5}
2.	M_3	85	25	1.75×10^{-5}

Accordingly, two series of trials were carried out. In the first series, three specimens were prepared, namely M_3C_1, M_3C_2 and M_3C_3 by adding 3%, 5% and 7% cement of dry weight of WRP and fly ash in mix M_3. In these mixes, cement was considered as a part of mix M_3 and therefore MDD-OMC values were kept the same as those of M_3. Furthermore, unconsolidated undrained triaxial compression tests and falling head permeability tests were carried out on specimens (at the MDD-OMC state and damp cured for seven days) and the results are tabulated in Table 16.3.

In the second series, two specimens were prepared namely M_2C_4 and M_2C_5 by adding 10% and 7% of clayey soil of dry weight of WRP and fly ash in mix M_2. Further standard Proctor compaction tests, unconsolidated undrained triaxial compression tests (at the MDD-OMC as well as the saturated state) and falling head permeability tests (at the MDD-OMC state) were carried out on specimens and the results are presented in Table 16.4.

From the laboratory investigations, it has been confirmed that mixes M_3C_2, M_3C_3, M_2C_4 and M_2C_5 conform to the desired permeability requirement and also possess considerable shear strength. However, a clay stabilised mix was preferred over a cement stabilised mix as during the triaxial compression tests, it was found that failure strain was in the range of 6%–8% for the cement stabilised mix as compared to 12%–14% for the clay stabilised mix. Also, the cement stabilised mix exhibited brittle

Table 16.3 Shear strength and permeability characteristics of different mixes

Sr. No.	Mix designation	Shear strength parameters (MDU-OMC)		Permeability (cm/s)
		Cohesion (kPa)	Angle of internal friction (°)	
1.	M_3C_1	25	36	2.39×10^{-6}
2.	M_3C_2	100	33	3.30×10^{-7}
3.	M_3C_3	120	30	2.38×10^{-7}

Table 16.4 Shear strength, compaction and permeability characteristics of different mixes

Sr. No.	Mix designation	Compaction parameters		Shear strength parameters (MDU-OMC)		Shear strength parameters (saturation)		Permeability (cm/s)
		MDU (kN/m^3)	OMC (%)	Cohesion (kPa)	Angle of internal friction (°)	Cohesion (kPa)	Angle of internal friction (°)	
1.	M_2C_4	17.82	15.80	60	35	25	29	2.36×10^{-7}
2.	M_2C_5	18.04	16.40	80	30	30	25	2.04×10^{-7}

behaviour as compared to the ductile behaviour of the clay stabilised mix. The cost of cement is also high as compared to clay. So, the mix M_2C_4 was finally selected for the construction of the dyke section considering the economy and maximum utilisation of the waste. Finally, two sections, namely, Section-I and Section-II, were suggested for the construction of the ash pond dyke based on the results of the laboratory investigation (Choudhary et al., 2009).

In Section-I, the major portion of the dyke section was proposed to be consisting of WRP, fly ash and clay while the right portion on the upstream side consists of local soils having gravels, soil and silt. Second, Section-II was proposed to be consisting of a thin, central impervious core of puddle (an intimate mixture of stiff clay, sand and gravel thoroughly tamped into the place) with upstream and downstream shells consisting of WRP, fly ash and clayey soil. Finally, Section-I was selected for implementation keeping not just the economic viability of construction but also other advantages such as saving in construction time and creating an additional storage capacity in the ash pond by using locally available soil borrowed from the ash pond itself.

16.9   GEOENVIRONMENTAL IMPACT ON SOILS AND ROCKS

Improper storage, handling, transportation, treatment and disposal of waste result in adverse impacts on the ecosystem, including the human environment. When discharged on land, heavy metals and certain phytotoxic organic compounds can adversely affect soil productivity and aquifers. Most of the industries of Jharkhand are situated in Jamshedpur, Bokaro, Ranchi and Dhanbad. In between them, heavy industries are mainly located in Jamshedpur (except for the SAIL steel plant in Bokaro). In Jamshedpur, the TISCO (a major subsidiary of Tata Steel Limited) has a well-equipped solid management team. At present, the daily MSW quantity produced at the Jamshedpur city is 442.32 metric tonnes as per the Global Tech Enviro Experts Private Limited (2015). The TISCO had done an excellent job by changing its previous damping yards to ECO-PARK. Now they are planning to adopt a most advanced integrated solid waste management system at Ramgamatiya village, 22 km from Jamshedpur by installing the RDF plant, aerobic–anaerobic decomposition plant, incineration, gasification unit and so on. As a result, Jamshedpur has bagged the top position in India having a population under 10 lakhs as the cleanest city, according to the central government cleanliness survey (2019). The effectiveness of solid waste management is not the same for the entire state. The Bokaro Municipal Corporation is also planning to set up a waste treatment plant unit of cost 225 crores with the help of some private companies. The proposed location for this unit is in Sector 11 of Bokaro city. Ranchi is going to introduce the radio-frequency identification tag smart bucket system to improve the effectiveness of solid waste management.

16.10 CONCLUDING REMARKS

The details, as presented in the previous sections, may be concluded as follows:

1. The soil of Jharkhand can be broadly classified into red soil, sandy soil, black soil, laterite soil and red micaceous soil.
2. Typical borehole data confirmed the presence of varied moisture content in soil resulting in different shear strengths of soil across Jharkhand.
3. The topsoil mainly consists of red coloured moorum in many parts of Jharkhand. Also, there is a significant water table variation across the state.
4. Implementation of suitable ground improvement techniques has always helped the industry rich state in mitigating the effects of environmental and industrial activities on the existing soil.
5. The idea of using locally available fly ash generated from various thermal power plants and steel industries, as raw materials in construction activities, is noteworthy.

REFERENCES

CGWB (2013a). Ground Water Information Booklet of Ranchi District, *Report of Central Ground Water Board*, New Delhi.
CGWB (2013b). Ground Water Information Booklet of West Singhbhum District, *Report of Central Ground Water Board*, New Delhi.
CGWB (2013c). Ground Water Information Booklet of East Singhbhum District, *Report of Central Ground Water Board*, New Delhi.
CGWB (2013d). Ground Water Information Booklet of Simdega District, *Report of Central Ground Water Board*, New Delhi.
CGWB (2013e). Ground Water Information Booklet of Dhanbad District, *Report of Central Ground Water Board*, New Delhi.
CGWB (2013f). Ground Water Information Booklet of Dumka District, *Report of Central Ground Water Board*, New Delhi.
CGWB (2013g). Ground Water Information Booklet of Palamu District, *Report of Central Ground Water Board*, New Delhi.
Choudhary, A.K., Jha, J.N. and Verma, B.P. (2009). Construction of an ash pond with waste recycled product, fly ash and locally available soil – A case study. *Indian Geotechnical Conference* 2009, Guntur, India.
Dwivedi, A. and Jain, M.K. (2013). Fly ash – Waste management and overview: A review. *Recent Research in Science and Technology*, 6(1): 30–35.
Global Tech Enviro Experts Private Limited (2015). Environmental Impact Assessment report of Integrated Municipal Solid Waste Management and Handling Facility in Hata, for Tata Steel Limited, East Singhbhum, Jharkhand.

Chapter 17

Karnataka

C.R. Parthasarathy

Sarathy Geotech & Engineering Services Pvt. Ltd.

CONTENTS

17.1 INTRODUCTION

This chapter describes the classification of the soils in Karnataka. Based on the geotechnical formation and engineering applications, the soils can be mainly classified into three types as shown in Figure 17.1. They are observed during geological and geotechnical site investigations and are mainly residual deposits (south interior region), black cotton soil/expansive soil (north interior region) and lithomargic clay/shedi Soil (coastal region).

17.2 MAJOR TYPES OF SOILS AND ROCKS

The residual deposits are found in the southern interior region (e.g., Bangalore). They broadly consist of an overburden layer (red, brown, yellow or grey in colour) and can be clay, sandy clay, clayey sand, sandy silt, silty sand and moorum (Figure 17.2), weathered rocks/refusal stratum (Figure 17.3) and bedrock (Figure 17.4). In general, these

DOI: 10.1201/9781003177159-17

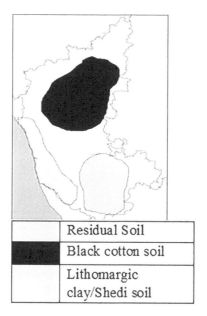

Figure 17.1 Approximate geology of Karnataka.

Figure 17.2 Residual deposits (overburden soils).

residual deposits do not pose any geotechnical challenges to the shallow foundation in terms of bearing capacity and settlement, but due to the high strength of bedrock, socketing of deep foundation (pile foundation) in these hard formations poses difficulty and is always debatable (Murthy and Pujar, 2009). The intermediate deposits, that is, the refusal stratum between overburden and bedrock, pose excavation challenges if not quantified properly.

Black cotton soil (expansive clay) predominantly exists in the northern interior of Karnataka. These expansive soils undergo volumetric changes with changes in the moisture content. An increase in moisture content causes swelling of the soil and loss

Figure 17.3 Residual deposit (excavation in weathered rock refusal stratum).

Figure 17.4 Residual deposit (excavation in bedrock).

in strength, while a decrease in moisture content causes soil shrinkage. Swelling and shrinkage of expansive soil cause differential movements resulting in severe damage to the foundations, buildings, roads, retaining structures, canal linings and so on. These expansive clays are one of the most problematic soils for civil engineers and exist around the world in the USA, Saudi Arabia, Jordan, South Africa, Kenya, Egypt, Australia and so on.

Lithomargic clay is the product of laterisation and is termed shedi soil in the local language. These are highly dispersive and highly erosive soils, and thus they pose problems to foundation, subgrades and stability of slopes. The natural consistency varies from soft to stiff. These soils are found in abundance in the coastal region.

17.3   PROPERTIES OF SOILS AND ROCKS

Prior to the design and construction of any structure, it is imperative to conduct a thorough site investigation. The following methods are customarily adopted.

- Trial pits – For physical examination and collection of samples.
- Hand-operated manual auger (Figure 17.5) – Exploratory boreholes are made by manual augers for shallow depth of investigation (say up to about 6 m). A standard penetration test (SPT) is conducted at regular intervals and undisturbed samples are collected. The depth of boreholes is normally about 6 m or refusal whichever is earlier. This method is very popular and commonly adopted for lightly loaded structures (e.g., residential buildings and pile lines).
- Rotary drilling (Figure 17.6) – This method utilises a non-standard rotary drilling machine (calyx type). It is useful for overburden, conducting SPT, collecting undisturbed samples and drilling in bedrock using a single tube core barrel. However, this has a very poor or non-performance while using the double tube/triple tube core barrel. This type of drilling rig is extensively used in India.
- Rotary hydraulic drilling (Figure 17.7) – This helps in conducting the SPT, collecting undisturbed soils and diamond drilling of rocks by using all types using single/

Figure 17.5 Hand-operated auger.

Figure 17.6 Rotary drilling using conventional calyx-type machine.

Figure 17.7 Rotary hydraulic drilling for good quality diamond drilling.

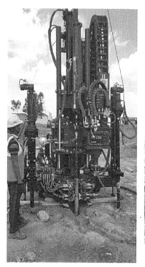

Figure 17.8 Multipurpose – SPT, CPT and wireline rotary core test drill rig.

double and triple tube core barrels. It is very useful in retrieving the core samples in weathered formation/refusal strata.

*   Multipurpose drilling rig (Figure 17.8) – This can be utilised to conduct the cone penetration test (CPT), SPT, auger drilling and core sampling by both rotary and wireline methods in all types of soils/rocks. The machine is equipped with automatic hammers for conducting the SPT, which imparts high energy to the drill rods (85%–95%). These are very useful for core sampling using the wireline methods.

17.3.1   Properties of residual deposits – overburden (south interior region, e.g., Bangalore)

Table 17.1 presents the range of various parameters of residual deposits (south interior region) as collected from an extensive database from Bangalore and neighbouring cities. Generally, kaolinite is the mineralogical composition of these soils and thus they are inactive. The thickness of the overburden varies widely from zero to about 20 m.

17.3.2   Properties of residual deposits – weathered rock/refusal stratum (south interior region, e.g., Bangalore)

It is very difficult to retrieve the core samples in the weathered formation (refusal stratum) using conventional rotary drilling machines (e.g., calyx type), where the rotary drill rods will wobble rather than rotating at an axis. Although the use of a single tube core barrel is successful in these machines for retrieving cores in bedrock, it is extremely difficult to sample these formations since the drilling fluid is in direct contact with the samples. The core recovery and RQD are very poor. However, a rotary hydraulic rig can be utilised with a double/triple tube core barrel to recover the core

Table 17.1 Physical and engineering properties of residual deposits/overburden

Soil parameters		Different types of soil				
		Silty sand	Clayey sand	Sandy clay	Sandy silt	Clayey silt
NMC (%)		7–35	8–22	10–28	9–42	19–35
Liquid limit (%)		18–47	24–40	36–61	48–85	50–60
Plastic limit (%)		15–20	12–23	16–26	23–34	25–40
Bulk unit weight (kN/m³)		18–20	18–20	17–20	16–21	18–20
Specific gravity		2.6–2.66	2.66–2.7	2.66–2.7	2.64–2.67	2.62–2.67
Grain Size	Gravel	Nil–22	1–20	Nil–17	Nil–6	-
Distribution	Sand	42–81	40–72	11–45	17–56	8–13
(%)	% passing 75 μm	20–64	23–54	54–88	45–83	87–92
	Silt	44–45	11–18	18–30	43–67	72–82
	Clay	5–10	24–31	32–46	7–13	9–11
Angle of friction (°)		29–43	30–36	30–35	25–32	23–35
Cohesion (kPa)		0–15	8–12	6–15	0–12	3–11

Table 17.2 Physical and engineering properties of residual deposits (weathered rock/refusal stratum)

Rock parameters	Highly weathered rock/refusal stratum
Water content (%)	0.2–1.25
Water absorption (%)	1.0–3.1
Porosity (%)	2.1–6.4
Dry density (g/cc)	2.1–2.73
Specific gravity	2.4–2.59
Unconfined compression strength (MPa)	1.5–12
Point load (MPa)	0.1–4.8
Brazilian tensile strength (MPa)	0.9–1.7
Core recovery (%)	20–90[a]
Rock quality designation (%)	10–85[a]

[a] With hydraulic rig and triple tube core barrel.

samples in these weathered formations, where the drilling fluid is not in direct contact with the core samples (with high core recovery and good to excellent RQD). Table 17.2 presents the range of physical and engineering parameters of the weathered rock/refusal stratum. The thickness of these formations varies from about 2 to 30 m.

17.3.3   Properties of residual deposits – bedrock (south interior region, e.g., Bangalore)

Below the overburden and/or weathered rock formation, bedrock is encountered in the southern interior region of Karnataka. These rock masses can be intact (sheet-rock) and/or highly fractured/jointed at some locations. The unconfined compressive strength of these intact rocks can be more than 200 MPa. Table 17.3 presents the range of physical and engineering parameters of bedrock.

Table 17.3 Physical and engineering properties of residual
deposits/bedrock

Rock parameters	Bedrock
Water content (%)	0.01–0.5
Water absorption (%)	0.03–0.9
Porosity (%)	0.03–2.3
Dry density (g/cc)	2.53–3.01
Specific gravity	2.54–3.02
Unconfined compression strength (MPa)	30–212
Point load index (MPa)	6–9
Brazilian tensile strength (MPa)	5–12
Core recovery (%)	30–100
Rock quality designation (%)	50–100

17.3.4 Properties of black cotton soil (expansive clay)

The term expansive soil is associated with soils that are sensitive to the changes in the
moisture regime. In general, expansive soils refer to soils that contain active clay minerals, especially montmorillonite. Their formation is associated with basalts but their
occurrence on granite, gneiss, shales, sandstones, slates and limestone has also been
recognised.

Black cotton soils are inorganic clays of medium to high compressibility and form
a major soil group in Karnataka. They are characterised by high shrinkage and swelling properties. This type of expansive soil occurs mostly in the northern parts and
covers approximately 20% of the total area of Karnataka. The districts that have the
black cotton soil are Belgaum, Davangeri, Bagalkot. Bijapur, Gulbarga and Bidar,
also part of Raichur, Chitradurga and Bellary. These soils have been a challenge to
the construction field because of their high swelling and shrinkage characteristics.
The black cotton soil is very hard when dry but loses its strength significantly under
wet conditions. It is observed that on drying, the black cotton soil develops cracks of
varying depth. As a result of the wetting and drying process, vertical movement takes
place in the soil mass. All these movements lead to failure of pavement, building and
similar lightly loaded structures, in the form of settlement, heavy depression, cracking
and unevenness. About 40%–60% of the black cotton soil has a size less than 0.001
mm. At the liquid limit, the volume change is of the order of 200%–300% and results
in a swelling pressure as high as 8–10 kg/cm^2. As such, black cotton soil has very low
bearing capacity and high swelling and shrinkage characteristics. It forms a very poor
foundation material for the construction due to its peculiar characteristics. Table 17.4
presents the summary of a range of parameters of expansive soils from different locations of Karnataka.

17.3.5 Properties of lithomargic clay/shedi

Shedi soil is a local term for the lithomargic soils formed due to the laterisation process.
This material is abundantly available throughout the west coast of India from Malabar
(Kerala) to Ratnagiri in Maharashtra and covers the coastal region of Karnataka in

Table 17.4 Physical parameters of black cotton soil

Districts \ Soil parameters			DAVANAGERE	BAGALKOT	HUBLI
Water content (%)			23–28	25–31	19–36
Liquid limit (%)			68–76	60–74	67–80
Plastic limit (%)			29–32	24–30	22–37
Shrinkage limit (%)			6–8	8–12	7–8
Plastic index (%)			37–44	36–44	39–45
Shrinkage index (%)			61–69	48–66	60–73
Bulk unit weight (kN/m^3)			14–16	12–13	14–17
Specific gravity			2.73	2.8	2.75
Grain size	Gravel		1.5–3	2.5	3
distribution	Sand	Coarse	2.5–4	2–6	1–4
(%)		Medium	5–7	6–10	4–8
		Fine	10–11	4.5–9	10–14
	Silt and Clay		78–81	78–84	75–87

Source: After Parthasarathy (2002).

Table 17.5 Physical and engineering properties of lithomargic clay/shedi soil

Soil parameters		Values
Water content (%)		10–30
Liquid limit (%)		30–62
Plastic limit (%)		18–42
Shrinkage limit (%)		11–36
Plastic index (%)		3–30
Bulk unit weight (KN/m^3)		15–18
Maximum dry density (kN/m^3)		15–22
OMC (%) (standard proctor)		16%–26%
Specific gravity		2.5–2.8
Grain size	Gravel	0.5–36
distribution (%)	Sand	20–80
	Silt	14–64
	Clay	0–45
California bearing	Unsoaked	3–25
ratio (%)	Soaked	1–9
Angle of friction (°)		6–40
Cohesion (kPa)		0.5–30

Source: After Shivashankar et al. (2015).

the west. Shedi soils are termed as treacherous soil by geotechnical engineers due to their low strength and unpredictable behaviour. These are highly dispersive and have high erosion potential. This nature of the soil in the slope hastens the slope failures. Table 17.5 presents the range of engineering properties of shedi soils abundantly found in the west coast of Karnataka.

17.4   USE OF SOILS AND ROCKS AS CONSTRUCTION MATERIALS

As far as the soils and rocks of Karnataka are concerned, the residual deposits are extensively used as fill materials for the embankments, increasing the ground level, construction of pavements for road/airfields/rail and so on. These overburdens in residual soils are inactive and thus do not need any additive to alter the natural properties. Perhaps, use of geosynthetic materials is also very common. The rocks encountered in the southern interior of Karnataka are used as construction materials and ornamental pieces, in railway ballast, in foundations and so on.

Black cotton soils are expansive, and montmorillonite is the primary mineralogy. These soils have a high free swell index and exhibit extreme volume changes due to the change in moisture content. In fact, the foundations placed below the active zone do not experience volume change, because the foundations within the active zone are detrimental to their safety. Thus, black cotton soil is not used as a construction material without treatments, such as mixing with lime, cement, fly ash, chemicals or with any other combination. The parent material is mostly basalt and weathering of basalt can also lead to brown/yellow expansive clays.

Shedi soils are extensively used in the construction of flyovers, pavement for rail and road. These soils are dispersive and erosive; thus, they pose greater challenges for the stability of slopes.

17.5   FOUNDATIONS AND OTHER GEOTECHNICAL STRUCTURES

The observed N-values in residual deposits vary from 15 to greater than 50 (Table 17.6a). Thus, most of the normal structures (residential, industrial commercial and so on) are built on shallow open foundations in the southern interior region of Karnataka. Typical allowable bearing pressure/safe bearing capacity is presented in Table 17.6b. For heavily loaded high-rise structures (without deep excavation) or infrastructure projects

Table 17.6a Range of measured N-values in residual deposits

Description	Observed SPT (N) values
Sandy clay	20–40
Sandy silt	20–50
Silty sand	15–R
Clayey sand	15–40
Clayey silt	30–60
Silty clay	10–30

Table 17.6b Typical bearing capacity values

Description	Allowable bearing pressure/ safe bearing capacity (t/m²)
Overburden layer	15–40
Weathered rock /refusal strata	70–120
Bedrock	150–300

(e.g., metro rail/flyover), it is a practice that if the bedrock (Figure 17.9a) is available less than 4.0 m, the open foundation is adopted. If the bedrock is not available less than 25 m, the foundations are designed as friction piles. If the bedrock is available anywhere between 4 and 25 m, the foundations are designed as end-bearing piles (piles socketed into rock). For deep excavation, many times, excavations are unprotected as shown in Figure 17.9b, which are detrimental to safety under monsoon. There are many excavation support methods, as shown in Figure 17.9c. For a shallow depth of excavation, up to about 5.0 to 6.0 m (vertical cut) with the soil nailing technique is adopted by

Figure 17.9 (a) Samples of bedrock, (b) unprotected excavation in residual deposits and (c) excavation support methods.

driving closely spaced bars (two to four nails per square metre). The diameter of the driven nails varies between 16 and 25 mm with a relatively limited length of about 4–8 m. For deep excavation (beyond say 6 m), a touch pile system which is a bored reinforced concrete pile with small gaps or secant piles (without gaps) are adopted. These piles can be free-standing cantilevers or with tieback/strut supports. One of the most commonly adopted temporary support systems is with a micropile, waler beam and grouted anchor system. This system has been successfully adopted in several projects in Bangalore for an excavation depth of about 18 m in the author's experience.

The principles of construction of foundation in black cotton soil have been summarised by Sitharam and Babu (1998) below:

- Foundation load is limited to 5.0 t/m^2 if there is accessibility of water to the foundation, otherwise load may be up to 10 t/m^2.
- Foundation excavation is extended beyond the depth of cracks observed in the soil.
- Foundation trench is made wider than required and the extra width is filled with granular materials to prevent intimate contact of black cotton soil with the foundation structure.
- If the thickness of black cotton soil strata is not more than 500 mm, it is advisable to remove it completely.
- In case of important buildings, RCC raft foundation is recommended.
- For ordinary buildings, any of the following techniques may be adopted according to the suitability: removal of black cotton soil and filling with granular material, underreamed piles, stabilisation with cement, lime, etc.

The lithomargic clay/shedi soils are located mainly in coastal Karnataka in western peninsular India comprising three districts, namely Uttara Kannada, Udupi and Dakshina Kannada. This area has quite a few rivers that flow westwards and exit into the Arabian Sea. The area has laterites and lateritic soils and a large number of sporadic lateritic hillocks. The top laterites are used as bricks for construction purposes in this area. Lithomargic clay is a product of laterisation and underlies the top hard and porous lateritic crust. The lithomargic clay behaves like dispersive soil and is present between the weathered laterite at the top and the hard granitic gneiss underneath. Lithomargic clays are also used for construction purposes and for backfilling purposes in low lying areas. A lot of engineering problems are being faced due to the presence of this lithomargic clay soil, either naturally or due to backfilling, and the fluctuating water levels. The problems of low bearing capacity, low subgrade strength and large settlement are very common. Low line agricultural and marshy lands around the Mangalore metropolis are being quickly converted into estates by filling with the locally silty soil lithomargic. These soils are merely dumped and usually not properly compacted. The other scenario is that, very often in areas where the lateritic crust is intact, existing mounds are levelled and the top laterite layer is removed for use as bricks in building constructions, thereby exposing the underlying lithomargic clays. In either case, the infrastructure constructions are invariably supported on the poor lithomargic clay. The problem of such lowlands adjacent to water bodies is low bearing capacities and settlements. Furthermore, the sensitivity of this lithomargic clay to moisture variations, from the point of view of their strength, is also of grave concern for highway engineers and foundation engineers, unless these soils are suitably stabilised. For other important and major constructions, such as multistoreyed constructions, pile foundations are commonly recommended and adopted.

The problems of slope stability and erosion are generally found in Mangalore and surrounding places on a sloping terrain. Slope failure is one of the most common geotechnical failures observed in these regions. These may be failures of natural slopes (viz., landslides or landslips), mainly because of changing drainage patterns, numerous human activities in the name of development or other manmade slope failures and wrong or poor geotechnical engineering practices. Slopes with lithomargic clays, which are generally stable during dry seasons, fail during rainy seasons.

Several soil stabilisation techniques are available in geotechnical engineering practice. Some of the possible soil stabilisation techniques that were attempted to improve the behaviour of lithomargic clays are cement stabilisation, stabilisation with sand, coir and/or geosynthetic material. In the literature, we may find several ground improvement techniques for the improvement of shedi soils, including the use of geosynthetics, lime, fly ash and coconut shells.

Vetiver is a very fast-growing grass and until very recently a relatively unknown plant in this area. It possesses some unique features of both grasses and trees by having a profusely grown, deep penetrating root system. The roots of vetiver grass can offer both erosion prevention and control of the shallow movement of surficial earth mass. Vetiver grassroots are very strong with an average tensile strength of 75 MPa or about one-sixth of the ultimate strength of mild steel. In addition to its unique morphological characteristics, vetiver is also highly tolerant to adverse growing conditions such as extreme soil pH, temperatures and heavy metal toxicities. The massive root system also increases the shear strength of soil, thereby enhancing slope stability appreciably. The effect of vegetation is beneficial in controlling surface erosion and improving the stability of slopes.

17.6   OTHER GEOMATERIALS

Fly ash is one of the wastes from thermal power plants generating millions of tonnes annually. There is a great demand for fly ash from factories making cement, bricks and tiles, and it is used as an additive for ground improvement of black cotton soils, municipal solid wastes and so on (Naveen et al., 2014). Fly ash can also be used as a geopolymer and an alternative binder to the Portland cement in the manufacturing of mortars and concrete (Rao and Acharya, 2014).

Slag is a by-product generated during the manufacturing of pig iron and steel. It is produced by the action of various fluxes upon gangue materials within the iron ore during the process of pig iron making in blast furnace and steel manufacturing in steel melting shop. Primarily, slag consists of calcium, magnesium, manganese and aluminium silicates and oxides in various combinations. The cooling process of slag is responsible mainly for generating different types of slags required for various end-use consumers. Although the chemical composition of slag may remain unchanged, physical properties vary widely with the changing process of cooling. There are millions of tonnes of these slags generated every year from steel industries.

Slag has different uses based on its types. The air-cooled blast furnace (BF) slag is crushed, screened and used mainly as road metal and bases, asphalt paving, track ballast, landfills and concrete aggregates. Steel slag has found use as a barrier material remedy for waste sites where heavy metals tend to leach into the surrounding environment. Steel slag forces the heavy metals to drop out of solution in water runoff because of its high oxide mineral content. Steel slag has been used successfully to treat acidic water discharges from abandoned mines. The BF slag in India is used mainly in cement manufacturing and in other unorganised work, such as landfills and railway ballast. A small quantity is also used by the glass industry for making slag wool fibres. Cement plants in the country producing slag cement require BF slag in granulated form. Slag is used as a substitute for clinker. This slag otherwise would have been a waste and used as a filler material. Slag, if used properly, will conserve valuable limestone deposits required for the production of cement. Currently, the Government of India has plans to introduce a fourth type of cement in the market, Pozzolana slag cement, which will contain both steel slag and fly ash. The potentiality for consumption of slag in cement manufacturing is bound for a substantial rise in the near future. Tables 17.7 and 17.8 present some physicochemical properties of steel slag tested in the author's laboratory. The chemical properties are like the one published in Indian Minerals Yearbook (2018).

Table 17.7 Physical properties of slag in Karnataka

Sample ID (Slag)	Specific gravity	Minimum unit weight	Maximum unit weight	Free swell index	Particle size analysis						
					Gravel		Sand			Fines	
		kN/m^3	kN/m^3	%	Coarse (%)	Fine (%)	Coarse (%)	Med (%)	Fine (%)	Silt (%)	Clay (%)
G	2.80	12.1	15.0	0	-	-	5	88	7	-	
G-SMS	2.88	14.32	17.64	0	-	3	7	82	8	-	
SMS-slag	3.48				-	-	-	-	-	-	
Crushed slag	3.18				-	-	-	-	-	-	
SMS crushed	3.49				-	38	38	24	-	-	

Table 17.8 Chemical properties of slag in Karnataka

Sl. No.	Parameters tested	Test results (% by mass)	
		G (steel slag)	G-SMS (iron slag)
1	Moisture content	6.85	7.08
2	Insoluble residue (IR)	0.65	1.12
3	Silica (as SiO_2)	32.00	30.38
4	Alumina (as Al_2O_3)	19.38	20.87
5	Iron oxide (as Fe_2O_3)	1.08	2.12

17.7 NATURAL HAZARDS

According to the seismic map of the Indian meteorological department, Karnataka is situated in the least seismically active zone and thus, effects of an earthquake on foundations are very minimal. The Deccan Plateau is in Zone 2, while the coastal region is in Zone 3.

There is a possibility of floods due to incessant rain and blocked storm water drains. The foundation and other structures are always designed for the submerged conditions for them to be safe even during floods.

17.8 CASE STUDIES AND FIELD TESTS

Despite the residual deposits not posing any geotechnical challenges, improper and/ or no geotechnical investigation leads to improper design of foundation (both size and depth). Even lightly loaded structures experience settlement and this is not an uncommon phenomenon. One of the remedial measures is to underpin the foundation by driving micro-piles that are widely adopted to prevent further settlements and restore foundations. Thereafter, internal and external walls are stitched to close the cracks to restore the foundations and structures monitored. In the author's experience, this method of treatment has been successfully adopted in nearly about 40 projects.

For a multistoreyed building project in Bangalore (Domlur layout), investigations were carried out up to 30 m depth. The ground profile consists of overburden (silty sand) of up to 10.5 m followed by weathered rock/refusal stratum. From 10.5 to 21 m, the rock was highly weathered and conventional single or even double tube core barrels did not result in any core sample. However, below 21 m depth, rock was moderately weathered and conventional coring resulted in some core samples. The borehole details and test results are presented in Table 17.9a.

With the uncertainty in the quantification of weathered rock for a thickness of about 10 m, it was decided to use a triple tube core barrel for sampling in an adjacent borehole with a rotary hydraulic machine. The triple tube core barrel has three tubes that are placed concentrically. The outer barrel is for cutting, the middle one is for finer cutting of the core and the inner one is for retrieving the sample without disturbance or damage. Thus, it was possible to obtain rock cores in the transition zone from 12 to 21 m. This bump over the borehole was terminated at 30 m. The boring log details and test results are shown in Table 17.9b. A comparison of both the tables regarding the core samples and their unconfined compressive strengths brings out the efficacy of using a triple tube core barrel. In the transition zone between 12 and 21 m, while the conventional core barrels yielded no cores, the triple tube barrel yielded rock core samples in the refusal stratum (Figure 17.10a) with unconfined compressive strengths varying from 3 to 6 MPa. Beyond 21 m and up to 30 m, the triple tube core barrel very consistently resulted in better unconfined compressive strengths. The most striking feature is just at the bedrock. The core qualities in both the cases were 'fair'. But the UCC strengths were 6.85 and 46.6 MPa, clearly demonstrating how even double tube core barrels can underestimate the rock strength. In addition, it also demonstrates the inadequacy of RQD in describing the core quality. Figure 17.10b presents core samples in weathered and bedrock.

Table 17.9a Results from the conventional core barrel

Depth (m)	Description	N value	Run (m) From	Run (m) To	CR (%)	RQD (%)	Core quality	UCS (MPa)
1.5	Silty sand	5						
3.0		12						
4.5		14						
6.0		22						
7.5		20						
9.0		31						
10.5	Highly weathered	>50	10.5	12.0	Nil	Nil		
12	rock	>50	12.0	13.5				
13.5		>50	13.5	15.0				
15		>50	15.0	16.5				
16.5		>50	16.5	18.0				
18		>50	18.0	19.5				
19.5		>50	19.5	21.0				
21	Moderately		21.0	22.5	33	15	Very	3.00
22.5	weathered rock		22.5	24.0	39	22	poor	3.42
24			24.0	25.5	44	26	Poor	4.28
25.5			25.5	27.0	Nil	Nil	Very poor	5.14
27			27.0	28.5	56	26	Poor	5.14
28.5			28.5	30.0	90	55	Fair	6.85

Borehole terminated at 30.0 m and water table met at 7.5 m below the existing ground level.

Table 17.9b Results from a triple tube core barrel with a rotary hydraulic drilling machine

Depth (m)	Description	N value	Run (m) From	Run (m) To	CR (%)	RQD (%)	Core quality	UCS (MPa)
1.5	Silty sand	12						
3.0		8						
4.5		27						
6.0		17						
7.5		33						
9.0		24						
10.5		26						
12	Weathered	>50	12.0	13.5	20	10	Very poor	3.05
13.5	rock		13.5	15.0	55	28	Poor	3.05
15			15.0	16.5	60	40		3.05
16.5			16.5	18.0	83	77	Good	3.48
18			18.0	19.5	38	32	Poor	5.14
19.5			19.5	21.0	36	13		6.62
21			21.0	22.5	80	74	Fair	6.11
22.5			22.5	24.0	90	50		7.47
24			24.0	25.5	96	85	Good	8.42
25.5			25.5	27.0	80	60	Fair	12.49
27			27.0	28.5	100	94	Excellent	22.00
28.5	Bedrock		28.5	30.0	83	70	Fair	46.61

Borehole terminated at 30.0 m and water table met at 7.4 m below the existing ground level.

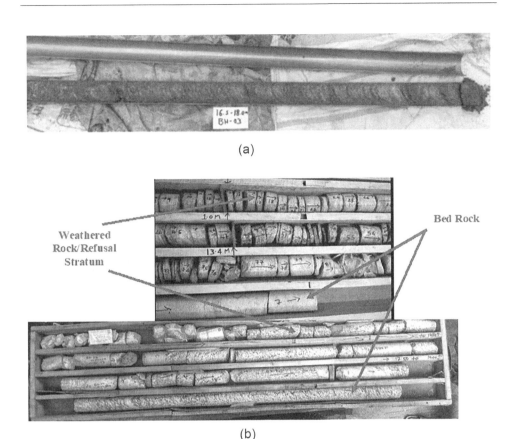

(a)

Weathered
Rock/Refusal
Stratum

Bed Rock

(b)

Figure 17.10 (a) Rock cores using a triple tube core barrel in refusal stratum and (b) rock
core samples (weathered and bed rock).

17.9 GEO-ENVIRONMENTAL IMPACTS ON SOILS AND ROCKS

Generally, the residual deposits of the southern interior region of Karnataka do not
pose any challenge to the environmental/climatic changes. However, rapid urbanisation leads to the accumulation of large quantities of municipal, industrial waste and
sewage sludge. The composition varies but more than half of municipal waste may be
paper, with the rest made up of food scraps, metals, glass and ash. Wastes from industrial plants may contain acids, alkalis, oils, metals and other noxious substances. Because of the presence of large amounts of decomposable organic substances collected
by the municipal authority, intense microbial activity occurs in the buried waste. From
being initially aerobic, the waste becomes anaerobic, a condition that can last for several years. Microbial decomposition results in the loss of organic materials, which
causes settlement of the site of up to about 25% of the depth of the landfill. Sewage
sludge is the organic material produced from domestic and industrial wastewater and
direct run-offs from roads. The sludge contains high concentrations of metals, which
are toxic to plants or animals. For example, chromium and nickel are released by the

iron and steel industry, cadmium and lead from the manufacture of batteries and zinc from zinc plating factories. Zinc and copper predominate in domestic waste but are usually present in lower concentrations than in sewage that contains waste from industrial processes.

In Karnataka, Bangalore and other cities are facing problems due to old municipal solid waste dump yards situated close to the expanding city as they are in the expansion zone of a growing city. Generally, these wastes include newspaper, junk mail, meal scraps, pieces of bread, roti, waste rice, raked leaves, dust grass clippings, broken furniture, abandoned materials, animals manure, sewage sludge, industrial refuse or street sweepings and are dumped in open sites in the form of a heap. These dump yards need to be reclaimed for infrastructure developments such as roads, buildings and other needs. Fly ash as one of the means to stabilise landfills is reported by Naveen et al. (2014).

Black cotton soils are problematic for engineers everywhere in the world, and more so in tropical countries like India because of wide temperature variations and because of distinct dry and wet seasons, leading to wide variations in the moisture content of soils. The following problems generally occur in black cotton soil: high compressibility, swelling and shrinkage. Black cotton soil cannot damage the structure if there is no change in its moisture content. The damages are caused by upheaving or settlement due to a change in moisture content of the black cotton soil. It is the water that creates the uplift pressure, which causes upheaving and then again it is the water that on removal from the soil causes settlement due to drying. Thus, without water, even the black cotton soil cannot cause damages to the buildings. There are two ways to prevent damage to structures due to black cotton soils, namely replacement of black cotton soil and preventing water from encountering black cotton soil. Hence, from the above points, it is essential that the water coming on the surface and percolating into the soil below foundation or floors is taken care of in such a way that either water does not enter the black cotton soil or the flow path of the water is lengthened in a way that water does not enter the active zone.

Shivashankar (2013) narrates a case study of a landslide due to a changed drainage pattern due to human activities. Mangalore and surrounding areas are generally on a sloping terrain. The large number of geotechnical failures that one observes in this area are slope failures or landslips due to changed drainage patterns because of numerous human activities in the name of development. More so, manmade slope failures are more common too. These are due to steep slopes or vertical cuts or near-vertical cuts excavated. A classic example of landslip due to changed drainage patterns in that area is Kethikal slope failure, reported by Bhat et al. (2008).

17.10   CONCLUDING REMARKS

The details of soils and rocks of Karnataka may be summarised as follows:

- Based on the geotechnical formation and engineering applications, the Karnataka soils can be mainly classified into three types: residual deposits, black cotton/expansive clays and lithomargic clays/shedi soils.

- Residual deposits consist predominantly of overburden, weathered rock/refusal stratum and bedrock/sheetrock. In general, these residual deposits do not pose any geotechnical challenges to the shallow foundation in terms of bearing capacity and settlement.
- Bedrock in Karnataka is very hard where the unconfined compressive strength can range from 50 to excess of 200 MPa. This poses challenges to the socketing of piles in the bedrock stratum.
- Black cotton soils are inorganic clays of medium to high compressibility and form a major soil group in Karnataka. They are characterised by high shrinkage and swelling properties and occurs mostly in the northern parts and covers approximately 20% of the total area of Karnataka.
- Shedi soil is a local term for the lithomargic soils formed due to the laterisation process. This material is abundantly available throughout the west coast of India from Malabar (Kerala) to Ratnagiri in Maharashtra and covers the coastal region of Karnataka in the West. These soils pose problems of low bearing capacity, low subgrade strength, large settlements and problems of slope stability and erosion.
- Apart from natural soils/rock, fly ash and slags from steel plants are other geomaterials that are abundantly available as wastes and pose challenges for disposals. Both these materials have found application in civil engineering.

ACKNOWLEDGEMENTS

The author sincerely acknowledges Jestin Joseph, Aksa John, Anirudh Singh Rathore, Prof K.S. Subba Rao, Prof R. Shivashankar, Ms Minnu KP and Dr Prashanth Talkad for material collection, drafting, content and advice. Special thanks to Ms Amala Krishnan for supporting at all stages, finalizing the manuscript and making it possible to bring out this chapter.

REFERENCES

Bhat, K.A., Shivashankar, R. and Yaji, R. (2008). Case study of landslide in NH-13 at Kethikal near Mangalore-India. *Proceedings of 6th International Conference on Case Histories in Geotechnical Engineering*, Arlington, Virginia, Paper No. 2.69, pp. 1–12.

Indian Minerals Yearbook (2018). *Slag – Iron & Steel*, Chapter 16, 57th Edition, India Bureau of Mines, Government of India, pp. 1–9.

Murthy, B.R.S. and Pujar, K.L. (2009). Socketing of bored piles in rock. *Proceedings of Indian Geotechnical Conference*, Guntur, India, pp. 678–681.

Naveen, B.P., Sivapulliah, P.V., Sitharam, T.G. and Sharma, A.K. (2014). Stabilization of waste dump using fly ash. *Proceedings of National Conference on Beneficial Use of Fly Ash in Construction Industry & Agriculture*, Bangalore, pp. 217–224.

Parthasarathy, C.R. (2002). Prediction of engineering properties of fine-grained soils. PhD Thesis, Indian Institute of Science, Bangalore, India.

Rao, S.M. and Acharya, I.P. (2014). Synthesis and characterization of fly ash geopolymer sand. *Journal of Materials in Civil Engineering*, ASCE, Vol. 26, No. 5, pp. 912–917.

Shivashankar, R. (2013). Role of case histories in geotechnical engineering teaching and practice. *Proceedings of 7th International Conference on Case Histories in Geotechnical Engineering*, Chicago, USA, pp. 1–10.

Shivashankar, R., Shankar, A.U.R. and Jayamohan, J. (2015). Some studies on engineering properties, problems, stabilization and ground improvement of Lithomargic clays. *Geotechnical Engineering Journal of the SEAGS & AGSSEA*, Vol. 46, No. 4, pp. 68–80.

Sitharam, T.G. and Babu, G.L.S (1998). *Workshop on Engineering Practice in Black Cotton Soils.* Indian Institute of Science, Bangalore, India.

Chapter 18

Kerala

Anil Joseph
Geostructurals (P) Ltd.

Jayamohan J.
LBS Institute of Technology for Women

Sreevalsa Kolathayar
National Institute of Technology Karnataka

CONTENTS

18.1 INTRODUCTION

Kerala is located between 8° 17′ 30″N and 12° 47′ 40″N latitudes and between 74° 27′ 47″E and 77° 37′ 12″E longitudes. The midlands and highlands of Kerala have been subdivided into Central Sahyadri, Nilgiris and Southern Sahyadri (Krishnan et al., 1996). Sporadic uplift of the Western Ghats during the Miocene–Pliocene periods is thought to be responsible for the development of the recent landscape (Chandran et al., 2005). The state spreads over about 38,864 km² comprising 1.18% of the whole land of India. Kerala experiences a tropical humid climate with alternative dry and wet periods, which is mainly related to two types of monsoons, that is, southwest monsoon and northeast monsoon. The state receives an average rainfall of 3,107 mm. Figure 18.1 shows the location map of Kerala along with the districts. There are 41

DOI: 10.1201/9781003177159-18

west-flowing rivers and 3 east-flowing rivers, which pass across Kerala. These rivers are monsoon-fed and hence may turn into rivulets in summer. The salient features of Kerala are given in Table 18.1.

Figure 18.1 Location map of Kerala along with its districts. (Source: https://www.vecteezy.com/free-vector/india-map, accessed 6 February 2021.)

Table 18.1 Salient features of Kerala

Capital of Kerala	*Thiruvananthapuram*
Language	Malayalam
Area	38,863 km^2
Number of districts	14
Population	**35,122,966**
Density of population	860 persons per square kilometre
Location of Kerala	8° 18′12° 48′ north latitude and 74° 52′77° 24′ east longitude
Rivers of Kerala	44 (41 west-flowing and three east-flowing)
Kerala climate	Southwest monsoons (June–September), northeast monsoons (October–November), winter (December–February), summer (March–May)
Forest area	11,125.59 km^2
Kerala economy	Agricultural
Major agricultural produce	Spices, rubber, coconut, beverages like coffee, tea

Table 18.2 Types of soils in different districts of Kerala

District	Type of soil
Thiruvananthapuram	Alluvium soil, Riverine alluvium, Lateritic soil, Brown hydromorphic soil, Forest soil
Kollam	Alluvium soil, Riverine alluvium, Lateritic soil, Brown hydromorphic soil, Forest soil, Onattukara soils, Coastal alluvial
Pathanamthitta	Forest soil, Lateritic soil, Brown hydromorphic soil, Riverine alluvium, Onattukara soils
Alappuzha	Coastal alluvium, Lateritic soil, Brown hydromorphic soil, Acid saline alluvium, Riverine alluvium, Kari soil
Kottayam	Forest soil, Lateritic soil, Brown hydromorphic soil, Riverine alluvium, Kari soil
Idukki	Alluvium soil, Lateritic soil, Brown hydromorphic soil, Forest soil, Hill soil
Ernakulam	Coastal alluvium, Lateritic soil, Brown hydromorphic soil, Riverine alluvium
Thrissur	Coastal alluvium, Lateritic soil, Brown hydromorphic soil, Riverine alluvium, Forest soil
Palakkad	Alluvium soil, Lateritic soil, Black cotton soil, Forest soil
Kozhikode	Coastal alluvium, Lateritic soil, Riverine alluvium, Forest soil
Wayanad	Lateritic soil, Brown hydromorphic soil, Riverine alluvium, Forest soil, Hill soil
Malappuram	Lateritic soil, Brown hydromorphic soil, Riverine alluvium, Forest soil
Kannur	Coastal alluvium, Lateritic soil, Brown hydromorphic soil, Riverine alluvium, Forest soil
Kasaragod	Lateritic soil, Brown hydromorphic soil, Riverine alluvium, Forest soil

18.2 MAJOR TYPES OF SOILS AND ROCKS AND THEIR PROPERTIES

The soil formation mainly depends on the variation in climatic conditions, topography and hydrological condition. The major soils found in Kerala are alluvial soil, Kuttanad alluvium/acid saline soil, kari soil, red soil, hill soil, forest soil, black cotton soil and brown hydromorphic soils. Types of soils in different districts of Kerala are presented in Table 18.2.

18.2.1 Soils

Three main types of alluvial soils are identified in the state. They are designated as the coastal alluvium, riverine alluvium and the Onattukara alluvium.

Coastal alluviums are greyish brown to reddish-brown and yellowish-red in colour. These soils have a sand to loamy sand texture. These soils mainly occur on gently sloping to level plains as narrow strips along the western coast of Kerala with an average width of about 10 km. The soil mainly comprises sand (80% of sand and 15% of clay) and is observed to have low cation exchange capacity (CEC) and organic content. Due to the presence of a higher fraction of sand, the permeability of soil is high. Since the soil has low CEC, the water holding capacity of the soils is also less.

Riverine alluviums are light brown to dark grey in colour. They have a sandy loam to clayey loam texture. These soils are developed from sediments of lacustrine and riverine sediments or their combinations by the fluvial process. Fine sand fractions are predominant. They consist of a considerable amount of organic contents and are acidic.

Onattukara alluviums are grey in colour. They have a sandy texture. They occur as marine deposits and extend to the interiors up to the laterite belt. They consist of sand without the mixture of laterite soil or silt and have high permeability and low CEC.

Kuttanad clay is very soft, highly compressible and highly organic. These soils are dark brown in colour. They have a clayey texture. They are susceptible to seasonal entry of saline water as a result of tidal inflow from the sea. They are considered to be poor foundation materials due to their low shear strength. The soil is generally acidic. Due to the presence of organic matter, the soil has low specific gravity and a high liquid limit.

Kari soils are marshy soil formed due to waterlogging and anaerobic conditions which leads to partial decomposition of organic matter. These soils are black in colour. They are commonly found in a high humid area. The soil is composed of high contents of iron and organic matter.

Laterites are the most extensive soil group which covers about 65% of the total area of Kerala. Lateritic soils are reddish-brown to yellowish-red in colour. They have a gravelly loam to gravelly clay loam texture. They have attracted the attention of earth scientists all over the world because of their importance in industry and agriculture (Schellman, 1981; Ollier and Galloway, 1990; Aleva, 1994). Early works on the mineralogy of these soils indicate both kaolinite and gibbsite as the dominant minerals, with some 2:1 minerals and quartz (Satyanarayana and Thomas, 1962; Gowaikar, 1972; Sahu and Krishna Murthi, 1984; Bronger and Bruehn, 1989). These were formed due to the leaching process under a tropical humid climate. Laterite soils have low CEC and are acidic. They are subjected to alternate cycles of wetting and drying resulting in lowering of liquid limit and plasticity limit, lower strength and higher permeability.

Red soils are formed due to weathering of igneous and metamorphic rocks. These soils are reddish in colour. They have a sandy clay loam to clay loam texture. The reddish colour of the soil is mainly due to the diffusion of iron in crystalline and metamorphic rocks rather than due to a high percentage of iron. Research works indicate the predominant clay mineral as kaolinite. Red soils are highly porous, friable and low in organic matter.

Hill soils are found in hilly, elongated ridges, rocky cliffs and narrow valleys. These soils are reddish-brown to yellowish-red in colour. They have a loam to clay loam texture. The soil comprises 10% to 50% of gravel. At greater depths, the amount of clay content increases. These soils are friable and highly susceptible to soil erosion.

Forest soils are formed due to weathering of rock under forest cover. These soils are dark reddish-brown in colour. They have a silty loam/clay loam texture. Forest soils are subjected to a slow weathering process and hence they have immature soil profiles. The depth of soil varies with the amount of erosion and vegetative cover. The soils are acidic, well drained and have a high organic content.

Black cotton soils are residual deposits that are formed from weathering of basalt or trap-rocks. These soils are dark reddish-brown in colour. They have a clay loam to clay texture. Montmorillonite is the predominant clay mineral. Black cotton soils have a low shrinkage limit and a high liquid limit. They are hard under dry conditions but lose strength under wet conditions. The swell–shrink characteristics of soils result in the differential settlement of the foundation.

Brown hydromorphic soils are commonly found in valley bottom and low-lying areas of the coastal strips. They were formed due to transportation and sedimentation of materials from adjacent hill slopes and deposition of the river. Brown hydromorphic soils are formed by the action of gravity and contain low to medium organic content. These soils are brown in colour. They have a sandy loam to clay texture. They are acidic and have low permeability.

18.2.2 Rocks

Geologically 88% of the state is underlain by crystalline rocks of the Archean age, which is a part of the peninsular shield. The crystalline complex of Kerala is composed of charnockite, gneisses, schists, migmatites and rocks of the Wayanad supracrustal. The major rocks found in Kerala are Wayanad supracrustal rocks, layered ultrabasic-basic rocks, the rocks of peninsular gneissic complex, khondalite group, quartzo-feldspathic, gneiss, garnet-biotite gneiss, hornblende gneiss, hornblende-biotite gneiss, quartz-mica gneiss and charnockite.

Wayanad supracrustal rocks are high-grade metasedimentary and ultramafic rocks that mainly occur in the Wayanad district. These rocks occur as a narrow ac-curate belt. The high-grade schist consists of talc-tremolite schist, fuchsite quartz-ite, calc granulite, quartz sericite schist, kyanite quartzite, garnet-sillimanite gneiss/schist, magnetite quartzite and kyanite mica schist.

Layered ultrabasic-basic rock group consists of peridotite, dunite, dolerite, anorthosite, gabbro and pyroxenite rocks. Anthrosite rocks are found in Wayanad, regions of Attapady, Palakkad and minor bodies of this rock are found in Kasaragod. Dunite is reported in the regions of the Punalur mica belt. Dismembered layered ig-neous complex consisting of alternate layers of peridotite and pyroxenite within char-nockite can be traced around the Panathadi area of Kannur District.

The rocks of the peninsular gneissic complex group comprise foliated granite, hornblende gneiss, pink granite gneiss and biotite gneiss. Granite gneiss is exposed along the boundary of the Palakkad and Idukki districts. Pink granite formation is well developed in the regions of Idukki. Hornblende gneiss can be traced from Manan-toody in Wayanad to the west coast. Biotite gneiss is formed around the regions of Mahe and Thalasseri.

The khondalite group of rocks are classified into calc granulite rocks and quartz-ite. Calc granulite rocks formation is developed in the eastern part of Kollam and Thiruvananthapuram, northeast of Munnar in the Idukki district and in parts of the Palakkad district. The rock is generally medium to coarse-grained, inequigranular and granoblastic in texture. Quartzite is usually formed as linear bands enclosed be-tween khondalite gneiss, charnockite and migmatitic gneisses. The rocks are white in colour with a brownish coating along the surface. The liner bands are found in the region of Pathanamthitta and Ernakulam.

Quartzo-feldspathic gneiss is formed due to stress-induced injection of acid mate-rials into the host rocks. They are found along the contact zone between garnet-biotite gneiss and garnet-sillimanite gneiss of the Thiruvananthapuram area.

Garnet-biotite gneiss is formed by retrogression and migmatisation of the khondalite group. These are well developed in the northeastern parts of Kollam, Thiruvananthapuram and Palakkad.

Hornblende gneiss, hornblende-biotite gneiss and quartz-mica gneiss have been formed due to migmatisation and associated retrograded within the charnockite group rocks and hence these are highly deformed. The naming of the rock is based on the predominant mineral present in the rock. They are medium grained and comprise alternate layers of rocks rich in hornblende or biotite. This rock formation is exposed in the Periyar valley area, east of Idukki.

Charnockites characterised by orthopyroxene-bearing granitic mineral assemblages are a common constituent of granulite facies metamorphic terrains. However, the relative importance of igneous versus metamorphic processes involved in their origin is debated (Rajesh and Santosh, 2004). They are either granitic rocks metamorphosed to the granulite facies (metamorphic charnockites; e.g., Newton et al. 1980) or rocks whose pyroxene crystallised directly from magma (igneous charnockites; e.g., Wendlandt, 1981). They are mainly exposed in the southern areas of Palakkad. These rocks are usually medium to coarse grained. This rock group consists of quartz, feldspar, pyroxenes, garnet and graphite with accessories like biotite, zircon, apatite and monazite. Table 18.3 gives the various types of rocks in different districts of Kerala.

Table 18.3 Types of rocks in different districts of Kerala

District	Major types of rocks
Thiruvananthapuram	Laterite, Sandstone, Gneiss, Khondalite
Kollam	Sandstone, Shale, Charnockite, Granite, Gneiss, Khondalite
Pathanamthitta	Laterite, Sandstone, Shale, Charnockite, Granite, Gneiss, Khondalite
Alappuzha	Charnockite, Khondalite
Kottayam	Laterite, Gneiss, Granite, Calc granulite, Magnetite quartzite
Idukki	Granite, Calc granulite, Magnetite quartzite, Granite gneiss, Charnockite, Pyroxene granulite, Hornblende-biotite gneiss, Dolerite
Ernakulam	Calc gneiss, Magnetite quartzite, Quartzite, Charnockite, Pink and grey gneiss, Laterite, Biotite gneiss, Granite gneiss
Thrissur	Biotite gneiss, Hornblende-biotite gneiss, Pyroxene granulite, Charnockite, Calc granulite, Pink granulite
Palakkad	Dolerite, Gabbro, Garnet-biotite gneiss, Hornblende-biotite gneiss, Quartzo-feldspathic gneiss, Magnetite quartzite, Pyroxenite, Charnockite, Granite gneiss
Kozhikode	Granite gneiss, Magnetite quartzite, Quartzo-feldspathic gneiss, Charnockite, Hornblende-biotite gneiss, Laterite
Wayanad	High-grade schist, Granite, Khondalite, Charnockite
Malappuram	Dolerite, Laterite, Charnockite, Hornblende-biotite gneiss, Quartzo-feldspathic gneiss, Pyroxenite, Magnetite quartzite, Hornblende granulite, Granite gneiss
Kannur	Dolerite, Laterite, Charnockite, quartz mica schist, Quartzo-feldspathic gneiss, Quartzite
Kasargod	Dolerite, Laterite, Granite, Hornblende-biotite gneiss, Fuchsite quartzite

18.3 GEOTECHNICAL PROPERTIES OF SOILS AND ROCKS

The state of Kerala has diverse soil types in different regions. Most part of the state is covered with laterites. The laterite soil shows good foundation characteristics with compressive strength ranging from low to medium. The compressive strength

Table 18.4(a) Subsurface profile of a typical borehole from a site in Northern Kerala

Soil strata

z_T (m)	z_B (m)	H (m)	Description and classification	SPT N_c	Unit weight (kN/m3)	c (kPa)	ϕ'	c_u (kPa)	u
0	3.2	3.2	Loose lateritic clayey silty sand with gravel (SM)	6	17	–	28	–	–
3.2	4.7	1.5	Medium dense lateritic sandy silty clay with gravel (GM-SM)	13	18	–	31	–	–
4.7	5.4	0.7	Medium dense lateritic silty sand with gravel (GM-SM)	19	18	–	33	–	–
5.4	6.3	0.9	Clayey silty sand with gravel	UDS	–	–	–	–	–
6.3	8	1.7	Medium dense silty sand with gravel (GM-SM)	22	18	–	34	–	–
8	8.7	0.7	Medium dense clayey silty sand with gravel (SM)	17	18	–	32	–	–
8.7	10.5	1.8	Loose silty sand with weathered rock (SM)	7	17	–	28	–	–

Rock strata

z_T (m)	z_B (m)	H (m)	Description	Least Q_u (kPa)
10.5	13.5	3	Granitic hard rock	1966

Table 18.4(b) Subsurface profile of a typical borehole from a site in Southern Kerala

Soil/IGM strata

z_T (m)	z_B (m)	H (m)	Description and classification	SPT N_c	γ (kN/m^3)	c' (kPa)	ϕ'	c_u (kPa)	ϕ_u
0.00	8.50	8.5	Hard residual soil (CL/ML)	100	20	10	28	250	0
8.50	20.0	11.5	Dense residual soil (SM)	35	20	0	35	–	–
20.0	30.0	10.0	Very dense residual soil (SM)	60	20	0	38	-	-

Rock strata

z_T (m)	z_B (m)	H (m)	Description	γ (kN/m^3)	CR	RQD	UCS (MPa)
30.0	33.0	3.0	Moderately weak Khondalite	23	35	20	10
33.0	35.0	2.0	Moderately strong Khondalite	25	65	65	20

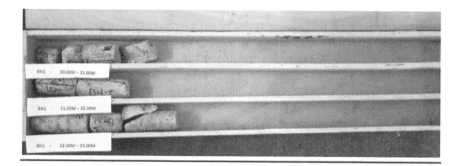

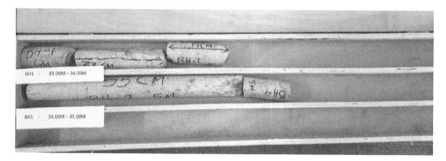

Figure 18.2 Rock core samples from a site in Thiruvananthapuram, Southern Kerala.

of coastal sediments ranges between 1 and 2 kg/cm² indicating that their foundation characteristics are poor.

Most of the coastal regions, including Kochi, consist of very soft marine clay. These marine clays have high Atterberg's limits and their liquidity indices range from 0.46 to 0.87. Marine clays of Kochi have equal fractions of clay and silt size with a sand content around 20%. These soils are highly compressible and have very low undrained shear strength. Hence deep foundation is recommended for sites with marine clays (Jose et al., 1988).

Table 18.4a shows an idealised subsurface profile and design parameters at a typical site from the Kozhikode district of Northern Kerala. Table 18.4b shows a typical subsurface profile from a site at Thiruvananthapuram district in Southern Kerala. Figure 18.2 shows rock cores obtained from the same site in Thiruvananthapuram.

18.4   USE OF SOILS AND ROCKS AS CONSTRUCTION MATERIALS

Traditionally, locally available materials such as rubble, laterite, wood and mud are extensively used for the construction of buildings in Kerala. Stones and laterites are mainly used for foundation and basement. The superstructure is commonly made out of timber structures. The roof structure is supported by wooden trusses, covered with thatch or clay tiles. Nowadays modern high rise RCC and steel structures are coming up extensively especially in the cities of Trivandrum, Cochin and Calicut. The five widely used materials for masonry are laterite blocks, stone blocks, burnt clay bricks, soil–cement blocks and hollow concrete blocks (Alex and Kasthurba, 2016).

Laterite is a tropical weathered rock and is generally used as a load-bearing masonry material, which is widely used in the Malabar area of Kerala. Laterite is a rock that hardens when exposed to the atmosphere. They are usually obtained in a size of about 330 mm × 200 mm × 200 mm. Laterite blocks are heavy (about 35 to 40 kg) and this makes it difficult for masons to work with laterite. They contain mineral oxides of iron and aluminium. Laterite is generally reddish in colour. As it is cut into the required shape and size manually, the thermal energy requirement for the manufacturing of blocks is negligible.

Natural stones are used as masonry and foundation materials, which are generally obtained by breaking huge natural stones into smaller sizes. Stone blocks are usually available in sizes of 180 mm × 180 mm × 180 mm. The expenditure for the production of stone blocks includes the quarry land cost, labour and machinery for breaking huge natural stones and also labour for shaping and sizing.

Burnt clay bricks are the most common masonry materials used and are available in different sizes like 230 mm × 105 mm × 70 mm, 200 mm × 90 mm × 90 mm. They are manufactured from red clay. The manufacturing process involves high thermal energy for the burning of bricks. The expenditure for the production of burnt clay bricks includes the cost of clay and water, labour for mixing and moulding and cost of fuel and labour for high temperature burning of a mix.

Soil–cement blocks are manufactured by pressurised pressing of a wet soil-cement mixture in the mould of required size. Mechanised pressing units and manual pressing units are available. Soil–cement blocks are produced by employing manually operated machines in a highly decentralised fashion. Load-bearing soil–cement blocks will have a cement content of about 6%–8%. The expenditure for the production of soil–cement blocks includes the cost of materials, which includes soil, cement and water, labour and machinery cost for mixing and moulding and curing charges.

18.5 FOUNDATIONS AND OTHER GEOTECHNICAL STRUCTURES

Kerala is known for its unique settlement pattern with independent houses on individual plots scattered across the habitable areas with different varieties of housing typologies of vernacular, traditional and modern types.

The different types of foundations normally used in house construction in Kerala are primarily categorised into rubble foundation, isolated foundation, strip/raft foundation and under-reamed/direct mud circulation piles.

Large rock pieces are used in a stepped manner to make the rubble foundation. To construct it, a minimum of 600 mm depth trench is made on the ground and rubble (hard rock) is skillfully placed there. The rubble is tightly packed to minimise the gaps in between. A rubble foundation runs directly beneath the layout of the main walls of the house. To make the rubble foundation stronger, sand filling or cementing is done in-between the rubble.

An isolated foundation is ideal for concentrated stress load release and can be applied even for high rise buildings, depending on the strength characteristics of the soil. In this type of foundation, columns are provided on the corners of the walls and footing (made of concrete).

For loose soil or clay layers, a strip/raft foundation is used. If the soil is very weak, prior to the foundation work, the soil beneath is strengthened. Depending on the type of soil, various ground improvement methods/systems, such as sand piling, stone column, preloading, PVD and geotextiles may be adopted.

Pile foundation is used when the soil in the immediate vicinity is very soft/loose and the load has to be transferred to better strata beneath. Commonly adopted pile foundations are of two types – under-reamed piles and direct mud circulation (DMC) piles.

Under-reamed piles use the friction of soil and hence are also called the friction piles. For friction piles, holes are dug and the side walls are strengthened with bentonite powder to prevent collapse. Depending upon the type of soil and load-carrying requirement, bulbs are formed using suitable under-reamer tools. Reinforcement cages are inserted into the borehole and concreting is done using the tremie method. The friction produced by the side walls of the borehole and the bulb provides strength to the pile. Generally, under-reamed piles are limited to 12 m depth and 500 mm diameter.

In DMC piles, the piles are taken to the deeper strata. It can be either a frictional pile, end-bearing pile or a combination of both. The depth of a pile for most of the major structures in the Cochin and Kuttanad areas goes up to 45 to 65 m. The piles are usually terminated in the dense sand strata encountered. If rocks are available at a reasonable depth, the DMC piles are provided with seating in hard rock.

18.6   NATURAL HAZARDS

The coastline of Kerala extends to a length of 580 km along the Arabian Sea coastline and about 322 km of this coast is prone to severe erosion. The state has the second-highest population density in the country with a density of 860 person/km^2 and it is also one of the most densely populated coastlines in the country, which adds to its vulnerability.

18.6.1   Floods

The frequency of great floods and extreme precipitation events has substantially increased under the warming climate. Frequent extreme precipitation exceeding the adsorptive capacity of soil and the flow capacity of river causes watercourses to overflow their banks onto flood plains. The great floods of 1924 and 2018 are considered to be the worst floods faced by the state in the past. In 2018, because of floods and landslides, about 5.4 million people were affected and 433 persons lost their lives. Table 18.5 provides details of flood-prone areas in Kerala.

Table 18.5 Flood-prone areas in Kerala

District	Area (km^2)	Area (%)	District	Area (km^2)	Area (%)
Thiruvananthapuram	268.09	12.23	Idukki	38.78	0.89
Kollam	283.62	11.41	Thrissur	688.44	22.65
Alappuzha	762.57	53.77	Malappuram	601.67	16.93
Pathanamthitta	212.76	8	Palakkad	567.16	12.66
Kottayam	461.33	20.95	Kozhikkode	288.83	12.3
Ernakulam	718.94	23.5	Wayanad	215.39	10.11
Kannur	339.18	11.45	Kasargod	198.79	9.99

Table 18.6 Landslide-prone areas in each district

District	Area (km²)	Area (%)	Area (km²)	Area (%)
	High		Low	
Thiruvananthapuram	45.59	2.08	114.9	5.24
Kollam	75.61	3.04	191.07	7.69
Pathanamthitta	170.28	6.41	426.25	16.04
Alappuzha	0	0	0	0
Kottayam	61.78	2.81	190.5	8.65
Idukki	388.32	8.9	873.71	20.02
Ernakulam	61.42	2.01	229.05	7.49
Thrissur	108.15	3.56	217.4	7.15
Palakkad	324.62	7.25	366.88	8.19
Malappuram	198.34	5.58	267.56	7.53
Kozhikode	109	4.64	206.71	8.8
Wayanad	102.56	4.82	196.57	9.23
Kannur	168.64	5.69	272.55	9.2
Kasargod	33.67	1.69	205.9	10.35
Total for the state	1847.98	4.75	3759.07	9.67

18.6.2 Landslides

Landslides mainly occur in the Western Ghats region where the slope is steep and the soil is over-saturated as a result of prolonged rainfall. About 40% of the state lies in the highland region forming the western slopes of Western Ghats except the coastal district of Alleppey. The west-facing Western Ghats scarps are more prone to landslides. The scrap faces of Western Ghats are heavily influenced by the intensity of rainfall and anthropogenic activities such as deforestation. When scrap faces are subjected to prolonged and intense rainfall or a combination of the two, variations of pore pressure result, leading to landslides. The slope in the Western Ghats region is generally steep to very steep with plateau edges highly indented having greater than 25° slopes. Table 18.6 presents landslide-prone areas in every district of Kerala.

18.6.3 Coastal erosion

Out of 14 districts in the state, 9 districts are bordering the Arabian Sea. These nine districts are, namely, Kasargod, Kannur, Kozhikode, Malappuram, Thrissur, Ernakulum, Alappuzha, Kollam and Thiruvananthapuram. Coastal erosion can occur due to natural causes such as winds, tides, nearshore currents, storm surges and sea-level rise or human activities such as the construction of harbours, jetties, mining and dredging.

18.6.4 Earthquakes

In the past, the state has experienced only mild tremors. None of them caused major significant damages to the structures. However, pockets of higher ground acceleration have been identified in central Kerala. According to the seismic hazard map of India (2002) by the Bureau of Indian Standards, all the districts in the state lay in seismic zone III. The maximum intensity of an earthquake in zone III is 5.6 M on the

Richter scale. Higher levels of seismicity are seen in Kottayam–Idukki and Thrissur–Palakkad districts.

18.6.5   Reasons and remedies for hazards

The state is prone to several natural as well as human-induced disasters due to high population density, geographical location of the state and varying climatic conditions. This disaster needs to be counteracted by adopting suitable preventive and remediation methods. Kerala is frequently ravaged by the disastrous consequences of numerous hazards. Hence it is named a multi-hazard prone state. The remedial measures adopted for various hazards in Kerala are listed below:

For floods

*   Drainage improvement should be implemented in the water management system of urban areas mainly focusing on flood control.
*   Important structures should be located in non-flood prone areas.
*   The amount of free-flowing water should be reduced by the construction of water holding structures.
*   Flood defence infrastructures such as embankments, dam and storm surge barriers should be constructed.

For landslides

*   Suitable dewatering techniques should be adopted.
*   Trench and interceptor drains should be constructed.
*   Vertical wells and pumps, horizontal or directional drilling, bioengineering solutions, geotextile matting and slope grading should be employed.

For coastal erosion

*   Infrastructures should be constructed on stilts or earth mounds.

For earthquakes

*   Structures should be designed and constructed to resist any type of ground movement.
*   The type of soil should be analysed and construction of buildings on soft soils, which can undergo liquefaction due to earthquake, should be avoided.
*   Restriction in height and density of building should be considered.

18.7   CASE STUDY

Kochi lies at the northern end of a narrow neck of land, about 19 km long and less than 1.6 km wide in many places and is separated from the mainland by inlets from the Arabian Sea and by the estuaries of rivers draining from the Western Ghats. As a result, Kochi is a natural harbour. Much of Kochi lies at the sea level, and the entire city spans an area of 87.5 km². The city has a seacoast of about 50 km. Willingdon island which is a large artificial island, created by dredging the Vembanad lake under the direction of Lord Willingdon, lies on the opposite shore of the Cochin Shipyard berthing facility. Cochin Shipyard Limited (CSL) is one of the major shipbuilding and maintenance facilities in India. It is a part of a line of maritime-related facilities in the port-city of Kochi, in the state of Kerala, India. Some of the services provided by the shipyard are building platform supply vessels and double-hulled oil tankers.

As a part of expansion, CSL has planned to develop a new dry dock facility at Cochin having a size of approximately 320 m × 75 m × 13 m. The consultant for the project is Royal Haskoning DHV, an international engineering consultancy firm with headquarters in Amersfoort, Netherlands. The geotechnical investigations comprise land and marine boreholes, conducting cone penetration tests at the proposed locations, field testing, laboratory testing and factual reports to provide the necessary input parameters for the design of different facilities. The proposed new dry dock site is located on the Ernakulam channel. The shipyard has an area of 170 acres of land, out of which about 30 acres are set aside for future expansion. The work of geotechnical investigation was awarded to Engineers Diagnostic Centre (P) Ltd., Cochin in 2015.

The area proposed for the new dry dock was found to be partially inside the shipyard premises and partly on the premises of the shipyard where 'Dolphin club' and staff quarters were situated. The premises inside the shipyard were used as a scrapyard. Ten numbers of land boreholes, five numbers of marine boreholes, eleven numbers of land static cone penetration tests and three numbers of marine static cone penetration tests were taken for the design of the proposed dry dock. Figures 18.3 and 18.4 show the locations of boreholes and locations of SCPTs, respectively.

Out of the ten land boreholes, five were taken along and outside the retaining wall alignment and the rest of them were in the bed area of the proposed dry dock. Out of the five marine boreholes, three were taken along and outside the retaining wall alignment of the dry dock, one in the area of the dry dock gate and the rest in the bed area of the dry dock. The boreholes were penetrated through clayey/sandy layers and terminated at 50 to 50.50 m depth. The reduced level of ground was noted to be +3.203 in the borehole LBH 06 during the time of the investigation. In LBH 06, the top 3.60 m comprises medium dense sand having an SPT value of 11 to 18. Below this, medium dense sand with silty clay having an SPT value of 11 was noted extending up to a depth of 5.20 m. This was followed by dense sand with silty clay having an SPT value of 40 extending up to a depth of 6.30 m. From 6.30 to 12.60 m, medium-stiff clay with sand having an SPT value of 5 to 7 was noted. Below this, stiff silty clay having an SPT value of 12 to 13 was noted extending up to a depth of 16.70 m. This was followed by stiff clay having an SPT value of 8 to 12 extending up to a depth of 24.50 m. From 24.50 to 25.65 m, very stiff clay having an SPT value of 27 was noted. This was followed by dense sand having an SPT value of 33 extending up to a depth of 28.00 m. From 28.00 to 32.30 m, very stiff clay having an SPT value of 21 was noted. Below this, medium dense sand with silty clay having an SPT value of 29 to 30 was noted extending up to a depth of 35.20 m. This was followed by hard clay with sand having an SPT value of 37 extending up to a depth of 36.40 m. From 36.40 to 40.00 m, dense to very dense sand having an SPT value of 30 to greater than 50 was noted. This was followed by medium dense to dense sand with silty clay having an SPT value of 26 to 43 extending up to a depth of 44.00 m. Below this, very dense sand having an SPT value of greater than 50 was noted extending up to a depth of 49.00 m. From 49.00 to 50.00 m, hard clay having an SPT value of greater than 50 was noted and LBH 06 was terminated at this depth. The water table was noted at a depth of 0.60 m from the ground level during the time of investigation. Figure 18.5 shows the boring log of LBH 06.

Figure 18.3 Location of boreholes.

Figure 18.4 Location of SCPTs.

PROJECT : GEOTECHNICAL INVESTIGATION FOR DRY DOCK FACILITY AT COCHIN SHIPYARD, THEVARA

ENGINEERS DIAGNOSTIC CENTRE (P) LTD	Project	: DRY DOCK FACILITY	Boring Started	:	06.11.2015
	Bore Hole No	: LBH-06	Boring Completed	:	10.11.2015
	Type of Boring	: ROTARY	Ground water table	:	0.60 m
	Termination Depth	: 50.00 m	Reduced Level	:	3.203 m

DEPTH W.R.T GL	DEPTH W.R.T RL	PROFILE	DESCRIPTION OF STRATA	THICKNESS OF STRATA	DEPTH In m	Test Depth	BLOWS/15cm			SPT "N"	CORRECTED SPT "N"	Remarks
							15cm	15cm	15cm			
			MEDIUM DENSE SAND (GREY)	3.60	1.00	1.00-1.45	1	5	6	11	11	UDS 1
					2.00	2.00-2.45	-	-	-	-	-	
3.60	-0.39				3.00	3.00-3.45	4	1	17	18	17	
5.20	-1.99		MEDIUM DENSE SAND WITH SILTY CLAY (GREY)	1.60	4.50	4.50-4.95	7	8	3	11	11	
6.30	-3.09		DENSE SAND WITH SILTY CLAY (GREY)	1.10	6.00	6.00-6.45	4	4	36	40	40	
			MEDIUM STIFF CLAY WITH SAND (BLACK)	6.30	7.50	7.50-7.95	1	2	3	5	5	UDS 2
					9.00	9.00-9.45	-	-	-	-	-	
					10.50	10.50-10.95	2	4	3	7	7	
12.60	-9.39				12.00	12.00-12.45	-	-	-	-	-	UDS 3
			STIFF SILTY CLAY (BLACK)	4.10	13.50	13.50-13.95	2	5	7	12	12	
					14.00	FIELD VANE SHEAR, Cu= 5.22 T/m2						
					15.00	15.00-15.45	4	6	7	13	13	UDS 4
16.70	-13.49				16.50	16.50-16.95	-	-	-	-	-	
					17.00	FIELD VANE SHEAR, Cu= 11.59 T/m2						
			STIFF CLAY (BLACKISH GREY)	4.30	18.50	18.50-18.95	2	3	5	8	8	UDS 5
					19.50	19.50-19.95	-	-	-	-	-	
21.00	-17.79				20.00	20.00-20.45	2	5	6	11	11	

Note : UDS- Undisturbed Sample   SPT "N"-Standard Penetration Test "N"

Figure 18.5 Bore log of LBH 06.

(Continued)

PROJECT : GEOTECHNICAL INVESTIGATION FOR DRY DOCK FACILITY AT COCHIN SHIPYARD, THEVARA

	ENGINEERS DIAGNOSTIC CENTRE (P) LTD		
Project	: DRY DOCK FACILITY	Boring Started	: 06.11.2015
Bore Hole No	: LBH -06	Boring Completed	: 10.11.2015
Type of Boring	: ROTARY	Ground water table	: 0.60 m
Termination Depth	: 50.00 m	Reduced level	: 3.203 m

DEPTH W.R.T GL	DEPTH W.R.T RL	PROFILE	DESCRIPTION OF STRATA	THICKNESS OF STRATA	DEPTH In m	Test Depth	BLOWS/15cm			SPT "N"	CORRECTED SPT "N"	Remarks
							15cm	15cm	15cm			
					21.50	21.50-21.95	2	4	7	11	11	
					22.50	22.50-22.95	-	-	-	-	-	UDS 6
			STIFF CLAY (BLACKISH GREY)	3.50	23.00	23.00-23.45	3	6	6	12	12	
					24.00	24.00-24.45	-	-	-	-	-	UDS 7
24.50	-21.29											
25.65	-22.44		VERY STIFF CLAY (BROWNISH GREY)	1.15	25.50	25.50-25.95	6	16	11	27	27	
			DENSE SAND (GREY)	2.35	27.00	27.00-27.45	9	16	17	33	24	
28.00	-24.79				28.50	28.50-28.95	-	-	-	-	-	UDS 8
			VERY STIFF CLAY (GREY)	4.30	30.00	30.00-30.45	6	9	12	21	21	
					31.50	31.50-31.95	-	-	-	-	-	UDS 9
32.30	-29.09											
			MEDIUM DENSE SAND WITH SILTY CLAY(GREY)	2.90	33.00	33.00-33.45	11	16	14	30	30	
					34.50	34.50-34.95	6	13	16	29	29	
35.20	-31.99											
36.40	-33.19		HARD CLAY WITH SAND (GREY)	1.20	36.00	36.00-36.45	12	21	16	37	37	
			DENSE TO VERY DENSE SAND (GREY)	3.60	37.50	37.50-37.95	36	>50	-	>50	>50	
					39.00	39.00-39.45	10	14	16	30	23	
40.00	-36.79				40.50	40.50-40.95	10	11	15	26	26	
			MEDIUM DENSE TO DENSE SAND WITH SILTY CLAY (GREY)	2.00								
42.00	-38.79				42.00	42.00-42.45	15	25	18	43	43	

Note : UDS- Undisturbed Sample   SPT "N"-Standard Penetration Test "N"

Figure 18.5 (Continued) Bore log of LBH 06.

(Continued)

PROJECT : GEOTECHNICAL INVESTIGATION FOR DRY DOCK FACILITY AT COCHIN SHIPYARD, THEVARA

ENGINEERS DIAGNOSTIC CENTRE (P) LTD

Project	: DRY DOCK FACILITY	Boring Started	: 06.11.2015
Bore Hole No	: LBH-06	Boring Completed :	10.11.2015
Type of Boring	: ROTARY	Ground water table :	0.60 m
Termination Depth :	50.00 m	Reduced level :	3.203 m

DEPTH W.R.T GL	DEPTH W.R.T RL	PROFILE	DESCRIPTION OF STRATA	THICKNESS OF STRATA	DEPTH In m	Test Depth	BLOWS/15cm			SPT "N"	CORRECTED SPT "N"	Remarks
							15cm	15cm	15cm			
44.00	-40.79		MEDIUM DENSE TO DENSE SAND WITH SILTY CLAY (GREY)	2.00	43.50	43.50-43.95	12	15	22	37	37	
			VERY DENSE SAND (GREY)	5.00	45.00	45.00-45.45	>50	-	-	>50	>50	BALANCE =33 CM
					46.50	46.50-46.95	37	>50	-	>50	>50	BALANCE =28 CM
					48.00	48.00-48.45	17	>50	-	>50	>50	BALANCE =21 CM
49.00	-45.79											
50.00	-46.79		HARD CLAY (GREY)	1.00	49.50	49.50-49.95	14	19	26	45	45	

BORE HOLE TERMINATED AT 50.00m DEPTH

Note : UDS- Undisturbed Sample

SPT "N"-Standard Penetration Test "N"

Figure 18.5 (Continued) Bore log of LBH 06.

Boreholes MBH5, LBH1, LBH3, LBH4, LBH7, LCPT 7 and LCPT 8 were taken in the floor bed area of the proposed dry dock. From the study of the boreholes and CPT data, it is noted that the soil profile has an erratic variation. Loading on the dock floor can be of three categories: (i) upward reaction from the ground and the groundwater pressure. Loading transmitted from the dock walls should also be considered; (ii) loading from the water in the dock, with a dock full of water, and (iii) loading from the ships, including tank testing loads and isolated loads from Jacketing and so on. In all the boreholes taken in these areas, very dense sand was noted at a depth of about 45 m from the ground level. For the proposed dry dock floor to overcome the huge uplift forces likely and to cater for the vertical axial forces, raft foundation supported by bored cast in situ DMC piles end bearing in the very dense sand strata at a depth of about 45 m with adequate anchorage may be adopted.

Boreholes MBH1, MBH4, LBH2, LBH6, LBH8, LBH10, LBH5, LCPT6, LCPT 11, LCPT 12, LCPT13, LCPT14, LCPT9, LCPT5 and MCPT3 were taken along the retaining wall alignment of the proposed dry dock and LBH9, MBH2, LCPT 1, LCPT 10, MCPT1 and MCPT2 were taken in the area outside the retaining wall alignment of the proposed dry dock. For the proposed retaining wall for the dry dock, tangent piles or sheet pile cut off walls with adequate back anchorage systems may be adopted. The back anchorage system may be obtained using suitable systems such as prestressed active or passive anchors, soil nailing, dead man and sheet pile walls. The use of a reinforced concrete relieving platform can be considered for reducing the lateral earth pressure on the retaining system. The termination depth of the main retaining system can be two to three times the depth of the retaining structure, based on the external stability analysis criteria.

Borehole MBH3 was taken in the location proposed for the dock gate area. Dry dock gates are highly variable in principle and in design. The choice is governed by different features, which may be required, and the different conditions under which the gates must operate. The factors affecting the choice of the dock gate are the (i) width of entrance, (ii) head of water to be retained, (iii) speed of operation, (iv) cost of construction, (v) ability to open against a head, (vi) depth available outside the dock and so on. For the proposed dock gate, an end-bearing pile foundation may be adopted in the very dense sand strata at a depth of 39 m with adequate anchorage.

18.8 GEOENVIRONMENTAL IMPACT ON SOILS AND ROCKS

Factors such as population growth, urbanisation, industrialisation and globalisation have led to significant changes in land use. Unscientific agricultural methods resulted in land degradation and natural disasters such as floods, droughts, landslides, soil erosion, stream bank erosion, sea erosion and salinity.

Mining of natural resources is a process involving intervention in the land environment, the magnitude and intensity of which vary based on the type of mining and environmental fragility of location. The major mining activity in the state is confined to the beach placers and China clay deposits. The demand for river sand is very high and extensive removal of this sand from the river has led to declining of aggregate grade sand in the rivers and mining of sand from river is currently banned. Exposed rock/laterites are also mined for construction purposes. The major impacts of mining are modifications of landscape and land stability and soil loss. The artificial ponds that are formed due to illegal mining led to land stability problems in the adjoining areas.

The steep soil slopes facilitate quick run off of water at a high velocity, hence the amount of soil displacement in these regions is more. The major portion of the state consists of lateritic soil and these soils are prone to erosion. This is mainly due to the high porous and coarse texture nature of the soil. The rate of erodability increases with an increase in the silt content.

18.9 CONCLUDING REMARKS

This chapter attempted to provide a brief description about the soils, rocks and other relevant geological and geotechnical information related to Kerala. A case study on geotechnical investigation for the construction of a dry dock at Cochin Shipyard is also included in this chapter. Major observations about geological and geotechnical aspects of soils and rocks are summarised below:

- There are mainly nine types of soils and eight types of rocks in Kerala.
- The major soils found in Kerala are alluvial soil, Kuttanad alluvium/acid saline soil, kari soil, red soil, hill soil, forest soil, black cotton soil and brown hydromorphic soils.
- All the soils have entirely different properties, which vary from place to place. The same soil type found at different locations shows drastic variation in properties.
- Lateritic soils are seen in all the districts of Kerala.
- The major types of rocks in Kerala are Wayanad supracrustal rocks, layered ultrabasic-basic rocks, the rocks of peninsular gneissic complex, khondalite group, quartzo-feldspathic gneiss, garnet-biotite gneiss, hornblende gneiss, hornblende-biotite gneiss, quartz-mica gneiss and charnockite.
- Kerala is vulnerable to natural hazards such as landslides, floods, earthquakes and coastal erosion. Remedial measures for natural hazards are also appended in this chapter.
- There is a greater scope for a detailed study of the soils and related issues of the behaviour of soils in Kerala. New alternatives and solutions for the existing foundation and the method of ground improvements needed can be developed after a detailed study of each soil type found in the state and based on the load transference criteria.

REFERENCES

Aleva, G.J.J. (1994). Laterites: concepts, geology, morphology and chemistry, Educational Department of Cruetzberg, *International Soil Reference and Information Centre (ISRIC), Wageningen*.

Alex, J., and Kasthurba, A.K. (2016). Masonry materials for the future: comparison on sustainability factors of a few popular masonry materials, *International Journal of Innovative Research in Science, Engineering and Technology*, Volume 5, Special Issue 14, 200–205.

Bronger, A., and Bruehn, N.K. (1989). Relict and recent features in tropical Alfisols from south India, *CATENA*, 16, 107–128.

Chandran, P., Ray, S.K., Bhattacharyya, T., Srivastava, P., Krishnan, P., and Pal, D.K. (2005). Lateritic soils of Kerala, India: their mineralogy, genesis and taxonomy, *Australian Journal of Soil Research*, 43, 839–852.

Gowaikar, A.S. (1972). Influence of moisture regime on the genesis of laterite soils in south India. II. Clay composition and mineralogy, *Journal of the Indian Society of Soil Science*, 20, 59–66.

Jose, B.T., Sridharan, A., and Abraham, B.M. (1988). A study of geotechnical properties of Cochin marine clays. *Marine Georesources & Geotechnology*, 7(3), 189–209.

Krishnan, P., Venugopal, K.R., and Sehgal, J.L. (1996). Soil resources of Kerala for land use planning, NBSS Publication, 48b. Soils of India series 10., *National Bureau of Soil Survey and Land Use Planning, Nagpur, India*.

Newton, R.C., Smith, J.V., and Windley, B.F. (1980) Carbonic metamorphism granulites and crustal growth, *Nature*, 288, 45–50.

Ollier, C.D., and Galloway, R.W. (1990). The laterite profiles, ferricrete and unconformity, *CATENA*, 17, 97–109.

Rajesh, H.M., and Santosh, M. (2004). Charnockitic magmatism in southern India, *Proceedings of Indian Academy of Science (Earth Planet Science)*, 113(4), 565–585.

Sahu, D., and Krishna Murthi, G.S.R. (1984). Clay mineralogy of a few laterites of Kerala, Southwest, India, *Clay Research*, 3, 81–88.

Satyanarayana, K.V.S., and Thomas, P.K. (1962). Studies on laterites and associated soils II. Chemical composition of laterite profiles, *Journal of the Indian Society of Soil Science*, 10(3), 211–222.

Schellman, W. (1981). Consideration on the definition and classification of laterites in laterisation process, *Proceedings of International Seminar on Laterisation Processes, Trivandrum, India*, pp. 1–10, Geological Survey of India: Calcutta.

Wendlandt, R.F. (1981). Influence of CO_2 on melting of model granulite facies assemblages: a model for the genesis of charnockites, *American Mineralogist*, 66, 1164–1174.

Chapter 19

Ladakh

Chandan Ghosh
National Institute of Disaster Management, Ministry of Home Affairs

Nitin Joshi, Prabhat Kumar, and Avinash Dubey
Indian Institute of Technology Jammu

CONTENTS

19.1 INTRODUCTION

Ladakh, a newly constituted union territory of India, is blessed with mountain beauty and distinct culture. It is bounded by Pakistan in the northwest, Jammu and Kashmir, another recently constituted union territory of India, in the west, Tibet in the north and eastern part and Himachal Pradesh in the southeast. It has an arctic desert climate and the main features of this 'Cold Desert' are having intense solar radiation (up to 6–7 kWh/mm) (CGWB, 2009), wide daily and seasonal variation in temperature with as low as −40°C in winter and as high as +30°C in summers (Guru et al., 2017). Ladakh used to be the gateway to Tibet and it was once adored as the nerve centre of Polyculture, including Indian, Chinese and Islamic cultures. The land area consists of unconsolidated loose deposits, sandy, silty, loamy and porous, and it has the least green cover. Soils vary in physicochemical properties as well as microbiological properties (Gupta and Arora, 2017). They vary widely in colour, texture, porosity and permeability. Under the predominant influence of physical weathering, these soils usually consist of coarse-grained materials and are often called skeletal soils with admixtures of distinguished rock fragments and colluvium materials. The soils of the Ladakh region vary in their morphological and physicochemical properties, thereby distinguishing them in the genesis and classification.

    The average soil cover loss in India is 16 tonnes/ha/year and Ladakh is next to Rajasthan, where it is the highest (18%). Usually in Ladakh, rainfall occurs twice a

DOI: 10.1201/9781003177159-19

year, June to September and October to May, and snowfall is quite extensive, which is the major source of making eight more sanctioned run-of-the-river hydroelectric power plants with a collective capacity of 144 MW along the Indus river and its five tributaries. The 114 MW capacity Nimo-Bazgo hydroelectric project has been successfully commissioned in the year 2014. Due to wide variation in day and night temperatures, erosion and weathering are rampant in Ladakh (Bhan et al., 2015). At this high altitude with little rainfall as well as with prolonged cold and arid climate, it is known as the country's High-Altitude Cold Desert (Thakur, 1981; Tandup, 2014).

Leh and Srinagar are connected by a national highway via the Kargil district. It is having land route connectivity only during summer whereas during winter because of heavy snowfall, the region remains mostly cut off from the rest of the country. Recently an all-weather tunnel of 14.5 km along the Zojila Pass is under construction. Leh is also connected by the Leh–Manali highway via the Rohtang Pass (NH-3, 479 km), where a new tunnel (named Atal Tunnel, 8.8 km) has recently started operation. Many strategic roads (Figure 19.1), tunnels and airstrips are being constructed using modern construction materials and techniques. Traditionally, Ladakhi houses are made of clay-mud, wood, stones and tree trucks with proper foundations both in plains and in slopes.

However, due to the promotion of tourism, modern living habitats with RCC and other modern constructions are mushrooming in the Leh town, where annual tourist counts exceed by 7–10 times the town residents. Flash floods (Ghosh and Parkash, 2010; Gupta et al., 2012), snow avalanche, earthquake, drought, locust attack and road blockade due to landslides are very common in Ladakh. Some thematic maps consisting of soil depth, texture, water capacity, geothermal potentiality, erosion, pH and so on are available. The soil profiles are shallow and they vary in colour, texture, structure and permeability. Therefore, using them alone as local construction materials for roads and buildings is limited. Due to advancement of civil engineering construction

Figure 19.1 Road construction in unestablished cut slopes in Ladakh.

materials, various admixtures as binding agents are tried increasingly. However, traditional buildings, using local mud, have monumental signatures in Leh and the surrounding habitat. Some details of the disaster resilient buildings and their foundation shall be described later.

19.2   PHYSIOGRAPHY

Ladakh is the second largest UT of India (about 59,146 km^2), after Jammu and Kashmir (about 222,236 km^2), with strategic mountains of varying heights between 4,000 and 7,000 m above the MSL. It is located between 32° 15′ and 36°N latitude and 75° 15′ and 80° 15′E longitude. It mainly consists of the barren treeless tract, which is intricately interspersed by high vertical cliffs of black grey and brown shimmering sand. Despite the changing topography and cold desert climate, the local habitants have made their harsh living intact for about 8,800 years (Miehe et al., 2009). With many modern interventions, like making strategic roads for enhancing border area surveillance, the risks are changing from the bad to worse. Ladakh is mainly attributed to the Indus river basin and its tributaries. The Leh district is in the northern part, Skardu is in the west and the Kargil district is in the southern half. The important mountain systems are Kun-Lun, Karakorum, Kailash, Aghil, Zanskar and the Himalayan range. These unique ranges are aligned in the northwest to southwest direction. Zanskar, Indus and Shyok are the main valleys and Nubra, Dras, Pangong (lake) and Suru. Nubra and Indus valleys are the side valleys. The K 2 (Mt. Godwin Austen, 8,611 m) lies in the Karakoram ranges and Saser Kangri (7,680 m) is having numerous glaciers. The Ladakh range has no major peaks though its average height is 6,000 m and some of its famous passes are of 4,800 m. The major river of Ladakh is Indus and it enters India from Tibet at Demchok. The main tributaries of the Indus river are Zanskar, Shmgo, Shyok, Nubra and their river valleys where the main areas of habitation lie. The Indus river has deep and narrow gorges, which form the alluvial flats, river terraces, alluvial fans and the main Leh town; currently, the capital of Ladakh lies in the alluvial fans. There are several lakes such as Pangong Tso, which is 150 km long and 4 km wide and it is at the height of 4,300 m above the MSL. A pearl shaped lake, named Tso Morari, and Tso Kar are the brackish water lakes and Yaye Tso, Kiun Tso and Amtitla are some of the important freshwater lakes in Ladakh. The UT of Ladakh has lots of glaciers. Notable among them are the 74-km long strategically important Siachen located in the Karakoram range, which is the largest in the Himalayas–Karakoram region, Hari Parbat, Chong Kmdan, Machoi, Drang-Drung, Nubra and Nun Kun.

19.3   GEOLOGY

The Leh area of Ladakh is made up of granites and alluvial tills. The slopes around are covered with loose, unconsolidated deposits and large fans of loose sediments. Million years of erosion can be seen along the Indus river banks. These sediments become loose and slide down when the surface layer becomes saturated due to a small amount of rains but with a huge amount of snow melt water (Thakur, 1981; Hussain et al., 2015). The snow cover in the Dras area often exceeds 6 m. The Ladakh area has (a) Continental passive margin sediments of the northern Indian margin, (b) Indus

Suture Zone, (c) Trans Himalayan Batholith representing the subduction-related calc-alkaline magmatism and (d) Shyok Suture Zone (Shah et al., 1976; Sinclair and Jaffey, 2001; Srikantia and Razdan, 1980). Ladakh is mostly a desert of granite dust with sandstones, shales and conglomerates. Crystalline rocks and Schists occupy large areas of the northwest Himalayas forming the core of Zanskar and ranges beyond in Ladakh and Baltistan. The Pir-Panjal is regarded only as a minor offshoot. The south of Ladakh is a band of cretaceous with ophiolites comprising the flysch facies as well as massif and ultra-massif volcanic rocks of the Dras volcanic type.

19.4 LAND COVER LAND USE

Agriculture occupies a greater part of the Ladakh economy, where people work as cultivators, agricultural labourers and livestock householders. The Leh town has a wide stretch of fertile land for growing barley, wheat, fodder, pulses, cereals, mustard and vegetables. Traditionally, people are conscious about the conservations of vegetation and greeneries (Angchok and Singh, 2006). In the Leh town, Willow and Poplar plantation are very high, especially utilising the wasteland. Human settlements are mostly found around glacial streams of the river Indus and its tributaries. Such settlements are mostly influenced by the local adaptability to diverting water from the streams to canals and towards deserts to grow crops and trees. Leh has a lot of local attractions like the hall of fame, Leh Palace, Shanti Stupa and Gompas that were maturely built up several centuries ago and ensuring features for earthquake resistance. The driving force for the shifts in the land use of the Leh town and many such upcoming sub-towns (Figure 19.2;

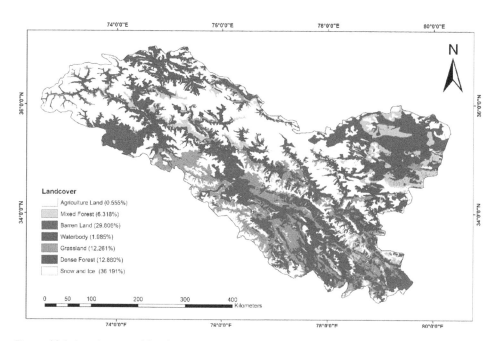

Figure 19.2 Land use and land cover map of Ladakh.

Table 19.1 Land cover areas in Ladakh

Land cover	Percent area	Area (km^2)
Agricultural land	0.56	924.71
Mixed forest	6.32	10,534.06
Barren land	29.81	49,688.38
Waterbody	1.99	3,268.92
Grassland	12.26	20,442.52
Dense forest	12.88	21,473.27
Snow and ice	36.18	60,314.80
Total area	100	166,646.66

Table 19.1) is its surging tourism industry. Eventually living in Leh depends on imports from neighbouring states amid shrinking natural resource depletion and environmental deterioration. In addition, haphazard urban proliferation and intense resources exploitation have been causing threat to environmental planning and proper use management.

19.5   MAJOR TYPES OF SOILS AND ROCKS

Soils in Ladakh are neutral to moderately alkaline except for soils on terraces and slopes. Mainly three types of rocks are found in Zanskar; they are granite, anorthosite, limestone and sandstone. The soil is clayey, loamy rich and light and it is mainly of alluvial origin but is found to be quite fertile (Walia et al. 1999). In the semi-mountainous tracts, the soil is coarse. The underlying rocks in this area are loose boulders.

Because of the absence of sufficient water action, there is less leaching effect but that has produced an abundance of detrital product in dry uplands and valleys forming a peculiar kind of mantle rock or regolith from the fresh undecomposed rock fragments. Thus, most of the morphological characteristics like colour, texture and permeability seem to be inherited from the so formed regolith due to the very slow rate of the soil-forming process (Table 19.2, Figure 19.3). The soils of Ladakh are coarse textured, shallow and sandy type which is derived from weathered debris of the rocks. The subsoil profile of the Leh city is shown in two borelog profiles (Figures 19.4 and 19.5). They are subjected to wind erosion and have usually boulders with high permeability and low water holding capacity. Some of the areas stand affected by salinity and sodicity (Gupta and Arora, 2017; Tundup et al., 2017).

Table 19.2 Major soil class in Ladakh

Soil class	Percent area	Area (km^2)
Glaciers and rock outcrops	30.21	48,428.74
Sandy soil	10.96	17,363.19
Rock outcrops	47.89	77,306.60
Glaciers	6.64	10,529.89
Water bodies	1.08	1,526.01
Loamy soil	3.22	5,053.99

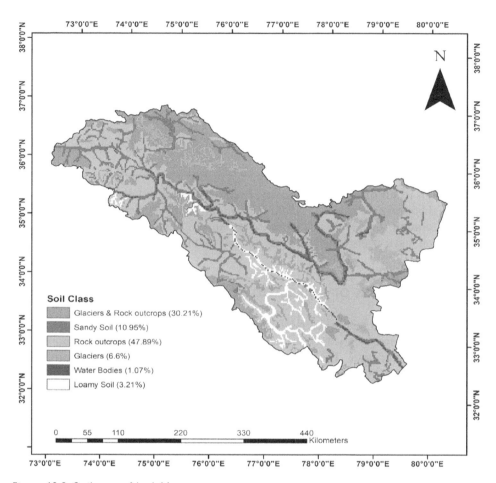

Figure 19.3 Soil map of Ladakh.

The physicochemical properties indicate that the soils developed on steep to very steep slopes are characterised by loamy sand to sandy in texture with a sand content from about 84% to 91% and silt and clay contents from 2% to 7% and 6% to 7%, respectively. The soils are alkaline in nature, with the pH ranging from 8.3 to 8.6; they are low in organic carbon (0.16% to 0.58%) and have cation exchange capacity (2.6 to 3.6 cmol/kg) with dominance of exchangeable Ca^{2+}. The soils are calcareous in nature. Although the soils formed at moderate slopes (10%–15%) are loamy sand in texture up to 51 cm depth, the clay content is higher as compared to the soils developed over steep to very steep and steep slopes (Gupta and Arora, 2017). The pH of these soils is in the alkaline range (7.9–8.8) with more status of organic carbon (0.22%–0.88%). Pebbles and gravels are also found at variable depths in the profile (16.8%–40.4%).

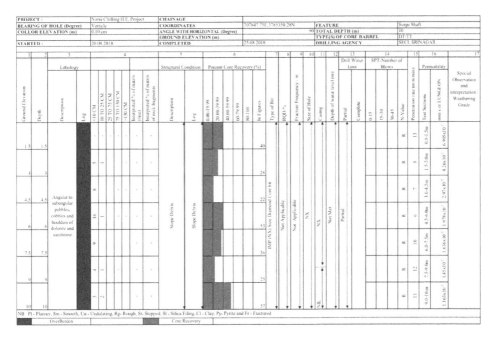

Figure 19.4 Borelog profile of Nimo at Leh with rocky strata. (Courtesy of M/s Space
Engineers Pvt. Ltd, Srinagar.)

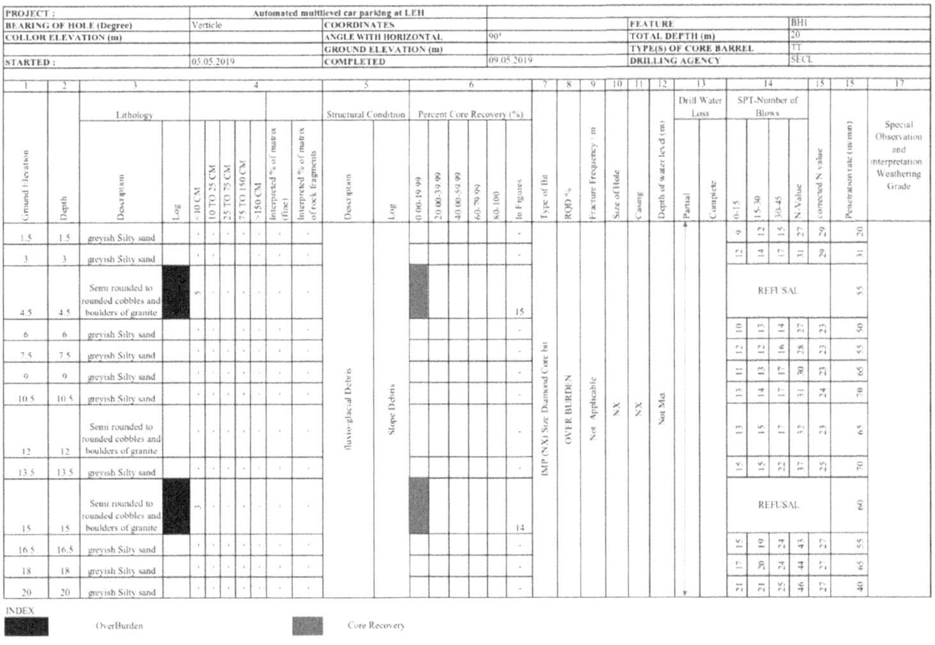

Figure 19.5 Borelog profile of Leh with soil strata. (Data courtesy, M/s Space Engineers
Pvt. Ltd, Srinagar.)

19.6 CONSTRUCTION MATERIALS

A traditional Ladakhi house is made from rammed earth, stone and brick walls and the roofs are made with poplar wooden beams and willow twigs; sometimes even with the dung of cows, donkeys or horses to ensure stiffness and durability. Inhabitants of Ladakh have harvested the sunlight together with the employment of the vernacular architecture exhibiting the passive measures for climatic control. The buildings have a unique spatial configuration to deal with the climatic conditions apart from using native building materials and techniques. The use of vernacular materials like mud-bricks, quartzite stones, poplar grass, willow fascines and timber in the construction helps in maintaining resilient climatic conditions.

19.6.1 Building materials and construction techniques

The dependency on sunlight to maintain comfortable climatic conditions is complemented by the traditional building materials and construction techniques. The primary building materials are local soils and timbers. Both earth and timber offer climatic comfort in wide diurnal range of Ladakh. It is due to their high thermal insulating properties that they are ideally used for centuries to construct not just houses but also the magnificent monasteries and palaces in the region. Quartzite stone is used for the construction of forts and palaces and the lower storeys provide additional strength.

Sundried earth blocks are the primary unit of masonry construction. Being usually made in the sizes of $300 \times 150 \times 150$ mm, they are employed in the construction of walls of 300–450 mm thickness. These earth blocks are made from alluvial soil along the banks of the Indus. A majority of earth blocks are made in Shey about 15 km away from Leh. In some cases, stone blocks are used in the lower courses of the walls for added strength and protection against water, especially in low-lying areas. Finally, the wall is finished in mud plaster. Roofs are constructed in flat spans by using the trunk of the local poplar tree as beams placed about 500–600 mm apart. The diameter of the trunks is 150 mm on an average while the length of the trunk is 3–4 m. These beams are covered by using poplar willows spread in the other direction. The usual thickness of the willows is 20–30 mm. A 150–200 mm layer of dry grass, hay, and so on is spread over the layer of willows and finally finished with plaster of clayey mud. Floors of a lower storey on the ground are made of mud while the upper storey is made of timber. Timber floors offer better thermal comfort along with the furnishing of carpets.

19.6.2 Other geomaterials

In the Ladakh region, the houses are built at two levels. The ground level is usually kept for animals, wood and fodder storage for winters, whereas the upper for the habitable spaces. Houses have a single large room with an oven in the corner which is used for cooling as well as heating. The houses are made entirely of mud, sometimes reinforced with horizontally placed timber members (Sharma and Sharma, 2016). The walls are either made of sundried bricks or rammed earth. Initially, the walls are of mud plaster while flooring is either in mud or wood. The ceiling height is low to provide the required insulation in all the areas. Every possible care is taken to trap the heat and maintain the temperature inside for a conducive living. Thus, the houses in Ladakh are essentially utilitarian. The descriptions of some of the materials used in the region are as follows (Sharma and Sharma, 2016; Joshi, 2013; Khan, 2013):

Mudbrick

Mudbricks are composites chiefly made of clay, silt and sand. Due to high sand contents, the surface water absorption of the brick reduces. The presence of clay and silt helps to bind the brick.

Foundations

Earth easily absorbs water which lowers its load-bearing capacity. Thus, lime is used as a binding material in the foundations to provide its strength.

Walls

Walls are built upon foundations raising above the ground level. The top of the foundation should not be wider than the bottom of the wall to avoid any structural damage in case of earthquakes. Walls are generally thick at the bottom and taper gradually as they rise.

Roof details

As Ladakh sustains under cold and dry climatic conditions, the ceiling is mainly built in mud and wood due to their insulating properties and easy availability.

Mud plaster

Mud plaster is applied wet with 15–25 mm thickness. Most mud plasters have to be repaired annually when used externally but adhere well to earth walls.

Insulation in the ceiling, outer walls and floors

During the daytime, the natural heat collected is retained by the insulation. The wood waste produced during the construction is filled in the ceiling to stop the heat loss through the roof.

Thermal mass in walls and floors

The foundations of buildings are laid about 1 m more below the ground. The building benefits from the stability of the earth's temperature at that depth, which is relatively warm in winter and cool in summer. It also helps in procuring the building material on site.

19.7 NATURAL HAZARDS

Several parts of Ladakh face wind blizzards, thunderstorms, cloud bursts, hailstorms, heavy snowing, human epidemics, livestock epidemics and so on from time to time; a few of those occasionally convert into disaster. The notable impact of locusts was observed in locust attacks in 2011 and 2015 in the Zanskar region, where 95% of people are dependent on agriculture. Ladakh is a fitting example of a highly evolved traditional set-up adjusting with the extreme climatic conditions through many vernacular architectures.

19.7.1   Potential glacial outburst

The high-altitude regions of the Himalayan Mountains have the highest concentration of glaciers outside the Polar Regions. Glaciers in the northwestern Himalayas are one of the most fragile and least studied. Glaciers monitored in the Karakoram region show different results; heterogeneous trends have been observed, some are losing mass and some are either advancing or in the equilibrium stage. These diverse results are due to the unique behaviours and dynamics of glaciers in the Himalayan regions. Ganjoo et al. (2014) investigated the glaciers of Nubra valley, Karakoram (Ladakh) for different time periods from 1969 to 2001. Out of 114 glaciers they monitored, 39 glaciers (34%) showed a gain in the area, 43 glaciers (38%) lost mass, whereas 32 glaciers (28%) showed no change in the area. A maximum area of 56 km^2 has been lost between 1969 and 1989 and only 4 km^2 between 1989 and 2001 suggesting the slowing down of glacier retreat in the Nubra valley. Ganjoo et al. (2014) mentioned the complexity and heterogeneous results of the glaciers studies in the valley and found that it is premature to draw any conclusion based on limited understanding. The complexity and variedness have been related to various factors such as topography, micro-climate and glacier type; by glacier retreat, several proglacial lakes have been formed, and most of them are dammed by ice-filled moraines. The potential risk of these lakes is shown by recent reports on glacial lake outburst flood in the villages Nidder in October 2010 and Gya in August 2014. Many of these glacial lakes are almost always frozen; a monitoring system based on high-resolution images and field surveys is required to detect potential lake outburst hazards. Scientists have noted that 146 outburst floods have taken place in the Karakoram range due to the obstruction of rivers passing through the valleys by advancing glaciers. Continuous monitoring of the satellite data to check where all rivers are being blocked due to surging glaciers would help the local administration for appropriate action.

19.7.2   Leh cloud burst (2010)

Cloud bursts are weather-related events causing highly concentrated rainfall over a small area lasting for a short time and thus lead to flash floods/landslides, house collapses, dislocation of traffic and human casualties (Figure 19.6). Though the exact mechanism of these systems is not yet known, they are mainly caused by intense vortices on a small scale. The heavy rainfall over Nimoo-Basgo and Leh on 5–6 August 2010 caused mudslides that travelled downslope towards the Indus river (Kumar et al., 2012). Leh district town faced heavy damages due to the rapid movement of huge volumes of debris with boulders, trees and building parts. At Choglamsar, the debris flow travelled approximately 10 km from the epicentre of the cloudburst, spreading up to 2 km (Figure 19.7). In Leh, the debris flow travelled about 3 km, from an elevation of 3,800 to 3,410 m, confined to the catchments of

Figure 19.6 Cloud burst induced debris flow affected area in the Leh town with heavy infrastructural damage.

Figure 19.7 Roads broken along the Leh–Manali highway at Upashi-Rumatse at km 10.10 due to cloud burst in 5–6 August 2010.

Shaksaling stream. The flow destroyed settlements, the bus stand and the BSNL mobile communications hub, and damaged the Sonam Norboo Memorial Hospital and the radio station. The worst affected areas including Leh town are Choglamsar village, Tashi Gyatsal area of Choglamsar, Saboo village, Taru, Nimoo, Basgoo, Stakna, Shey, Arzoo Thiksay Kungam, Anlay, Nidder, Achinathang Lungba, Skurbuchan, Rezong Ulley, Tia Temisgam and Tyakshi in Turtuk area, where about 233 human lives were lost, 424 people were injured and about 79 people were still missing (SEEDS, 2010).

19.8 CONCLUSIONS

Based on the soil types explained earlier, it is emphasised to study the vernacular construction typology, local building materials, stabilised rammed earth, stone masonry vis-a-vis checking/monitoring construction quality for planned sustainable development. With the onrush of an accelerated pace of construction of strategic all-weather roads/tunnels, administrative as well as institutional set-up after the formation of this union territory of India, it is emphasised to use soils as the local resources. Along with the soil formation, geology and future potential to grow up with Ladakhi culture, it is important for site development, design and construction technique. The use of local building materials/masonry/mortar and indigenous skills with inclusion of the concept of green cum thermally efficient buildings as well as harnessing the extreme potential of geothermal/solar/wind energy are to be explored.

REFERENCES

Angchok, D. and Singh, S. (2006). Traditional irrigation and water distribution system in Ladakh. *Indian Journal of Traditional Knowledge*, 5(3), 397–402.

Bhan, S.C., Devrani, A.K. and Sinha, V. (2015). An Analysis of monthly rainfall and meteorological conditions associated with cloudbursts over the dry region of Leh (Ladakh), India. *MAUSAM*, 1(66), 107–122.

CGWB (2009). Ground water information booklet of Leh District, Jammu and Kashmir State. http://cgwb.gov.in/District_Profile/JandK/Leh.pdf, accessed 12 March 2021.

Ganjoo, R.K., Koul, M.N., Bahuguna, I.M. and Ajai. (2014). The complex phenomenon of glaciers of Nubra valley, Karakorum (Ladakh), India. *Natural Science*, 6, 733–740. http://dx.doi.org/10.4236/ns.2014.610073

Ghosh, C. and Parkash, S. (2010). Cloud burst induced debris flows on vulnerable establishments in Leh. *Journal of Indian Landslides*, 3(2), 1–6. https://www.researchgate.net/publication/296561558_Cloud_burst_induced_debris_flows_on_vulnerable_establishments_in_Leh

Gupta P., Khanna A. and Majumdar S. (2012). Disaster management in flash floods in Leh (Ladakh): a case study. *Indian Journal of Community Medicine*, 37(3), 185–190.

Gupta, R. and Arora, S. (2017). Characteristics of the soils of Ladakh region of Jammu and Kashmir. *Journal of Soil and Water Conservation*, 16, 260–266.

Guru, B., Seshan, K. and Bera, S. (2017). Frequency ratio model for groundwater potential mapping and its sustainable management in cold desert, India. *Journal of King Saud University – Science*, 29(3), 333–347.

Hussain, G., Singh, Y. and Bhat, G. (2015). Geotechnical investigation of slopes along the National Highway (NH-1D) from Kargil to Leh, Jammu and Kashmir (India). *Geomaterials*, 5, 56–67. doi: 10.4236/gm.2015.52006.

Joshi, N. (2013). Efforts to resurrect and adapt earth building and passive solar techniques in Ladakh, India. *Vernacular Heritage and Earthen Architecture: Contributions for Sustainable Development*. Correia, M., Carlos, G., Rocha, S. C. S., CRC Press, London. 868.

Khan, N. (2013). Vernacular architecture and climatic control in the extreme conditions of Ladakh. *National Conference on Advancements in Sustainable Practices and Innovations in Renewable Energy*, Patiala, Punjab.

Kumar, S.M., Shekhar, M.S., Rama Krishna, S.S.V.S. et al. (2012). Numerical simulation of cloud burst event on August 05, 2010, over Leh using WRF mesoscale model. *Natural Hazards*, 62, 1261–1271. https://doi.org/10.1007/s11069-012-0145-1

Miehe, G., Miehe, S., Kaiser, K., Reudenbach, C., Behrendes, L., Duo, L. and Schlütz, F. (2009). How old is pastoralism in Tibet? An ecological approach to the making of a Tibetan landscape. *Palaeogeography, Palaeoclimatology, Palaeoecology*, 276(1), 130–147.

SEEDS (2010). Shelter strategy – Leh floods report by the sustainable environment and ecological development society. www.seedsindia.org, accessed 3 March 2021.

Shah, S.K., Sharma, M.L., Gergan, J.T. and Tara, C.S. (1976). Stratigraphy and structure of western part of Indus suture belt, Ladakh North-Western Himalaya. *Himalayan Geology*, 6, 534–556.

Sharma, A. and Sharma, S. (2016). Vernacular architecture in cold & dry climate: Ladakh – a case study. *International Journal for Scientific Research & Development*, 3(12), 767–769.

Sinclair, H.D. and Jaffey, N. (2001). Sedimentology of the Indus Group, Ladakh, Northern India implications for the timing of initiation of the Palaeo-Indus river. *Journal of the Geological Society, London*, 158, 151–162. http://dx.doi.org/10.1144/jgs.158.1.151

Srikantia, S.V. and Razdan, M.L. (1980). Geology of part of central Ladakh Himalaya with particular reference to the indus tectonic zone. *Journal of the Geological Society of India*, 21, 523–545.

Tandup, C. (2014). Natural Environment of cold desert region zanskar (Ladakh). *International Journal of Scientific and Research Publications*, 4(8), 1–8.

Thakur, V.C. (1981). Regional framework and geodynamic evolution of Indus Tsangpo Suture zone in Ladakh Himalayas. *Transactions of the Royal Society of Edinburgh: Earth Sciences*, 72, 89–97. http://dx.doi.org/10.1017/S0263593300009925

Tundup, P., Wani, M.A., Dawa, S. and Dorjay, N. (2017). Soil conservation methods and revegetation of trans-Himalayan cold desert region-Ladakh. *International Journal of Current Microbiology and Applied Sciences*, 6(3), 1996–2001. https://doi.org/10.20546/ijcmas.2017.603.227

Walia, C.S., Rana, K.P.C., Sidhu, G.S., Mahapatra, S.K. and Lal, T. (1999). Characterisation and classification of some soils of Ladakh region for land use. *Agropedology*, 9, 16–21.

Chapter 20

Lakshadweep

Sreevalsa Kolathayar
National Institute of Technology Karnataka

Nizar P.K.
Calicut Central Subdivision, CPWD

Anil Joseph
Geostructurals (P) Ltd.

CONTENTS

20.1 INTRODUCTION

The name Lakshadweep means 'one lakh islands' in the Sanskrit language. Lakshadweep, formerly (1956–73) called the Laccadive, Minicoy and Amindivi islands, is a union territory of India. It is a group of 27 islands, three reefs and six submerged sandbanks scattered over 78,000 km^2 of the Arabian Sea off the south-western coast of India, between north latitudes 8° 00′ and 12° 13′N and east longitudes 71° 00′ and 74° 00′E. Each island is surrounded by coral sands and has vast, shallow, calm lagoon on the western side, which separates the island from incoming surges of the outer sea by the reef of massive coral boulders. The easternmost island lies about 300 km from

DOI: 10.1201/9781003177159-20

the coast of the state of Kerala. Ten of the islands are inhabited. The inhabited islands are Agatti, Amini, Androth, Bangaram, Bitra, Chetlat, Kadmat, Kalpeni, Kavaratti and Minicoy. One island (Bangaram) has a tourist resort only. Table 20.1 presents the geographical details of these islands. Androth, having an area of 4.84 km^2, is the largest island, whereas Bitra, with an area of 0.1 km^2 is the smallest. The administrative centre is Kavaratti. The total land area of Lakshadweep is 32 km^2, and it has a total

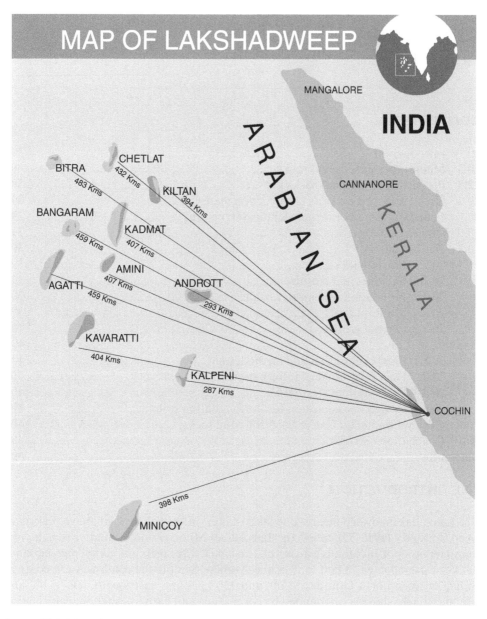

Figure 20.1 Map of Lakshadweep. (Source: https://lakshadweep.gov.in/about-lakshadweep/lakshadweep-map/, accessed 5 January 2021.)

Table 20.1 Geographical details of the Lakshadweep islands

S. no.	Islands	Area (km^2)	Latitude	Longitude
1	Agatti	2.71	10° 48′–10° 53′N	72° 09′–72° 13′E
2	Amini	2.59	11° 06′–11° 07′N	72° 42′–72° 45′E
3	Andrott	4.84	10° 48′–10° 50′N	73° 38′–73° 42′E
4	Bangaram	0.58	10° 55′–10° 58′N	72° 17′–72° 20′E
5	Bitra	0.1	11° 36′N	72° 10′E
6	Chetlat	1.04	11° 41′–11° 43′N	72° 41′–72° 43′E
7	Kadmat	3.13	11° 11′–11° 16′N	72° 45′–72° 48′E
8	Kalpeni	228	10° 03′–10° 06′N	73° 37′–73° 39′E
9	Kavaratti	3.63	10° 32′–10° 35′N	72° 37′–72° 40′E
10	Kiltan	1.63	11° 28′–11° 30′N	73° 0′
11	Minicoy	4.37	8° 15′–8° 19′N	73° 01′–73° 05′E

Table 20.2 Basic details about the union territory of Lakshadweep

Capital of the union territory of Lakshadweep	Kavaratti
Language	Malayalam, Mahl
Area	32 km^2
Number of islands	36 (10 inhabited)
Population	64,473
Density of population	2,013 persons per square kilometre
Location	8° 18′ and 12° 48′ north latitude and 74° 52′ and 77° 24′ east longitude, north latitude 8° 00′ and 12° 13′N and east longitude 71° 00′ and 74° 00′E
Water resources	No freshwater streams or ponds
Economy	Agricultural, fishing
Major agricultural produce	Coconut

population of 64,429 (Census of India, 2011). Table 20.2 presents the basic information on the union territory of Lakshadweep. Figure 20.1 shows a map of the Lakshadweep islands. The islands have a coral reef stretching from the north to the south. The land height above the mean sea level varies from 1 to 2 m. The Lakshadweep islands are nearly level and have mainly sandy soils formed by physical weathering of coral limestone. Figure 20.2 presents the cross-sectional profile of the soil hill inside the Kadmath island.

20.2 MAJOR TYPES OF SOILS AND ROCKS

The soils in all islands of Lakshadweep were formed from the fragmentation of coral limestone (Krishnan, 1997). Figure 20.3 shows the limestone deposit near the beach, which leads to the formation of island sand. Figure 20.4 shows the beach sand free from the clay content, which is a powdered form of coral lime. The soils in the islands are porous, extremely drained and poor in water-holding capacity. As the soils are highly permeable, the rainwater infiltrates quickly vertically down to the underlying substratum of coral limestone. Hence there is no surface runoff except in some

Figure 20.2 Cross-sectional profile of soil hill inside the Kadmath island.

Figure 20.3 Limestone near the beach which leads to the formation of island sand.

Figure 20.4 Beach sand free from the clay content in powdered form of coral lime at the Kavaratti island.

Figure 20.5 Coral rocks with a high lime/calcium content.

Figure 20.6 Excavated soil at a construction site.

compacted regions. There are no streams or major surface freshwater bodies on the island. In some areas along the coast and lagoon, the soil is less permeable, and water is impounded for a short period after the rainfall. Brackish water ponds exist at Bangaram and Minicoy islands. Figure 20.5 shows coral rocks with a high lime/calcium content. Figure 20.6 shows excavated soil at a construction site in the Kavaratti island.

20.3 PROPERTIES OF SOILS AND ROCKS

The Lakshadweep islands comprise coral reefs and materials derived from them. The soils are formed due to the fragmentation of coral limestone and sedimentary rocks formed from the sediments deposited by wind and water. The hard coral limestone can be seen along the coast during low tides. The profile below the seafloor has calcareous sediments of Upper Paleocene to Pleistocene age (CGWB). Darwin (1889) proposed that the submergence of volcanic islands resulted in reef formation and an atoll was formed encircling the lagoon. The islands are located on the eastern side of the reef and on the western side are the lagoons between the islands and reef. The Andrott island is oriented east-west and has a lagoon on the southern side. Tinnakara islands are surrounded by lagoons all around.

Alcock (1992) reported that Lakshadweep islands are remains of eroded atolls raised only a few metres above the mean sea level and are formed entirely of coral rock and coral sand. Gardiner (1903) proposed that Maldives and Lakshadweep islands were moulded on a large bank of an ancient land which sank into the sea, and some islands are remnants of mountains that existed in the sunken land. Stoddart (1973) stated that the reefs of Lakshadweep were built in Tertiary and Quaternary eras on volcanic structures, and the present-day surface features of the reefs are the results of erosional and depositional consequence. Vadivelu (1998) studied the characteristics of soil profiles in the islands of Minicoy, Kavaratti and Kadmat through investigation of the soils. He reported that the orientation of the islands inside the ring reefs is the result of sea surface circulation and wave action, which follows seasonal monsoon winds. The age of Lakshadweep islands is less than 5,000 years (Wood, 1983;

Table 20.3 Properties of soil at different depths in a typical borehole at Kavaratti

N-value	Depth (m)	Sample type	Soil description	Clay %	Silt %	Sand %	Gravel %	IS classification	G_s
72	1.00	DS	Medium sand (white)	5	10	83	2	SP	2.62
33	2.00	DS	Medium sand (white)	3	8	73	16	SP	2.62
45	3.00	DS	Weathered rock (white)	5	9	65	21	SP	2.63
34	4.00	DS	Weathered rock (white)	2	7	42	49	SP	2.63
37	5.00	DS	Weathered rock (white)	7	9	42	42	SP	2.63
28	6.00	DS	Weathered rock (white)	5	8	73	14	SP	2.64
28	7.00	DS	Weathered rock (white)	8	12	59	21	SP	2.64
	8.00	DS	Soft rock						
	9.00	DS	Soft rock						
	10.00	DS	Soft rock						
	11.00	DS	Soft rock						
	12.00	DS	Soft rock						
	13.00	DS	Soft rock						
	14.00	DS	Soft rock						

Vadivelu, 1998). The soil on the islands is sand. The clay content increased to about 10% towards central and eastern parts from about 2% to 4% in the west coast of these islands which can be attributed to the increase in organic carbon in the eastern and central areas (Gregorich et al., 1988; Vadivelu, 1998). The soil along the seashore is white in colour. The soil in the interior of islands is ash in colour due to the remnants of plants and animals.

CPWD, Government of India, has carried out detailed site investigations wherein six boreholes were drilled at different locations in the Kavaratti island. Figure 20.7 presents a typical boring log from a site at the Kavaratti island. It can be observed that the top 2.2 m is dense sand, followed by 7.2 m thick weathered coral rock and further soft rock. Similar soil profiles were observed in each borehole with slight variations in the thickness of layers.

The graphical representation of N-values is given in Figure 20.8. It was found that the N-values for the top layer has large variations at different locations of the island. The typical SPT results as shown in Figure 20.8 are a larger N-value for the top sand layer but, in some of the locations, low N-values ranging from 25 to 30 for top 2 m. Table 20.3 presents the laboratory experimental results of the disturbed soil sample collected from various depths of the borehole. Grain size analysis shows that the topsoil is sand with very less clay or silt content.

DEPTH	PROFILE	DESCRIPTION OF STRATA	THICKNESS OF STRATA	DEPTH in mm	Test Depth	BLOWS/15cm			SPT 'N'	Corrected SPT 'N'	Remarks
						15cm	15cm	15cm			
		VERY DENSE SAND (WHITE)	2.20	1.00	1.00-1.45	31	34	38	72	72	
2.20				2.00	2.00-2.45	15	16	17	33	33	
				3.00	3.00-3.45	18	21	24	45	45	
				4.00	4.00-4.45	10	15	19	34	34	
		WEATHERED CORAL ROCK	5.00	5.00	5.00-5.45	13	17	20	37	37	
				6.00	6.00-6.45	10	12	16	28	28	
7.20				7.00	7.00-7.45	6	9	19	28	28	TOTAL CORE RECOVERY =8% WAS NOTED FROM 7.20M TO 14.56M
		SOFT CORAL ROCK	7.36	8.00	8.00-8.45	1ST LIFT, CORE RECOVERY =10CM AND RQD NIL					
				9.00	9.00-9.45	2ND LIFT, CORE RECOVERY =8CM AND RQD NIL					
				10.00	10.00-10.45	3RD LIFT, CORE RECOVERY =5CM AND RQD NIL					
				11.00	11.00-11.45	4TH LIFT, CORE RECOVERY =8CM AND RQD NIL					
				12.00	12.00-12.45	5TH LIFT, CORE RECOVERY =8CM AND RQD NIL					
				13.00	13.00-13.45	6TH LIFT, CORE RECOVERY =7CM AND RQD NIL					
17.56				14.56	14.56-15.01	7TH LIFT, CORE RECOVERY =10CM AND RQD NIL					
BOREHOLE WAS TERMINATED AT 14.56M DEPTH											

Figure 20.7 Boring log of a typical borehole from a site at the Kavaratti island.

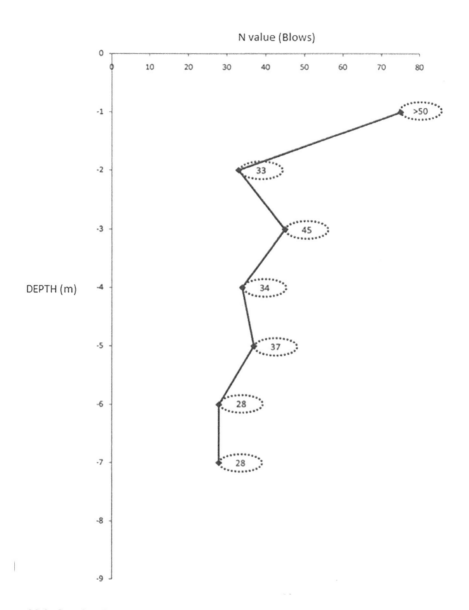

Figure 20.8 Graphical representation of *N*-values at a typical site in Kavaratti.

20.4   USE OF SOILS AND ROCKS AS CONSTRUCTION MATERIALS

Since ancient days, soils and rocks have been used for the construction of structures in the Lakshadweep islands. Figure 20.9 shows typical structures in the Kavaratti island constructed purely using naturally available materials in the locality. Figure 20.10 shows an olden-day house with sandstone blocks in the Kadmath island.

(a)   (b)

(c)

Figure 20.9 Different views of an ancient house built with calcium stones with lime mortar in the Kavaratti island.

20.5   FOUNDATIONS AND OTHER GEOTECHNICAL STRUCTURES

The locally available soils and rocks have been used as foundations of structures in the olden days. In modern-day constructions, the foundation types used in general for structures in Lakshadweep are rubble foundation, isolated foundation, strip/raft foundation and direct mud circulation (DMC) piling. Table 20.4 presents the test results of a plate-load test carried out at the Kavaratti island.

Figure 20.10 Old-days house with sandstone blocks in the Kadmath island.

Table 20.4 Pressure versus settlement values from the plate-load test at the Kavaratti island

Pressure (t/m^2)	9.262	18.524	27.786	37.048	46.31	55.567	64.833	74.10	83.358
Settlement (mm)	0.345	1.375	2.175	3.085	3.835	4.55	5.36	6.235	7.22

From the load settlement curves for various foundation widths, it is found that settlement does not exceed 25 mm. Hence it is inferred that the bearing capacity of the ground is very high. The bearing capacity based on shear failure and settlement criteria is high.

20.6   OTHER GEOMATERIALS

The economy of the Lakshadweep islands depends mainly on coconut cultivation. Hence, the coir fibre is plentifully available in the islands. Coir fibres can be efficiently used to reinforce the soils in the form of fibres, geogrids and geocells (Rao and Balan, 1997; Subaida et al., 2009; Vinod et al., 2009; Kolathayar et al., 2020; Chitrachedu and Kolathayar, 2020). The coir ropes can be used to reinforce soil to use as a foundation medium as well as subgrade for pavements. As the soil in the islands is relatively strong, these fibres have not been finding popular applications in construction. However, the coir geotextiles are popular in protecting the shores from beach erosion. Lakshadweep has got a total shore length of 132 km. A total coastal length of 59 km has been protected using concrete blocks which have created a concrete jungle along the beaches. The Lakshadweep Action Plan on Climate Change report (LAPCC, 2012) recommends implementing an eco-friendly scheme for coastal protection. The coir fibres are abundantly available in the islands, and coir geotextiles can be effectively used for coastal protection along with other long-term environmentally friendly measures. Protection of coral reefs and retting of coconut husks along the banks will aid in reducing the coastal erosion.

20.7 NATURAL HAZARDS

The Lakshadweep islands are susceptible to various natural hazards like storms, cyclones, heavy rains, floods, inundation, tsunami and coastal erosion due to their geographical position.

20.7.1 Earthquakes

No major earthquakes have been reported so far in the region. It is classified as the seismic zone III as per IS 1893 (2002) with moderate seismicity. However, the faults running parallel to the west coast of the mainland of India have the potential of generating large magnitude earthquakes, causing ground shaking in the islands.

20.7.2 Cyclones and storms

The islands were hit by a great storm in 1847. A violent storm struck in 1891 upon the Kavaratti island, Agatti and Amindivi group of islands. Major storms have hit the Kalpeni island in 1922, Kavaratti in 1941, Andrott in 1963, and Andrott and Kalpeni in 1965 and 1977. A cyclone in 2004 affected the Kavaratti, Amini, Kiltan and Agatti islands. The storms and cyclones affect the coconut trees in the islands and also fishing activities. During storms, the coconut trees get uprooted, and other vegetation and properties are also damaged. There were very few cyclones that have crossed near Lakshadweep. Andrott, Kalpeni and Minicoy lie in a cyclone belt.

20.7.3 Inundation of seawater

Lakshadweep islands are among the low-lying group of islands in the world. Considering the geographical position of the islands, the islands face the risk of inundation of seawater due to anticipated sea-level rise, storm surges and tsunami waves. The 2004 tsunami with a wave direction propagating east from Sumatra had not affected the Lakshadweep islands.

20.7.4 Sea-level rise

Sea-level rise, though a long-term threat, is an important natural hazard concerning the islands due to global warming and climate change. The low level of the islands of Lakshadweep makes them very vulnerable to sea-level rise, and hence a major future threat to these island chains is potential global climate change. Low-lying islands are at greater risk from the sea-level rise.

20.7.5   Coastal erosion

Coastal erosion is a serious problem faced by the Lakshadweep islands. Erosion takes place due to wave action as well as due to the destruction of coral reefs. The high-speed wind and huge waves hitting the seashore also lead to erosion, reducing the size of the islands. During the southwest monsoon season, many low-lying regions of the islands get washed out due to erosion. There is a need to review the existing coastal protection schemes considering plantation of woody, herbaceous vegetation and creepers.

20.8   CASE STUDY

The details of geotechnical investigation for expansion of the Agatti airport are presented here. Agatti island is a 5.6 km long island located about 459 km off Kochi in the mainland and 7 km to the southwest of Bangaram, the nearest island. Agatti's total land area is approximately 2.7 km². The small island of Kalpati is located at the southern end of the same reef. The Agatti airport is spread over 18.56 ha. It has one asphalt runway, oriented 04/22, 1,204 m long and 30 m wide, while its terminal building can handle 50 passengers during peak hours. It is operated by the Airports Authority of India (AAI). It was proposed to extend the runway by 336 m in the south-west direction over the lagoon towards the Kalpati island without interconnecting the two islands.

   The work of geotechnical investigation for the proposed project was carried out by M/s Engineers Diagnostic Centre (P) Ltd, Cochin. The field works for the geotechnical investigation started in December 2013 and ended in March 2014. The scope of work for this investigation included performing a site survey, drilling and sampling of boreholes, and conducting geotechnical field and laboratory testing. The primary purpose of these activities was to collect subsurface information at the site for subsequent preparation of geotechnical recommendations for the design of foundations for various structures like the terminal building, apron and extended airstrip of the proposed project.

20.8.1   Field exploration and methodology

Boring was done in accordance with the provisions of IS 1892 (1979), using the rotary calyx rig technique, which is mechanically operated. The borehole locations are shown in Figure 20.11. The borehole was penetrated through the sandy and clayey layers and advanced up to the hard rock strata. While drilling through the topsoil layers, sodium bentonite slurry was circulated to prevent the sides from caving. Disturbed samples were collected in plastic bags for visual inspection and classification of strata from all the layers as recorded in log sheets of the boreholes. Undisturbed tube samples were collected at specified intervals. The disturbed samples were collected as per the specification and kept in the plastic container with a proper level identification chart on it.

   Boring in water was carried out on a special platform mounted on country boats. The boring units, including tripod stand, rig and mud circulation pump, were erected on the platform with a sufficient number of anchoring so that the lateral movement of the barge was completely arrested. To avoid the oscillation of the boat, that is, vertical movement, GI pipes of 50 mm diameter were tightly fixed at the sides of the boat to

Figure 20.11 Borehole locations for the site investigation for expansion of the Agatti airport.

cater for the tidal effect. To ascertain the vertical alignment of the borehole, the casing guard was fixed at the centre of the barge. Casing pipes of 150 mm diameter were introduced from the boat floor level to a sufficiently hard stratum in the subsoil to withstand the movement of the boat as well as to protect the sides.

20.8.2 General geological information

Agatti is one of the 36 islands of the union territory of Lakshadweep and is a coral island and one of the 12 inhabited islands among them. Located about 459 km off Kochi in the Arabian Sea (Lakshadweep Sea), it is a 5.6 km long narrow island with an elevated centre and not much of morphological features. The Agatti airport spread over 18.56 ha is 1.2 km long and 30 m wide and is operated by the AAI.

Basic volcanic rocks (basalt) form the basement rocks (with an average depth of 40 m from the surface) over which the corals have grown to form coralline limestone, coralline grits and gritty conglomerates exposed on the beaches as wave-cut terraces. The sediments of the lagoon and interior of the islands are mainly coralline materials of pebble to fine sand size, broken and remodelled by wave action. The lithology varies from coralline sands and coralline fragments on the top grading to weathered coralline limestone and then towards depth, karstic coralline limestone showing voids and solution cavities. Coralline limestone is a rock consisting of the calcareous skeletons of corals often cemented by calcium carbonate. Solution/replacement, cementation, compaction and recrystallisation processes are analogous to syntaxial overgrowth and compaction involved in sediment diagenesis. Study of the subsurface geological sections has revealed three episodes of diagenic lithification identified at depth ranges of 8–12 m, 14–20 m and 22–32 m below the present level, which fluctuate within individual boreholes.

20.8.3 Evaluation of stratum

Twenty-four boreholes were drilled to evaluate the soil stratum. BH01, N1197002-E190541, was located in the sea in-between Agatti and Kalpati islands. In this area, the airstrip extension was proposed. In this borehole, 2.15 m water was noted during high tide, and 0.70 m was noted during low tide. In this borehole, the top 1.10 m from the bed level comprises white medium sand. From 1.10 to 7.00 m soft/medium-strong karstic coralline limestone was noted. This was followed by medium-strong karstic coralline limestone extending up to the depth of termination of the borehole at 15.00 m. A core recovery of 12% was noted from 7.00 to 8.00 m, 15% was noted from 8.00 to 9.00 m, 15% was noted from 9.00 to 10.00 m, 11% was noted from 10.00 to 11.00 m, 3% was noted from 11.00 to 12.00 m, 5% was noted from 12.00 to 13.00 m, 8% was noted from 13.00 to 14.00 m and 10% was noted from 14.00 to 15.00 m in the medium-strong karstic coralline limestone strata.

20.8.4 Design consideration for the foundation system

From the study of the boreholes taken for the proposed borehole locations, bored cast-in-situ piles end bearing with adequate anchorage on hard bedrock may be adopted as the foundation system. If any weathered rock is noted during boring operation, then the piles should fully penetrate the weathered rock and should be seated on the hard bedrock by 2.00 m. A load test is essential for the accurate determination of pile capacity. Hard rock was met from a depth of 3.00 to 10.30 m from the ground level. It was noted that soft rock was present from a depth of 0.90 to 8.30 m. Recommendations are based on the assumption that the soil profile/rock found in the boreholes tested is indicative of the entire area. Any deviation in the soil profile other than that observed in the borehole tested should immediately be referred to the consultant, and proper modification should be implemented.

In all the boreholes, rock was encountered at shallow depth from the bed level varying from 0.9 to 4.3 m. For the proposed bridge structure for airstrip, bored cast-in-situ piles end bearing with 2.00 m anchorage in hard rock was recommended. If any weathered rock/soft rock is noted during the boring operations, then the piles should fully penetrate the weathered rock and should be seated on hard rock. For boring in water, the permanent liner has to be installed until refusal strata. The hard rock was observed at around 9.00 m in the area. The minimum depth of the pile shall be 3.50 m. The investigation was carried out in the area of the proposed terminal building. A 3.55 m depth of water was noted during high tide, and 2.30 m was noted during low tide. The top 1.10 m comprises white medium sand. From 1.10 to 10.00 m, soft/medium-strong karstic coralline limestone was noted. This was followed by medium-strong karstic coralline limestone extending up to the depth of termination of borehole at 16.00 m. The rock was encountered at a very shallow depth of around 1.00 m from the bed level. For the proposed terminal building, bored cast-in-situ piles end bearing with 2.00 m anchorage in hard rock were recommended. If any weathered rock/soft rock is noted during the boring operations, then the piles should fully penetrate the weathered rock and should be seated on hard rock. For boring in water, the permanent liner has to be installed until refusal strata. The hard rock was observed at around 11.00 m in the area. The minimum depth of the pile shall be 3.50 m. The pile capacity shall be verified by a pile load test.

The existing fencing was found to be made with a precast-concrete post and wire fence resting on a solid concrete block as the foundation. For compound wall/fencing, a shallow foundation may be provided in the medium dense sand strata. A safe bearing capacity of 12 t/m^2 may be adopted for a footing of width 1 m at a depth of 1.20 m from the ground level. Depending upon the intensity of loading, wall footing, isolated foundation, strip footing or raft foundation may be adopted. Scouring of soil due to waves needs to be considered while designing the foundation system. Sufficient counterweight to resist the wave and scouring needs to be provided while designing the foundation of the compound wall. For heavy structures coming in the area, bored cast-in-situ piles end bearing with adequate anchorage on hard rock may be adopted. If any weathered rock is noted during the boring operations, then the piles should fully penetrate the weathered rock and should be seated at 2.00 m in hard rock. The investigation was carried out along the area proposed for runway expansion at the north-east side. The top 8.25 m comprises dense white sand with shell dust. This was followed by medium-strong karstic coralline limestone extending up to the depth of termination of the borehole at 15.00 m. The water table was noted at a depth of 3.25 m below the ground level during the time of investigation. A *CBR* value of 3.8 was obtained in the area. In the area, pavement design could be done based on the *CBR* value obtained.

20.9 GEOENVIRONMENTAL CONCERNS ON SOILS AND ROCKS

The open sea coral islands of Lakshadweep are one of the low-lying small groups of islands in the world. The low level of the islands of Lakshadweep makes them very sensitive to sea-level rise, and therefore, the foremost future threat to these island chains is potential global climate change. The IPCC Report by Pachauri and Reisinger (2007) predicts a global sea-level rise of at least 40 cm by 2100 that shall inundate vast areas on the coast, and up to 88% of the coral reefs, termed the "rainforests of the ocean", may be lost. Researchers have warned that in India, the region most vulnerable to inundation from accelerated sea-level rise is the Lakshadweep archipelago.

20.10 CONCLUDING REMARKS

This chapter presented an overview of geological and geotechnical information of Lakshadweep, a union territory of India. A case study of geotechnical investigation for the construction of the Agatti airport in one of the islands was also presented. The important observations on the soils and rocks of the islands are summarised below:

1. The soils in all the islands of Lakshadweep were formed from the fragmentation of coral limestone and sedimentary rocks formed from the sediments deposited by wind and water.
2. The major type of soil in all islands is sand.
3. The soils in the islands are porous, extremely drained and poor in water-holding capacity. As the soils are highly permeable, the rainwater infiltrates quickly vertically down to the underlying substratum of coral limestone.

4. The sand is dense, and the bearing capacity of the soil is relatively high.
5. The locally available soils and rocks were used as foundations of structures in the olden days. In modern-day constructions, the foundation types used in general for structures in Lakshadweep are rubble foundation, isolated foundation, strip/raft foundation and DMC piling.
6. The islands are exposed to different natural hazards, especially cyclones, sea-level rise and coastal erosion.

The islands are blessed with coral reefs which protect the islands naturally. It is recommended that further human interventions on the coast should be limited, and the developmental activities in the islands should be environment-friendly and sustainable. Having a long coast of 132 km, coastal protection using natural materials is the need of the hour, and more research studies need to be conducted on the same using locally available natural materials.

REFERENCES

Alcock, A. (1992). *A naturalist in Indian seas; or four years with the Royal Indian Marine Survey Ship, "Investigator".* John Murray, London.
Census of India. (2011). Census of India 2011 provisional population totals. Office of the Registrar General and Census Commissioner, New Delhi.
Chitrachedu, R.K. and Kolathayar, S. (2020). Performance evaluation of coir geocells as soil retention system under dry and wet conditions. *Geotechnical and Geological Engineering*, 38(6), 6393–6406.
Darwin, C. (1889). *The structure and distribution of coral reefs* (Vol. 15). D. Appleton, Boston.
Gardiner, J. S. (1903). *The fauna and geography of the Maldive and Laccadive Archipelagoes.* Cambridge University Press, Cambridge.
Gregorich, E.G., Kachanoshki, R.G. and Voroney, R.P. (1988). Ultrasonic dispersion of aggregates: distribution of organic matter in size fractions. *Canadian Journal of Soil Science*, 68, 395–403.
https://lakshadweep.gov.in/about-lakshadweep/lakshadweep-map/, accessed 5 January 2021.
IS 1892 (1979). *Bureau of Indian Standard – code of practice for subsurface investigations for foundations.* Bureau of Indian Standards, New Delhi.
IS 1893-Part 1 (2002). *Criteria for earthquake resistant design of structures – general provisions and buildings.* Bureau of Indian Standards, New Delhi.
Kolathayar, S., Saayinath, N., Rizfana, K. and Sitharam, T.G (2020). Performance of footing on clay bed reinforced with coir cell networks. *International Journal of Geomechanics (ASCE).* DOI: 10.1061/(ASCE)GM.1943-5622.0001719.
Krishnan, P. (1997). Soils of Lakshadweep: their kinds, distribution, characterization, and interpretations for optimising land use. *National Bureau of Soil Survey & Land Use Planning*, Vol. 70, Indian Council of Agricultural Research.
LAPCC (2012). Lakshadweep Action Plan on Climate Change. Department of Environment and Forestry, Union Territory of Lakshadweep.
Pachauri, R.K. and Reisinger, A. (2007). *IPCC fourth assessment report.* IPCC, Geneva.
Rao, G.V. and Balan, K. (1997). Reinforcing sand with coir fibre. *Geosynthetic Asia*, 97.
Stoddart, D.R. (1973). Coral reefs of the Indian ocean. In: *Biology and geology of coral reefs*, (Ed. Jones, O.A. & Endean. R.) Geology. 1, 51–92.

Subaida, E.A., Chandrakaran, S. and Sankar, N. (2009). Laboratory performance of un-paved roads reinforced with woven coir geotextiles. *Geotextiles and Geomembranes*, 27(3), 204–210.

Vadivelu, S. (1998). A Hypothesis on the formation of Lakshadweep islands from Pedogenetic Standpoint. *Agropedology*, 8, 1–9.

Vinod, P., Ajitha, B. and Sreehari, S. (2009). Behaviour of a square model footing on loose sand reinforced with braided coir rope. *Geotextiles and Geomembranes*, 27, 464–474.

Wood, E.M. (1983). *Corals of the world*. T.F.H. Publications. Inc., Ltd., Surrey.

Chapter 21

Madhya Pradesh

Lalit Borana
IIT Indore

Sanjoy Bhowmik and V.K. Panwar
Engineers India Limited

Manish Kumar Goyal and Vikas Poonia
IIT Indore

CONTENTS

21.1 INTRODUCTION

Madhya Pradesh is the second-largest state of India in terms of the geographical area, which is about $308,245\,km^2$ (Som and Mishra, 2016). This state lies over a transitional area between the Indo-Gangetic plain in the north and the Deccan plateau in the south. The major rivers of the states are namely Narmada, Tapti (Tapi), Mahanadi and so on. The Narmada River runs across the state from the east to the west between the Vindhya and Satpura ranges. Generally, fertile black soil/alluvial soil is found in the Malwa Plateau, Narmada valley and parts of the Satpura Range (Central Ground Water Board, 2013). Figure 21.1 presents the district map of Madhya Pradesh.

Geology-wise, in Madhya Pradesh (MP), the oldest group of rocks comprise Archaeans and Proterozoic formations constituting nearly 45% area of the state. The next younger formation of Carboniferous to lower Cretaceous comprising the Gondwana group covers 10% area while the formation of Cretaceous to Paleocene comprising mostly Deccan Trap basalt constitutes 38% area of the state. By referring to the Seismo-tectonic atlas (Dasgupta et al., 2000), it is inferred that the site generally falls under the Central Indian Tectonic Zone. Most of the areas, as noticed from previous records, are semi-consolidated formations, which are classified as Tertiary, Mesozoic, upper Palaeozoic, claystone, grit, sandstone, shale, conglomerate and limestone,

DOI: 10.1201/9781003177159-21

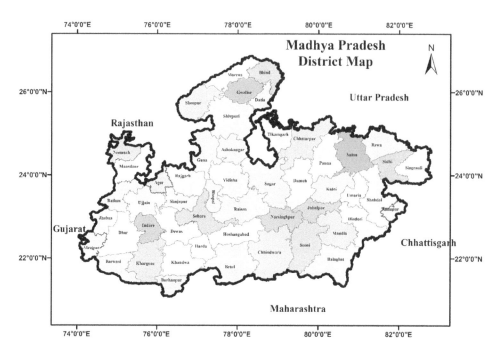

Figure 21.1 District map of Madhya Pradesh. (Source: Watershed Atlas of India. It was made using ArcGIS Desktop (Version 10.6.8321).)

including noticed intrusive. Most of the area under consideration is occupied by sandstone, limestone and basalt of Upper Vindhyan and Lower Vindhyan of the Cretaceous age.

21.2   MAJOR TYPES OF SOILS AND ROCKS

In the state, the overburden soil can be divided broadly into four categories:

1. Alluvial soil
2. Black soil
3. Clayey soil
4. Red and yellow soils

From the borehole log collected from different parts of the state, it is found that northern, western and southern portions of the state are covered with the medium to stiff black silty clay/clayey strata. The topsoil of other parts (central and eastern part) of the state is composed of very stiff to hard silty clay/clayey silt/alluvium soil strata and the thickness of overburden soil is comparatively less. However, at some locations, rock outcrops are observed but the rocky strata are highly weathered to slightly weathered in nature.

Below the overburden soil, sandstone, limestone, basalt and granitic rocks are generally found in this area. The top layer of the rock is highly weathered to weathered in nature. However, with depth, the weathering of rock is less.

21.3   PROPERTIES OF SOILS AND ROCKS

In Kailaras, which is located in the northern part of MP, the top layer of soil consists of very stiff to hard silty clay of medium plasticity, as seen in Table 21.1. The standard penetration test (SPT) N value is generally found to be more than 15, at a depth

Table 21.1 Boring log at northern MP Kailaras location, near Gwalior

Field observation: Kailaras

Project:	Geotechnical investigation	G.W.L. - not encountered
Borehole No: 1	Final Depth: 10.50 m B.G.L.	Date of start: 06-03-2009
Chainage: 228.036 m		Date of completion: 08-03-2009
Method of boring: Rotary wash		

Casing φ	Depth (m)	R.L (m)	Thickness of layer	Visual description of strata	Field test SPT				N'	Type of Sample	Wash watercolour
					15	15	15	15			
100 mm φ	1.00 1.60 2.00				03	08	10	12	18	D1 SPT1 WS1	Brownish
75 mm φ	2.60 3.00 3.60 4.00 4.60 5.00 5.60 6.00 6.60 7.00 7.60 8.00		7.60	Dark brown clayey sandy silt	04 10 15 16 16 11	10 15 18 15 14 13	16 26 20 18 21 17	15 27 20 20 22 21	26 41 37 33 35 30	SPT2 WS2 SPT3 WS3 SPT4 WS4 SPT5 WS5 SPT6 WS6 SPT7 WS7	
INXDTCB	8.60 9.00 9.60 10.50		2.90	Yellow/dark brown sandy silty clay	14 18	18 18	19 22	26 24	37 40	SPT8 WS8 SPT9 WS9	

Source: After EIL (2021).

Notes:
- Borehole is terminated as per instructions of the engineer-in-charge.
- The strata described above are specific to the borehole location.
- GWL is not encountered at this borehole location.

D – Distributed sample	SPT – Standard penetration test	Scale: V: 1:100
WS – Wash sample	U – Undistributed sample	H: N.T.S.
WO – Water sample	CP – Core pieces SP – Small pieces	Date: 10-03-2009
		DRN BY: SUPY
		CHKD BY: SMZ

of about 1.50 m below the existing ground level. The top layer of soil consists of dark brown clayey sandy silt having a thickness of about 7.60 m, followed by low plasticity hard to very hard clayey silt/silty clay with an SPT N value of more than 30 and continued up to the termination depth of the borehole (15.0 m). Furthermore, the groundwater table is not encountered up to 15.0 m depth in the borehole location.

Additionally, in the northern part of MP, Sheopur, geotechnical investigation was conducted for three areas, i.e., intake area (Table 21.2), switchyard area (Table 21.3) and tank area (Table 21.4) to determine the surface and subsurface profiles and to

Table 21.2 Boring log (Borehole # 1) at a location in northern MP, Sheopur

Part – A									
Field observation: Sheopur									

Borehole #1									
Method of boring: Machine							Dia. of borehole: 150 mm		
B.H. location: Intake area							Water level: 4.0 m		

Soil description	Symbol	Type of symbol	Depth/Run (m) From	To	Length (m)	SPT N value			
						0-15 (N_1)	15–30 (N_2)	30–45 (N_3)	N Value
Black soil	CI	DS	0	1.50	1.50	-	-	-	-
	CI	UDS/SPT	1.50	1.95	0.45	2	2	3	5
	CI	DS	1.95	3.00	1.05	-	-	-	-
	CI	UDS/SPT	3.00	3.45	0.45	3	3	4	7
	CI	DS	3.45	4.00	0.55	-	-	-	-
Yellow soil	CL	DS	4.00	5.00	1.00	-	-	-	-
of low	CL	UDS/SPT	5.00	5.45	0.45	5	7	9	16
plasticity									
Yellow soil	CL	DS	5.45	6.00	0.55	-	-	-	-
with fine	CL	UDS/SPT	6.00	6.45	0.45	5	9	11	20
gravels	CL	DS	6.45	7.00	0.55	-	-	-	-
of low	CL	UDS/SPT	7.00	7.45	0.45	6	8	10	18
plasticity	CL	DS	7.45	8.00	0.55	-	-	-	-
	CL	UDS/SPT	8.00	8.45	0.45	7	9	12	21
	CL	DS	8.45	10.00	1.55	-	-	-	-

Part – B						
Field test results by the SPT						

S. No.	Depth (m)	Corrected N value	Unconfined compressive strength $q_u = N/8$ (kg/cm²)	Shear strength $c_u = q_u/2$ (kg/cm²)	Allowable load $c_u \times N_c / 3$ (kg/cm²)	Net bearing capacity (t/m²)
1.	5.00	16.00	2.00	1.00	1.713	17.13
2.	6.00	20.00	2.50	1.25	2.142	21.42
3.	7.00	18.00	2.25	1.125	1.928	19.28
4.	8.00	21.00	2.625	1.3125	2.249	22.49

Source: After EIL (2021).
Note: SPT – Standard penetration test, DS – Disturbed sample, UDS – Undisturbed sample.

Table 21.3 Boring log (Borehole # 2) at a location in Sheopur, northern

Part - A

Field observation: Sheopur

Borehole #2
Method of boring: Machine
B.H. location: Switchyard area

Dia. of borehole: 150 mm
Water level: 4.0 m

Soil description	Symbol	Type of symbol	Depth/Run (m)		Length (m)	SPT N value			
			From	To		0-15 N_1)	15–30 (N_2)	30–45 (N_3)	N value
Black soil	CI	DS	0	1.50	1.50	-	-	-	-
	CI	UDS/SPT	1.50	1.95	0.45	2	3	5	8
	CI	DS	1.95	3.00	1.05	-	-	-	-
	CI	UDS/SPT	3.00	3.45	0.45	3	4	5	9
	CI	DS	3.45	4.00	0.55	-	-	-	-
Yellow soil	CL	UDS/SPT	4.00	4.45	0.45	5	7	8	15
of low plasticity	CL	DS	4.45	5.00	0.55	-	-	-	-
Yellow soil	CL	UDS/SPT	5.00	5.45	0.45	6	8	11	19
with fine	CL	DS	5.45	6.00	0.55	-	-	-	-
gravels	CL	UDS/SPT	6.00	6.45	0.45	8	9	12	21
of low	CL	DS	6.45	7.00	0.55	-	-	-	-
plasticity	CL	UDS/SPT	7.00	7.45	0.45	5	8	10	18
	CL	DS	7.45	9.00	1.55	-	-	-	-
	CL	UDS/SPT	9.00	9.45	0.45	7	8	12	20
	CL	DS	9.45	10.00	0.55	-	-	-	-

Part - B

Field test results by SPT

S. No.	Depth (m)	Corrected N value	Unconfined compressive strength $q_u = N/8$ (kg/cm^2)	Shear strength c = $q_u/2$ (kg/cm^2)	Allowable load $c_u \times N_c /3$ (kg/cm^2)	Net bearing capacity (t/m^2)
1.	4.00	15.00	1.875	0.937	1.606	16.06
2.	5.00	19.00	2.38	1.19	2.035	20.35
3.	6.00	21.00	2.63	1.31	2.249	22.49
4.	7.00	18.00	2.25	1.125	1.928	19.28

Source: After EIL (2021).
Note: SPT- Standard penetration test, DS – Disturbed sample, UDS- Undisturbed sample.

determine the parameters of soil. The site investigation picture is presented in Figure 21.2. The drilling was made using machine up to 10.0 m depth and an SPT was conducted at a regular interval. During investigations, disturbed and undisturbed samples were collected from different depths. The top layer of soil consists (Table 21.2) of black soil up to a depth of about 4.0 m below the existing ground level. This black soil layer is followed by yellow soil of low plasticity up to a thickness of about 1.0 m (4.0–5.0 m). Finally, yellow soil with fine gravels having low plasticity is encountered

Table 21.4 Boring log (Borehole # 3) at a location in Sheopur, northern MP

Part - A

Field observation: Sheopur

Borehole #3
Method of boring: Machine
B.H. location: Tank area

Dia. of borehole: 150 mm
Water level: Nil

Soil description	Symbol	Type of symbol	Depth/Run (m)		Length (m)	SPT N value			
			From	To		$0-15$ N_1)	$15-30$ (N_2)	$30-45$ (N_3)	N value
Black soil	CI	DS	0.0	1.50	1.50	-	-	-	-
	CI	UDS/SPT	1.50	1.95	0.45	2	2	4	6
	CI	DS	1.95	3.00	1.05	-	-	-	-
	CI	UDS/SPT	3.00	3.45	0.45	2	3	4	7
	CI	DS	3.45	3.50	0.05	-	-	-	-
Yellow soil	CL	DS	3.50	4.00	0.50	-	-	-	-
of low plasticity	CL	UDS/SPT	4.00	4.45	0.45	5	7	7	14
Yellow soil	CL	DS	4.45	5.00	0.55	-	-	-	-
with fine	CL	UDS/SPT	5.00	5.45	0.45	5	7	11	18
gravels	CL	DS	5.45	6.00	0.55	-	-	-	-
of low	CL	UDS/SPT	6.00	6.45	0.45	7	8	12	20
plasticity	CL	DS	6.45	7.00	0.55	-	-	-	-
	CL	UDS/SPT	7.00	7.45	0.45	7	9	13	22
	CL	DS	7.45	9.00	1.55	-	-	-	-
	CL	UDS/SPT	9.00	9.45	0.45	8	9	13	22
	CL	DS	9.45	10.00	0.55	-	-	-	-

Part - B

Field test results by SPT

S. No.	Depth (m)	Corrected N value	Unconfined compressive strength $q_u = N/8$ (kg/cm^2)	Shear strength $C_u = q_u/2$ (kg/cm^2)	Allowable load $C_u \times N_c/3$ (kg/cm^2)	Net bearing capacity (t/m^2)
1.	4.00	14.00	1.75	0.875	1.499	14.99
2.	5.00	18.00	2.25	1.125	1.928	19.28
3.	6.00	20.00	2.50	1.250	2.142	21.42
4.	7.00	22.00	2.75	1.375	2.356	23.56

Source: After EIL (2021).
Note: SPT- Standard penetration test, DS – Disturbed sample, UDS- Undisturbed sample.

and it continues up to the termination of the borehole (10.0 m). It is important to note that the groundwater table is observed about 4.0 m below the existing ground level in two boreholes, whereas in one borehole, the groundwater table is not observed up to 10.0 m depth. Moreover, the SPT reveals that at 5.0 m depth, the SPT N value is 16; however, it increases to 21 up to the depth of 8.0 m. A similar subsoil profile is found for Borehole #2 and #3.

Figure 21.2 A view of site investigation of a location in the Sheopur region.

Hoshangabad city (in the southern part of MP) is situated on the south bank of the Narmada River, where the geotechnical investigation was performed to examine the soil behaviour and soil characterisation at a different stratum. The relevant data regarding subsoil profile are obtained for all kinds of strata at the investigation locations by drilling boreholes at three different locations up to 10.0 m depths. The top layer of soil as detailed in Table 21.5 consists of silty clay up to a depth of 3.50 m below the existing ground level and the site investigation photograph is presented in Figure 21.3. Furthermore, the silty clay with sand is encountered and it continues up to the termination of the borehole (10.0 m). It is important to note that the groundwater table at this location was observed at a shallow depth (1.0–2.0 m below the existing ground level). Boreholes #2 and #3 also show similarly the subsoil profile (Tables 21.6 and 21.7).

As we proceed towards the central parts from the northern regions of MP (i.e. Guna and Shajapur), it is observed that the thickness of overburden soil is very less compared to the northern and southern parts. The overburden strata are gravelly mixed with clay as a binder and this layer is very hard in nature (with an SPT N value more than 50). This layer is followed by highly weathered to moderately weathered sandstone/basalt. The RQD of this rock layer varies from 0% to 30%. However, with depth, the rock layer gets harder with an RQD of more than 60%. Details of the subsoil profiles are given in Tables 21.8 and 21.9. During the soil investigation, natural groundwater table was not encountered at these locations.

In Dhar, which is located in the western region of MP, the soil profile at a borehole log (Table 21.10) shows 2.00 m of filled-up soil mixed with gravels, pebbles and boulders, pieces with clay as binder from the existing ground level. Black cotton soil of 0.7 m thickness is observed below the filled-up soil, which indicates that at this location, the top natural soil is expansive. From 2.70 to 6.50 m, light yellow soil is observed,

Table 21.5 Boring log (Borehole # 1) at a location at Hoshangabad, southern

Field observation: Hoshangabad								

Borehole #1
Method of boring: Rotary drilling　　　　　　　　　　　　　　　*Dia. of borehole: 150 mm*
Location: Gharghoda aqueduct　　　　　　　　　　　　　　　　*Water level: 2.00 m*

Depth below GL (m)			Engineering classification	Type of sampling	SPT N Value			
From	To	Run			0–15 (N_1)	15–30 (N_2)	30–45 (N_3)	N value ($N_2 + N_3$)
0.00	1.50	1.50	Silty clay	WS				
1.50	1.95	0.45		SPT-1	03	05	06	11
1.95	3.00	1.05		WS				
3.00	3.50	0.50		UDS-1				
3.50	4.50	1.00	Silty clay with sand	WS				
4.50	4.95	0.45		SPT-2	08	10	13	23
4.95	6.00	1.05		WS				
6.00	6.50	0.50		UDS-2				
6.50	7.50	1.00		WS				
7.50	7.95	0.45		SPT-3	14	19	31	50
7.95	9.00	1.05		WS				
9.00	10.00	1.00		WS				

Source: After EIL (2021).
Note: SPT – Standard penetration test, DS – Disturbed sample, UDS – Undisturbed sample.

Figure 21.3 A view of site inspection of a location in the Hoshangabad region.

Table 21.6 Boring log (Borehole # 2) at a location at Hoshangabad, southern

Field observation: Hoshangabad

Borehole #2
Method of boring: Rotary drilling
Location: Gharghoda aqueduct

Dia. of borehole: 150 mm
Water level: 1.00 m

Depth below GL (m)			Engineering classification	Type of Sampling	SPT N Value			
From	To	Run			$0-15$ (N_1)	$15-30$ (N_2)	$30-45$ (N_3)	N value (N_2+N_3)
0.00	1.50	1.50	Silty clay	WS				
1.50	1.95	0.45		SPT-1	03	05	08	13
1.95	3.00	1.05		WS				
3.00	3.50	0.50	Silty clay with sand	UDS-1				
3.50	4.50	1.00		WS				
4.50	4.95	0.45		SPT-2	09	13	17	30
4.95	6.00	1.05		WS				
6.00	6.50	0.50		UDS-2				
6.50	7.50	1.00		WS				
7.50	9.00	1.50		WS				
9.00	10.00	1.00		WS				

Source: After EIL (2021).
Note: SPT – Standard penetration test, DS – Disturbed sample, UDS – Undisturbed sample.

Table 21.7 Boring log (Borehole # 3) at a location at Hoshangabad, southern

Field observation: Hoshangabad

Borehole #3
Method of boring: Rotary drilling
Location: Gharghoda aqueduct

Dia. of borehole: 150 mm
Water level: 1.00 m

Depth below GL (m)			Engineering classification	Type of sampling	SPT N value			
From	To	Run			$0-15$ (N_1)	$15-30$ (N_2)	$30-45$ (N_3)	N value (N_2+N_3)
0.00	1.50	1.50	Silty clay	WS				
1.50	1.95	0.45		SPT-1	03	05	06	11
1.95	3.00	1.05		WS				
3.00	3.45	0.45	Silty clay with sand	SPT-2	07	12	15	27
3.45	4.50	1.05		WS				
4.50	6.00	1.50		WS				
6.00	6.50	0.50		UDS				
6.50	7.50	1.00		WS				
7.50	7.95	0.45		SPT-3	11	18	30	48
7.95	9.00	1.05		WS				
9.00	10.00	1.00		WS				

Source: After EIL (2021).
Note: SPT – Standard penetration test, DS – Disturbed sample, UDS – Undisturbed sample.

Table 21.8 Boring log at a location at Shajapur, Central MP

Field observation: Sandavata, Shajapur Termination depth (m) = 10.0					Groundwater table: Not encountered
S. No.	Depth (m)	N value	Type of sample	Specific gravity	Description
1.	0.00	-	DS	-	Dark brown coloured, completely weathered & fractured soft rock
2.	1.50	-	DS	-	Dark brown coloured, completely weathered & fractured soft rock
3.	2.00	-	DS	-	Dark brown coloured, completely weathered & fractured soft rock
4.	2.50	-	SPT	-	Dark brown coloured, completely weathered & fractured soft rock
5.	3.00	-	DS	-	Blackish coloured, highly weathered basalt
6.	3.50	-	SPT	-	Blackish coloured, highly weathered basalt
7.	4.00	-	DS	-	Blackish coloured, highly weathered basalt (CR = 32%, RQD = NIL)
8.	5.50	-	DS	-	Blackish coloured, moderately weathered basalt (CR = 76%, RQD = 66%)
9.	7.00	-	DS	-	Yellowish grey coloured, highly weathered basalt CR = 30%, RQD = 16%
10.	8.50	-	DS	-	Yellowish grey coloured, highly weathered basalt (CR = 24%, RQD = 9%)
11.	10.00	-	DS	-	Blackish coloured, highly weathered rock Basalt CR = 34% RQD = 8%

Source: After EIL (2021).

which is followed by yellowish silty soil. Another borehole log of the Dhar location shows the black cotton soil from the existing ground level up to 2.7 m depth followed by yellowish soil. This yellowish silty stratum continues up to the termination depth of the borehole. At this location, the thickness of black cotton soil is more compared to the Borehole #1 location. The detailed subsoil profile is given in Table 21.10 and the site view is shown in Figure 21.4. Borehole #2 also shows a similar subsoil profile (Table 21.11).

The detailed soil profile at the Jabalpur location, the western part of MP, is given in Table 21.12. Another city in the western part of MP, the Jhabua location shows that from the ground level to the top 3.0 m of soil consists of filled-up soil mixed with gravel and pebbles (Table 21.13). This layer is followed by severely weathered rocky strata; the strength of rocky strata improves with the increase in depth.

In the eastern part of MP (Singrauli), the subsoil stratum is similar to the central part of the state as seen in Table 21.14. The thickness of the overburden soil layer in this part of the area is also less compared to the northern and southern parts. This soil layer is followed by highly weathered to moderately weathered granite rocky strata. The RQD of this rock layer varies from 0% to 5% in the top 4.0 m depth; however, the RQD value increases with depth.

Table 21.9 Boring log at a location at Bina, Central MP

Field observation: Bina

Constell Consultants Pvt. Ltd.
Borehole # NB-5
Location: N 50290; E 48310
Method of Boring/ Drilling: Rotary
Boring/Drilling equipment: G.O. -6
Casing Lowered: NX-5.50 m

Project: Soil investigation works
Ground elevation: +399.101 m
Water Level (Static): Not encountered
Diameter of boring/drilling: NX
Date: From 12.09.06 To 19.09.06

Date (dd/mm)	Elevation (m)	Depth/Run (m) From	To	Length (m)	Nature of sampling	SPT: No. of blows 0–15 cm	15–30 cm	30–45 cm	45–60 cm	N value	Time taken (min)	Total length of Core Pieces (m)	Core recovery (%)	RQD (%)	Description
12/09	+399.101 +398.901	0.15	0.2	0.05	P	71	-	-	-	>100	-	-	-	-	Reddish-brown silty clay
		Drilling started from 0.2 m depth													
		0.20	1.00	0.8	C	-	-	-	-	-	35	0.15	19	NIL	Highly weathered reddish-brown coloured ferruginous sandstone highly fractured, cores are broken into several pieces non-jointed
13/09		1.00	2.00	1.00	C	-	-	-	-	-	45	0.20	20	NIL	-do-
		2.00	3.00	1.00	C	-	-	-	-	-	60	0.36	36	NIL	Moderately weathered (one set joint, joint plane smooth)
		3.00	4.00	1.00	C	-	-	-	-	-	80	0.45	45	15	-do-(fractured at top and bottom at the run)
		4.00	5.00	1.00	C	-	-	-	-	-	160	0.50	50	NIL	-do-(highly fractured along bedding plane)
14/09		5.00	6.00	1.00	C	-	-	-	-	-	190	0.80	80	NIL	Moderately weathered reddish-brown coloured ferruginous sandstone, fractured along bedding plane
15/09		6.00	7.00	1.00	C	-	-	-	-	-	150	0.69	69	NIL	Do (one set of joint at depth 6.2 m, joint plane rough)

(Continued)

Table 21.9 (Continued) Boring log at a location at Bina, Central MP.

Field observation: Bina

Constell Consultants Pvt. Ltd.
Borehole # NB-5
Location: N 50290; E 48310
Method of Boring/ Drilling: Rotary
Boring/Drilling equipment: G.O. -6
Casing Lowered: NX-5.50 m

Project: Soil investigation works
Ground elevation: +399.101 m
Water Level (Static): Not encountered
Diameter of boring/drilling: NX
Date: From 12.09.06 To 19.09.06

Date (dd/ mm)	Elevation (m)	Depth/Run (m) From	To	Length (m)	Nature of sampling	SPT: No. of blows 0– 15 cm	15– 30 cm	30– 45 cm	45– 60 cm	N value	Time taken (min)	Total Length of Core Pieces (m)	Core recovery (%)	RQD (%)	Description
16/09	+384.101	7.00	8.00	1.00	C	-	-	-	-	-	140	0.68	68	21	-do- (highly fractured)
17/09		8.00	9.00	1.00	C	-	-	-	-	-	150	0.59	59	NIL	-do- (highly fractured along bedding planes)
		9.00	10.00	1.00	C	-	-	-	-	-	180	0.60	60	20	-do- (non-jointed)
17/09		10.00	11.00	1.00	C	-	-	-	-	-	160	0.75	75	NIL	-do-
18/09		11.00	12.00	1.00	C	-	-	-	-	-	180	0.79	79	16	-do-
		12.00	13.00	1.00	C	-	-	-	-	-	150	0.65	65	10	-do-
19/09		13.00	14.00	1.00	C	-	-	-	-	-	190	0.80	80	40	-do-
		14.00	15.00	1.00	C	-	-	-	-	-	180	0.75	75	13	-do-

Source: After EIL (2021).

NOTES
1. Abbreviation Used: U – Undisturbed sample, C – Core sample, D – Disturbed sample, P – Standard penetration test
2. Level at which Artesian condition experienced and its pressure, if any: Not encountered
3. Water loss with depth, if any: No perceptible quantity
4. Colour of water during drilling: Reddish-brown.

Table 21.10 Boring log (Borehole #1) at a location in Dhar, western

Borehole #1
Field observation: Dhar

S. No.	Depth/Run (m)		Length (m)	Description
	From	To		
1	0.00	2.00	2.00	Gravels, pebbles and boulders, pieces with clay as binder
2	2.00	2.70	0.70	Black cotton soil
3	2.70	6.50	3.80	Light yellow soil
4	6.50	13.0	6.5	Yellow silty soil

Source: After EIL (2021).

Table 21.11 Boring log (Borehole #2) at a location in Dhar, western

Borehole #2
Field observation: Dhar

S. No.	Depth/Run (m)		Length (m)	Description
	From	To		
1	0.00	2.70	2.70	Black cotton soil
2	2.70	6.50	3.80	Light yellow soil
3	6.50	13.0	6.5	Yellow silty soil

Source: After EIL (2021).

Figure 21.4 A view of site investigation at Dhar.

21.4 USE OF SOILS AND ROCKS AS CONSTRUCTION MATERIALS

Construction materials such as sand, gravel, wood and so on are sufficiently available in the state of MP. However, to check the suitability of these materials as construction materials, requisite testing has to be carried out as per the Indian and other

Table 21.12 Boring log at a location at Jabalpur, western

	Field observation: Jabalpur			
S.No.	Location	Longitude	Latitude	Group name
1	Gorakhpur	79°55′59″ E	23°9′14″ N	Clayey Sand
2	Commercial Complex at Victoria Hospital	79°56′14″ E	23°10′20″ N	Lean Clay
3	Reliance Exchange Building Vijay Nagar	79°54′41″ E	23°11′32″N	Fat Clay
4	Medical College	79°52′52″ E	23°8′59″ N	Clayey Sand
5	Womens Polytechnic College	79°56′59″ E	23°10′28″ N	Clayey Sand
6	Government Science College	79°58′8″ E	23°9′49″ N	Lean Clay
7	Mahanadda Road	79°55′9″ E	23°9′6″ N	Clay of Medium Plastic
8	MGRGVS Adhartal	79°56′46″ E	23°11′58″ N	Clayey Sand
9	Krishi Mandi Karyalaya Damoh Road	79°54′57″ E	23°11′16″ N	Fat Clay
10	MP High Court	79°57′30″ E	23°9′43″ N	Clayey Sand

Source: After EIL (2021).

relevant codal provisions. Based on the suitability, existing subsoils can be effectively used as construction materials for granular sub-base layers on roads and highways. By effective utilisation of locally available materials in construction, it is possible to significantly reduce the cost and construction time of the project. But, as mentioned earlier, black/brownish clayey strata are also observed in different parts of the state, which is expansive/swelling in nature. Therefore, such soils cannot be directly used below the foundations, road, pavements and so on. The properties of these materials need to be improved based on the functional aspects of the engineering structures. For instance, soil stabilisation can be done by blending or mixing with a variety of materials such as fly ash, construction wastes, industrial wastes, stone dust, lime, additives, admixtures and so on. It is prudent to carefully examine the engineering characteristics of the soil and rocks especially in the case of regions consisting of an expansive subsoil stratum.

21.5 FOUNDATIONS AND OTHER GEOTECHNICAL STRUCTURES

Considering the subsoil profile stratification and loading requirements, both shallow and deep foundations are feasible. As rocky strata are found at shallow depths, socketed piles are preferred in this region. From the borehole logs, it is observed that the topsoil is quite capable to carry the desired load for minor or lightly loaded structures. Primarily, bearing capacity and settlement are not an issue for such types of subsoil strata. Pile and shallow foundations are quite common for refineries, petrochemical complexes and compressor stations in this state. In addition to industrial plants, huge reservoirs and dams are also found in this state. The existing dams are mainly earthen, gravity, concrete and masonry dams.

Table 21.13 Boring log at Western MP Jhabua location

Field observation: Jhabua

N value	Reduced level (m)	Depth (m)	Sample no.	Soil description	Grain size analysis				Atterberg's limit			Index properties				Triaxial test		
					Gravel %	Sand %	Silt %	Clay %	Liquid limit %	Plastic limit %	Plasticity index %	Specific gravity	Bulk density	Dry density	Moisture content	Confining pressure (kg/cm²)	Cohesion (kg/cm²)	Angle Of friction
	323.66	0.00	DS1	Filling: Brown Silty Sand with gravels and pebbles														
	323.16	0.50																
62	323.16	0.50	SPT1															
	322.71	0.95																
102/15cm	322.16	1.50	SPT2															
	321.86	1.80																
100/10cm	320.66	3.00	SPT2	Refusal met on rock, drilling through rock formation adopted														
	320.41	3.25																

(Continued)

Table 21.13 (Continued) Boring log at Western MP Jhabua location.

Field observation: Jhabua

Location: Compressor house | Type of boring: Rotary | Size of hole: NX | Return water | Rg: Voltas 12B

Reduced level (m)	Depth (m)	Drill run (m)	Sample no.	SPT value	Serial Casing no. of lowered cores	Description	Percent recovery (%)	(RQD) (%)	Rock mass rating (RMR)	Penetration rate (minute/cm)	Hydraulic Pressure (kg/sq. cm)	Colour	Loss	Bits used	Remarks
323.66	3.00	3.00	SPT2	100/10 cm	NX	Very weak to moderately weak grey quartzite, jointed and fractured very severely weathered.	0	0	15	0.2	1.5			32 CT impregnated diamond bit	
320.66	3.50	3.25					0	0							
320.16	4.00		WS1												
319.66	4.50	4.50	SPT4	102/15 cm											
319.16	5.00	4.80													
318.66	5.50					Very weak, very severely weathered, disintegrated, 3.0 to 6.0 m	0	0							
318.16	6.00	6.00	WS2												
317.66	6.50	6.20	SPT5				33	0	20	0.3	2.0				
317.16	7.00														
316.66	7.50	7.50	RK1	Ref/5 cm	1-6										
316.16	8.00														
315.66	8.50		RK2		7-14	Moderately weak, severely weathered, 6.0–10.5 m	35	0				Grey	Partial		
315.16	9.00	9.00													
314.66	9.50														
314.16	10.00		RK3		15-21		35	0							
313.66	10.50	10.50													
313.16															

Source: After EIL (2021).

Table 21.14 Boring log at Eastern MP Singrauli location

Field observation: Singrauli

Elevation (m)	Depth of the sample below the existing ground surface (m)	Run number	Visual description of substrata	Depth (m)	Run number	Penetration rate (cm/h)	Core recovery (%)	RQD (%)	Water loss (%)	Dry density (g/cc)	Natural moisture content (%)	Water absorption (%)	Porosity (%)	Specific gravity	Unconfined compressive strength (kg/cm²)
396.490	0.0		Soil overburden	0						–	–	–	–	–	–
394.540	1.95	1	Grey medium to coarse-grained highly weathered disintegrated rock mass	1.95	1	315	NIL	NIL	NIL	–	–	–	–	–	–
	3.00	2			2	300	NIL	NIL	NIL	–	–	–	–	–	–
	4.00	3			3	264	NIL	NIL	NIL	–	–	–	–	–	–
	5.10	4			4	200	NIL	NIL	5	–	–	–	–	–	–
390.390	6.10	5	Grey coarse-grained slightly weathered to fresh granitic gneiss	6.10	5	42	87.0	87.0	10	–	–	–	–	–	–
	7.00	6			6	43	100.0	100.0	10	2.68	0.07	0.25	0.67	2.81	259.8
388.490	8.00			8.00											

Source: After EIL (2021).

For shallow foundations, roads, pavements and so on, the major concerns related to design are swelling and shrinkage characteristics of the soil under the influence of water. Therefore, it is very much essential to carry out tests for swelling pressure, free swell index and Atterberg's limits to evaluate the expansive potentiality of the soil and such potentiality must be taken care of during the design of any foundation/structures.

21.6   NATURAL HAZARDS

Based on IS 1893 (Part-1) (IS:1893, 2002), seismic zoning map of India, most of the MP state falls under seismic Zone-II, whereas a little portion of southern and southern-eastern parts of the state falls under seismic zone-III. Based on the zonal map, it is observed that Khargone, Hoshangabad, Jabalpur, Seoni, Mandla, Chhindwara, Annupur and Sidhi fall under earthquake zone-III. Based on the earthquake history of the state, it is found that most of the earthquakes occurred in zone-III locations. Therefore, earthquake forces and hazards must be kept in mind during the design of structures/foundations.

In general, due to heavy rainfall, the central part of the state witnesses flood in the monsoon period. Therefore, suitable drainage and protection systems should be adopted while designing to protect the structures from floods/erosion. Furthermore, it is suggested to adopt the riverbank protection systems and river training works to reduce the hazards related to flood and slope failures.

21.7   CASE STUDIES AND FIELD TESTS

In this section, safe bearing pressure values from the SPT, pile load test and direct shear test are discussed.

An SPT was carried out at Hoshangabad to evaluate the allowable bearing capacity. The depth-wise safe bearing capacity is given in Table 21.15 for different sizes of foundations.

At a site in Singrauli, due to project requirements, the site was graded to a higher level by filling. Considering the recent filled-up soil strata of significant depth, pile foundations are adopted for foundations. The pile was designed to be socketed into slightly moderate to hard rock and the design vertical and lateral capacity of the pile are verified by the initial pile load tests.

21.8   EFFECTS OF ENVIRONMENTAL CHANGES ON SOILS
AND ROCKS

Climate change is emphasizing ecosystems including sea levels, agriculture (Das et al., 2020; Poonia et al., 2021a) and natural products, flora and fauna, water cycles, drought (Kumar et al., 2021; Poonia et al., 2021b, 2021c) and soil properties (Simonovic, 2017). Climate change affects the soil properties (Biswas et al., 2018; Brevik, 2012; Gelybó et al., 2018). Moisture, CO_2 and temperature are the major drivers of climate change and have various effects on several soil properties and processes (Borana et al., 2017a,

Table 21.15 Bearing capacity values at Hoshangabad

Safe bearing capacity, Hoshangabad

Borehole #1

S. No.	Depth (m)	Width (m)	N value	Settlement consideration (40 mm) (t/m²)	Settlement consideration (75 mm) (t/m²)
1.	5.00	2.0	16.0	23.81	18.0
2.	6.00	2.0	20.0	31.14	22.6
3.	7.00	2.0	18.0	27.47	24.3
4.	8.00	2.0	21.0	32.97	25.6

Borehole #2

S. No.	Depth (m)	Width (m)	N value	Settlement consideration (40 mm) (t/m²)	Settlement consideration (75 mm) (t/m²)
1.	4.00	2.0	15.0	21.98	19.1
2.	5.00	2.0	19.0	29.31	21.2
3.	6.00	2.0	21.0	32.97	25.6
4.	7.00	2.0	18.0	27.47	20.3

Borehole #3

S. No.	Depth (m)	Width (m)	N value	Settlement consideration (40 mm) (t/m²)	Settlement consideration (75 mm) (t/m²)
1.	5.00	2.0	14.0	20.15	16.9
2.	6.00	2.0	18.0	27.47	20.3
3.	7.00	2.0	20.0	31.14	24.3
4.	8.00	2.0	22.0	34.80	26.5

Source: After EIL (2021).

2017b; Chen et al., 2020; Smith et al., 2005). Long-term engineering characteristics of soil greatly vary depending on the available moisture content of the soil. Yin and Graham built the EVP model, by considering the viscous, plastic and elastic behaviour of clayey soil (Chen et al., 2020; Yin and Graham, 1989, 1994). Singh et al. (2020) examined the long-term consolidation behaviour of such expansive black cotton soils of the Indore region in the MP state. The increase or variation in temperature may have a direct consequence in the physical and engineering aspects of the soil (Karmakar et al., 2016), such as the reduction in the moisture content, loss of soil structure, loss of soil organic matter, increase in soil respiration rate, increase in mineralisation rate, reduction in the labile pool of SOM and so on. Similarly, an increase in CO_2 may cause an increase in soil organic matter, accelerate the nutrient cycle and result in increased availability of carbon to soil microorganisms (Trost et al., 2013). Reduction in rainfall may also have some effects like soil salinisation, reduction in nutrient availability and reduction in soil organic matter (Pareek, 2017).

21.9   CONCLUDING REMARKS

Based on the study of the available subsoil data from different locations and the geology of the MP state, the following can be concluded:

- The soil strata vary from one location to another. Moreover, the thickness of the overburden strata also varies from one location to another. Therefore, it is essential to have a detailed soil investigation work prior to any construction works. Based on the geology and available information, the designer has to judiciously decide the detailed soil investigation test programme to be suitable for a particular location of the state. This will economise the soil investigation works in terms of time and costs and will help determine the required subsoil parameters for the design and selection of the construction materials.
- For lightly loaded structures, the bearing capacity and settlement criteria are not a concern for the design of shallow foundations in this state. For shallow foundations, the primary concern is swelling, free swell index, swelling pressure and shrinkage characteristics of the topsoil. Therefore, an extensive investigation/ study is required to know about these parameters area-wise. Based on the study, a guideline has to be formulated about the construction methodology for such areas along with the suitability of the construction materials.
- For pile foundations, a bored cast-in-situ rock socketed pile is the most preferable pile in this area. However, during the execution of piling works, a suitable length of socketing should be ensured based on the availability of rock type.
- Considering seismicity, seismic analysis of the foundation system is compulsory for a safe and stable structure.
- In some parts of the state, the groundwater table is close to the existing ground surface. Hence, for designing shallow foundations and deep excavations, the designer should judiciously study the design considerations at such locations.

REFERENCES

Biswas, B., Qi, F., Biswas, J.K., Wijayawardena, A., Khan, M.A.I., Naidu, R., 2018. The fate of chemical pollutants with soil properties and processes in the climate change paradigm – A review. *Soil Syst.* 2, 1–20. https://doi.org/10.3390/soilsystems2030051

Borana, L., Yin, J.H., Singh, D.N., Shukla, S.K., 2017a. Influence of matric suction and counterface roughness on shearing behavior of completely decomposed granitic soil and steel interface. *Indian Geotech. J.* 47, 150–160. https://doi.org/10.1007/s40098-016-0205-7

Borana, L., Yin, J.H., Singh, D.N., Shukla, S.K., Pei, H.F., 2017b. Influences of initial water content and roughness on skin friction of piles using FBG technique. *Int. J. Geomech.* 17, 04016097. https://doi.org/10.1061/(asce)gm.1943-5622.0000794

Brevik, E.C., 2012. Soils and climate change: Gas fluxes and soil processes. *Soil Horizons* 53, 12. https://doi.org/10.2136/sh12-04-0012

Central Ground Water Board, 2013. District groundwater information booklet, Bhopal District (Madhya Pradesh), Ministry of Water Resources, Central Ground Water Board, North Central Region, Bhopal, India, 1–26.

Chen, W.B., Liu, K., Feng, W.Q., Borana, L., Yin, J.H., 2020. Influence of matric suction on nonlinear time-dependent compression behavior of a granular fill material. *Acta Geotech.* 15, 615–633. https://doi.org/10.1007/s11440-018-00761-y

Das, J., Poonia, V., Jha, S., Goyal, M.K., 2020. Understanding the climate change impact on crop yield over Eastern Himalayan Region: Ascertaining GCM and scenario uncertainty. *Theor. Appl. Climatol.* 142, 467–482. https://doi.org/10.1007/s00704-020-03332-y

Dasgupta, S., Pande, P., Ganguly, D., Gupta, H.K., 2000. *Seismotectonic Atlas of India and Its Environs*, Special Pu. ed. Geological Survey of India.

EIL, 2021. Engineers India Limited (EIL), New Delhi, India.

Gelybó, G., Tóth, E., Farkas, C., Horel, Kása, I., Bakacsi, Z., 2018. Potential impacts of climate change on soil properties. *Agrokem. es Talajt.* 67, 121–141. https://doi.org/10.1556/0088.2018.67.1.9

IS:1893, 2002. Criteria for earthquake resistant design of structures - General provisions and buildings Part-1. *Bur. Indian Stand. New Delhi Part* 1, 1–39.

Jian-Hua Yin, Graham, J., 1994. Equivalent times and one-dimensional elastic viscoplastic modelling of time-dependent stress-strain behaviour of clays. *Can. Geotech. J.* 31, 42–52. https://doi.org/10.1139/t94-005

Karmakar, R., Das, I., Dutta, D., Rakshit, A., 2016. Potential effects of climate change on soil properties: A review. *Sci. Int.* 4, 51–73. https://doi.org/10.17311/sciintl.2016.51.73

Kumar, N., Poonia, V., Gupta, B.B., Goyal, M.K., 2021. A novel framework for risk assessment and resilience of critical infrastructure towards climate change. *Technol. Forecast. Soc. Change* 165, 120532. https://doi.org/10.1016/j.techfore.2020.120532

Pareek, N., 2017. Climate change impact on soils: Adaptation and mitigation. *MOJ Ecol. Environ. Sci.* 2, 136–139. https://doi.org/10.15406/mojes.2017.02.00026

Poonia, V., Das, J., Goyal, M.K., 2021a. Impact of climate change on crop water and irrigation requirements over eastern Himalayan region. *Stoch. Environ. Res.* Risk Assess. 6. https://doi.org/10.1007/s00477-020-01942-6

Poonia, V., Goyal, M.K., Gupta, B.B., Gupta, A.K., Jha, S., Das, J., 2021b. Drought occurrence in Different River Basins of India and blockchain technology based framework for disaster management. *J. Clean. Prod.* 312, 127737. https://doi.org/10.1016/j.jclepro.2021.127737

Poonia, V., Jha, S., Goyal, M.K., 2021c. Copula based analysis of meteorological, hydrological and agricultural drought characteristics across Indian river basins. *Int. J. Climatol.* joc.7091. https://doi.org/10.1002/joc.7091

Simonovic, S.P., 2017. Bringing future climatic change into water resources management practice today. *Water Resour. Manag.* 31, 2933–2950. https://doi.org/10.1007/s11269-017-1704-8

Singh, M.J., Weiqiang, F., Dong-Sheng, X., Borana, L., 2020. Experimental study of compression behavior of indian black cotton soil in oedometer condition. *Int. J. Geosynth. Gr. Eng.* 6, 1–13. https://doi.org/10.1007/s40891-020-00207-0

Smith, K.L., Colls, J.J., Steven, M.D., 2005. A facility to investigate effects of elevated soil gas concentration on vegetation. Water. *Air. Soil Pollut.* 161, 75–96.

Som, K.S., Mishra, R.P., 2016. Fertility and rch-status in Madhya Pradesh: A district level analysis. *Trans. Inst. Indian Geogr.* 38, 281–290.

Trost, B., Prochnow, A., Drastig, K., Meyer-Aurich, A., Ellmer, F., Baumecker, M., 2013. Irrigation, soil organic carbon and N_2O emissions. A review. *Agron. Sustain. Dev.* 33, 733–749. https://doi.org/10.1007/s13593-013-0134-0

Yin, J.H., Graham, J., 1989. Viscous-elastic-plastic modelling of one-dimensional time-dependent behaviour of clays. *Can. Geotech. J.* 26, 199–209. https://doi.org/10.1139/t89-029

Chapter 22

Maharashtra

Hemant S. Chore and S. Rupali
Dr. B.R. Ambedkar National Institute of Technology Jalandhar

Gaurav Kumar
SNF India

CONTENTS

22.1 INTRODUCTION

Maharashtra is one of the largest states, in the western peninsular region of India, occupying a substantial portion of the Deccan Plateau. Spread over an area of 3,07,713 km², it is the third-largest state in the country by area. The state shares the borders with Karnataka to the south, Goa to the southwest, Madhya Pradesh to the north and Chhattisgarh to the east along with Telangana to the southeast. Furthermore, Gujarat lies to the northwest, with the Union territory of Dadra and Nagar Haveli adjoining both the state borders. Maharashtra occupies the western and the central part of the

DOI: 10.1201/9781003177159-22

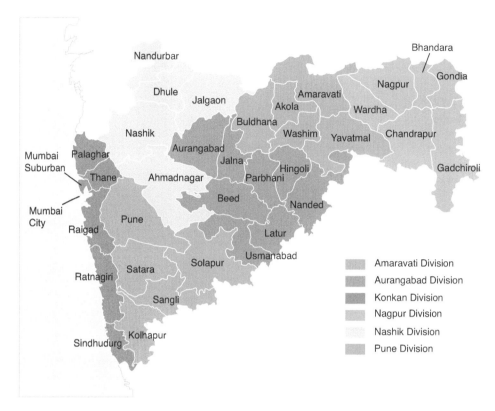

Figure 22.1 Political map of the state of Maharashtra. (After https://www.en.wikipedia. org/wiki/List_of_districts_of_Maharashtra.)

country and has a long coastline stretching 720 km along the Arabian Sea. The geographical location of Maharashtra renders it to be a key economic contributor and industrial region of the country and makes the state one of the wealthiest and most developed amongst other states. The coordinates of the state are latitude 19° 39′ 47″ north and longitude 75°18′ 1″. Figure 22.1 shows the general political map of the state.

The state occupies between the western and central parts of the country. One of the more prominent physical features of the state is the Deccan Plateau, which is separated from the Konkan coastline by 'Ghats'. The Ghats are a succession of steep hills, periodically bisected by narrow roads. Most of the famous hill stations of the state are at the Ghats. The Western Ghats, commonly called the Sahyadri Mountain range, provide a physical backbone to the state on the west while the Satpura Hills along the north and Bhamragad-Chiroli-Gaikhuri ranges on the east serve as its natural borders.

Maharashtra is divided into six administrative divisions and currently has 36 districts. The Western Ghats are hill ranges running parallel to the coast, at an average elevation of 1,200 m. Kalsubai, a peak in the Sahyadris, near the Nashik city is the highest elevated point in Maharashtra. To the west of Sahyadris lie the Konkan coastal plains, 50–80 km in width with an elevation of below 200 m while to the east lies the flat Deccan Plateau. Around 17% of the total area of the state comprises the forest with a majority of the forest falling in the eastern and Sahyadri regions of the state. Krishna, Bhima, Godavari, Tapti, Purna, Wardha and Wainganga form the main rivers of the state. Since the central parts of the state receive low rainfall, many dams have been constructed across the rivers flowing in this region.

Maharashtra is divided into five geographic regions. Konkan is the western coastal region, located between the Western Ghats and the sea. Khandesh is the northwestern region lying in the valley of the Tapti River. Nashik, Jalgaon, Dhule and Bhusawal are some of the major cities of this region. Desh is in the centre of the state. Marathwada is located in the southeastern part of the state. Aurangabad and Nanded are the main cities of the region. Vidarbha is the easternmost region of the state. Nagpur, where the winter session of the state assembly is held, and Akola, Amravati, Wardha and Chandrapur are the main cities in the region.

The annual average rainfall, being fairly high (2,000–5,000 mm) in Maharashtra, the process of mechanical and chemical weathering and rapid erosion of the soils and bedrocks appear significant. Brief classifications and characteristics of the soils and rocks of Maharashtra are given in the next section.

22.2   MAJOR TYPES OF SOILS AND ROCKS

The soils of Maharashtra are mainly residual, derived from the igneous basalts. Different types of soils are found in various parts of the state. These soils include red soil, lateritic soil, black and black cotton soil, yellowish soil, alluvial soil and marine clay. The soils in the river basins of Godavari, Bhima, Krishna and Tapti have a deep layer of fertile black basalt soil, which is rich in humus. This soil is the best suited for cotton production and is often called black cotton soil. The rest of the semi-dry plateau of Deccan has a medium layer of black regur soil, which is clayey and moisture retentive, rich in iron but poor in nitrogen and organic matter. The higher plateau region has moorum soils, which contain more gravel. The peaks of the Sahyadri Mountains, the districts of Ratnagiri and Sindhudurg and the western regions of Kolhapur and Satara have a reddish lateritic soil, which is locally called Jambhi. The Konkan coast has sandy loam soil. Farther away towards the east in the Vidarbha region, with a better mixture of lime, the morand soils form the ideal Kharip zone. North Konkan and eastern Vidarbha regions like Bhandara, Gondia and Gadchiroli districts have reddish and yellowish soils. The soil map of the state indicating the distribution of different types of soils in various parts of the state is available in the technical bulletin (Sehgal, 1990). Brief information on the soil found in different categories of the region is provided below.

22.2.1   Soils in Konkan

Konkan has an enormous amount of basalt with a minor share of granite laterites and gneisses. The relief features of the region are essentially the product of its geological past carrying the signatures of climate change from humid to semi-arid. The soils in the region are mainly red and black which shows lateritic nature on exposure. The soils of the region show varying crop productivity, which in turn, depends upon various properties of the soil that include clay contents, carbon, clay cation exchange capacity and a host of physical, chemical and microbiological properties including the mineralogical parameters. The region receives heavy rainfall many times and therefore, there is better vegetation that preserves the organic carbon in these soils for a considerable period as against those in semi-arid and tropical parts of the country.

22.2.1.1   Soils in the Western Ghats

Soils found in the Western Ghats are rich in red soil. Unlike the soils in the northeastern part of the country and the lateritic soils in Kerala, the red soils of Konkan and the Western Ghats are extremely rich in nature and are Alfisols and not Ultisols. Few research studies have indicated the presence of zeolites even under a humid climate. This renders the soil calcium rich. This suggested that the loss of bases during the leaching of soils is regularly replenished by bases from these calcium-rich zeolites. The soils in the Sindhudurg district are classified into Inceptisols, Alfisols and Entisols with Inceptisols being the primary type dominating the region. In the Ratnagiri district, the soil is classified into Entisols, Alfisols and Inceptisols with shallow Entisols indicating a huge loss of surface soils due to water erosion. The Raigad district comprises four soil orders, i.e., Entisols, Inceptisols, Vertisols and Alfisols. Most soils of the district, which are moderately deep to deep, fall under the categories of Vertisols and Alphisols. In the Thane district including the newly formed Palghar district of the region, Inceptisols, Entisols and Vertisols are identified while Alfisols have not been reported in this district (Bhattacharya et al., 1999; Gunnel and Radhakrishna, 2001).

22.2.1.2   Soils in Upper Maharashtra

The soils in the upper region of the state are dominated by the black and yellowish lateritic soil formed due to erosion. In the Vidarbha region, located on the east side of the state, a better mixture of lime, mortar and soils forms the ideal zone for Kharif crops. Northern Konkan and Bhandara, Gondia and Gadchiroli districts of the Vidarbha region comprise reddish and yellowish soil.

22.2.1.3   Deccan Trap

Around 78% of the total area of the Maharashtra State is covered by the lava flows of Deccan Trap. The other significant formations include the Archean and Proterozoic, Gondwanas and alluviums, which occupy 11%, 2.5% and 4.5% area, respectively. These rocks are a vast formation of basaltic rocks strikingly uniform in their lateral composition and formed as a result of eruption on the surface of the earth from the upper mantle. These basaltic flows are termed plateau basalts and in geological literature, they are known as flood basalts. The Deccan Plateau ranks as the world's fourth-largest (volumetrically) subcontinental outpouring of plateau basalt lavas extruded at unstable margins during active plate tectonic movements. In the Deccan basaltic plateau, the compound flows are exposed around Nashik and Igatpuri elliptical region and simple flows are dominant in the peripheral region (Bhattacharya et al., 2006; Mundhe et al., 2008).

22.2.1.4 Alluvial soil

They are depositional soils, transported and deposited by rivers and streams. The alluvial soils vary in nature from sandy loam to clay. They are generally rich in potash but poor in phosphorous. The colour of the alluvial soils varies from light grey to ash grey. Its shades depend on the depth of the deposition, the texture of the materials and the time taken for attaining maturity. Alluvial soils are intensively cultivated and mainly, rice crop is cultivated in this soil. This type of soil is found in the lowland region of the Konkan coast. It is also called Bhabar soil. This type of soil is found in the area parallel to the Arabian Sea in the north–south direction. The rivers flowing in Konkan deposit clay brought with them in this region. Similarly, this type of soil is produced in estuaries and creeks due to the accumulation of mud and scum. The rice crop is mainly cultivated in this soil (Patil, 2015).

22.2.1.5 Red and yellowish soil

Red soil develops on crystalline igneous rocks in areas of low rainfall in the eastern and southern parts of the Deccan Plateau. Along the piedmont zone of the Western Ghats, a long stretch of area is occupied by red loamy soil. The soil develops a reddish colour due to a wide diffusion of iron in crystalline and metamorphic rocks. It looks yellow when it occurs in a hydrated form. The fine-grained red and yellow soils are normally fertile, whereas coarse-grained soils found in dry upland areas are poor in fertility. They are generally poor in nitrogen, phosphorous and humus. This type of soil is observed in limited regions in Maharashtra. This soil is produced in the area of Sahyadri Mountain, especially in north Konkan. The yellowish soil is found in the basins of Wardha and Wainganga rivers in the north Vidarbha. This soil is made from the erosion of Granite and Gneiss rocks. It is with limestone and sand and is red coloured due to iron peroxide. Commonly, this is a loam type of soil. It has phosphoric acid, humus and potash to a low extent. There is no permanent structure, colour, depth and chemical substance. The soil of this type is brown, yellow and ash coloured. This is suitable for cultivation of cereals and pulses (Patil, 2015).

22.2.1.6 Lateritic soils

Lateritic soil is rich in aluminium and iron, formed in wet and hot tropical areas. Almost all lateritic soils are red due to the presence of iron oxides. It is formed by the prolonged and rigorous weathering of the parent rock. Lateralisation or tropical weathering is a long drawn out process of chemical and mechanical weathering, which results in large variations in the chemistry, grade, thickness and ore mineralogy of the ensuing soils. Laterite soils are pregnant with aluminium and iron oxides but are deficient in potash, phosphoric acid, lime and nitrogen. Lateritic soils are habitually poor and can hold only scrub forests and pastures, and they are mainly found in the southern regions of the Western Ghats, including the adjoining coastal region of the state, getting incredibly heavy rains. High rainfall encourages the leaching of soil where lime and silica are leached away and soil rich in oxides of aluminium predominates and the abundance laterite is called bauxite. Due to the presence of iron oxides, the colour of laterite soils is red. This soil is poor in lime contents and hence, it is acidic (Anisur Rahman, 1986; Sheth and Chore, 2015).

22.2.1.7   Black soil

Black soil covers most of the Deccan Plateau of the state in the upper reaches of Godavari and Krishna river, which includes parts of Maharashtra, Madhya Pradesh, Gujarat, Andhra Pradesh and some parts of Tamil Nadu. In the upper reaches of the Godavari and the Krishna, and the northwestern part of the Deccan Plateau, the black soil is very deep. These soils are also commonly known as the 'regur soil' or the 'black cotton soil'. The black soils are generally clayey, deep and impermeable. They swell and become sticky when wet and shrink when dried. Therefore, during the dry season, these soils develop wide cracks. Thus, there occurs a kind of 'self-ploughing'. Owing to such characteristics of slow absorption and loss of moisture, the black soil retains the moisture for a considerable time, which helps the crops, especially, the rain-fed ones, to sustain even during the dry season. From a chemical composition viewpoint, the black soils are rich in lime, iron, magnesia and alumina. They also contain potash, but they lack phosphorous, nitrogen and organic matter. The colour of the soil ranges from deep black to grey (Modak et al., 2012; Patil, 2015).

22.2.1.8   Marine clay

Marine clay is a type of clay found in coastal regions around the state. Marine clay is a particle of soil that is dedicated to a particle size class; this is usually associated with USDA's classification with sand at 0.05 mm, silt at 0.05–0.002 mm and clay being less than 0.002 mm in diameter. Paired with the fact, this size of particle was deposited within a marine system involving the erosion and transportation of the clay into the ocean. Soil particles become suspended in a solution with water, with sand being affected by the force of gravity first with suspended silt and clay still floating in the solution. This is also known as turbidity, in which floating soil particles create a murky brown colour in a water solution. These clay particles are then transferred to the abyssal plain in which they are deposited in high percentages of clay. Once the clay is deposited on the ocean floor, it can change its structure through a process known as flocculation. Clays can also be aggregated or shifted in their structure besides being flocculated.

When clay is deposited in the ocean, the presence of excess ions in seawater causes a loose, open structure of the clay particles to form, a process known as flocculation. Once stranded and dried by ancient changing ocean levels, this open framework means that such clay is open to water infiltration. Construction in marine clays thus presents a geotechnical engineering challenge. Where clay overlies peat, a lateral movement of the coastline is indicated and shows a rise in the relative sea level. Swelling of marine clay has the potential to destroy building foundations in only a few years. Due to the changes in climatic conditions on the construction site, the pavement constructed on

the marine clay (as subgrade) will have less durability and requires a lot of maintenance cost. Some simple precautions, however, can reduce the hazard significantly.
Geotechnical problems posed by marine clay can be handled by various ground improvement techniques. Marine clay can be densified by mixing it with cement or a
similar binding material in specific proportions. Marine clay can be stabilised using
wastes of various industries like the porcelain industry and tree-cutting industries.
This method is usually adopted in highways where marine clay is used as a subgrade
soil (Manjula Devi and Chore, 2021).

22.2.1.9   Rock formation

Maharashtra has stratigraphic sequences of rocks ranging from Archaean to Quaternary. The oldest Archaean Gneisses, variously designated as Amgaon Gneiss, Bengpal
Gneiss, Tirodi Gneiss and Peninsular Gneiss, can be seen in different parts of the state.
The 3000 m thick unfossiliferrous plate formal cover sediments exposed along the Godavari valley is classified into Pakhal, Penganga and Sullavai groups. In the Sindhudurg district, basic and ultra-basic rocks exposed are considered to be an extension of
the Chitradurga group of rocks of Karnataka and Goa. The Palaeoproterozoic acidic
volcanic and volcanoclastics of the Nandgaon group exposed in the easternmost Gondia and Gadchiroli districts is intruded by Dongargarh granite. Khairagarh Group of
rocks of Palaeo- to Mesoproterozoic age and comprise bimodal volcano-sedimentaries
unconformably overly the Dongargarh granite. The rocks of the Sausar group occur in
Nagpur and Bhandara districts and are exposed in a 75 km wide and 210 km long belt
extending into the neighbouring state of Madhya Pradesh. The late Palaeozoic to the
Mesozoic fluvial sedimentary sequence of the Gondwana supergroup is exposed in the
Wardha Valley region of Chandrapur, Yavatmal and Gadchiroli districts extending
northward in Kampti and Umrer areas of the Nagpur District. A minor proportion of
fossiliferous sedimentary rocks of the marine bagh formation and freshwater lameta
formation are exposed below the Deccan flood basalt sequences (DCFB). Tholeiitic
basalts of DCFB cover a major part of Maharashtra (Upper Cretaceous to Palaeocene). The DCFB and associated intertrappean sediments are, further, classified as the
Sahyadri Group and the Satpura Group.

Quaternary deposits of the state are mainly confined in Tapti, Purna and Godavari valleys while fluviomarine deposits occur along the west coast. Laterite capping
covers almost all the rock formations, but mostly over the Deccan Trap basalts. The
eastern and southwestern parts of the state are occupied by stratigraphically older
rock formations forming a repository of vast mineral wealth including manganese,
coal, sillimanite, kyanite, copper, gold, tungsten, dolomite, limestone, bauxite and
iron ores. Geological Survey of India (GSI) is involved in the mineral investigations
for gold, tin-tungsten and base metals and the established resources. Presently, the activities of GSI are more focused on mineral exploration for base metals and platinum
group of elements in the Chandrapur and Gadchiroli districts of the state. Geothermal
energy has gained popularity in the western reason. Figure 22.2a shows the columnar
basalt found in the Deccan Trap area of the state while the fractured basalt rock found
in the western part of the state is shown in Figure 22.2b.

(a)   (b)

Figure 22.2 Different types of basalt in various parts of the state: (a) columnar basalt (Deccan Trap) and (b) fractured basalt (Western Maharashtra).

22.2.2   Mineral resources

Maharashtra is rich with respect to natural mineral resources. The state is considered the second-largest producer of kyanite and the third-largest producer of manganese ore in the country. Minerals are generally found in the Vidarbha and Konkan areas of the state. Important minerals found in the state are bauxite in the region of Satara, Sindhudurg, Kolhapur, Raigad, Ratnagiri and Thane districts; china clay in areas of Amravati, Bhandara, Chandrapur, Nagpur, Sindhudurg and Thane districts; chromite in Chandrapur, Nagpur and Sindhudurg, Bhandara districts; coal in Chandrapur, Yavatmal and Nagpur districts. Dolomite and fireclay are seen in Amravati, Chandrapur, Nagpur and Ratnagiri districts; iron ore (hematite) in Chandrapur, Gadchiroli and Sindhudurg districts. The kyanite is found in the parts of the Bhandara and Nagpur districts. The laterite is found in abundance in the Kolhapur district. Similarly, minerals such as limestone, manganese ores, corundum, pyrophyllite quartz and silica sand quartzite are found in Nagpur and Gondia districts. Sillimanite is found mainly in the Chandrapur district. Other minerals that are found in the state include barytes, felspar, gold; granite districts; graphite and mica; lead-zinc and tungsten marble ochre; silver and vanadium, steatite and titanium. Apart from the aforementioned minerals, the state has huge deposits of petroleum in its offshore area in the Arabian Sea.

22.3   PROPERTIES OF SOILS AND ROCKS

The borehole lithology of Maharashtra has a very high variation, which can be seen from the boring log data at few locations in different regions of the state (Figures 22.3–22.6). The boring log data in respect of various soils giving description, consistency and colour of each stratum are given in these figures. The standard penetration test results are

Borehole No: 7									Date of commencement: 13.06.19			
Location: Pier No: 1									Date of completion: 22.06.19			
Ground level: 208.584 m												

Ground level (RL) (m)	Depth (m)	Log	Type of soil	Sl. no of sample	SPT				Core recovery (%)	RQD	Frac of frequency	Sl. no. of core
					0-150	150-300	300-450	N Value				
208.584	0.00											
207.084	1.50		Sand mix gravels with pebbles	S1	4	10	8	18				
205.584	3.00			S2	8	13	11	24				
204.084	4.50			S3	5	8	10	18				
203.784	4.80											
202.584	6.00		D.I. rock	S4	11	16	19	35				
201.084	7.50			S5	10	17	23	40				
199.584	9.00			S6	13	19	21	40				
198.084	10.50			S7	17	23	28	51				
196.584	12.00			S8	27	33	13 cm in 100 blows $N > 100$					
195.084	13.50		Soft rock (Non-coring)	S9	31	39	12 cm in 100 blows $N > 100$					
193.584	15.00			S10	43	10 cm in 100 blows N > 100						
192.984	15.60								76	20	11	1-12
191.984	16.60								78	30	8	13-20
190.984	17.60								82	45	7	21-27
189.984	18.60		Hard rock						83	55	7	28-34
188.984	19.60								89	89	3	35-37
187.984	20.60								96	88	4	38-41
186.984	21.60								92	92	4	42-45
185.984	22.60											

Figure 22.3 Geotechnical boring log at Shivaji Nagar, Pune, India.

also given as N-values in these boreholes. The boring logs taken at different locations in various regions show a complete variation.

Figure 22.3 shows the boring log details of one of the locations at Shivaji Nagar in Pune. Shivaji Nagar, a locality in Pune City, is at the centre of the city and home to several educational institutions. The city is a part of the Deccan Plateau consisting of basaltic rocks as is evident from the boring log summary. The borehole was drilled to a depth of 22.6 m and it was observed that up to a depth of 4.5 m, sand mixed with gravel along with the presence of pebbles was prevalent with an average N-value of 20 (N-value of 18 at 1.5 m depth, 24 at 3 m depth and 18 at 4.5 m depth). The sand is followed by a disintegrated rock to a depth of 12 m having an average N-value of 41. The soft rock is seen from depths of 12–15.6 m and the N-value is found to be greater than 100. The hard rock is encountered from 15.6 to 22.6 m with the rock quality designation (RQD) varying from 20 to 92. It is also observed that the core recovery is in the range

Borehole number	Chainage (m)	Ground RL (m)	Groundwater table depth (m)	Depth of investigation (m)			RQD between 10 m to 20 m depth (%)
				In soil	In rock (soft/hard)	Total	
BH 1	-271	2.9	2.5	4.6	15.4	20	54-82
BH 2	-151	2.61	2.5	5.5	14.5	20	64-83
BH-1 RLY	949	3.361	2.5	4.8	9.3	14.1	
BH-2 RLY	1708	3.747	2.5	8.6	4.4	13	
BH-3 RLY	1913	3.767	2.5	4.9	8.3	8.3	
BH-4 RLY	2543	3.871	2.5	3.6	21.8	25.4	
BH-5 RLY	2904	3.731	2.5	3.6	2.8	6.4	
BH-6 RLY	3271	4.877	2.5	3.6	2.8	6.4	
BH 3	3523	5.48	3.5	5	9.3	14.3	20-34
BH-7 RLY	3928	6.96	3	9.5	2	11.5	
BH 4	4622	5.77	4.1	8.8	10.4	19.2	35-87
BH 5	5122	5.674	5.8	6.95	6.8	13.75	92-99
BH 6	5566	5.19	6.5	8.5	6.25	14.75	98-100
BH 7	5949	5.016	7.1	9.6	11.4	21	15-86
BH 8	6433	5.743	4.75	11	12	23	16-60
BH 9	6446	3.75	4.1	9.45	17.05	26.5	14-66
BH 10	6958	2.123	4.5	9	15.4	24.4	46-81
BH 11	7863	2.23	4.2	4	13.7	17.7	36-92
BH 12	8444	1.787	4	4	14.1	18.1	20-100
BH 13	9090	1.99	2	6	18.55	24.55	12-75
BH 14	9449	2.451	2	2.4	14.1	16.5	48-95
BH-8 RLY	10064	3.418	2.35	6.31	6.31	12.62	
BH-9 RLY	10086	2.347	2.12	7.5	7.5	15	
BH 15	11110	2.328	2.75	5.5	6.8	12.3	60-90
BH 16	11993	2.079	3	7.5	17.05	24.55	0-65
BH 17	12307	2.14	4	3.5	14	17.5	10-90
BH 18	12753	2.301	3.5	4	19.5	23.5	10-34
BH 19	13162	3.244	3	5	15	20	42-85
BH-10 RLY	13931	4.258	4.2	8.5	10.5	19	
BH 20	14256	7.197	2.5	8.5	13.5	22	14-68
BH 21	14553	3.8	2	10	7.7	17.7	15-89
BH 22	15058	3.334	1.8	8.5	8.9	17.4	10-69
BH 23	15568	4.31	2	12.8	5.9	18.7	30-100
BH 24	16041	4.535	1	11.5	6.3	17.8	32-75
BH 25	16551	4.895	2	11.25	6.45	17.7	40-98
BH 26	17008	4.656	2	7.5	6.5	14	72-100
BH 27	17519	4.809	2.5	11.6	5	16.6	22-92
BH 28	17986	5.216	2.5	11	5.85	16.85	22-97
BH 29	18389	5.537	1.5	13	9	22	20-90
BH 30	18845	5.634	4	6.5	13.5	20	15-93
BH-11 RLY	19198	4.428	4	8	8	16	

Figure 22.4 Geotechnical boring log at Mahim Station, Mumbai, India.

of 76%–92% which signifies the quality of the rock as that of good quality. The RQD value up to 18.6 m demonstrates poor and weathered rock; but beyond 18.6 m, the RQD is 88%–92%, which demonstrates good and fresh quality rock.

The boring log of the site near the suburban railway station in the western suburbs of Mumbai is indicated in Figure 22.4. Mumbai is the capital of Maharashtra and the business capital of India. Mahim is the most preferred locality in Mumbai.

Project: Construction of AIIMS at Mihan (Maharashtra)						Location: As per location map										
Client: Deputy General Manager (Civil), HSCC (I) Ltd.						Water table depth			1.50 m							
Sample				Subsoil profile		Atterberg limits					Density					
S. No.	Depth from NSL (m)	Type of sample	Description of strata	Observed DCPT value	Depth v/s DCPT N-value	Liquid limit (%)	Plastic limit (%)	Plastic index (%)	Dia of core (cm)	qc* (kg/cm2)	Bulk density (g/cc)	Dry density (g/cc)	Moisture content (%)	Void ratio	Porosity (%)	Sp. gravity
1	0.00 0.75	UDS		24.00	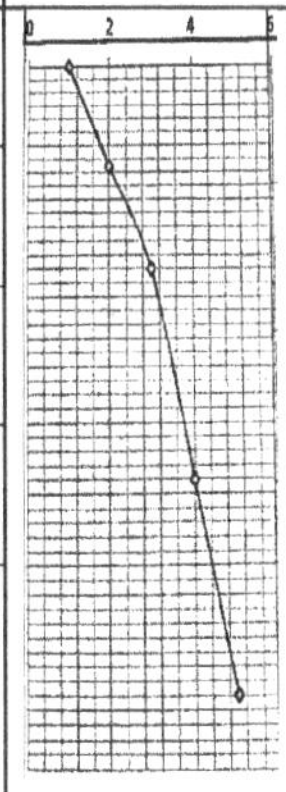	NP	NP	NP								
2	0.75 1.50	UDS	Salty claystone	31.00		NP	NP	NP			2.4	2.2	3.9	0.1	12	2.5
3	1.50 3.00	UDS		42.00		NP	NP	NP	5.2	131						
4	3.00 4.50	UDS		N>50		NP	NP	NP			2.4	2.3	4.7	0.1	12	2.6
5	4.50 6.00	UDS	Hard quartzite rock	N>50		NP	NP	NP	5.2	308						

Figure 22.5 Geotechnical boring log at AIIMS Nagpur, India.

The geological formation of the city indicates the presence of basaltic and spilitic rocks and further, sandy soil is predominant. A typical boring log, as seen in Figure 22.4, shows an RQD between 20% and 93% with initial strata of sandy soil followed by weak to hard rocks.

Nagpur, the winter capital of the state, is at the centre of the country. The zero milestone of the country is located in Sitabuldi, an old locality of the city. The city is also famous for oranges in the country and is also popularly called the orange city. In the recent past, the Government of India decided to open the state-of-the-art medical institute in every state and one such Institute, called the All India Institute of Medical Sciences (AIIMS), is being founded at Nagpur in the state. The site for this institute is located in Mihan, a new part of the city, developed on the outskirts of the city. The soil in the Nagpur region comprises black cotton soil, loamy alluvial soils, shallow sandy soils and red clayey soils. The geology of the region shows the presence of Deccan Trap basalt. A typical boring log (Figure 22.5) for the site of AIIMS at Nagpur demonstrates soft clay stone up to a depth of 3 m with a strength of 131 kg/cm^2 and hard rock up to a depth of 5 m with a strength of 309 kg/cm^2.

Figure 22.6 shows a typical boring log of 13.5 m from one site in Kolhapur city. The major soil in and around Kolhapur comprises lateritic soils, brown and black soils, and clayey soils. The boring log demonstrates the presence of low plasticity clay up to a depth of 3 m followed by sandy silts and sandy clay up to a depth of 13.5 m.

Project: Soil investigation work for proposed construction of building (2 basements + Ground floor + 6 stories) at Survey No. 276/2

Opp. Court. Dist. Kolhapur, State. Maharashtra

Water table below ground level (m): Not encountered

Termination depth (m): 25.00

Depth (m)	Sample type	SPT N value	Sp. gravity	In situ bulk unit weight (g/cc)	In situ water content %	Gravel (%)	Coarse	Medium	Fine	SILT & CLAY %	LIQUID LIMIT (%)	PLASTIC LIMIT (%)	PLASTICITY INDEX (%)	IS Soil Classification	Type of shear test	Cohesion (kg/cm²)	Angle of internal friction (degrees)	Compression index	Shrinkage limit (%)	Free swell index (%)	Swelling pressure (kg/cm²)	Unconfined compressive strength (kg/cm²)
0.00	DS	..	..	..	..	0	0	2	28	70	29	14	15	CL	..	..	..	..	..	..	..	..
1.50	SPT	15	..	..	..	15	3	2	22	58	28	15	13	CL	..	..	..	..	..	10	..	..
3.00	UDS	..	2.67	1.77	10.7	0	0	6	26	68	34	15	19	CL	TUU	0.5	16	..	..	25	..	..
4.50	SPT	19	..	..	..	0	0	4	64	32	..	..	NP	SM	..	..	..	..	..	..	..	..
6.00	SPT	22	..	..	..	0	1	7	45	47	..	..	NP	SM	..	..	..	..	..	..	..	..
7.50	UDS	..	2.66	1.84	11.9	1	1	8	44	46	33	16	17	SC	TUU	0.12	28.5	..	..	..	..	..
9.00	SPT	27	..	..	..	0	0	2	50	48	..	..	NP	SM	..	..	..	..	..	..	..	..
10.50	UDS	..	..	1.75	6.1	4	3	5	44	44	..	..	NP	SM	DCD	0	32	..	..	..	..	..
12.00	SPT	31	..	..	..	0	0	6	58	36	..	..	NP	SM	..	..	..	..	..	..	..	..
13.50	UDS	..	..	1.82	6.9	5	2	6	45	42	..	..	NP	SM	..	..	..	..	..	..	..	..

Figure 22.6 Geotechnical boring log for a construction project in Kolhapur, India.

22.4 CONSTRUCTION MATERIALS

In Maharashtra, the mineral resources found are mainly laterite soil-based bricks, limestone and basalt rock. All these materials are used for construction to a variable degree. There have been different types of soils found in the state of Maharashtra. Lateritic soils are continuously being developed with the use of different types of additives. These soils can be used for the construction of roads and production of compressed earth blocks. This type of soil is readily available, and the cost of its procurement is relatively low. Lateritic soil has many advantages over other materials, which makes it potentially a very good and appropriate material for construction. For the construction of rural structures in developing countries, it acts as a vital construction material. These merits include little or no specialised skilled labour required for laterised compressed blocks/bricks production and for its use in other construction works. Apart from this, moorum found in various parts of the state is also used as a material for the construction of embankments, subbase courses of most of the roads and base courses of low-volume roads. The clayey soils are also used in the construction of earthen dams, especially as the core walls. Normally, the soils found in the nearby area are used for construction in that vicinity. In case, if the soil is found unsuitable from an application viewpoint for a particular type of construction, then the stabilisation technique is normally resorted to (Sheth and Chore, 2015).

Basalt, being strong, helps in laying the foundation for any construction. Moreover, basalt is extensively used for making smaller aggregates for construction. Furthermore, recently, there has been a huge demand for basalt fibres. Basalt fibres are made up primarily of minerals like plagioclase, olivine and pyroxene. It resembles very close to carbon fibre and fibreglass and possesses better physico-mechanical properties than fibreglass and is comparatively cheaper than carbon fibres. These products are used as fireproof textiles in the aerospace and automotive industries and have various other applications. Basalt fibres have many unique advantages over other similar construction materials in the same segment. These fibres are ideal for fire protection and insulative applications. Basalt fibres are 100% natural and inert. They have been proven to be non-carcinogenic and non-toxic. On count of durability, they are tough and long-lasting (Singh, 2012).

22.5 FOUNDATIONS AND GEOTECHNICAL STRUCTURES

Many old cities and towns of the state nowadays have taken the shape of metropolitan cities and cities. There has been growth of the infrastructure facilities in these cities. Moreover, with the paucity of land in the nearby surrounding of these cities, multi-storeyed buildings, high rise towers and large malls are being constructed. In addition to this, flyovers, bridges, tunnels and so on are being constructed in many cities to ease out traffic congestion. Such structures necessitate different types of foundations, namely, shallow foundations that include isolated footings and raft or mat foundation and deep foundation, which include pile and piled raft foundation depending upon the type of structure and the nature of sub-soil. Even for simple two-three storeyed buildings, framed structures are preferred constructed using isolated RCC footing from the earthquake consideration as the state falls under the medium to high seismic zone. However, in the interior parts of the rural area, still people follow the conventional method of providing foundation such as spread footing constructed using locally available stones owing to the economy in the construction. However, a devastating earthquake that had hit Killari in the year 1993 has created awareness even in the rural parts of the state regarding safe geotechnical construction practices.

Various geotechnical structures have been constructed in the state over the years. Most of the dams constructed across various rivers in the state are earthen (earth fill) dams. Besides, the construction of the expressways connecting prominent cities in the state such as Mumbai, Pune and Nagpur and the upgradation of the existing state highways and national highways to superior standards has seen the construction of various structures using soils. These structures include the embankment, earth retaining structures. Another geotechnical structure that can be seen prominently is the network of canals constructed in various parts of the state. Similarly, in the interior of the rural parts of the state, roads catering to the low traffic volume are constructed using locally available soils. In case the locally available soil is not suitable to be used in the construction of the roads, it is rendered suitable by resorting to the suitable soil stabilisation technique.

22.6   OTHER GEOMATERIALS

The industrial development in the state has seen the establishment of many steel industries, which produce waste materials (to be referred to as industrial waste) as the by-product on a large scale. It includes mostly the granulated blast furnace slag. Even the thermal power plants also produce different types of ash such as fly ash, pond ash and coal bottom ash as a by-products. The safe disposal of such waste material is a concern from an environmental and ecological viewpoint. These materials can also be used as a substitute either in full or in parts to the conventional soil. They find application in the construction of various geotechnical structures including embankments, slopes, roads and so on. They can also be used for the ground improvement.

The quality construction materials are not readily available in many locations and are difficult to transport over long distances, which renders the construction costly. On this backdrop, over the last few years, environmental and economic issues have stimulated interest in the development of many alternative materials that can fulfil design specifications. To utilise fly ash in bulk quantities, ways and means are being explored all over the world to use it for the construction of embankments and roads as fly ash satisfies major design requirements of strength and compressibility except for its susceptibility to erosion and possible liquefaction under extreme conditions. Along similar lines, the ground granulated blast furnace slag, a by-product of the steel industries, is also being used either as a substitute to fly ash or in conjunction with fly ash as the stabilisation agent (Patil, 2016).

In Maharashtra, there are several sugar factories. The disposal of sugarcane bagasse poses a serious environmental concern. The ash generated from the burning of such bagasse can also be used similar to that of other industrial waste materials (such as fly ash and ground granulated blast furnace slag) containing Pozzolanic properties (Manjula Devi and Chore, 2020).

Geosynthetics are artificial fabrics used in conjunction with soil or rock as an integral part of the manmade project. The term geosynthetics is the generic name given to the manmade products used in geotechnical construction and it includes geotextile, geogrid, geonet, geomembrane, geocomposites, geocell, geofoam, geomesh, geopipe, geospacer and so on (Shukla, 2002).

For any given application of a geosynthetic, there can be one or more functions that the geosynthetic is expected to serve during the performance and the design life. They have numerous application areas in civil engineering. They always perform one or more of the basic functions that include reinforcement, separation and filtration, drainage, fluid barrier and protection when used in contact with soil, rock and/or any other civil engineering-related materials. Basically, geosynthetics are referred to as ground improvement techniques. The other ground improvement techniques include soil nailing, gabions, sand drain, band drains, stone columns and so on (Shukla and Yin, 2006).

The soil nailing and gabions have been used in various stretches, especially the ghats sections, of Mumbai–Pune expressway, as the landslide mitigation measures. These stretches have witnessed landslides in the past. Similarly, gabions are being preferred because of their free draining nature and speed in construction nowadays in various stretches of the new expressways being constructed in the state and the stretches of the old highways being upgraded to superior standards. The sand drains were used

Figure 22.7 Laying of nonwoven geotextile for a road project in Maharashtra.

as the ground improvement technique for the approaches of Airoli–Mulund Creek Bridge, constructed in the 1990s. The widening work of the eastern express highway in Mumbai passing through the belt comprising marine clayey soil in Mumbai has entailed the use of geotextiles in the decade of 2000. The widening of Jogeshwari–Vikhroli Link Road which took place around the same time has seen the application of band drains as the ground improvement measure. Apart from these examples, geo-synthetics and other ground improvement measures have been implemented in several construction works in other parts of the state. The performance of the structures/facilities is reported to be satisfactory. Figure 22.7 shows the laying of nonwoven geotextiles for a road project in the state.

Apart from the conventional geosynthetics, nowadays, the discarded and waste tyres of the vehicles are being used as the rubber grid (similar to the geogrid) and rubber cell (similar to that of the geocell). The research in this regard has shown the superior performance of such grids and shells when compared with geogrid and geocell. This may pave the way for the utilisation of discarded tyre wastes as sustainable construction materials (Patil, 2016; Manjula Devi and Chore, 2019).

22.7 NATURAL HAZARDS

Maharashtra being a coastal state is prone to various types of natural calamities throughout the year. Moreover, the state falls in a region of moderate to high seismic hazards. The state has suffered a lot due to several reasons such as earthquakes, landslides and floods.

Different parts of the state have experienced seismic activities, However, the earthquake with an intensity of 6.6 on the Richter scale experienced at Koyana Nagar in 1967 and that with an intensity of 6.2 in Latur-Killari along with Osmanabad in the Marathwada region of the state are regarded the most devastating ones. The Koyana Nagar earthquake occurred near the site of the Koyana dam claiming around 200 lives and leaving more than 2,000 injured. Around 80% of the houses constructed in Koyana Nagar Township had suffered damage. The earthquake did not cause any major damage to the gravity dam constructed across the Koyana river at the site barring few cracks in the dam. However, they were quickly repaired. The epicentre was found to be within 5 km of the site of the Koyana dam. The depth of the focus of the earthquake was estimated to be around 30 km. The earthquake was reported to have caused 10–15 cms wide fissure in the ground having spread over a length of 25 km. Tremors were experienced strongly in many towns and cities in Western Maharashtra, including, Mumbai, Pune, Goa and other parts of western and southern India. There have been several opinions with regard to its occurrence. According to few geologists, the earthquake resulted from the movements along the fault, possibly a satellite fault along the Malabar coast while others claimed that it was due to the reservoir-triggered seismic activity. The region of Koyana Nagar has been experiencing tremors of smaller magnitude of the earthquake since then frequently. However, no damage has been reported after 1967.

Another devastating earthquake, the state has witnessed, was the one that occurred at Killari near Latur and in the surrounding areas of Ausa and Omerga falling into the nearby Osmanadabad district in September 1993. The earthquake of intensity 6.2 had claimed more than 10,000 lives and left more than 30,000 people injured. However, the damage that occurred at Killari was much more as the epicentre of the earthquake was reported to have been at Killari. The hypocentre was around 10 km below the surface. Relatively shallow hypocentre was attributed to have caused more damage. It was also observed that most of the houses were constructed of stone masonry and this was also one of the reasons for an increased toll. There had been different opinions as to what led to the earthquake that occurred in this region, which did not lie on the plate boundary. It is believed that the Indian subcontinent crumples as it pushes against Asia and the pressure is released. The release of this pressure along the fault lines in this region could have been the reason for the earthquake. However, according to some of the researchers, the construction of the reservoir across the Terna river could be responsible for mounting pressure on the fault lines, the analogy similar to the one that applied in the context of the Koyana Nagar earthquake.

Maharashtra experiences a high amount of floods every year in different parts of the state causing casualty and property loss. Mumbai, the capital city of the state and the financial capital of the country, has been experiencing floods almost every year with waterlogging and submerging of the low-lying parts of the region, bringing some part of the city to a standstill position. However, the floods that occurred in the year 2005 were the most devastating in the history of the city. The Mumbai Metropolitan Region, commonly called the MMR, comprising the island city, suburbs, and surrounding region in the Thane district was struck by a heavy storm on 26th July 2005 and the subsequent deluge with the highest rainfall recorded to be 944.2 mm for 24 hours. This was the eighth highest rainfall in the city. This lashed the city and MMR region on that day and intermittently continued for the next day. The people were stranded on roads, lost their homes while many had to walk long distances back home from work on the evening of 26th July. Thousands of school children were stranded due to flooding and could not reach home for the next 24 hours. The runway and apron part of the international airport of the city was under flood. The tracks of long-distance and suburban trains were under the floods in several parts of the state and some parts of the tracks had suffered considerable damage, especially, in between Kalyan and Kasara. The rail links were disrupted, and this necessitated the cancellation of the long-distance trains for a period of 1 month or so. The loss due to this flood was unprecedented and it, further, caused a stoppage of the entire trading and industrial activities for few days. The antiquated drainage system in the city, uncontrolled and unplanned development in northern suburbs and destruction of the existing mangroves ecosystems along the Mithi River and Mahim Creek because of the increased construction activities were reported to have been the main causes for the disaster.

Apart from the Mumbai floods, the state has witnessed devastating floods in several other parts. The deluge of the Wardha river in the year 1991 devastated Mowad, a small town located in the Nagpur District and along the banks of the Wardha river, completely. The collapse of the embankment constructed along the riverbank was believed to have been one of the reasons for this havoc. According to the residents, part of the embankment was repaired a week before the rains wrought havoc. However, several trees were uprooted in the process and that further weakened the barrier.

The heavy rains in various parts of the state, especially, in the hilly region causes landslides many times. The state has witnessed many devastating landslides. Mumbai witnessed a landslide in the Ghatkopar suburb of the city in the year 2000 killing 78 people leaving several injured. The landslide was caused by land erosion, followed by heavy rains and subsequent flooding that coincided with a high tide in the Arabian Sea. The heavy downpour of 26th July 2005 impelled the closure of the Mumbai–Pune expressway for the first time in its history due to landslides though the expressway witnessed a number of landslides before. The most tragic landslide occurred in the year 2014 in Malin, a small village located in Ambegaon Tahsil of Pune district. The village was completely hit by a landslide on 30th July 2014 claiming at least 150 casualities. There were no traces of the existence of the village in the post landslide scenario. This landslide was reportedly attributed to a burst of heavy rainfall. Apart from the heavy downpour that continued throughout the following day, there were many causes for the destruction, which included negligence of the geological aspect before undertaking any developmental processes, deforestation in the area, a shift in agricultural practices and construction of the dam a few years back and quarrying in the nearby vicinity.

22.8 CASE STUDIES AND FIELD TESTS

This section presents the description site investigation and geotechnical data for the site of the Mumbai Metro Line 3 (UEIA Report, 2017). The entire Mumbai region is prevalent by Deccan basalt flows and the associated pyroclastic and the plutonic rocks of upper cretaceous to Palaeogene age. The sandy soil is observed in Mumbai city whereas the suburbs of Mumbai comprises alluvial and loamy soils. A total of six soil samples were collected and tested for the quality of soil in the vicinity of the site. The test result shows that soil texture is sandy silt with medium contents of nitrogen, phosphorus and potassium. The city falls under Seismic Zone III of the specifications brought out by the Bureau of Indian Standards (BIS) indicating that the city is at moderate risk. The shallow water levels between 2 and 5 m ground level are observed in the southern part, whereas moderate water levels in the range of 5–10 m ground level are observed in the northern part of the area during pre-monsoon. The water levels during post-monsoon in the major part of the district range between 2 and 5 m ground level. To test the quality of the water, five samples from different locations along the metro alignment were collected and analysed. The test suggested that the water at the location of Mahim Creek contained TDS more than the permissible limit. At the rest of the locations, the parameters were within the limit.

The proposed route of the metro has many heritage sites and the route passes through a densely populated area. The annual rainfall over the district varies from about 1,500 to 2,500 mm. This can cause flooding which would disturb the complete workability of the project. The operation of the Tunnel Boring Machine (TBM) during construction of a tunnel and rolling stock during operation of Metro Rail causes vibration in the nearby vicinity. To know the impact of vibration due to TBM operation and metro train operation, the study was conducted at six locations by selecting the sensitive area (structures) falling on the proposed metro line alignment. The hard rock structure was considered while predicting the vibration impact.

Some of the strata along the route of the Metro comprise old and heavy weathered rocks, which are susceptible to the landslide in wake of the heavy rainfall affecting the city. This aspect was also considered during the design and construction. Taking into consideration the in-situ conditions (sample disturbance), confinement aspect and in light of the recovery patterns, visual inspection of samples covering texture, fracture and weathering aspect, the bored cast-in-situ piles were designed for the weathered rock stratum. The lengths of the piles were kept between 8 and 10 m. The piles are to be socketed in rock. The lengths of the piles were decided based on the considerations of the adverse environmental consequences, economy and taking into consideration the type of the proposed structure.

The link road connecting the Airoli suburb of the city of Navi Mumbai and the Mulund suburb of Mumbai, via a bridge constructed across Thane Creek near Airoli, was made operational in the year 1999. However, within a span of 45 years, the road started showing signs of distress and in the year 2007, most of the portion of around 5 km long road had undergone complete distress. The 2 km portion of the road emanating from the Thane–Belapur Road to Sector 9 of Airoli was severely distressed with the formation of potholes and depressions, washing out of the base course in one or two lanes of the dual carriageway. The remaining portion was found to suffer medium to considerable damage, which included potholes, shear failure, settlement and alligator cracking. Extensive investigations of the distress were carried out. It was observed that primarily under-estimation of the repetitions of heavy commercial vehicles, overloading of the commercial vehicles and inadequate design of the crust were the major factors affecting the poor performance of the road. Furthermore, the use of inferior materials and inadequate quality control during construction also contributed to the distresses. In addition to this, poor drainage of the surface water was found to be the root cause of the distress in the pavement. On one side of the dual carriageway, the drainage was not provided at all while in some portion, although the drainage was provided on either side, the poor maintenance of the drainage resulted in the clogging of the drains and blockage of the gratings as a result of which the surface water could not be removed from the surface of the asphaltic pavement. This also formed the major cause of distress of the pavement and allied components of the flexible pavement (Chore and Manjula Devi, 2009). Figure 22.8 shows the photographs taken of the different locations of the study stretch.

Figure 22.8 (a) Pothole, damaged base course and accumulation of surface water due to choking of drainage near Sector-5 and Sector-17, (b) completely washed out base course in a stretch located near Sector-9, (c) heavily damaged carriageway portion of extreme right land central lane near bridge approach at Sector-10 and (d) accumulation of rainwater due to the absence of side drains near Sector 9.

22.9   EFFECTS OF ENVIRONMENTAL CHANGES ON SOILS

Climate change has played a significant role in causing large-scale floods across central India, especially the flood that occurred in the Mumbai Metropolitan Region in 2005. The increased number of extreme rains is attributed to an increase in the fluctuations of the monsoon westerly winds due to increased warming of the Arabian Sea. This causes occasional surges of moisture transport from the Arabian Sea to the subcontinent, resulting in heavy rains that last normally for 2–3 days. The floods of Mumbai in 2005 were also the outcome of such a moisture surge from the Arabian Sea. However, the heavy rains at that time did not remain confined to the Mumbai Metropolitan Region but also extended over a large region of the state.

The encroachment for any developmental activities on the soils having vegetation leads to the fiddle with the original composition of the soil. It also affects significantly the environment. The environmental destruction results in landslides. It has also been observed from various catastrophic incidences that the negligence of the geological facts before undertaking any developmental activities has led to the failure of many structures and occurrences of landslides. In an effort of undertaking large developmental activities, nowadays, natural vegetation is being removed. This is called deforestation. This deforestation removes not only trees but also root structures that hold the soil together thereby rendering the soil in the surrounding piece of land. Under heavy rains, such loose soil causes landslides. Deforestation is regarded as the primary underlying anthropogenic cause for the landslide. Furthermore, changing agricultural practices is also causing nowadays soil erosion leading to landslides. Many times, the farmers shift cultivation of few crops such as rice to wheat, which requires levelling of the steep or hilly areas. In this exercise, the stability of the hills gets endangered. Further, in hilly areas, construction activities are carried out without any careful investigation of the environmental consequences and they further result in the instability of the hillside. Similarly, stone quarrying is also responsible for the instability of hillsides. The investigations after the occurrences of the many disasters including landslides at Ghatkopar and Malin and the flood of Wardha river that had washed away the entire tiny town Mowad in the Nagpur District, have emphasised on the preservation of the environment so that the disasters can be avoided.

The mangrove ecosystem exists along the Mithi river and the Mahim Creek in Mumbai city was destroyed to be replaced with the construction. More than hundreds of hectares of swamps in Mahim Creek were reclaimed for construction activities. Some portion is reported to have been destroyed by the slum dwellers. The Bandra-Kurla Complex in the city was created by removing such swamps. The sewage and garbage dumps have also added to the destruction of the mangroves. The mangrove ecosystem serves as the buffer between the piece of land and sea. The destruction of such buffer was also attributed to the flood disaster that occurred in the year 2005 in Mumbai along with several other causes.

22.10 CONCLUDING REMARKS

- Maharashtra, the third-largest state of the country, is rich in terms of forest and mineral resources. The state is located on the Deccan Plateau with 78% of the area of the state having been covered by the Deccan Trap. Besides, there are Ghats and coast along the Arabian Sea. While basalt rock is predominantly found in the state, granite also exists in the Gondia and Gadchiroli District of the state.
- Different soils including alluvial soils, red and yellowish soils, lateritic soils, black soils and marine clay soil are found to exist in different parts of the state. While alluvial soil is found normally in the low land coastal region of Konkan, red and yellowish soils are seen in the limited parts of the state. The region of north Konkan surrounded by the ranges of Sahyadri comprises red soil, the basins of Wardha and Wainganga river in north Vidabha show the presence of yellow soil. Lateritic soils are mainly found in the southern part of the state including the adjoining Konkan region. Black soils are found to cover most of the Deccan Plateau of the state in the upper reaches of the rivers such as Godavari and Krishna.
- Each soil has got its own peculiarities. Alluvial soil is suitable for agriculture in general and rice production. The red and yellow soil suits for the cultivation of cereals and pulses. Black soils are referred to as the most suitable for the cultivation

of cotton. The soils and rocks prevailing in the state are found to have a deep impact on the social economy of the state. The presence of hard rocks to a considerable extent in many regions of the state increases the construction cost. At the same time, fertile types of soils such as alluvial, red and black found in various regions support the farming community in the concerned regions of the state.

•   The by-products of the various thermal power plants and industrial units are being used nowadays as an alternative construction material in lieu of the conventional construction material. This shift in the utilisation of such waste materials in civil engineering construction helps in preserving the natural resources, maintaining the ecology and eliminating the environmental concern. The state has witnessed many natural disasters and calamities arising out of earthquakes, floods and cyclones. These disasters resulted in a large number of casualties and loss of properties. Although natural disasters cannot be completely ruled out, the instances have underscored the necessity of proper construction practices and preservation of nature by maintaining the ecosystem. This would cause minimum damage.

ACKNOWLEDGEMENTS

Acknowledgement is due to Head, PG Department of geology, Nowrosji Wadia College, affiliated to Savitribai Phule Pune University for providing useful information in respect of geological formation of the state, Mr. S.M. Das, Geotechnical Consultant for providing the geotechnical information in the context of various locations within the state and Dr.Ajit Vartak of the University of Pune for providing useful information and few photographs of geological formation. Thanks are also due to Dr. (Mrs). Vidya Patil, Associate Professor, AISSMS's College of Engineering, Pune and Dr. (Mrs.) Smita Patil, Associate Professor and Mrs. B. Manjula Devi, Assistant Professor, Department of Civil Engineering, Datta Meghe College of Engineering, Airoli, Navi Mumbai for providing the information in respect of some of the soils and other geomaterials within the state which has been the part of their research work carried out under the supervision of the lead author. The help provided by Mr. Bhupati Kannur, research scholar in the Department of Civil Engineering of Dr. B.R. Ambedkar National Institute of Technology, Jalandhar in retracing some of the figures and tables pertaining to the bore log data, is also deeply acknowledged.

REFERENCES

Anisur Rahman, M.D. (1986). The potentials of some stabilizers for the use of lateritic soil in construction. *Build. Environ.* Vol. 21, No. 1, pp. 57–61.

Bhattacharya, T., Pal, D.K., Lal, S., Chandran, P. And Ray, S.K. (2006). Formation and persistance of Mollisols on zeolitic Deccan basalt of humid tropical India. *Geoderma.* Vol. 136, No. 3, pp. 609–620.

Bhattacharya, T., Pal, D.K. and Srivastava, P. (1999). Role of zeolites in persistence of high altitude ferruginous Alfisols of the Western Ghats, India. *Geoderma.* Vol. 90, No. 3–4, pp. 263–276.

Chore, H.S. and Manjula Devi, B. (2009). Diagnosis of distress in Airoli-Mulund Link Road in New Mumbai. *Proceedings International Conferences Advances in Mechanical and Building Sciences (ICAMB-2009)*, VIT Vellore (Dec. 14–16), pp. 1596–1601.

Gunnel, Y. and Radhakrishna, B.P. (2001). Environmental changes in Western Ghats during Quaternary. In *Sahayadri: The Great Escarpment of the Indian Subcontinent (Memoir Geological Society of India)*, pp. 817–832.

https://en.wikipedia.org/wiki/Maharashtra, Wikipedia of Maharashtra, accessed April 3, 2021.

Manjula Devi, B. and Chore, H.S. (2019). Use of reclaimed rubber cell in highway pavement: an experimental research. *Int. J. Innovative Tech. Exploring Eng.* Vol. 8, No. 6 (S4), pp. 1047–1052.

Manjula Devi, B. and Chore, H.S. (2020). Feasibility study on bagasse ash as light weight material for road construction. *J. Mat. Today Proc.* Vol. 27, No. 2, pp. 1668–1673.

Manjula Devi, B. and Chore, H.S. (2021). Strength behaviour of marine clay stabilized with marble dust and quarry dust. In *Ground Improvement Techniques* (Springer, Singapore) (Eds. Sitharam, T.G., Parthasarathy, C.R. and Kolatyhayar). pp. 99–109.

Modak, P.R., Nangare, P.B., Nagrale, S.D., Nalavade, R.D. and Chavan, V.S. (2012). Stabilization of black cotton soils using admixtures. *Int. J. Eng. Innovative Tech. (IJEIT).* Vol. 1, No. 5, pp. 1–3.

Mundhe, M.S., Pandhare, V.B., Methekar, N.M. and Vaijapurkar, S.R. (2008). Case history compilation of engineering properties of common rocks in Maharashtra, India, for database (1982–2002). *Proceedings of Sixth International Conference Case Histories in Geotechnical Engineering, Missouri University of Science and Technology*, Arlington, VA (Aug. 11–16), pp. 1–10.

UEIA Report. (2017). Updated Environmental Impact Assessment (UEIA), Part-I, Mumbai Metro Line -3. Mumbai Metro Rail Corporation, Mumbai (Unpublished).

Sehgal, J. (1990). Agro-ecological regions of India. *Technical Bulletin, National Bureau of Soil Survey and Land Use Planning (Indian Council of Agricultural Research*, NBSS PUB 24, India.

Sheth, S.S. and Chore, H.S. (2015). Evaluation of strength of lateritic soil stabilized with fly ash and cement. *Proceedings of National Conference on Smarter Cities-India: Smarter Solutions for Better Tomorrow (SCI-2015)*, Sinhagad College of Engineering, Pune, Maharashtra (June 8–9), pp. 7–13.

Shukla, S.K. (2002). Fundamentals of Geosynthetics. In *Geosynthetics and Their Applications*, Thomas Telford, London.

Shukla, S.K. and yin, J.H. (2006). *Fundamentals of Geosynthetic Engineering.* Taylor & Francis, London.

Singh, K. (2012). A short review on basalt fibers". *Int. J. Textile Sci.* Vol. 1, No. 4, pp. 19–28.

Patil, S.B. (2015). *Numerical Modeling of Contaminant Transport through Porous Media.* Ph.D. Thesis, University of Mumbai, Mumbai, India.

Patil, V.N. (2016). *Experimental Investigations on Strength and Stability of Reinforced Embankments Made Up from Pozzolanic Materials.* Ph.D. Thesis, University of Mumbai, Mumbai, India.

Patil, V.N. and Chore, H.S. (2016). Bearing capacity behaviour of strip footing on model slopes made up of fly ash and furnace slag. *Int. J. Geotech. Eng.* (Taylor & Francis). Vol. 11, No. 5, pp. 431–440.

Chapter 23

Manipur

Sangeeta Shougrakpam and Ashutosh Trivedi
Delhi Technological University

Arunkumar Yendrembam
Manipur Institute of Technology

CONTENTS

23.1   INTRODUCTION

Manipur is one of the border states in the northeastern (NE) region of India, bounded by Nagaland in the north, Assam in the west and Mizoram in the southwestern sides. Along the east, it shares a 398 km long international border with Myanmar (Figure 23.1). Thus, it significantly influences an important strategic corridor (Trans-Asian International Highway) for India's 'Look East Policy' (Oinam, 2007). The total geographical area of Manipur is 22,327 km^2, extending approximately between 23°50′N and 25°42′N latitudes and between 93°00′E and 94°45′E longitudes. Manipur is a hilly state constituting about nine-tenths of the hills, which surround the remaining one-tenth valley. Imphal city (24°48′N, 93°56′E) is the capital of the Manipur state, which forms the central part of the oval-shaped Imphal valley extending over an area of about 2,238 km^2. The Imphal valley, also known as the Manipur valley, has an NNW-SSE orientation with a gentle tilting slope towards the south measuring 798 and 746 m above the mean sea level (MSL) at the extreme north and the south, respectively. The saucer-like Imphal valley is surrounded by a series of mountain chains (Patkai hills on the north and northeast and the Manipur hills, Lushai hills, Naga hills and Chin hills on the south) running from north to south, forming an integral part of the Indo–Burma ranges (Laiba, 1992; Soibam, 2000). There are also many lakes in the

DOI: 10.1201/9781003177159-23

(a)

(b)

Figure 23.2 (a) Cutting of hillsides for a new road constructed along the hilly terrain in the Ukhrul district. (b) A facet of a hill protruding above the Imphal valley.

Figure 23.3 A plain terrain surrounded by a chain of hill ranges at a nearby place of Sekmai river in Imphal-West district of Manipur (along NH-39) from where sands and pebbles are commonly collected for various construction works.

thrust (which partly forms the tectonic contact between the Disangs and Berails) on the western side bound the valley. Churachandpur-Mao thrust (CMT) is situated west of the Indo–Myanmar subduction zone of the high seismic zone. Due to the proximity of the Indo–Burma Range (IBR), Shillong Plateau (SP) and Naga Disang Thrust, the vulnerability of the Imphal city to damaging earthquakes has increased. The structural and tectonic patterns near Imphal valley are transitional between the NE-SW trending pattern of Naga-Patkai hills and the N-S trend of Mizoram and Chin hills (Brunnschweiler, 1974). These hills comprise geologically young rock formations due to tertiary upward displacement of the Himalayas from the shallow bed of the Tethys sea. The rocks are mainly of Tertiary and Cretaceous sediments with minor igneous and metamorphic rocks. Major river basins surround the Imphal valley. Major rivers originate from the surrounding hills and deposit their sediment load in the Loktak Lake (Laiba, 1992), located in the south of the valley.

Pallav et al. (2010, 2012) show that faults surround almost the entire region of Imphal city. These faults contribute to the tectonic activities in the region, mainly due to the interaction between the north-south convergence along the Himalayan boundary and the east-west convergence along the IBR. The studies of seismic activity in NE India show that the Kopili fault and Indo–Burma arc (IBA) are highly active and could be the sources for large magnitude earthquakes (Nandy, 2001; Kayal, 2008). Moreover, as evident from the damage during the past earthquakes, most parts of Imphal city are found vulnerable to liquefaction-induced failure.

The Indo–Myanmar ranges extend between 93°E and 95°E longitudes and between 19°N and 27°30′ N latitudes. The margin of the eastern foothills is supposed to be aligned with the Myanmar plate boundary overriding above the Indian plate. The rocks are made up of Neogene molasses, Tertiary and Cretaceous flysch sediments with minor igneous and metamorphic rocks of older age. The whole range is longitudinally divided into four major tectonostratigraphic units, namely, Metamorphic Belt, Ophiolite Belt (Ophiolite Mélange Zone), Central Flysch Belt and Molasses Belt.

The metamorphic complex on the extreme eastern part comprises low to medium grade phyletic schist, quartzite, marble, granite-gneiss and is the oldest in the stratigraphic sequence. The eastern ranges contain the Ophiolite Mélange zone, approximately 200 km long Ophiolite suite, chiefly made up of basic intrusive and extrusive, associated conglomerates, volcanic pillows, coarse-grained exotic sandstones, pelagic sediments like shales, chert and limestones. This Ophiolite belt in the western side of the metamorphic belt is thrusting over the central flysch belt of Disangs and Barail sediments. Thus, Ophiolite rocks, mainly peridotite ultramafic, with serpentinites and associated pelagic sediments like limestones, chert, shales and exotic coarse-grained blocks of sandstones, together constituted the Mélange zone.

The Disang group of rocks are overlain with the Barail group of rocks of the Cenozoic age. The Barail are usually light-brownish grey, fine- to medium-grained sandstone often interbedded with shales. The contact between Disangs and Barails runs more or less parallel to NH-39 of the Senapati–Mao sector. Within the flysch belt, the older Disang formation overlies the younger Barail sediments along with the CMT; however, synclinal remains of the same Barail beds are seen frequently on top of the ranges.

23.3   PROPERTIES OF SOILS AND ROCKS

There are alluvial soils near the river banks of Imphal valley. The clayey soils in the valley region are rich in humus, highly compressible and susceptible to settlement. In the low-lying areas of Imphal city, viz., Lamphelpat, Takyelpat, Porompat and Kakwapat (pat means lake in the Manipuri language), virgin soils like clayey loam, dark clayey and boggy-type soils are found (Laiba, 1992). Detailed soil investigation data have been obtained from a few construction sites, using the standard penetration test (SPT) at different locations of Lamphelpat, Porompat, Langol, Thangal Bazar, Tarung and Canchipur, which are located nearby national highways in Imphal city. The N value was measured at a 1.5 m depth interval up to 20–30 m. Based on the SPT data, the soil deposit in the specified location of Imphal comprises primarily silty clay and clayey soil (i.e., clay, silt, clayey sand, silty sand, clayey silt, sandy clay, silty clay and so on) with organic matters deposited in different layers. These are typical alluvial/fluvial deposits that are generally loose in density. Indeed, the SPT data indicate that the N values even at 10–20 m depth below ground surface are mainly of the order of 10–20.

The soil generally contains small rock fragments, sand and sandy clay and is of different varieties. The topsoil on the steep slopes is fragile. In the plain areas, especially in flood plains and deltas, the soil layer is of considerable thickness. The soil on the steep hill slopes is subjected to high erosion resulting in the formation of sheets and gullies and barren rock slopes. The shifting cultivation in the hills leads to deforestation, soil erosion in the hills, silting of the rivers and lake beds, causing floods in the plain. Terrace farming, landslides, flash floods and mass movement of soil are additional man-made problems of the region.

Landslides and related natural disasters frequently disturb the hilly routes of Manipur, especially the national highways NH-39 (now NH-2) and NH-53 (now NH-37), which connect the state with other parts of the country during heavy and prolonged rainfall. The main factor contributing to landslides is the fragile nature of the litho units, resulted from the rock types and structures characterising these rocks. However,

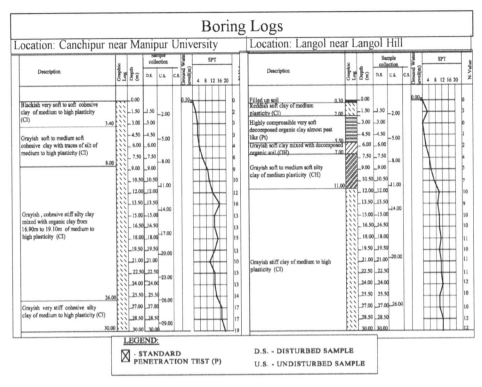

Figure 23.4 The subsurface soil conditions and SPT-N profile at two typical boreholes located at Canchipur near Manipur University and the other at Langol to the east of Langol Hill.

certain places are found to be covered with thick columns of soil, rendering instability conditions.

The subsurface soil conditions with SPT-N profiles at two typical boreholes located in Canchipur near Manipur University and the other at Langol to the east of Langol Hill are presented in Figure 23.4.

23.4   CONSTRUCTION MATERIALS

The use of proper sources for stones and aggregates has become a significant issue in the state. Small quarries on steep slopes are often enlarged by blasting or excavation at the base. Blast is dangerous and can cause slope failures. Sand and gravel are obtained from river deposits of the Sekmai river, Thoubal river and so on. The silt content in the river sand from various sources is in the range of 5%–10%, which is on the higher limit. Hence, the strength of the concrete and other construction materials is not up to the specifications as per the Indian Standards. For important structures to meet the specifications, sand is brought from neighbouring states, which escalates the overall cost of the projects. Stone dust and crushed stones in different sizes are also used in construction works; however, these are very limited in quantity. With the upcoming various infrastructures of national importance (e.g., Trans-Asian-International highways, rail and head connectivity up to the state capital Imphal), the local availability of suitable construction materials is always a significant issue in the state. New construction materials and advanced technologies are to be investigated to meet the present challenges. The construction wastes and excavated hill soils are used to fill the quarry and borrow pits. Further, the excavated hillslopes may be filled up to resemble an original ground surface. Figure 23.5a shows a site view during construction of pile-raft pier

(a)

(b)

Figure 23.5 (a) A site view during construction of pile-raft pier foundations for a bridge over Imphal river at Heigang located in Imphal-East district during the dry season of December 2020. (b) Pile-raft foundations constructed in a low-lying waterlogged area to reduce differential settlement located in the RIMS campus at Lamphelpat in Imphal-West district.

foundations for a bridge over Imphal river at Heigang located in Imphal-East district during the dry season of December 2020, while Figure 23.5b shows pile-raft foundations constructed in a low-lying waterlogged area to reduce total and differential settlement located in the Regional Institute of Medical Sciences (RIMS) campus at Lamphelpat in Imphal-West district.

23.5 FOUNDATIONS AND OTHER GEOTECHNICAL STRUCTURES

Other than Imphal, the capital city, elsewhere in Manipur, varied locations have witnessed a significant spurt in the construction activities. Figure 23.6 shows the site of a foundation to construct a structure for National Sports University at Kangpokpi District, Manipur and a hillside cutting for road expansion at Tengnoupal district, Manipur, where soil type is of red ferruginous.

The Jiribam–Tupul–Imphal new broad gauge (BG) rail line of Indian Railways has the world's tallest pier railway bridge across the river Ijai near Noney with a height of 141 m. It is constructed over a gorge with an overall length of about 703 m and is a part of the 111 km long Jiribam–Tupul–Imphal BG rail line in Manipur by the Northeast Frontier Railway (NFR) (PTI, 2014). It is one of the 148 bridges along the route, which also has 47 tunnels through steep hills to connect the capital of Manipur with the BG network of Indian Railways. The alignment of the railway line passes through steep rolling hills of the Patkai region, an eastern trail of the Himalayas. Jiribam is situated 37 m above the MSL, while Imphal is located at 780 m above the MSL. Therefore, the alignment has to traverse through several deep gorges and over several rivers flowing at low ground levels, necessitating 47 tunnels measuring 61.028 km in length and tall bridges to maintain a suitable gradient for the efficient operation of the railway. The longest tunnel is tunnel #12, with a length of 10.28 km between the Tupul–Imphal section. While the high mountains are penetrated by constructing tunnels, the deep river gorges between the mountain ridges are connected by tall bridges.

The main superstructures of the tallest bridge, Bridge #164, have been scheduled to be steel open web through type girders of 103.5 m span (centre to centre of bearings). The pillars are reinforced cement concrete (RCC) hollow type, with the tallest pier being 141 m high, as shown in Figure 23.7a.

According to the Indian Railways, the general soil profile consisting of shale is undesirable for such bridges. The area receives very heavy rainfall for over 6 months in a year. Besides, the bridge is located in seismic zone V, making it highly vulnerable to earthquakes. The piers of the bridge are constructed using hydraulic augers. The tall piers needed a specially designed 'slip-form technique' to ensure efficient and continual construction. The steel girders are pre-fabricated in a workshop, transported in segments and erected at the site by a cantilever launching scheme. The self-erecting electric lifts are used at each pier for the safe and speedy conveyance of men and materials to the top.

The NF Railway Construction Organisation has constructed the first rail-girder bridge, Bridge #44, with over 100 m tall pier and 555 m long across the valley of river Makru in Tamenglong district of Manipur. The bridge is also a part of the Jiribam–Tupul–Imphal new BG rail line project. Figure 23.7b shows the launching of the steel girders by the cantilever erection method.

(a)

(b)

Figure 23.6 Red ferruginous soil in the hilly regions: (a) site of a foundation to construct a hostel building for National Sports University at the Kangpokpi District, Manipur. (b) A hillside cutting along a road for road expansion at the Tengnoupal district, Manipur.

(a)   (b)

Figure 23.7 (a) Construction of bridges along the Jiribam–Tupul–Imphal new BG rail line
project (111 km long) of the Indian railway network: the 141 m tall pier constructed for Bridge #164 over river Ijai in Noney district and (b) the launching
of the steel girders by the cantilever erection method of Bridge #44 of 555 m
long with a pier height of 100 across the valley of river Makru in the Tamenglong district of Manipur.

Figure 23.8a shows the rock type of tunnel #12 portal, and Figure 23.8b shows the
view of the tunnel excavation. It is dominantly a sedimentary origin rock. It is light
to dark-grey, fine- to medium-grained sandstone with siltstone interlayer, and in some
places intercalated with shale and intersected by shear seam containing clay and disintegrated rock fragments, weak to medium-strong strength as observed in Figure 23.8.

The rock mass is characterised by prominent 2+ random joints, also, the 3 and
3+random joint sets (moderately to highly fractured), which are rough and irregular
planes with overconsolidated, non-softening clay mineral fillings and zone or bands
of clay. Rockmass is characterised by dripping to flowing water conditions. Based
on the rating of importance of rock mass quality parameters, the rock mass quality
(Q) ranges from 4.0 to 10 (poor to fair rock mass) according to the Barton criterion
consisting of rock quality designation (RQD). The subsurface investigations to interpret engineering properties and RQD of jointed rocks, which provide the scope of the
evaluation of deformation modulus (E_{mr}) and compressive strength of rock masses
(σ_{mr}) based upon the interpretation of rock mass rating (RMR), GSI and in situ RQD
values. The strength and modulus ratio of rock masses may be presented as

$$\sigma_{mr} = E_{mr} \exp\left[\alpha\ C_{hs} \exp(\beta RQD)\right] \tag{23.1}$$

where C_{hs} is a plasticity parameter for the rock masses, while α and β depend upon the
relationship of J_{fg} (joint factor of rock masses) and RQD (Trivedi, 2015, 2017).

The construction of a single line BG tunnel, tunnel #12, of about 10.28 km long
in connection with Jiribam–Tupul–Imphal new BG line is a project of the NFR of
Indian Railways. The construction work at Portal 2 of tunnel #12 of Tupul–Imphal
section line was started at Atom Khuman, Haorang Keirel, Imphal-West district. It is

(a)

(b)

Figure 23.8 (a) Sedimentary rock types found at tunnel #12 site of the Jiribam–Tupul–Imphal rail line. (b) A view of the tunnel excavation.

the second-longest railway tunnel in India. It also has a 9.3 km long escape tunnel. In addition to that, there are two Adits in this tunnel, Adit1 and Adit1-A. Adit1 is 1.25 km, and Adit 1-A measures 1.4 km long. Adit has been planned for speeding the work during construction. After completion, they will also be helpful in emergencies. An escape or safety tunnel will be made parallel to tunnel #12, which will have a length of 9.35 km. A cross-passage connects the main tunnel and safety tunnels at every 500 m at a distance of 50 m. The safety tunnel is made as per the international guidelines, namely UIC (International Union of Railways) and TSI/SRT (Safety in Railway Tunnels – Europe Railway Agency) for the safety of the passengers during the operational time. A state-of-the-art NATM (New Austrian tunnelling method) is being used in all the tunnels for tunnelling work. The stone pitching was also constructed as a part of the slope protection along the hill slopes of the rail line. Slope protection in and around the tunnels was done using cement grouting.

The oldest rocks in Manipur include shales, slates, siltstone, sandstone and quartzites. Geologically, Manipur can belong to the recent formation, which has implications on mineral exploitation. The surface rock is loose and soft and, therefore, vulnerable to the weathering process. This peculiar characteristic also accentuates erosion, silting and sedimentation. The MN-06 road section is a part of tranche 2 of the Asian Development Bank as Tupul–Kasom Khullen road comprising two roads, namely, Tupul to Bishenpur and Thoubal to Kasom Khullen, covering a total length of 97.925 km. This project will improve the quality of life for the rural population in the project influence area (six districts and provide connectivity to three national highways, i.e., NH-39, NH-53 and NH-150). The project road passes through plain as well as hilly terrain. The terrain in the hillside is undulating in horizontal geometry, resulting in a frequent zig-zag configuration in the horizontal alignment necessitating protective works to retain earth fill in stretches and retain unstable hillside cut, particularly in soil and soil mixed with boulders and soft rock. The innovative and state-of-the-art slope protection and stabilisation measures by geo-strengthening the soil have been adopted to make the road safer and sustainable. The methods adopted are soil reinforcement using geosynthetics, soil anchorage, soil nailing and erosion control on hillslopes using geonets. The stability of road formation and road traffic safety is ensured by providing protective works in hilly terrain, as made by providing retaining walls, breast walls and parapet walls/guideposts/railings/edgestones at critical sections. The pavement profile will have 285, 250, 60 and 40 mm thicknesses for GSBC, WMM, DBM and BC, respectively.

23.6   OTHER GEOMATERIALS

Slope protection measures such as retaining walls and bioengineering measures are suitable for slope protection in hill roads. The following steps are suggested as bioengineering measures for slope protection in hill roads for the Asian Development Bank road project in Manipur:

- Turfing of slopes through rough grassing
- Tree plantation along the hill section (slopes) of the roads to control soil erosion

Tree roots provide an important stabilising influence on forest soils and operate to reinforce the ground in three ways:

- Attaching potentially unstable regolith to stable substrates
- Providing a matted network that offers lateral attachment near the surface
- Providing a localised zone of reinforcement associated with individual trees, similar to the arching restraint provided by piles

Recently, a PMGSY (Prime Minister Gram Sadak Yojana) rural road of 2.4 km long between Hiyanglam to Hiranmei in Kakching District was constructed using the jute geotextiles (JGTs) as an environmentally friendly biodegradable material. The use of JGT is one of the essential diversified jute products. It can be applied in many areas of civil engineering, including soil erosion control, road pavement construction and the protection of river banks. Non-woven geotextiles are also being used for pavement construction of a few major road sections along the NH-150.

Microbially induced calcite precipitation (MICP) is considered a sustainable and environmentally friendly technique of soil treatment to improve the engineering properties in loose soils(Shougrakpam and Trivedi, 2020). Calcite ($CaCO_3$) as a biocement can be used as an innovative geomaterial that requires less energy to produce, unlike cement that requires high energy consumption and generates a substantial amount of CO_2, that is, about $0.95 \ t \ CO_2/t$ Portland Cement (Yi et al., 2013). However, the improvements varied with treatment conditions, soil types, calcium source, type of soil bacteria and other environmental factors such as pH (≥ 7) and temperature of the local environment. MICP is triggered in the soil due to a biomineralised reaction product of ureolytic bacteria such as *Sporosarcina pasteurii* (a non-pathogenic soil bacterium) and a cementation reagent solution containing calcium and urea in soils. In the MICP process, urease enzymes released by the bacteria play a significant role in urea hydrolysis. The enzymatic reaction hydrolyses urea to ammonium [NH_4^+] and carbonate ions in the presence of water; as a result, the pH level increases. The hydroxide ions combine with CO_2 to produce bicarbonate ions, which react with Ca^{2+} present in the cementation solution to precipitate $CaCO_3$. The carbonate ion reacts with the calcium ions from a calcium source such as $CaCl_2$ to precipitate insoluble calcite grains ($CaCO_3$) that bind the sand particles together or fill the pores. The precipitates act as a bonding material to increase the shear strength of soil or as a pore-filling material to reduce the permeability of soil (Dejong et al., 2013; Ivanov and Chu, 2008). Higher compressive strength and calcite content have been obtained in the soil cured for 14 days (Shougrakpam and Trivedi, 2018). The judicious use of reagent concentration has been found economical and reduces the hazardous effects of chemicals on the environment. The calcite grain formation has been found to reduce the pore throat size, causing a reduction in pore pressure and hence hydraulic conductivity of the soil. This method would reduce liquefaction and settlement problems of structures constructed in loose sandy soils. The soil structure depends on the interaction between minerals and organic matter present in it. The soil also provides a spatially heterogeneous habitat for microorganisms with varying substrate needs, nutrients, oxygen concentration and pH levels, thus affecting bacterial diversity and structure. Thus, the achievable cementation level, uniformity, durability and other engineering properties in the sand formations are challenging but not unpredictable.

In the future, the MICP technique would help solve several geotechnical problems. This technique can be used to reduce the liquefaction potential in sand formations, reduce the swelling potential of clayey silt, mitigate wind erosion potential of loose sand deposits, pre-treatment of the subsurface to reduce settlement of highway structures and in slope stabilisation. Despite the challenges faced in field applications at present, this method has tremendous potential in the future. This technique would serve as a cost-effective and environmentally friendly approach in improving the engineering properties of silty soils. Besides, calcite may provide a green construction material as a sealing agent for filling the gaps, cracks and fissures for civil engineering structures. It is also possible to form a hard and thin water-impermeable crust on the sand surface by applying 0.6 g of Ca/cm^2 of the sand surface as cementing material to bind the sand particles to reduce the permeability of the biocemented sand from 10^{-4} to 1.6×10^{-7} m/s (Stabnikov et al., 2011). Studies have been carried out to use biocement for soil erosion control, slope protection, mitigation of internal erosion or liquefaction, reducing fluid flow in soil and sealing the leakage of construction landfill (Dejong et al., 2013; Ivanov and Chu, 2008; Xiao et al., 2018) and to allow repair of cracks formed in the water-barrier layer (Yang et al., 2019). Seepage control is a common construction process for many infrastructure projects such as reservoirs, earth dams, tunnels and other underground constructions. Seepage leads to the largest single loss of water for aquaculture ponds, which may also diminish the nutrients contained in the pond in addition to the water loss. Thus, the soil treatment using the MICP method on loose barren soils of the terrace or Jhum cultivated hillslopes of the state will effectively mitigate the mass movement of soil, landslides and soil erosion. Cement grouting for slope protection is commonly adopted in important civil engineering structures, which may cause soil pollution, and thus, it is not sustainable. Cement is not locally available and is being transported from neighbouring states of Nagaland and Assam, which increases the project cost. Therefore, systematic use of the MICP process to trigger calcite will be an alternative soil stabilisation method. The calcite will bridge the soil particles to increase the shear strength and stiffness of the soil matrix. Besides, calcite materials may clog the soil pores to reduce pore pressure and permeability of the soil. Hence, the precipitation of calcite materials in soil formations may replace cement in soil stabilisation, and slope protection works in the future.

23.7   NATURAL HAZARDS

Landslides and earthquakes are unpredictable natural phenomena and cause damage to property and loss of life in Manipur. These are generally isolated processes that individually may not be very large but can occur with high frequency. Non-consideration of geological and geotechnical factors during construction considerably increases the incidences of landslides along the roads and seismic hazards to property and life. These two hazards are discussed here.

23.7.1   Earthquakes

Manipur is in the zone of the most severe seismic hazard (i.e., zone V). The past seismicity data show that the NE region of India has experienced more than 2,000 earthquakes of magnitude, $M_w \geq 4.0$, spanning a period of 154 years (1866–2020) years. These earthquakes of both crustal and the subduction zone origin are close to Imphal city (<300 km). Almost the entire region around Imphal city is surrounded by faults and contributes to tectonic activity in the region. The earliest documented earthquake was on 10 January 1869, referred to as the '1869 Cachar earthquake' (Oldham, 1882) was of magnitude $M_w = 7.5$ and caused widespread damage in Imphal city. The details on the seismicity of NE India and damages incurred due to past earthquakes in Imphal city have been reported in the literature (Raghukanth et al., 2009; Pallav et al., 2010, 2012, 2015). They have predicted ground motions of Imphal city using a systematic seismic microzonation of Imphal city. The data obtained can be used directly in the seismic design of new engineering constructions. They will also facilitate the evaluation of the seismic risk of the existing industrial and infrastructural facilities. The seismic hazard within the Imphal varies with variation in soil layering and bedrock depth (Raghukanth et al., 2008).

The Manipur earthquake on 4 January 2016, measuring a magnitude, $M_w = 6.7$, on the Richter scale, epicentre at 24°49'48.46"N and 93°39'35.88"E with a focal depth of 54 km, rocked the state. The ground motion felt during the earthquake was strong to very strong with the potential threat of light to medium damage to the engineering structures. The ground motion could have reached a peak acceleration of about $0.18g$ (g=acceleration due to gravity), resulting in a velocity of 15 cm/s, especially in the alluvium of Imphal valley, implying possible major damage in the valley region. The earthquake caused damage only to civil structures, such as Ima Market (Women market at Khwairamband Bazar) and Khuman Lampak ISBT buildings. Investigating the damage of these public buildings indicates major damage to the columns of the structures, many of which burst during the earthquake. The primary cause of failure appears to be vertical loading, as evident from the well-developed conjugate shear fractures in some of the columns.

23.7.2 Landslides

The hilly region on the Senapati–Mao sector is highly landslide-prone. The landslides frequently blocked the NH-39 connecting Assam–Manipur (India) to Myanmar (Kumar and Sanoujam, 2007). Disang and Barail groups of rock represent the area. Considering the importance of NH-39, landslide susceptibility zonation studies along NH-39 between Karong and Mao have been conducted (Chanam et al., 2011). Landslide susceptibility zonation (LSZ) map reveals that 57% of the facets belong to low susceptible zone, 33% to moderate susceptible zone and 9% to high susceptible zone, and 1% to a very low susceptible zone. All the facets of the high susceptible zone are concentrated along the road section of NH-39, especially the Maram–Mao sector. The LSZ map gives an overall picture of the stability condition of the hill slopes. Such a map can be used as a base map for planning any developmental scheme or maintenance of the existing infrastructure.

Proper maintenance of the hillside drain is essential in the hilly terrain. Due care should be given to the extensive slope cutting for the road either during the cutting by following proper scientific slope stability norm or after the cutting by constructing a support wall/structure. As the entire study area shows a fragile vegetative cover, planting more trees will be a means to improve the slope stability condition.

In 2004, a hazardous landslide/mudflow occurred at Phikomei (Mao) and Gopibung along the hilly terrain of NH-39. The landslide at Phikomei destroyed at least 50 houses, and more than 100 houses were severely affected. The mudflow at Gopibung destroyed 10 houses and many hectares of paddy fields. Debris materials of the mudflow consist of clay particles to boulder along with tree trunks. Rocks found along the NH-39 (Imphal–Mao) road section are highly disturbed and fragile due to proximity to the CMT zone. Several joints and fractures also develop in the rocks, and mineralisation zones are also observed along this zone, resulting in a thrust fault. Most of the joints have 50–100 mm spacing having the potential of producing rock fragments. With slight disturbance on these rocky slopes, there is a possibility of slides. Due to this structural complicacy, the rocks are weakened, and a thick column of soil and rock fragments are developed. The heavy rainfall and alternate wetting and drying also lead to developing a thick soil column on the slopes through mechanical and chemical weathering processes. Therefore, the slopes along this highway are prone to landslide and related phenomena, further aggravated by other environmental and anthropogenic factors. The slopes of the study area are found to be near equilibrium condition and more precisely inclined towards unstable condition as safety factor values come slightly below unity. Hence, the majority of the slopes along the road section could be prone to landslides.

Two river systems are draining between Imphal and Mao, namely the Barak River and the Manipur river. Barak River originates from Liyai, a village in the Senapati district, and it confluences with Irang, Makru and Jiri rivers. Due to fluvial activities, such as fluvial down cutting, toe erosion by the river systems, slope cutting for building constructions, quarrying and road construction on the slopes adjacent to NH-39 make the route vulnerable to failure at different points. Practically the moisture contents (33%–35%) of the slope material along NH-39 are not only higher than the plastic limits of the soils but also remain for a prolonged duration during the monsoon period (May–October). When the moisture contents exceed the plastic limit, the soil begins to deform. The deformation is augmented by other factors such as prolonged saturation, slope morphology, environment factor and anthropogenic intervention. As a result, the slope materials will initiate deformation, destabilising the slope to initiate landslide and other phenomena. Such landslides are quite common along the NH-39 (Imphal–Mao) road section during the monsoon period.

The geotechnical properties of the rocks as well as the soils on the slope at Nung Dolan (24°46'8.2"N and 93°18'1.3"E) situated along the national highway, NH-37, were considered to investigate the instability conditions of the slope masses (Khuraijam, 2018). Geological formations exposed in the study area are highly fragile, jointed and structurally complex, belonging to the Surma group mainly made up of sandstone and shale with minor conglomerates. The compressive strength (approx. 90 MPa) of the rock specimen perpendicular to the beddings gives higher strength values (approx. 65 MPa) than parallel beddings. The tensile strength of the rock specimen perpendicular to the beddings (approx. 8 MPa) is more than that of the parallel beddings specimens (approx. 6.50 MPa). The type of rock is found to be massive to thickly bedded sandstone. The RMR developed by Bieniawski (1989) is found to be approximately 58 of fair category belonging to class III (RMR is between 41 and 60) and the slope mass rating (SMR) after Romana (1985) as around 57, which falls under partially unstable condition. The kinematic analysis based on the discontinuity data suggests a wedge type of failure of the landslides.

A geotechnical assessment of landslides along the Jiribam–Imphal rail line, between Barak and Tupul, was undertaken by the Geological Survey of India-North East Region. It has been found that a sudden uncontrolled increase in the discharge of major rivers, including Ijai had caused severe erosion of the rivers and their tributaries, which would pose a danger to nearby villages.

23.8   CASE STUDIES

The physical properties of soil and its implication to slope stability of Nungbi Khunou along NH-150 between Ukhrul and Jessami, Manipur, has been reported by Heisnam et al. (2017). The details are discussed here as a case study. The site is a part of the Mélange zone, which belongs to the Indo–Myanmar Range (IMR). The study area is situated 38 km from Ukhrul town and belongs to the UDF comprising rhythmic intercalation of fine-grained Barail sandstones, Disang shales and siltstones. Rocks are soft, fragile and highly weathered, thereby converted to the soil. A thick column of soil covers the area. As it is a thrusting zone, the rock of the region is highly deformed, jointed and prone to weathering.

During heavy monsoon, soft lithology like shale and mudstone become mud and silt and susceptible to slide. This study determined the physical property of soil and its implication to landslide occurrence. Water plays an important role in triggering landslides and slope failures. An increase in water content reduces the stability of the slope. When the moisture content exceeds plastic limits, the slope begins to deform. Three soil samples were collected from the study area, and the average bulk density, moisture content and specific gravity were found to be 1.577 g/cc, 37% and 2.5, respectively. Atterberg limits are the most distinctive property of fine-grained sediments and may distinguish silts from clays. Plastic limit (w_P), liquid limit (w_L)and shrinkage limit (w_S) values of Nungbi Khunou were 26%, 48% and 9%, respectively. Plasticity index (I_P), consistency index (I_C) and liquidity index (I_L) values were 21%, 0.38% and 60% respectively. From index properties values, the soil is highly plastic, stiff, and semi-solid. The soil sample falls under the CI group in the plasticity chart, indicating organic silt and clay soil with medium compressibility and plasticity. Phase determination and particle size distribution result in very high porosity (52%) and highly saturated (80%) soils that are well graded in nature. A slope and aspect map was prepared from DEM using ArcGIS. The slope is an important contributory factor to landslides, and the slope reported from the sampling area indicates a gentle slope. The aspect refers to the direction of the terrain faces, which is influenced by a component like vegetation, settlement, agriculture, precipitation, wind and so on. The factor of safety ($F_s = 0.405$) calculated using shear stress data is less than unity, indicating an unstable slope. The findings of the study indicate that the area may result in sudden and unpredictable failure due to volumetric changes in the soil.

The soil at the site is characterised by low clay particle size content with medium compressibility and plasticity. Its moisture content exceeds the plastic and shrinkage limits, and it indicates that the soil is in a semi-solid state and tends to deform under its weight. From the particle size distribution curve, the soil is well-graded sand (coefficient of uniformity, $C_u = 8.33$ and coefficient of the curvature, $C_c = 1.763$), and its factor of safety ($F_s = 0.405$) indicates the studied area as prone to landslides. Generally, the moderate slope has a higher frequency of landslide, though the landslide reported from the area indicates a gentle slope. Hence, it can be concluded that the slopes are not the major controlling factor; lithology and structure are the controlling factors for the landslide. Terrace cultivation and resultant waterlogging in the study area cause slide and damage to the road during the monsoons. This study will help undertake mitigation measures and other development activities in the area. Further analyses, including preparation of landslide hazard zonation map, landslide susceptibility map, RMR, SMR and so on, are to be carried out to know the exact problem and provide adequate mitigation measures.

23.9   EFFECTS OF ENVIRONMENTAL CHANGES ON SOILS AND ROCKS

The groundwater table is generally high in different valley areas of Manipur. Saturated macro-pore flow is the dominant hydrological process in tropical and subtropical hilly watersheds of northeast India. The infiltration process into saturated macro-porous soils is primarily controlled by the size, network, density, connectivity, saturation of surrounding soil matrix and depth-wise distribution of macro-pores (Shougrakpam et al., 2010). The Jhum cultivated plots showed disconnected subsoil macro-pores due to frequent soil disturbance during land preparation for cultivation. Figure 23.9 shows a view of a terrace/shifting cultivation site at Leimakhong in the Senapati district of Manipur, as it is being practised on a mass scale in all the hilly terrains of the state.

The Jhum cultivation referred to as the terrace/shifting cultivation is practised in places of convenient slope grades on the medium hill ranges in regular cycles of 5–10 years. It has been seriously affecting the ecological balance of the area. The shifting cultivation and deforestation in gently sloping narrow valleys and strongly sloping

Figure 23.9 A view of a terrace/shifting cultivation site at Leimakhong in the Senapati district of Manipur.

hills cause severe erosion, landslides, flash floods in rivers and mass movement of soil, leading to severe mudflow during heavy rainfalls in monsoon season. The changes in vegetable cover influence the stability of hill slopes. Deforestation leads to desiccation cracking, loss of cohesion and subsurface erosion. During the rainy season, infiltrated water usually saturates the slope materials, decreasing the shear strength and increasing the driving force. Infiltration is further enhanced by the various land use practices such as shifting cultivation on slopes. The shifting cultivation makes the soil more porous that may initiate preferential flows. Some of the highways are characterised by the regular phenomenon of subsidence, which is intimately associated with terrace cultivation (paddy cultivation) in the adjoining areas. Water is retained for about 60–90 days for paddy plantations, which helps in more or less complete saturation of soil column, aggravating the problems of landslide and subsidence. The collection of detailed data of soil macro-porosity from different parts of northeast India may help assess the impacts of different land use, land cover and local crop management practices on soil structures. Different land use and land cover data should give a better understanding of macro-pore dominated flow and transport processes occurring in the region to prevent the probable hazards during monsoon season in Manipur.

23.10 CONCLUDING REMARKS

Based on the details presented in the previous sections, the following general conclusions may be made:

- The Imphal valley, also known as Manipur valley, has an NNW-SSE orientation with a gentle tilting slope towards the south measuring 798 and 746 m above the MSL at the extreme north and the south, respectively.
- The Inceptisols are the dominant soils followed by Ultisols, Entisols and Alfisols, occupying about 38.4%, 36.4% and 23.1% of the total geographical area of Manipur.
- The predominant sediments are clay and a mixture of clay and silt, while silt and sand form the lensoids and bands within the clay. The average thickness of alluvium is at least 150–200 m.
- The nearby roadside area of the Imphal valley mainly consists of residual soils and rocks, whereas ferruginous red soils in hilly regions occupy nine-tenths of the total area of Manipur. Consequently, the CBR value of the valley regions varies between 5% and 8%, while that of the hill soils is in the range of 10%–15%.
- Sand and gravels obtained from river deposits are mainly used for construction works, and the silt content in the sand is considerably high in the range of 5%–10%.
- The rocks are dominant of sedimentary origin. They are light to dark-grey, fine- to medium-grained sandstone interlayer and in some portions, intercalated with shale and intersected by shear seam containing clay and disintegrated rock fragments.
- The natural hazard potential of Manipur has shown the probability of liquefaction as interpreted from the geotechnical investigation and microzonation of Imphal city.
- The problems of working on rugged terrains through deep gorges, high hills and valleys areas for construction of tall-pier bridges and long tunnels to connect

Jiribam (MSL of 37 m) and Imphal (MSL 780 m) of new BG rail line has been highlighted to show the examples of engineering solutions.

- Fresh investigations call for slope protection to prevent slope failures, landslides, mudflow hazards and soil erosion using new sustainable and innovative materials to strengthen and stabilise the soils and rocks in the hilly regions.
- Studies on RMR, rock structure rating (RSR), rock mass quality (Q), RQD, SMR and so on, for all the landslide-prone areas along national highways, show usefulness for the design of infrastructural developments and to avoid slope failure by providing proper protection works.

REFERENCES

Bieniawski, Z.T. (1989).*Engineering Rock Mass Classifications*. Wiley, New York.

Brunnschweiler, R.O. (1974). Indo-Burman ranges. In A.M. Spencer (ed.), *Mesozoic-Cenozoic Orogenic Belts: Data for Orogenic Studies*, Special Publ., Vol. 4, Scottish Academic Press Ltd for the Geol. Soc., London, pp. 279–299.

Chanam, D.S., Kallola, K.B. and Wungmarong, S.R. (2011). Landslide susceptibility along NH-39 between Karong and Mao, Senapati district, Manipur. *J. Geol. Soc. India*. Vol. 78, pp. 559–570.

DeJong, J., Soga, K., Kavazanjian, E., Burns, S., Van Paassen, L., Al Qabany, A., Aydilek, A., Bang, S., Burbank, M. and Caslake, L.F. (2013). Biogeochemical processes and geotechnical applications: Progress, opportunities and challenges. *Geotechnique*. Vol. 63, No. 4, pp. 287–301.

Das gupta, A.B. and Biswas, A.K. (2000). Geology of Assam. *Geological Society of India*, Bangalore, p. 169.

Heisnam, B., Kushwaha, R. S., Mairembam, C. and Moirangthem, O. (2017). Physical properties of soil and its implication to slope stability of NungbiKhunou, NH-150, Manipur. *Int. J. Geosci*. Vol. 8, pp. 1332–1343.

Ivanov, V. and Chu, J. (2008). Applications of microorganisms to geotechnical engineering for bioclogging and biocementation of soil in situ. *Rev. Environ. Sci. Bio/Technol*. Vol. 7, No. 2, pp. 139–153.

Kayal, J. R. (2008). *Microearthquake Seismology and Seismotectonics of South Asia*. Capital Publishing Company, New Delhi.

Kesari, G.K. (2011). Geology and mineral resources of Manipur, Mizoram, Nagaland and Tripura. Geological Survey of India, *Misc. Pub*. Vol. 1, No. 30, Part IV, p. 103.

Khuraijam, M.S. (2018). Slope stability analysis of Nung Dolan Landslide, NH-37, Manipur, NE-India: Geotechnical approach. *Int. J. Eng. Tech. Sci. and Research*. Vol. 5, No. 1, pp. 6–16.

Kumar, A. and Sanoujam, M. (2007). Landslide studies along the national highway (NH-39) in Manipur. *Nat. Hazards*. Vol. 40, pp. 603–614.

Laiba, M.T. (1992). *The Geography of Manipur*. Public Book Store, Imphal.

Nandy, D.R. (2001). *Geodynamics of Northeastern India and the Adjoining Region*. ACB Publications, Kolkata.

Oinam, A. (2007). *Look East Policy and Manipur*. The Sangai Express (27 December 2003), Imphal.

Oldham, T. (1882). The Cachar earthquake of 10 January, 1869, *Memoirs. Geol. Surv. India*. Vol. 19, No. 1, pp. 1–88, Eds. R.D. Oldham.

Pallav, K., Raghukanth, S.T.G. and Konjengbam, D. S. (2010). Surface level ground motion estimation for 1869 Cachar earthquake (M_w 7.5) at Imphal city. *J. Geophys. Eng*. Vol. 7, No. 3, pp. 321–331.

Pallav, K., Raghukanth, S.T.G. and Konjengbam, D. S. (2012). Probabilistic seismic hazard estimation of Manipur, India. *J. Geophys. Eng.* Vol. 9, No. 5, pp. 516–533.

Pallav, K., Raghukanth, S.T.G. and Konjengbam, D. S. (2015). Estimation of seismic site coefficient and seismic microzonation of Imphal City, India, using the probabilistic approach. *Acta Geophys.* Vol. 63, pp. 1339–1367.

PTI (2014). *Construction of world's tallest railway bridge begins in Manipur*, Press Trust of India, https://www.indiatvnews.com/news/india/world-tallest-railway-bridge-manipur-39804.html, accessed 29 July 2014.

Raghukanth, S.T.G., Konjengbam, D.S. and Pallav, K. (2009). Deterministic seismic scenarios of Imphal City. *Pure Appl. Geophys.* Vol. 166, pp. 641–672.

Raghukanth, S.T.G., Sreelatha, S. and Dash, S.K. (2008). Ground motion estimation at Guwahati city for a M_w 8.1 earthquake in the Shillong plateau. *Tectonophysics*. Vol. 448, pp. 98–114.

Romana, M. (1985). New adjustment ratings for application of Bieniawski classification to slopes. In *Proceedings of the International Symposium on the Role of Rock Mechanics in Excavations for Mining and Civil Works*. International Society of Rock Mechanics, Zacatecas, pp. 49–53.

Shougrakpam, S., Sarkar, R. and Dutta, S. (2010). An experimental investigation to characterise soil macroporosity under different land use and land covers of northeast India. *J. Earth Syst. Sci.* Vol. 119, pp. 655–674.

Shougrakpam, S. and Trivedi, A. (2018). Formation of biomineralized calcium carbonate precipitation and its potential to strengthen loose sandy soils. In *Proceedings of China-Europe Conference on Geotechnical Engineering*, 2018. Springer Series in *Geomech. and Geoeng.* Springer Nature, Switzerland, pp. 830–833.

Shougrakpam, S. and Trivedi, A. (2020). Engineering properties of bacterially induced calcite formations. *Curr. Sci.* Vol. 118, No. 7, pp. 1060–1068.

Soibam, I. (2000). Structural and tectonic framework of Manipur. In *Manipur Science Congress*, Manipur University, Imphal, India.

Stabnikov, V., Naeimi, M., Ivanov, V. and Chu, J. (2011) Formation of water-impermeable crust on sand surface using biocement. *Cem. Concr. Res.* Vol. 41, No. 11, pp. 1143–1149.

Trivedi, A. (2015). Computing in-situ strength of rock masses based upon RQD and modified joint factor: Using pressure and damage sensitive constitutive relationship. *J. Rock Mech. Geotech. Eng.* Vol. 7, No. 5, pp. 540–565.

Trivedi, A. (2017). Comments on modulus ratio and joint factor concepts to predict rock mass response. *Rock Mech. Rock Eng.* Vol. 50, No. 5, pp. 1357–1362.

Xiao, P., Liu, H., Xiao, Y., Stuedlein, A.W. and Evans, T.M. (2018). Liquefaction resistance of bio-cemented calcareous sand. *Soil Dyn. Earthquake Eng.* Vol. 107, pp. 9–19.

Yang, Y., Chu, J., Yang, X., Liu, H. and Cheng, L. (2019). Seepage control in sand using bioslurry. *Constr. Build. Mater.* Vol. 212, pp. 342–349.

Yi, Y., Liska, M., Unluer, C. and Al-Tabbaa, A. (2013) Carbonating magnesia for soil stabilization. *Can. Geotech. J.* Vol. 50, No. 8, pp. 899–905.

Chapter 24

Meghalaya

Smrutirekha Sahoo
National Institute of Technology Meghalaya

CONTENTS

24.1   INTRODUCTION

Meghalaya, 'The abode of clouds', was created as an autonomous state within the State of Assam on 2 April 1970. The state lies between 25°4′ N and 26°10′ N latitudes and between 89°48′ E and 92°50′ E longitudes with a total geographical area of 22,429 km^2 (Singh and Tiwari, 2008). The plateau with rolling grasslands interspersed by river valleys forms the main physical feature of Meghalaya. The state can be divided into three natural sectors, the central plateau, the southern border area and the northern border area. The central plateau forms the highest region of the state. It lies between 1,230 and 1,850 m above the mean sea level. It comprises mainly the highlands of Khasi and Jaintia Hills and is more or less situated centrally. The central plateau is the source of all the big rivers of the state. The southern border area has a highly irregular feature. This region begins where the central plateau ends. It is more or less a continuation of the central plateau up to a few kilometres with interruptions here and there caused by sudden drops and depressions. As it recedes further from the central plateau and moves closer to the border of Bangladesh, the sudden drops and depressions become more prominent till they abruptly end in sheer precipices. The northern border area lies to the north of the state and merges with the border districts of Assam. The northern border area continues in features more or less similar to the central plateau in its gradual downward move till it merges with the border of Assam. The plateau of Garo Hills slopes down to the Brahmaputra valley in the north and drops down towards Bangladesh in the south and west. The whole area of the state is full of scenic beauty.

DOI: 10.1201/9781003177159-24

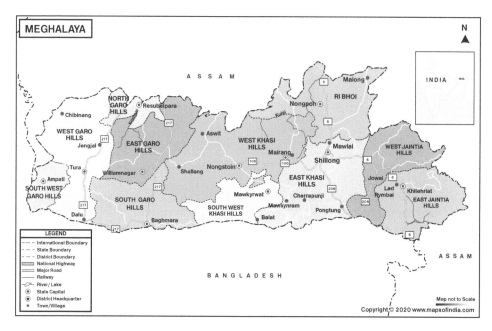

Figure 24.1 General map of Meghalaya indicating districts with key towns. (https://www.mapsofindia.com/maps/meghalaya/, accessed 2 February 2021.)

Waterfalls, lakes, peaks and hills, meadows, valleys and rushing rivers combine to make a rich panorama.

Meghalaya has a monsoon type of climate and is directly influenced by the southwest monsoon originating from the Bay of Bengal and the Arabian Sea. But there are some variations in the climatic variables from place to place depending upon altitude and physiographic differences of the landmass.

The general map of Meghalaya indicating all the districts with key towns and the locations where the detailed subsoil investigation was carried out at Mawlai, Mawiongrim, Shillong is shown in Figure 24.1. The site plan showing the layout of boreholes is shown in Figure 24.2.

. About 90% of the total geographical area of Meghalaya is hilly. Plain land with fertile alluvial soil is located in river valleys in the form of narrow strips in the fringes of the state, that is, in the lower altitude areas to the north, west and south of the plateau region. The topographical view of the typical sites is shown in Figure 24.3.

24.2   MAJOR TYPES OF SOILS AND ROCKS

Meghalaya represents the remnant of the ancient plateau of the Precambrian Indian peninsula. It forms a prominent geomorphic unit stretching across the Garo Hills, Khasi Hills and Jaintia Hills in the east-west direction. Meghalaya consists of five geological formations: (i) Archean gneissic complex with acidic and basic intrusive; (ii) Shillong group of rocks mostly quartzites, usually friable, phyllites, schists, and

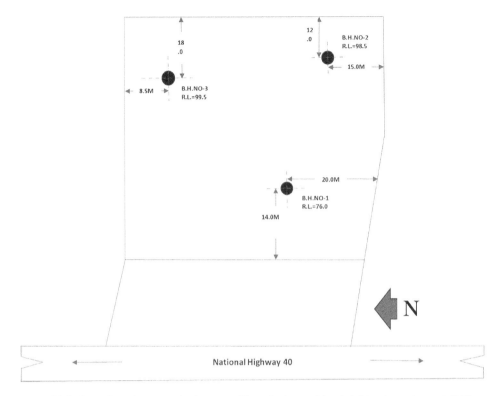

Figure 24.2 Site plan showing the layout of boreholes at Mawlai, Mawiongrim and Shillong (not to scale).

(a)   (b)

Figure 24.3 Topographical views of two typical sites in Meghalaya at (a) Smit and (b) Laitlum.

conglomerates; (iii) Lower Gondwana rocks; (iv) Sylhet traps; and (v) Cretaceous tertiary sediments (Tripathi et al., 1996).

The soil of Meghalaya varies from dark brown to dark reddish-brown in colour.
The depth of soil varies from 50 to 200 cm in different parts of the state with texture
ranging from loamy to fine loamy. The soils are rich in organic carbon with high nitrogen supplying potential but deficient in phosphorus and potassium. Soil reaction varies from acidic (pH 5.0–6.0) to strongly acidic (pH 4.5–5.0) (Singh and Tiwari, 2008)).
Most of the soils occurring on higher altitudes under high rainfall belts are strongly
acidic due to intense leaching. The red loamy soils occupy the entire central part of
Garo Hills and central uplands of Khasi and Jaintia Hills from west to east except for
the valley part of the Simsang River. The soils are generally loamy and red in colour.
Red Loamy soils are the result of weathering of rocks like granite, gneisses, diorites
and others. The Red and Yellow soils are extended parallelly west to east along with
the southern slope of red loamy soils. The soils are generally found in the grade of
fine-textured, ranging from loam to silty loam. The laterite soils are extended from
west to east in the northern part of the state. Most parts of this belt fall under the rain
shadow area. The alluvial soils are found all along the northern, western and southern
fringe of the state. The soil textures in this region vary from sandy to clayey loam with
varying degrees of nitrogen and are very much acidic in character.

24.3   PROPERTIES OF SOILS AND ROCKS

A fairly accurate assessment of the characteristics and engineering properties of the
soils at a site is essential for the proper design and successful construction of any structure at the site. The field and laboratory investigations required to obtain the necessary
data for the soils for this purpose are collectively carried out.

The required geotechnical properties for all the three boreholes located at Mawlai, Mawiongrim, Shillong as shown in Figures 24.1 and 24.2 are depicted in Figures
24.4–24.6. Site exploration programmes with standard penetration tests along with
the required geotechnical investigations reveal a generalised pattern in subsoil strata.
In all these fields as well as laboratory tests, provisions set by the relevant bureau
of Indian standard codes of practice were strictly adhered to. The subsoil in general
consists of a medium-stiff to stiff layer of sandy silty clay with gravel ranging from 1.5
to 3.0 m from the existing ground level and beyond that hard layer, sandy silty clay/
weathered rock/rock fragment and so on are present up to the explored depth. Generally, the consistency of subsoil significantly increases from 3.0 m in the boreholes.
Top weathered rock/gravel at these locations is to be chipped off and properly levelled.

For the proposed extension of the Annexe Building of the Meghalaya High Court,
Shillong, the subsoil investigation consisted of drilling of exploratory boreholes at
three different locations covering the entire area under consideration. The soil samples
were collected from different depths of all three boreholes for laboratory investigations by conducting standard penetration tests. Field investigation at the site consists
of vertical boring in three different locations up to the depth where refusal state occurred ($N > 50$). During this operation, changes in strata, ground water level, visual
identification of soil such as colour and nature were recorded. The disturbed and undisturbed samples were collected at different depths and were properly packed and
sealed after collection.

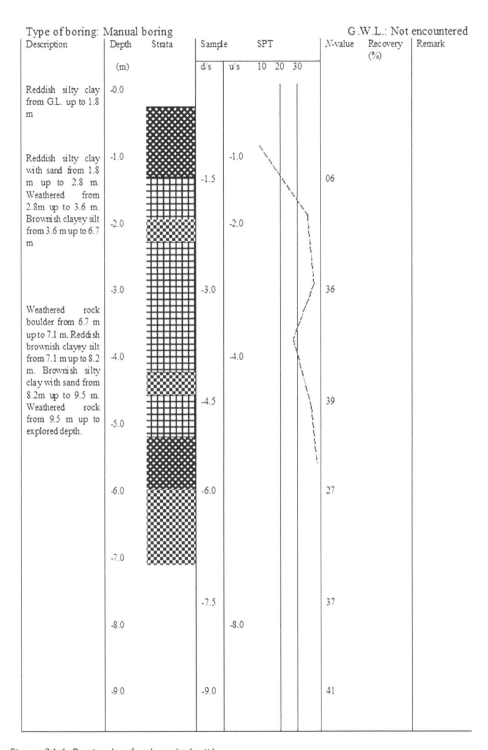

Figure 24.4 Boring log for borehole #1.

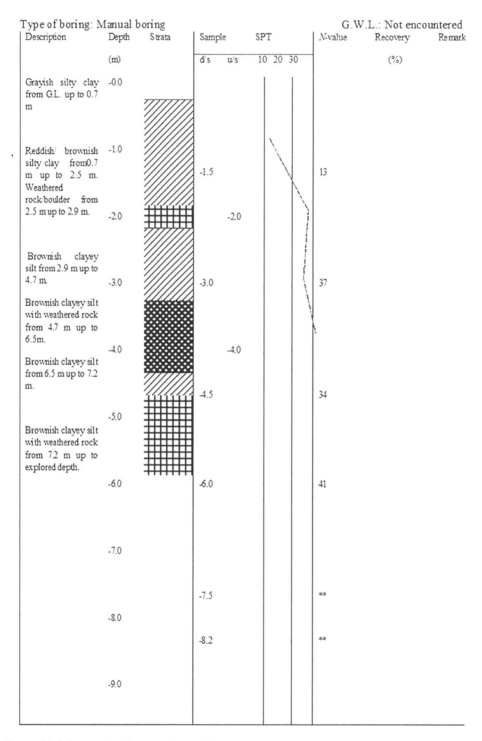

Figure 24.5 Boring log for borehole #2.

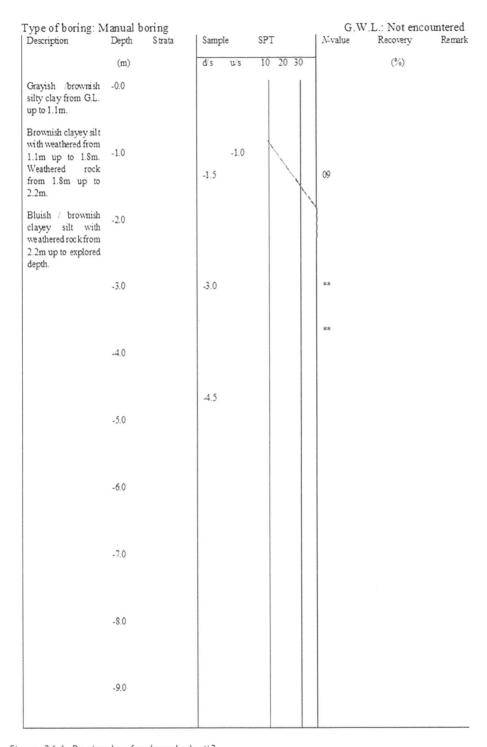

Figure 24.6 Boring log for borehole #3.

Whenever significant changes in soil stratum were encountered, disturbed representative soil samples were collected for laboratory investigation to determine the physical properties of soil such as grain size distribution and specific gravity. The borehole layout of the proposed extension of Annexe Building of the Meghalaya High Court, Shillong is shown schematically in Figure 24.7.

The field investigation was done on 25 October 2017. The field investigation and laboratory test results are reported in Table 24.1. The ground water table has not been encountered during the field investigation. The entire area under consideration was covered with filled-up soil up to a depth of around 3.00 m. The layer of weathered sandy rock is encountered at shallow depth. Hence the foundation was suggested to be designed on normal isolated footing considering a minimum grip length of 1.00 m in the weathered sandy rock layer. All the laboratory investigations were carried out as per the relevant Bureau of Indian Standards.

The soil of Shillong is usually yellowish to deep red in colour except for the soil in polo ground and the area adjoining the riverbanks of Umkhen and Umkhra. The soils are clayey to silty clay in nature. The thickness of the soil varies from less than 15 cm to over 3 m depending upon the slope of the terrain. The soil is derived from the weathering of Shillong Groups of rocks, which generally lack vital elements like potash and phosphorous. The thick vegetative cover incorporates a high percentage of organic matter into the soil. The soil is acidic in nature. Analysis of soil for heavy metals indicates iron content at 18% to the north of Mawiong and northwest of Laitumkhrah.

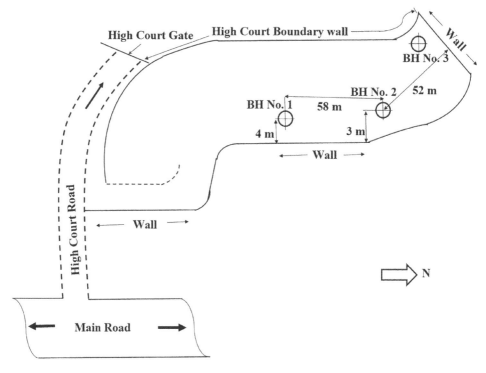

Figure 24.7 Site plan showing the layout of boreholes at the proposed extension of Annexe Building site of the Meghalaya High Court, Shillong (not to scale).

Table 24.1 Soil properties at the proposed extension of Annexe Building of the Meghalaya High Court, Shillong

BH no.	Depth (m)		Observation from SPT test
	From	*To*	
01	0.00	2.00	$N = 02$, Filled-up soil with Gravel, Boulder from GL
	2.00	2.80	$N > 50$ (Refusal), Black/bluish silty clay with little sand
	2.80	$\rightarrow$	Weathered sandy rock
02	0.00	3.80	$N > 50$ (Refusal), Filled-up soil with Gravel, Boulder from GL
	3.80	4.30	$N > 50$ (Refusal), Brownish silty sand
	4.30	$\rightarrow$	Weathered sandy rock
03	0.00	5.00	Filled-up soil with Gravel, Boulder from GL
	5.00	$\rightarrow$	Weathered sandy rock

24.4 CONSTRUCTION MATERIALS

The state is rich in timber resources. The Forest Survey of India Report (FSI, 1990) has identified six types of forest, such as Khasi pine, teak, sal, hardwood mixed with conifers, upland hardwood and miscellaneous based on the availability of economically important tree species. Timber is widely used for the construction of houses and for making furniture. It finds varied uses as construction material. The state is also endowed with abundant sources of granite and other crystallised rocks of different colours and shed (viz. black, pink, grey, etc.).

24.5 FOUNDATIONS AND OTHER GEOTECHNICAL STRUCTURES

Mostly isolated footings are being used as building foundations throughout the state if no special conditions are required and the soil at the site is favourable. This is because predominantly unweathered/weathered rocks are usually encountered at shallow depth in most parts of Meghalaya. The construction works at two of the typical sites viz., one, at Mawlai-Mawiong area at the side of Guwahati–Shillong road, Meghalaya and the other, near Umiam, Meghalaya have been shown in Figures 24.8 and 24.9, respectively.

24.6 OTHER GEOMATERIALS

The State of Meghalaya is a store house of economic minerals. The major minerals which are presently being mined are coal, limestone, sillimanite, clay and kaolin, glass sand, quartz and feldspar. Deposits of these minerals are spread throughout the state. Recently, the presence of uranium deposits was discovered in the southern part of West Khasi Hills and this discovery brings Meghalaya into the uranium map of India. According to the Directorate of Mineral Resources, Government of Meghalaya,

(a)   (b)

Figure 24.8 A typical construction site in the Mawlai-Mawiong area at the side of Guwahati–Shillong road, Meghalaya: (a) view #1 and (b) view #2.

Figure 24.9 A typical construction site beside a landslide area near Umiam, Meghalaya.

maximum limestone reserves are present in the Khasi Hills District, while maximum coal reserves are present in Garo Hills District, whereas the extraction is more in Jaintia Hills District. Jaintia Hills District alone contributes more than 70% of the total coal production of the state. The quality of limestone found in the state varies from cement grade to chemical grade. These minerals are utilised in several mineral-based industries in the state as well as in the country.

Limestone is a major mineral that occurs in an extensive belt (approx. 200 km. long) along the southern border of Meghalaya. Limestone found in the state is fine-grained, hard, compact and fossiliferous with little variation in colour from light to dark grey. The total inferred reserve of limestone in the state is about 5,000 million tonnes. The CaO content in the limestone of Meghalaya may be found up to 53% and can be of great use to the steel, fertiliser and chemical industries.

Coal deposits can be found in all districts and particularly in the southern slopes of the State. The coal of Meghalaya is characterised by low ash content, high calorific value (ranges between 6,500 and 7,500 K Cal/kg) and high sulphur content. The total estimated inferred reserve of coal in the state is about 563.5 million tonnes. The coal is mainly of sub-bituminous type and can be utilised for various purposes. The coalfields of the Jaintia Hills are small and spread out in different patches.

Jaintia Hills district has a total coal deposit of about 40 million tonnes, which is only % of the total coal deposits of the state but produces more than 70% of the total coal produced in the state.

Clay of various types such as kaolin (China clay), white clay and fire clay is also found in various parts of the state. The clay found in the state is suitable for ceramic, paper, rubber and refractory industries. It has been estimated that there are a few hundred million tonnes of clay reserves in the state.

Sillimanite is one of the important minerals found in the West Khasi Hills district of Meghalaya. This mineral is one of the best in the world. The state is the leading producer of this mineral and 95% of India's Silimanite comes from the state. It occurs predominantly in the Mawshynrut (Sonapahar) region of the West Khasi Hills District. In addition to these, other economically viable minerals like uranium, glass sand, quartz and feldspar are also found in the state. It is to be noted that in terms of the size of the estimated reserves, the most important mineral of the state is limestone, followed by coal, clay, kaolin, glass sand, feldspar and sillimanite. One of the essential functions for the assessment of geological and natural resources is a continuous process of exploration which would include geological mapping, core drilling and exploratory mining. This work is mainly being carried out by the Geological Survey of India, Government of India and the Directorate of Mineral Resources, Government of Meghalaya.

24.7 NATURAL HAZARDS

Seismically, the Shillong plateau falls in zone V of the 'Seismic Zoning' map of India. In the Shillong plateau, two large earthquakes of magnitude≥7.0 and 25 earthquakes of magnitude≥4.5 have occurred during the last 100 years.

Rainfall is one of the most important factors which trigger landslides in the region. Almost every year during rainy seasons, the NH-44 gets blocked for some time due to heavy slides and disturb road communications. Two of the most recent landslide pictures both adjacent to Shillong bypass road are shown in Figure 24.10. Figure 24.11 presents the remedial measures undertaken with the help of wire mesh with nailing to cover up the falling debris because of landslides adjacent to the Guwahati–Shillong road.

A major landslide occurred in 1988 popularly known as the Sonapur landslide (Rao and Singh, 2008). The Sonapur landslide is an old active, huge rock-cum-debris slide, which affected the traffic movement due to the accumulation of slide debris on the road brought by a steeply descending nala from hill slope along with rainwater.

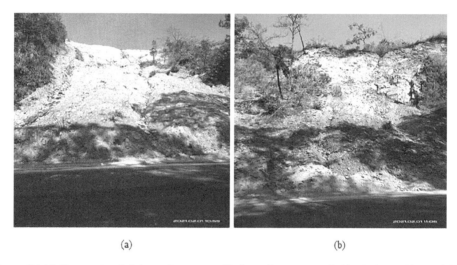

(a)   (b)

Figure 24.10 Recent landslides adjacent to Shillong bypass road, Meghalaya: (a) landslide
site #1 and (b) landslide site #2.

(a)   (b)

Figure 24.11 Remedial measures are undertaken to cover up the falling debris because of
landslides adjacent to Guwahati–Shillong road, Meghalaya: (a) view #1 and
(b) view #2.

The slope washed debris is composed of blocks of heterogeneous material viz. shale, siltstone and sandstone which blocked the road to a height of 15 m at the slide zone. But during the monsoon of 1999, the slide became active and caused major disruption to road communication thereafter every year.

A huge landslide again occurred in June 2001 and the slide debris washed away a bus, a truck and an oil tanker into the Lubha River. The history of Sonapur land-slide occurrences indicates that almost all landslides took place between June and September during the peak of the monsoon period. Various visual observations of the Sonapur landslide from road level are presented in the study of Rao and Singh (2008).

The main causative factors for the slope instability in this area are weak geological formation, discontinuity pattern of joints and their spacing, fractures and faults, steep slope of fie hill and heavy rainfall. Efforts have been made in the past to control the landslides by adopting suitable remedial measures.

24.8   CASE STUDIES AND FIELD TESTS

24.8.1   Case study #1

Geotechnical evaluation of the surface and subsurface data has been carried out for an FCI Godown area located south of the Dakopgre–Arai Mile public road and on a southerly sloping ground with elevation varying from EL±343 to EL±325 m (Sharma, 2011). It was constructed at the proposed site by slightly shifting towards the western side to avoid disturbed soil in the eastern end. A wall surrounding the building was constructed to prevent the erosion and movement of the foundation soil with time as the ground mass was of sloping nature. A suitable drainage arrangement has been provided to drain out the seepage water without the removal of the foundation soil.

Geotechnical investigation of the land area, including its slope stability and foundation for the proposed construction of 2,500 metric tonne capacity food grain godown in the FCI Complex, Tura was carried out between 23 June and 25 June 2010. In the course of the geosurvey work, including preparation of contour plan in 1:500 scale of the complex was done with GSI surveyor. The entire FCI Godown area is lo-cated south of the Dakopgre–Arai Mile public road and on a southerly sloping ground with elevation varying from EL±343 to EL±325 m. The site is about 240 km by tar-topped road from Guwahati and about 102 km from Krishnai on NH 37. The area has an undulated topography. The investigation area (N 25° 30′ 48.8″ /E 90° 11′ 19.5″) forms the top portion of the southeastern slope of a hillock. It is about 101 km to Tura, the district H.Q. of West Garo Hills District, Meghalaya along NH 51 from its crossing with NH 37. The area falls in seismic zone V as per the GSI seismotectonic atlas. In and around the proposed godown site, no rock exposure was seen. The structure was to be constructed on soil/overburdened material. The soil in the area is composed of highly oxidised reddish-brown clayey silt at the top followed by cherry red angular cobble to pebble size sandstone pieces. However, exposures of medium-grained, ferruginous sub-horizontal sandstone beds belonging to the Tura Formation of the Jaintia Group are encountered along road cuttings some 800 m away.

During the ground survey, demarcation of the disturbed area was made upon sub-sidence of soil leading to failure of earlier retaining walls. During the field work, areas

Table 24.2 Geotechnical properties of collected soil

Depth (m)	Visual identification	Sp. Gr	Clay (%) < 0.002 mm	Silt (%) from 0.002 to 0.075 mm	Sand (%) 0.075– 4.75 mm	Gravel (%) > 4.75 mm	Angle of internal friction (°)
0.30	Disturbed	2.67	25.80	1.97	1.566	0.34	10
0.60	Disturbed	2.6555	28.30	1.94	1.512	0.28	9
0.90	Disturbed	2.65	27.47	1.94	1.522	0.29	9

of stable soil and unstable soil were recorded and subsequently depicted on the geological plan. It was also ascertained where in-situ soil lies and the area where slopes have been filled/dumped with extraneous soil. The land mass comprises highly oxidised clayey silt with sand. Depending on the stability, the area was divided into (i) disturbed and filled soil and (ii) undisturbed soil masses. To have an idea of the nature and condition of the soil, it was pertinent to have analytical values of its properties. Hence, disturbed soil samples were scooped out (from the wall of the pit dug) from three horizons at 30–50, 60–80 and 90–110 cm from the proposed FCI Godown construction site and packed in double polythene bags, sealed (airtight) and labelled for analysis. These samples were analysed by the project authorities from M/s North East Engineers, Guwahati for laboratory tests. The firm opined that as density properties cannot be derived from loose (disturbed) soil samples, they remoulded them for shear tests taking natural moisture content and moderate compaction. The geotechnical properties of the collected soil samples are presented in Table 24.2.

24.8.2   Case study #2

The Ganol Stage-I Small Hydroelectric Project is located about 7 km north of Tura, West Garo Hills District (Kumar, 2014). The project envisages the construction of a 35 m high concrete gravity dam across the Ganol River to generate 22.5 (7.5 X 3) MW of power potential by utilising a head of 148 m. It has a surface powerhouse and a 2.06 km long headrace tunnel. The area forms the western part of the Meghalaya plateau and comprises low hills drained by small to medium rain-fed rivers. The project site is located in seismic zone V and records of past seismic events (Period: 1901–1993) as available in SEISAT, 2000 map indicate that earthquakes up to 5.9 magnitudes (with epicentre up to 40.0 km depth) have occurred in the 50 km and 100 km radii around the project site. Hence all the project components were designed considering the seismicity of the area. The project is located within the Archaean gneissic complex comprising granite gneiss, migmatites frequented by Ptygmatic folding, intrusives of K-feldspar-quartz veinlets, K-feldspar granites, irregular intrusive amphibolites. Mostly the rock outcrops are exposed in river sections and the hill slopes are occupied by residual soil, the debris material and colluviums supporting thick forest growth. Detailed geotechnical investigations in this project have been done by the GSI (2009). Two geological L sections: (i) 672 m long along the entire length of penstock alignment on 1:600 and (ii) the second 175 m long section was prepared for finalisation of surge shaft location. Since the depth of overburden along penstock slopes was higher, of the order of more than 20 m, therefore, anchor blocks (pillars) were properly founded on firm ground, after determining the soil properties. Each pillar was erected after analysing slope stability and with a broader foundation, narrowing down step by step to the desired dimension near the surface. The finalised surge shaft has a lateral cover of 55.53 m and a vertical cover of 25 m at the centre of the surge shaft.

24.9   GEO-ENVIRONMENTAL IMPACT ON SOILS AND ROCKS

Soils on soft rock and loose-mantled hill slopes are inherently weakly structured and at risk of large-scale erosion, including slope failure. Vegetation that grows on hill slopes plays an important role in binding the soil particles, minimising soil erosion and hence stabilises the hill slope. The mechanical stabilisation of soil slopes by means of tree roots depends largely on the strength properties of the roots and their growth pattern within the soil. Soil particles on hill slopes tend to be unstable and become easily mobilised when disturbed or exposed. But there are very few cases of erosion and landslide in the state as compared to other regions. This may be attributed to the protective role of forest in the stabilisation of the hill slope and controlling the loss of soil. Most of the landslides and erosion-prone areas in the state are due to human activities like unscientific cutting of hill slopes for various developmental activities, mining, deforestation and shifting cultivation.

24.10   CONCLUDING REMARKS

The terrain of Meghalaya is mostly hilly. This northeastern part of India has a monsoon type of climate and receives the highest rainfall in the country. Seismically, the entire state falls in zone V of the 'Seismic Zoning' map of India. Based on the observations made throughout the chapter, it can be concluded with the following remarks as follows:

i.   The soils of the state are generally loamy, red in colour and fine-textured. Some part of Meghalaya (i.e., from west to east in the northern part of the state) consists of the laterite soils and a part of the state (i.e., the northern, western and southern fringe of the state) is covered with the alluvial soils.

ii.   Meghalaya comprises five different geological formations consisting mostly of gneisses, quartzites, phyllites, schists, conglomerates, lower Gondwana rocks, Sylhet traps and Cretaceous tertiary sediments.

iii. The subsoil in general consists of a medium-stiff to stiff layer of sandy silty clay with gravel ranging from 1.5 to 3.0 m from the existing ground level and beyond that hard layer, sandy, silty clay/weathered rock/rock fragments and so on are present up to a great depth.

iv.   One should take utmost care while planning, designing and/or implementing civil engineering facilities in this zone considering both seismic and heaviest rainfall conditions. The civil engineering structures must be made flexible enough to resist the seismic forces and at the same time, a proper drainage system should be provided to all facilities.

v. Suitable remedial measures may be taken by using locally available materials such as treated timber resources and/or bamboo as these are sufficiently available in this area.

vi. To strengthen the unstable slopes and to minimise the soil erosion to avoid landslides, vegetation having good root strength and required growth pattern may be grown on hill slopes. Suitable adaptive measures may be taken by wrapping up the loose-mantled hill slopes which are inherently weakly structured and at risk of large-scale erosion, including slope failure so that they will not be directly exposed to heavy rainfall. Unscientific cutting of hill slopes for various developmental activities, mining, deforestation, shifting cultivation and so on must strictly be avoided.

REFERENCES

FSI (1990). *Forest Resources Survey of Meghalaya: State-Inventory Results*. Forest Survey of India, New Delhi, India.

GSI (2009). Geology and mineral resources of Meghalaya, Miscellaneous publication, No. 30, Part IV, Vol. 2(ii), Geological Survey of India, North Eastern Region, Shillong, India.

https://www.mapsofindia.com/maps/meghalaya/, Maps of India, accessed 2 February 2021.

Kumar, V. (2014). Report on construction stage geotechnical investigations of Ganol stage I small hydroelectric project, Tura, West Garo Hills, Meghalaya, Engineering Geology Division, Geological Survey of India, Mission- IV, North Eastern Region, Shillong, India.

Rao, K.S. and Singh, C.D. (2008). Final report on site specific studies of Sonapur landslide, Jaintia Hills District, Meghalaya, T.S. No. 83 C/8, Item Code No. Env/rhq/2004/011, Acc. No. 2950, Landslide Hazard Zonation Project, North-Eastern Region, Shillong, India.

Sharma, S. (2011). Geotechnical report on the assessment of slope stability and foundation for construction of a 2500 metric food grain godown at FCI complex, Tura, West, Garo Hills District, Meghalaya, Engineering Geology Division, Geological Survey of India, North Eastern Region, Shillong, India.

Singh, O.P. and Tiwari, B.K. (2008). Environmental accounting of natural resources of Meghalaya: Land and forest resources, Central Statistical Organisation, Ministry of Statistics & Programme Implementation, Government of India, New Delhi.

Tripathi, R.S., Pandey, H.N. and Tiwari, B.K. (1996). *State of Environment of Meghalaya*. North Eastern Hill University, Shillong, India.

Chapter 25

Mizoram

Naveen Kumar, Prashant Navalakha, and Ashish D. Gharpure
Genstru Consultants Pvt. Ltd.

CONTENTS

25.1 INTRODUCTION

Mizoram is one of the eight sister states of northeast India. It is located in the southern part of northeast India and has international borders with Myanmar on the east and Bangladesh on the west. The neighbouring Indian states sharing borders with Mizoram are Manipur and Assam on the north and Tripura on the west. Mizoram has eight districts with Aizawl being the state capital. A map of the state has been presented in Figure 25.1.

The state has a hilly terrain and is geologically young. The state has approximately N-S trending steep mostly anticlinal, parallel to subparallel hill ranges and narrow adjoining synclinal valleys with parallel hillocks or topographic highs (Figure 25.2). They have been developed as a result of east-west compression due to the movement of the Indian plate towards the Burmese plate. The hill ranges are gentler and peaks are of lower elevation on the northern part of the state and become steeper and higher towards the southern and eastern parts. All major rivers in the state are north-south flowing.

The hill state in general has relatively thin soil cover. The top stratum in hill slope is majorly composed of a thin layer of top soil followed by decomposed completely weathered rock. The parent rock in the region is a sedimentary rock and the commonly found types are sandstone, siltstone and shale. The parent sedimentary rock is highly fragile, young and prone to quick weathering. Since the hill slope consists of weak sedimentary rock and due to heavy rainfall in the region, frequent landslides occur in the state. The occurrence of landslides is getting aggravated due to human activities, which include unscientific constructional activities, jhum cultivation and deforestation. The region receives heavy rainfall (>2,000 mm/year) and falls in the seismic zone V.

DOI: 10.1201/9781003177159-25

Figure 25.1 Map of Mizoram (SI 2020).

25.2 MAJOR TYPES OF SOILS AND ROCKS

The soil cover in the region is in general shallow. In the major part of the region, a rocky stratum has been found near the ground surface. The major soil types in the region are sand, silt and clay, which are formed by in-place weathering of parent rock, i.e. sandstone, siltstone and shale.

Figure 25.2 Drone image showing hilly terrain of Mizoram. (Courtesy of Genstru Consultants Pvt. Ltd., Pune, India.)

The rocks present in the state are sedimentary. The major lithological unit exposed in the state is rocks of Bhuban and Bokabil formation, both of which fall under the Surma group. The Barail group of rocks is present in some parts of eastern Mizoram. The Bhuban formation mainly comprises argillaceous rock while the Bokabil formation comprises mainly arenaceous rocks. The upper layers of rock mass are dominantly arenaceous, while the lower layers are dominantly argillaceous. The rock mass in general is present in an interbedded form as seen in Figure 25.3, that is, different rock types, sandstone, siltstone and shale are present in layered form. Siltstone and shale are also found in intercalated/interlaminated form within sandstone mass. Sandstone is mostly found to be consisting of fine-grained sand particles along with a small percentage of silt and clay.

Figure 25.3 Image showing interbedded rock strata. (Courtesy of Genstru Consultants Pvt. Ltd., Pune, India.)

The soil in the region is in general yellowish to brownish in colour. The rock mass in the upper layers is khaki, buff, yellowish grey to light grey in colour and light grey to dark grey in lower layers. The discolouration in the top layers is interpreted to be due to weathering.

The bedding plane of the sedimentary rock in most parts of the region dips towards ENE. At few locations, the dip direction of the bedding plane is also noted towards WSW. The dip amount of the bedding plane in general is observed to vary between 15° and 40°. The dip amount of the bedding plane at some places was even observed to be as steep as 70° as noticed during the field survey carried out by the authors for projects in the state. The joints in sandstone rock mass are generally not much visible. The bedding joints in siltstone and shale are clear and usually closely spaced. In some layers, which are probably shear zones, very thin sheets with appreciably thick clayey infilling have been observed in this rockmass. The rock mass rating values evaluated for the rockmass for tunnelling majorly fall under the poor to very poor category (NFR 2016a, 2016b).

25.3   PROPERTIES OF SOILS AND ROCKS

Major types of soils found in the state of Mizoram are silty sand, sandy silt and silty clay. The sandy soils are found to consist of fine to medium-grained particles. The sandy silt soils have appreciable fine sand content and are non-plastic to low plastic. Soils with major clay content are observed to be of mostly medium to high plastic. Some of the properties of soil are given in Table 25.1. The soil in the region has majorly formed as a result of weathering of parent rocks, which are sandstone, siltstone and shales. Thus, the soil in the region is mostly a mixture of fine sand, silt and clay. The values of properties presented in this section have been collected from the geotechnical investigation reports conducted for bridges and tunnels for the new Broad Gauge rail line from Bairabi to Sairang (NFR 2012, 2014, 2015, 2016a–2016d) and those conducted by the Geological Survey of India (GSI) for two of the hydropower projects and dams in Mizoram (GSI 2000, 2005).

As discussed in the earlier section, the rock types present in the region are sandstone, siltstone and shale, which are all sedimentary rocks. These rocks are soft and weak in nature. The average unconfined compressive strength of slightly weathered to fresh rock is in the range of 10–15 MPa. The range of compressive test values for different rocks in the region is presented in Table 25.2. Due to this weak nature of rock, it is difficult to get the core recovery in highly to moderately weathered rock mass.

Table 25.1 Range of soil properties

Type of soil	Gradation		Atterberg limit		Shear strength (DS/ Triaxial)	
	% Coarse	% Fines	LL	PL	c (kPa)	ϕ (°)
Silty sand	60–85	15–40	NP		10–20	30–32
Sandy silt	30–40	60–70	NP–35	NP–20	10–20	26–30
Silty clay	20–30	70–80	35–50	15–25	30–60	5–10

Table 25.2 Range of compressive test values for rock

Type of rock	UCS (MPa)	Point load index (MPa)
Sandstone	1–17	0.1–0.5
Siltstone	1–8	0.1–0.3
Shale	1–19	0.1–0.6

Moreover, due to closely spaced bedding plane joints, the rock quality designation obtained in this rockmass is generally low, even with a high core recovery. In some cases, the rock cores get sheared off along the bedding plane with slight hand pressure. In general, it is observed that slightly weathered to fresh rock is encountered in most of the locations from a depth of about 20 m from the ground level. Based on the grain size distribution, the rocks vary as silty sandstone, shaly siltstone and silty shales. The maximum depth of borehole available is up to 162.5 m, which has been carried out for one of the railway tunnels (T1 P1).

Permeability tests have been carried out by the double packer method for an average thickness of 2 m. The tests have been carried out at the depth from the ground level varying from 10 to 50 m. A range of permeability in different rock types in the state are presented in Table 25.3. In the field, it has been observed from the subsurface flow seeping out from the face of cut slopes that the flow is majorly through sandstone layers. The subsurface flow through siltstone and shale has rarely been observed. The subsurface flow generally emanates out from the interface of sandstone and siltstone/shale layer. The permeability of rock mass is affected by the joint conditions as well and hence it can be interpreted from the test results that the siltstone and shale in the location of tests are having open joints. Another factor that would have affected the permeability test is the possibility of the presence of sandstone layers within the siltstone and shale beds, which contributes to a higher flow through the rock mass.

The shale rock in general falls under the category of weak and soft rock. Shales undergo quick weathering on exposure to the atmosphere and water. Thin flakes and chips can be observed on the exposed surface of excavated cut slopes in shale as seen in Figure 25.4. The amount of weathering differs from one location to another. During rainfall, the surface runoff brings down the weathered chips deposited on the surface and deposits at the toe of the cut slopes. This has caused hindrance to smooth operation of many of the infrastructures in northeast India, particularly railways. It has been reported from various studies across the world that the weathering depends upon the amount of clay content and type of minerals in soft rocks such as marls, shales, claystones and flysch. One such study is by Miscevic and Vlastelica (2014), wherein weathering in marl rock and its effect on slope stability are discussed. Weathering also

Table 25.3 Range of field permeability test values for rock

Type of rock	Permeability (Lugeon)
Sandstone	5–20
Siltstone & Shale	10–25

Figure 25.4 A view of cut slope along new BG line under construction. (Courtesy of Genstru Consultants Pvt. Ltd., Pune, India.)

Table 25.4 Range of slake durability test values for rock

Type of rock	Slake durability	
	Index (%)	Classification
Sandstone	60–90	medium to high
Siltstone	70–85	medium to high
Shale	70–80	medium to high

occurs due to repeated wetting and drying processes. A slake durability test is carried out to measure the durability of rock against weathering. From the available data as presented in Table 25.4, it can be seen that the rocks in the region in general fall under medium to highly durable.

25.4 USE OF SOILS AND ROCKS AS CONSTRUCTION MATERIALS

The soil in the region consists of a high percentage of fines as it is formed by weathering of the fine-grained parent rock. The rock available is of low strength and hardness. Thus, the soil and rock available in the state are mostly not suitable to be used as construction materials. Construction materials such as coarse aggregates for concrete and metal for the road are brought from other neighbouring states. Good quality aggregates are brought from the state of Meghalaya. Sand is brought either from Meghalaya or from the Barak basin in lower Assam. Pakur sand from Meghalaya is known to be of better quality among the available sands in the region. Moderately hard sandstone and siltstone or other rock intrusions available in small pockets are quarried in the state for local use such as the construction of masonry walls, local roads, low-rise buildings and so on.

25.5 NATURAL HAZARDS

One of the major hazards in the state is landslides. Most of the reported landslides are caused by anthropogenic activities like the construction of buildings and hill cutting for the development of infrastructures. One of the major landslides that occurred is in Aizawl, the capital town of the state. The images of the Laiputlang landslide that occurred in September 2014 in Aizawl are shown in Figures 25.5 and 25.6. It has been reported that the area consisted of massive sandstone and shale beds with sandstone layers overlying the shale rock mass (Verma 2014). The dip of the layers varied from 40° to 70°. Subvertical joints were also observed in the rock mass. Some of the buildings in this location were founded on shale rock and up to five-storeyed buildings were being constructed on this hill slope. The major causes of landslides have been noted as heavy rainfall during September and load on the shale rock due to the buildings on the hill slope.

Figure 25.5 A view from toe of the landslide. (Courtesy of Genstru Consultants Pvt. Ltd., Pune, India.)

Figure 25.6 A view of a landslide from a farther distance. (Courtesy of Genstru Consultants Pvt. Ltd., Pune, India.)

25.6   CONCLUDING REMARKS

In this chapter, the major soil and rock types present in the state of Mizoram and their properties have been highlighted. The lithology mainly comprises young sedimentary rock formations, which are weak and soft in nature. The rock types present, particularly shale, is highly prone to quick weathering on exposure to water and the atmosphere. Most of the landslides and foundation issues have reported the presence of shales as one of the causative factors. Thus, a careful study needs to be carried out while designing foundation and cut slopes in the hilly region, particularly when there is a presence of shale.

REFERENCES

GSI (2000). Progress Report No. 4 on The Preconstruction Stage Geotechnical Investigation of Turial Hydro Electric Project, Aizawl District, Mizoram, NER Shillong, Geological Survey of India.

GSI (2005). A Report on The Preliminary Stage Geotechnical Investigation of Kolodyne Hydroelectric Project (Stage II), Chimtuipui District, Mizoram, NER Shillong, Geological Survey of India.

Miscevic, P. and Vlastelica, G. (2014). Impact of weathering on slope stability in soft rock mass. *Journal of Rock Mechanics and Geotechnical Engineering*, Vol. 6, Issue 3, pp. 240–250.

NFR (2012). Report on Geotechnical Investigation for Bridge No. 21 of New BG Line from Bairabi to Sairang, Office of Dy. Chief Engineer Silchar/CON/BS, Northeast Frontier Railway (NFR), Silchar, Assam.

NFR (2014). Report on Geotechnical Investigation Work for Bridge No. 196 at Ch. 48.85 km in between Kawnpui – Sairang Section of NF Railway in Mizoram, Office of Dy. Chief Engineer Silchar/CON/BS, Northeast Frontier Railway, Silchar, Assam.

NFR (2015). A Report on Geotechnical and Geological Investigation for Slope Stability Analysis of Slope in Embankment between 0 – 15.3 km in Connection with Bairabi – Sairang New BG Line Project, Office of Dy. Chief Engineer Silchar/CON/BS, Northeast Frontier Railway, Silchar, Assam.

NFR (2016a). Report on Geotechnical Investigation Reports for Tunnels T1 to T5 between Ch. 0-30 km for Bairabai – Sairang New BG Line Project in Mizoram, Northeast Frontier Railway, Office of Dy. Chief Engineer Silchar/CON/BS, Northeast Frontier Railway, Silchar, Assam.

NFR (2016b). Report on Geotechnical Investigation Reports for Tunnels T18 to T23 between Ch. 30-51.3 km for Bairabai – Sairang New BG Line Project in Mizoram, Northeast Frontier Railway, Office of Dy. Chief Engineer Aizawl/CON/BS, Northeast Frontier Railway, Aizawl, Mizoram.

NFR (2016c). Report on Geotechnical Investigation Work for Bridge No. 104 at Ch. 33172.290 km in Connection with Proposed Construction of New BG Rail Line between Bairabi – Sairang Section of NF Railway in Mizoram, Office of Dy. Chief Engineer Silchar/CON/BS, Northeast Frontier Railway, Silchar, Assam.

NFR (2016d). Report on Geotechnical Investigation Work for Bridge No. 115 in Connection with Proposed Construction of New BG Rail Line between Bairabi – Sairang Section of NF Railway in Mizoram, Office of Dy. Chief Engineer Silchar/CON/BS, Northeast Frontier Railway, Silchar, Assam.

SI (2020). Map of Mizoram, Survey of India, Department of Science and Technology, Government of India. http://www.surveyofindia.gov.in/pages/political-map-of-india, accessed 29 Dec 2020.

Verma, R. (2014 June). Landslide hazard in Mizoram: Case study of Laipuitlang landslide, Aizawl. *International Journal of Science and Research (IJSR)*, Vol. 3, Issue 6, pp. 2262–2266.

Chapter 26

Nagaland

Raju Sarkar, Devashish Gupta, and Aman Pawar
Delhi Technological University

Sunil Saha
University of Gour Banga

Chandan Ghosh
National Institute of Disaster Management, Ministry of Home Affairs New Delhi

CONTENTS

26.1 INTRODUCTION

Nagaland is a state of northeast India. It is a part of the NER (northeastern region), a region that constitutes a total of eight northeastern states (Figure 26.1). The majority of people in the state belong to the Naga tribe, with diverse dialects and cultural characteristics. It is a predominantly rural state, with 82.26% of the people residing in small villages, therefore conserving the natural flora and fauna of the area. Geographically Nagaland extends between 93°15′ E and 95° 6′E longitudes and between 25°10′ N and 27°4′ N latitudes. Its international borders are shared with Myanmar in the eastern part and are bounded by the Indian states, namely, Arunachal Pradesh in the northeast, Manipur in the south and Assam in the west. The capital of Nagaland is Kohima, located in the southern part of Nagaland. Dimapur, the most densely populated city of the state, lies 70 km northwest of the capital city Kohima. It is also known as the commercial centre of Nagaland. There are a total of ten national highways in the state that

DOI: 10.1201/9781003177159-26

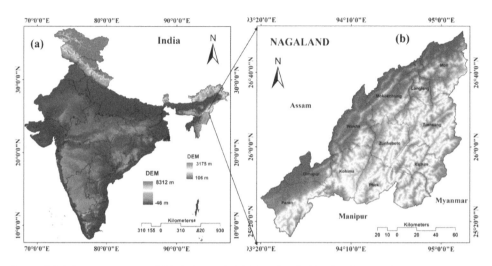

Figure 26.1 Location map of (a) India and (b) Nagaland. (Author's own figure.)

covers a road length of 1,150 km. The annual rainfall ranges from 1,800 to 2,500 mm and the annual temperature varies from 21°C to 40°C. Barring the foothills bordering the Assam plains, the state is almost exclusively hilly in nature. As far as the physiography is concerned, Nagaland lies in the Indian subcontinent's Purvanchal zone.

The terrain can be broadly grouped into four topographic units, from a geomorphological point of view, which includes the alluvial plains which are 150–200 m, low to moderate linear hills which are 200–500 m, moderate hills which are 500–800 m and high hills which are 800 m above the MSL (mean sea level). There are five main rivers that flow through the state and they are Doyang, Dhansiri, Melak, Dikhu and Tizu. There are narrow valleys of the rivers and the adjoining streams, the varying climate and the dense forest cover in the state nurture a plethora of habitats, supporting rich flora and fauna, which are indigenous to the region. The soils of Nagaland are formed from tertiary rocks belonging to the sequence Barail and Disang. People's economy is entirely dependent on agriculture but soil productivity and its resulting agricultural production are alarmingly decreased due to the misuse of land. The practice of shifting agriculture and deforestation is one of the main causes of soil degradation in the state.

26.2 MAJOR TYPES OF SOILS AND ROCKS

Inceptisol, Ultisols, Entisols and Alfisols are the four major soil orders to be observed in the state as per the National Bureau of Soil Survey and Land Use Planning. Their details are as follows (source citation as used):

1. Inceptisols – These cover 66% of the land area of the state. Fine clayey, clayey loamy and fine loamy clay are the textural classes of Inceptisols soil. Inceptisols are Predominant near river beds.

2. Ultisols – These soils cover 23.8% of the land area of the state and are clayey in texture, predominant in regions that receive high rainfall, mainly forest areas.
3. Entisols – They cover 7.3% of the land area of the state. These soils are fine loamy in texture and predominant in the north and NERs.
4. Alfisols – They cover 2.9% of the land area of the state. These soils are predominant near the border regions with Assam. These soils are fine loamy in texture, of light colour and mineral rich.

Nagaland can broadly be divided into four zones based on its geological index.

1. Belt of Schuppen – The Schuppen Belt is said to follow the boundary of the Assam valley. In Naga Hills, it covers the portion between the Naga and Disang thrusts. It is a priority area for petroleum exploration.
2. Disang Belt – Disang is a tributary of Dihing (Bhramaputra River) in its southern bank. Hence, it is not very geologically active in terms of mineral decomposition. It is generally considered economically barren.
3. Ophiolite Belt – In the Naga Hills, the Ophiolite Belt is relatively wider in the northern Nagaland sector. Metals, limestones and marbles are primarily found in this area.
4. Low-grade Metamorphic Belt – This belt has high-grade limestone, marble and noble metals as the main rocks to be found.

26.3   PROPERTIES OF SOILS AND ROCKS

The mechanical compositions of soil based on the land use effect on soil erodibility parameters in four villages, viz., Tsiesema, Riisoma, Chiephobozou and Botsa in the Kohima district of Nagaland, are presented by Dutta et al. (2017). The dispersion ratio of soil under different land uses for the four villages as mentioned in the above paragraph is also proposed by Dutta et al. (2017).

As per the criterion of Middleton (1930), those soils that have a dispersion ratio value of more than 15 are said to be erosive in nature. Dispersion ratio of less than 5 was categorised as 'very stable', similarly a ratio between 6 and 10 is categorised as 'stable', that between 11 and 15 as 'fairly stable' and as the ratio further increases stability starts to decrease comparatively with greater than 30 being 'very unstable'. Only 1 was found to be stable, 5 fairly stable, 15 somewhat unstable, 10 unstable and 1 very unstable, out of 32 samples. Usually in all land use types, soils that were found to have a lower dispersion ratio were from the forests as compared to those of cultivated lands.

The standard penetration test (SPT) values for settlement analysis for shallow foundations on Dimapur to the Kohima section of NH-39 stretch located in Kohima district, Nagaland are obtained by the state govt. of Nagaland during soil investigation works for minor bridge structures (Four Laning of Dimapur to Kohima Road, 2017). From the work, it has been found that the N value indicates the relative density and consistency of non-cohesive and cohesive soils respectively. The value basically represents the energy required to penetrate the soil by 300 mm, in this case via a spoon sampler.

26.4 CONSTRUCTION MATERIALS

The primary construction materials used are timber, bamboo and thatch. There are three types of houses found in Nagaland. These are principally categorised as earth floor, earth and raised floor, solely raised floor.

26.5 FOUNDATIONS AND OTHER GEOTECHNICAL STRUCTURES

Due to active tectonism, Nagaland is one of the most disturbed states of the country. These disturbances are observed in the layers of rocks, mainly in the geological form of rocks that includes folds, joints, fractures and faults. These bedding, joints, fractures, faults and so on are the planes of weakness and largely determine the slope stability of a structure. In the rocks of the study area faults, joints, shear zones and fractures are common and play a major role in promoting instability, either individually or in their varying combinations (Balasubramanian, 2013). One of the important criteria for the determination of slope stability is the altitude of the bedding or joint planes in relation to the slope. Anbalagan (1992) has given the three relationships based on which ratings are assigned to each of the facets.

1. First is the extent of parallelism in between the directions of the discontinuity or between the lines of intersection of these discontinuities with the slope.
2. Of the two discontinuities, it tells us about the plunge of line of intersection or about the steepness of the dip.
3. The last one is the difference in the dip of the discontinuity or the plunge of the line of intersection of the two discontinuities to the inclination of the slope.

In northeast India, the major structural features are likely due to the collision of the Indian subcontinent with Eurasia, anciently during the rift of Gondwana land, as the eastern margin of the Indian Peninsula was positioned at 50°S latitude and was oriented towards an E-W direction (Chatterjee and Hotton, 1986); further in the Cretaceous period, the Indian plate was moving northward and the eastern continental passive margin rotated 20° (Gordon et al., 1990) until it collided with Eurasia during the Late Eocene, in a clockwise direction. The palaeomagnetic studies of the Indian rocks (McElhinny, 1973) and oceanic magnetic anomalies are in favour of the above-mentioned theories (McKenzie and Sclater, 1971). Large negative isostatic anomalies to the east of Arakan-Yoma along with the relations among high seismicity and depth of foci suggest that the subduction processes are still continuing (Verma, 1985). The major structural units of Nagaland owe their origin to the above stresses.

Nagaland being a "Natural Hazard Prone Area" is likely to have one or more of these undermentioned hazards that include moderate to high intensity of earthquake, cyclonic storm, significant flood flow or inundation and landslides/mudflows/avalanches. Hence there are special provisions provided by PWD Nagaland for the construction of foundations for structures of Nagaland which include:

1. For Earthquake Protection: (xiv) IS: 1893–2002 "Criteria for Earthquake Resistant Design of Structures (Fifth Revision)" (xv) IS:13920-1993 "Ductile Detailing of Reinforced Concrete Structures subjected to Seismic Forces – Code of Practice".

(xvi) IS:4326-1993 "Earthquake Resistant Design and Construction of Buildings – Code of Practice (Second Revision)". (xvii) IS:13828-1993 "Improving Earthquake Resistance of Low Strength Masonry Buildings – Guidelines". (xviii) IS:13827-1993 "Improving Earthquake Resistance of Earthen Buildings – Guidelines",

2. For Protection from the Landslide Hazard: (xx) IS 14458 (Part 1): 1998 Guidelines for retaining wall for hill area: Part 1 Selection of type of wall. (xxi) IS 14458 (Part 2): 1997 Guidelines for retaining wall for hill area: Part 2 Design of retaining/breast walls. (xxii) IS 14458 (Part 3): 1998 Guidelines for retaining wall for hill area: Part 3 Construction of dry stone walls. (xxiii) IS 14496 (Part 2): 1998 Guidelines for preparation of landslide – Hazard zonation maps in mountainous terrains: Part 2 Macro-zonation.

On 7th July 2020, Nagaland Railways had got its first rail tunnel with an estimated distance of 83 km towards the Dhansiri Railway Station of the North-East Frontier Railway in Zubza in Nagaland; it was to combat the complications that came during construction; stabilisation was done for Nagaland's hilly tracks prone to landslides. Another complication included the presence of sale (type of clay rock which expands when contracted with water resulting in a sluggish and muddy texture). It created problems as it weakens the structure of the tunnel; to counter this problem, hydraulic drilling and controlled blasting was used. It was also reported that aquifers also hampered the work progress while digging as water flowed through such layers, which turns the tunnel road soggy making it difficult for workers and vehicles to move about.

26.6   OTHER GEOMATERIALS

The soil of Nagaland is rich in mineral contents like limestone coal, copper, zinc, nickel, cobalt, magnetite and chromium, and in recent years, geologists have discovered platinum, petroleum and natural gas as the other major minerals available in Nagaland. The state still has many areas that are unutilised and unexploited and includes reserves of granite, limestone, marble, natural gas and petroleum. Coal (the black combustible mineral) is found in Borjan, Nazira and Teru valley of Mon district and limestone (the calcareous mineral) is found at Satuza in the Phek district, Wazeho and Nimi Bert in the Tuensang district. It is grey to whitish-grey in colour. Ores of chromite-magnetite and nickel-ferrous occur in the Ultra Basic Beltat Pokhpur in the Tuensang district. Nagaland has huge reserves of natural oil that are yet to be fully explored. In the western portion of the state, hydrocarbons are generally found. The metallic and non-metallic minerals are located in the parts bordering Myanmar. These are ideal for export to the South-East Asian region hence giving an opportunity to exploit the mineral wealth of the state for trade and commerce both within the country and outside with South-East and East Asian Countries.

Nagaland is rich in mineral resources, mainly, the following mineral reserves are found in the state:

1. Petroleum and natural gas
2. Nickel–cobalt–chromium bearing magnetite
3. Marble, dimensional decorative stones
4. Coal

Eastern Nagaland has enormous reserves of up to 1,000 million tonnes of high chemical grade limestone, which is a major prospect for setting up industries such as:

1. Bleaching powder
2. Cement
3. Calcium carbide
4. Hydrated lime
5. White and green marble mining and polishing
6. Ceramic glazed tiles
7. Ceramic insulators
8. Ceramic crockery
9. Slate for building materials
10. Coal
11. Granite
12. Glass sand

26.7 NATURAL GEOHAZARDS

The state of Nagaland is vulnerable to all sorts of natural hazards such as earthquakes, flash floods, landslides and forest fires. The state experiences massive property damages per year due to disasters. Two major natural disasters i.e. earthquakes and landslides are discussed below.

26.7.1 Landslides

Landslides are one of the most dangerous natural calamities that are unpredictable and come under geohazards, damaging vital infrastructures, having an adverse effect on mankind. Generally, landslides occur due to extreme natural events that include cloudburst rainfalls, earthquakes, volcanic eruptions and so on. The distinctive combinations of active and diverse tectonic settings along with high rates of weathering including abundant rainfall aggravated by human interference in the form of rapid development of infrastructure and urbanisation and deforestation adversely affect the sensitive and fragile ecosystems of the state. Moreover, hilly terrains are subjected to slope failure, which is due to usually more than one of the above-mentioned geological causes. Landslides are the rapid mass movements of the pre-consolidated masses of soils and pose huge hazards in mountainous terrains (Sharma et al., 1996). Slope failures could also be triggered by a combination of external and internal forces such as intense rainfall, ground vibrations due to earthquakes and changes in the subsurface water table, (Dai et al., 2002). The vibrations and loads due to heavy vehicles along with weak slopes might have a minor role in leading to slope failures (Aier et al., 2005, 2009).

Due to failure along curved and planar surfaces, landslides occur. Landslide is defined as the down-slope movement of rock debris and earth masses due to gravity when a material loses its shearing strengths, may or may not be because of the existing saturation levels of the soils and water tables of the specified areas. A landslide occurs when the forces of mobilisation of earthly particles exceed the resisting forces.

Sometimes, the shearing strength of the soil can sufficiently resist these forces and then the slope becomes stable and safe from the hazard of landslides. It may occur abruptly or through a series of time, which may be with or without any apparent warning or provocation. The factors that influence landslides are namely structure, drainage, slope angle, relief of the area, groundwater conditions, lithology and so on. They are divided into two sub-divisions, viz. natural and anthropogenic as per their mode and genesis. Naturally, landslides are caused by neo-tectonic activities e.g. earthquakes and reactivation of thrusts, faults, water action and denudational processes acting on the upper surface of the tectonic plates. Man-induced landslides are possibly due to human intervention associated with activities such as urbanisation, deforestation, mining and public utility activities.

A landslide that occurs as a result of changes in landforms might also be triggered due to gravitational forces, which are constantly acting on soils and are aided by increased buoyancy of the soils, which are formerly stable on the steep slopes and then become less stable and can potentially slide. Gray (1973), Swanston and Dyrness (1975) and Swanston and Swanson (1977) claimed that forest destruction and road construction are the most basic causes initiating landslides. Bhandari (1987) claimed that human interference in the ecosystem has a great role in accelerating the occurrence of landslides. Valdiya (1987) blames road construction without proper planning and scientific knowhow as causing destructive landslides. It is said that weak lithology, unfavourable slopes, poor vegetative cover and abnormal rainfall together are the most probable causes of landslides. The thick deposits of unconsolidated soils and other associated materials including vegetationless areas on steep hill slopes, adverse hydro-geological conditions, lithological and anthropogenic activities such as road cutting and construction of heavy structures are responsible for the majority of landslides (Kumar et al., 1995). Cruden and Varnes (1996) said that landslides are caused by geological, geomorphological and human processes. Petley and Reid (1999) indicated that landslides are inevitable where mountain chains are being uplifted. For managing the landslides in the state, some strong measures should be taken mainly for infrastructural development that is accelerating the occurrences of landslides.

26.7.2   Seismicity

India is a highly vulnerable area to seismic hazards because of its high population density and extensive developmental investments that are not properly planned for sustainability and focus on short-term goals. The NER of India had been shaken by more than 40 earthquakes with magnitudes greater than 6 in the previous century, of which two earthquakes with greater than M8 have occurred in the years 1897 and 1950 respectively. Verma (1985) said that repeated occurrences of earthquakes in the northeastern areas are caused by intermittent tectonic stress releases that indicate the orogenic movements are still in progress. Froehlich et al. (1992) mentioned that the high density of joints of rocks is probably connected with the high seismicity of any region. This large scale mass wasting shows the support to the above mentioned as Nagaland falls in the Zone-V category (the maximum magnitude of a possible earthquake is greater than 8). It is hence crucial to consider seismicity while studying about landslides. Although, in context to the development on the topic, we still lack the data pertaining to earthquakes and their effects on surface instability. Slopes are weakened by vibrations in the earth thereby causing failure by reducing the factor of safety of the slope material. The intensity and duration of shaking, its geotechnical characteristics, local geological details, the type of slope, slope geometry and existing pore-water pressure due to seepage, impounded water and other conditions are the factors causing slope failure. As a triggering mechanism, the role of seismicity must be studied for historic landslide events (Thigale, 1999). Usually, earthquakes are responsible for triggering landslides (Schuster and Highland, 2001). Varnes (1978) and Brabb (1985) assumptions required in the verification of landslide susceptibility calculation models are related to spatial factors such as land cover, topography and geology. Also, future landslides would be triggered by a specific impact e.g. seismic shock or heavy rainfall. Malamud et al. (2004) opined that an earthquake, a large storm, a rapid snowmelt or a volcanic eruption triggers are associated with landslide events. A landslide event may include a single landslide or many thousands and can be quantified by the frequency-area distribution of the triggered landslide. However, on its own, ground surface acceleration is not a good measure of the effect of shaking on slope stability (National Institute of Disaster Management, 2009). Considering the same acceleration, an earthquake may not trigger landslide movement though it may still result in the collapse of or damage to unreinforced stone buildings.

26.8   CASE STUDY AND FIELD TESTS

Works within the field of landslides studies and Geology of Nagaland are quite insufficient as a consequence of its inaccessibility, remote location and socio-political issues. As discussed by Oldham (1883), one of the earliest pieces of literature was established on the geology of Kohima and parts of northern Manipur. Works of Evans (1932) reported a tertiary succession of the tectonics of Nagaland. Mathur and Evans (1964) studied Tertiary sediments of upper Assam and Nagaland and concluded about the stratigraphy, structure and conditions of deposition of the state. Sarmah (1989) studied about the Disang and Barail sediments of the Kohima district. Agarwal and Shukla (1996), in their investigations, told that drainage is mainly controlled structurally as the majority of rivers flow along a multitude of lineaments trending NE-SW and NW-SE of the ophiolite complex. Pillai et al. (2008) told about the identification, significance and distribution of clay minerals in the shale of Kohima Disang and concluded that the Disang flysch comprises clastic sediments that were derived from complex sources. As a result of severe weathering and rapid erosion from orogenic sources, the sediments were possibly derived where climatic conditions were humid. Aier et al. (2011) and Imtiwapang (2011) gave an in-depth account of the Medziphema intermountain basin and Quaternary deformation within the Schuppen Belt of Nagaland from where the sediments were derived. The geology of the Kohima and Dimapur districts is discussed below.

26.8.1   Geology of Kohima

The rock types found in Kohima include Disang Formation, Barial Group, Tipam Group and the alluvial and terrace deposits; the details are mentioned below.

a. Disang Formation is the oldest and lowermost part of this formation, seen along Dzuu and Sidhu River flowing below Chakhabama. They incorporate dark grey to almost black phyllites and shale. Evidence suggests that there have been deep marine sediments that had undergone high-grade burial diagenesis during plate subduction. Salt spring in the shale is found in the lower regions of Viswema, Khuzama and Jakhama.

b. Barial Group is the topmost part of the Barial seen from the cliffs and peak of Mt Japfu and in the upper regions of Viswema, Jakhama, Kigwema, Phesama, Jotsoma, Khonoma and Mezoma. They are multistoreyed massive sandstones measuring almost 2,000 m thick. In the Jenam Formation, there are thick carbonaceous shale beds with subordinate siltstones and sandstones that are seen in Phekerukriema, Tsosinyu, New Remisheyu and Honku. There is gas and oil seepage in the stream bed located 2 km south of Honku. Rocks of the lowermost Laisong Formation are found in Botsa, Tseminyu, Sentenyu, Khamyu and along the Kohima–Imphal highway.

c. Surma and Tipam Group rocks are found in and around Rengmapani, Henimi, Hanomi and Longwesunyu, in the western side of the district bordering Assam. They are also found in the Dzukou valley and also in the southern parts of Viswema and Khuzama. In the Dzukou valley, Tipam stone and Girujan clay beds measuring up to 25 m thickness lie over the Surma rocks. The Peak of Mt Tempfu (2994 m), which is the second highest peak, is constituted by horizontally bedded Surma rocks.

d. In Rengmapani and adjoining areas, deposits of alluvium and terrace deposits are found along the Nagaland and Assam border. It includes the pebbly Dihing Formations of unconsolidated alluvium of the Holocene age and the Plio Pleistocene age. Unconsolidated terrace deposits of pebbles, boulders, clay, silt and sand of the Holocene age are exposed along the upper parts of the course of many rivers and streams.

26.8.2 Geology of Dimapur

Dimapur is known for its Barail, Tipam, Surma, Namsang and Dihing rocks and it is rich in alluvium and terrace deposits. A few km below Piphema, towards the national highway, areas are exposed to Renji formation, carbonaceous shale, thin-bedded siltstone and stone of Jenam Formation are seen along the national highway between Piphema and Pherima. Laisong formations are present on the roadside between Pherima and Medziphema. The complete landscape between Piphema and Medziphema mostly consists of burial rocks.

Stone, siltstone, shale, clay, pebble and conglomerate beds that are the characteristics of the Surma group are found in Chumukedima, Seithekema and Niul. Along the Charge River in Chumukedima, there is a magnificent exposure of Surma rocks. Multistoreyed ferruginous stones with thin interbeds of siltstone and shale being features of the Tipam group are present in Chumukedima, Seithekema, Kukidolong, Jharnapani, Ruzhiephema, Medziphema and Pherima. Excellent exposure of mottled clay, clay of Girujan and clay formations are present at the Chathe riverbank in New Chumukedima. Namsang and Dihing Formations are also found in Medziphema, Jharnapati, New

Chumukedima and Ruzhiephema. Alluvium and terrace deposits are found along the entire western side of the Dimapur district. These areas include Manglumukh, Honour, Dhansiripar, Rangapahar, Chumukedima, seithekema and Niul. They are represented by loose as well as compacted sandy silt, clay, pebbles and boulders.

26.9 EFFECT OF ENVIRONMENTAL CHANGES ON SOILS AND ROCKS

Three main factors which control the development of soil are climate, parent material and vegetation type. The primary effects of climate change on soil development are alteration in soil moisture content and an increase in temperature and CO_2 levels. The direct climatic effects on soil moisture content are via phenomena like precipitation, the effect of temperature on evaporation or by induced changes in the climate. This can be done via vegetation, increasing plant growth rates, rates of soil water extraction by plants and the effect of enhanced CO_2 levels on transpiration in plants. Changes in the proportion of water content of the soil can be directly related to the climate itself and may lead to drought-like conditions by a decrease in available moisture, altering patterns in circulation and an increase in air temperatures.

Talking about soil development, the effect of climate is significantly visible in weathering of rocks and minerals. The various stages of weathering in these rocks and minerals are governed by a change in temperature and rainfall patterns which are primarily associated with climate change. These result in a change in the soil-forming rocks. The mentioned changes are chemical and mineralogical. An increase in rainfall is said to accelerate weathering. The above-mentioned variations in the various climatic conditions give rise to different secondary minerals from primary minerals owing to the different conditions of weathering. Thus, rock types that are similar may give rise to distinctive soil profiles. This could have both positive and negative effects on the respective soil profiles.

The distribution of soils is uneven around the whole state. The inherent constraints that limit the productivity of the soil for various uses are properties of the parent material, climate, relief features, vegetation cover and time that determine soil formation. Most soil constraints are dependent on each other.

26.10 CONCLUDING REMARKS

Nagaland is known for its vernacular architecture. The primary construction materials used are timber, bamboo and thatch. Climate change can have a substantial influence on the properties of the soils and rocks. Soils are unevenly distributed around the state and as a consequence have various uses as per their properties.

1. The most prominent orders of soils found in the state are Inceptisols, Ultisols, Entisols and Alfisols. The most dominantly found variety is Inceptisols (66% land area of the state) with fine clayey, clayey loamy and fine loamy clay textures.
2. On the basis of the geo-tectonic and morpho-tectonic features, Nagahills can be divided longitudinally into a number of belts. These are the Disang Belt, Ophiolite Belt, Low-Grade Metamorphic Belt and the Schuppen Belt.

3. The forest soils were found to have a lower dispersion ratio as compared to those of cultivated lands, and the soil of Nagaland is rich in minerals. It has great areas of unutilised and unexploited mineral reserves. These reserves are mainly of limestone, marble, granite, petroleum and natural gas.

4. Landslides and earthquakes are the most prominent natural disasters in Nagaland. Nagaland falls in the Zone-V category (with a maximum expected magnitude of a possible earthquake greater than 8). Landslides are a common occurrence in Nagaland especially during the monsoon season due to heavy precipitation.

The data pertaining to earthquakes and their effects on the instability of the surfaces are lacking, hence the direct impact of an earthquake in triggering/inducing a landslide cannot be established beyond a certain degree. Some suggested measures to stop this natural calamity are retaining walls with tiebacks and binding of soil through artificial vegetation.

REFERENCES

Agarwal, N.K. and Shukla, R.C. (1996). Kohima urban area. *Contri. Environ. Geol. Geol. Surv. India Sp. Publ.*, vol. 43, p. 150.

Aier, I., Walling, T. and Thong, G.T. (2005). Lalmati slide: Causes and mitigation measures. *Naga. Univ. Res. J.*, vol. 3, pp. 44–47.

Aier, I., Supongtemjen, Khalo M. and Thong G.T. (2009). Geotechnical assessment of the Mehrülietsa slide (179 km) along NH 39, Kohima, Nagaland. In Kumar, A., Kushwaha, R.A.S., Thakur, B. (Eds.) *Earth System Science*, Concept Publishing Company, New Delhi, pp. 81–88.

Aier, I., Pradipchandra, M., Thong, G.T. and Soibam, I. (2011). Instability analyses of Merhülietsa Slide, Kohima, Nagaland. *Nat. Haz.*, vol. 60, pp. 1347–1363.

Anbalagan, R. (1992). Landslide hazard evaluation and zonation mapping in mountainous terrain Engineering. *Geology*, vol. 32, pp. 269–277.

Balasubramanian, A. (2013). *Welcome to the state of Nagaland (A Video Documentary Script).* Centre for Advanced Studies in Earth Science, University of Mysore, Mysore.

Bhandari, R.K. (1987). Slope instability in the fragile Himalaya and strategy for development. *Ninth Annual Lecture. Indian Geotechnical J.*, vol. 17, pp. 1–77.

Brabb, E.E. (1985). Innovative approaches to landslide hazard and risk mapping. In *International Landslide Symposium Proceedings*, Toronto, Canada (Vol. 1, pp. 17–22).

Chatterjee, S. and Hotton, N. (1986). The palaeoposition of India. *J. Southeast Asian Earth Sci.*, vol. 1, no. 3, pp. 145–189, ill, maps.

Cruden, D.M. and Varnes, D.J. (1996). Slope movement types and processes. In *Landslides: Investigation and Mitigation. Spl. Rep. Natl. Acad.*, Press, Washington, vol. 247, pp. 36–75.

Dai, F.C., Lee, C.F. and Ngai, Y.Y. (2002). Landslide risk assessment and management: An overview. *Eng. Geol.*, vol. 64, no. 1, pp. 65–87.

Dutta, M., Mezhii, R., Kichu, R. and Ram, S. (2017). Erodibility status of soils under different land uses in Chiephobozou sub-division soils of Kohima, Nagaland. *Asian J. Bio. Sci.*, vol. 12, no. 2, pp. 248–253.

Evans, P. (1932). Tertiary succession in Assam. *Trans. Min. Geol. Inst. Ind.*, vol. 27, pp. 155–260.

Four Laning of Dimapur to Kohima Road from existing KM 156.000 to existing KM 172.900. On EPC Mode of Contract (Package 3). (2017). The state of Nagaland, India-soil investigation works for minor bridge structures. https://www.scribd.com/document/376659571/Soil-Investigation-Report-Package-3.

Froehlich, W., Starkel, L. and Kasza, I. (1992). Ambootia landslide valley in the Darjeeling hills, Sikkim Himalaya, active since 1968. *J. Himalayan Geol.*, vol. 3, no. 1, pp. 79–90.

Gordon, R.G., DeMets, C. and Argus, D.F. (1990). Kinematic constraints on distributed lithospheric deformation in the equatorial Indian Ocean from present motion between the Australian and Indian plates. *Tectonics*, vol. 9, no. 3, pp. 409–422.

Gray, D.H. (1973). Effects of forest clear-cutting on the stability of natural slopes: results of field studies, Ann Arbor, MI: College of Engineering, Department of Civil Engineering, University of Michigan; Interim report, DRDA project 002790. p. 119.

Imtiwapang, A., Thong, G.T. and Supongtemjen. (2011). Geological evaluation of surface instability along NH 39 (180 km), west of Raj Bhavan, Kohima, Nagaland. In Singh, T.N. and Sharma, Y.C. (Eds.) *Slope Stability: Natural and Man Made Slope*, Vayu Education of India, New Delhi, pp. 192–201.

Kumar, B., Virdi, N.S., Sahe, M.P., Bartarya, S.K. and Gupta, V. (1995). Landslide hazard zonation between Rampur and Wangtu. H.P. *Symp. Rec. Adv. Geol. Studies on NE Himalayas*, Lucknow, pp. 324–326.

Malamud, B.D., Turcotte, D.L., Guzzetti, F. and Reichenbach, P. (2004). Landslides, earthquakes, and erosion. *Earth Planet. Sc. Let.*, vol. 229. pp. 45–59

Mathur, L.P. and Evans, P. (1964). Oil in India. *22nd International Geological Congress*, New Delhi, pp. 7–52.

McElhinny, M.W. (1973). *Paleomagnetism and Plate Tectonics*. Cambridge University Press, Cambridge.

McKenzie, D. and Sclater, J.G. (1971). The evolution of the Indian Ocean since the Late Cretaceous. *Geophys. J. Int.*, vol. 24, no. 5, pp. 437–528.

Middleton, H.E. (1930). Properties of soils which influence soil erosion. USDA Technical Bulletin. United States Department of Agriculture, Washington, DC.

National Disaster Management Policy, National Institute of Disaster Management (2009). https://ndma.gov.in/images/guidelines/national-dm-policy2009.pdf

Oldham, R.D. (1883). Report on the geology of parts of Manipur and Naga Hills. *Geol. Surv. India Mem.*, vol. 14, pt. 4.

Petley, D.N. and Reid, S. (1999). Landscape sensitivity and change at Taroko, Eastern Taiwan. In Smith, B.J., Whalley, W.B. and Warke, P.A. (Eds.) Spl. Pub., *Geological Society London*, vol. 162, no. 1, pp. 169, 179–195, London, UK.

Pillai, R. Vineetha., Thong, G.T. and Aier, I. (2008). Identification, distribution and significance of clay minerals in the Disang shale of Kohima, Nagaland. *Naga. Univ. Res. J.*, Spl. Pub. pp. 24–32.

Rothacker, L., Dosseto, A., Francke, A. et al. (2018). Impact of climate change and human activity on soil landscapes over the past 12,300 years. *Sci. Rep.*, vol. 8, p. 247, doi: 10.1038/s41598-017-18603-4

Sarmah, R.N. (1989). Clay minerals in Disang-Barail groups of sediments from Kohima, Nagaland. *Bull. Ind. Geol. Assoc.*, vol. 22, pp. 107–111.

Schuster, R.L. and Highland, L.M. (2001). Impact of landslides and innovative landslide mitigation measures on the natural environment. USGS Open File Rep. 01-0276, pp. 29–36.

Sharma, V.K., Sharma, A. and Attre, J.K. (1996). Slope mass rating (SMR) technique in landslide susceptibility evaluation in parts of Nainital area, Kumaon Himalaya. *J. Eng. Geol.*, vol. 25, pp. 289–295.

State of Environment Nagaland (2005). Nagaland Pollution Control Board. https://npcb.nagaland.gov.in/wp-content/uploads/soe.pdf

Swanson, F.J. and Dyrness, C.T. (1975). Impact of clear-cutting and road construction on soil erosion by landslides in the Western Cascade Range, Oregon. *Geology*, vol. 3, no. 7, pp. 393–396.

Swanston, D.N. and Swanson, F.J. (1977). Timber harvesting, mass erosion and steep land forest geomorphology in the Pacific Northwest. In Coates, D.R. (Ed.) *Geomorphology and Engineering*. Hutchinson and Ross, Stroudsburg, pp. 199–221.

Terzaghi, K. and Peck, R.B. (1993). Mekanika Tanah dalam Praktek Rekayasa, Penerbit Erlangga, Jakarta.

The Sema Naga Traditional Dwelling, Nagaland – April, 2016, ENVIS Center on Human Settlement's Hosted by School of Planning and Architecture, Delhi Sponsored by Ministry of Environment, Forests & Climate Change, Government of India. https://www.slideshare.net/sanjanaaggarwal2/nagalandjpg, accessed 2nd April 2020.

Thigale, S.S. (1999). Recommendation on landslide and water scarcity problems of Nagaland with special reference to Kohima. *Report*.

Tomlinson, M.J. (1999). *Foundation Design and Construction*. 6th ed., Pearson Education Limited, Edinburgh.

Valdiya, K.S. (1987). *Environmental Geology: Indian Context*. Tata-McGraw Hill, New Delhi, 583 p.

Varnes, D.J. (1978). Slope movement types and processes, landslides analysis, and control. *Spl. Rep., 176*. Transportation Research Board, Washington, DC, pp. 11–80.

Verma, R.K. (1985). *Gravity Field, Seismicity, and Tectonics of the Indian Peninsula and the Himalaya*. Allied Publishers, New Delhi, pp. 155–189.

Xu Qiang, Chen Jianyun, Li Jing, Zhao Chunfeng and Yuan Chenyang. (2015). Study on the constitutive model for jointed rock mass. *PLoS ONE*, vol. 10, no. 4, pp. 1–20. DOI:10.1371/journal.pone.0121850.

Zaman, F. and Bezbaruah, D. (2019). Morphotectonic aspects in a part of Naga-Schuppen belt, Assam Nagaland region, North-East India. *Sci. Vision*, vol. 19, pp. 6–11.

Chapter 27

Odisha

Manas Chandan Mishra
IIT Bhubaneswar

Swagatika Senapati
SOA University

Bendadi Hanumantha Rao
IIT Bhubaneswar

CONTENTS

27.1 INTRODUCTION

Odisha, the ancient state, historically known as Kalinga, is the eighth largest state (by area) of the country, situated in the eastern part of peninsular India. It expands amid the latitudes of 17° 49′N and 22° 34′N parallels and the longitudes of 81° 27′E and 87° 29′E meridians, spanning over an area of 1,55,707 km^2 and contributing to over 450 km of coastline to the country's maritime border (https://en.wikipedia.org/wiki/Odisha#Geography). It is surrounded by four states: Jharkhand towards the north, West Bengal towards the northeast, Chhattisgarh towards the west, Andhra Pradesh towards the south and the Bay of Bengal towards the east. It consists of 30 districts housing various major rivers such as Mahanadi, Subarnarekha, Budhabalanga, Baitarani, Brahmani and Rushikulya along with the largest salt-water lake of India, namely Chilika.

The climate of the Odisha state can be considered as tropical with mild winters, hot-humid summers and a high annual rainfall of around 1,450 mm. The state is subjected to frequent natural calamities of drought, cyclone, flood and moisture stress over the recorded history. The geographical area of Odisha can be divided into five key expanses: coastal plains, central plateaus, mountains and highlands, major flood plains and rolling uplands, which are further subdivided into ten agro-climatic regions. The major soil types show drastic variations along these five expanses. The coastal plains mostly comprise fertile alluvial deltaic soils, whereas the plateaus are covered with

DOI: 10.1201/9781003177159-27

mixed red and black soils in the centre, red and yellow soils with low fertility in the north and red, black-brown forest soils in the east. The pH values of these soils vary from highly acidic to slightly alkaline with soil types ranging from loose sandy to stiff clays with around 2.5% of the total land area exposed to frequent saline inundation, 2.3% to flooding and 0.5% to waterlogging, particularly in the coastal deltaic regions.

The forest coverage in Odisha (Figure 27.1) is around 35% of its total area, while the total cultivated area of the state is around 40%. The majority of the cultivation is based on paddy farming of many varieties. The tropical forests in the state are of two

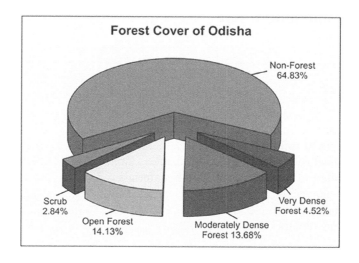

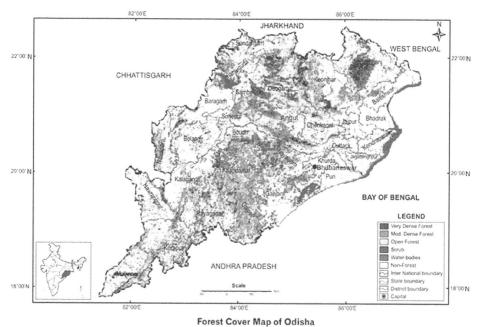

Figure 27.1 Distribution of forests in Odisha (ENVIS, 2021).

types: tropical moist deciduous type in the northeast part of the state and tropical dry deciduous type in the southwest region. These forest areas are prohibited and protected for the home to a wide variety of species and the protected areas are 4.1% of the total geographical area.

Table 27.1 presents the history of the geological evolution of the Odisha coast. During the upper Cretaceous period, the Odisha coast was said to be first developed witnessing extensive transgressions and regressions as depicted in Figure 27.2.

The lithological classification of the Odisha coast is given in Table 27.2. The geologic history of Odisha starting from Archean to Quaternary reveals the typical lithologies representing the preservation of different rock masses that depict the evidence of a multiphase tectonothermal history (Mahalik, 1996). The geological history of Odisha confesses the presence of rich mineralogy such as coal, iron ore, limestone, gemstone, bauxite, chromite, manganese ore, dolomite, graphite and so on (according to the Department of Steel and Mines, Govt. of Odisha).

The crustal blocks, i.e., Cratonic blocks (2 nos) and mobile belt (1 no.), are known as North Odisha Craton (NOC), where low-grade banded iron formation and granite

Table 27.1 Evolution of the geological age of Odisha coast ENVIS Centre, Odisha (ENVIS, 2021)

S. no.	Geological age	Geological events
1	Holocene	Flandrian transgression and formation of the Holocene coast with subsequent regression to the present coast, development of all deltas, Chilika lake, Bhitarkanikaand so on
2	Upper Pleistocene	Transgression followed by regression up to the last glacial maximum (LGM)
3	Pleistocene	Gunz-Mindel glaciation, regression of the Mio-Pliocene sea. Formation of laterites
4	Mio-Pliocene	Transgression of Mio-Pliocene sea and formation of Baripada Bed in north Orissa and Tolapada Bed in south Orissa
5	Oligoene	Non-deposition in shelf areas. Regression of the sea into further offshore
6	Upper Cretaceous Paleocene and Eocene	Sedimentation in early formed marine basin as observed at the Konark–Chandbali sector and shelf areas
7	Upper Jurassic to Cretaceous	Rifting (R_3) and volcanic eruption, breaking of the Gondwana assembly – formation of a marine basin over the Gondwana basin. Marine sedimentation started in the Upper Cretaceous Period
8	Triassic – Jurassic	No marine basin was found during this time. Continuation of Upper Gondwana sedimentation in Athagarh and Salepur depressions
9	Upper Carboniferous to Permian	Lower Gondwana sedimentary sequence with coal beds in Talcher basin
10	Carboniferous	Rifting within Gondwanaland (R_2) giving rise to the Mahanadi rift basin which extended into Antarctica as Lambert Rift
11	Precambrian to Carboniferous	Unified East Gondwanaland: Antarctica–Australia–India formed a stable land mass. Rifting in the Eastern Ghats mobile belt (R_1)

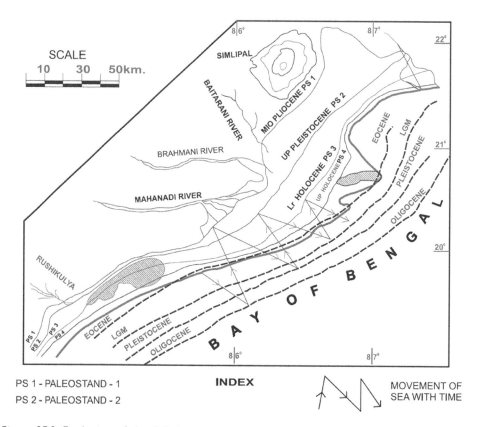

Figure 27.2 Evolution of the Odisha coast (ENVIS, 2021).

intrusive occurs, West Odisha Craton, which is underlain by Archean granites and limestone bearing sediments, and Eastern Ghats Granulite Belt (EGB), which consists of high-grade granulites such as khondalites, migmatites, charnockites and augen gneisses. Each of these blocks is separated by deep-seated regional fault boundaries, i.e., North Odisha Boundary Fault, passing across the Mahanadi valley also known as the Mahanadi rift and West Odisha Boundary Fault. The occurrences of laterites, alluvial plains, lagoons, swamps, water bodies and sandy beaches are found across the western hilly terrain on the onshore coast (Mahalik, 2000, 2006). The maximum thickness of sediments is found in the deltaic region surrounded by Mahanadi–Brahmani–Baitarani rivers. The lithological study of the Mahanadi basin predicts the formation of sand, clay and silt within 200–600 m depth and sand and clay formations within 200–700 m depth during the Pleistocene age. The formation of claystone, siltstone, sandstone and fossiliferous patchy limestones during the Miocene age from 600 to 1,900 m thickness is also observed.

The formation of charnockite followed by granite (leptynite) and pegmatite in the Eastern Ghats terrain lies in the middle Proterozoic time (Halden et al., 1982; Park and Dash, 1984). Charnockite formation is attributed to carbonic fluids derived from the upper mantle that contains carbon dioxide (CO_2)-rich basic magma (Frost and Frost, 1987).

Table 27.2 Lithological classification of the Odisha offshore region (Fuloria et al., 1992; Bharali et al., 1987; Faruque and Lahiri, 2002; ENVIS, 2021)

Age	Lithology	Environment	Thickness
Holocene	Clay, sand, silt, grainstone	Marine, shelf, transgressive phase	30 m
Pleistocene Pliocene	Clay, fine to medium sandstone, claystone with few interbeds of sands and silt	Marine and prograding deltaic deposits on basin margin	4,200 m
Upper-middle-lower Miocene	Claystone, siltstone calcareous claystone with minor sand in the upper parts, sandstone, dolomite and limestone in the lower part	Deep outer shelf to progressional detail. Tectonic uplift, shallowing of the depositional basin upward	1,903 m
Unconformity Oligocene	Claystone, siltstone, shale with interbedded sandstone, fossiliferous argillaceous limestone	Abrupt deepening of the basin due to subsidence and tilt in the basin floor	640 m
Unconformity Eocene	Fossiliferous dark grey to buff massive limestone, dolomite and thin interbeds of calcareous sandstone, limestone and claystone	Gently sloping stable shelf cut by numerous gullies, mild uplift of the source area	772 m
Late Paleocene	Fine-grained argillaceous limestone with interbeds of shale/sandstone	Shallow supratidal to the deltaic middle shelf, sloping shelf with classic sediments due to intense uplift	198 m
Early Paleocene	Grey calcareous shale with interbeds of glauconite sandstone, calcareous silt, shale and sandstone	Deltaic shallow marine	
Unconformity Upper Cretaceous	Fine-grained sandstones, shales, basalts with layers of volcanic tuff, clays and carbonaceous shale/coal	Marginal marine to mid-shelf, rift-related volcanism, lacustrine sedimentation (carbonaceous)	293 m
Early Cretaceous	Tuffs and volcanic with minor interbeds of coal and carbonaceous shale and siltstone (Upper Gondwana)	Rift-related volcanism, lacustrine sedimentation during volcanic quiescence	858 m
Unconformity Precambrian	Metamorphic basement		

The geological features of Odisha offshore are similar to those of the Mahanadi basin of Odisha. The offshore Odisha coast, lying in the Krishna–Godavari basin, is rich in hydrocarbons. Reliance Industries Limited deeply investigated the occurrence of hydrocarbons and gas along the Odisha offshore coast. These hydrocarbons are deposited deep offshore during the Eocene to Pliocene age. The geological, geophysical and geochemical studies by the Geological Survey of India, Oil and Natural Gas Commission, Utkal University and many other sources reveal the presence of a petroleum system in the Mahanadi basin of Odisha (both onshore and offshore) through geophysical investigation and extensive drilling. Late Paleocene, Eocene and Basal Miocene-type rocks are found in the catagenetic stage and have the potential for generating oils (ENVIS, 2021).

27.2 MAJOR TYPES OF SOILS AND ROCKS

Soils of Odisha can be classified broadly into eight groups: red soils, mixed red and yellow soils, black soils, laterite soils, deltaic alluvial soils, coastal saline soils, brown forest soils and mixed red and black soils. These soils can be grouped into four families as per their taxonomy, namely Inceptisols, Alfisols, Entisols and Vertisols arranged as per prominence. The majority of soils in Odisha are acidic (69%), about a quarter are neutral and 6% are saline.

Soils available in the eastern coast of India are mostly alluvial, red soils and lateritic. Red soils are predominantly available in Odisha as they are decomposed from granites and gneisses and many other rocks that are rich in iron and magnesium. The prevalent hot and humid climate conditions had led to alternate dry and wet seasons, which in turn resulted in the leaching of siliceous matter and lime from the rocks, forming lateritic soils that are rich in iron and aluminium oxide. Marine clay deposits are generally encountered near the coastal areas of the east coast of India (Goswami et al., 2013). Gravelly and lateritic soils are mostly found in the northeastern hilly region of the state. Table 27.3 shows the broad classification of soils in the major districts of Odisha separated with different agro-climatic zones.

Table 27.3 Soil classification in the major districts of Odisha

S. no.	Zone	District	Soil type
1	Northwestern plateau	Sundargarh	Mixed red and yellow
2	North-central plateau	Keonjhar	Red
3	Northeastern coastal plain	Balasore	Coastal alluvial
4	East and southeastern coastal plain	Puri and Cuttack	Deltaic alluvial and lateritic
5	Northeastern Ghats	Phulbani	Red loam and brown forest
6	Eastern Ghats highland	Koraput	Red and laterite
7	Southeastern Ghats	Koraput	Red
8	Western undulating	Kalahandi	Red and black
9	West-central table	Sambalpur and Bolangir	Mixed red and black
10	Mid-central table land	Dhenkanal	Red and laterite

27.3 PROPERTIES OF SOILS AND ROCKS

Soil samples from boreholes were collected to study the geotechnical properties of the soils of Odisha. The drilling equipment employed is able to make a borehole of 150 mm size. The samples were collected at a depth interval of 1.50 m. The summary of field investigation results of boreholes for the Ganjam district is given in Table 27.4.

Borehole #1 was explored up to a depth of 10 m. The layer starts with strong strata from 0 to 10 m followed by weak strata for the remaining depth. The core recovery and RQD (rock quality designation) at 1.50 m is 12% and nil, at 3.0 m is 13% and nil, at 4.5 m is 10% and nil, at 6.0 m is 11% and nil, at 7.50 m is 12% and nil, at 9.0 m is 10% and nil and at 10.0 m is 13% and nil respectively. The engineering classification of the rock from 0 to 10 m depth is 'moderately weak' type. Unconfined compressive strength (UCS) (in MPa) was found in the range from 1.25 to 5 MPa, as per IRC:78 (2014).

In Borehole #2, only boulder pieces were collected, which were not suitable for any testing purpose. For Borehole #3 and #4, up to 10 m (layer 1) from the ground level, boulder pieces were obtained, which again were found not suitable for the testing purpose. The test report of layer 2 starting from 10 to 12 m shows moderately strong strata with UCS of 12.5 to 50 MPa for Borehole #3 and moderately weak strata with UCS of 5 to 12.5 MPa for Borehole #4. The UCS values were determined as per the procedure defined by IRC:78 (2014).

Similarly, subsoil investigation test results of Bhabanipatna Bridge over Pipal Nallah on Iorepada to the Railway Station Road shows the presence of poorly graded clayey sand followed by weak strata, strong strata and very strong strata. A very strong stratum was found from a depth of 7.5 to 12.5 m with an RQD of 86%–89% and UCS (in MPa) of 100 to 200.

A detailed summary of soil properties that were commonly found in different areas of Odisha is provided in Table 27.5. The various parameters presented in this table clearly highlight the distinct nature of soils location-wise. For a vivid and better understanding of soil types of Odisha, profiles of soils obtained from boreholes of different sites are presented in Table 27.6. Such elaborate soil details could help in deciding the type of foundation system based on the site conditions. Generally, foundation system selection includes understanding visual observations of soil samples at the site, field bore log records, local geology of the area, type of samples received, laboratory classification of soil samples and experience of a geotechnical expert. The selection of the type of foundation is governed primarily by the magnitude of the maximum and permanent loading conditions and the type of structure as well. As such, the geotechnical parameters presented in Table 27.5 are of great help in deciding the foundation type.

Table 27.4 Exploration details of boreholes made at the Ganjam district of Odisha

Borehole reference	Terminated depth of borehole (m)	Ground water table (m)
BH-1	10.00	3.00
BH-2	10.00	4.00
BH-3	12.00	6.00
BH-4	12.00	3.00

Table 27.5 Details of soil properties obtained for different sites across Odisha

Site	Borehole depth (m)	Ground water table (m)	Liquid limit (%)	Plastic limit (%)	Engineering classification of soil (USCS)	DFS (%)	Specific gravity	Seismic zone
Brajarajnagar	10	NIL	18	Not possible (pebbles)	GP	0	2.67–2.7	
Berhampur, Ganjam	10 to 12	3 to 6	NP	NP	Moderately weak and strong rock strata with boulder pieces	NP	–	II
Bhabanipatna Pipal Nallah on lorepada to Railway Station Road	12.5	NIL	35	18	SC	34–45	2.45–2.47	II
Kendrapada Nahanga	10.5	1.75	–	–	SC up to 1.5 m CH (1.5 to 10.5 m)	–	–	II
Kesinga (Kalahandi dist.)	21.5	No WT up to 21.5 m at some places and it was found below 5.29 m at other places	18–30	NP	Mixed soil layer with GW, SC, SP, weak strata with pebbles and gravel, fractured rock and so on	0–15	2.68–2.7	II
Rayagada	17	NIL	–	–	Rock pieces found. Classified as weak strata	–	2.65	II
Balasore Baragarh	15 15	– 4.8	50–53 22–28	23–26 15 up to 6 m and not possible beyond this depth	CH 0–6m: SC 6–10.5: SM 10.5–15: weak sedimentary rocks	45–47 10 up to 6.5 m and 0 after 6.5 m	2.35–2.36 2.42–2.62	III III
Dhenkanal (Brewery Plant)	15	3.7–4.2	18–26	13 up to 0.5 m	3–12 m laterite pieces Weak strata after 12 m	10 up to 0.5 m and 0 after 0.5 m	2.62–2.69	III
Cuttack (Jagatpur)	10	3.8–4	20–42	For the first few layers not possible and after that 20–22	SP and CI	–	2.35 to 2.36	III

Table 27.6 Boring logs of different sites in Odisha

(a) Berhampur

| Name of the project | Soil investigation test at Berhampur for SWM projects | | Water table: 3 m | | | | | | | | | |

Type of soil	Soil profile	Depth (m)	SPT value	Gravel (%)	Coarse sand (%)	Medium sand (%)	Fine sand (%)	Silty clay (%)	DFS (%)	Liquid limit (%)	Plastic limit (%)	Classification of soil
Poorly graded gravel		1.5	26	62.3	15.3	10.32	7.42	4.66	0	18	Not possible	GP
Pebbles		3										
		4.5										
			>50	Boulders are collected and not suitable for test								
		6										
Poorly graded gravel		7.5	>50	Boulders are collected and not suitable for test								

(b) Kesinga, Kalahandi

Name of the project	Bridge over the Roul river at Kesinga, Kalahandi	Water table: NIL m

Type of soil	Soil profile	Depth (m)	SPT value	Gravel (%)	Coarse sand (%)	Medium sand (%)	Fine sand (%)	Silty clay (%)	DFS (%)	Liquid limit (%)	Plastic limit (%)	Classification of soil
Well-graded gravel		1.5	–	52.63	7.58	18.69	19.63	1.47	0	18	Not possible	GW
		2	42	54.23	15.6	12.75	15.69	1.73	0	20	Not possible	GW
Poorly graded sand		3.5	>50	20.36	15.42	22.46	37.86	3.90	0	20	Not possible	SP
Pebbles		5 6.5 7.5				Pebbles are collected						

(c) Bhabanipatna

Name of the project	Bridge over Pipal Nallah at Bhabanipatna	Water table: Nil m

Type of soil	Soil profile	Depth (m)	SPT value	Gravel (%)	Coarse sand (%)	Medium sand (%)	Fine sand (%)	Silty clay (%)	DFS (%)	Liquid limit (%)	Plastic limit (%)	Classification of soil
Poorly graded clayey sand		1.5	24	0	1.2	50.22	13.39	35.19	45	35	18	SC
GSB filling		3	50									
		4.5	>50			GSB filling material Not suitable for the test						
		5	>50									

(d) Balasore

| Name of the project | 60 men's barrack SO's and GO's mess single storeyed at Balasore | | | | | | | | | | Water table: 1 m | | |

Type of soil	Soil profile	Depth (m)	SPT value	Gravel (%)	Coarse sand (%)	Medium sand (%)	Fine sand (%)	Silty clay (%)	DFS (%)	Liquid limit (%)	Plastic limit (%)	Classification of soil
High plastic silty clay		1.5	5	0.00	0.00	0.00	4.32	95.68	46	52	26	CH
		3.0		0.00	0.00	0.36	2.36	97.28	46	51	25	CH
		4.5	10	0.00	0.00	0.42	3.12	96.46	45	53	26	CH
		6.0		0.00	0.00	0.52	3.46	96.02	45	50	23	CH
		7.5	17	0.00	0.00	0.63	2.48	96.89	47	52	25	CH
		9.0	16	0.00	0.00	1.23	2.63	96.14	47	52	25	CH
		10.5	21	0.00	0.00	2.13	4.78	93.09	45	50	25	CH
		12.0	21	0.00	0.00	1.42	3.56	95.02	45	50	25	CH
		13.5	26	0.00	0.00	1.75	3.48	94.77	46	52	26	CH
		15.0	30	0.00	0.00	1.32	4.96	93.72	45	52	26	CH

(e) Bargarh

| Name of the project | Commercial cum office complex for MARKFED at Bargarh | | | | | | | | | | Water table: 4.8 m | | |

Type of soil	Soil profile	Depth (m)	SPT value	Gravel (%)	Coarse sand (%)	Medium sand (%)	Fine sand (%)	Silty clay (%)	DFS (%)	Liquid limit (%)	Plastic limit (%)	Classification of soil
High plastic silty clay		1.5	10	0.00	12.45	25.13	25.63	36.79	10	28	15	SC
		3.0	22	0.00	13.69	30.15	25.63	30.53	10	28	15	SC
		4.5	28	0.00	10.45	13.42	31.25	44.88	10	28	15	SC
		6.0	31	0.00	5.36	29.45	34.69	30.50	0	22	NP	SM
		7.5	>50	0.00	4.25	28.61	41.63	25.51	0	22	NP	SM
		9.0	>50	0.00	7.85	28.42	43.96	20.04	0	22	NP	SM
		10.5	>50	0.00	9.42	25.85	32.69	32.04	0	22	NP	SM
		12.0	>50									
		13.5	>50									

Sedimentary rock

| | | 15.0 | >50 | | | | | | | | | |

(f) Jagatpur

| Name of the project | 30" diameter mainline pipe at Jagatpur | | | | | | | | | | Water table: 3 m | | |

Type of soil	Soil profile	Depth (m)	SPT value	Gravel (%)	Coarse sand (%)	Medium sand (%)	Fine sand (%)	Silty clay (%)	DFS (%)	Liquid limit (%)	Plastic limit (%)	Classification of soil
		1.5	20	0.00	21.63	63.42	10.42	4.53	0	20	NP	SP
		3.0	26	0.00	26.42	45.63	25.96	1.99	0	20	NP	SP
		4.5	35	0.00	45.69	25.63	26.5	2.18	0	20	NP	SP
		6.0	40	0.00	46.90	30.69	18.42	3.99	0	20	NP	SP
		7.5	2	0.00	5.40	3.56	4.75	86.29	38	40	18	CL
		9.0	5	0.00	6.42	5.42	6.23	81.93	38	40	20	CL
		10.5	7	0.00	2.31	6.13	4.79	86.77	36	40	20	CL
		12.0	12	0.00	2.16	5.14	4.62	88.08	36	42	20	CL
		13.5	12	0.00	2.71	4.78	4.85	87.66	38	42	22	CL
		15.0	8	0.00	3.62	6.32	6.12	83.94	38	42	20	CL

(g) Jharsuguda

| Name of the project | Commercial cum office complex for MARKFED at Jharsuguda | | | | | | | | | | Water table: 2.15 m | | |

Type of soil	Soil profile	Depth (m)	SPT value	Gravel (%)	Coarse sand (%)	Medium sand (%)	Fine sand (%)	Silty clay (%)	DFS (%)	Liquid limit (%)	Plastic limit (%)	Classification of soil
Sand		1.5	7	0.00	12.36	38.36	44.25	4.76	0	20	NP	SP
		3.0	20	0.00	26.89	29.53	38.63	4.95	0	20	NP	SP
		4.5	45	0.00	25.86	35.89	35.42	2.83	0	15	NP	SP
		6.0	>50									
		7.5	>50			Pebbles						
		9.0	>50									

Red soils cover about 49% of the land in the state and are the highest coverage of all the soil groups of the state, extending to the districts of Koraput, Rayagada, Nabarangpur, Malkangiri, Keonjhar, Ganjam, Kalahandi, Nuapada, Bolangir, Dhenkanal and Mayurbhanj. The soils are strongly and moderately acidic, with low to medium organic matter status and poor water retentive capacity. These soils are deficient in soil nutrients such as nitrogen and phosphorous. Micronutrients like boron and molybdenum are also highly deficient in these soils, which usually have a low cation exchange capacity (CEC). Mixed red and yellow soils occupy around 35% of the land being the second highest in the state. These soils are eminently found in the districts of Sambalpur, Bargarh, Deogarh, Jharsuguda and Sundargarh. The upland soils are moderately acidic, whereas the lowland soils are slightly acidic. The lowland soils are formed

(h) Keonjhar

Name of the project	H.L. Bridge over river Salandi, Keonjhar											Water table: 4.8 m
Type of soil	Soil profile	Depth (m)	SPT value	Gravel (%)	Coarse sand (%)	Medium sand (%)	Fine sand (%)	Silty clay (%)	DFS (%)	Liquid limit (%)	Plastic limit (%)	Classification of soil
Clayey sand		1.5	15	4.80	10.37	29.69	17.78	37.36	13	30	15	SC
		3.0	35	15.57	12.49	29.03	14.08	28.83	14	32	15	SC
		4.5	42	5.77	9.36	24.20	21.14	39.53	13	30	15	SC
Clayey gravel		6.0	>50	45.63	10.43	7.63	2.48	33.83	10	28	13	GC
		7.5	>50	48.75	7.42	8.24	1.36	34.23	10	28	14	GC
		9.0	>50	52.15	8.42	5.63	1.59	32.21	10	28	14	GC
Pebbles		10.5	>50	50.63	6.35	0.69	6.42	35.91	10	28	16	GC
		12.0	>50									
		13.5	>50									

Sedimentary rock

(i) Khordha

Name of the project	D.A.V. Public School, Khordha											Water table: 3.8 m
Type of soil	Soil profile	Depth (m)	SPT value	Gravel (%)	Coarse sand (%)	Medium sand (%)	Fine sand (%)	Silty clay (%)	DFS (%)	Liquid limit (%)	Plastic limit (%)	Classification of soil
Clayey sand		1.5	41	35.69	10.25	12.63	8.25	33.18	8	26	14	GC
		3.0	>50	42.63	15.42	05.89	5.69	30.57	10	28	15	GC
		4.5	>50	58.42	31.25	06.30	6.30	01.45	0	20	NP	GP
Clayey gravel		6.0	>50	55.36	22.31	08.69	8.69	04.01	0	20	NP	GP
		7.5	>50									
		9.0	>50						Rock pieces				
Pebbles		10.5	>50									

mainly by colluvial deposition. The upland soils are low in nitrogen and phosphorous, whereas low land soils are medium in phosphorous and high in potassium contents. Upland light textured soils have a low boron content and lowland soils, which are mostly used for paddy farming, are deficient in zinc.

There is no systematic manifestation of black soils in the state. These soils are sporadically distributed in the districts of Puri, Ganjam, Malkangiri, Kalahandi, Nuapada, Bolangir, Sonepur, Khurda, Boudh, Sambalpur, Bargarh and Angul, covering an area of approximately 6% of the land. The black colour of soil is mostly due to the presence of titaniferous magnetite. The soil pH is neutral to alkaline having

(j) Paradeep

Name of the project				30" diameter mainline pipe at Paradeep								Water table: 3 m	
Type of soil	Soil profile	Depth in m	SPT value	Gravel (%)	Coarse sand (%)	Medium sand (%)	Fine sand (%)	Silty clay (%)	DFS (%)	Liquid limit (%)	Plastic limit (%)	Classification of soil	
		1.5	–	0.00	0.00	03.26	01.58	95.16	42	48	22	CL	
		3.0	3	0.00	0.00	03.26	05.23	91.51	43	49	23	CL	
		4.5	6	0.00	32.56	33.15	31.42	02.87	0	23	NP	SW	
		6.0	7	0.00	33.00	01.42	01.86	63.72	0	22	NP	SW	
		7.5	12	0.00	33.42	32.56	30.15	03.87	0	23	NP	SW	
		9.0	19	0.00	33.42	32.15	33.10	01.33	0	22	NP	SW	
		10.5	22	0.00	32.15	32.63	30.42	04.80	0	23	NP	SW	
		12.0	27	0.00	42.56	25.63	27.42	04.39	0	22	NP	SP	
		13.5	5	0.00	13.75	52.46	32.78	01.01	0	40	20	CL	
		15.0	10	0.00	32.56	48.96	13.56	04.92	0	22	NP	SP	
		16.5	13	0.00	28.42	49.52	17.42	04.64	0	20	NP	SP	
		18.0	20	0.00	23.63	32.16	42.36	01.85	0	20	NP	SP	
		19.5	25	0.00	25.14	28.42	42.63	03.81	0	22	NP	SP	
		21.0	29	0.00	07.63	50.23	38.56	03.58	0	22	NP	SP	
		22.5	27	0.00	38.52	19.52	39.63	02.33	0	23	NP	SP	
		24.0	37	0.00	48.96	30.12	16.75	04.17	0	23	NP	SP	
		25.5	35	0.00	30.26	32.45	34.86	02.43	0	22	NP	SP	
		27.0	33	0.00	38.76	33.86	25.13	02.25	0	23	NP	SP	
		28.5	38	0.00	38.75	37.63	22.15	01.47	0	22	NP	SP	
		30.0	42	0.00	42.96	37.65	17.42	01.97	0	23	NP	SP	

GP: poorly graded gravel, SC: poorly graded clayey sand, SM: silty sand, SP: poorly graded sand, CI: silty and clay with medium compressibility.

free calcium carbonate nodules in the soil profile. The soil is rich in calcium but deficient in phosphorous, zinc and boron. Laterite soils occupy 4.5% of the land of the state, mostly in the districts of Puri, Khurda, Nayagarh, Cuttack, Dhenkanal, Keonjhar, Mayurbhanj and Sambalpur. These laterite soils are mostly acidic, with acidity varying from slight to strong and pH ranging from 4.5 to 5.8. These soils are less fertile with low organic matter. Available nitrogen and phosphorous are very low and potassium abundance is medium. Most of the available nitrogen is lost due to leaching, and the available phosphate becomes unavailable due to fixation in the presence of aluminium and iron oxides. The CEC of these soils is low. Deltaic alluvial soils cover around 4% of the land and occur in the deltaic regions of the rivers Mahanadi, Subarnarekha, Brahmani, Rushikulya and Baitarani in the districts of Balasore, Bhadrak, Jajpur, Kendrapada, Jagatsinghpur, Cuttack, Puri, Gajapati and Ganjam. Usually, these soils are fertile; however, the soil fertility in these regions decreases if the soil is not recharged regularly due to flooding. The pH values of these soils range from acidic to neutral. The coarse-textured soils are deficient in nitrogen, phosphorous, potassium and sulphur. Alluvial soils with high dissolved salts are mostly found along the coast of the Bay of Bengal in a narrow strip of less than 25 km. The salinity found in these regions can be attributed to the littoral deposits of estuarial intrusion of brackish water from the sea through small creeks. Nearly 1.6% of the state's area is covered with saline soils, which are distributed in the districts of Balasore, Bhadrak, Jagatsingpur, Kendrapada, Puri, Khurda and Ganjam. Saline soils are rich in soluble salts of chlorides and sulphates in conjunction with sodium and magnesium. The pH of these soils varies from 6.0 to 8.0. The electrical conductivity is generally high in summer as these are generally rich in sodium and potassium but low in phosphorous. Brown forest soils, rich in organic content, are distributed mostly in the districts of Kandhamal, Rayagada, Koraput and parts of Ganjam and Nayagarh and cover about 1% of the total area. They are brown to grey-brown in colour, light in texture and acidic in nature. Organic matter and nitrogen contents in these soils are medium to high, with medium phosphorous and potassium contents. These soils are generally adequate in micronutrients. Mixed red and black soils mostly occur in association with both red and black soils together in which black soils occur in patches within the predominant red soils. The red and black soils are so intermixed that red soils are found in the upper ridges, whereas black soils occur in the lower ridges. These types of soils occupy about 1% land of the state in the western districts of Sambalpur, Bargarh, Sonepur and Bolangir. These soils are light to medium textured having a neutral pH. Black soils are rich in calcium and red soils are abundant in iron. These soils extend to deeper depths with medium to high fertility.

27.4   USE OF SOILS AND ROCKS AS CONSTRUCTION MATERIALS

Laterite soils, predominant in the eastern districts of Odisha, are commonly extensively used for construction purposes. Moreover, most of the medieval architectures built in the state are extensively based on lateritic stone carving. The construction industry, especially infrastructure development related, is on the rise in the state, which emphasises the necessity for adopting advanced practices. However, the majority of the brick industry in the state is clay based. On similar notes, it is important to note that river sand is the major source of fine aggregates to be used in the construction industry. The coarse aggregates used in local construction are generally quarried from nearby locations, indicating the availability of strong rocks at shallow depths at various parts of the state.

Odisha houses various major dams on different rivers, which calls for not only an in-depth study of the local geology and geotechnology but also constant monitoring for structural safety and proper functioning. The research fraternity is frequently posed with various technological and technical challenges for such major constructions on versatile soil types of Odisha. However, it is also essential to note that in addition to water reservoirs, modern-day geotechnology also has to adhere to the design of impoundments for storage and maintenance of chemical effluents in all three forms including liquid, semi-solid and solid to prevent possible accidents. The state houses numerous thermal power plants and aluminium plants leading to the requirement of large impoundments designed for the disposal and storage of fly ash and bauxite residue. These impoundments need to be specifically designed as per the quality of the waste, solid–liquid ratio, drainage and permeability conditions, volume requirements and so on, demanding meticulous execution and maintenance.

27.5   OTHER GEOMATERIALS

The state is abundant in various natural resources such as minerals, fossil fuels and ores along with forests and water bodies. The availability of minerals and ores has attracted massive industrialisation in various parts of the state. There are numerous major thermal power plants, aluminium refineries, steel plants and so on. The wastes generated by these industries such as fly ash, blast furnace slag, phosphogypsum, steel slag and red mud include a few geomaterials worth mentioning. The state pollution control board, along with national bodies, emphasises on minimisation and utilisation of these wastes. The state has witnessed a drastic improvement in the utilisation of fly ash and blast furnace slag in various forms such as filling material, cement manufacturing, concrete blocks and tiles and brick manufacturing. However, there is a dearth of technological advancement for promoting utilisation of several other industrial solid wastes such as red mud and phosphogypsum.

27.6   NATURAL HAZARDS

Odisha has faced 17 major natural calamities, including four severe to very severe cyclones, since 2000 along with droughts, floods, hailstorms, tornadoes and unseasonal rains. According to statistics published by the Meteorological Department of India, the state has been affected by 35% of all the cyclonic and severe cyclonic storms that have crossed the eastern coast of the country. A lot of geomaterials and other resources have been spent by the state government on the restoration and rebuilding of the infrastructure post calamities, which otherwise could have been spent on the developmental front. According to the report, apart from losses to life and property, natural disasters also lead to crop failure, decline in surface and groundwater levels and inundation and salination of the affected land masses. Major rivers like Mahanadi and its tributaries have the potential to cause severe floods (out of a total geographical area of $1,55,707\,\mathrm{km}^2$, around 5% is very flood prone). Recent trends in the past few years have demonstrated that the frequency, intensity and extent of droughts in the state are gradually on the rise leading to severe negative impacts on the agricultural sector. In addition, Odisha is also affected by disasters like heat waves, pest attacks, moisture stresses and forest fires.

27.7   CONCLUDING REMARKS

The chapter deals with the geological features and geotechnical properties of soils and rocks available in different districts of Odisha. River sands are the major source for fine aggregates and strong rock quarried at shallow depths are used as coarse aggregates. The state is rich in various natural resources such as minerals, fossil fuels and ores along with forests and water bodies. Following are the major facts pointed out on the soil types of Odisha based on various case studies and field test results:

- Red soils are decomposed from granites and gneisses and many other rocks, which are rich in iron and magnesium. They are strong and moderately acidic, containing low to medium organic matter and have highest coverage among all the soil groups of the state. Mixed red and yellow soils occupy around 35% of land, being the second highest in the state. The black soils in the state cover an area of approximately 6% of the land with the pH varying from neutral to alkaline.

- Laterite soils, formed by leaching of siliceous matter and lime from rocks, are mostly acidic with acidity varying from slight to strong and pH ranging from 4.5 to 5.8. These soils occupy 4.5% of the land of the state and are prevalently used for construction purposes.
- Deltaic alluvial soils cover around 4% of the land and occur in deltaic regions of the rivers Mahanadi, Subarnarekha, Brahmani, Rushikulya and Baitarani. Saline soils are distributed in the districts of Balasore, Bhadrak, Jagatsingpur, Kendrapada, Puri, Khurda and Ganjam, which are spread over nearly 1.6% of the state land.
- Soils available in the east coast are mostly alluvial, red in colour and lateritic in nature. Marine clay deposits are generally encountered near the coastal areas. Gravelly and lateritic soils are mostly found in the northeastern hilly region of the state.

REFERENCES

Bharali, B., Srivastava, S.K. and Rath, S. (1987). Exploration for oil and gas in Orissa offshore areas. Proc. SGAT Sem. Development in mineral exploration techniques, pp. 29–39.

ENVIS (2021). *Distribution of forests in Odisha*, http://orienvis.nic.in/, accessed 16 January 2021.

Faruque, B.M. and Lahiri, A. (2002). Post Wisconsin glaciation sea level stands off parts of east coast, Andhra Pradesh. In *2002 In-House Workshop on Quaternary Geology of Coromandel Coast and Drainage Basins of Andhra Pradesh* (pp. 7–s). Hyderabad: Geological Survey of India.

Frost, B.R. and Frost, C.D. (1987). CO_2, melts and granulite metamorphism. *Nature*, Vol. 327, pp. 503–506

Fuloria, R.C., Pandey, R.N., Bharali, B.R. and Mishra, J.K. (1992). Stratigraphy, structure and tectonics of Mahanadi Offshore Basin. In *Recent Geoscientific Studies in the Bay of Bengal and Andaman Sea*. Geol Surv Ind, Spl. Publ., No. 29, pp. 255–265.

Goswami, R., Ghosh, M. and Saga, D. (2013). Report on the soil fauna of Bhadrak and Balasore district, Orissa. *Records of the Zoological Survey of India*, Vol. 113, Part-4, pp. 213–227.

Halden, N.M., Bowes, D.R. and Dash, B. (1982). Structural evolution of migmatites in granulite facies terrane: Precambrian crystalline complex of Angul, Orissa, India. *Earth and Environmental Science Transactions*, Vol. 73, No. 2, pp. 109–118.

IRC:78 (2014). Standard Specifications & Code of Practice for Road Bridges–Section VII: Foundation & Substructure, Indian Roads Congress, New Delhi.

Mahalik, N.K. (1996). Lithology and tectonothermal history of the Precambrian rocks of Orissa along the eastern coast of India. *Journal of Southeast Asian Earth Sciences*, Vol. 14, No. 314, pp. 209–219.

Mahalik, N.K. (2000). *Mahanadi Delta, Geology, Resources & Biodiversity*. AIT Alumni Assn Publ, New Delhi, 169p.

Mahalik, N.K. (2006). A study of morphological features and borehole cuttings in understanding the evolution of and geologic processes in Mahanadi Delta, *East Coast of India. Journal of the Geological Society of India*, Vol. 67, pp. 595–603.

Park, A.F. and Dash, B. (1984). Charnockite and related neosome development in the Eastern Ghats, Orissa, India: petrographic evidence. *Earth and Environmental Science Transactions*, Vol. 75, No. 3, pp. 341–352.

Chapter 28

Puducherry

J. Saravanan, T. Kaviarasu, and N.J.L. Ramesh
Annamalai University

R. Premkumar
Pondicherry University

CONTENTS

28.1 INTRODUCTION

Puducherry, formerly known as Pondicherry until 2006, is one of the union territories of India. It covers an area of 483 km^2, consisting of four territories of former French India, namely Pondicherry (293 km^2), Karaikal (161 km^2), Mahe (9 km^2) and Yanam (20 km^2), as shown in Figure 28.1. Puducherry lies in the southern part of the Indian Peninsula. The areas of Puducherry and Karaikal districts are bound by the state of Tamil Nadu, while Yanam and Mahe districts are enclosed by the states of Andhra Pradesh and Kerala, respectively. Topographically, the places spread over the southern and eastern parts of India, Chandernagore in Bengal, Yanam at the mouth of the Godavari River, Pondicherry and Karaikal on the Coromandel Coast and Mahe on the Malabar Coast. Puducherry and Karaikal have the largest areas and population, and are both enclaves of Tamil Nadu. Yanam and Mahe are enclaves of Andhra Pradesh and Kerala, respectively. Its population, as per the 2011 Census, is 1,244,464.

The city of Puducherry is divided into two parts by a canal, and all the main streets, running parallel to one another, lead to the open roadstead offshore. The port of Puducherry does not have a harbour, and ships are forced to lie about 1.5 to 3 km offshore, but its roadstead was once considered the best on the Coromandel Coast. There are a promenade, a landing place for cargo, and a pier. In and around the city are artesian wells that supply a large quantity of water for irrigation. The chief local crops are rice, sugarcane, cotton, and peanuts (groundnuts). Puducherry has about 300 villages and hamlets. The Karaikal region, south of the Puducherry sector, is in the fertile Kaveri River delta, one of the most important rice-producing areas of India. The exceptional fertility of the region is to some extent reflected in the unusually high density of its rural population. The town is on the Mayiladuthurai–Peralam route, a branch line of the southern railway. The Mahe sector consists of two parts, the quaint

DOI: 10.1201/9781003177159-28

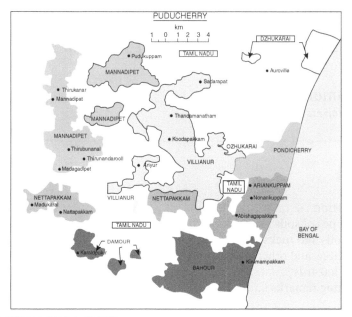

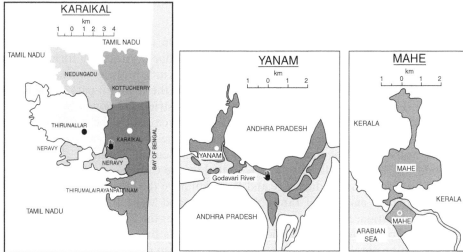

Figure 28.1 Map showing the districts of Puducherry (redrawn).

picturesque town of Mahe, with its buildings situated on the left bank of the Mahe River close to its mouth, and the isolated tract known as Naluthrara, on the right bank. Yanam is a small town on the bank of a branch of the Godavari River, about 650 km north of the city of Chennai (Madras), near Kakinada. All four regions of Puducherry are located in the coastal region. Five rivers in the Puducherry district, seven in the Karaikal district, two in the Mahe district and one in the Yanam district drain into the sea, but none originates within the territory.

28.2   MAJOR TYPES OF SOILS AND ROCKS
WITH THEIR PROPERTIES

Subsoil investigations have been carried out at several locations of this union territory. Some of the boring logs of selected locations such as Chinna Kalapet, Thattanchavadi Industrial Estate and Mannadipet are presented in table form with graphical representations as shown in Figures 28.2, 28.3 and 28.4, respectively, with properties of soils.

The east region of Puducherry consists of red earth along with sandy loam soil and this area is well drained. The west region of Puducherry is an industrial area consisting of full of red earth along with pebble stones. This red earth is being excavated and supplied to the construction projects. The ground water table is found at a depth of 9–10 m. The north region of Puducherry consists of red earth along with silty sand and the area with no water table within the top 10 m. The south region of Puducherry consists of red earth along with clay and silty sand soils. The ground water table is available at 1 m depth in this region. The Karaikal region completely consists of silty sand along with very thin clayey soils in between. The ground water table is available at 1 m depth throughout the region. The Mahe region consists of laterite sand up to 1 m followed by hard rocks. The water table available is found at a depth of 8 m.

28.3   USE OF SOILS AND ROCKS AS CONSTRUCTION MATERIALS

Gravel is a loose aggregation of rock fragments and is used for filling and construction of roads and railway tracks. The red earth deposits found in the places are fertile and fine to medium grained in nature, as shown in Figure 28.5. The red earth is suitable for the use of garden filling and vegetation purposes. The red earth has more water-retaining capacity. These characters are very much good enough for planting vegetation. In some areas, the red earth is also used for brick manufacturing.

Well-rounded pebbles are available in varying sizes (0.5 to 7 cm) as shown in Figure 28.6 and are used for making the exterior decoration, particularly elevation work. They are quarried and graded with different sizes and sold at a small-scale level. They are also used for making special concrete works, deep boreholes as a filter medium, and landscaping.

Clay is used for the brick manufacturing process on a large scale as shown in Figure 28.7. After moulding and drying, the raw bricks are burnt by temporary or permanent kilns as shown in Figure 28.8. The burnt bricks are widely used for masonry wall construction in all types of construction works. It fulfils the construction needs for the entire Puducherry and Karaikal regions.

28.4   FOUNDATIONS AND OTHER GEOTECHNICAL STRUCTURES

The Puducherry region faces coastal area inundation problems due to which the shore side is protected against erosion by construction of revetments consisting of locally available soil and brick materials. The seawater intrusion badly affects the deep foundation structures, including affecting the reinforcing elements. Special effort is required to manage the deep foundation construction works.

Project No:	AAL.1224/2018	Date:	05.05.2018	Location:	Chinna Kalapet,Puducherry																
Borehole #	1	GWT(m)	Not encountered	Graphical representation of standard penetration resistance (*N*)					Index properties (%)						Shear strength parameters		Gradation properties (%)				
Depth of boring(m):		7.50													Direct shear test		Sieve analysis				
Depth below GL	Soil stratum	Classification of soil	Thickness of layer(m)	Depth of sampling(m) UDS	DS	*N* value	10 20 30 40 50	Description/consistency	Natural moisture content	Density (t/m³)	Liquid limit	Plastic limit	Plasticity index	Consistency index	Free swell index	c(kg/cm²)	Φ(degrees)	Gravel	Coarse sand	Medium sand	Fine sand	Silt and clay
		Red earth silty sand (SM-SW)			0.5																
1.0					1.0	29															
					1.5			Medium	7	1.90	Non-plastic				Nil	0.030	30°45'	0.00	0.00	62.56	37.44	0.00
2.0					2.0																	
					2.5	50		Dense	9	1.47	Non-plastic				Nil	0.119	31°10'	20.11	3.45	52.30	24.14	0.00
3.0					3.0																	
					3.5	35		Dense	10	1.66	Non-plastic				Nil	0.116	32°17'	4.55	7.58	59.09	28.79	0.00
4.0					4.0																	
					4.5	40		Dense														
5.0					5.0																	
					5.5	Rebound		Very	11	1.84	Non-plastic				Nil	0.125	32°39'	8.08	16.67	53.03	22.22	0.00
6.0					6.0																	
					6.5	Rebound		Very														
7.0					7.0																	
					7.5	Rebound		Very	13	1.59	Non-plastic				Nil	0.174	33°25	4.04	2.02	40.40	53.54	0.00

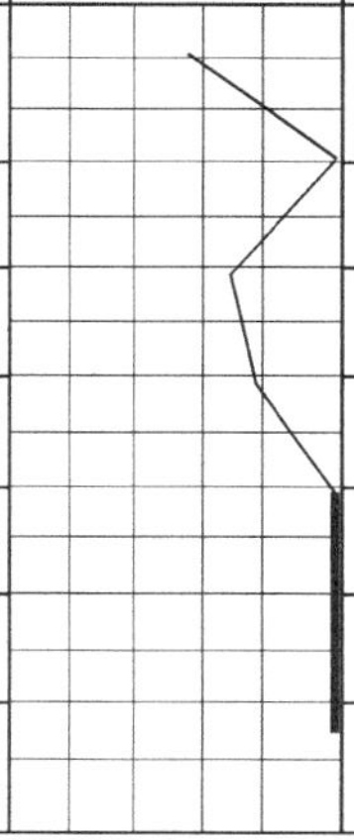

Figure 28.2 Boring log with properties of soil at Chinna Kalapet.

Project No:	AAL1186/2018	Date:	05.03.2016	Location	Industrial Estate, Thattanchavady
Borehole #	1	GWT(m):	5.00		
Depth of boring(m):		10.0m			

Depth below GL	Soil stratum	Classification of soil	Thickness of layer(m)	Depth of sampling (m) UDS	DS	N value	Graphical representation of standard penetration resistance (N)	Description/consistency	Natural moisture content	Density (t/m³)	Liquid limit	Plastic limit	Plasticity index	Consistency index	Free swell index	c (kg/cm²)	Φ (degrees)	Gravel	Coarse sand	Medium sand	Fine sand	Silt and clay
		Filled-up earth	0.50		0.5																	
1.0		Silty sand (SM-SW)			1.0	4		Very Loose	15	1.94		Non-Plastic			Nil	0.186	30°45'	1.21	1.52	50.61	44.55	2.12
			1.50		1.5																	
2.0					2.0			Very Loose														
					2.5	4																
3.0					3.0																	
		Sand (SW)			3.5	10		Loose	25	1.71		Non-Plastic			Nil	0.119	31°32'	0.00	0.00	65.05	34.47	0.49
4.0					4.0																	
					4.5	10		Loose														
5.0			6.00		5.0																	
					5.5	15		Medium	16	1.90		Non-Plastic			Nil	0.064	31°10'	0.00	0.39	43.41	29.46	26.74
6.0					6.0																	
					6.5	16		Medium														
7.0					7.0																	
					7.5	16		Medium	12	1.42		Non-Plastic			Nil	0.125	32°39'	0.00	0.00	42.78	30.48	26.74
8.0					8.0																	
		Sand (SW) with pebbles			8.5	30		Medium														
9.0			2.00		9.0																	
					9.5	33		Dense	10	1.98		Non-Plastic			Nil	0.091	33°46'	15.16	5.74	36.89	27.87	14.34
10.0					10.0																	

Figure 28.3 Boring log with properties of soil at the Thattanchavadi Industrial Estate.

Project No:	AAL.1132/2017	Date:	25.12.2017	Location:	Mannadipet, Puducherry
Borehole #	1	GWT(m)	11.50		
Depth of boring (m):		21.50			

Depth Below GL	Soil stratum / Classification of soil	Thickness of Layer(m)	Depth of sampling(m) UDS	Depth of sampling(m) DS	N value	Graphical representation of standard penetration resistance (N) 10 20 30 40 50	Description/consistency	Natural moisture content	Density (t/m³)	Index properties (%) Liquid limit	Plastic limit	Plasticity index	Consistency index	Free swell index	Shear strength parameters, Direct shear test c (kg/cm²)	Φ(degrees)	Gradation properties (%), Sieve analysis Gravel	Coarse sand	Medium sand	Fine sand	Silt and clay
	Brown silty sand (SM-SW)			0.5																	
1.0		1.00		1.0	11																
	Brown clayey sand (SC)			1.5			Medium	10	1.66	Slightly plastic				40	0.122	28°48'	0.50	1.97	21.67	33.50	42.36
2.0		0.80		2.0	13		Medium														
	Brown silty sand (SM-SW)	1.40		2.5				19	1.59	Non-plastic				Nil	0.113	29°12'	0.00	0.00	23.56	35.09	41.35
3.0				3.0																	
	Dark grey clay (CH)			3.5	9		Very stiff	27	1.63	60	33	27	1.22	70	0.385	9°52'	0.00	0.00	0.00	10.00	90.00
4.0		1.30		4.0																	
				4.5	9		Very stiff														
5.0	Light brown silty sand (SM-SW)			5.0																	
				5.5	21		Medium	14	1.71	Non-plastic				Nil	0.106	32°17'	0.00	0.00	14.93	47.06	38.01
6.0		1.50		6.0																	
	Dark grey clay (CH)			6.5	15		Very stiff														
7.0				7.0																	
				7.5	12		Very stiff	28	1.46	63	31	32	1.09	70	0.192	10°52'	0.00	0.00	0.00	8.00	92.00
8.0				8.0																	
				8.5	13		Very stiff														
9.0		4.00		9.0																	
				9.5	41		Hard	19	1.60	66	35	31	1.52	71	0.287	10°22'	0.00	0.00	0.00	5.00	95.00
10.0				10.0																	

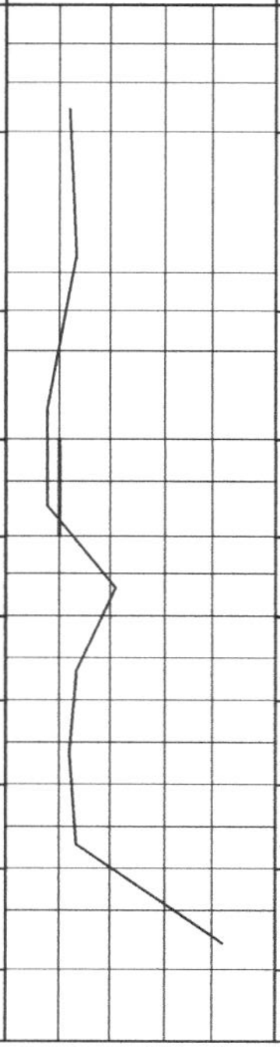

Figure 28.4 Boring log with properties of soil at Mannadipet.

Figure 28.5 Red earth.

Figure 28.6 Pebbles.

28.5 NATURAL HAZARDS

In the past, Tsunami not only affected the livelihoods of the fishing community of Puducherry and Karaikal regions in particular and coastal communities in general, but also to some extent exposed their pre-existing vulnerability from pre-Tsunami times. Flash floods and cyclones often hit Puducherry, Karaikal, Yanam and Mahe regions with heavy to very heavy rainfall. The flashing floods affect the entire union territory.

Figure 28.7 Preparation of clay bricks.

Figure 28.8 Country bricks kiln.

28.6   CONCLUDING REMARKS

The east region of Puducherry has mainly red earth along with sandy loam soil, while the west region of Puducherry consists of red earth along with pebble stones. The north region of Puducherry consists of red earth along with silty sand and the south region of Puducherry consists of red earth along with clay and silty sand soils. The Karaikal region completely consists of silty sand along with very thin clayey soils. The Mahe region consists of laterite sand up to 1 m followed by hard rocks. The depth of the groundwater table varies from 1 m to greater than 10 m.

Chapter 29

Punjab

Harvinder Singh and K.S. Gill
Guru Nanak Dev Engineering College

J.N. Jha
Formerly Muzaffarpur Institute of Technology

CONTENTS

29.1 INTRODUCTION

The state of Punjab is a vast monotonous fertile, alluvial flood plain of major five rivers originating from the Himalayas and lies in northwestern India between the latitudes of 29° 32′ and 32° 32′ N and longitudes of 73° 52′ and 76° 56′ E. The state gets its name

DOI: 10.1201/9781003177159-29

'Punjab' from these five rivers passing through it and the name is a combination of two words *Punj* (five) and *aab* (rivers). It has a total geographical area of 50,362 km² and is bounded by Pakistan on its west, Jammu and Kashmir on the north, Himachal Pradesh on the northeast and Haryana and Rajasthan on the south. A belt of undulating land (known as Shiwalik hills) extends along the northeastern part of the state at the foot of the Himalayas. The average elevation of the state is 300 m above the mean sea level, with a range from 180 m in the southwest to an elevation of more than 500 m around the northeast border (ENVIS, 2019). The southwest of the state is semi-arid that eventually merges into the *Thar* desert in the Rajasthan. Punjab falls under three different seismic zones, namely II, III and IV.

It receives annually south-easterly current of the summer monsoon coming over the Bay of Bengal. The mean annual rainfall in the state varies from less than 300 to about 1,400 mm per annum; a major portion (70%) of which is received in the months of July to September as the summer monsoon. Then, dry conditions prevail from October to the end of June, except for some few light showers received from the westerly depression during the months of December to January. The summer period is extremely hot and winter period is slightly cool. The temperature variations, especially the mean annual temperature over the whole region, are moderate and vary from 23.2°C (Pathankot) to 25.8°C (Abohar). The mean monthly maximum temperature in the month of June is as high as 42.2°C while the mean monthly minimum temperature during the month of January is as low as 2.3°C (ENVIS 2019).

The soil deposit in the state is alluvial, mostly having stratified soils of silt and sand layers. In a few pockets of southern and western districts, saline soils are also found. The land to population ratio of Punjab is very low; therefore, it is not practically feasible to change the allotted site due to poor soil conditions. Moreover, the chances of liquefaction are also present in some regions of the state due to the easy availability of soil moisture in cohesionless soil strata and the possibility of severe seismic activity. Therefore, to enhance the safety of people, engineering intervention can play a major role in the development of infrastructure. Second, optimum utilisation of locally available resources, such as blending of locally available soils and gravels, fly ash and rice husk ash and so on, is also of paramount interest. Not only materials, but also the judicious use of land is very important, as the cost of land is very high in Punjab and there is a widespread network of canals passing through many major cities of the state. The knowledge of subsoil characteristics, i.e., its index and engineering properties along with the location of the groundwater table is a prerequisite before the commencement of any project, but unfortunately, this component has been ignored due to the lack of awareness or financial reasons. In this chapter, an effort has been made to present the data pertaining to soil deposits and other materials available in the state.

29.2   MAJOR TYPES OF SOILS AND ROCKS

The plain land area in the state is an outcome of the alluvial deposits of *Indo-Gangetic* river systems originating from the Himalaya range. The deposition of this alluvium is supposed to be commenced after the final upheaval of the Himalayas (*Siwaliks*) and continued all through the Pleistocene up to the present. The soil characteristics in various regions of the state are the outcome of the changes that took place due to the prevailing climatic conditions, the presence of different rivers flowing through the state, the position of the water table, vegetation and the type of parent rock. A rocky stratum is generally not present at shallower depths. However, gravels, pebbles and small stones (*Kankar*) are found mixed in the soils of some of the districts, mainly along the border with Himachal Pradesh and Jammu & Kashmir. Punjab is divided into three distinct regions based on soil types: southwestern, central and eastern.

The alluvium is recognised as '*Bhangar*' – old alluvium that occupies a relatively higher elevation in the northeastern region, and it contains '*Kankar*'; and '*Khadar*' mixed in the sand and or silty deposits. The newer alluvium occupies the lower ground levels towards the southern region of the state; it contains clayey stratum along with silty and/or sandy layers and '*Kankar*' in smaller quantities. The foothills are composed of layers of weakly consolidated, medium to coarse grained sandstones, siltstones, conglomerates, shales and so on with the occasional organic remains (Agarwal, 1961). The following sections briefly outline the nature of soils available in different regions of the state.

29.2.1   Bet soils

These are *Khadar* soils of the periodically flooded or flood plain areas of various rivers, streams or *choes* in the state. These are found in the form of elongated belts on both sides of the rivers, such as those of Sutlej, Ravi, Beas and Ghagghar. They are pale to yellowish-brown in colour. The soils are well drained, vary in texture and possess organic matter in some quantity. Depending upon the source of alluvium, the soils are calcareous or non-calcareous.

29.2.2   Loamy soils

It is the most fertile and productive soil group of the state and covers about 25% of its area. It is important to note that this type of soil is available in the top depths only and subsequently, sandy and/or silty deposits are present from a depth of 3 m onwards (Singh, 2019). It is found mainly in the Doaba region of the state comprising districts of Nawanshahr, Jalandhar, Phagwara and central parts of the Kapurthala district and is intensively cultivated for wheat and paddy crops. In the Malwa region, this type of soil exists in western Patiala, Nabha area, Sangrur area, southern Moga district, some patches in Muktsar area and Bathinda district. The soils become clayey towards the northwest in Amritsar and Gurdaspur districts.

29.2.3   Sandy soils

This type of soil is found mainly in the southwestern and south-central Punjab, covering the districts of Bathinda, Mansa, southern parts of Ferozepur and Muktsar districts, a larger part of Sangrur, south-central parts of the Patiala district and some patches of the Ludhiana district. It is yellowish to grey in colour and has pH values ranging from 7.8 to 8.5 (ENVIS, 2019). The overall grey colour reflects the deficiency of organic matter. However, from a depth of 3 m onwards, it is present throughout the state albeit with different densities.

29.2.4   Desert soils

These soils cover more than 11% of the total area of the state. They have developed under arid, hot climatic conditions with thin vegetations prevailing in the southwestern parts of the state, mainly comprising Abohar, Zira, Mukatsar and large parts of Bathinda and Mansa districts.

29.2.5   Kandi soils

This group of soils consists of a mixture of sand, sandy loam, silt loam, clay-silt and gravels. It is found mainly in the areas of Pathankot, Hoshiarpur, Nawanshahr and Roopnagar (Ropar) districts; mainly regions along the state border with Himachal Pradesh and Jammu & Kashmir. The texture becomes coarser eastward of the *Sivalik* Hills where gravel, pebbles and conglomerates predominate because of the presence of numerous *choes* coming from the *Sivalik* Hills.

29.2.6   Forest soils

These soils are stony, gravelly and sandy available in the forest area of some blocks of Gurdaspur districts and *Sivalik* Hill zone in Hoshiarpur, Nawanshahr and Ropar districts. These soils have developed under shrubs and forests, steep slopes and rugged topographical conditions. These soils are reddish-brown to olive-brown in colour.

29.3   PROPERTIES OF SOILS AND ROCKS

Main soil investigation data pertaining to various locations in the Punjab state were analysed. Figure 29.1 depicts the positions of various test points considered in the analysis along with the position of ground water level as of January 2014. The study reveals that the soil in the state is mostly cohesionless in nature, consisting of layers of silt, sand and/or mixture of these soils at different depths. In some regions of the state, clayey soil is also present along with silts or sands in mixed form. Mostly, the top 2–3 m layer of the soil is highly variable, region to region, and manifests in different forms as is outlined in the previous section. It mainly possesses low plasticity.

Sand is the most dominant soil type beyond a depth of 3 m and is present almost throughout the state with some exceptions in some of the districts where silt or sandy silt or silty sand is found. Silt along with Kankar is also found at lower depths in some of the areas. The area near the northern border with Jammu & Kashmir (such as upper areas of the Pathankot district) possesses a mixture of silt and gravel. It is important to mention that the soil strata being alluvium of flood plains is highly unpredictable in the state; many loose pockets are present in many places that exhibit very low N-values. Based upon the analysis of bore log data collected from various regions of the state, Tables 29.1 to 29.4 present a summary of the soil classification and their geotechnical properties for a ready reference.

The state of Punjab lies in three seismic zones; almost 50% area of Punjab falls under seismic zone IV, the remaining area falls under zone III and only a very small

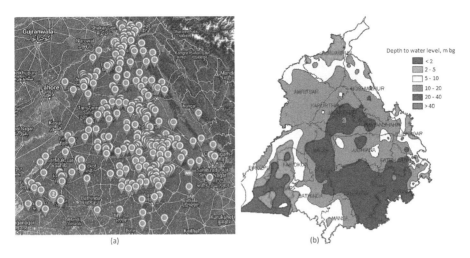

Figure 29.1 Map of Punjab: (a) position of the test points considered and (b) depth of the water levels below the NGL. (Source: Author.)

Table 29.1 Soil properties (Muktsar, Ferozepur, Faridkot, Bathinda and Mansa districts)

Soil properties as observed in the region (distance from the natural ground level)			Remarks
Depth	Up to 8 m	From 8 m onwards	
Soil type	Soil of low plasticity (ML/CL) or silty sand (SM)	Poorly graded fine loose sand (SP)	• In some pockets where the water table is very high, standard penetration resistance is negligible
Bulk unit weight, kN/m^3	16–17	16–18	• In this region, the ground water table rose almost to the ground surface level in the late eighties
N – value	0–6	3–15	
'c', kPa	2–8	0–2	• The ground water table in this region is saline in nature; therefore, proper care needs to be exercised while designing any substructure components in such types of aggressive soils
'Φ' value	22°–28°	28°–30°	
Consistency limits	LL = 19%–25% PI = 2%–4%	Non-plastic	• For the analysis purpose, only two layers are considered representing average values of the soil parameters in that layer

area of Abohar comes under the seismic zone II. The easy availability of the ground moisture coupled with the cohesionless nature of the soils, existing in most parts of the state makes some areas therein highly susceptible to liquefaction in the event of any potential seismic activity. The severity index in the case of the liquefaction occurring under a PGA of 0.24g is depicted in Figure 29.2. The approach proposed by Seed and

Table 29.2 Soil properties (Ludhiana, Jalandhar, Kapurthala, Amritsar, Sangrur, Patiala and Hoshiarpur districts)

Soil properties as observed in the region (distance from the natural ground level)			Remarks
Depth	Up to 4 m	From 4 m onwards	
Soil type	Low plasticity with occasional nodules classified as ML or CL	Poorly graded sand is encountered classified as SP. Seams of cohesive soil may also be encountered at some of the locations	• The ground water table is at great depth varying between 20 and 40 m. However, in some regions along the rivers and unlined canals, it is available at shallow depths between 2 and 10 m
Bulk unit weight, kN/m^3	15.5–17.5	16.5–19	
N – value	3–10	6–28	• For analysis purposes, only two layers are considered representing the average values of the soil parameters in that layer
'c', kPa	2–10	0–2	
'Φ' value	22°–27°	29°–32°	
Consistency limits	LL = 19%–27% PI = 3%–5%	Non-plastic	

Table 29.3 Soil properties (belt along the Sivalik Hills in Ropar, Gurdaspur and Hoshiarpur districts)

Soil properties as observed in the region (distance from the natural ground level)			Remarks
Depth	Up to 3 m	From 3 m onwards	
Soil type	Soil of low plasticity with occasional small pebbles classified as ML or CL	Silty sand along with gravels of medium size is encountered making the soil quite hard to penetrate with a split spoon sampler. Classified as SW	• Ground water table in these areas is available at shallow depths and lies between 2 and 15 m • DCPT or other test methods become mandatory in many parts of the area beyond a depth of 3 m
Bulk unit weight, kN/m^3	16.5–18.5	18.5–21.5	• For analysis purposes, only two layers are considered representing average values of the soil parameters in that layer
N – value	3–10	20 to 'the point of refusal'	
'c', kPa	2–8	0	
'Φ' value	24°–29°	33°–39°	
Consistency limits	LL = 29%–27% PI = 3%–5%	Non-plastic	

Table 29.4 Soil properties (Patiala, Mohali and Fatehgarh Sahib districts)

Soil Properties as observed in the region (distance from the natural ground level)			Remarks
Depth	Up to 6 m	From 6 m onwards	
Soil type	Soil of low plasticity with occasional small pebbles classified as ML or CL	Cohesive soils classified as CL or CI	• Soils in these areas are cohesive up to great depths, but the seams of sandy soil may be encountered at some intervals
Bulk unit weight, kN/m³	15.5–17.5	16–18	• For analysis purposes, only two layers are considered representing average values of the soil parameters in that layer
N value	5–14	7–25	
'c' value, kPa	7–14	15–25	
'Φ' value	20°–24°	15°–21°	
Consistency limits	LL = 24%–32% PI = 5%–7%	LL = 30%–42% PI = 8%–12%	

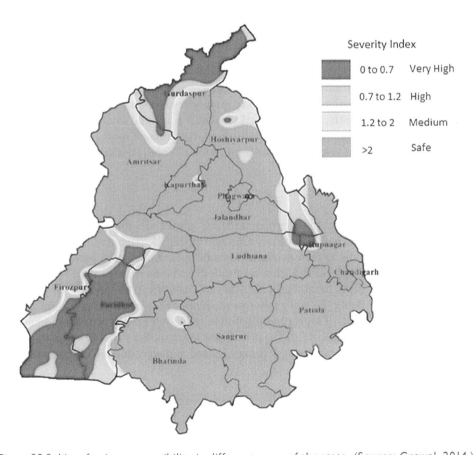

Figure 29.2 Liquefaction susceptibility in different areas of the state. (Source: Grewal, 2014.)

Idriss (1971) and Seed et al. (1983) has been used to assess the severity of the liquefaction at various locations presented in the figure.

The results show that the southwestern area of the state comprising the districts of Abohar, Fazilka, Faridkot, Gurdaspur, Muktsar and some areas of Ropar and Pathankot are highly vulnerable to liquefaction problems. For a PGA of 0.24g, around 15% of the state area shows a factor of safety of less than unity and around 25% of the area has this value between 1 and 2. This area with a factor of safety of less than unity increases to 33% in the case of any seismic activity having a PGA of 0.34%. Therefore, proper planning and design of buildings and other infrastructural projects falling under these locations are recommended. Soils along the riverbanks are also highly prone to liquefaction, but it is not shown in the figure due to insufficient data available along these stretches of the land.

29.4   USE OF SOILS AND ROCKS AS CONSTRUCTION MATERIALS

The description of natural construction materials available in Punjab is given below:

29.4.1   Fine sand

Fine sand is available as the riverbed material deposited in beds, along the banks of various rivers flowing through the state. It is also available in the lower layers of the ground, normally found at depths varying from 3 m onwards in various regions of the state. The fineness modulus of this type of sand varies from 1.5 to 2.5. There are around 50 sand mines all over the state, which are mainly controlled by the state government (DMGP, 2019). Currently, the local market rate of fine sand is about Rs. 22,000 per 1,000 cubic feet and it is commonly used for preparing cement mortar, plastering work, flooring and so on. It is also used in the production of concrete along with coarse sand.

29.4.2   Coarse sand

Coarse sand is coarser than the riverbed sand (fine sand) and is mainly available from the two sources – from the bed of the river Ghaggar (locally known as *Ghaggar* sand) and from Pathankot (locally known as *Pathankot* sand). *Ghaggar* is the perennial river descending from the Himalayas in Himachal Pradesh and it finally enters the state from the Mohali side; on the other hand, Pathankot gets this sand from the beds of the Ravi river and various *chows*. Being the entry point to the plains, the sand deposits found at these two locations are coarser. This is generally found to be well graded coarse sand having a fineness modulus of more than 2.5. The price in the local market varies between Rs. 27,000 to Rs. 32,000 per 1,000 cubic metre. It is mainly used in the production of concrete.

29.4.3   Coarse aggregates

Coarse aggregates of different sizes are available as natural stones/gravels/pebbles as a river bed material, especially from the Pathankot district and also as crushed stone aggregates from various quarries located in the Ropar and Pathankot districts (foothills of the Himalayan range). The sizes of the natural aggregates vary from 10 mm to as high as 40 mm; for crushed stones, it varies from 10 to 20 mm and is available as single sized particles or as graded particles with nominal size varying from 10, 12 and 16 to 20 mm. The rate of these aggregates varies between Rs. 22,000 and Rs. 28,000 per 1,000 cubic feet, depending upon the distance of the site from the quarry. Recycled materials like demolition waste from the construction industry and furnace slag from various furnaces at Ludhiana and Mandi Gobindgarh are also being used in construction activities as a partly replacement of coarse aggregates.

29.4.4 Fine-grained soil

Soil is being used as the backfill material in various projects. It also finds applications in the construction of embankments, reinforced earth walls and so on. The soil available in most of the areas of the state is good for highway construction; the plasticity index of the soil ranges from 0% to 6% with a liquid limit of less than 25% and a swelling index of less than 10%. Due to the higher land prices in the state, the cost of soil ranges from Rs. 300 to Rs. 350 per cubic metre.

29.4.5 Granular subbase material

The CBR value of subgrade soil in various regions of the state varies from 3% to 8%. The low CBR values coupled with a manifold increase in the commercial traffic on state and national highways passing through the state have necessitated requirements of huge quantities of granular subbase material (GSB); the crust thickness required for these pavements normally go up to 1 m. The continuous mining of natural GSB is creating a lot of ecological imbalance and has resulted in a manifold increase in the cost over the past few years; at present, the GSB normally costs around Rs. 12,000 per 1,000 cubic feet. The recent technological advancements have paved the way for more sustainable design of pavements by encouraging the use of Geosynthetics to improve the sub-bases; it has been employed in some projects in addition to the use of other ground improvements/stabilisation techniques to reduce the crust thickness and other associated costs.

29.5 FOUNDATIONS AND OTHER GEOTECHNICAL STRUCTURES

Being alluvium deposits, the soil in the state consists of different strata of silty, sandy or a mixture of these two types of soils. As described in the previous sections, layers, gravels/pebbles are also found mixed in the sand and/or silty deposits, especially in the northeastern regions of the state. The water table is also available at shallower depths in this part of the state. The seams of clayey and/or sandy-clayey deposits are found in the southern regions.

The geotechnical investigation test data (Singh, 2019) across the state reveal that in much of the area, the soil is of 'soft' type with N-values reported to be less than or around 10. Table 29.5 presents an overview of soil strata available at various depths in different districts. In the belt along the border with Himachal Pradesh and

Table 29.5 A typical set of N-values observed in various districts

District		Depth below NGL., m						
		Up to 3	6	9	12	15	18	21
Pathankot	Upper areabordering the J&K	17	20	25	27	30	32	35
	Middle area	8	18	21	25	27	30	32
	Lower area	5	7	8	9	11	13	15
Gurdaspur		5	12	14	12	13	14	15
Amritsar		7	13	16	15	16	20	28
Tarntaran		6	12	16	18	20	22	23
Kapurthala		10	16	18	20	22	23	25
Jalandhar		8	10	18	20	22	24	25
Hoshiarpur		4	14	16	20	24	25	26
Roopnagar		4	8	6	6	7	8	9
Nawanshehar		5	12	14	12	13	14	15
Ludhiana		7	12	15	18	20	21	23
Fathegarh Sahib		6	8	14	15	16	18	21
Patiala		7	15	15	17	20	21	23
Ferozepur		7	14	18	20	21	22	23
Faridkot		8	10	12	13	14	15	16
Sangrur		10	11	14	13	14	15	16
Barnala		9	16	18	20	20	21	22
Mohali		6	12	15	18	20	22	24
Mansa		10	12	14	18	20	24	27
Abohar		4	6	14	20	24	25	27
Bathinda		11	12	18	20	22	24	26
Moga		10	12	20	22	25	28	29
Muktsar Sahib		3	6	8	12	15	17	19

Jammu & Kashmir, the soil type changes to 'Medium' with N-values ranging from 12 to 18. Accordingly, the safe-net bearing capacity of the soil (for normal footing sizes of 2 m × 2 m, for a 40 mm settlement) ranges from 90 to 110 kPa at a shallow depth of 1.5 m or so; the settlement criterion mainly controls the bearing capacity values in most of the land area in the state. However, at deeper levels of 3 m or higher, the net soil-bearing capacity is found to be in the range of 130 to 250 kPa where the shear criterion starts controlling the design values. Nevertheless, there are many local loose pockets of soils across the state where the bearing capacity values are found to be as low as zero. This mainly happens because of the existence of high-water table or filled-up areas (as in the case of filled-up ponds, wells and so on), loose soil pockets, areas along the unlined canals, rivers, drains and so on. The silt factor for most of the soil available in the riverbeds generally varies from 0.7 to 0.9 due to thick deposits of poorly graded sands.

Isolated footings, strip/wall footings and combined footings are the common footing types that are in practice for the construction of low-rise building structures (up to ground floor +3 storeyes). The raft foundations with or without piles and the pile foundations (both under reamed and straight bored concrete piles) are used for the high-rise buildings or in cases where the water table is available at shallow depths. On the other hand, the good foundations and the piled raft are the most common in the case of heavily loaded structural systems, such as bridges, buildings with long spans and/or heavy loads. Because of very high scour depths in the riverbeds, the bridge piers are supported over the good foundations. Stabilised sub-grade, use of stone columns and the cellular confinement system in the form of a geocell mat with or without a geomembrane are other alternatives that are often used by engineers in areas of problematic soils having very low bearing capacities or high compressibility.

29.6 OTHER GEOMATERIALS

Besides the natural soil, many other materials are nowadays available that can be easily used in construction activities. Mostly, these are industrial by-products that possess certain properties that make them competing for materials to be used in certain engineering applications. This section describes two such materials.

29.6.1 Fly and pond ashes

This material is mainly produced as a waste product from various thermal power plants operating in the state. These are producing approximately 6 million tonnes of fly ash annually. Figure 29.3 depicts loading of pond ash into trucks from one such thermal plant. Out of the total available quantity in the state, only 12% to 15% finds its use in various engineering applications, such as reclamation of low-lying areas, as re-inforced fly ash in the construction of roads and flyovers (Gayathri and Kaniraj, 2012). Cement plants are another bulk user of this waste material. The source of the coal used in the plants decides the type of the fly ash; it is mainly C-type in the state with mostly tan colour. The typical silt-sized spherical particles are 10 and 100 μm in size. The small-sized spherical particles improve the fluidity and workability of the fresh concrete if used in the production. Geotechnical properties of these two forms of the material are quite comparable with those of natural sand, albeit it comes with lesser unit weights. However, proper compaction to its maximum dry density at the OMC

Figure 29.3 Pond ash being loaded in trucks from dykes of the Ropar Thermal Plant. (Source: Author.)

produces strength parameters very similar to those exhibited by the natural sand. Because of this, the compacted pond ash finds its use as a potential substitute for natural sand material in many construction activities.

29.6.2   Rice husk ash

Paddy is the major summer crop of the state, though it is not a native crop of the state. Punjab accounts for 11–12 million tonnes of rice annually or more than a tenth of India's output of the staple grain. Rice husk is a major by-product of the paddy. During the milling of paddy, around one ton of rice husk is produced for every four tonnes of paddy going for the milling. Rice husk finds its use in various industries as a fuel to run the boilers; this process leaves behind around 15% to 20% ash called rice husk ash (Jha and Gill, 2006). It contains an enormous amount of silica; Table 29.6 shows a typical chemical composition of the rice husk ash available in the state.

Despite being a good candidate for use as an alternative material for stabilising the problematic soils and as an additive in concrete production, it is being used in the reclamation of low-lying areas; Figure 29.4 depicts one such dumping operation. These days, however, many industries have started commercialising the material, which is being marketed under the trade name 'Agro Silica'.

Table 29.6 Chemical Composition of rice husk ash

Component	Content (%)
SiO_2	93.2
Al_2O_3	0.50
F_2O_3	0.22
CaO	0.51
MgO	0.41
Loss on ignition	1.91

Figure 29.4 Rice husk ash dumped for reclamation of low-lying areas. (Source: Author.)

29.7   NATURAL HAZARDS

Punjab, being the flood plains of five rivers, has alluvial deposits of cohesionless soil. However, the texture and soil type vary from region to region depending upon the prevailing climatic conditions. The depth of the water level is also highly erratic; it varies from as low as less than 2 m to as high as more than 40 m in some areas; it is graphically depicted in Figure 29.1. These two factors, coupled with the fact that most of the area fall in the seismic zone IV leads to a very high probability of possible liquefaction, especially in the southwestern area of the state and along the banks of different rivers passing through it.

Flooding of different areas adjoining the riverbanks during the monsoon season is another major natural hazard. The heavy rainfall in the catchment area of the rivers compels the authorities to open the flood gates of various dams built thereon that many times leads to breach in the river and/or canal embankments causing huge damage worth billions of rupees. In the year 2019, there was a severe flood in the river *Sutlej* that affected thousands of square kilometres of area falling in Ludhiana, Jalandhar, Kapurthala and Ferozepur districts of the state. Excessive flooding and/or release of excess water from the dams also lead to severe erosion/scour of soils along the riverbanks and riverbeds. In the past, it has caused a non-repairable damage to many structures, such as bridges and high mast built thereon. Figure 29.5 depicts two such cases of a very severe scour at the tail of the river Sutlej in the southwestern area of the state, which eroded the underlying soil to the extent that the culvert constructed across the river has to be demolished.

The groundwater level rise is another problem that had affected the number of buildings and other infrastructures in the state, especially in the southwestern districts of the state in the late eighties. The water table in this part of the state is at shallow depths of around 1 m and in some pockets, it has come over the ground, which has led to severe damage to the roads, the other housing infrastructure and buildings. Due to this, settlements as high as 1 m were observed in many parts of the region, especially in case of the non-engineered structures such as masonry buildings.

The other problem that is prevalent in the southwestern region of the state is the relatively fast deterioration of reinforced concrete structural components. This mainly exists because of the presence of harmful salts in the soil and water in larger quantities.

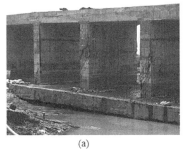  

(a)   (b)   (c)

Figure 29.5 Scour of riverbed: (a) damaged culvert, (b) dismantling operations of a damaged culvert due to more than anticipated scouring in the bed and (c) exposed footing of a high mast pole installed on the riverbank. (Source: Author.)

Either this region of the state is the tail-point of all canals that supply water to the region or they pass through this region in the adjoining states of Haryana and Rajasthan. The continuous running of these canals has resulted in the accumulation of a large concentration of salts in the soil and the water being used in the construction that finally reduces the active life of the structural components constructed in reinforced concrete. Though this is the result of man-made activities, its existence over the decades has made it at par with other natural hazards that must be adequately addressed at the very start of the projects to optimise the resources. Alternatively, the construction practice can be shifted to the time-tested masonry type, especially in this region of the state.

29.8   CASE STUDIES AND FIELD TESTS

The soil strata being alluvium deposits in the state is highly erratic, having a huge variation in the soil properties. This happens very often in the top 2 m soil depths. Many times, damage to buildings and other projects constructed over such soils has been reported in the past. This section presents a set of three typical such problems.

29.8.1   Pavement failure

The state has a well-connected network of canals. Due to very high land costs, the state government has started using the canal embankments for the construction of highways. One such road pavement was constructed on the bank of the Sirhind canal connecting Roopnagar district with the Doraha town in Ludhiana. Unfortunately, this important road immediately started showing signs of distress at many locations along its stretch (Singh et al., 2018). Figure 29.6 shows one such typical failure noticed in the pavement.

To identify the possible reasons for this failure, around 20 test locations were selected at various embankment heights along the road alignment. In-situ and laboratory tests were conducted to evaluate the stability of embankment slopes. The strength of the subgrade was also evaluated using LFWD. Subsequently, the slopes were analysed using the test data. The analysis results tabulated in Table 29.7 indicate an inadequate

Figure 29.6 A typical pavement failure of a road over the embankment. (Source: Author.)

Table 29.7 Results of the slope stability analysis

Chainage, km + m	Test location No.	Embankment height, m	Factor of safety (FoS) with different methods			Average value of the FoS	Remarks
			Morganstern–price	Spencer	Sharma		
1 + 600	1	16.0	0.908	0.910	0.877	0.9	<1.30
3 + 630	2	16.0	0.912	0.911	0.873	0.9	<1.30
5 + 580	3	11.5	1.001	0.990	0.993	0.99	<1.30
7 + 965	4	12.5	1.145	1.120	1.07	1.11	<1.30
10 + 410	5	12.0	1.372	1.360	1.322	1.35	OK
13 + 180	6	8.5	1.914	1.909	1.91	1.91	OK
15 + 195	7	9.0	2.000	1.989	1.993	1.99	OK
18 + 105	8	13.5	1.358	1.350	1.349	1.35	OK
20 + 650	9	13.5	1.502	1.495	1.493	1.5	OK
22 + 950	10	15.5	1.487	1.490	1.487	1.49	OK
24 + 700	11	15.5	1.393	1.393	1.391	1.39	OK
27 + 800	12	13.5	1.618	1.614	1.608	1.61	OK
31 + 375	13	8.0	2.686	2.670	2.685	2.68	OK
34 + 760	14	7.5	1.696	1.687	1.68	1.687	OK
38 + 100	15	8.0	1.453	1.446	1.435	1.44	OK
41 + 150	16	8.5	1.392	1.396	1.395	1.39	OK
15 + 500	17	8.5	1.333	1.333	1.334	1.33	OK
47 + 100	18	8.5	1.353	1.343	1.273	1.32	OK
48 + 650	19	8.5	1.358	1.343	1.361	1.35	OK
50 + 600	20	12.5	0.810	0.830	0.850	0.83	<1.30

factor of safety at many locations on the road. It was mainly found at the points where the embankment was of more than 10 m height. The phreatic line at all such locations was also found to be crossing through the embankment height, thereby causing changes in the in-situ soil properties from their design values. It led to the appearance of rutting and other cracks in the pavement.

29.8.2 Cracking and uneven settlements in buildings

This case study pertains to the *Mugal Dhikky, Qusi Muhalla* and *Kingra Gate* area of Jalandhar city where many houses developed cracks in their walls and floors. Some of these houses were very old while many others were constructed in the last 5–10 years. Most of the houses were observed to be constructed in brick masonry with the cement mortar and seemed highly distressed leading to a widespread cracking in the walls and uneven floor settlements. A typical set of cracking developed in the walls and floors of the building is depicted in Figure 29.7. The slabs, stairs and projections in almost all houses were constructed using the reinforced concrete. The area was visited on 5 May 2018 to see the possible cause of distress exhibited by various buildings (Singh, 2019).

The locality visited has properly laid concrete roads, and it seems that houses constructed therein are situated over some raised ground/mound, not naturally found in this region of the state; some houses were G + 1 on their front side road and G + 3 on the backside road, thereby indicating a sloppy ground. The dampness was also observed in the brick walls of many houses. As informed by the area residents, earlier, the sewage/

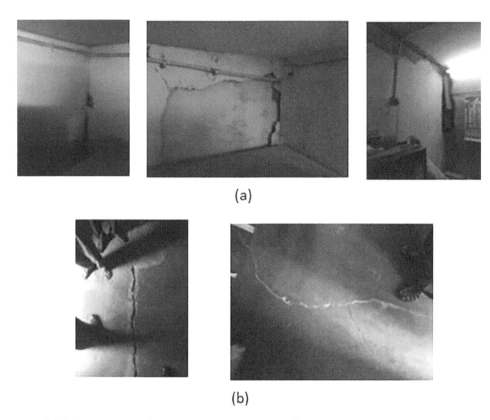

Figure 29.7 A typical set of cracks appeared due to differential settlement: (a) in walls; and (b) in floors. (Source: Author.)

wastewater was carried through a network of open drains laid along the streets. However, underground sewer lines were laid a few years ago in the streets in place of the open drains. Mostly the walls near the newly laid sewer line in the street were showing more settlement in comparison to the other walls of the building. Some of the walls were displaced/tilted sideways as high as 40 to 75 mm near the roof levels. In some of the houses, the floors were looking titled/sloppy with angular displacements as large as 25–50 mm.

To ascertain the reasons for this settlement, geotechnical investigations were conducted in November 2018 and it was found that there was earth filling up to a depth of 7.5 m from the point of the testing over the existing ground level. The penetration resistance of the soil in the area was found to vary from zero to three up to a depth of 6 m, thereby indicating loose soil/weak soil/fill-up soil deposits in the area. Settlements were observed to be controlling the soil bearing capacity. Although the load from the building floors was not too high to cause any distress in the soil when in a dry state, the seepage from the sewer lines, laid underground in the streets, was found to cause a reduction in the soil bearing capacity leading to the uneven ground settlements. Loosening of the soil strata in already loose soil near the building foundations at the time of construction of the sewer lines was observed to be another major factor for these differential settlements between the walls on the roadside and the walls opposite to the roadside. Therefore, the municipal corporation officials were advised to check and stop the seepage/leaking of water into the ground by suitable means or start using the open drains as was the earlier practice till the defects are contained. Retrofitting/strengthening of the damaged buildings was suggested and some buildings, which have developed very wide cracks or have wall inclination more than the safe values, were advised to be vacated for the safety of the occupants.

Such types of loose soil pockets, either filled-up wells and ponds or raised grounds, exist in many areas of the state. Careful soil explorations are mandatory in all such areas to avoid such large-scale damage to the buildings. The people living in the nearby areas can be consulted to inquire about the existence of any such loose pockets and investigations must be planned accordingly to quantify the soil response.

29.8.3   Tilting of an overhead service reservoir

One more interesting case was reported wherein an overhead service reservoir (OHSR) tilted sideways in a few months after its construction. It was constructed at Alt 262 m, N 30.72998, E 76.36983. The site was visited on December 24, 2019 (Singh, 2019). Figure 29.8 shows a typical photograph of the tilted OHSR. A standard penetration test on the two opposite sides of the foundation of the OHSR (where the settlement is reported to be maximum and minimum) was conducted to check the soil strata.

Figure 29.8 The OHSR that tilted due to large differential settlements. (Source: Author.)

Table 29.8 Comparison of N-values measured at the site

Depth (m)	N-Value available in an available soil investigation report	N-Value observed during the day of testing
1.67	06	01
3.0	08	04
4.5	10	06
6.0	10	06
7.5	12	10
9.0	13	12
10.5	13	12
12.0	15	13

Table 29.8 shows a comparison of the N-values available from the records (the BC report dated December 12, 2002) for this project and those observed during the day of the testing. No water table was encountered during the investigations even up to a depth of 12 m from the ground level. The current tilt of the OHSR was also measured which translates to a horizontal displacement of 610 mm at the OHSR balcony level.

Soil exploration revealed that the foundation of the OHSR has been located at 2.1 m below the existing ground level. Up to a depth of 3 m, filled-up soil was observed at the site with very low N-values (around 1). Because of the very low N-values, the settlement criterion was found to be a controlling parameter for the foundation design in this case. The bearing capacity at the site of the OHSR was determined to be 60 kPa. This value is observed to be less than the soil pressure taken in the design of the OHSR footing (=87.5 kPa). The unequal soil pressure caused by the lateral loads seems to be one of the major reasons for this tilt being exhibited by the OHSR; the reported soil pressures in the structural design report at the footing extreme edges are 43.1 and 103.38 kPa for the wind loading case and 31.62 and 103.38 kPa for the seismic loading case. The overstressing of the underlying soil is obvious, which may be one of the dominant factors that led to the differential settlements and the subsequent tilting of the staging. In addition, initial non-verticality of the OHSR staging at the time of its construction may be another probable cause for this large drift exhibited by it.

The subsequent investigations revealed the shifting of the OHSR site from its initial planned place; it was done due to some administrative reasons, but the design was not modified and adjusted to the new site conditions. Unfortunately, the OHSR after its shifting was founded over some loose pocket generally found in many places in the alluvium strata that led to the problem. This case study shows the importance of a thorough soil exploration right at the place of the construction. It may therefore become unsafe and financially unsound to use the site test data of some nearby areas for any new construction site.

29.9   GEOENVIRONMENTAL IMPACT ON SOILS AND ROCKS

Ludhiana and Jalandhar are two major industrial cities of Punjab having a large number of tanneries and dying industries. The subsoil in these cities is sandy at shallow depths (at about 2 to 3 m depth), which causes a quick spread of contaminated water into the soils. The percolation of untreated effluent from these industries in the ground has been reported in electronic and print media repeatedly. This unethical and illegal practice not only polluted the underground water, but also led to a drastic deterioration of the soil, altering nearly all its characteristics. Recent studies have reported a change in the pH value of soils and the presence of salts (Singh, 2019). The change in the pH value and excessive percolation of salts, such as sulfates and chlorides into the soil, has increased the chemical aggressivity, thereby causing a relatively faster deterioration of the substructure components of buildings, bridges and buried structures. Despite the best efforts of the law enforcing agencies, the problem of disposal of untreated effluent is quite common. Therefore, this aspect needs careful attention while designing the substructure in such areas. A thorough soil exploration should be carried out to ascertain the geotechnical properties of the soil, including its chemical aggressivity in the context of the presence of harmful salts.

29.10 CONCLUDING REMARKS

Intensive soil explorations at the site of any construction project are the backbone of any subsequent structural design process. Inadequate soil investigations or change of project location at the construction sites without carrying out the studies at new places had resulted in considerable damage to buildings and other built facilities in the past. Some of these failures have been highlighted in this chapter. It is believed that this chapter may be of great help, especially during the initial/planning stage of the projects to decide the detailed soil exploration. Some of the salient points that may be of some help to engineers and planners in this regard are presented below:

- The soil in the state is primarily an alluvial deposit from flood plains of five rivers. It consists of various layers of different soil types, such as silt, sand, clay and a mixture of soil and gravels.
- The top layer up to a depth of 2 m or so is highly variable region to region, formed by the local prevailing climatic conditions and its use over a long period of time. Mostly, a sandy stratum is present 3 m onwards below the NGL.
- The level of the water table in the state is highly variable; it varies from as deep as 40 m in some areas and causes waterlogging in some other parts of the state. Special care is needed in areas of shallow water table depth.
- The presence of sand and/or silty deposits and the position of the water table/soil moisture at shallow depths in some areas of the state may lead to liquefaction in the event of any seismic activity; some areas are prone to liquefaction, even for a PGA of $0.18g$, especially in southwestern areas and that fall along the banks of the rivers and unlined canals.
- There exist many loose soil pockets in some areas; their identification must be among the main objectives of any soil exploration programme.
- At the routine founding levels, the settlement criterion mainly dictates the safe-bearing capacity of the soil stratum across the state, except for some areas along the boundary with Himachal Pradesh and Jammu & Kashmir and at larger depths (more than 3 m) where the shear criterion may dominate due to the high in-situ N-values.

ACKNOWLEDGEMENT

The authors are highly indebted to the Testing & Consultancy Cell of Guru Nanak Dev Engineering College, Ludhiana who provided access to the vast soil exploration data in the state during the preparation of this chapter.

REFERENCES

Agarwal, R.R. (1961). Soil classification of the alluvium derived soils of the Indian Gangetic plains in U.P. *Journal of Indian Society of Soil Science*, Vol. 9, pp. 219–231.

DMGP (2019). Online monitoring portal, Department of Mines and Geology, Punjab. https://www.minesandgeology.punjab.gov.in/, accessed 19 November 2019.

ENVIS (2019). Punjab status of environment and related issues. http://punenvis.nic.in/, accessed 19 November 2019.

Grewal, G.K. (2014). Liquefaction Potential Mapping of Punjab. M. Tech. Thesis submitted to Guru Nanak Dev Engineering College, Ludhiana, India.

Gayathri, V. and Kaniraj, S.R. (2012). Geotechnical reuse of waste material-special focus to stabilised fly ash as pavement base course material. *SDPProceeding, Ground Improvement and Ground Control including Waste Containment with Geosynthetics, Ludhiana, India*, pp. 192–212.

Jha, J.N. and Gill, K.S. (2006). Effect of rice husk ash on lime stabilization of soil. *Journal of Institute of Engineers India*, Vol. 87, pp. 33–39.

Seed, H.B. and Idriss, I.M. (1971). Simplified procedure for evaluating soil liquefaction potential. *Journal of Soil Mechanics and Foundation Division (ASCE)*, Vol. 107, No. SM9, pp. 1249–1274.

Seed, H.B., Idriss, I.M. and Arango, I. (1983). Evaluation of liquefaction potential using field performance data. *Journal of Geotechnical Engineering*, Vol. 109, No. 3, pp. 458–482.

Singh, H. (2019). Personal communication.

Singh, D., Jha, J.N. and Gill, K.S. (2018). Evaluation of soil subgrade characteristics in field application" PhD thesis submitted in Punjab Technical University Jalandhar, India.

Chapter 30

Rajasthan

Anil Dixit, Nirbhay Mathur, and Harsh Chittora
Landmark Material Testing and Research Laboratory Pvt. Ltd.

CONTENTS

30.1 INTRODUCTION

Rajasthan, the largest state of India, covers around 10.50% of the land of the country and is situated 23° 3′ to 30° 12′ north and 69° 30′ to 78° 17′ east with the Tropic of Cancer passing from its southernmost end. Rajasthan is the most diversified state in the context of geographic formations. These vary from deserts of windblown sand dunes and lofty hills to dense forests along sides of rocky riverine. Wide-ranging soils from silty dune sand to black cotton expansive clay and rocks, from hard basalt to weathered rock of bentonite deposits make the geology of the state most diversified (Figure 30.1).

General topography suggests a long cyclic process of stripping and erosion. Under harsh desert-like conditions, the soils of plains are generally characterised as sandy

DOI: 10.1201/9781003177159-30

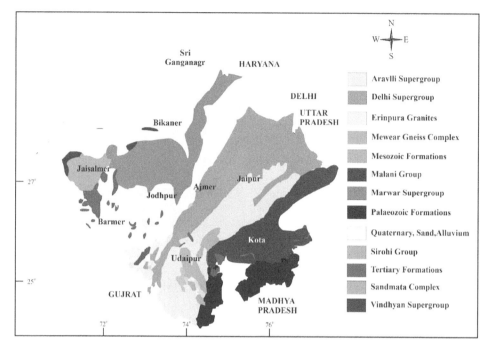

Figure 30.1 Geological map of Rajasthan.

(cohesionless) and clayey (cohesive) soils. The northwest part of the arid Great Indian Desert covers a larger part of the Thar Desert while the southeast part of the state is covered by a north-western extension of the peninsular block. Broadly, the soil physiology can be divided into four major parts as detailed below.

30.1.1   Western desert region

This region spreads right from the stepped land of west of Aravali hills to the extreme of Thar desert and holds almost all the sand dunes found in Rajasthan. However, at some places, rocky outcrops may also be seen. The most important river flowing in the region is the Luni, which originates from Aravali, southwest of Ajmer, and flows towards the southwest and finally fall into the Rann of Kuchchh. The districts in this area are Bikaner, Barmer, Churu, Jodhpur, Jaisalmer, Nagaur, Hanumangarh, Sriganganagar, Pali, Sirohi, Sikar and Jhunjhunu. Figure 30.2 shows the sand dune at Jaiselmer.

30.1.2   Aravali hill region

The Bhilwara district and the major parts of Udaipur district, Chittaurgarh district, Sirohi district and the tract of Aravali hills form a part of the inter-mountain plateau, consisting of dark-lava hill soils along with different types of rocks.

Figure 30.2 Sand dunes.

30.1.3 Eastern plain region

The eastern plain is flown by Chambal, Banas and Mahi rivers and depositions of these three river systems have made the eastern plain containing black cotton clay and loamy sand. Kota, Jhalawar, Sawai Madhopur, Baran, Bharatpur Alwar and Bundi are major districts falling in this area.

30.1.4 Southeast plateau

This high tableland has a very diverse topography consisting of more or less sandy uplands, broad depressions and level stretches of deep black cotton soil to red and yellow loamy sand. Bhilwara, Banswara, Dungarpur and partly Udaipur are the districts belonging to this area.

However, Rajasthan is predominantly known for its deserts; it is also endowed with a wide variety of rocks. The Aravali mountain range, which starts from Delhi and extends up to Gujarat bifurcates the land into two geological regions. The older group of rocks is in the east and the younger ones in the west. Rajasthan is the state that has almost all the eras of the geological time scale. The oldest sediments in southeastern parts of the state around Bhilwara, Udaipur, Ajmer, Nathdwara and Rajsamand have granites and marble deposits.

While some of the Aravali sediments have numerous ductile shear zones and brittle faults, deposits of limestone and sandstone are found in these areas. The land consists of a wide range of metamorphic rocks like schist, quartzite, marble and gneisses while sedimentary deposits of lime and sandstones like Kota stone, Karauli Stone and Dholpur stone are also abundantly available. Basaltic flows of Deccan traps are found in the southernmost region of the state.

Several mineral deposits and renowned building stones of commercial importance occur in association with these rock units. Milk white Makrana marble popularly known as *"Sangmarmar"* which was used in the construction of the world-class monument Tajmahal of Agra is produced by Rajasthan.

Rajasthan has impressive supremacy in the manufacturing of major minerals, like lead, zinc, wollastonite, gypsum, calcite, ochre and rock phosphate, contributing around 90% of national production. Massive reserves of crude oil, lignite, heavy oil, lean gas and bitumen further add to the state's mineral strength. Rajasthan contributes appreciably to the production of lead, copper and zinc.

Regular production of these metals in Udaipur, Bhilwara, Rajsamand and Jhunjhunu has created large tailings of zinc, lead and copper ores. The tailings are treated to neutralise the waste for reclamation purposes through the use of calcium hydroxide. Efforts have also been made to retrieve low-grade zinc and lead found in tailings through bioleaching.

Large-scale activities of thermal power productions in Rajasthan have generated an abundance of fly ash and pond ash, which are now widely being used in the manufacturing of building bricks and production of concrete. The availability of pond ash is a favourite choice for reinforced earth wall construction in areas where a suitable reinforced fill material is not available.

In this chapter, physiological features of the land of Rajasthan are presented and various issues related to natural soil formations and their effect on the built environment is briefed. Major types of available soils and subsoils, rocks and minerals and their peculiar properties, uses and problems associated are also discussed based on the data obtained from laboratory testing of the material collected from boreholes at various field locations.

30.2 MAJOR TYPES OF SOILS AND ROCKS

The soils of Rajasthan exhibit wide variations in chemical composition, texture, structure, physical properties, available water content and strength parameters. Soils and rocks available in Rajasthan can be classified based on various aspects such as geological formations, composition, weathering conditions and engineering applications. Here, the classifications are mainly based on the properties obtained during the geotechnical site investigations for both surface and subsurface exploration and laboratory testing. These studies are based on many field tests such as standard penetration tests, plate load tests, field permeability test and in-situ density conducted at various locations of Rajasthan at different times. Figure 30.3 shows a standard penetration test arrangement at Sriganganagar. Data, collected from boreholes drilled up to 35.0 m depth and further based on laboratory results, are analysed for the determination of index and engineering properties. The geological formations of soil and rock can be classified broadly into the following groups (Government of Rajasthan, 2021).

30.2.1 Desert soils

These soils are chiefly found in areas of desert climate, where arid conditions prevail almost year-round. Lack of water for binding and huge cyclic range of exfoliation due to varying temperatures have contributed richly to this type of soil. Soil contains a

Figure 30.3 Conducting the standard penetration test (SPT) in Sriganganagar.

high percentage of soluble salt and has a high pH value. Districts of Nagaur, Jodhpur, Jalore Barmer, Churu, Jhunjhunu and Sikar are covered by these types of soils.

30.2.2 Dune soils

These soils are deposited sand and liable to be blown off with the air; hence, they are grouped separately from the desert soils. They are of frequent occurrence with sand dunes in western Rajasthan. The texture of these kinds of soil varies from loamy fine sand to coarse sand.

30.2.3 Brown soils

They are found in major parts of Tonk, Bundi, Sawai Madhopur, Bhilwara and Chittorgarh districts. These soils vary from sandy loam to clay loam and are rich in calcium salts but have poor organic matter.

30.2.4 Black cotton soils

These soils are found in Kota, Jhalawar, Bundi, Baran and some parts of Sawai Madhopur, Dholpur, Bharatpur and Alwar. Well-irrigated flood plains of the Chambal River have highly clayey soils over a hard bed of sandstone strata in these areas.

30.2.5 Sierozemics

On both sides of Aravali hills, these soils are found. Pali, Nagaur, Ajmer, Jaipur, Alwar and Dausa districts have an abundance of such soils. Having good permeability and

Figure 30.4 Vertical cut in Alwar.

mostly found in yellowish-brown colour, these are sandy loam to sandy clay loam in texture. Figure 30.4 shows a vertical cut in this soil deposit in Alwar.

30.2.6   Red soils

The sandstone of the Vidhyan rock group in red and yellow colour is the parent material of these soils. These soils are sandy loam to loam in texture with a granular structure and are also well drained. They are gravelly or sandy on hilltops and deep reddish loam in the valleys. They are predominantly found in areas of Sikar, Jhunjhunu and some areas of Dholpur.

30.2.7   Hill soils

These soils are found on and at the foothills of the main hills and hill ranges of Rajasthan. The colours of soils vary from reddish to yellowish-red to yellow-brown. Hill soils are mainly found in Sirohi, Pali, Nagaur, Udaipur, Rajsamand, Chittorgarh and Ajmer.

30.2.8   Saline-sodic soils

These soils are found in the natural depressions, mainly in Panchpadra, Didwana, Sambhar, Ranns of Jalore and Barmer districts. Besides these, saline-sodic soils are also found in the far flood plain of Ghaggar and in parts of the Luni Basin.

Western and northwestern parts of Rajasthan are covered with dune sand and other areas are having a wide variety of rocks. The boring log studies reveal wide variation in rock properties along the depth. Geologically, Rajasthan is among the very few regions in the world, which has well-preserved records of a continuous geological history for over 3,500 million years. Archean, Proterozoic, Paleozoic, Mesozoic, Cenozoic and Quaternary have constituted the Rajasthan geology.

The oldest formation of the Archean age belongs to the Bhilwara supergroup. These are exposed in Udaipur, Dungarpur and Chittorgarh. Varieties of hard granites, colourful marbles and sediments of sandstones comprise the rock formation of this area. The geological setup of Ajmer, Pali and Bhilwara is represented by various igneous and meta-sedimentary rocks.

The Bhilwara supergroup of the Archean age comprises shale, phyllite, slate, limestone, marble, schist, quartzite and so on. The Marwar supergroup also occurs in the northern part of this area and is represented by limestone, dolomite, sandstone and shale. The geological configuration is quite complex as it is composed of a heterogeneous assemblage of different litho-units of igneous, sedimentary and metamorphic origin. Figure 30.5 shows core samples in a box obtained from core drilling at the site in Udaipur at 15.0 m depth.

Most parts of the Kota, Bundi, Baran, Jhalawar and Chittorgarh districts are occupied by the Vindhyan supergroup. The occurrence of sandstone at different stratigraphic horizons indicates fluctuation of the sea level due to transgression and regression of the sea several times during the Vindhyan period.

Physiography of areas covered by the Karauli, Dausa, Bharatpur and Sawai Madhopur districts is characterised by the great Vindhyan plateau. Primarily the terrain comprises pre-Cambrian metamorphic, igneous and sedimentary rocks belonging to pre-Aravali Vindhyas.

Figure 30.5 Rock core samples in Udaipur.

30.3 PROPERTIES OF SOILS AND ROCKS

Rajasthan is characterised by a wide range of soils and rocks having varied engineering properties. Though the compilation of geotechnical properties of all Rajasthan cannot be done in this literature, a brief representative description of different soils and rocks is summarised here.

During the past few years, extensive geotechnical investigations were carried out at different locations all over the state for the determination of various parameters. Boreholes were drilled up to 35 m depth at some places and the results were analysed. A standard penetration test (SPT) and plate load tests were carried out for safe-bearing capacity and settlement determination. Field permeability tests, electrical resistivity test and geophysical surveys were also conducted.

Various laboratory tests were carried out on many soil and rock samples collected from different depths to determine index and engineering properties such as moisture content, density, specific gravity, grain size, Atterberg's limit, shear strength parameters and consolidation parameters. In the case of rocks, core samples as obtained from fields were further analysed in a laboratory to determine unconfined compressive strength, point load index, water absorption, specific gravity, shear strength parameters and modulus of elasticity. Few photographs of field test observations are illustrated here for a better understanding of geotechnical properties and their variation at different locations and depths (see Figures 30.6–30.9).

The data compiled in Table 30.1 are the true depiction of variation in the properties of soils at various locations. The areas dominating the SM group have large deposits of silty sand with very little to negligible clay content. The permeability in these

Figure 30.6 Core recovery and RQD in Nasirabad.

Figure 30.7 Core recovery and RQD in Jhalawar.

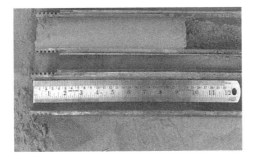

Figure 30.8 SPT sample of silty sand in Nagour.

Figure 30.9 SPT sample of clayey silt in Bharatpur.

areas is good and rapid percolation of water is observed; the water table is found at around 80 to 180 m depth. No serious foundation problem is observed except for low bearing capacity. However, soil erosion of slopes in cut and fill needs special treatment for stabilisation due to no cohesion. More cohesive and expansive soils of CH and CI groups are predominantly found in the southern and eastern areas of the state. The clay content varies from 40% to 80% in these areas. Poor drainage and expansion on wetting are the main issues that they need geotechnical intervention. In many parts of western Rajasthan, underground deposits of highly expansive bentonite soils are found. The swelling pressure of such soils ranges around 250 kPa, which is a major concern for subsoil foundation structures.

In most of the southern parts of Rajasthan, rocks are predominant geological features. Outcrops of granites, quartz and basalt type good quality hard rock can be easily seen there. Not many geotechnical problems are encountered in these areas due to the presence of rock under sound conditions. Cutting and excavating slopes and tunnels sometimes pose the challenges of weak zones of enormous joints and faults. Figure 30.10 shows the bore log profile of Jhalawar.

While in eastern to southeastern parts, sedimentary deposits of limestone and sandstone are chiefly found, these stones are very popular for architectural and aesthetic use in building construction. Roofing, wall cladding and flooring stones are the preferred choice from the stones available in these areas. A brief description of location-wise rock properties is given in Table 30.2.

Table 30.1 Location-wise engineering properties of soils

Location	Depth (m)	Soil group	N values	% Finer than 75 μm	Cohesion (kg/cm²)	Angle of internal friction (°)	UCS (kg/cm²)	LL (%)	PL (%)	PI
Jaipur, Jhunjhunu, Sikar, Churu	1.5	SM	5–16	5–25	0	26–29	—	21–24	Nil	NP
	4.5	SM	10–35		0	28–30	—	22–25	Nil	NP
	10	SM	25–50		0	29–32	—	22–25	Nil	NP
Jalore, Sirohi	1.5	SM	10–20	10–25	0	26–28	—	21–23	Nil	NP
	4.5	SM	18–35		0	27–29	—	22–26	Nil	NP
	10	SM	30–50		0	30–32	—	22–26	Nil	NP
Kota, Bundi, Baran	1.5	CH	10–22	85–95	1.5–2.0	0–5	30–40	51–65	28–35	23–30
	4.5	CI	20–35	70–90	0.8–1.5	2–8	20–30	36–50	22–28	14–22
	10	CI	30–50	70–90	0.8–1.5	2–8	20–30	36–50	22–28	14–22
Jhalawar	1.5	CI	20–50	70–80	0.8–1.5	5–8	16–30	36–50	22–28	14–22
	4.5	Rock	—	—	—	—	—	—	—	—
Swai Madhopur, Tonk	1.5	CI	10–20	70–85	0.45–0.6	8–12	9–12	36–45	20–25	16–20
	4.5	CL	18–30	50–70	0.3–0.5	12–16	6–10	25–35	15–22	10–13
	10	CL	25–40	50–70	0.3–0.5	12–16	6–10	25–35	15–22	10–13
Dausa, Alwar, Karauli, Bharatpur, Dholpur	1.5	CL	8–18	50–75	0.4–0.6	10–15	8–12	25–30	15–18	10–12
	4.5	CL	18–30	50–70	0.3–0.4	12–16	6–8	27–32	17–20	10–12
	10	CL	25–40	50–70	0.3–0.5	12–16	6–8	30–35	19–22	11–13
Sri Ganganagar, Bikaner, Hanumangarh, Barmer, Jaisalmer	1.5	ML	9–15	50–60	0.05–0.1	20–22	—	22–26	Nil	NP
	4.5	ML	16–30	50–60	0.05–0.1	22–24	—	22–28	Nil	NP
	10	ML	25–50	50–60	0.05–0.1	24–26	—	22–28	Nil	NP
Barmer	4.5	CH	15–30	92–97	2.0–3.0	2–5	40–60	150–280	40–60	110–220
	10	CH	25–45	92–97	2.0–3.0	2–5	40–60	150–280	40–60	110–220
Ajmer, Bhilwara	1.5	CL	15–35	50–75	0.3–0.4	10–14	6–8	26–30	14–17	12–13
	4.5	CL	30–50	50–70	0.2–0.3	15–18	4–60	29–34	17–20	12–14
	10	Rock	—	—	—	—	—	—	—	—
Udaipur, Banswara, Chittorgarh, Dungarpur, Pratapgarh, Rajsamand	1.5	CL	12–30	50–65	0.2–0.4	10–15	4–8	26–35	14–20	12–15
	4.5	Rock	—	—	—	—	—	—	—	—
	10	Rock	—	—	—	—	—	—	—	—
Nagaur, Jodhpur, Pali	1.5	CI	12–18	68–80	0.4–0.6	8–11	8–12	36–45	20–25	16–20
	4.5	CL	20–35	50–65	0.2–0.4	12–18	4–8	25–30	15–18	10–12
	10	CL	35–50	50–65	0.2–0.4	12–18	4–8	31–35	19–22	12–13

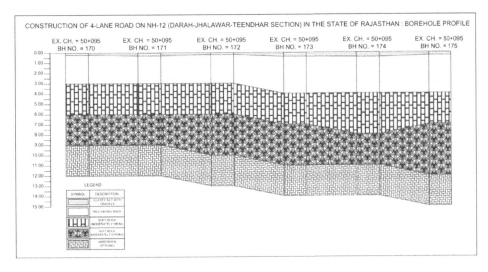

Figure 30.10 Bore log profile of Jhalawar.

Table 30.2 Location-wise engineering properties of rocks

Location	Classification	Specific gravity	Water absorption (%)	UCS (MPa)	Shear strength parameter	
					Cohesion (kg/cm^2)	Angle of shearing resistance
Kota, Bundi, Baran, Jhalawar	Strong	2.72–2.85	0.15–0.30	50–100	9.50–31.0	45–62
Ajmer	Moderately strong	2.70–2.80	0.20–0.45	30–50	10.50–32.80	40–50
Udaipur, Banswara, Chittorgarh, Dungarpur, Pratapgarh, Rajsamand, Bhilwara	Moderately strong	2.69–2.80	0.22–0.45	30–50	10.50–32.80	40–50
Karauli, Bharatpur	Moderately week	2.63–2.69	0.40–0.70	5–12.5	11.80–37.0	35–45

30.4 USE OF SOILS AND ROCKS AS CONSTRUCTION MATERIALS

Soil and rock along with other natural products are among the oldest and most common construction materials for human civilisation. Right from the adobe of mud mortar to modern ceramic tiles and further to ultra-modern development of mechanically stabilised reinforced earth walls, the journey is very long and continued. The abundance of wide varieties of soils from sticky plastic clay to loamy silty sand, various types of marbles, granite, sandstones, basalt, slate, pebbles, lime and other useful minerals makes the land of Rajasthan a rich treasure of construction materials.

Adobe is one of the oldest construction materials of western rural Rajasthan. It is basically a moistened mixture of highly plastic clay, fine sand, cow dung and chopped straw as a reinforcement. Panels or bricks made up of adobe are dried in bright sunlight and then used as building blocks for construction. Having good insulation properties, the adobe is still a preferred choice in the construction of *kuchcha* (temporary) houses in western Rajasthan where large temperature variations are observed.

For cement and lime-based permanent construction of the modern era, the river sand is a very important ingredient. Banas and Masi river sands exhibit very favourable properties for concrete and mortar production. Silica is the main ingredient of river sand. Particles are in spherical to angular shape, chemically and biologically inert, possess negligible proportions of soft and deleterious matters and have considerable hardness and resistance to weathering. But due to serious environmental problems, excessive mining of river sand is banned in many parts of the state now. This has encouraged the use of dune sand as a replacement for finer sand in some specific areas.

For other relevant properties of fine aggregates required for the production of concrete and mortar, quarry dust, crushed stone sand and artificially prepared sand popularly known as M-sand (manufactured sand) are becoming an effective alternative. M-sand is manufactured from locally available plentiful materials such as granite, quartzite, basalt and sandstone. It is produced by crushing these hard rocks into sizes of 4.75 mm and lower. The crushed material is sieved, washed and dried to make it usable for different purposes. Even in mechanically stabilised earth (MSE) walls, the M-sand is being used as a blending agent for property enhancement of reinforced fill materials.

In MSE walls, the selection of a suitable fill material that meets the requirements of the standard is a difficult task. As in some situations, the naturally available silty soil may not be as coarser as required. However, the silty sand, abundantly available in many parts of the state confirms the other requirements of fill. Such as for a plasticity index lesser than 6, almost no cohesion, angle of internal friction more than 30°, substantially free from shale, organic or other soft material, but in most of the cases, the percentage finer does not fall in the satisfying range. To overcome this deficiency, such soils are blended with M-sand and the desired ranges of grain size are achieved with minimum extra expenditure. This blending technique reduces the higher costs of importing a suitable granular material that may not be available in nearby locations. The results of one such blending of M-sand with naturally available silty sand are elaborated in Table 30.3. The data presented here belong to the MSE wall construction project for the railway overbridge on the Jaipur Sikar rail section near Jaipur.

Basalt, which is one of the hardest and most durable construction stones, is prominently available in many parts of the state, mainly around the Kota district. Good quality aggregate made up of basalt is used in pavement construction and for railway ballast. Other than basalt, quartz, granite and some other sandstones are also being used in aggregate manufacturing for infrastructure development works.

The hardness and durability are not the only qualities of Rajasthan stones but many of these are amalgamated with ornamental features such as enchanting colours, rich texture, smooth surface finishes and ease to process and install. Rajasthan stones

Table 30.3 Property enhancement of soil by blending

	Parameter and test method	Original soil	M-sand	Soil blended with 30% M-sand
1	Liquid limit (%)	23.5	–	22.0
2	Plastic limit (%)	Nil	NP	Nil
3	Optimum moisture content (%)	10.0	4.50	9.00
4	Maximum dry density (g/cc)	1.82	1.68	1.79
5	Direct shear test			
	Angle of shearing resistance (°)	29	32	30
	Cohesion intercept kg/cm^2	0.00	0.00	0.00
6	Grain size analysis			
	(i) 75 mm	100	100	100
	(ii) 0.425 mm	70.25	25.11	55.20
	(iii) 0.075 mm	17.30	1.52	11.57

Table 30.4 Different stones available in Rajasthan

Types	Density (kg/m^3)	Unconfined compressive strength (dry) (MPa)	Colour	Origin
Marble – A metamorphic sedimentary limestone or dolomite	2,600–2,800	80–150	White	Makrana, Rajsamand, Alwar
			Green	Udaipur, Sirohi
			Pink	Pali, Makrana
			Black	Udaipur, Jaipur, Banswara
			Grey and banded	Ajmer, Bhilwara, Nagaur
Granite – Igneous granular rock of plutonic origin	2,600–2,750	100–175	White and cream	Pali
			Pink	Jalore, Jhunjhunu
			Red	Jaisalmer
			Blue	Sawaimadhopur
			Black	Rajsamand
			Green	Jaisalmer
			Yellow and cream	Barmer, Alwar
			Grey and brown	Tonk
Sandstone – Sedimentary rock having fine grains of quartz and silica	2,000–2,600	70–180	Red and pink	Jodhpur, Nagour
			Grey, green	Kota, Bundi
			Beige, pink, red	Dholpur, Karauli
			Brown	Nagour
Limestone – Sedimentary rock of lime and silica	2,500–2,700	70–170	Blue, green, brown	Kota
			Yellow	Jaisalmer

such as granites, marble, sandstone, slate and schist and other dimensional limestones have always been a favoured choice of architects and engineers for the beautification of buildings along with durability and other functional benefits. So many world-class monuments all over the country are constructed using Rajasthan stones. A typical list of major products is given in Table 30.4.

Besides the above, there are many other useful construction materials available in plenty. Limestone available in Gotan, a place near Jodhpur, is whitish and well matched in qualities required for white cement production. Many white cement production plants are set up in nearby locations. 'Gotan Lime' is a well-known term in the construction industry due to its superior properties.

Slate and schist are available in Tonk, Bhilwara and Ajmer in multicolours such as copper yellow, ocean green, golden, black, silver-grey, terra red and zeera green. They are widely used as decorative stones and for roofing purposes. Low water absorption and good thermal resistance make the slate an ideal choice of roofing and elevation cladding worldwide. It has good resistance to damage against freezing and thawing and hence it is exported to many countries.

30.5   FOUNDATIONS AND OTHER GEOTECHNICAL STRUCTURES

In engineering, foundation is the part of a structure that connects the superstructure to the ground and transmits the load to the ground. Generally, foundations rest on soil or rock. Engineering aspects of foundation primarily depend on the type of soil or rock. In a state like Rajasthan where the variation in geotechnical properties is extremely large, a detailed engineering analysis is obligatory for geotechnical engineers and many times, special treatments are required to be provided for foundation strata to achieve the desired stability.

The aeolian sands of Thar deserts are metastable and represent the most relevant image of Rajasthan soils. These windblown collapsible types of soil particles are radically rearranged and experience a loss of volume on wetting without application of pressure. The SPT N value ranges from 20 to 35 at 5.0 m depth for stable dunes while from 10 to 14 for unstable dunes. The safe-bearing capacity (SBC) values of dune sands are on the lower side due to the collapsible nature of particle arrangement. Due to high compressibility, more settlement is observed on the application of load as compared to other such soils. The SBC improvement techniques, which are commonly adopted, are not found suitable in such soils. An increase in settlement of about 2.8 times under saturated conditions as compared to dry conditions was observed in a plate load test done for both saturated and dry conditions. Therefore, a water table or saturation correction factor of 0.30 is adopted in dune sand generally (Gupta and Sundaram, 2017). Skirted footing in which vertical walls surround sides of the soil mass beneath the footing is a well-known bearing capacity improvement technique in such situations. Construction of vertical skirts at the base of the footing, confining the underlying soil, generates a soil resistance on skirt sides that helps the footing to resist sliding. Other than skirted footings, pile foundations are also an effective and commonly adopted solution in cases of foundations to cater to heavy loads of superstructures. But for structures of larger spread such as pavements and canals linings, stabilisation with lime and cement is done normally. Injection of a silicate solution that gels with time is a relatively newer technique with growing applicability. The thin film of silicate formaldehyde grout envelopes every particle of sand and makes a firm bond that remains stable in water and results in an increased bearing capacity, lesser settlement of structure and decreasing permeability of dune sand (Tiwari and Sharma, 2013).

Table 30.5 Properties of expansive soil obtained from different locations

Properties	Jaisalmer	Barmer	Jodhpur	Pali	Kolayat
Plasticity index (%)	30.00	41.0	55.0	50.0	27.0
Shrinkage limit (%)	16.0	18.40	21.80	20.0	15.0
Silt: clay	15:85	25:75	20:80	10:90	30:70
Free swell index (%)	200	260	340	230	140
Swelling pressure range (kPa)	145–220	160–255	205–310	190–275	110–150

Variation in volume due to the presence of water is not the problem only with dune sand. A different kind of behaviour is observed with expansive soil deposits of bentonite where the soil swells upon getting wet and exerts significant pressure on the foundation system. The presence of underneath expansive soil of high swelling potential first came into picture when the Indira Gandhi Canal lining work was going on during the 1970s in western Rajasthan and cracks emerged. The expansive soils are found generally from shallow depths in most areas of Jaisalmer, Barmer, Nagaur, Pali and Jodhpur districts. Table 30.5 provides variation in engineering properties of expansive soils obtained from different locations. Damages sustained by the occurrence of such expansive soil may include deformation, tilting and cracking of foundation systems, grade beams, floors and even pavements. In the case of buildings, the best ways to prevent structures from the ill effect of expansion are to make deeper foundations beyond the zone of variation in water content and by checking surface infiltration of water around foundations. Methods that are more suitable for pavement and other such infrastructure constructions are lime stabilisation, fly ash blending and replacement of expansive soil layers by non-expansive soils.

However, in some areas of Kota, Baran, Chittorgarh, Pratpgarh, Banswara and Udaipur, the extent of expansive clay is more severe due to very soft subsoil conditions. The replacement of clayey soil up to a considerable depth by any other suitable non-expansive cohesionless material such as silt or sand is an acceptable solution. Sometimes blending of existing soil with fly ash, lime slurry or any other slurry waste is also being used in view of cost effectiveness and to check environmental hazards due to indiscriminate disposal of such waste.

In situations where the use of such traditional practices is not feasible, a more engineered solution of application of geosynthetics is found very successful. However, in Rajasthan, the use of geosynthetics as soil reinforcement for ground improvement is in an emerging stage and yet to gain popularity, but the initial response of these techniques has been very encouraging.

Since Rajasthan has so many rocky and mountainous terrains, the development and construction of safe rail and road networks through such mountains have always been a challenging task. Construction of open cuts and tunnels on Rajasthan's topography is rapidly gaining acceptance after the availability of modern techniques of rock slope stability and rockfall protection. Though tunnels are constructed after a long and thorough geotechnical investigation, sometimes failures occur due to uncertainties in rock mass composition and due to the geometry of the slope surface, material profile and their spatial distribution, material properties viz. geo-mechanical, structural, phreatic surface, and local climatic conditions, and so on. One such failure occurred at the highway from Udaipur to Pindwara due to landslides and rockfall and consequent to this, the twin tunnel was closed temporarily. This twin tunnel was excavated through the structurally complex rocks comprising soil, phyllite and quartz of highly weathered condition with enormous random joints and folds. The engineering investigation and design were properly done for 85 m rock height over the tunnel crown. But the cuts at both the entrance and exit of tunnels were not treated properly and sliding had taken place. The erratic phreatic surface above the floor of the tunnel and piezometric surface higher than the natural ground level at some places were also among the reasons. Moreover, the reason includes very close joint spacing of approx. 1.0 m with a discontinuity of 10.0 m approx. Rock mass of tunnel was categorised as fair to poor due to variation in material composition. Based on further studies and software-based analysis, preventive measures were suggested including application of geotextile, side design modification, lowering of phreatic surface and prevention of rainwater percolation in the sensitive rock mass.

Besides those described above, many other conventional and modern ground improvement and stabilisation techniques are practised in Rajasthan. In Jaipur, which has experienced severe erosion of silty sand in 1981 during unprecedented floods, many gabion structures were made in the flood path and stone pitching covered with wire mesh was done on silty soil slopes, prone to slide.

Growing infrastructural development in some cities has pushed the developers to go for deep excavation to build multilevel basements. Non-cohesive silty sand in many areas poses a tough challenge of collapse and sliding for such construction. Shoring of continuous RCC piles is becoming a solution to this but has proved a very costly alternative for deep construction in smaller areas.

Geotechnical engineers in many parts of Kota and Jodhpur are facing a very different situation for underground basement construction. Rocky strata in both areas are allowing rapid seepage of subsoil water, which is sourced from Chambal River and Kaylana Lake. Subsoil structures are struggling with the continuous flow of water and a threat to subsoil constructed areas and foundation systems.

There are different forms of geotechnical problems and at the same time, solutions are there. It is a regular cycle of learning and implementation ideas to resolve the snag. Research and development are on the way to find out most sustainable and cost-effective solutions for all such problems.

30.6   OTHER GEOMATERIALS

Rajasthan has many coal-based thermal power plants. These plants use coal as fuel and fly ash is a by-product of coal combustion. Calcium fly ash is stockpiled or landscaped and causes serious ecological hazards. The use of fly ash in the construction industry is rapidly catching the attraction due to its easy availability. The most prominent use of fly ash is in the production of building bricks and in the production of pozzolanic Portland cement. Industries of bricks and cement are consuming the largest chunk of fly ash produced in Rajasthan.

Besides these commercial uses, fly ash is also being used for ground improvement. In many areas where the SBC is considerably low and soil is highly expansive, fly ash as a blending material has been proved to be a good option due to its pozzolanic action. The shear strength and bearing capacity of the soil can be considerably increased

Table 30.6 Properties of the original soil

District	Section	Soil type	Soil group	Max. dry density (g/cc)	OMC (%)	CBR (%)
Baran	Khyawda to Haripura	Clayey silt with sand and gravel	CI	1.84	15.0	2
Bundi	Madhorajpura to Jakhron	Clayey silt	CH	1.77	16.0	1
Jhalawar	Chaumahla to Piplai	Clayey silt with sand	CH	1.71	18.0	5
Kota	A/R to Marjhana	Clayey silt	CI	1.76	16.0	1

Table 30.7 Properties of blended soil

District	Section	Blending percentage		Max. dry density (g/cc)	OMC (%)	CBR (%)
		Industrial waste (%)	Fly ash (%)			
Baran	Khyawda to Haripura	5	15	1.75	20.0	19.0
Bundi	Madhorajpura to Jakhron	5	15	1.64	24.0	15.0
Jhalawar	Chaumahla to Piplai	5	15	1.62	26.0	14.0
Kota	A/R to Marjhana	5	15	1.70	22.0	16.0

by stabilising it with fly ash. Fly ash reduces the plasticity index and shrinkage limit, which has a potential impact on the engineering properties of fine-grained soil.

An experimental study conducted on black cotton soil of the Kota region demonstrated that blending of mineral slurry obtained from DCM Fertilizers Ltd. of Kota and fly ash with natural soil improved the soil properties significantly. The outcome of this study was used for property enhancement of subgrade soil by Public Works Dept. (Govt of Rajasthan) for construction of various roads. Tables 30.6 and 30.7 indicate property enhancement for each district in which this work was carried out.

The most fine-grained soil was effectively stabilised with 5%–11% of waste obtained from DCM Fertilizer Ltd., which was a mineral slurry and with fly ash obtained from Kota and Chhabda thermal power plants. The blending was used extensively to improve the engineering properties of fine-grained expansive soils. It was most effective in treating the plastic clays capable of holding large amounts of water. The addition of mineral slurry to a fine-grained expansive soil in the presence of water initiates several chemical reactions, which lead to improvements in its swelling properties and plasticity.

30.7 NATURAL HAZARDS

Rajasthan due to its peculiar type of geology and ecological system is not very much prone to suffer natural hazards. Since most of the parts of the state fall in Zone-II of the seismic map of India as per IS:1893 (2002), no serious loss of any structure is reported due to earthquakes in the past. Flood water-borne calamities are prone in

Figure 30.11 Damages to the highway near Jaipur.

areas where the drainage is not proper due to typical geological strata. Otherwise, in many areas where silty sand is prominently available, the permeability is fairly good and percolation of excess runoff takes place without much loss. In rare instances, the foundation soil is eroded due to heavy flow caused by the extraordinary deluge.

However, the presence of loose silty and sandy strata over the impervious layers of bentonite and gypsum in some areas of Barmer and Jodhpur caused havoc and damaged the rail and road network in the year 2007. These areas, which are known for drought, witnessed the devastating loss of infrastructure due to heavy rains and further due to typical geological conditions. Negligible permeability of underneath layers did not allow the huge quantity of runoff to percolate and loose sand washed out below the railway tracks and highways.

Despite heavy rains during this time, Barmer remained thirsty for the upcoming time due to the destruction of water harvesting structures like the *tankas* (UG water collection tanks) and anicuts. The villagers had to build all of them again.

One such damage due to a downpour occurred in embankment slopes of filled silty soil for highway construction near Jaipur. Photos taken after the failure reveal poor compaction and probably the wrong design of chute drains. Runoff caused severe soil erosion of adjacent slopes and heavy damage to the upper paved surface. Figure 30.11 shows the damages to this highway near Jaipur due to heavy rains. Provision of a reinforced earth slope could have averted the erosion and damage to the pavement. Damages to geotechnical structures reported so far are more due to manmade blunders or inadequacy in the design and execution of work rather than natural calamities in Rajasthan.

30.8   CASE STUDY

Rapidly growing transport infrastructure across the cities of Rajasthan has emerged with the need for smooth and hassle-free traffic movement, which cannot be accomplished without the construction of flyovers and railways over bridges. Mechanically stabilised reinforced earth walls are becoming a preferred choice as an alternative due to their cost effectiveness and ease of implementation.

These walls are the perfect example of a geotechnical structure in which apart from RCC panels and geosynthetics, the soil itself is a structural element. The properties of soil to be used as reinforced fill, the methodology of filling and compaction, protection of soil from seepage, workflow cycle and so on are important aspects for durability and serviceability.

This case study includes a recent failure that has occurred to a mechanically stabilised reinforced earth wall for a railway overbridge in eastern Rajasthan. The RCC panels were started falling just after two years of construction. The geosynthetic-reinforced soil wall was constructed using polyester geogrids and silty sand as reinforced fill with precast concrete facia panels as per the MORT&H specifications and designed as per AASHTO (2010). When this ROB was closed for investigation and maintenance operation, few panels had been fallen down, corner panels had been moved out, bulging occurred in the wall and the close wall was in a highly distressed state as there was relative movement both in the horizontal planes and vertical planes as seen in Figure 30.12.

Possible reasons for the failure as identified were the percolation of rainwater into the reinforced fill and lack of compaction. Differential settlement in a close wall due to a poorly compacted foundation portion was also observed due to which connections of panels with geogrid got broken and allowed the panels to bulge.

The structure of MSE walls is of such built that any surface treatment applied from outside cannot perfectly restore the panels in position once the loosening of earth and detachment of panels from reinforcement has occurred. Dismantling the whole system and redoing the work, with recoverable material plus additional materials, are the only choice for repairing.

In this case, following a thorough investigation, it was decided to provide soil nails through RCC panels in such a way that all acting loads were assumed to be taken by soil nails. This method was adopted to save considerable costs that could have been spent on the removal and reconstruction of the whole structure including existing pavements and other accessories. Nails were designed for length, diameter and number against active earth pressure using the ultimate bond strength of nails in soil. The diameter of the nail was worked out as 25 mm for lengths ranging between 5.0 and

Figure 30.12 MSE wall failure.

Figure 30.13 Soil nailed MSE wall.

6.50 m depending on the load the spacing of nails was calculated. This arrangement was checked for internal and external stability and found satisfactory. Figure 30.13 shows the completed soil nailed structure. The work of soil nailing was executed by an expert agency. After completion of this rehabilitation exercise, the bridge was opened for traffic and the treatment was found successful.

30.9   GEOENVIRONMENTAL IMPACTS ON SOILS AND ROCKS

There is a strong scientific belief that the earth's climate is changing rapidly and the concentration of greenhouse gases in the atmosphere is continuing to increase. Many studies have reported that climate change affects the hydrological cycle or water cycle components, especially precipitation, evapotranspiration, temperature, streamflow, groundwater and surface runoff. Global warming causes unequal distribution of rainfall and some regions experience greater rainfall and flooding while others become more prone to droughts. The flood in drought-prone Barmer due to unprecedented deluge is an example of such global warming. The deluge in Barmer was caused by a low-pressure zone over the area, itself a result of extreme heat conditions. Extreme weather events occur in cycles. But worldwide, the frequency of such extreme events is increasing because of climate change.

Droughts in Rajasthan bring a different sequence of consequences for geotechnical engineers. Low rainfall for a longer duration causes depletion of vegetation and deforestation that further lead to soil erosion. The low dry density of soil mass with low bonding of vegetation becomes a significant cause of slope failure in Rajasthan where silty sand is prominently used for such construction.

Apart from global warming, the environmental changes in Rajasthan are observed due to excessive mining as well. The main impacts on the environment by mining are degradation of land and forest, soil contamination, surface and groundwater pollution, noise and vibrations and deterioration of natural drainage systems. The excess mining leads to rearrangement of plate tectonics and chances of earthquake increases. In Rajasthan, extensive mining of sandstone, marble and other minerals has converted the Aravali into a rocky wasteland at some places. Soil erosion is rampant, natural recharging of groundwater has been affected and riverbeds have been flooded with coarse sand. Dams filled with sand need frequent de-silting operations.

The poor disposal management of mining waste along with industrial and municipal wastes is also posing a threat to not only the ecological system but also the geotechnical properties of land. Open dumping, indiscriminate disposal of waste materials and garbage landfills, is badly affecting the engineering properties of the underlying soil system. Plasticity index and permeability index are the parameters that are affected at the initial stage and further with due passage of time, a loss of strength is observed. Some experimental studies conducted for shear strength parameters and CBR values indicate a loss in values of these parameters. The erratic behaviour of soil due to the presence of waste material perforce the geotechnical engineers to adopt deep and heavy foundation systems. The worst affected parameter is electrical resistivity, which in turn poses difficulties in the installation of railway tracks and electrical transmission systems. However, it is difficult to quantify the exact effect of environmental changes on geotechnical structures, still, the stability and serviceability of geotechnical structures are analysed for several combinations of failure mechanisms. These also include loads on account of climate change. These are based on previous historical data in general but sometimes some unprecedented situation does not allow the technology to win over the nature. It is a continuous long chase between technology and nature.

30.10 CONCLUDING REMARKS

Rajasthan, the place of most diversified geology, has wide variations in its geotechnical features. An attempt is made to collect and concise different aspects of engineering properties and uses of soil and rocks available in Rajasthan. Though a thorough description of all available soils and rocks is beyond the limits of this script, the chapter presents an overview of common geotechnical aspects of the area. Issues related to low bearing capacity and expansive soils have been tackled well in the state from ancient times and now with the development of technology, solutions are relatively of low cost and more sustainable. But rapid urbanisation of Rajasthan cities poses challenges to geotechnical engineers and still, a lot more has to be done.

The presence of extreme saline soils with a high concentration of soluble salts in some parts of western Rajasthan is a challenge for the sustainability of many subsoil constructions. Inconsistent rainfall and high temperatures lead to high concentrations of sodium and calcium chlorides in soils. The geotechnical behaviour of soil is significantly changed due to chemical reactions between salts and soil minerals. Decreases in plasticity index, shear strength parameters and hydraulic conductivity were observed in some studies carried out on saline soils. The major impact of such soils is on underground infrastructures. Corrosion of buried pipelines and reinforcement, efflorescence in bricks, blistering of plaster and spalling of concrete are the other major points of concern. However, the use of marine-grade cement in such a situation may bring some relief to construction engineers.

Large-scale marble and granite processing units in the state are producing an abundance of marble slurry. The utilisation of slurry in concrete production to partly replace fine sand is under a trial state. Initial results are satisfying, and it may get acceptance for commercial production, which will not only reduce the environmental hazards due to slurry landfills but also reduce sand mining.

Providing sustainable and cost-effective solutions for the low bearing capacity of soft and compressible soils, slope failure analyses and mitigation measures, and identification and utilisation of waste and marginal materials are new points of consideration in the state.

Developments are on the way to use other marginal materials such as pond ash, zinc and copper slag and other mining waste and tailings as geomaterials. For ground improvement, soil stabilisation and reinforced fill of MSE walls, these materials have shown a sign of success in initial research and gaining popularity.

In the field of geotechnical engineering of Rajasthan, the applications of geosynthetics are becoming a popular choice to solve prolonged geotechnical problems and a very encouraging response is being received from all parts of infrastructure development.

REFERENCES

AASHTO (2010). *AASHTO LRFD Bridge design specification*, Fifth edition, Washington, DC.

Government of Rajasthan (2021). Various district survey reports – Government of Rajasthan, http://environmentclearance.nic.in, accessed 28 November 2020.

Gupta, S. and Sundaram, R. (2017). Geotechnical characterization of dune sands of the Thar Desert, Technical report, Cengrs Geotechnica Pvt. Ltd., New Delhi, India.

IS:1893 (2002). Criteria for earthquake resistant design structures, Part 1, Bureau of Indian Standards, New Delhi.

Tiwari, S.K. and Sharma, J.P. (2013). Characterization and strength improvement techniques for collapsible dune sand. *Electronic Journal of Geotechnical Engineering*, Vol. 18, pp. 4291–4302.

Chapter 31

Sikkim

Kaushik Bandyopadhyay, Abhipriya Halder, Saptarshi Nandi, Bappaditya Koley, and Subhajit Saraswati
Jadavpur University

CONTENTS

31.1 INTRODUCTION

Sikkim is a state in northeastern India. It borders Tibet in the north and northeast, Bhutan in the east, Nepal in the west and West Bengal in the south. The Kingdom of Sikkim was founded by the Namgyal dynasty in the 17th century. It was ruled by a Buddhist priest-king known as the Chogyal. It became a princely state of British India in 1890. After 1947, Sikkim continued its protectorate status with the Republic of India. It enjoyed the highest literacy rate and per capita income among the Himalayan states. In 1973, anti-royalist riots took place in front of the Chogyal's palace. In 1975, the monarchy was deposed by the people. A referendum in 1975 led to Sikkim joining

DOI: 10.1201/9781003177159-31

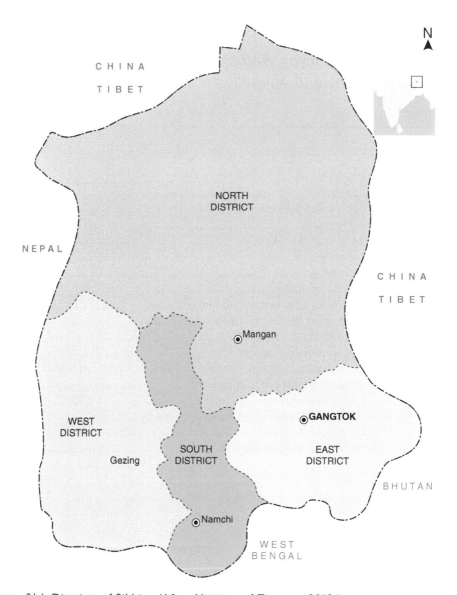

Figure 31.1 Districts of Sikkim. (After Ministry of Tourism, 2012.)

India as its 22nd state. Sikkim is also located close to India's Siliguri Corridor near Bangladesh. All the districts of Sikkim are shown in Figure 31.1.

31.1.1   Physiography of Sikkim

Sikkim is the 22nd state and the second smallest state of the Indian Union. The state has a total area of 7,096 km^2 and is situated between 27°46′ and 28°7′48″ N latitude and between 88°0′5″ and 88°55′25″E longitude in the eastern Himalayas. As an important transboundary landscape, it shares borders with countries like Bhutan (approx. 30 km long) represented by the Pangolakha range in the east, the Tibetan Autonomous Region of China (approx. 220 km long) represented by the Chola range in the east, Trans-Himalayan region in the north and Nepal in the west (approx. 100 km long) represented by the Singhalila range. Towards the south, Sikkim shares a border of approx. 80 km with the Darjeeling district of West Bengal, which is segregated by three river systems, viz., Teestachu, Rangitchu and Rongpochu. It stretches approximately 64 km from the east to the west and 114 km from the north to the south. The altitude in Sikkim ranges between 300 and 8,586 m (Basnett et al., 2013).

There are 84 glaciers in Sikkim, e.g., Zemu glacier (largest glacier in Sikkim), Rathong glacier, Zum-thul Phuk glacier, Onglathang glacier, Tasha Khang glacier and Teesta Khangse glacier. The water from these glaciers is discharged into two major river systems in Sikkim, namely Teesta and Rangit, which are considered the lifeline of Sikkim.

31.1.2 Physical features

Sikkim has a very rugged topography and formidable physical features. The whole state is enclosed on three sides by lofty ranges and spurs of Greater Himalayas with varying heights. In the north, the Greater Himalayas is stretched in a convex form while in the west, the Singalila range, which is a spur of the great Himalayas, is extended from the north to the south. The Donkya range, forming the eastern boundary of Sikkim, is much segregated with only two gaps, Nathu la and Jalepa la, which provide trade routes between Sikkim and China. The crowning glory of the state is the world's third-highest mountain Mt. Kangchenjunga (8,586 m).

The northern portion of the state, particularly beyond Chungthang, is the highest region of the state and is cut into a deep escarpment. This region has no populated area except Lachen and Lachung valleys. Southern Sikkim is low and more open and fairly cultivated in patches. It is subjected to erosion by River Teesta and its tributaries.

Broadly, it can be physiographically divided into the following zones (Forest and Environment Department, 2007):

- Lower hills – It stands between an altitude from 300 to 1,800 m and has hilly topography with flat cultivated lands in patches.
- Upper hills – The altitude of this area is from 1,800 to 3,000 m. Major forest areas are found in this zone.
- Alpine zone – The area between 3,000 and 4,500 m is termed the Alpine zone. It is covered with scrubs and grassland.
- Snow land – The area above 4,500 m is perpetually snow-covered and is without vegetation.

The general slope of the state is from the north to the south. However, the degree of slope varies from place to place. The slope in the whole north district, except Teesta valley below Chungthang and the northeastern part of the east district is 600 m/km. Towards the south, Teesta valley below Chungthang and the area around Rabongla in the south district, the slope is between 300 and 600 m/km. The rest of the state consisting of the whole west district, southern portion of the south district and extreme southwest part of the north district have a slope of 150–300 m/km (Forest and Environment department, 2007).

Glaciers are the important physiographic features of the state. They are mostly found in the north district. The most important one is Zemu glacier, which is 26 km in length and is situated at the base of Mt. Kangchenjunga. Other glaciers like Rathong, Lonak, Tolung and Hidden are in the northwestern part of the state. Some are also situated in the northeastern part of the state and are sources of important rivers of the state.

Teesta is the largest river in Sikkim. It flows essentially north-south across the length of Sikkim and divides the state into two parts.

The narrow and serpentine Teesta in its upper part becomes swollen, swift and muddy during monsoon and is full of rocks and hence is not navigable. Teesta and its tributaries receive the water from snow melting on the mountains as well as rain that accumulates during the monsoon. River Teesta and its tributaries provide huge surface water resources for the production of hydroelectricity in the state. Apart from small tributaries, Teesta receives the water of Zemu and Rangyong rivers on its right bank and from Lachung (Sebojung), Dik Chhu, Rongni and Rangpo rivers on its left bank. The Great Rangit, which is the most important right-hand tributary of Teesta, is another important river of the state. Its major tributaries are Rathang, Kalej, Reshi and Rangbhong. Based on the NCIWRD (1999) report, nine subriver basins/catchments have been demarcated on the map. These subbasins may be very useful for land-use planning particularly in a state like Sikkim.

31.2   MAJOR TYPES OF SOILS AND ROCKS

This section describes the major types of soils and rocks found in this region.

31.2.1   Types of soils

The Precambrian gneissic and Daling group of rocks with some intermediatories cover the major portion of the state. The gneissic group constitutes mainly the Himalayas. The Daling group consists of predominantly phyllites and schist. The slopes on these rocks are highly susceptible to weathering and are prone to erosion and landslides. As a major part of Sikkim lies on Darjeeling gneiss, the soil developed from this rock is brown clay, generally shallow and poor in lime, magnesia, phosphorous and nitrogen. However, it is quite rich in potassium. The texture of the soil is loamy sand to silty clay loam. The depth of the soil varies from 30 to 150 cm and in some cases, even more than 150 cm. The soils are typically coarse with poor organic mineral nutrients.

The soils of Sikkim are characterised by their variations in texture, colour and depth from one place to another owing to variations in topography, vegetation and climatic conditions. Broadly speaking, the soils are gritty to gravel from highly micaceous to sandy loam or pebbly and at places soft and chalky to stony in texture. The colour varies from grey to dark black or dark yellow. The colour, texture and depth of the soils have been influenced by the nature of the parent rocks.

31.2.2   Types of rocks

Sikkim is situated in the eastern part of the Indian Himalayan region and is covered by Precambrian rock assumed to be much younger. This state is considered as a young mountain system, still in the process of formation, with highly folded and faulted rock strata at several places. The mountain system in the state generally runs in the east–west direction while the chief ridges run in the north–south direction. The northern portion of the state is deeply cut into escarpment while the southern portion is comparatively open, which is basically due to the direction of the main drainage (southern). The northern, eastern and western portions of the state consist of hard massive gneissose rocks, which are capable of resisting denudation; however, the central and southern portions consist of soft, thin slate and half-schistose rocks, which are highly susceptible to weathering. In many places, the soil is good for the cultivation of crops; it is sandy alluvial in the lower regions and black, white or red in the higher hills.

Typical engineering geological information and geotechnical properties of soils/rocks as observed in some boreholes are summarised in Table 31.1. Variations in the soil profile other than these are also common and may be found in some locations.

The geological information and soil type of main districts of the state are shown in Table 31.2 and Figure 31.2.

31.3 PROPERTIES OF SOILS AND ROCKS

Schematic diagrams showing different soil layers obtained from boring logs of some of the soil exploration works and photographs of the work in progress are furnished in Figures 31.3–31.8. Deviations from these layers are also not uncommon at some locations.

31.4 USE OF SOILS AND ROCKS AS CONSTRUCTION MATERIALS

By and large, the soil is fairly homogenous in nature and character. The soil types vary from silty clay to sandy clay of medium plasticity, with the plasticity index varying from 7 to 18. The soaked CBR value ranges from 5 to 7.

The rock deposits are available all along the state with varying overburden soil. Besides, cobbles, pebbles and sand deposits are available in the rivers or streams within the state. Construction materials for granular subbase, cross-drainage and masonry RE wall works and so on are available at local quarry within the project corridor and wet mix macadam, dense bituminous macadam and bituminous concrete materials from Teesta River and LANCO tunnel excavated mug within the project corridor. The water absorption and aggregate impact value of these quarries are within the limit of the ministry's specifications. Bitumen, steel and cement are generally taken from Siliguri.

31.5 FOUNDATIONS AND OTHER GEOTECHNICAL STRUCTURES

Foundations and other geotechnical structures are decided based on soil strength, i.e., bearing capacity and settlement of soil.

31.5.1 Dams and roads

Sikkim is rich in hydropower potential despite its small area and is drained by many perennial rivers. Several hydroelectric power plants are currently in operation in Sikkim over these rivers and many more are being planned. There are two hydroelectric power

Table 31.1 Properties of soils and rocks in different districts

Sl no.	Name of the district and location	Depth of overburden soil and rock below the EGL	Properties of soils and rocks
1	North Sikkim (Mangan)	Overburden soil is present up to 2.10 m depth from the EGL and rock starts from 2.10 m depth	

(Continued)

Table 31.1 (Continued) Properties of soils and rocks in different districts

Properties of soils and rocks

Sl no.	Name of the district and location	Depth of overburden soil and rock below the EGL	Properties of soils and rocks
2	East Sikkim (Makha)	Overburden soil is present up to 2.20 m depth from the EGL and rock starts from 2.20 m depth	

(Continued)

Table 31.1 (Continued) Properties of soils and rocks in different districts

Sl no.	Name of the district and location	Depth of overburden soil and rock below the EGL	Properties of soils and rocks
3	East Sikkim (Signtham)	Overburden soil is present up to 2.50 m depth from the EGL and rock starts from 2.50 m depth	

(Continued)

Table 31.1 (Continued) Properties of soils and rocks in different districts

Properties of soils and rocks

Sl no.	Name of the district and location	Depth of overburden soil and rock below the EGL
4	South Sikkim (Namchi)	Overburden soil is present up to 4.50 m depth from the EGL and rock starts from 4.50 m depth

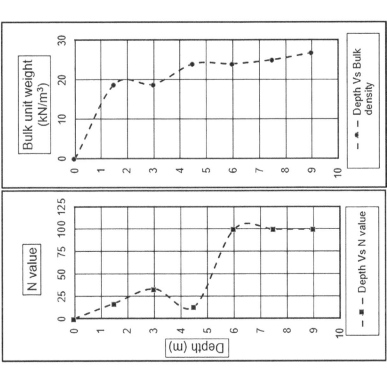

(Continued)

Table 31.1 (Continued) Properties of soils and rocks in different districts

Properties of soils and rocks

SI no.	Name of the district and location	Depth of overburden soil and rock below the EGL	Properties of soils and rocks
5	West Sikkim (Pelling)	Overburden soil is present up to 15.00 m, i.e., termination depth from the EGL	

Table 31.2 Geological and soil information of seismically high villages in Sikkim

Station	Latitude (°N)	Longitude (°E)	Geological information	Soil type
Singtam	27.15	88.29	Daling Metapelite	Coarse-grained soil with rock fragments. Typical Hapludolls to typic udorthents.
Gezing	27.17	88.15	Daling Metapelite	Excessively drained coarse to fine loamy soil with slight surface stoniness. Entic Hapludolls and typichamplumbrets.
Jorethang	27.21	88.18	Daling Metapelite	Coarse-grained soil with rock fragments. Typical Hapludolls to typic udorthents.
Gangtok	27.21	88.35	Lingtse granitoid gneiss	Excessively drained coarse to fine loamy soil with slight surface stoniness. Entic Hapludolls and typichamplumbrets.
Mangan	27.30	88.31	Pelitic Schist	Excessively drained coarse to fine loamy soil with slight surface stoniness. Entic Hapludolls and typichamplumbrets.
Chunthang	27.35	88.38	Pelitic migmatite	Excessively drained coarse to fine loamy soil with slight surface stoniness. Entic Hapludolls and typichamplumbrets.
Lachen	27.43	88.32	Politic migmatite	Excessively drained coarse to fine loamy soil with slight surface stoniness. Entic Hapludolls and typichamplumbrets.

Source: After Nath et al. (2000).

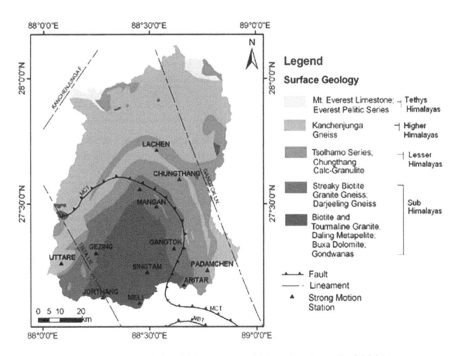

Figure 31.2 Surface geology of the Sikkim state. (After Nath et al., 2008.)

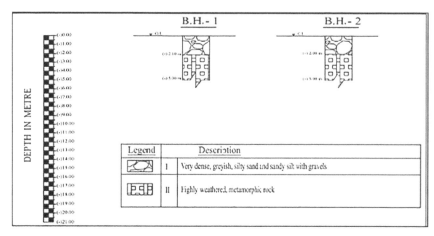

Legend		Description
▨	I	Very dense, greyish, silty sand and sandy silt with gravels
▦	II	Highly weathered, metamorphic rock

Figure 31.3 Typical soil profile and photographs showing boring work in progress for the project on 66 kV Mangan (extn.) Substation at north Sikkim, Sikkim.

plants, one of 513 MW near Dikchu and the other over Rangit River (60 MW) near Rangit Nagar (Figure 31.9a and b). No damage due to earthquake shaking or landslide was observed in the body of any of these dams. The power stations also performed extremely well; the only visible damage was minor cracking in masonry infill walls at various locations in the power stations. Severe landslides and ground deformations were observed near both the dam sites that resulted in the accumulation of excessive debris and silt in the reservoir and on downstream of the dam (Figure 31.9c). The 2011 Sikkim earthquake will be known for the severe damage caused to properties and lifelines by hundreds of landslides triggered due to the shaking and subsequent rainfall in the region. Subsidence was observed at Ranipool near Gangtok that resulted in the settlement of a part of the national highway by about 180 mm.

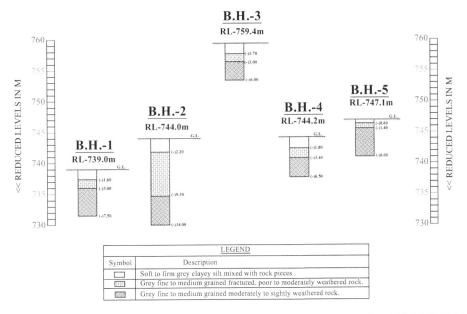

SUB-SOIL PROFILE
SITE: MAKHA, SIKKIM

B.H.-3
RL-759.4m

B.H.-2
RL-744.0m

B.H.-4
RL-744.2m

B.H.-5
RL-747.1m

B.H.-1
RL-739.0m

LEGEND	
Symbol	Description
☐	Soft to firm grey clayey silt mixed with rock pieces .
▦	Grey fine to medium grained fractured, poor to moderately weathered rock.
▨	Grey fine to medium grained moderately to sightly weathered rock.

Figure 31.4 Typical soil profile for the project on soil investigation for 66/11 kV Makha (New) substation under PGCIL at East Sikkim, Sikkim.

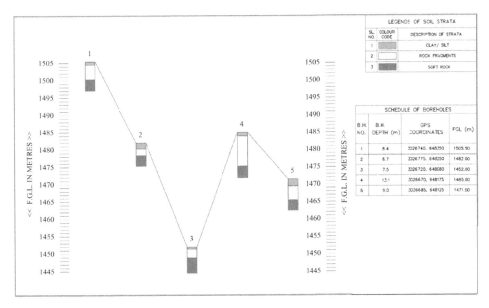

	LEGENDS OF SOIL STRATA	
SL NO.	COLOUR CODE	DESCRIPTION OF STRATA
1		CLAY/ SILT
2		ROCK FRAGMENTS
3		SOFT ROCK

	SCHEDULE OF BOREHOLES		
B.H. NO.	B.H. DEPTH (m)	GPS COORDINATES	FGL (m)
1	8.4	3326740, 648350	1505.50
2	6.7	3326775, 648250	1482.00
3	7.5	3326720, 648080	1452.00
4	13.1	3026670, 648175	1485.00
5	9.0	3026685, 648125	1471.50

Figure 31.5 Typical soil profile for the Project: 132/66/11 Dikchu substation under PGCIL East Sikkim, Sikkim.

SUB-SOIL PROFILE

SITE: MAMRING(EXTENSION)SUB STATION UNDER PGCIL AT SIKKIM

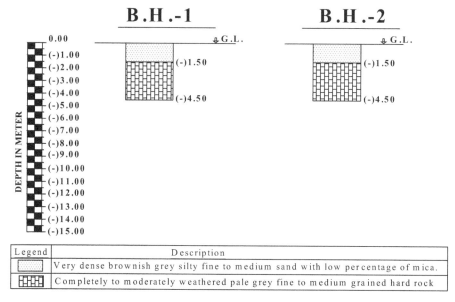

Figure 31.6 Typical soil profile for Project 66/11 kV Mamring (extension) substation under PGCIL at South Sikkim, Sikkim.

31.5.2 Vernacular housing systems

Primarily two types of vernacular housing systems which are mostly non-engineered are found to have been constructed in Sikkim, mostly wooden houses and masonry houses. Most traditional houses in Sikkim constructed privately are typically single storey bamboo houses known locally as the *Ikra* houses. *Ikra*-type wooden housing is commonly constructed in Sikkim due to its natural advantages related to lightweight, local availability of materials and so on. Performance of traditional *Ikra* houses was extremely good in both the earthquake shakings of 2006 and 2011. The only damage observed in *Ikra* constructions due to earthquake shaking alone (not due to landslides) was those of the additional classrooms constructed on the third storey of the new building of Government Secondary School at Sichey in East Sikkim. This observation presents a case for encouragement to the construction of vernacular housing systems, especially in seismically active regions. After the 2011 earthquake, one reinforced concrete (RC) column in the first storey of a new building of the Rumtek-Lingdum monastery suffered severe damage. Most of these monasteries reportedly suffered minor damage during the 2006 shaking and subsequently minor repair works, mostly cosmetic, were carried out (Kaushik et al., 2006).

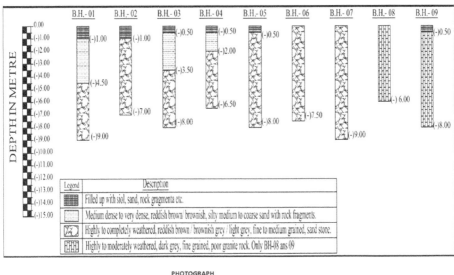

PHOTOGRAPH

Figure 31.7 Typical soil profile and pictures showing boring work in progress for Project: 132/66 kV new substation at Namchi (new), at South Sikkim, Sikkim.

31.5.3 Performance of residential, commercial and Government buildings

It was observed during the 2006 earthquake that Government buildings generally performed quite well and small repair or retrofitting work essential for few buildings was quickly carried out after the shaking. On the other hand, during the 2011 shaking, several Government buildings responded extremely poorly and had to be abandoned. A four-storey RC building at Dikchu bazaar constructed about 10 years back on the rear side of

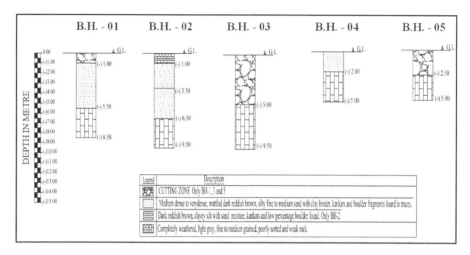

Figure 31.8 Typical soil profile for Project: soil investigation for 66/11 kV substation under
PGCIL at Rinchengpong, West Sikkim, Sikkim.

(a)   (b)   (c)

Figure 31.9 Performance of dams during the 2011 Sikkim earthquake: (a) Project Teesta-V,
(b) Project Rangit and (c) accumulation of excessive debris and silt in reservoirs
due to landslides.

an old wooden house collapsed completely after two days of the mainshock. The building
was constructed on a slope over an RC retaining wall and the slope started failing after the
main shock. This resulted in severe tilting of the building along the slope prompting the
occupants to vacate the building before it collapsed completely during an aftershock two
days later. Poor construction practice, inadequate reinforcement detailing and inadequate
measures to arrest slope failures appeared to be the primary reasons for the collapse.

Though the performance of a large number of RC buildings in Sikkim was below
par, several RC buildings performed exceptionally well during earthquake shaking.
A three-storey RC building in front of the Tashiling secretariat building hosts many
Government offices. It performed exceptionally well during the earthquake shaking,
though the performance of the secretariat building constructed right across the road
was quite poor. One of the major sports stadiums in Gangtok, Paljor Stadium, was constructed in 2005 and the performance of the RC and steel supporting structure of the
main stadium and ancillary buildings was very good during the earthquake shaking.

31.5.4   Geotechnical aspects of damages

The geotechnical damages include landslides, slope failures, ground failures, settlement of soils, failure of retaining walls, failure of foundations and damage to roads. In this section, the pattern of some of the geotechnical damages is presented.

31.5.4.1   Failure of retaining walls

In hilly areas, normally toe lines of multi-storey buildings are founded on loose soil, which is supported by a flexible retaining wall. Due to the earthquake, many of these retaining walls were damaged, which led to the failure of supporting foundations. Failure due to collapse of retaining wall in Ranipol, Gangtok damaged a building badly. It was observed that the retaining wall on the backside of the building had wide cracks with a gap of more than 15 cm. The cracked portion of the retaining wall had moved away from the building, creating a wide gap. Due to the earthquake, the staircase got separated from the wall of the building by about 15 cm. It was observed that this failure of the retaining wall was up to a considerable distance on both sides of the building.

31.5.4.2   Ground failure and damage to roads

In Lachung, wide cracks with a gap of about 20 cm were observed on the road, which can be attributed to the failure of the backfill on the riverside. Also, the road was damaged due to the falling of heavy rock pieces. This ground failure in Lachung indicated the high intensity of shaking (VIII+).

31.5.4.3   Foundation failures

The damage survey was carried out in Jorthang (West Sikkim) which is located on the terrace of Rangit River. Jorthang has an epicentral distance of about 66 km and the intensity assigned is VIII. One of the buildings in Jorthang was totally damaged of which two of the floors (G+1) were collapsed. The ground floor of the building was completely collapsed, perhaps due to foundation failure, which might be due to excessive settlement caused by shaking. However, no information on the foundation of the building could be available.

31.5.5   Structural aspects of damages

The structural aspects of damages include failures of buildings, monasteries, Government buildings and bridges. Both brick masonry and RCC frame buildings were damaged during the earthquake. Two of the buildings near the entry point of Sikkim, namely Rangpo, which were constructed on the slope of the mountain downside of the road towards the Teesta River, were damaged due to the earthquake. One of the buildings was the Godown used to store the waste material. These buildings were found to have minor cracks.

31.6 OTHER GEOMATERIALS

This section includes different types of wastes, which are utilised or can be utilised in geotechnical and other construction applications.

The handmade paper unit was established as an important section of the Directorate of Handicrafts & Handloom in the year 1957 to process/convert used waste into new products to prevent wastage of potentially useful materials, reduce consumption of fresh raw materials, reduce energy usage and reduce air pollution and water pollution by minimising the need for conventional waste disposal as compared to virgin production. Recycling is a key component of modern waste management. Further old files and documents of offices are restricted from being disposed of in the open market due to many reasons, confidentiality being the main one. There are other sources from which waste paper and raw materials are gathered, e.g., printing press and Argali (Edgeworthia gardenia) bark.

Handmade paper and its attractive products are becoming popular in the world market due to their eco-friendly attribute, artistic appeal and durability. As the handmade paper is made entirely from the recycling of waste materials, its use helps in saving trees and this contributes to a healthier environment. The Government in furtherance of its vision to make Sikkim a clean green state is committed to ensuring that trees are planted regularly, banning the use of plastic bags and encouraging the use of eco-friendly articles in a day-to-day life.

The main sources of solid waste in Namchi under Namchi Municipal Corporation are summarised below:

- Municipal sources of waste – Waste from households, different organisations, schools, colleges, other institutions, restaurants and other public places, excess foods, used plastic bottles, polythene bags, old clothes, broken furniture and used paper.
- Medical sources of waste – Waste from different health care institutes like hospitals, nursing homes, waste materials from expired medicine, used needles and syringes, blood, used bandages and so on.
- Waste from automobiles – The various waste materials produced from these sectors are old broken vehicles and different parts of a car.
- Construction sources – The waste materials mainly produced from the construction of roads and buildings are concrete waste, plastic bags of building materials and cement and so on.
- Electronic sources of waste – Different electronic wastes are computers, broken TV, damaged electronic parts and so on.
- Tourism linkage – Almost 1 lakh tourists come to this place every year from various parts of India. As a result, a huge amount of solid waste is generated.

31.7 NATURAL HAZARDS

This section includes details of earthquakes, landslides/erosion, floods and so on focussing on how they affect foundations and other geotechnical structures.

Due to its climatic conditions and physiographic conditions in India, the state is considered to be one of the most earthquake-prone areas in the world. It has experienced numerous earthquakes with 58% of land under threat to seismic hazards. India has been a victim to six major earthquakes among the most disastrous events which occurred in the world. Among the various earthquake zones, zone V is the most seismically active region. The division of earthquake-prone regions according to the MM intensity scale includes Zone II with VI intensity, Zone III with VII intensity, Zone IV with VIII intensity and Zone V with IX intensity (Nath et al., 2005; Pal et al., 2008). The areas with moderate to low risk are relatively safe from earthquakes, but due to non-engineering foundations and structures, the regions are susceptible to earthquakes.

India is seismically active with low to high earthquakes frequently, but comparatively the northern part of the country is very active. Sikkim is a small state in the northeastern part of India, with irregular topography and with geology more prone to earthquake-induced landslides.

The state is full of mountain terrains with elevations ranging from 280 to 8,586 m, with the existence of the world's third-highest peak Kangchenjunga. According to the seismic zonation map of India, the state of Sikkim comes under Zone IV (IS 1893:2002), but the north zone of Sikkim comes under zone V (Chakraborty et al., 2011). It has been affected by great earthquakes of the world with a magnitude of 8 or more. As it has uneven topography, Sikkim experiences both minor and major earthquakes.

Sikkim is in a very high zone prone to many disasters due to its favourable terrain conditions, mainly earthquakes and landslides, which account for the loss of property and life, network systems, agricultural lands and human settlements. The Sikkim region is vulnerable to natural disasters but the continuation of the occurrence of vulnerability may increase or decrease by human action. The perennial disturbances of earthquakes and accidents like landslides are very high in Sikkim.

Several types and sizes of landslides occur from the top to the bottom of the hills of Sikkim during an earthquake. Besides earthquakes, the rapid growth and huge structural developments also make Sikkim more vulnerable to landslides. Furthermore, this earthquake is enhanced due to the topographical features or due to complex tectonics. The state consists of gneissose, migmatite and schistose rocks and its stratigraphic features include the Buxa and Gondwana groups of rocks (Nath et al., 2000). These rocks are covered with unconsolidated screened deposits, which are favourable conditions for failure of slopes during the earthquake.

All four districts of Sikkim have witnessed massive landslides due to earthquakes. It has been reported that earthquakes triggered more than 300 landslides. The extremely high occurrence of landslides is attributed to the geology of the region and intense rainfall. The high intensity of rainfall contributes to rapid erosion and weathering of rock mass. The increase in water level causes instability in natural slopes. Furthermore, human activities such as excavation work for buildings, roads and embankments add to further instability (Mehrotra et al., 1996). Earthquake provides just a triggering action to the already unstable or marginally stable slopes. As per the landslide hazard zonation map of India, Sikkim lies in the second-highest zone categorised as "high zone". This spells the risk the state inherits for landslides. A recent earthquake just validated this unfortunate fact.

Due to incessant rain followed by an earthquake, a water stream changed its path and created a new stream. This phenomenon is known as landslide lake outburst flood. In this stream, water had flown at such a high velocity that debris of rock had also flown with it and created mud mountains at both the sides of this stream. This may be attributed to the reduction in strength of the rock material due to saturation

(Kramer, 1996) and then it slid due to the earthquake force carrying huge boulders and rock pieces with it. The locals were trying to divert the path of this new stream as it poses threat to the buildings downstream. There were massive landslides in other parts of Sikkim too.

31.7.1   Seismicity aspects of Sikkim

The northeastern part of the country is most vulnerable and is a hot bed of earthquakes due to its different interring plate stresses. The compression forces generated due to the tectonic movement of the Indian plate and its subsequent plate abruptly releasing forces reaching the threshold are the primary causes of earthquakes. Sikkim is part of northeastern India and is situated in the eastern Himalayas, with one-third of its area being covered by heavy forests without a flat land. The high mountains in the state are raised towards the north and are affected by high tectonic events from the past two decades.

Some of the severe earthquakes in Sikkim and surrounding areas are Cachar 1869 (7.5), Shillong 1897 (8.7), Dhubri 1930 (7.1), Bihar–Nepal 1934 (8.3), Arunachal Pradesh 1950 (8.5), Gangtok 1980 (6.1), Nepal–India 1988 (6.4), Sikkim 2006 (5.7), Bhutan 2009 (6.2) and Sikkim 2011 (6.9) (SSDMA, 2012). The state is divided into four regions, namely, northern Sikkim, southern Sikkim, western Sikkim and eastern Sikkim. The entire state is covered with beautiful rivers like Rangit and Teesta and 180 perennial lakes, which pass through to deep valleys with the location of epicentre at 68 km from north-west Sikkim according to the meteorological department.

Each portion of the state is different from the other with different geological conditions. The northern portion is covered with steep escarpments. The eastern, western and northern parts of the state are covered with massive gneissose rocks that can resist denudation. Southern Sikkim is covered with flattened lands with soft, thin and half-schistose rocks, which are used in cultivation.

As the state is seismically active, various techniques are applied to know the state's geology, subsoil conditions, the topography of bedrock, amplification of earthquake ground motion and so on. Using site response studies, geology, soil type, computing response characteristics, slopes, resonance frequency and analysis of the seismic hazard, a microzonation map of Sikkim Himalayas was developed by Nath et al. (2005) and is presented in Figure 31.10. Six levels of hazards from very low to severe have been identified.

31.7.2   Landslides

As already mentioned, the Tethys and Higher Himalayas are dominating the lesser and sub-Himalayas, as the Tethys and Higher Himalayas are above the Main Central Thrust dominating lesser Himalayas and exhibit low hazard than lesser Himalayas. The report of "Sikkim earthquake of 18th September, 2011" by the Disaster Mitigation and Management Centre of Uttarakhand states that the Sikkim region consists of Daling, Gondwana, Chungthang and crystalline gneiss group of rocks from the north to the south.

As we know that the state is mainly covered with Daling and Precambrian group of rocks, which consist of phyllites and schist, the occurrence of minerals such as

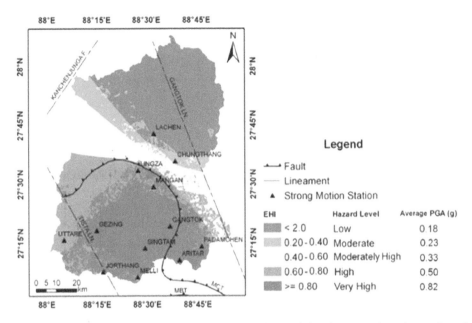

Figure 31.10 Microzonation map of Sikkim with hazard level covered with earthquake motion stations, faults and lineament. (After Nath et al., 2005.)

dolomite, graphite, copper and limestone is rather lesser known in different portions of the state. The slopes of Sikkim are prone to erosion and landslides due to weathering of these rocks. Loading and unloading shear distortions in the Daling group during orogenesis cause high landslide susceptibility. The mountain slopes turn unstable very easily during earthquakes and rainfalls, because gravity forces govern the slopes. Due to its geological conditions and seismic conditions, landslides in the Sikkim state are very high. These landslides are spread from the east to the west and from the north to the south. The districts in the east and south sides of Sikkim are more prone to landslides, especially the eastern part due to the Daling group of rocks, which are more prone to landslides, in comparison to the gneiss and schist deposits. These eastern and southern parts are covered with Daling group formation. These landslides are mainly controlled by seismicity, climate and non-tectonism. The eastern Himalayas including the Sikkim state is a hotbed for natural disasters mainly of landslides and earthquakes (SSDMA, 2012).

The slopes with 50% steepness are covered with excessively drained coarse loamy to fine loamy soils with little surface stoniness. If the slopes are greater than 30%, their surfaces are covered with coarse-grained soils with small rock fragments. The slope with 30% exhibits a thick soil cover. The moderately steep slopes with 15% exhibit well-drained fine loamy soils and also these are associated with coarse loamy soils (Nath et al., 2000). There are different types of landslides with different movements in Sikkim, which are shown in Table 31.3.

The main cause of landslides is triggering by many factors. The combinations of many factors will finally initiate the movements in slopes causing failure. Some factors,

Table 31.3 Different types of landslides with different movements in Sikkim Himalayas (SSDMA, 2012)

Landslide name	Type of landslide/Rocks type	Magnitude and intensity	Area in danger	Triggers	Warning	Strike time	Damage
AoKhola Rongli	Debris flow Overburden of gneiss and schist	Huge Fast Avalanche	Rongli Bazaar	Heavy rain Thick debris, Steep slope	No Warning	21/5/97 12.30 pm	07 persons dead, one injured
Gangtok & Vicinity	All type Rock/ Soil Materials	Widespread Fast Mud & Debris flow	Dev. Area Rongnek Syari	Heavy rain Overflow of drain water and so on	2–3days before the strike, Cracks in Roads and Subsidence	8–9.30 pm 08/06/97	43 persons dead, 300 house completely 1,000 partially
Deorali Gangtok	Mudslide (shallow) Rock/soil/ Construction Materials	Localised Fast flow & Avalanche	Kopibari School Area	Water supply Pipe burst and rainfall	No Warning	9.30 pm 05/9/95	32 people dead, 08 houses completely
Gyalshing Bazar	Translational (shallow) Rocks & soil debris	Localised Fast Avalanche	Road to Kyongsa, Legship and Houses	Wayward Rainwater	No warning	Morning 26/9/2000	05 person dead 28 families evacuated
Manzing	Complex	Massive fast Avalanche	22 km from Ravangla To Lingmoo	High rainfall steep slope week geology	Persistent rain subsidence	5 pm 24/9/2005	07 person dead 28 families evacuated

Table 31.4 Factors attributing the failure of slopes

A Geotechnical factors		B Morphological causes	
a.	Weak material	a.	Slope angle
b.	Sensitive materials	b.	Uplift
c.	Weathered materials	c.	Rebound
d.	Sheared materials	d.	Fluvial erosion
e.	Jointed or fissured materials	e.	Wave erosion
f.	Adversely oriented discontinuities	f.	Glacial erosion
g.	Permeability contrasts	g.	Erosion of lateral margins
h.	Material contrasts	h.	Subterranean erosion
		i.	Slope loading
		j.	Vegetation Change
C Physical causes		*D Human causes*	
a.	Intense rainfall	a.	Excavation
b.	Rapid snowmelt	b.	Loading
c.	Prolonged precipitation	c.	Drawdown
d.	Rapid drawdown	d.	Land-use change
e.	Earthquake	e.	Water management
f.	Volcanic eruption	f.	Mining
g.	Thawing	g.	Quarrying
h.	Freeze–thaw	h.	Vibration
i.	Shrink–swell	i.	Water leakage
j.	Groundwater changes		
k.	Other mass movements		

Source: After SSDMA (2012).

which attribute to failure, are presented in Table 31.4 (SSDMA, 2012). It is easy to determine the trigger after the occurrence of the landslide but it is difficult to find the exact factors that trigger the movement. If the slope is having these factors and is subsequently triggered with rainfall or earthquake, it finally leads to landslides.

31.8   CONCLUDING REMARKS

Sikkim, a part of the eastern Himalayas, is a state in the northeastern part of India. It is divided into four districts, namely east, west, north and south. The major types of soils and rocks of this state are summarised below:

*   The northern, eastern and western portions of the state have hard massive gneissose rocks.
*   The central and southern portions constitute soft, thin slate and schistose rocks.
*   Phyllites and schists are susceptible to landslides and erosion.
*   The depth of overburden soil varies in the entire state. In the high-altitude areas, it is less but in the lower ones, sometimes it extends up to a considerable depth. By and large, this depth varies from 30 to 150 cm.
*   The soils are typically coarse with poor organic mineral nutrients.
*   By and large, shallow foundations in the form of isolated/strip footings are suitable for the typical low-rise structures constructed in the state.

ACKNOWLEDGEMENT

The kind permission of M/S B.S. Geotech Pvt. Ltd., Hooghly to publish the different district-wise soil data and also contributions from Mr. Sumit Mukherjee and Mr. Santosh Kr. Dey (Directors of M/S B.S. Geotech Pvt. Ltd.) are highly acknowledged.

REFERENCES

Basnett, S., Kulkarni, A. and Bolch, T. (2013). The influence of debris cover and glacial lakes on the recession of glaciers in Sikkim Himalaya, India. *J. Glaciol.*, 59(218), 1035–1046. doi: 10.3189/2013JoG12J184.

Chakraborty, I., Ghosh, S., Debasish, B. and Bora, A. (2011). Earthquake-induced landslides in the Sikkim-Darjeeling Himalayas-An aftermath of the 18th September 2011 Sikkim earthquake. Geological Survey of India, Kolkata, 1–8.

Forest and Environment Department (2007). Report of Natural Resources of Sikkim. http://www.sikkimforest.gov.in/docs/Sikkim/Natural%20Resources%20of%20Sikkim%202007.pdf, accessed 3 February 2021.

IS1893 (Part I): 2002. *Indian Standard on Criteria for Earthquake Resistant Design of Structures*, Bureau of Indian Standards, New Delhi, India.

Kaushik, H.B., Dasgupta, K., Sahoo, D.R. and Kharel, G. (2006). Performance of structures during the Sikkim earthquake of 14 February 2006. *Current Sci.*, 91(4), 449–455.

Kramer, S.L. (1996). *Geotechnical Earthquake Engineering*, Prentice Hall, Inc., Upper Saddle River, NJ.

Mehrotra, G.S., Sarkar, S., Kanungo, D.P. and Mahadevaiah, K. (1996). Terrain analysis and spatial assessment of landslide hazards in parts of Sikkim Himalaya. *J. Geol. Soc. India*, 47, 491–498.

Ministry of Tourism (2012). Tourism Survey Report for the State of Sikkim, Ministry of Tourism, Government of India. p. 109. https://tourism.gov.in/sites/default/files/2020-04/Sikkim%20tourism%20Final%20Report%2019th%20July.pdf, accessed 9 February 2021.

Nath, S.K., Sengupta, P., Sengupta, S. and Chakrabarti, A. (2000). Site response estimation using strong motion network: A step towards microzonation of Sikkim Himalayas. Seismology 2000, *Curr. Sci.*, 79, 1316–1326.

Nath, S.K., Vyas, M., Pal, I. and Sengupta, P. (2005). A seismic hazard scenario in the Sikkim Himalaya from seismotectonics, spectral amplification, source parameterization, and spectral attenuation laws using strong motion seismometry. *J. Geophys. Res.*, 110, B01301.

Nath, S.K., Thingbaijam, K.K.S. and Raj, A. (2008). Earthquake hazard in Northeast India – A seismic microzonation approach with typical case studies from Sikkim Himalaya and Guwahati city. *J. Earth Syst. Sci.*, 117, 809–831. doi: 10.1007/s12040-008-0070-6

NCIWRD (1999). Water requirement for various uses. National Commission on Integrated Water Resources Development, Government of India. http://nwm.gov.in/?q=integrated-water-resource-management, accessed 9 February 2021.

Pal, I., Nath, S.K., Shukla, K., Pal, D.K., Raj, A., Thingbaijam, K.K.S. and Bansal, B.K. (2008). Earthquake hazard zonation of Sikkim Himalaya using a GIS platform. *Nat Hazards*, 45, 333–377. doi: 10.1007/s11069-007-9173-7

SSDMA (2012). Sikkim State Disaster Management Plan. Land Revenue and Disaster Management Department, Sikkim State District Disaster Management Authorities (SSDMA), Government of Sikkim, Sikkim, 1–180. http://www.sikkimlrdm.gov.in/downloads/publications/sdmp.pdf, accessed 10 February 2021.

Chapter 32

Tamil Nadu

K. Premalatha
Anna University

M. Muthukumar
Vellore Institute of Technology (VIT)

B. Arun
WRD, TNPWD Chennai

M. Dhanasekaran
Public Works Department Kottar
TNPWD

CONTENTS

DOI: 10.1201/9781003177159-32

32.1 INTRODUCTION

Tamil Nadu is located in the south-eastern part of the Indian peninsula between North Latitudes 08°00′ and 13°30′ and East Longitudes 76°15′ and 80°18′. The total area of the state is about 1,30,058 km². The state consists of 37 districts (Figure 32.1). The state is

Figure 32.1 Map of Tamil Nadu along with the districts.

bounded by the Bay of Bengal in the east, the Indian Ocean in the south, Kerala in the west and Karnataka and Andhra Pradesh in the north.

32.2   MAJOR TYPES OF SOIL AND ROCK

32.2.1   Major types of soil

The major types of soil found in Tamil Nadu are red loam, laterite, black cotton soils, alluvial, peat and marshy soils. Images of different types of soils that prevail in Tamil Nadu are shown in Figure 32.2.

32.2.1.1   Red soils

These soils are predominant and occupy a major part of the state. The colour of the soil is due to the presence of ferric oxide. The soils are derived from granite, gneiss and other metamorphic rocks. They are also slightly acidic to alkaline in nature and contain more percentage of sand and silt with moderate permeability. The soil away from the coasts are gravelly, sandy, porous and light coloured and are more susceptible to erosion. On the lower plains and valleys, they are dark coloured fertile loams. This soil occupies particularly in interior districts and along the coastal districts. It is found predominantly in the northern districts such as Kancheepuram, Vellore, Thirupathur, Cuddalore, Salem, Dharmapuri and in some southern districts such as Ramanathapuram, Coimbatore, Trichy, Pudukkottai, Thanjavur, Sivaganga, Virudunagar, Madurai, Dindigul, Nagapattinam, Thoothukudi, Tirunelveli and the Nilgiris.

32.2.1.2   Black cotton soils

Black soils are characterised by their dark brown to black colour. The soils contain 35%–60% of clay and are highly plastic in nature. These soils undergo high swelling and shrinkage. Deep polygonal cracks occur when the water evaporates during summer. Because of its high water retention capacity, rainfed crops like cotton, millets, pulses and so on are well grown on these soils. These soils are also known as black cotton soils as cotton is the major crop grown in these soils. It is found in parts of Coimbatore, Madurai, Dindigul, Thoothukudi and Tirunelveli and in patches in the districts of Kancheepuram, Vellore, Salem, Dharmapuri, Ramanathapuram, Virudunagar and the Nilgiris.

32.2.1.3   Laterites and lateritic soils

Lateritic soils are enriched with oxides of iron and aluminium under the conditions of high rainfall with alternate dry and wet periods. The silica present in these soils is leached during rainfall leaving the iron and aluminium oxides in the top layers. All lateritic soils are poor in calcium, magnesium, nitrogen, phosphorus and potash. They are usually pervious in nature. They are found in parts of Thanjavur, Nagapattinam, Nilgiris and Kanchipuram districts.

Figure 32.2 Different types of soils in Tamil Nadu: (a) black cotton soil, (b) red soil, (c) lateritic soil, (d) alluvial soil and (e) Teri soil (rare earth).

32.2.1.4   Alluvial soils

Alluvial soils cover the largest area along the river coast of Cauvery, Vaigai and Vennar basins of Tamil Nadu. The main features of alluvial soils have been derived as silt deposition laid down by the river systems. The rivers carry the products of weathering of rocks constituting the mountains and deposit them along their path as they flow down the plain land towards the sea. Deltaic areas of Thanjavur, Nagapattinam,

Tiruchirappalli, Cuddalore, Kancheepuram, Tirunelveli, Tuticorin, Kanyakumari, Ramanathapuram and Sivaganga districts have this kind of soil. From the agricultural point of view, alluvial soils are considered as most important soils.

32.2.1.5   Peat and marshy soils

Peat soils are formed due to the accumulation of organic matter in humid regions. The soils are black clayey and contain organic matter of up to 40%. Marshy soils occur on the southeast coast of Tamil Nadu.

In addition to the above soil types, special red sand dune soils are available which come under rare earth categories. This soil is also called Teri soil. This type of soil is found in the districts of Therikadu in Thuthukodi, Kanyakumari and Tirunelveli districts of Tamil Nadu. Teri soil is formed in the omega thermic and ustic regimes from geogenic sand deposits. The soils are dominated by heavy minerals like limonite and zircon and a little rutile, monasite and garnet and quartz in light mineral fractions. Ilmenite is used in the production of titanium dioxide pigment, which is used in paints, papers, plastics, textiles and so on. TiO_2 slag is further processed for producing TiO_2 pigments and titanium sponge, which finds application in aircraft bodies, artificial limbs and so on. Zircon is used in the production of ceramics, sanitary ware and so on to impart opacity to the end-products. Garnet is used as an abrasive in grinding wheels, as a medium for water filtration, for rough glass polishing and so on (Indian Rare Earth Ltd., 2020).

32.2.2   Major types of rock

Over 80% of the state is covered with crystalline rocks of Archaean to late Proterozoic age, while the remaining area is covered by Phanerozoic sedimentary rocks mainly along the coastal belt and in a few inland river valleys. The stratigraphic succession of Tamil Nadu is discussed here.

32.2.2.1   The Archaean rocks

Most of the Precambrian rocks of charnockite, khondalite, Sathyamangalam and Bhavani groups in Tamil Nadu belong to the Archaean age. The charnockite group of rocks are identified in the northeastern parts of the state comprising pyroxene granite, magnetite quartzite and pink coloured feldspathic granulite. They are found in places like Shevaroy, Chitteri, Kalrayan, Kollimalai, Pachaimalai and in the Nilgiris. The pyroxene granulites represent the basic volcanics; the banded magnetite quartzite indicates a volcanic exhalative origin, while the pink coloured granulites represent the acid volcanics. Charnokitic rocks also occur in the Dharmapuri and Erode districts comprising the hills in the west and northwest of Mettur.

The khondalite group belongs to metamorphose sedimentary rocks that include argillaceous, arenaceous and calcareous members (Gopalakrishnan et al., 1976). The khondalite consists of parent sedimentary rocks such as quartzites, calc-granulites, calc-gneiss, crystalline limestone, gametiferous quartzo-feldspathic gneisses and meta-pelitics (Sriramguru et al., 2002).

Sathyamangalam group of rocks consists of mica schist calc-granulites, para am-phibolites and crystalline limestone. The fuchsite and kyanite bearing characteristic assemblages are the key members of the Sathyamangalam group (Gopalakrishnan et al., 1976).

Bhavani group is a combination of gneisses of different compositions and tex-tures and is found to be like the peninsular gneisses of Karnataka. Rocks that are highly fissile including banded mica gneiss, quartzo-feldspathic gneiss, augen gneiss, hornblende gneiss, biotite gneiss and so on are found in this group. These rocks are well found in the central part of the state like Erode, Coimbatore, Salem, Namakkal, Tiruchirapalli and Perambalur districts (Hansen et al., 1988).

Geologists identified that the anorthosite rocks from Sithampoondi and Kun-namalai (a village in Namakkal district) have similar geological properties to that of the lunar composition (Venugopal et al., 2020). About 60 tonnes of rock from Sitham-poondi were excavated and pulverised to various sizes to match the chemical and me-chanical properties of the lunar soil to study the rover's movements on it in a simulated environment.

32.2.2.2   Paleozoic sediments

Along the coast of Tamil Nadu, the crystalline rocks are overlain by Phanerozoic sed-iments. The Talchir Formation composed of boulder bed, conglomerates, olive-green to brown shale with minimal sandstone of Lower Gondwana age (Murthy and Ahmed, 1977) is well developed in the Palar basin in the northeast.

32.2.2.3   Mesozoic rocks

Mesozoic rocks are along the Tamil Nadu coast, namely in Palar, Vriddhachalam, Tiruchirapalli and Sivaganga sub-basins. The Mesozoic rocks are represented by upper Gondwana formations and marine Cretaceous rocks. The upper Gondwana rocks occur in three sub-basins in Sivaganga, Perambalur (with marine Cretaceous sediments) and Tiruvallur districts. The Sivaganga formation comprises basal boul-der bed, conglomerate, micaceous sandstone and shale. Shale bears a fossil impres-sion. These forms indicate Lower Cretaceous and marine influence at the time of deposition.

32.2.2.4   Cenozoic

The sandstones of Paleocene age rest above the Cretaceous rocks. Rocks of Mio-Pliocene age (Cuddalore formation) occupy large areas and rest over the crys-talline basement. The formation comprises large quantities of fossil wood. National Fossil Wood Park has been established by the Geological Survey of India (2006) at Thiruvakkarai in Villupuram district.

Different types of soil, rock and water-bearing formations prevailing in some of the important cities/major different districts are summarised in Table 32.1.

Table 32.1 Types of soils and rocks and their occurrence in some districts of Tamil Nadu

Districts	Geomorphology (major physiographic units)	Major soil types	Predominant geological formations
Chennai	1. Fluvial landforms 2. Marine landforms 3. Erosional landforms	Beach sands, clay and alluvial soils	Alluvium, sandstones, clay, shale, siltstone, granites, gneisses and charnockite
Coimbatore	Upland plateau region with hill ranges, hillocks and undulating plain	Red calcareous soil, Red non-calcareous soil, black soil, alluvial and colluvial soil	Archaean crystallines and recent alluvial and colluvial formations
Madurai	Hilly region and plains of Madurai and Melur regions	Red soil, black soil and sandy soil	Granite gneiss, charnockites and alluvium as patches along the river
Salem	Upland plateau region bounded by the Eastern Ghats with many hill ranges, hillocks and undulating plain slopping towards the east	Red soil, black soil, brown soil, alluvial soil and mixed soil	Alluvium, colluvium, laterite, granite, dolerite, quartzite, charnockite and granite, gneiss
Thoothukudi	Coastal plain and upland	Black soil, red soil and sandy soil	Recent river alluvium and coastal sand, red teri sand, calcareous sandstone, pink granites, charnockites and Peninsular gneisses
Tirunelveli	Papanasam upper slopes, Kalakkadu upper slopes, Chittar plains, Tamarabarani Plains and Nanguneri Plains	Deep red soil, black cotton soil, red sandy soil, coastal alluvium and river alluvium	Recent alluvium, granite gneiss and charnockite
Vellore	(i) Hilly terrain in the eastern and southwestern parts. (ii) Plain regions in the eastern part.	Sandy soil, sandy loam, red loam, clay, clay loam, black cotton soil	Alluvium, granite, gneisses and charnockite

32.3 PROPERTIES OF SOILS AND ROCKS

32.3.1 Soil properties

The soil properties of Tamil Nadu range widely with different consistency and strength. The typical soil profile for the construction of a shopping mall in Cuddalore is presented in Figure 32.3. Cuddalore is located in the southern part of Tamil Nadu along the coastal region. Soil exploration is conducted up to a depth of 34 m. From the boring log details, the hard bearing stratum is not encountered even at a depth of 35 m. Alternate layers of silty sand, clayey sand and sandy clay with different colours and textures are found to exist up to a depth of 34 m with an N value in the range of 10–80. A similar type of soil profile exists along the coastal districts/regions of Tamil Nadu.

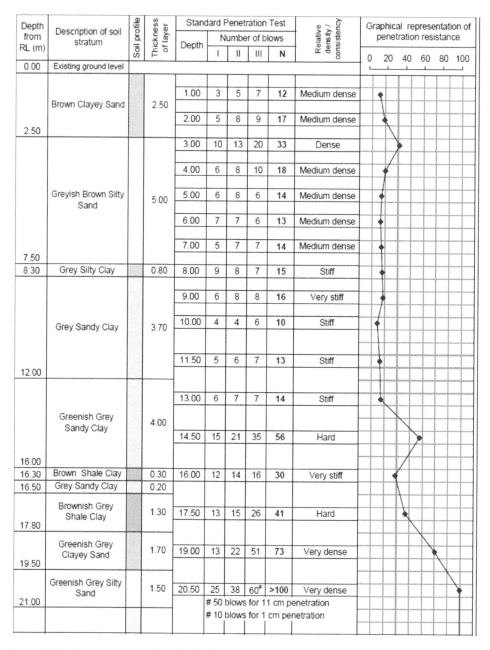

Figure 32.3 Typical soil profile (Cuddalore district).

(Continued)

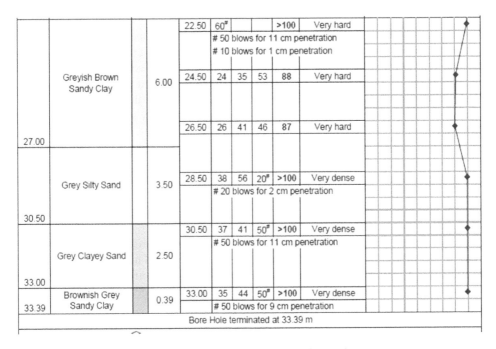

Figure 32.3 (Continued) Typical soil profile (Cuddalore district).

In the interior districts, the hard stratum is found to exist at shallow depths of less than 1 m, with topsoil comprising silty or clayey sand or gravel.

32.3.2   Properties of rock

Boominathan et al. (2004) conducted geotechnical and geophysical investigations in the East Coast of Tamil Nadu for the construction of a thermal power plant. The investigation was performed up to a depth of 120 m. The range of values on various properties of rock is reported in Table 32.2.

Table 32.2 Properties of rocks

Sl. No.	Property of rock	Values
1	Density (kg/m^3)	2,650–3,400
2	Specific gravity	2.75–3.5
3	Water absorption (%)	0–0.36
4	Porosity (%)	0.61–8.64
5	Point load index (MPa)	7.5–26
6	Unconfined compressive strength (MPa)	30–115
7	Young's modulus from UCC tests (MPa)	39,000–149,000
8	Tensile strength (MPa)	8–19
9	Compressive strength (MPa)	396–587
10	RQD	50–75

32.4 USE OF SOILS AND ROCKS AS CONSTRUCTION MATERIALS

In general, locally available materials such as mud, limestone and granite were extensively used in the construction of buildings. Heritage structures and temples were constructed mainly using granites. Stone masonry is mainly used for the construction of foundations and the basement and brick masonry is used for the construction of superstructures in earlier days. Lime mortars with natural ingredients like natural herbs, jiggery and so on were used for the construction of masonry structures in the olden days and cement mortars are used in recent years.

Red soil, which is the predominantly available soil in all parts of Tamil Nadu, is used for various applications. The red soil with plasticity ranges from medium plastic to highly plastic is used as a mortar for brick masonry, plastering works, utensils and so on. In Tamil Nadu, bricks are essentially manufactured using red soil with reasonably very good strength. Fibre-reinforced earthwork is also generally adopted using red soil because of its availability with desired characteristics.

Limestone is another important raw material used in construction. Limestone is available in abundant in Ariyalur, Salem, Tirunelveli and Thoothukudi districts of Tamil Nadu. The images of limestone available in Tamil Nadu are shown in Figure 32.4. Several cement factories have been set up in these districts, because of the availability of limestones.

Sand is another important material used in construction. River sand is generally used in all parts of Tamil Nadu and is also exported to other states because of its good quality. The sand obtained from Ennore and Thiruvallur districts also called Ennore sand possesses very good quality and can yield very good strength of concrete. This sand is recommended by Indian standards. The properties of Ennore sand is mentioned in Table 32.3.

Black granites, geologically known as dolerites, are found in many parts of Tamil Nadu. The black granite of Kunnam (taluk of the Perambalur district) is comparable to Ebony black of Sweden. Black granite in Paithur village, Salem district, with a brown background is also praised in the international market.

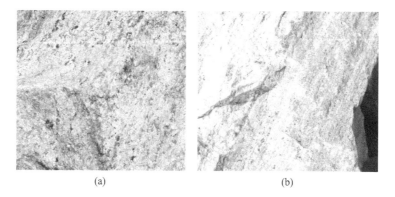

          (a)   (b)

Figure 32.4 Types of limestones: (a) limestone in Thirunelveli district, and (b) limestone in Thoothukudi district.

Table 32.3 Properties of Ennore sand

Properties	Ennore sand
Physical properties	
Colour	Greyish White
Specific gravity	2.64
Absorption in 24 hours	0.80%
Shape of grains	Sub-angular
Chemical analysis	
SiO_2	99.30%
Al_2O_3	
Fe_2O_3	0.10%
CaO	
Loss on extraction with hot HCl	0.11%
Petrographic analysis	
Quartz	97.40%
Feldspar	2.50%
Compressive strength	
3-day curing	160 kg/cm^2
7-day curing	220 kg/cm^2

32.5 FOUNDATIONS AND OTHER GEOTECHNICAL STRUCTURES

Different types of foundations are practised in Tamil Nadu based on the nature of the structures, as shown in Figure 32.5. Stone masonry foundations are generally adopted for load-bearing walls in the olden days. For heavy structures like dams and temple towers, compensated foundations were practised in Tamil Nadu, and a well foundation was adopted for bridges. In recent years, spread footings are adopted for lightly loaded structures, but for heavy structures, raft or pile foundations are generally adopted.

(a) (b)

Figure 32.5 (a) Spread footing for high-rise building and (b) well foundation (open caisson) for steel bridge at Thiruvalam, Vellore district.

The Pamban Railway Bridge (Figure 32.6) is the first sea bridge in India. This bridge spans 2.065 km between the Pamban island and Rameswaram. It is made of a 12 m steel girder, which spans 145 m and has also Scherzer navigational rolling lift span. The bridge was constructed in 1914 and is located in the second most corrosive environment in the world, leaving maintenance a challenging job. The location is also prone to high wind velocity.

The soil that prevails in Rameswaram is suitable for the cultivation of cotton. The Pamban bridge was constructed by the British government for trading cotton between Tondi and Devapattanam of Ramanathapuram district. The length between the two terminals is 20.05 miles, of which 7.19 miles is on the land with various islands and 12.86 miles in seawater. The section on the sea was constructed with a double row of reinforced driven concrete piles in the sand. These piles were braced together with ties, struts and chains. The bridge was 2 m above the high tide and the rails were constructed at that level. The bridge is still in use. The sea at the location of the bridge is only 1.5–2 m deep, and only in the navigation channel is the depth more than 6 m. There are 74

Figure 32.6 Pamban bridge, Rameswaram.

Figure 32.7 Grand anicut, Trichy.

open foundations for this bridge, of which 64 foundations lie in the sea. Construction proceeded successfully with the erection of cofferdams inside circular platforms.

Kallanai Dam, also known as Grand Anicut, is one of the oldest dams in the world and it is still in use (Figure 32.7). The dam was constructed by Karikala Cholan, King of the Chola Dynasty sometime around AD 190. The dam is 3.5 km long and 20 m wide. It was constructed to shift the flow of water from the Kaveri river to help irrigate 400,000 hectares of land around the Cauvery Delta.

32.6   OTHER GEOMATERIALS

32.6.1   Fly ash

Lignite in the Cauvery Basin of Tamil Nadu was first discovered by an agriculturist in 1934. It occurs at a depth ranging from 50 to 500 m below the ground level. The ONGC reported that the down-dip extension of this lignite zone is from 500 to 1,800 m depth in Tiruvarur, Kamalapuram, Kovilkalappal and Mayiladuthurai. At present, the Neyveli Lignite Corporation Limited mines lignite in two mines, Mine-I and Mine-II in the Neyveli region. Figure 32.8 shows the lignite mine in Neyveli, Cuddalore district. The lignite mined from these mines is used in two thermal powers, which have 600 and 1,470 MW generation capacity. These thermal power plants produce 1.5 million tonnes of fly ash annually. Fly ash is used in the construction of roads, embankments, structural fills, reclamation of low-lying areas and also in the manufacturing of construction products like bricks, blocks, tiles and so on. Fly ash-based bricks, blocks and tiles are like clay-based conventional building products.

32.6.2   Manufactured sand

Manufactured sand refers to crushed fine aggregates developed to use as fine aggregates in concrete instead of natural river sand. Manufactured sand also called M-sand has been developed worldwide to avoid the depletion of natural sand. The comparison of the various chemical properties of M-sand with river sand is tabulated in Table 32.4.

Figure 32.8 Lignite mine (Neyveli, Cuddalore district).

Table 32.4 Properties of M-sand

Parameters	River sand	M-sand
pH	6.76	8–10.77
Electrical conductivity (micromhos/cm)	55	125–1,575
Silica	78.5	53.66–67
Alumina	7.19	17.56–25.84
Iron oxide	0.15	2.12–12.71
Calcium oxide	0.77	4.28–10.9
Soluble salts	0.03	0.007–0.0993
Sulphates	0.006	0.006–0.3
Chlorides	0.0369	0.0739–0.44
Fluorides	-	0.00002

32.6.3 Vermiculite

Vermiculite is an interesting clay mineral available at Sevathur Village, Thirupathur Taluk, Vellore district. The vermiculite is a 2:1 phyllosilicate mineral comprising hydrated magnesium/aluminium sheet silicates containing water molecules within the layered structure (Mitchell and Soga, 2005). The internal structure of this vermiculite under exfoliation process, when heated at very high temperatures (870°C–1,100°C), changes and its volume increases manifold than that of the original material (Marcos and Rodríguez, 2014). This exfoliated vermiculite is light in weight and has got good thermo-insulating properties. Exfoliated vermiculite is used in the construction of lightweight concrete (Silva et al., 2010) and used as a thermal insulation and an adsorbent material in the removal of heavy metals because of its high citation exchange capacity (Baldev et al. 2020).

32.7 NATURAL HAZARDS

Tamil Nadu is prone to several hazards compared to other states and the vulnerability of the coastal region became more evident when Tsunami struck the coastal region of Tamil Nadu. Besides Tsunami, the coastal region also faces various disasters like cyclones, floods and so on frequently. The plain and hilly regions of the state also face other hazards like landslides, earthquakes and floods, where urban flooding is becoming a major concern in the state (Tamil Nadu Natural Disaster Management, 2020).

32.7.1 Cyclone

The coastal region of the state is more prone to cyclones and depressions. Cyclone forms in low-pressure zones in the Bay of Bengal. A severe cyclone causes high wind and torrential rain. The areas mostly affected along the coast are in between Mamallapuram and Puduppattinam zone, Marakkanam and Cuddalore zone and Tharangambadi, Nagapattinam and Vedaranyam zone.

32.7.2   Tsunami

The word Tsunami became familiar in India when an earthquake of 9.0 magnitude hit the Sumatra Coast, Indonesia in 2004. This resulted in Tsunami along the coast of Tamil Nadu. This Tsunami affected the coastline of Tamil Nadu ferociously. Thirteen districts along the coastline, including Chennai, Kancheepuram, Tiruvallur, Villupuram, Cuddalore, Nagapattinam, Tiruvarur, Thanjavur, Pudukkottai, Ramanathapuram, Thoothukkudi, Tirunelveli and Kanniyakumari were affected. In this catastrophe, several thousands lost their lives and lakhs of people became homeless.

32.7.3   Landslides

The Nilgiris district is located in the Western Ghats of the state and is surrounded by Coimbatore and Erode districts on the east, Kerala on the west and Karnataka on the north. Numerous landslides have occurred in Nilgiris in the past few decades. Some landslides caused severe damage to various infrastructures. The Nilgiris district receives heavy rainfall during SW and NE monsoons, which triggers the landslides there. It is also observed that cutting of slope at the toe, loading heavily on slope crest, blocking of the surface drainage system and weep holes in retaining structures and lack of proper remedial measures are the main reasons for most of the landslides in Nilgiris.

32.7.4   Urban flooding

Urban flooding has become more common in recent years because of urban spraling. The flood that hit the state during December 2015 has totally devastated the Chennai city, resulting in a loss of more than 400 lives with huge financial loss.

The torrential rainfall during November and December 2015 that inundated the coastal districts of Chennai, Kancheepuram and Tiruvallur affected more than 4 million people with economic damages that cost around 3 billion US dollars. There was a very heavy downpour on 2nd December 2015 and was reported as extremely heavy rainfall in Chennai of about 50 cm. Chennai city has been urbanising very rapidly over the last few decades. The progression of urbanisation and consequent land-use change in and around Chennai in recent years is the major cause of infiltration rate, which causes to increase the peak run-off discharge. Another consequence of urbanisation is the disappearance of many small and medium water bodies. These water bodies served as detention basins and resulted in a decrease in the peak discharge. Unplanned urbanisation also encroached on the waterways, which in turn reduces their vent way.

32.8   CASE STUDIES AND FIELD TESTS

Three case studies are discussed here. The first case study is about a study on the prevention of vibrations due to rail tracks for the nearby structure. The second study is on the pile integrity test for the construction of a grade separator at Chennai and the third case study is about the renovation of a temple structure in the Thoothukudi district.

Case study #1: Prevention of vibration due to the movement of loaded heavy-duty vehicles in the nearby vehicle test track for a nearby data centre at Velliveyalchavadi, near Minjur, Chennai, Tamil Nadu.

A data centre was proposed to be constructed at the technical centre in Vellivey-alchavadi, near Minjur, Chennai. The area selected for the construction of the data centre was located very close to the heavy-duty vehicle test track. For prevention of vibration from the heavy vehicle test track, the level had to be checked at the site as it houses the server and network systems. Through detailed subsoil investigation, the type of foundation was decided as the pile foundation. The necessary precautionary measure shall be employed for the foundation and superstructure of the proposed data centre. For finding out the vibration levels existing on the ground, some field tests were conducted. Vibrations were measured using acceleration sensors, which were sensitive for the frequency range of 1 Hz–5 kHz. The vibration characteristics of two types of vehicles were measured. The rms acceleration values were studied in different pits to understand the effect of vibration in the proposed site. The observed values were maximum observed vibration level in rms of $0.02563g$ and permissible vibration level as specified by Ashok Leyland in rms of $0.25g$.

The measured vibration level was compared with the permissible vibration levels from the relevant FTA code, DIN and Euro code, and it was found to be well within the permissible limits in terms of rms acceleration, velocity and displacement. The coefficient of attenuation at the foundation level of two pits was calculated and the observation was that the computed value was higher than the recommended value as per IS 5249 (1995). The high-value attenuation represents the decay rate of the wave amplitudes. The amplification of vibration for framed structures is 3.5 as specified in the literature. From this, the vibration level at the first floor was computed and found to be well within the permissible limits.

From the results of the detailed vibration characteristics analysis, it was concluded that theoretically no special precaution was required to screen the vibration for the foundations of the data centre. As an additional precautionary measure, considering the service of the structures, the vibration shall be mitigated using various vibration reduction measures. As the vibration levels were very less, the reduction measure of 'Building modifications' was recommended for the foundation of the data centre. The pile cap of the data centre building was recommended to support by elastomer pads of normally available thickness as used in bridges.

Case study #2: Construction of a grade separator in EVR Salai at the junction with Nelson Manickam Road and Anna Nagar III Avenue for Highways Department at EVR Salai, Nelson Manickam Road-Anna Nagar III Avenue junction, Chennai.

The construction of a grade separator in the EVR salai at the junction of the Nelson Manickam road and Anna Nagar III Avenue was in progress. Initially, a pile was cast on 19th November 2013. From the field boring log, it was observed that the clay deposit is present at different consistency up to a depth of 28.5 m underlain by weathered shale.

The pile was constructed in an abutment area and it was bored cast in-situ type of pile. The length and diameter of the pile were 30.7 m and 1,200 mm, respectively. The pile construction was taken up in the night on 19th November 2013 using the rotary rig method. The pile concreting was stopped after laying 9.50 m^3 of concrete. The balance depth was measured with chain on 27th November 2013 and the length was matched with the theoretical balance depth. The core was again rebored using a 600 mm diameter auger and washed continuously to flush out the disturbed soil substances. The balance pile installation process was continued and completed on 30th November 2015.

In continuation with the work after 8 months, the construction was again started. Therefore, to assess the integrity of the pile, a pile integrity test and high strain dynamic test were performed on the pile on 16 June 2016. The pile integrity testing is shown in Figure 32.9. The pile integrity test results indicate that the quality of concrete was good and matched with the Indian Standard guidelines. The reflection of the sonic velocity profile showed that there were no serious damages on the pile throughout the length. Poor toe reflection was observed at the tip of the pile. However, a change in reflection of velocity profile observed at 30.0 m showed that a full pile length was achieved. High strain pile integrity test results showed that the predicted capacity of the pile through the test was higher than the designed capacity of the pile. Based on the pile integrity test results and high strain dynamic test results, the pile was accepted for service. Since the pile was used in the abutment area, the stability of the pile had to be tested by considering the surcharge load due to an adjacent abutment.

Case study #3: Construction of Sri Sankaralingaswamy temple at Siruthondanallur, Tuticorin district.

A project has been proposed for the reconstruction and renovation of Sri Sankaralingaswamy temple at Tuticorin district. A detailed geotechnical investigation was carried out by conducting SPT at two locations up to N value more than 100 and up to rock level. The two boreholes were drilled at locations, one inside the Garbhagriha and another at the gopuram location. The water table was found at a depth of 5.25 m from the ground level. The water level fluctuates depending upon the supply channel present 1 km away and Thamirabharani river at 2 km distance.

There exists no hard stratum at shallow depths. So, it was advisable to go for a full raft instead of going for strip footing. Also, it is a monolithic structure that must not settle differentially. There were two options based on the bearing capacity assessment. That is to place a full raft for the entire structure or to place a raft for Garbhagriha alone and strip footing for the outer mandapam. The structures should be separated, and additional pillars must be placed.

Figure 32.9 Pile integrity test.

The entire area of the temple is around 210.29 m^2 for Center Garbhagriha, the load was estimated as 600 tonnes and the additional load due to different structures was assumed to be approximately 900 tonnes and hence a total load of 1,500 tonnes was assumed. The safe bearing capacity requirement was about 6.61 tonnes/m^2 considering the total temple area for a raft. The bearing capacity assessed for the size of raft 11.50 × 19.75 m was 10 tonnes/m^2 or 100 kN/m^2.

The base area of Garbhagriha is around 36.25 m^2. The load due to the structure as intimated by the client was 600 tonnes. The safe bearing capacity requirement was about 16.50 tonnes/m^2 or 165 kN/m^2. The bearing capacity assessed for the size of raft 5 × 7.25 was 16.50 tonnes/m^2 or 165 kN/m^2.

Based on the load assessment, the recommendations were furnished as follows:

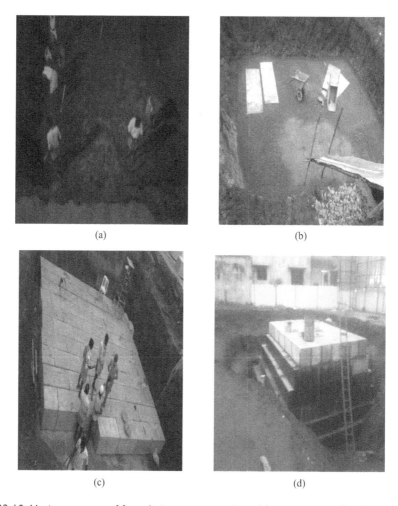

(a)　　　　　　　　　　　　　　(b)

(c)　　　　　　　　　　　　　　(d)

Figure 32.10 Various stages of foundation construction: (a) excavation of soil up to 2.25 m, (b) construction of raft foundation, (c) laying of granite stone above the raft and (d) completion of temple basement.

1. Provision of a full raft of size 11.50 m × 19.75 m at a depth of about 2.25 m from the existing lowest ground level was recommended as one of the choices of foundation. The anticipated settlement was 18 mm for 25 years.
2. Provision of a raft of size 5 m × 7.25 m at a depth of about 2.25 m from the existing lowest ground level was recommended for the centre Garbhagriha. The anticipated settlement due to the load intensity was 25 mm for 25 years as calculated based on the Schmertmann method. For the outer mandapam, a safe bearing capacity of 150 kN/m^2 for a raft of size 1.60 m × 11 m at a depth of about 2.25 m from the existing lowest ground level was recommended. The anticipated settlement due to the load intensity was 19 mm for 25 years.

Based on feasibility, one of the recommended foundations was selected for construction. Some of the pictures showing the construction progress are shown in Figure 32.10.

32.9   GEOENVIRONMENTAL IMPACT ON SOILS AND ROCKS

A majority of tannery industries are present in Ranipettai, Vellore, Tirupattur and Ranipet districts. More than 600 tannery industries are present in Ranipettai and Vellore districts (Sankaran et al., 2010). The tannery industry uses many chemicals for processing the tanneries. Most of the tannery industry do not have a waste effluent treatment and the untreated wastes are discharged directly into the drainage. The untreated waste contains heavy metals such as lead and chromium. These heavy metals contaminate the soil and groundwater. Another major source of soil contamination is the hazardous waste discharged from textile industries. A majority of the textile industries are located in Tirupur district and have nearly 750 dying units.

32.10   CONCLUDING REMARKS

There are many types of soils and rocks in Tamil Nadu. All the soils have entirely different properties, which vary from place to place. The same soil type seen at different locations shows drastic variations in its properties. A brief description of soils, rocks and other relevant information related to Tamil Nadu is presented. There is a greater scope of detailed study of the soils and related issues of the behaviour of soils in Tamil Nadu. New alternatives and waste construction materials must prevent the depletion of natural resources. Fly ash and other waste materials are generated in excess. The use of fly ash in the construction of embankments for laying railway lines has also significant potential for large-scale utilisation of fly ash and hence study has to be considered by researchers on these aspects.

REFERENCES

Baldev, D., Kumar, A, Chauhan, P., Muthukumar, M. and Shukla. S.K. (2020). Hydraulic and volume change characteristics of compacted clay liner blended with exfoliated vermiculite. *International Journal of Environment and Waste Management*, Vol. 25, No. 4, pp. 430–440, doi: 10.1504/IJEWM.2020.107563.

Boominathan, A., Gandhi, S.R., Elango, J. and Sivathanu, P. (2004). Evaluation of rock characteristics for a power plant site in India, *Fifth International Conference on Case Histories in Geotechnical Engineering*, New York, 13–17 April 2004, Paper No. 6.13, pp. 1–6.

Geological Survey of India. (2006). Geology and Mineral Resources of India, Miscellaneous Publication No. 30, Part IV. Published by Geological Survey of India, Kolkata, 17 January 2006.

Gopalakrishnan, K., Sugavanam, E.B. and Venkata Rao, V. (1975). Are these schistose rocks older than Dharwars? A reference to rocks in Tamil Nadu. *Journal of the Geological Society of India*, Vol. 16, No. 3, pp. 385–388.

Gopalakrishnan, K., Sugavanam, E.B. and Venkata Rao, V. (1976). Palaeo-volcanism in the granulite terrains of Tamil Nadu, *National Symposium on Palaeo-volcanism*, Ind. Acad. of Geoscience, Hyderabad, India.

Hansen, G.N., Krogstad, E.J. and Rajamani, V. (1988). Tectonic setting of Kolar schist belt, Karnataka, India. *Journal of the Geological Society of India*, Vol. 31, No. 1, pp. 40–41.

Indian Rare Earth Ltd. (2020). https://www.irel.co.in/, accessed 01 June 2020.

IS 5249 (1995). Indian Standard for the determination of dynamic properties of soil, Bureau of India Standards, New Delhi, India, Karnataka, India. *Journal of the Geological Society of India*, Vol. 31, No. 1, pp. 40–41.

Marcos, C. and Rodríguez, I. (2014). Some effects of trivalent chromium exchange of thermo-exfoliated commercial vermiculite. *Applied Clay Science*, Vol. 90, pp. 96–100.

Mitchell, J.K. and Soga, K. (2005). *Fundamentals of Soil Behaviour*. John Wiley & Sons, Hoboken, NJ.

Murthy, N.G.K. and Ahmed, M. (1977). Palaeogeographic significance of the Talchirs in the Palar basin near Madras (India), *Fourth International Gondwana Symposium*, Kolkata, India, pp. 515–521.

Sankaran, S., Rangarajan, R., Kumar, K.K., Rao, S.S. and Humbarde, S.V. (2010). Geophysical and tracer studies to detect subsurface chromium contamination and suitable site for waste disposal in Ranipet, Vellore district, Tamil Nadu, India. *Environmental Earth Sciences*, Vol. 60, No. 4, pp. 757–764.

Silva, L., Ribeiro, R., Labrincha, J. and Ferreira, V. (2010). Role of lightweight fillers on the properties of a mixed-binder mortar. *Cement and Concrete Composites*, Vol. 32, No. 1, pp. 19–24.

Sriramguru, K., Janardhan, A.S., Basava, S. and Basavalingu, B. (2002). Prismatine and Sapphirine bearing assemblages from Rajapalaiyam area, Tamil Nadu: Origin and metamorphic history. *Journal of the Geological Society of India*, Vol. 59, pp. 103–112.

Tamil Nadu Natural Disaster Management. (2020). https://tnsdma.tn.gov.in/, accessed 04 February 2020.

Venugopal, I., Muthukkumaran, K., Sriram, K.V., Anbazhagan, S., Prabu, T., Arivazhagan, S. and Shukla, S.K. (2020). Invention of Indian moon soil (lunar highland and soil simulant) for chandrayaan missions. *International Journal of Geosynthetics and Ground Engineering*, Vol. 6, No. 4, pp. 44.1–44.9, Switzerland, doi: 10.1007/s40891-020-00231-0.

Chapter 33

Telangana

Balunaini Umashankar
Indian Institute of Technology Hyderabad

Sravanam Sasanka Mouli
VNR VJIET Hyderabad

Chennarapu Hariprasad
Mahindra University

CONTENTS

33.1 INTRODUCTION

Telangana is the 29th and the 12th largest state of India. The geographical area of the state covers 1,12,077 km², and it lies between the coordinates NL 15° 48′ and 19° 54′ and EL 77° 12′ and 81° 50′. The state shares its boundaries with Andhra Pradesh on the east and south, Chhattisgarh in the east, Maharashtra on the north and west, and Karnataka on the west side (TSP, Portal, 2020). The entire state is divided into 31 districts, as seen in Figure 33.1. The state of Telangana is a semiarid area and has a predominantly hot and dry climate. Two major rivers flow across the state, with about 79% of the Godavari River catchment area and about 69% of the Krishna River catchment area, but most of the land is arid. Telangana is also drained by several minor rivers, such as Bhima, Manjeera and Musi. The annual rainfall is between 900

DOI: 10.1201/9781003177159-33

Figure 33.1 Political map of the state of Telangana (TSP, 2020).

and 1,500 mm in northern Telangana and 700–900 mm in southern Telangana, from the southwest monsoons. Various soil types abound, including chalks, red sandy soils, deep red loamy soils and very deep black cotton soils.

Telangana is characterised by a wide range of geological formations ranging from the Archaean age to recent age. Nearly 85% of the state is underlain by hard rocks (consolidated formations) belonging to the Peninsular gneissic complex, Dharwar and Eastern Ghats of Archaean to Middle Proterozoic age, Pakhal group of rocks belonging to Middle to Upper Proterozoic age, and Deccan traps. The rest of the state is underlain by semi-consolidated sedimentary zones comprising Gondwanas, tertiary group of formations and sub-recent to recent unconsolidated sediments (GSI, 2015). Historians discovered sonorous rocks in the borders of Jangaon and Siddipet Districts in Telangana; sonorous rocks are mysterious rocks that sing musically when struck.

33.2 MAJOR SOILS AND ROCK TYPES

Telangana comprises a variety of geologic and tectonic provinces, which are divided broadly into Eastern Dharwar Craton, Proterozoic Basins, Phanerozoic Gondwana Basin, Deccan Trap of the Cretaceous-Tertiary and the Quaternary cover of inland valleys (GSI, 2015). Each of these provinces is characterised by distinctive litho-assemblages, tectonic impress and metamorphic grades. In the state of Telangana, various minerals are available (refer to Table 33.1). Authors have found claystone formations at the Annaram irrigation site in the state of Telangana (Figure 33.2a). Sandstone (sedimentary formation) deposits were found at some places in the Medigadda barrage site located at the Medigadda Village, Mahadevpur Mandal, Jayashankar Bhupalpally district in the state. The sandstone was found to be intact without the presence of any fissures or joints (Figures 33.2b and c).

The soils in the state of Telangana are mostly red in colour (about 60%) and these soils are formed due to weathering of ancient metamorphic rocks. These soils cover a large part of Mahbubnagar, Nalgonda, Karimnagar, Khammam, Rangareddy, Nizamabad and Adilabad. Expansive soils (commonly termed *black cotton soils*) cover about 25% of the total area of Telangana and these are made up of volcanic rocks and lava flow (Figure 33.3). Expansive soils extend in the districts of Adilabad, Rangareddy, Nizamabad and Warangal, and fewer parts of Karimnagar, Nizamabad, Sangareddy and Medak. Lateritic soils cover about 7% of the area and these soils are formed due to intense leaching where high temperature and high rainfall occur. These soils are available in Medak and Khammam districts. Alluvial soils are formed by the deposition of sediments by rivers and these soils are present in riverbanks (Anitha et al., 2015; Biswas et al., 2015).

The state is also rich in coal mine deposits. The Singareni Collieries Company owned by the state government and Union government extends over six districts of the state. It contributes to 9.2% of the total domestic coal production (TSP, 2020). Major power plants of the National Thermal Power Corporation (NTPC) and Telangana State Power Generation Corporation Limited (TSGENCO), namely, Ramgundam and Kothagudem, are situated around these mines.

Table 33.1 List of major minerals found in the state of Telangana (DMG, 2020)

Major Minerals	Districts
Manganese ore	Adilabad, Jagityal
Garnet	Bhadradri-Kothagudem
Limestone	Vikarabad, Jogulamba-Gadwal, Wanaparthy, Mancherial, Nalgonda, Peddapalli, Suryapet, Komarm Bheem, Jagityal
Iron ore	Khammam, Mahabubabad, Peddapalli, Jagityal, Jayashankar
Coal	Khammam, Komaram Bheem, Jayashankar Bhupala, Bhadradri-Kothagudem, Mancherial, Peddapalli
Gold	Mahabubnagar, Nalgonda, Suryapet
Diamond	Mahabubnagar, Nalgonda, Suryapet
Stowing sand	Mancherial, Peddapalli, Bhadradri-Kothagudem, Jayashankar

(a)

(b)

(c)

Figure 33.2 Typical rock samples: (a) rock formation near raft level at the Annaram site, (b) rock formation at the Medigadda site and (c) intact rock present near the foundation level of the Medigadda site.

Figure 33.3 Typical soil samples: (a) red soil sample and (b) black cotton soil at stretches near IIT Hyderabad, Kandi, Sangareddy district.

33.3 PROPERTIES OF SOILS AND ROCKS

In Telangana, top loose soils up to a depth of 2–4 m are found in most of the places and the rock stratum underlies these deposits. This observation was made based on the available geotechnical investigations from different sites of the state. The properties of topsoil layers vary from loose to dense state with a representative Standard Penetration Test (SPT N) blow count as given in Figure 33.4.

Normal red soil found in the state has high shear strength parameters (angle of shear resistance ranging from 30° to 45° and cohesion intercept ranging from 0 to 10 kPa (Figure 33.5a)). It was observed that the expansive soil (black cotton soil) in the state possesses a varied swelling value with a free swelling index ranging from 50% to as high as 400% (Charan, 2019) (Figure 33.5b). However, the extent of expansive soil is relatively shallow, ranging between 1 and 3 m depth from the surface.

33.4 OTHER CONSTRUCTION MATERIALS

In the state of Telangana, various types of construction materials are available. Majorly, river sand, natural aggregates, limestone slabs, bricks, wood, cement and lime are found. In addition to this, the geomaterials as detailed below are used in the construction.

33.4.1 Stone dust

It is obtained from the crusher plants (Figure 33.6a). It has a potential replacement of river sand in concrete. These materials are available in the Nalgonda and Vikarabad districts of Telangana state.

33.4.2 Construction and demolition waste

Construction and demolition (C&D) materials are contributing majorly to the solid waste (Figure 33.6b). The C&D waste is generated from the renovation of buildings, road works, demolition, natural disasters and any other construction activities. These

Hyderabad			Mahabubnagar			Warangal		
Soil type	Depth (m)	N value	Soil type	Depth (m)	N value	Soil type	Depth (m)	N value
Clayey	0.00	25		0.00	10		0.00	45
sand	1.50			1.00			1.00	
Rock	3.00	Rebound	Silty sand	2.00	50	Silty	2.00	
	4.00			3.00		gravel	3.00	
	5.00		Sand	4.00			4.00	>50
	6.00			5.00	37		5.00	
Poorly	7.00			6.00			6.00	
graded	8.00			7.00		SDR	7.00	
sand	9.00		SDR	8.00	Rebound		8.00	
	10.00			9.00			9.00	
	11.00			10.00			10.00	

(a)

Karimnagar			Sircilla			Rangareddy		
Soil type	Depth (m)	N value	Soil type	Depth (m)	N value	Soil type	Depth (m)	N value
	0.00	14	Gravel	0.00			0.00	
Sandy	1.00			1.00		Clayey sand	1.00	25
soil	2.00			2.00			2.00	47
	3.00	16		3.00			3.00	Rebound
	4.00	Rebound		4.00			4.00	
	5.00			5.00	NA		5.00	
	6.00		SDR	6.00			6.00	
Rock	7.00			7.00		SDR	7.00	Rebound
	8.00			8.00			8.00	
	9.00			9.00			9.00	
	10.00			10.00			10.00	

(b)

Medak			Yadadri			Peddapalli		
Soil type	Depth (m)	N value	Soil type	Depth (m)	N value	Soil type	Depth (m)	N value
Reddish	0.00	23		0.00	NA		0.00	21
brown clay	1.00		Morrum	1.00		Lime mixed	1.50	
silty sand				2.00		with BC soil	2.50	Rebound
with Gravel	1.50			3.00			3.00	
Whitish	2.00	NA	SDR	4.00			4.00	
brown	3.00			5.00			5.00	
granite	4.00			5.50	Rebound		6.00	
based	5.00		Boulder	6.00		Lime with	7.00	Rebound
weathered	6.00			8.00		dense sand	8.00	
rock	7.00		SDR	9.00			9.00	
	8.00			10.00			10.00	

(c)

Figure 33.4 Boring logs at different districts of Telangana: (a) Hyderabad, Mahabubnagar and Warangal, (b) Karimnagar, Sircilla and Rangareddy, (c) Medak, Yadadri and Peddapalli.

(Continued)

Medak -BH1			Medak -BH2			Medak -BH3		
Soil type	Depth (m)	N value	Soil type	Depth (m)	N value	Soil type	Depth (m)	N value
Clayey sility sand with Gravel	0.00	27	Clayey sility sand with Gravel	0.00	48/10cms	Clayey sility sand with Gravel	0.00	44/7.5cms
	1.00			1.00			1.00	
	1.50			2.00			2.00	
Silty sand with Gravel	2.00	75/25cm		3.00			3.00	
	3.00		Granite based weatherd rock	4.00	45/7.5cm		4.00	
Granite based weatherd rock	4.00	NA		5.00			5.00	
	5.00			6.00		Granite based weatherd rock	6.00	45/7.5cm
	7.00			7.00			7.00	
	9.00			8.00			8.00	
	10.00			9.00			9.00	
				10.00			10.00	

(d)

Karim nagar- BH1			Karim nagar- BH2			Karim nagar- BH3		
Soil type	Depth (m)	N value	Soil type	Depth (m)	N value	Soil type	Depth (m)	N value
Sand	0.00	18	Filled up Soil	0.00	18	Sand	0.00	NA
	1.00			1.00			1.00	
	2.00			2.00			2.00	
	3.00	Rebound		3.00	Rebound		3.00	
Rock	4.00	NA	Rock	4.00	NA		4.00	
	5.00			5.00			5.00	
	6.00			6.00		Rock	6.00	
	7.00			7.00			7.00	
	8.00			8.00			8.00	
	9.00			9.00			9.00	
	10.00			10.00			10.00	

(e)

Ranga reddy -BH1			Hyderabad - Yadadri BH1			Hyderabad - Yadadri BH2		
Soil type	Depth (m)	N value	Soil type	Depth (m)	N value	Soil type	Depth (m)	N value
Silty sand	0.00	Rebound	Morrum	0.00	NA	Sandy	0.00	13
	0.50			1.00			1.00	
SDR	1.00			2.00			2.00	
	3.00			3.00	39		3.00	16
	4.00			4.00	Rebound		4.00	Rebound
	5.00		SDR	5.00			5.00	
	6.00			5.50		SDR	5.50	
	7.00			6.00			6.00	
	8.00			8.00			8.00	
	9.00			9.00			9.00	
	10.00			10.00		HDR	10.00	

(f)

Figure 33.4 (Continued) (d) Medak, (e) Karimnagar (f) Rangareddy and Hyderabad-Yadadri Highway.

(Continued)

Hyderabad - Yadadri BH3			Hyderabad - Yadadri BH4			Hyderabad - Medchal BH1		
Soil type	Depth (m)	N value	Soil type	Depth (m)	N value	Soil type	Depth (m)	N value
Chalky morrum	0.00	27	Sandy soil	0.00	13	Clayey sand	0.00	8
	1.00			1.00			1.00	
	2.00			2.00			2.00	11
SDR	3.00	31		3.00	16		3.00	15
	4.00	42		4.00			4.50	
	5.00			5.00		Silty sand	5.00	54
	6.00	Rebound		5.50	Rebound		6.00	
	7.00		SDR	6.00		Rock	7.50	Rebound
	8.00			8.00			8.00	
	9.00		HDR	9.00			9.00	
	10.00			10.00			10.00	

(g)

Hyderabad - Medchal BH2			Hyderabad - Medchal BH3			Hyderabad - Medchal BH4		
Soil type	Depth (m)	N value	Soil type	Depth (m)	N value	Soil type	Depth (m)	N value
Clayey sand	0.00	25	Silty sand	0.00	NA	Boulder	0.00	NA
	1.00			0.60			1.00	
	1.50		Rock	1.50			1.50	
Rock	2.00	Rebound		2.00			2.00	
	3.00			3.00			3.00	
Poorly graded sand	5.00		Boulder	5.00			5.00	
	6.00			6.00			6.50	
	7.00			7.00		Clayey sand	7.00	36
	8.00			8.00			8.50	
	9.00		Poorly graded sand	9.00		Rock	9.00	
Boulder	10.00			10.00			10.00	

(h)

Hyderabad - Medchal BH5			Sundilla (Kaleshwaram) BH1			Sundilla (Kaleshwaram) BH2		
Soil type	Depth (m)	N value	Soil type	Depth (m)	N value	Soil type	Depth (m)	N value
Boulder	0.00	NA	Morrum	0.00	14	Black cotton soil	0.00	28
Silty sand	1.00	12		1.50			1.50	
	1.50			2.50	16		2.50	29
Sand	2.00			3.00			3.00	
	3.00			4.00			4.00	
Rock	4.00			5.00			5.00	
	6.50	NA	Lime with dense sand	6.00	Rebound		6.00	NA
	7.00			7.00			7.00	
Sand	8.50			8.00		SDR	8.00	
	9.00			9.00			9.00	
	10.00			10.00			10.00	

(i)

Figure 33.4 (Continued) (g) Hyderabad- Yadadri and Hyderabad- Medchal Highways (h) Hyderabad- Medchal Highway (i) Hyderabad- Medchal Highway and Sundilla (Kaleshwaram).

(Continued)

Jadcherla -BH1			Jadcherla-BH2			Jadcherla -BH3		
Soil type	Depth (m)	N value	Soil type	Depth (m)	N value	Soil type	Depth (m)	N value
	0.00			0.00			0.00	
Sandy clay	1.00	6		1.00	33		1.00	6
	2.50			2.50		Clayey sand	2.50	
Clayey sand	3.00	23		3.00	46		3.00	19
	4.00		Dense silty sand	4.00	51		4.00	
Dense Sand	5.00	54		5.00			5.00	100
	6.00			6.00	100		6.00	
	7.00			7.00	100	Dense Sand	7.00	100
SDR	8.00	Rebound		8.00			8.00	
	9.00			9.00	100		9.00	Rebound
	10.00			10.00		SDR	10.00	

(j)

Mahabubnagar -BH1			Mahabubnagar -BH2			Mahabubnagar -BH3		
Soil type	Depth (m)	N value	Soil type	Depth (m)	N value	Soil type	Depth (m)	N value
	0.00			0.00		Medium	0.00	
	1.00	17	Dense Silty sand	1.00	41	Dense Clayey	1.00	14
	2.50			2.00		sand with	2.00	
Dense Clayey sand	3.00	16		3.00		Gravel	2.50	NA
	4.00			4.00			4.00	21
	5.00	100		5.00			5.00	32
	6.00		Rock	6.00	NA	Medium to	6.00	36
	7.00			7.00		dense calyey	7.00	39
Dense Sand	8.00	100		8.00		sand	8.00	49
	9.00			9.00			9.00	35
	10.00	100.00		10.00			10.00	42

(k)

Komaram Bheem			Jayashankar Bhupalapally -BH1			Jayashankar Bhupalapally- BH2		
Soil type	Depth (m)	N value	Soil type	Depth (m)	N value	Soil type	Depth (m)	N value
	0.00		Coarse sand with pebbles	0.00		clay	0.00	
Sandy Soil	1.00			1.00	22		1.00	
	2.00			2.00			2.00	
	3.00			3.00	22		3.00	
	4.00			4.00			4.00	
	5.00	NA		5.00	25		5.00	NA
	6.00		SDR	6.00			6.00	
SDR	7.00		Reddish silty stiff clay	7.00	31	Sand	7.00	
	8.00			8.00	29		8.00	
	9.00			9.00			9.00	
	10.00			10.00	27		10.00	

(l)

Figure 33.4 (Continued) (j) Jadcherla (k) Mahabubnagar (l) Komarram Bheem and Jayashankar Bhupalapally.

(Continued)

On River course- Jayashankar Bhupalapally -BH3			On River course-Jayashankar Bhupalapally -BH4			On River course-Jayashankar Bhupalapally -BH5		
Soil type	Depth (m)	N value	Soil type	Depth (m)	N value	Soil type	Depth (m)	N value
Silty Sand	0.00	23		0.00			0.00	
	1.50		Sand	1.00	27		1.00	16
sand with pebbles	2.00	23		2.00			2.00	
	3.00			3.00	24		3.00	15
	4.00	23		4.00	27		4.50	
	5.00	27		5.00		Sandy Soil	5.00	17
	6.00		Stiff clay	6.00	>100		6.00	19
Stiff Clay	7.00	31		6.50			7.50	
	8.00			8.00			8.00	23
	9.00	32	Rock	9.00	NA		9.00	24
	10.00			10.00			10.00	

(m)

On River course- Jayashankar Bhupalapally BH6			On River course-Jayashankar Bhupalapally-BH7			On River course-Jayashankar Bhupalapally-BH8		
Soil type	Depth (m)	N value	Soil type	Depth (m)	N value	Soil type	Depth (m)	N value
Silty Sand	0.00	23		0.00			0.00	
	1.50		Sand	1.00	20		1.00	19
sand with pebbles	2.00	23		2.00			2.00	
	3.00		Pebbles	3.00	20		3.00	22
	4.00	23		4.50			4.50	
	5.00	27		5.00		Sandy Soil	5.00	20
	6.00		SDR	6.00	>100		6.00	23
Stiff Clay	7.00	31		6.50			7.50	
	8.00			8.00			8.00	19
	9.00	32	Rock	9.00	NA		9.00	21
	10.00			10.00			10.00	

(n)

Nizamabad -BH1			Nizamabad -BH2			Nizamabad -BH3		
Soil type	Depth (m)	N value	Soil type	Depth (m)	N value	Soil type	Depth (m)	N value
Clayey sility sand with Gravel	0.00	18	Reddish (SDR)	0.00	NA	Reddish (SDR)	0.00	NA
	1.00			1.00			1.00	
	1.50			2.00			2.00	
Silty sand with Gravel	2.00	22		3.00			3.00	
	3.00		Granite based weatherd rock	4.00	NA		4.00	
Granite based weatherd rock	4.00	NA		5.00			5.00	
	5.00			6.00		Granite based weatherd rock	6.00	NA
	7.00			7.00			7.00	
	9.00			8.00			8.00	
	10.00			9.00			9.00	
				10.00			10.00	

(o)

Figure 33.4 (Continued) (m) along river at Jayashankar Bhupalapally (n) Jayashankar Bhupalapally (o) Nizamabad.

(Continued)

Adilabad- BH1			Adilabad- BH2			Adilabad- BH3		
Soil type	Depth (m)	N value	Soil type	Depth (m)	N value	Soil type	Depth (m)	N value
Sand	0.00	10	Filled up Soil	0.00	10	Sand	0.00	10.00
	1.50			1.00			1.50	
	2.00	12	SDR	2.00	NA	SDR	2.00	NA
	3.00			3.00			3.00	
	4.50			4.00			4.00	
Rock	5.00	NA		5.00			5.00	
	6.00			6.00			6.00	
	7.00			7.00			7.00	
	8.00		Rock	8.00			8.00	
	9.00			9.00			9.00	
	10.00			10.00			10.00	

(p)

Figure 33.4 (Continued) (p) Adilabad.

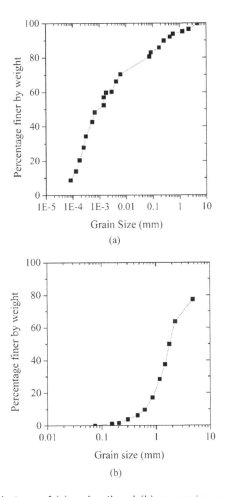

(a)

(b)

Figure 33.5 Typical gradations of (a) red soil and (b) expansive soil.

Figure 33.6 Photographs of construction materials: (a) stone dust, (b) C&D waste, (c) rice husk ash and (d) pond ash.

materials will be collected from various places and transported to a treatment plant. The C&D waste materials undergo the process of recycling/milling and crushing. After the recycling process, the materials will be separated as fine sand, coarse sand, and aggregates. These materials can be used for construction activities in place of natural resources (Bhushan et al., 2019).

33.4.3   Rice husk ash

It is obtained by the burning of rice husk. Rice husk is the waste product of the rice industry (Figure 33.6c). As per the statistics given by the Central Government of India, Telangana is a rice bowl of India. Hence, the quantity of rice husk produced is high in the state.

33.4.4   Pond ash

Burnt coal ashes in thermal power plants are collected in the form of fine fly ash from electrostatic precipitators and in the form of heavy particles (bottom ash) near the boilers (Figure 33.6d). The collected ash particles are mixed with water and transported to ash ponds in a wet slurry form. During the slurry deposition process in ash ponds, the heavier particles settle down near the inflow point and the finer ash

(a)

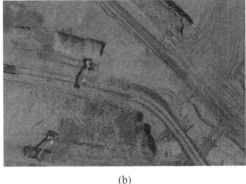

(b)

Figure 33.7 Ash pond at Ramagundam: (a) ash slurry deposition into a pond and (b) top view of deposited ash in the lagoon.

particles are carried to the outflow point (Figure 33.7). Approximately, 263 km^2 of land is occupied within the ash ponds in India alone; the expansion of existing ash ponds and an increase in the number of ash ponds across the country are severe environmental concerns (Karnam et al., 2019). The entire state has electricity production of about 6,700 MW thermal power (including all NTPC and TSGENCO plants). The upcoming thermal power production was estimated to be 11,600 MW. Hence, the expected production of pond ash is also high. The extent of ash pond for the power plant of NTPC Ramgundam itself is about 5.7 km^2 (Figure 33.7). The total ash pond is divided into four lagoons. Two of these lagoons are almost full to their capacities. There are two provisions of utilisation. One way is to utilise the pond ash for various construction purposes and the second way is to utilise the area of ash ponds for various construction activities after the pond is fully charged and abandoned.

33.4.5 Limestone slabs

Limestone slabs are used as flooring tiles and even roofing tiles in some parts of the state.

33.5   FOUNDATIONS OF A FEW IMPORTANT STRUCTURES

33.5.I   Kaleswaram project

This is the world's largest multi-stage lift irrigation project with 3 barrages, 19 pump houses and 1,832 km of the canal. The barrages were constructed with radial gates and rest on a raft foundation of a plan area of about 110 m wide × 10,000 m across the river Godavari (I&CAD, 2020).

33.5.2   Thousand pillars temple

This is a sculptural marvel of Telangana built 1,000 years ago. The temple had an interesting, engineered foundation. This temple is constructed by using a unique technique called the sandbox technique for strengthening the foundation. In this technique, a deep pit is dug and filled with sand and covered with huge rock beams. On this rock platform, the huge columns were raised and then the main structure was built. The skill of Kakatiya sculptors is also evident in the skilful craftsmanship and flawless ivory carving technique in their art (Baral et al., 2018).

33.5.3   Ramappa temple (built with floating bricks)

The roof (Garbhalayam) of the temple is built with bricks, which are so light that they can float on water. The temple is rich in its sculptural wealth too.

33.6   NATURAL HAZARDS

Telangana is one of the safest states in India. As it does not have a sea coastline, there is no possibility of a Tsunami. The entire state lies in Zone-II and Zone-III as per the earthquake zonation map as per IS 1893 (2016). Hence, it is safe from seismic hazards as well. Owing to this, the possibility of liquefaction is also low, with the depth of hard stratum encountered at shallow depths.

33.7   CONCLUDING REMARKS

Telangana is the youngest state of India. Some of the engineering properties of the available soils are detailed in this chapter. Construction materials available in the state include soils, wood, brick and so on. The following are keys facts about the soils in Telangana:

1. The state is covered majorly with red soils, black cotton soils, laterite soils and alluvial soils.
2. In a major part of the state, the hard rock stratum is found at a shallow depth at approximately 3–5 m, and in some places of Hyderabad, rock is found at 1 m depth.

3. Other than conventional materials, various other recycled materials such as C&D wastes, rice husk ash and pond ash are abundantly available.

4. Telangana state is one of the safest states against natural hazards.

ACKNOWLEDGEMENTS

The authors thank the various agencies for their help in providing the information reported in this chapter. They include Megha Engineering & Infrastructures Ltd. (MEIL), L&T Construction Ltd. and various departments of Telangana state, namely Irrigation & CAD Department, Ground Water Department and Mines and Geology Department.

REFERENCES

Anitha, G., Haseena, M. and Vinod, B. (2015). Soil investigation at Basar village, Telangana State. *Journal of Agroecology and Natural Resource Management*, Vol. 2, No. 5, pp. 325–327.

Baral, B., Divyadarshan C.S. and Rakshitha, M.D. (2018). Warangal fort and temple architecture. NID, Bengaluru. https://www.dsource.in/resource/warangal-fort-and-temple-architecture/thousand-pillar-temple, accessed 13 February 2018.

Bhushan, J.Y.V.S., Parhi, P.S. and Umashankar, B. (2019). Geotechnical characterization of Construction and Demolished (C&D) waste. In Stalin V., Muttharam M. (eds.) *Geotechnical Characterisation and Geoenvironmental Engineering. Lecture Notes in Civil Engineering*, vol. 16, Springer, Singapore.

Biswas, H., Raizada, A., Mandal, D., Kumar, S., Srinivas S. and Mishra, P.K. (2015). Identification of areas vulnerable to soil erosion risk in India using GIS methods. *Solid Earth*, Vol. 6, pp. 1247–1257.

Charan, K. (2019). *Stabilization of expansive subgrade using geogrid reinforcement-an experimental study.* M. Tech Thesis, IIT Hyderabad, India.

DMG (2020). Department of Mines and Geology (DMG), Government of Telangana, https://mines.telangana.gov.in/, accessed 21 August 2020.

GSI (2015). *Geology and Mineral Resources of Telangana*, Miscellaneous Publication No. 30, Part VIII A, First Edition, Geological Survey of India (GSI), Hyderabad.

I & CAD (2020). Irrigation and Catchment Area Development Department (I and CAD), of Telangana, https://irrigation.telangana.gov.in/icad/home/, accessed 21 August 2020.

IS 1893 (2016). *Criteria for earthquake resistant design of structures-Part 1: General provisions and buildings*, Bureau of Indian Standards, New Delhi, India.

Karnam, B.K.P., Guda, P.V. and Balunaini, U. (2019). Optimum mixing ratio and shear Strength of granulated rubber–fly ash mixtures. *Journal of Materials in Civil Engineering*, Vol. 31, No. 4, pp. 04019018-1–8.

TSP (2020). Telangana State Portal (TSP), https://www.telangana.gov.in/about/districts, accessed 22 August 2020.

Chapter 34

Tripura

Sima Ghosh
National Institute of Technology Agartala

J.R. Kayal
National Institute of Technology Agartala,
(Formerly Geological Survey of India)

CONTENTS

DOI: 10.1201/9781003177159-34

34.1   INTRODUCTION

Geographically, Tripura is one of the seven states in the North-East Region (*NER*) of India with a geographical area of 10,491 km^2. It is located in the south-west extreme corner of the *NER*, between latitudes 22°57′–24°33′ *N* and longitudes 91°10′–92°20′ *E*. The state is situated between the river valley of Myanmar and Bangladesh and is bounded by Bangladesh on the north, west, south and south-east, and in the east, it has a common boundary with Assam and Mizoram (Figure 34.1). Due to the high (2,000–3,000 mm) annual average rainfall in Tripura, the process of mechanical and chemical weathering and rapid erosion of the soils and bedrocks appear significant. Brief classifications and characteristics of the soils and rocks of Tripura are given in the following section.

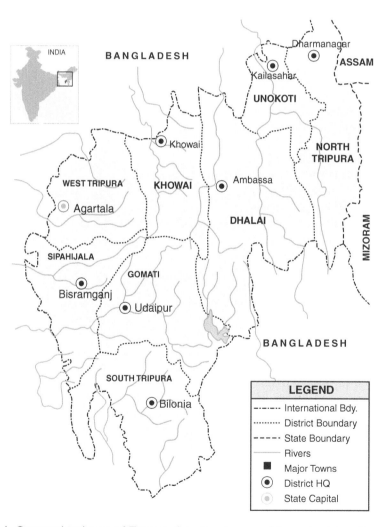

Figure 34.1 Geographical map of Tripura showing rivers and important locations.

34.2 SOIL TYPES

Das Gupta (1980) and Bhattacharyya et al. (2010) provided a detailed document of soils of the Tripura state; a short description for each group of soil is given below.

34.2.1 Reddish yellow brown sandy soil

This soil type covers nearly one third (~33%) of the total area of the state, distributed mostly along a north–south axis. Due to poor nutrients and leaching under heavy rainfall, this soil shows a resilience process through utilisation of ground biomass of leaf litters. The tropical evergreen forest of Tripura largely grows in this sandy soil. Deforestation, however, causes serious erosion problems to the soil resources.

34.2.2 Red loam and sandy loam

About 43% of the total area of Tripura is covered by the red loam and sandy loam soils, which are mostly associated with the forest ecosystem. This soil is rich in nutrients, but prone to heavy erosion, particularly in slope areas. Although *Jhum* cultivation is much popular, rubber, tea, coffee and pineapple plantations are also common in slope areas, which protect the soil erosion to a great extent. At least 14 different soil series are identified by the National Bureau of Soil Survey and Land Use Planning (NBSSLUP, 1997) in red loam and sandy loam soils of Tripura.

34.2.3 Older alluvial soil

The older alluvial soil, mostly located in river terraces and high plains, covers about 10% area of the state. This soil is rich in organic nutrients and is much suitable for farming. A large portion of the older soil remains under the tropical forest cover in the state. Vulnerability of erosion in uplands or in slopes and in river terraces demands soil conservation measures to protect the older alluvial soils.

34.2.4 Younger alluvial soil

The younger alluvial soil covers about 9% of the state, and it is mostly confined to the flood plains of the rivers like Khowai, Haora, Gumti and Muhari (Figure 34.2). This soil is composed of clay and loam. Due to the impact of annual river flooding, this soil is much rich and fertile. Cultivation of jute and paddy is assured in such soil. The erosion by lateral cutting and bank collapse, however, needs special attention to avoid wash out of the soil to the Bangladesh plains.

34.2.5 Lateritic soil

Lateritic soil can be recognised along the western boundary of Tripura; approximately 5% of the total land is identified as lateritic soil. This soil is coarse in texture and poor in nutrients; it, however, supports wild bushes. No area for farming or agroforestry is undertaken in this soil.

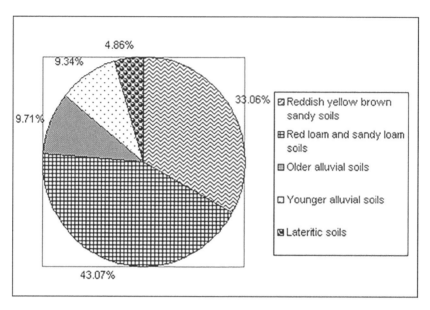

Figure 34.2 Pie chart of the soil types in Tripura. (After Das Gupta, 1980.)

34.3   OPTIMISING LAND USE

The NNRMS-ISRO (2014), based on the results of a detailed study, published a map of the state showing land use of the state Tripura. About 19% of the total area is categorised as Class-2, which is the main food grain-producing area in the south and west. Drainage, surface or subsurface, is a major limitation for irrigation. Drainage is much restricted due to the high groundwater table and poor permeability of the soil as it contains high clay and silt. About 12% of the land is grouped under Class-3, which is composed of narrow valleys. More than 20% of the land is categorised as Class-4 and has restricted access due to rough topography, but may be used for tree plantation, horticulture crops, spices and so on.

Soil erosion is a serious concern of environmental conservation. The soil erosion factors are categorised into eight groups. The maximum (26.2%) of the total area comes under Class 2-2 that suffers a middle level of erosion, while insignificant erosion is noted in Class 1-0 in 19.0% area, and the highest level of erosion is noted in Class 4-3 in 10.4% area. Thus, it is noted that about 36% of the total area falls under a severe level of erosion, and the maximum area falls under moderate to insignificant erosion. Deforestation is one of the major issues for severe erosion in the state.

In respect of soil degradation, four different categories are identified: *slight, moderate, strong and extensive.* An *extensive* degradation is noted in Tripura. The causes for degradation are normally attributed to water erosion, chemical decomposition and physical degradation. It is reported that only about 113,000 ha soil area is stable terrain under natural conditions, and about 508,000 ha, almost 50% of the total geographical area of the state, falls under strong degradation (NNRMS-ISRO, 2014).

34.4   ROCK FORMATIONS

The rock formations in the Tripura state range from the Lower Tertiary (40 my) to the Recent. The base of the sequence is not exposed. The sedimentary rocks from the Miocene basement to loosely consolidated sediments of Recent represent the geology of the state. The rocks are sandstone, siltstone and shale grading into clay. These rock types are repeated as layers, one above the other. Depending on their character and the presence of fossils, these sedimentary rock sequences are divided into the *Surma group*, the *Tipam Group* and the *Dupitila group*. From the nature of the grains and the texture imprinted on these rocks, it is inferred that originally the sediments were deposited in the sea, and later converted into rocks. The recent fluvial deposits occupy quite a large part of the South Tripura district. The sedimentary rocks are deformed and folded. The generalised geological succession shows four major groups of rocks, which are *Surma, Tipam, Dupitila* and *Recent* (GSI, 1974). The *Bhubans* and the younger *Bokabils* of the Surma Group are conformably overlain by the Tipam Group comprising Tipam sandstone. The Tipam Group is overlain by the Dupitilas with an erosional break upon which lies the Quaternary alluvium separated by an unconformity.

The major structural trend is in *N* 15°*W–S* 15°*E* to *N* 15° *E–S* 15°*W*. The structure, geologic and geomorphic pattern reveals that the ridges are anticlines and the valleys synclines, an early stage of morphotectonic evolution. The Tertiary folds, called *Tripura fold belts*, are characterised by open anticlines alternating with broad synclines (GSI, 2000). The fold movement progressively increases towards the east. Strike faults and oblique faults trending *NE-SW* to *ENE-WSW* are evident (GSI, 2000). About 76% of the total area is marked as of Tertiary origin and 24% to Quaternary. None of these contains any major or economic mineral resource, metallic or non-metallic, as reported by the GSI (1999) and Kesari et al. (2011). The Oil and Natural gas Corporation (ONGC), however, reported production of natural gas 1 million m^3/day.

34.5   PROPERTIES OF SOILS AND ROCKS

Borehole lithology shows various soils, mainly sand, silt and clay; an example is illustrated in Figure 34.3 giving description, consistency and colour of each stratum. Standard penetration test (SPT) results are given as *N*-values. The *N*-values are an important governing factor in the calculation of bearing capacity for civil engineering construction.

34.6   USE OF SOILS AND ROCKS AS CONSTRUCTION
MATERIALS

In Tripura, the soils are mainly sand, limestone, silts, plastic clay and so on; all of these materials are used for construction to a variable degree. The abundance of clayey soil is, however, suitable for making bricks, which are of maximum use for the construction of dwelling houses in Tripura.

Layer depth from G.L.(M)	Symbol	Description	Avg. S.P.T (N)	Bulk density (kN/m³)	LL	PL	Water content in soil (%)	Specific gravity
0 - 1.2		Brownish sandy silty clay with brick, kankars.						
1.2 - 5.0		Dark gray soft clayey silt with lenses of fine sand	6	16.6	30	18	31.12	2.66
5.0 - 7.5		Light gray clayey silty fine sand	16	18.9	-	-	23.4	2.67
7.5 - 9.4		Light gray soft firm clayey silt	8	17.6	52	31	36.49	2.66
9.4 - 12.4		Light gray fine to medium sand	44	19.2	-	-	71.2	2.66
12.40 - 18.3		Light gray soft to medium clayey silt	14	17.9	57	33	71.2	2.66
18.3 - 21.0		Light gray fine to medium sand	51	19.2	-	-	22.82	2.67

Figure 34.3 An example of subsoil stratum, Kalibari, Krishnanagar and Agartala. (After Das et al., 2018.)

34.6.1 Brick manufacturing

Pure soil or clay is required for brick manufacturing. The top layer of soil may contain impurities, and so the top 2 m soil is removed. After removal of the 2 m top layer, the clay is dug out and spread on the plain ground. The clay is cleaned of stones, vegetable matter and so on; the lumps are converted into powder with earth crushing rollers, then it is screened. The cleaned clay is exposed to the atmosphere for softening. Then the clay is blended and tampered adding water, pressed and mixed. The pressing may be done with feet for small-scale projects, or using a machine grinder for large-scale projects. The clay obtains a plastic nature when it is moulded to a rectangular shape by dices, or by machines. After moulding, the bricks need to be dried. The drying of raw bricks is done by a natural process under the Sun; bricks are laid in stacks on the plain ground. The period of drying may be 5–10 days depending on the weather condition. After drying, the dried bricks are burnt in kilns to gain appropriate strength. The burning temperature is maintained at 1,100°C; beyond this limiting temperature, the bricks become brittle and under this limiting temperature, the bricks do not get their full strength.

34.6.2 Building materials

The following building materials are common in Tripura:

i.   Soil – Soil for building making is collected at depth, below some 50 cm, to avoid the top organic matter, and any hard piece of rock is removed.
ii. Gravel – Small pieces of stone (sandstone or bricks) vary from the size of a pea to that of an egg. It is normally used with sand/cement for soling, flooring and so on.
iii. Sand – Course to medium-grained sand is used for making a wall, plastering and so on.
iv. Silt – It is like fine sand, so fine that individual grains are not observable; it is mainly used for filling.
v.   Clay – Soils that stick when wet, but hard when dry, may be used as a bonding material for wall making, particularly for low-cost village houses.

34.7   FOUNDATIONS FOR HOUSES AND GEOTECHNICAL STRUCTURES

In the villages and semi-urban areas, common people construct their residential houses without considering the seismic risk, rather people are ignorant about seismic design. They simply follow the thumb rule of masons who are equally or more ignorant about the seismic risk or seismic design. The Tripura region falls in zone V, the highest seismic risk zone in the country (BIS, 2002). In the past, villagers used to construct mud-wall houses. Even the local authorities used to distribute mud-wall houses to the villagers. Nowadays, these mud-wall houses are replaced by semi-permanent brick structured houses or *GCI* sheet-wall houses.

In the city, especially in Agartala, people started to construct their dwelling houses over pile foundations assuming that the pile foundation would be safe from the earthquake disaster. In shallow foundation, trench down to an optimum depth is dug and the trench floor is levelled with loose sand of nominal thickness. Over the loose sand, lean cement concrete of thickness 75–100 cm is provided. After 24 hours of the lean cement concrete, foundation base rods are placed at desired locations and concreting is done. In the case of deep foundation, precast RCC piles are driven. The depth of the pile foundations may vary from 8 to 10 m.

In geotechnical engineering structures, soils and rocks are also used as foundations for high-rise buildings, bridges, towers, retaining structures, embankments, slopes, dams, canals, tunnels and so on. Specific foundation problems, for any engineering structures like dams and tunnels must be investigated in detail and properly attended. For example, the detailed geological/lithological mapping should be made to ensure the depth of bedrock/engineering base, or geo-structural weakness, if any, like faults. Then, a site response study should be carried out to determine the predominant frequency at which the ground motion amplifies, and accordingly, the engineering structure should be designed.

34.8   NATURAL HAZARDS

Among several natural hazards, the Tripura state is much vulnerable to earthquakes, landslides and flood hazards.

34.8.1   Earthquakes

Tectonically, the entire Tripura region is in close link with the Indo-Burma subduction zone, where the Indian plate is subducting beneath the Indo-Burma ranges. The whole NER of India is in the seismic zone V, the highest seismic risk zone in the zoning map of India (BIS, 2002). The Tripura state region is affected by several large earthquakes ($Mw > 7.0$) that occurred in the NER, India and in adjoining Bangladesh (Figure 34.4a). Chronologically, these are the 1869 Cachar earthquake (Mw 7.4) on the Kopili fault in Assam, the 1918 Srimangal earthquake (Ms 7.6, revised to Mw 7.1) earthquake in Bangladesh and the 1930 Dhubri (Ms 7.1, revised to Mw 7.0) earthquake in Assam (Kayal, 2008). Most recently the 2016 Manipur earthquake (Mw 6.7) on the Kopili fault is also well felt in Tripura (Singh et al., 2017). The revised Mw for the past earthquakes is given after Ambraseys and Douglas (2004). The 1897 earthquake (Mw 8.1) with maximum intensity in MM scale X-XI (Figure 34.4b) and ground acceleration 1g in Shillong (Oldham, 1899) produced an east–west trending fissure ~152 m long and 45 m wide in Agartala, Tripura, and the ground acceleration in Agartala was equivalent to 0.13g. The 1918 Srimangal earthquake on the Sylhet fault, with maximum intensity X (RF Scale), caused the ground to sink at many places with the development of long fissures and sprouting of sand and water causing damage to several masonry structures in Tripura (Stuart, 1926). The 1930 Dhubri earthquake (Mw 7.0) struck in Assam, and the shock was felt all over the area with the movement of objects and damage to buildings (Gee, 1934). The most recent January 2017 moderate magnitude (Mw 5.7) shallow (depth ~30 km) earthquake in Tripura, about 75 km northeast of the Agartala city, caused damages to the mud-houses, fissures on the ground and liquefaction in the cultivated field (Das et al., 2018). The Tripura area is much close to the past large earthquake source zones like the Sylhet fault in Bangladesh and the Shillong plateau source zone in Meghalaya, Kopili fault zone in Assam valley and the Indo-Burma subduction zone to the east (Kayal, 2008).

With increasing population and mushrooming constructions of multistoreyed luxury apartments, high-rise office buildings, supermarkets, warehouses and so on with no seismic-design push people at high risk in this region of seismic zone V. Although some continuing research on earthquake occurrences and their source processes is carried out for the NER (e.g. Kayal, 2008 and references therein), no detailed work is done on seismic hazard microzonation, particularly in the Tripura region. Efforts are needed for seismic microzonation studies in Tripura, not only for engineering construction, like dams, highways and bridges, but also for high-rise buildings, supermarket complexes, hospitals, schools and so on.

34.8.1.1   Tectonic movement

Tripura area forms the western fringe of the north–south trending Indo-Burma Ranges flanking the Bengal Basin on the west. The region is tectonically much active since Miocene. The tectonic movements in the area can be dealt under the following groups:

34.8.1.2   Neogene movement

The late Pliocene orogenic movements folded the Neogene formations into a series of alternating open anticlinal ridges and synclinal valleys with sub-meridianal trends. Later the Quaternary fluvial sediments are unconformably deposited.

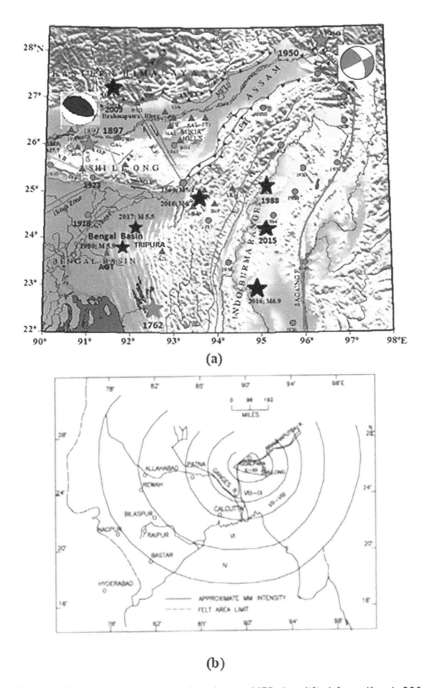

(a)

(b)

Figure 34.4 (a) Map showing large earthquakes in NER (modified from Kayal, 2008) and two recent felt earthquakes in Tripura (smaller back stars). Red stars indicate two great earthquakes (1897 and 1950) and red circles indicate large earthquakes in the region. Bigger black stars indicate recent strong earthquakes (*Mw* > 6.0). (b) Intensity map of the 1897 Shillong great earthquake (*Mw* 8.1), the maximum intensity X–XII was recorded in Shillong (after Oldham, 1899), and Tripura experienced intensity VII–VIII which was much damaging too.

34.8.1.3   Quaternary movements

Mahmood et al. (2003) recorded a series of enechelon faults, tilting, uplift and sub-sidence in the Sylhet basin. The Lalmai hills, commonly called the red banks near the town of Comilla, are an uplifted block (horst) of highly oxidised red Pleistocene sediments. The recent sediments of the Sylhet basin are traversed by several fault scars between the folded Tripura Hills and the Madhupur Jungle.

34.8.1.4   Contemporary movements

The Tripura state region is much affected by several large ($Mw > 7.0$) earthquakes in the recent past (Figure 34.4a), like those of the 1897 Shillong, 1918 Srimangal and 1930 Dhubri earthquakes as mentioned earlier. The ground-shaking/movement by these earthquakes caused long fissures, subsidence and uplift in the Tripura area. Most recently, a medium magnitude earthquake (Mw 5.7) in 2017, about 75 km northeast of Agartala city, caused landslides and fissures with liquefaction in the cultivated field (Das et al., 2018). Thus, a site-specific study should be made by microzonation approaches for any engineering construction in NER, India (Nath et al., 2008).

34.8.2   Floods

River and river systems are much important; the entire civilisation in the world grew around or near the rivers. Rivers and their basins, both as part of the natural ecosystem, have a variety of human uses, but the negative values are the destruction of properties and casualties due to river floods. It is one of the most devastating natural hazards in India.

   In Tripura, there are some ten important rivers (Figure 34.1). The main river Gumti runs almost through the centre of the state and is formed by the junction of two rivers, the Saroma and the Raima. Two rivers, Haora and Sonai Gang in West Tripura district, mostly get flood affected (Ganguly and De, 2015). A case study of flood-affected areas in the West Tripura district shows that about 60% of the area and 55% of the people are affected by flood in this district. Some 17,952 people were homeless and forced to take shelter in 61 relief camps in the 2019 monsoon flood as per the State Emergency Operations Centre Report. Monsoon rainfall intensity and duration of rainfall are the main causes of floods. Proper flood hazard management is necessary through proper land-use planning of the flood plains, embankment design and changing the cropping pattern.

   Floods have also occurred in areas, which were earlier not considered flood prone. Moreover, 80% of the precipitation takes place in the monsoon months from June to September. The rivers bring heavy sediment load from catchments. These sediments, coupled with an inadequate carrying capacity of rivers, are responsible for causing floods, drainage congestion and erosion of riverbanks. Flash floods also lead to huge losses at times. Continuing and large-scale loss of lives and damage to public and private properties due to floods indicate that there is still a need to develop an effective response and management to floods. The *NDMA*'s Executive Summary Guidelines may enable various implementing and stakeholder agencies to effectively address the critical areas for minimising the flood damage.

34.8.3 Landslides

Landslides are among the major hydro-geological hazards that affect large parts of India besides the Himalayas, the *NER* hill ranges, the Western Ghats, the Nilgiris, the Eastern Ghats and the Vindhyans, in that order, covering about 15% of the landmass (Ghosh and De, 2017). The *NER* is badly affected by landslide problems. The landslides in Tripura are reportedly increased ever since the road and other infrastructure developments have taken a greater pace in the hilly areas (Ghosh et al., 2019). Most of the landslides along the National Highway (NH-8) occurred in the high terrain/hill areas. During monsoons, occurrences of new landslides and reactivation of the old landslides create a frequent disturbance on the highway, which is the lifeline of Tripura (Ghosh et al., 2019).

Different factors, like lithology, slope, relative relief, rainfall, faults, land use/land cover and drainage systems, are among the several factors that cause landslides. The analysis revealed that the geological structure and human interference have more influence. It is found that the recent Ambassa landslide in Dhalai district, Tripura was located within 150 m buffer zone of a fault line (Ghosh et al., 2016). To mitigate disasters due to landslides, detailed geological and geophysical investigations should be carried out to delineate vulnerable weak zones or prone areas for landslides.

34.8.4 Cyclones

The word cyclone is derived from the Greek word *Cyclos* meaning the coils of a snake. The tropical storms in the Bay of Bengal and the Arabian Sea appear like coiled serpents of the sea, so the word *Cyclone* was coined by Henry Piddington in 1855 for such storms. Cyclones are caused by atmospheric disturbances around a low-pressure area distinguished by swift and often destructive air circulation. The air circulates inward in an anticlockwise direction in the northern hemisphere and clockwise in the southern hemisphere. Cyclones are classified as: (i) tropical cyclones and (ii) extratropical cyclones. Extratropical cyclones occur in temperate zones and high latitude regions, though they are known to originate in the polar regions. During the recent cyclones in May 2019, Tripura and adjoining *NER* states witnessed intermittent rains due to the impact of the cyclone, named *Fani* by the India Meteorological Department (IMD). Due to *Fani*, several houses were fully damaged and around 140 houses were partly damaged in three of the eight districts: Khowai, Gomati and Sepahijala (Figure 34.1). A school structure was partially damaged in Amarpur in the Gomati district when a tree fell on it. Many trees and electric poles were uprooted in different parts of the state disrupting vehicular movement and electricity supply as per the Tripura Disaster Control Centre.

34.9 CASE STUDY ON RIVERBANK EROSION

Riverbank erosion is one of the severe threats to environmental conservation in Tripura. A case study on river bank soil erosion in the West Tripura district given by Bhowmick et al. (2018) is briefly discussed here. Riverbank erosion is a common factor during monsoons along the Haora river and the Sonai Gang in the West Tripura district (Figure 34.1). Erosion occurs along~45 km length (about 96% of the total length) of the Haora river and ~20 km (90% of the total length) along the Sonai Gang. The riverbank erosion depends on the nature of bank material, its erodibility or resisting force. Samples from 12 sites at various depths from the top of the riverbank down to the water level were examined. The results revealed that the bank soils contain more than 90% sand and less percentage of silt and clay that make the soil non-cohesive and lead to erosion. At these sites, the riverbank erosion has a strong impact on flood plain dwellers, agricultural lands, tea gardens and National Highways. Erosion depends on bank materials, morphology and vegetation. The tractive force to carry load increases with channel slope, flow velocity and water depth. The erosive power of the river increases with the higher flow velocity. Bank erosion normally widens the river channel and leads to meandering, without change in the river size. This happens when erosion on one bank is compensated by deposition on the opposite bank to retain the same width of the channel.

At the observed sites, the riverbanks are composed of non-cohesive materials, i.e., bank soil with a higher percentage of sand, which leads to high erosion and eventually widening of the channel, At Golak Thakur Para and Trishghar sites, due to deeper depth and density of vegetation roots and moderate bank angle, the amount of bank erosion is, however, less. Bank stability depends on several factors such as temperature regimes, composition of bank materials, hydraulic forces and presence or absence of vegetation. From grain size analysis, it was reported that the samples were poorly graded, fine sand with a diameter size from 2.0 to 0.06 mm with the presence of least percentage of silt in the samples. The test result reveals the group name of the sample as silty sand (SM). According to the Unified Soil Classification System, the samples are under the group SP-SM, poorly graded silty sand. As the samples are dominant in sandy composition, they are under the cohesionless category.

Thus, it is concluded that the nature of bank material is the main causative factor behind bank erosion. A bank material containing more than 90% sand and less percentage of silt and clay makes the soil non-cohesive, which leads to maximum erosion and ultimately widens the channel. Bank erosion hazards need proper attention for providing effective policy decisions and execution of appropriate soil protection measures.

34.10 GEOENVIRONMENTAL IMPACT

34.10.1 Impact of river basin environmental changes

Growing urbanisation and human activities are major problems to the river systems (Arohunsoro et al., 2014). Urbanisation such as the construction of bridges, dams and other developments, though necessary for a better living society, has an adverse impact on the river environment or its morphology (Stump, 2006; James and Marcus, 2006; Kondolf, 1994) and on its natural characteristics (Khan and Islam, 2015).

Tripura is a riverine state, and among the ten major rivers, the Haora river is flowing through Agartala, the capital city of the state (Figure 34.1). This river basin was almost uninhabited before 1900. After the partition of India and East Pakistan (presently Bangladesh), huge immigrants entered the Tripura state and settled down along the Haora river in Agartala city. The dwellers, directly or indirectly, are not only polluting the river water but also hampering the morphology of the river. Although much work on pollution (Mili et al., 2013; Bandyopadhyay and De, 2018) and ecosystem (Duda and El-Ashry, 2000; Ellis and Marsalek, 1986) are published, there is much less work and only a few publications on the changes in physical configuration or morphology of the rivers due to human activities. To cope with the increasing water demand, the State Government and the local people are constructing several obstructions over the rivers, like bridge piers, causeways, sandbag filling for collecting water and so on. These activities make the rivers change their natural flow. Furthermore, human activities like land use, sand mining, water collection, solid waste disposal and agriculture channelling also cause a change of the river courses and its natural dynamics that gradually make the river sick. A case study on the Haora river was reported by Bandyopadhyay and De (2018), which is briefly given below.

The Haora river basin covers an area of 405.8 km² in India and Bangladesh and is in the southern part of the West Tripura District (Figure 34.1). It originates from the western flank of the Baramura range, flows through the hilly tracts, foothill sand through the plain until confluences with the Titas River in Bangladesh. The total length of the Haora river is 61.2 km, out of which 52 km is flowing in the Indian Territory. Physiographically, the major part of the river flows through moderately low undulating denudational topography i.e., tilla lands, having an elevation between 6 and 201 m. All the tilla lands in the Haora river basin possess the Dupitila group of rocks. The composition of the Dupitila group is sandy clay, clayey sandstone, ferruginous sandstone with pockets of plastic clay, silica and laterite. Most of the rocks of the Dupitila group are soft and fragile and are easily excavated. The Haora river is mainly a rain-fed river having a sinuous drainage pattern. From the hydrological data of the Central Water Commission (CWC), it is found that the total annual discharge of the Haora river is about 180 m³/s. The monthly rate of discharge widely fluctuates; maximum monthly discharge was recorded in 2004 followed by 2006 (about 140–165 m³/s) which are considered as intense flood years.

The course change of the Haora river between the years 1932 and 2005 is identified from the 1932 Survey of India (SOI) topographical maps and from the Google images. In a recent detailed study, Bandyopadhyay and De (2018) have shown that in the year 2010, a point bar of silt composition was found along the left bank of the river. In the year 2011, the earlier bar was eroded and a new bar of purely clay deposit was formed in its frontal part. In the year 2012, this newly formed bar was found to be enlarged. In 2014, the bar was found to be much enlarged, and the width of the channel narrowed by 0.6 m. The narrow channel failed to carry its water during the high discharge period, and it started to form a new course through the left side of the earlier silty point bar. This phenomenon has occurred in several places up to 1.5 km downstream. They further reported that two brick fields, namely R.B.I and Ramthakur, situated along the right bank of the Haora river were dumping their wastes along its bank. The brick fields collect mud from the surrounding tilla lands by axing here and there randomly along the tillas resulting in a change in their structures. The left out loose particles are transported by rain waters to the Haora river and create excessive sedimentation. These sediments are transported up to a long distance and are finally deposited in the lower course of the river, and being deposited randomly in the channel, they increase obstruction of the flow and cause channel migration.

34.10.2   Impact on agricultural lands

Bandyopadhyay and De (2018) reported anthropogenic effects on agricultural lands. It was observed that excavation for brick production damaged agricultural lands to a great extent. It was estimated that in the first two years (2009–2011), the brick fields have excavated 27,664.5 m³ and in the next three years (2011–2014), they have excavated 45,900 m³ of soil from the agricultural lands.

34.10.3   Impact due to road construction

The National Highway NH-8 is situated at the right bank of the Haora river, and most of the brick fields are on the left bank. The brick fields transport materials across the river. Along an 11.83 km stretch, between Champaknagar and Jirania, there are only two bridges. The trucks, carrying raw materials as well as the finished products of the brick fields need to travel a long distance, which increases the cost of production. So, the brick fields constructed some temporary roads across the river and even sometimes use the riverbed itself as their means of transportation by dumping brick bats. This dumping of brick bats on the riverbed makes the riverbed hard affecting the rugosity of the river.

34.10.4   Impact due to sand mining

The brick fields and some other construction industries collect sand from the Haora riverbed as well as from the bank. Such mining causes different types of hydrological changes within the river system (Saviour, 2012). Quarrying of sediments from the riverbed is commonly treated as a good practice as it reduces the sedimentation problem, but in this case, it makes the riverbed irregular and fragile and generates further sedimentation to the lower reaches of the river. Investigations show that near Mohanpur Bazar, the width of the river was about 27.75 m and the maximum depth was 0.51 m in the year 2010, but the width has come down to 24.6 m and the maximum depth increased to 0.65 m in 2011. The condition further deteriorated in 2012, when it became much narrower (21.05 m) and the maximum depth increased to 1.4 m. Continuous sand mining in the riverbed not only affects the bed, but also causes erosion.

34.10.5   Impact of bridge piers on the Haora river

Constructions of bridge piers have morphological impacts on rivers (Lane, 1955). There are several bridges over the Haora river. It is evidenced that when there was no bridge near Khayerpur in 2001, no bar was noticed, but the 2010 Google image indicates both bridge and bar. In 2014, the width of the Haora river became narrow (20.4–9 m), and active bank erosion is evidenced from 20 m downstream of the bridge. The effects of bridge piers vary with their size and composition; big and extended bars (22 m × 3.2 m) were formed near the concreted heavyweight bridge piers, whereas near the lightweight bridge piers, the length of the bar is less, and no bar is noticed near the hanging bridge located close to the international border.

Thus, it may be concluded that different anthropogenic activities, like sand collection, road construction, brick fields, bridge piers, tilla cutting and so on, along the Haora river are causing huge soil erosion and morphological changes of the river course (Bandyopadhyay et al., 2013). These aspects for soil conservation as well as environmental care need to be addressed for the societal benefit and its survival.

34.11   CONCLUDING REMARKS

Geologically, surface deposits in Tripura are dominated by alluvial soil. The Quaternary sediments and loose unconsolidated sand, silt and clay deposits cover a major part of the area. The soil characteristic features of Tripura vary with the topographical as well as morphological changes from hills to riverbeds. The climatic changes, weathering of rock materials and the local vegetation are also influencing factors. The soils in Tripura are mostly classified into five categories. About 43.07% of the total land area is occupied by red loamy soil and sandy soil. The reddish yellow brown sandy soil covers 33.06% of the land area. The three other types of soil are the older alluvial soil (9.71%), younger alluvial soil (9.34%) and the lateritic soil (4.86%). Rapid erosion by mechanical and chemical weathering occurs due to high annual rainfall. Another factor for the rapid erosion is the withdrawal of vegetation, which causes the removal of soil cover due to heavy wind. Furthermore, anthropogenic activities like sand collection, road construction, brick fields, bridge piers, tilla cutting and so on are causing huge soil erosion and morphological changes like narrowing of river courses.

Besides other natural disasters in Tripura like floods, landslides, cyclones and so on, the prominent disaster is caused by large earthquakes. The state falls in zone V in the seismic zoning map of India and has been much affected by several large earthquakes in the recent past. The ground-shaking by past large earthquakes in NER caused long fissures, subsidence and uplift in Tripura. Apart from taking serious measures for soil conservation, we should learn to live with natural earthquakes making earthquake-resistant construction. The old buildings should be retrofitted, and the new buildings should be built strictly following the building codes. A site-specific study should be made by microzonation for any engineering construction.

ACKNOWLEDGEMENT

We sincerely thank the Editor for kindly inviting and encouraging us to write this chapter. Our sincere thanks are to the Director, NIT Agartala for kindly permitting us to publish this work.

REFERENCES

Ambraseys, N. N. and Douglas, J. (2004). Magnitude calibration of north Indian earthquakes, *Geophys. J. Int.*, 159, 165–206.

Arohunsoro, S. J., Temitayo, O. J. and Omotoba, N. (2014).Watershed management and ecological hazards in an urban environment: The case of River Ajilosun in Ado Ekiti, Nigeria, *European J. Acad. Essays*, 1(2), 17–23, www.euroessays.org.

Bandyopadhyay, S. and De, S. K. (2018). Anthropogenic impacts on the morphology of the Haora River, Tripura, India, *Geomorphology*, 24, 151–166. doi: 10.4000/geomorphologie.12019

Bandyopadhyay, S., Ghosh, K., Saha, S., Chakravorti, S. and De, S. K. (2013). Status and impact of brick fields on the River Haora, West Tripura, *Inst. Indian Geographer Trans.*, 35. www.iigeo.org.in.

Bhattacharyya, T., Sarkar, D., Pal, D. K., Manda, C., Baruah, U., Telpand, B. and Vaidya, P. H. (2010). Soil information system for resource management–Tripura as a case study, *Curr. Sci.*, 99(9), 1208–1217.

Bhowmick, M., Das, M., Das, C., Ahmed, I. and Debnath, J. (2018). Bank material characteristics and its impact on river bank erosion, west Tripura district, Tripura, North-East India, *Curr. Sci.*, 115(8), 1571–1576. doi: 10.18520/cs/v115/i8/1567-1571

BIS (2002). Seismic Zoning Map of India, Bureau Indian Standard, New Delhi.

Das, S., Ghosh, S. and Kayal, J. R. (2018). Liquefaction potential of Agartala City in Northeast India using a GIS platform, *Bull. Eng. Geol. Environ.*, doi: 10.1007/s10064-018-1287-5

Das Gupta, S. P. (1980). Atlas of Agricultural Resources of India, *Ministry of Agriculture*, http://trpenvis.nic.in/lusoil.htm

Duda, A. M. and El-Ashry, M. T. (2000). Crises through integrated approaches to the management of land, water and ecological resources, *J. Water Int.*, 25, 115–126. doi: 10.1080/02508060008686803

Ellis, J. B. and Marsalek, J. (1986). Overview of urban drainage: Environmental impacts and concerns, means of mitigation and implementation policies, *J. Hydraulic Res.*, 34, 723–732, doi: 10.1080/00221689609498446

Ganguly, K. and De, S. K. (2015). Spatio-temporal analysis of flood and identification of flood hazard zone of West Tripura district Tripura, India using integrated geospatial technique, *Hill Geographer*, XXXI(1), 1–22. www.researchgate.net/publication/312947921

Gee, E. R. (1934). Dhubri earthquake of 3rd July 1930, Geol. Surv. India Mem., 65, No. 1, pp. 1–106.

Ghosh, K., Bandopadhyay, S. and De, S. K. (2016). Geophysical investigation and management plan of a shallow landslide along the NH-44 in Atharamura Hill, Tripura, India, *Int J. Geohaz. Environ.*, 2(3), 110–133, http://www.ijge@camdemia.ca

Ghosh, K. and De, S. K. (2017). Multi-scale modelling of landslide hazard and risk assessment in data scarce area – a case study on Dhalai district, Tripura, India, *19th EGU General Assembly, Vienna, Austria, Proceedings*, p. 17947.

Ghosh, K., Bandopadhyay, S. and De, S. K. (2019). A comparative evaluation of Weight-Rating and Analytical Hierarchical (AHP) for landslide susceptibility mapping in Dhalai district, Tripura. In Hazra, S., et al., *Environment and Earth Observations*, Springer, doi: 10.1007/978-3-319-46010-9-12

GSI (1974). Geological mapping and related studies on geomorphological and fluvial processes in Gumti basin (Progress Report for 1973–1974), by Sarkar, K. and Dasgupta, S. (Unpublished Report), https://www.portal.gsi.gov.in

GSI (1999). Appraisal of mineral/ natural resource potential for rural development on cadastral map base in outh Tripura district. (Proj: DOVEMAP), by Mukherjee, D., Bhattacharya, A. K. and Srivastava, S. C. (Unpublished Report), https://www.portal.gsi.gov.in

GSI (2000). Seismotectonic Atlas of India and its environs, *Geol. Surv. India, Sp. Pub.*, Narula, P. L., Acharya, S. K. and Banerjee, J. (eds.), 86 p, https://www.portal.gsi.gov.in

James, A. and Marcus, W. A. (2006). The human role in changing fluvial systems: Retrospect, inventory and prospect, *Geomorphology*, 79(3), 152–171, doi: 10.1016/j.geomorph. 2006.06.017

Kayal, J. R. (2008). *Microearthquake Seismology and Seismotectonics of South Asia*, Springer, The Netherlands and Capital. Publisher, India, 503p.

Kesari, G. K., Das Gupta, G., Prakash, H. S., Mohanty, B. K., Lahiri, S., Ray, J. N. and Behara, U. K. (2011). Geology and mineral resources of Manipur, Mizoram, Nagaland and Tripura, *Misc. Publications*, No. 30 Part IV, Vol. 1(Part 2), *Geol. Surv. New Delhi, India*.

Khan, M. S. S. and Islam, A. M. T. (2015). Anthropogenic impact on morphology of Teesta River in Northern Bangladesh: An exploratory study, *J. Geosci. Geomatics*, 3(3), 50–55, doi: 10.12691/jgg-3-3-1

Kondolf, G. M. (1994). Geomorphic and environmental effects of instream gravel mining, *Land. and Urb. Plan.*, 28, 225–243, doi: 10.1016/0169-2046 (94)90010-8

Lane, E. W. (1955). The importance of fluvial morphology in hydraulic engineering, *Proceedings, ASCE*, 81(7), 1–17.

Mahmood, A., Alam, M. M., Curray, J. R., Chowdhury, M. L. R. and Gani, M. R. (2003). An overview of the sedimentary geology of the Bengal Basin in relation to the regional tectonic framework and basin-fill history, *Sed. Geol.*, 155(3–4), 179–208. doi: 10.1016/S0037-0738(02)00180-X

Mili, N., Acharjee, S. and Konwar, M. (2013). Impact of flood and river bank erosion on socio-economy: A case study of Golaghat Revenue Circle of Golaghat District, Assam, *Int. J. of Geology, Earth Env. Sciences*, 3(3), 180–185, http://www.cibtech.org/jgee.htm2013

Nath, S. K., Raj, A., Sharma, J., Thingbaijam, K. K. S., Kumar, A., Nandy, D. R., Yadav, M. K., Dasgupta, S., Majumdar, K., Kayal, J. R., Shukla, A. K., Deb, S. K., Pathak, J, Hazarika P. J., Paul, D. K. and Bansal B. K. (2008). Site amplification, Qs and source parameterization in Guwahati region from seismic and geotechnical analysis, *Seis. Res. Lett.*, 79, 526–539.doi: 10.1785/gssrl.79.4.526

NBSSLUP (1997). Tripura state, 1:1:250,000 scale from SRM database, www.nbsslup.net.in

NNRMS-ISRO (2014). Land use and land cover map of Tripura, https://www.isro.gov.in/earth-observation/land-use-cover

Oldham, R. D. (1899). *Report on the Great Earthquake of the 12th June 1897, Geol. Surv. India Mem.*, 29: Reprinted, 1981, Geological Survey of India, Calcutta, 379 p.

Saviour, M. N. (2012). Environmental impact of soil and sand mining: A review, *Int J. Sci., Environ Tech.*, 1(3), 125–134, www.ijset.net.in

Singh, A. P., Rao, P. C. N., Ravi Kumar, M., Hsieh, M. C. and Zhao, L. (2017). Role of the Kopili fault in deformation tectonics of the Indo-Burmese Arc inferred from the rupture process of the 3 January 2016 Mw 6.7 Imphal earthquake, *Bull. Seism. Soc. Am.*, doi: 10.1785/0120160276

Stuart, M. (1926). The Srimangal earthquake of 8th July, 1918, *Geol. Surv. India Mem.*, 46, pt.1, 1–70.

Stump, G. (2006). *Inflectional Morphology: A Theory of Paradigm Structure*, Cambridge University Press, ISBN: 978-0521780476, doi: 10.1017/CBO9780511486333

Chapter 35

Uttarakhand

S.S. Gupta
G.B. Pant University of Agriculture and Technology

V.A. Sawant
Indian Institute of Technology Roorkee

CONTENTS

35.1 INTRODUCTION

Soil is the most basic and vital natural resource essential for agricultural production. Maintenance of soil fertility and productivity is key to achieve sustainability in agriculture. A larger portion of Uttarakhand is hilly terrain (Figure 35.1). Due to the presence of active tectonic movements, the terrain is characterised by fractured and weathered rock mass of variable origins. Uttarakhand has various types of soil and their geotechnical properties are entirely different. In the north, the soil ranges from gravel (debris from glaciers) to stiff clay. Brown forest soil often being shallow, gravelly and rich in organic content is found farther to the south. The Bhabar area is characterised by soils that are coarse-textured, sandy to gravelly, highly porous and largely infertile. In the extreme south-eastern part of the state, the Tarai soils are mostly rich, clayey loams mixed to varying degrees with fine sand and humus. There are mainly five types of soil found in the state, as detailed below:

1. Tertiary soil – found in Shivalik and Doon valley; this soil is suitable for the production of tea.
2. Cord soil – contains shell cysts and quartz; this soil is light and unproductive and is found in the Nainital district.

DOI: 10.1201/9781003177159-35

Figure 35.1 All district map of Uttarakhand. (https://www.uttarakhand-tourism.com/map/uttarakhand-map.php, accessed 4 February 2021.)

3. Volcanic soil – found in the mountain slopes.
4. Alluvial soil – found in lower slopes of the Shivalik range and Dun valley containing lime, iron and biological remains and is very suitable for agriculture.
5. Grey soil – contains lime but it is short of the productive element; found in Nainital, Mussoorie and Chakarata.

35.2   MAJOR TYPES OF SOILS AND ROCKS

Very steep to steep hills and glacio-fluvial valleys are dominantly occupied by very shallow to moderately shallow excessively drained, sandy-skeletal to loamy skeletal, neutral to slightly acidic with low available water capacity soils. They have been classified as Lithic/Typic Cryorthents. The Lesser Himalayan range is mainly composed of highly compressed and altered rocks like granite, phyllites and quartzite and a major part of it is under forest. The broader valley slopes dominantly have deep, well-drained, fine-loamy, moderately acidic and slightly stony structure.

35.2.1   Lesser Himalaya

The Lesser Himalayan formations occupy almost one-third area of the state. These formations comprise dominantly unfossiliferous metasedimentary sequences along with low to medium grade metamorphic ranging in age from Precambrian to Palaeogene. The main rock types are granite, granodiorite, phyllites, slates, quartzites, schists and gneiss. The Krol and Blaini formations comprise mainly sandstones, limestones and quartzites.

35.2.2   Outer Himalayan foothill zone

This zone can be classified into the Lower Shivalik, Middle Shivalik and Upper Shivalik. The Lower Shivalik is characterised by hard, massive, grey to brownish-grey sandstones interbedded with grey to maroon clays. They form the outermost zone in the Nainital Himalayas and occasionally exhibit local structural discontinuities. The dip is usually northwards. The middle Shivalik is characterised by massive light grey micaceous sandstones. They exhibit sporadic patterns of cementation at different stratigraphic intervals. The Upper Shivalik is constituted by pebbles, cobbles, boulders, conglomerates and clay lenses. The pebbles and boulders are mostly quartzitic. Thin lenses of grey to light green colour clays are common. Outcrops of Upper Shivalik are exposed in the western part between Kaladhungi and Ramnagar.

The area between Ganga and Yamuna rivers (Siwalik formation) forms a distinctive geomorphic unit in the Garhwal Himalaya. Drainage characteristics are associated with spring-fed perennial rivers. The formation of top soil cover is associated with four glacial and interglacial stages of the Himalayan glaciations. The gravel deposits have been formed by the superimposition of the alternate erosional and depositional phases caused by climate and crustal movements. Geomorphologically, the morpho-units comprise hills of structural origin, hills of denudation origin and units of fluvial origin. The soil matrix in Shivalik ranges comprises boulders, cobbles, pebbles, sand, silt and clay minerals (Figure 35.2).

35.3   PROPERTIES OF SOILS AND ROCKS

Various geotechnical parameters were determined for soil samples collected from different parts of the Kumaon region of Uttarakhand. Tables 35.1–35.3 present the typical geotechnical parameters at different locations in the state. A typical boring log represented in Table 35.2 shows general characteristics of alluvial deposits observed in the plains. The sandy strata with intermittent clay layers can be noticed. Even boulders of variable sizes are also noticed at shallow depths in the vicinity of streams.

A typical hill terrain shows the presence of rock outcrops as well as overburden was observed. The overburden is generally observed to be less than 2.0 m in thickness. However, at few locations, it may go effectively up to 5 m including weathered rock formation. The overburden material includes a fine-grained silty matrix with

Figure 35.2 Visible rock outcrop at the Joshimath site.

angular fragments of meta-basics. The meta-basics rock and phyllites are exposed in the Garhwal region showing signs of moderate to heavy weathering. These were observed to be moderately to highly jointed, thinly to medium bedded and dipping towards the northeast at moderate to steep angles. The rock mass is slumped at many places due to the fractured and jointed nature of rocks and the influence of seepage of water.

35.4 USE OF SOILS AND ROCKS AS CONSTRUCTION MATERIALS

A detailed description of construction materials available in the Kumaon region of Uttarakhand is presented here.

Stonecrete blocks are generally $300 \times 200 \times 150$ mm in size. As a result, they produce walls that are 200 mm thick. The stones can be fully encased in the concrete or they can be so placed as to be exposed on the long face. Such blocks with the exposed stone could be used to build a wall that has the appearance of a stone wall if not plastered. In short, stonecrete block is made using moulds of appropriate size using cement, sand, aggregates and stone. It is generally viable where stones of 100–150 mm size are easily available like in Almora, Pithoragarh, Ramnagar and so on.

A subgrade is made up of native soil that has been compacted to withstand the loads above it. It is a layer required in many structures such as pavements and slabs, although it needs to have certain characteristics. A subgrade might need special drainage structures to let water if it is composed of impermeable soil, and it should be graded to within plus or minus 1.5 inches of the specified elevation.

When the subgrade material is not adequate to support the necessary loads, then additional measures should be adopted to make the material suitable for the construction. Normally, properties of a subgrade material are improved by stabilising the subgrade mixing with other materials including the use of geotextiles resulting in improved geotechnical properties of soil making the subgrade material suitable for pavement.

Bedrock aggregate is aggregate cut from solid rock that is usually crushed and used for construction purposes. Depending on the use, bedrock aggregate can contain

Table 35.1 Characteristics of foundation soils in Uttarakhand

Depth (m)	Classification	Fine (%)	Sand (%)	Gravel (%)	γ (t/m³)	w (%)	G	LL	PL	c (kPa)	φ (°)	Cc	UCS (MPa)
	Udham Singh Nagar												
1.0	CL	87.7	12.3	-	1.54	14.6	2.57	25.8	12.4	47.5	12	0.122	
3.5	SM	79.9	20.1	-	1.56	14.9	2.58	NP	NP		29.5		
	Almora												
0.1	Soil-rock matrix				1.67			NP	NP		34		
1.5	Weathered rock with mica				1.8			NP	NP		37		5.38
	Pithoragarh												
1.2	Soil-rock matrix				1.72						34		
4.0	Hard weathered rock				1.86						41		
	Rudrapur												
2.0	CL	91.2	8.8		1.55	13.1	2.57	26.6	11.1	60	10	0.160	
4.5	CL	84.7	15.3		1.56	14.5	2.60	23.2	14.9	42.5	15	0.116	
	Bhimtal												
0.5	Soil-rock matrix										34		
1.0	Sandy silt ML	80.2	19.8		1.63	14.1	2.60				28		
4.5	Soil-rock matrix				1.76						37		
	Jaspur												
1.3	Sandy silt	81.5	18.5		1.55	14.2	2.58	26.3	11.1	47.5	28.5	0.150	
3.5	Clayey silt	90.2	9.8		1.54	16.3	2.56				11		
	Pantnagar												
0.6	Sandy silt	89.0	11.0		1.56	14.1	2.60	NP	NP		29		
2.0	SP-SM	8.2	87.9	3.9	1.59	13.7	2.61	NP	NP		31		
6.0	SP	5.8	88.7	5.5	1.63	13.2	2.61	NP	NP		33		

Sources: After Gupta and Kumar (2002, 2009, 2011, 2015, 2016, 2019a–c)

Table 35.2 Subsoil boring log at the borehole location IIT Roorkee Campus

Depth (m)	Classification	Grain size analysis (%)			LL (%)	PL (%)	NMC (%)
		Gravel	Sand	Fines			
1.5	SP-SM	0.0	89.25	10.75			3.6
3.0	SP	0.0	95.27	4.73			3.6
4.5	SM	0.0	83.75	16.25			24.5
6.0	CI	0.0	2.75	97.25	41	24.8	40.6
7.5	SM	0	67	33			22.5
9.0	SP-SM	0	90.75	9.25			24.9
10.5	SP-SM	0	91.75	8.25			27.5
12.0	SP-SM	0	93	7			28.8
15.0	CI	0	47.25	52.75			27.3
16.5	SM	1	80.25	18.75			30.4
19.5	CI	12	3.5	84.5	38.3	22.3	33.6
21.0	CL	5.75	8.5	85.75	31.7	22.8	31.5
22.5	CL	1	28	71	22.3		29.4
24.0	SP-SM	3	86.75	10.25			26.9
27.0	SP	0	96	4			29.7
30.0	SP	0	98.25	1.75			27.5
33.0	SP-SM	3.75	91	5.25			29.5
36.0	SP	0	98.25	1.75			30.6

Table 35.3 Subsoil boring log at borehole locations at Dhak–Joshimath

PLT Location	Depth (m)	I.S. Classification	Grain size analysis			Liquid limit (%)	Plastic limit (%)	Natural moisture content (%)	Unit weight (t/m³)
			Gravels (%)	Sand (%)	Fines (%)				
BH 1	1.0	SM (NP)	17.1	35.1	47.8	Non-plastic		1.8	-
	2.0	SM (NP)	25.6	40.0	34.4			1.84	
	3.0	SM (NP)	36.2	36.9	26.9			-	
	4.0	ML (NP)	2.4	20.7	76.9			-	
BH 2	1.0	GP-GM	69.1	22.8	8.1	Non-plastic		0.5	-
	1.5	GM (NP)	53.4	30.8	15.8			1.91	
	2.0	GP-GM	71.9	18.6	9.5			-	
	3.0	SM (NP)	29.4	38.8	31.8			-	
	4.0	GM (NP)	40.7	31.2	28.2			-	
BH 3	1.0	GM (NP)	48.6	26.7	24.7	Non-plastic		1.7	-
	1.5	ML (NP)	5.7	7.6	86.7			1.70	
	2.0	GM (NP)	46.0	27.1	26.9			-	
	3.0		42.0	35.4	22.6			-	
	4.0		45.2	34.4	20.4			-	

(Continued)

Table 35.3 (Continued) Subsoil boring log at borehole locations at Dhak–Joshimath

PLT Location	Depth (m)	I.S. Classification	Grain size analysis			Liquid limit (%)	Plastic limit (%)	Natural moisture content (%)	Unit weight (t/m³)
			Gravels (%)	Sand (%)	Fines (%)				
BH 4	1.0	ML-CL	22.1	37.6	40.3	27.6	22.8	2.4	-
	2.0	SM (NP)	29.2	30.6	40.2	Non-plastic		4.3	1.81
	3.0	GM (NP)	46.9	34.4	18.7			-	
	4.0	GM (NP)	52.1	34.8	13.1			-	
BH 5	1.0	ML-CL	25.4	34.4	40.2	27.4	22.5	2.4	-
	1.5	SM (NP)	33.6	39.3	27.1	Non-plastic		1.8	1.74
	2.0	SM (NP)	26.6	35.0	38.4			-	
	3.0	GP-GM	55.1	35.7	9.2			-	
	4.0		59.4	32.3	8.3			-	

Source: After Viladkar et al. (2009).

different kinds of fractions that can vary in size from fine grains to large boulders. Typical uses of bedrock aggregate include structure layers of roads and railroads and making concrete. It is becoming increasingly difficult to get high-quality esker aggregates, and it has led to increased use of bedrock aggregates. The utilisation rate of replacement rock materials and secondary rock materials is also increasing. They include, for instance, recycled rock materials, demolition waste from the construction industry and large volumes of gangue from the excavation of natural stones.

35.5 FOUNDATIONS AND OTHER GEOTECHNICAL STRUCTURES

Based on the height and loading, different types of foundations are adopted. Residential buildings up to three storeys are provided with shallow foundations. For multi-storey and heavy industrial buildings, raft/pile foundation/piled-raft foundations are adopted. Well foundations are provided to support bridges.

Several dams have been constructed in Uttarakhand. Sitarganj is an industrial city in the state of Uttarakhand. Sitarganj city is located between three major water reservoirs, Baigul, Dhora and Nanak Sagar which are used mainly for fisheries.

The Baigul Dam has an impressive height of 13.7 m and a length of 15,300 m. The water of this dam is used for irrigation purposes. The nearby areas are Rudrapur, Pilibhit and Haldwani.

Nanak Sagar Dam has been constructed on river Saryu or Deoha at Nanak Matta forming Nanak Sagar, which adds up to the beauty of Nanakmatta, which is a nearby town to Sitarganj. The length of the Dam is 19,700 m. The volume of the dam is 3,833 × 10³ m³. The irrigation potential of this dam is 39,200 hectares.

The astounding Dhora Dam is situated in Udham Singh Nagar district of the Uttarakhand state. This type of TE Dam was built on the Kichha River. The basin area of this dam is the holy Ganga River. The Dhora Dam has a height of 14.63 m and a length of 9,610 m. The dam was built with the purpose of irrigating the fields. The construction work of the Dhora Dam was completed in 1960.

The Baur Dam is a distinguished dam situated in Bajpur in the Udham Singh Nagar district of the Uttarakhand state. It is a type TE Dam, having a height of 17.98 m and a length of 9,500 m. The basin of this dam is the Ganga River. The water of this dam is used for irrigating the farmlands. Kichha in Udham Singh Nagar is the nearest city to this dam.

One of the major projects in recent times is the Dobra–Chanti bridge on Tehri Lake. It has the longest suspension bridge in the country of length 725 m. Slope protection works were carried out with bolts and shotcrete.

35.6   NATURAL HAZARDS

The Kumaon region of Uttarakhand has witnessed several earthquakes, landslides, glacier avalanches and floods in the past. While the occurrence of these incidents cannot be precisely predicted, their impacts are well understood and can be managed effectively through a comprehensive programme of hazard mitigation planning.

Ongoing changes in climate patterns around the world may alter the behaviour of hydro-meteorological phenomena within our lifetimes. The frequency and severity of floods, storms, droughts and other weather-related disasters are expected to increase. The details of some past natural hazards are provided here.

35.6.1   Earthquakes

According to the official website of Disaster Mitigation and Management Centre (https://dmmc.uk.gov.in, 2021), the first of the major earthquakes to strike the Indian subcontinent in the 1990s, the quake devastated Garhwal, especially Uttarkashi, Tehri and Chamoli districts on Dusshera night. The official death toll was 769, though unofficial estimates put the toll at a higher level. Major cities in the area, including Dehradun and Almora, were affected while the tremors were felt as far away as Pune in the Deccan. It recorded 58 aftershocks till the 28th of November 1991. The epicentre of the earthquake was located 10.5 km east of Pilang, now in Uttarakhand. This was the deadliest earthquake in the Himalayan region since 1950. While most of the deaths were in Uttarakhand, then part of Uttar Pradesh, Himachal Pradesh reported casualties too. According to the UP Government, 1,819 villages were affected by this earthquake. An estimated population of 4.22 lakhs was directly affected. About 90,000 houses have been damaged of which about 20,000 fully and 70,000 partially. The quake also attracted attention due to its proximity to the high Tehri dam.

35.6.2   Landslides

Landslides are common in this Himalayan state of Uttarakhand, especially following heavy monsoon rains. Over 5,300 people were killed in landslides since 2000 when the mountain state was formed. Besides rains, scientists blame weak topsoil for the frequent occurrence of landslides. Environmentalists, however, blame developmental activities and poor vegetation for these phenomena. Recently, the public works department (PWD) identified 39 zones across the state that are prone to landslides. Nine of these zones are being undertaken by the union roads and transport ministry for mitigation while the state PWD is working on 21 sites with funds from the Asian Development Bank. The detailed project report of the remaining nine zones is underway. To avoid landslides in these areas, various measures like using gabion walls, wire mesh, rock bolting and other steps are taken depending on their location. Despite all these efforts, landslides – big or small – are reported every now and then. A typical case of the Balianala landslide, which underlines tentative causes, is discussed here (Figure 35.3).

The Balianala landslide near Nainital is 300 m in length, 223 m in width and 240 m in height. The slide is located on the right bank of the Balianala. The slide is classified as a rock cum debris slide (meaning rocks and debris together form the slide). It is retrogressive (that is, it propagates in the up direction). The crown of the slide is located at about elevation 1,865 m and is marked by a prominent scarp whereas the toe lies at about elevation 1,632 m near the Balianala bed. The slide faces towards the east with an average slope of about 55°. The slide is barren and rocky. The crown portion of the slide is made up of dolomite whereas the toe part consists of greenish-grey shale of the Krol Formation. The slide material consists of rock fragments of dolomite and shale mixed with the argillaceous matrix. Intense weathering is observed on the upper surface of the rocky slope. Many faults/shear zones were observed in the area. Numbers of slide zones were also observed on both the banks of the Balianala up to the Brewery Bridge. All these landslides are very old, active and are retrogressive. Two seepages are located in the slide face and both the seepage areas are located at El ± 1,782 m. The seepages are located near the contact of dolomite and greenish-grey shale, a potentially weak zone. The Balianala Fault passes through the toe of the slide following the nala channel. The Bareilly–Nainital road is located 140 m from the crown and the area between the crown and the road is thickly populated. The evidence of subsidence in the area is manifested by the cracks developed in the houses.

The presence of the Balianala Fault associated with several minor sympathetic faults in the area worsens the quality of the rock mass resulting in very poor quality of the slope forming material. Balianala Fault is an extension of a regional scale Lake

Figure 35.3 Balianala landslide.

Fault that passes through the area that caused juxtaposition of the rocks of Middle and Upper Krol with the rocks of Lower Krol formation (Pant and Kandpal, 1990). Due to the presence of the Balianala Fault, the rocks in the area are intensely crushed, jointed and fractured. Numbers of faults are present in the Balianala section. In addition, four sets of joints are recorded in the Balianala section and such joints exhibit planar and wedge failure at a number of places (Figure 35.3). The disposition of the joint sets present in the rock mass facilitates planar and wedge failure in the area. The shale is overlain by dolomite in the area. The specific gravity of dolomite is higher than the shale. The incompetent shale does not seem to bear the load of the overlying dolomite and creates instability of the slope. Moreover, the habitations in the crown portion (founded on dolomite) also add load on the friable shale having a low bearing capacity. The spring points and the flow of water along undefined channels saturate the slope forming material and facilitate reduction in the shear parameters of the slope forming material and in turn slope instability. Balianala flows with a steep gradient (> 40°) causing intense erosion of the slopes on both the banks as well as the nala bed due to the high velocity of the water. Shale on the nala bed and lower portion of the slopes on either bank gets eroded more rapidly and intensely due to its soft nature. Consequently, the remedial measures provided in the recent past like retaining walls get redundant due to scouring of the nala bed. The slope particularly on the right bank of the Balianala below the GIC ground is very steep (>55°) and the toe of the slope, which is also being eroded by the nala, also adds to the slope instability problem.

Landslide and slope failure are common problems associated with the hilly area of Uttarakhand. Providing transportation facilities across such a hilly terrain is a major challenge. To provide enough width for access roads, many slope-stabilisation measures are invoked. Breast wall, retaining wall and gabions are very common features employed for stability. Recently, soil nailing techniques have been used solely or in combination with gabion walls. The advantages of the soil nailing technique are to facilitate nearly vertical slope to get more land-use area for roads. Figure 35.4 demonstrates soil nailing employed at the Nainital site.

Figure 35.4 Soil nailing for slope stabilisation at Nainital. (Courtesy of Professor S. Mittal.)

35.6.3   Flooding

The main reason behind the disaster was the widespread heavy rains that led to flash floods in all the major river valleys in Uttarakhand. The heavy rains also caused landslides at several locations. According to a report of the National Institute of Disaster Management, the Kedarnath area was the worst affected region, where heavy rains led to the collapse of the Chorabari lake resulting in the release of a large volume of water that caused another flash flood in the Kedarnath town leading to further devastation in downstream areas.

More than 9 million people were affected due to the flash floods. The most affected districts were Bageshwar, Chamoli, Pithoragarh, Rudraprayag and Uttarkashi. The region has one of the most important pilgrimage circuits in India. Kedarnath disaster took place in 2013 during the peak tourist and the pilgrimage season. A similar situation happened on 7 February 2021.

35.7   FIELD TESTS

This section presents the sample data of the initial pile load test and plate load test data. A routine pile load test was conducted on the working pile for a railway bridge in the Kichha–Pantnagar Section (Gupta, 2009). The pile was having a length of 9 m and a diameter of 1 m. The safe capacity was estimated to be 93 t. A pile load test was conducted up to double of safe capacity. Under the maximum load, the gross settlement was 6.01 mm and the net settlement was 2.93 mm (Figure 35.5). Results indicated a higher safe capacity than predicted by formulae.

Conducting plate load tests in many situations is not easy considering the presence of gravels and boulders in the strata. The proximity of a nearby slope is an additional challenge. Figure 35.6 shows the location of the plate load test in Pithoragarh near the slope. In many situations, bridges are to be designed connecting two hills at the lowest possible location. The design of bridge foundations is based on the estimation of allowable bearing pressure. The soil matrix with the presence of gravels and boulders makes it

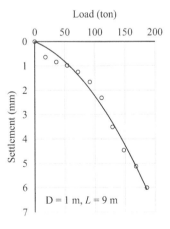

Figure 35.5 Pile load test at the Railway Bridge Kichha–Pantnagar section.

Figure 35.6 Plate load test location at Pithoragarh. (Courtesy of Professor S. Mittal.)

difficult to extend the borehole beyond 3–5 m. In that case, the plate load test is the only reliable option. But conventional plate load tests on rocky strata are not recommended as tests are unable to comment on ultimate load for hard rocks. In such cases, cyclic plate load tests are performed. The elastic subgrade modulus is computed based on the load versus elastic settlement plot, which can be used as the foundation design parameter.

Considering the nature of strata, conducting the standard penetration test (SPT) after a certain depth is impossible. Excavation pits are used as a measure to collect soil information. Figure 35.7 shows excavated pits at the Uttarakhand border near Ponta-Sahib. The SPT was attempted after having an excavation of about 2 m depth.

35.8   GEOENVIRONMENTAL IMPACT ON SOILS AND ROCKS

Climate/environmental change has a significant role in the process of weathering of minerals forming soils and rocks. The main three variables of climate change, namely temperature, pressure and water/rainfall, govern various stage phases of weathering of rocks and minerals resulting in chemical and mineralogical changes in the formation

Figure 35.7 Excavation pits at the Uttarakhand border near Ponta-Sahib:(a) showing big boulder, and (b) soil-boulder matrix (Courtesy of Professor S. Mittal.)

of soils and rocks. Water is also essential for chemical weathering, and therefore, an increase in rainfall accelerated weathering. This whole process of weathering is responsible for the formation of the secondary mineral. In the Kumaon region of Uttarakhand, different environmental changes have been responsible for giving rise to district soil profiles from similar rock types.

35.9 CONCLUDING REMARKS

In Uttarakhand, soil and rock conditions are quite different due to geographic conditions, as summarised below:

1. The major types of soil present in the Kumaon region of Uttarakhand are clayey silt (CL-ML), silty sand (SM) and poorly graded sand (SP) and in the Tarai area are silty sands (SM) and gravely sand (SW).
2. The sandy strata with intermittent clay layers can be noticed in the plains. Even boulders of variable sizes are also noticed at shallow depths in the vicinity of streams.
3. Due to the presence of active tectonic movements, the terrain is characterised by fractured and weathered rock mass of variable origins. These were observed to be moderately to highly jointed, thinly to medium bedded and dipping towards the northeast at moderate to steep angles.

4. Shale, shale with mica and quartzite are abundant in weathered soft and hard rock beds. There is a need for extensive study to understand the behaviour of these soils/rocks under earthquakes.

5. Landslides are very common in this region and hence simple devices/methods need to be developed for improving the geotechnical properties as well as for protective measures.

6. A lot of work must be carried out regarding understanding the behaviour of these soils for the construction of megastructures like dams.

REFERENCES

Gupta, S.S. (2009). Initial pile load test report for railway bridge no. 47, Kitcha, Pantnagar Section, Uttarakhand. Personal communication.

Gupta, S.S. and Kumar, A. (2002). Subsoil investigation report of guest house site, Kumaon Engineering College, Dwarahat (Uttarakhnad). Personal communication.

Gupta, S.S. and Kumar, A. (2009). Subsoil investigation report for Pithoragarh. Personal communication.

Gupta, S.S. and Kumar, A. (2011). Subsoil investigation report for Almora. Personal communication.

Gupta, S.S. and Kumar, A. (2015). Subsoil investigation report for Bhimtal. Personal communication.

Gupta, S.S. and Kumar, A. (2016). Subsoil investigation report for Pantnagar. Personal communication.

Gupta, S.S. and Kumar, A. (2019a). Subsoil investigation report for U.S Nagar. Personal communication.

Gupta, S.S. and Kumar, A. (2019b). Subsoil investigation report for Rudrapur. Personal communication.

Gupta, S.S. and Kumar, A. (2019c). Subsoil investigation report for Jaspur. Personal communication.

https://dmmc.uk.gov.in, Official Website of Disaster Mitigation and Management Centre, accessed 30 September 2021.

http://dmmc.uk.gov.in/pages/display/95-earthquake-zone, Disaster Mitigation and Mitigation Centre, Government of Uttarakhand, accessed 8 February 2021.

https://www.slideshare.net/actionaidindia/uttarkashi-flash-flood-in-pictures. Uttarkashi flash flood in pictures, accessed 4 February 2021.

https://www.uttarakhand-tourism.com/map/uttarakhand-map.php. Uttarakhand Map, accessed 4 February 2021.

Pant, G. and Kandpal, G. C. (1990). A report on the evaluation of instability along Balianala and adjoining areas of Nainital, U.P. *Geological Survey of India* (unpublished report).

Viladkar, M.N., Singh, M., Samadhiya, N.K., Sawant, V.A. and Maheshwari, P. (2009). Geotechnical investigations for proposed NTPC township at Dhak-Joshimath (unpublished report).

Chapter 36

Uttar Pradesh

Ashish Gupta
BIET Jhansi

Vinay Bhushan Chauhan
MMMUT Gorakhpur

Vikrant Patel
BIET Jhansi

CONTENTS

DOI: 10.1201/9781003177159-36

36.1   INTRODUCTION

Geographically, Uttar Pradesh is the fourth-largest state in terms of area (7.34% of the total area of the country) and the most populated state of India. It is situated in the north of the country with 75 districts and 18 divisions. With a geographical area of $2,40,928\,km^2$, Uttar Pradesh is positioned between 23° 52′ N and 31° 28′ N latitudes and between 77° 3′ E and 84° 39′ E longitudes (FARMECH, 2021). The major rivers of the state are the Ganga and Yamuna with prominent rivers Gomti and Varuna. The state is bounded by Rajasthan to the west, Himachal Pradesh, Haryana and Delhi to the northwest, Uttarakhand and an international border with Nepal to the north, Bihar to the east, Madhya Pradesh to the south and touches the state of Chhattisgarh and Jharkhand to the southeast. The annual average rainfall is fairly high (650–1,000 mm) with humid subtropical climate temperatures ranging from 0°C to 50°C.

The process of mechanical and chemical weathering and speedy erosion of the soils due to various agencies and bedrocks seem noteworthy. Brief classifications and characteristics of the soils and rocks of Uttar Pradesh are provided in the next section.

36.2   SOIL CLASSIFICATIONS

As Uttar Pradesh is located in the north-central Gangetic plains of India, the state features highly rugged terrain and elevations above the mean sea level range from 300 to 5,000 m. In geophysical terms, the Siwalik range of the Himalayas in the north, the river Yamuna and the Vindhyas in the west, south-west and the south and the Gandak river in the east define the terrain of Uttar Pradesh. Topographically, Uttar Pradesh consists of Terai-Bhabar, alluvial plains and the southern plateau. The Siwalik range in Uttar Pradesh falls to the Bhabar area, which is a porous bed of coarse pebbles and boulders brought down by the rivers flowing down the Siwalik hills. The river streams tend to sink in these porous beds of sediments. The Bhabar tract lies along the periphery of Siwalik foothills and it changes into the Terai region gradually. This transition belt is termed the 'Terai and Bhabar area' and is indicated by the rich forests and various river streams. A great area of the state of Uttar Pradesh is covered by a deep alluvial layer spread by the slow-flowing rivers of the Ganges system. Those extremely fertile alluvial soils range from sandy to clayey loam. The soils in the southern part of the state are generally mixed red and black or red-to-yellow.

Geomorphologically, Uttar Pradesh can be divided into three major topographical regions: (i) Siwalik foothills of the Himalayas and the Terai region border Uttar Pradesh on the north; (ii) Gangetic plains constitute the major central portion of the state; and (iii) Vindhya Range and plateau lie in a relatively smaller part of southern Uttar Pradesh. Based on its physical features, the state can be divided into three broad regions such as the sub-mountainous region, the Ganga plain region and the Trans-Yamuna region. The sub-mountainous region also known as Terai-Bhabar is mostly covered with forests. Immediately below the foothills, it is characterised by damp sub-soil punctuated by marshy land, thick jungles and tall grasses. The Ganga plain claims more than half of the state's area, extending from northwest to southeast. The whole plain is further divided into five sub-regions viz., the Ganga-Yamuna Doab, Ganga-Gomti Interfluve, Gomti-Ghaghara Interfluve, Trans-Ghaghara and Rohilkhand region. The Ganga plain, the most fertile region, is an alluvial soil that is traversed by large numbers of rivers, almost running parallel to one another. The Trans-Yamuna region also known as the Bundelkhand region is the least fertile in the entire state due to deficiency of rainfall.

36.2.1   Siwalik hills and Terai region

Siwalik range forms the southern foothills of the Himalayas, bordering Uttar Pradesh in the north. The lowland area dispersed with marshes, thick forests, swamps rich in clay that runs parallel to the Bhabar tract is called the Terai region. This area is composed of fine alluvium primarily sand, clay, silt and gravel. As the rivers flow down the slopes of Bhabar and course through the relatively planar areas of Terai, the sediments are deposited in shallow beds, and the sunken river streams of Bhabar re-emerge on the surface. Major soil-forming rocks are granite, Schists, Gneiss, Shales, Sandstones, Phyllite, Quartzite and so on. The soils in the Greater Himalayas, the Lesser Himalayas and Shiwaliks vary from sandy to loamy, are slightly acidic and of low Available Water Capacity (AWC) type.

36.2.2   Gangetic plains

The soils developed from the alluvium beds are deposited by Ganga, Yamuna and their tributaries in the vast Gangetic plains. The soils in this region are coarse loamy/fine silty (calcareous and non-calcareous). These features slight alkalinity and have a deeper soil depth as well as a high content of organic matter and plant nutrients. Such soils have good water holding capacity and are well-drained. The Gangetic plains are characterised by flat topography and highly fertile alluvial soil. The two-river system called the Ganges, which includes Ganga and Yamuna and their tributaries, that flows down the Himalayas, is responsible for laying in alluvial deposits. These plains cover about three-fourths of the total area of the state, stretching from east to west and covering most of its central portion. The Gangetic plains consist of the Ganga–Yamuna Doab, the Ganges plains, Terai and the Ghaghara plains, and this entire stretch of alluvial terrain is divided into three sub-regions, namely the eastern tract, central tract and western tract.

36.2.3   Vindhyas range and plateau region

Vindhyas is a discontinuous range of hilly terrain and mountain region that exhibits arid conditions. The southernmost stratum of Gangetic plains in Uttar Pradesh is rendered by hard and varied topography of hills, highlands and plateaus. The soils in this region are generally developed from Vindhyan rocks that include gneiss, granites, sandstone, quartzite, limestone, dolomite and so on. The soil type is fine loamy with stones and gravel. The soils of this region have mixed red and black hues that is slightly alkaline with low AWC.

The following are the four divisions that fall under the canopy of this region: Bundelkhand plateau; Tehsils of Prayagraj district; Mirzapur district and Chakia Tehsil of the Varanasi district.

In the western districts of Uttar Pradesh, the soil is generally deep brown and loamy in some places, also mixed with sand. The soil is gravely and full of stones – being generally acidic. Further eastwards, the soil becomes loamy, in some places the soil becomes acidic, while the rest has some alkaline properties. The soil in the central regions is loamy and sandy loam.

The eastern part of the state contains two varieties of soil, which are locally known as 'Bhat' and 'Banjar'. The alluvial soil is called 'Dhur' or 'Dhuh'. The one described as 'Mant' is loamy sandy calcareous, comparatively. In some districts of the eastern part of the state, the soil found in drier areas (affected by the presence of salt) are known as 'Usar' and 'Reh'.

In the southern part of the state (Jhansi Division of the state, Mirzapur and Sonbhadra and some parts of Prayagraj), there are mixed red and black soils that are sticky, calcareous, fertile and have expensive properties. On the upper plateau of these districts, the soil is red and consists of two types of slightly sandy or sandy loam and the alkaline loam known locally as 'Parwa' and 'Rackar', respectively.

36.3 OPTIMISING LAND USE

The NNRMS-ISRO (2019), based on the results of a detailed study, identified two major land classes and seven subclasses. Out of the total area of 2,40,926 km^2 of the state, approximately 80.7% of the total area is classified as agricultural land (includes cropland, current shifting cultivation, fallow and plantation), which is the main growing area for food grain in the entire state. Good drainage, surface, or subsurface water is an important source of irrigation throughout the state. More than 6.4% of the area is covered with forest, which consists of evergreen/semi-evergreen deciduous trees, forest plantations and scrub forest swamps/mangroves.

A very small area, which is nearly 0.05% of the total area of the state, is covered with grass or the area covered for grazing for herds with no land covered with snow and glacier. Build-up area including rural and urban development and mining area covers more than 5% area of the state. Around 3% area of the state is barren/unculturable/wasteland, which includes Barren Rocky Gullied/Ravenous Land, Rann Salt Affected Land and Sandy Area Scrub Land.

The land is facing serious threats of deterioration due to unrelenting human pressure and utilisation incompatible with its capacity. The information on land degradation is needed for a variety of purposes like planning reclamation programmes, rational land use planning, bringing additional areas into cultivation, and also improving productivity levels in degraded lands. Assessment of salt-affected and waterlogged areas is an important prerequisite for planning reclamation and improving land productivity. A few patches of seasonal waterlogging have been traced near the eastern boundary of the state, however seasonal waterlogging is associated with salinity/sodicity in form of ridges in a smaller part and connected as a line in the northwestern region.

The largest category of the land of the state is affected by water and wind erosion, which accounts for 80% followed by salinisation/alkalisation and waterlogging. Nearly 4.4% area of the state is wetlands/water bodies, which include Inland Wetland, Coastal Wetland, River/Stream/Canals Waterbodies. The waterlogged area may result in various types of soil degradation like physical degradation or chemical degradation or salinity. The salt-affected area of the state is classified by NNRMS-ISRO (2012). Salt-affected areas are one of the most important degraded areas where soil productivity is reduced due to either salinisation (Electrical Conductivity, EC > 4 dS/m) or sodicity (Exchangeable Sodium Percentage, ESP > 15) or both. However, dense patches of slight to moderate sodic and saline-sodic are observed affecting the many regions in the state.

There can be rather serious effects in terms of soil erosion, loss of soil fertility and thus reduced plant growth or crop productivity, clogging up of rivers and drainage systems, extensive floods and water shortages. Assessment of soil erosion status is an important prerequisite for land resources and conservation planning. Water and wind erosion is the mainland degradation process that occurs on the surface of the earth. Rainfall, soil physical properties, terrain slope, land cover, and management practices play a very significant role in soil erosion. The displacement of soil material by water can result in either loss of topsoil or deformation of the terrain, or both. These water-induced erosions are divided into sub-categories named water erosion, sheet erosion, rill erosion, gully erosion and ravines based on their extent of soil erosion.

The erosion maps should be used widely for the purposes such as soil conservation and regional planning; watershed management; agricultural productivity improvement planning; and scientific research involving carbon cycle, hydrologic cycle, energy budget studies, weather/climate prediction and so on.

36.4 ROCK FORMATIONS

Rocks ranging in age from Archean to Holocene are the oldest crystalline rocks. These rocks are confined to the southern and south-eastern parts of the state and the extensions of rock sequences exposed are in the neighbouring state of Madhya Pradesh, Bihar and Jharkhand. The stratigraphic sequences of Uttar Pradesh are summarised in Table 36.1 (GSI, 2012).

These rock groups are dominantly represented by basement gneisses, older metamorphites with younger granitoids as intrusives; metasedimentaries; sedimentaries and the alluvial. Various types of schist, quartzite, marble, and gneiss constitute the metamorphites present in the state. The metasedimentaries present in the state generally belong to Palaeo and Palaeo-Mesoproterozoic periods. The sedimentaries represent Meso to Neoproterozoic period comprise the rocks of the Vindhyan Supergroup and Gondwana Supergroup of the Late Palaeozoic period (GSI, 2018).

Gneisses with metasedimentary enclaves along with intrusive granite and other igneous rocks constitute the dominant lithology and represent the oldest suite of rocks confined to the southern part of the state and exposed in the Bundelkhand and Sonbhadra regions. In the Bundelkhand region, granite gneiss and granitoid constitute the dominant rock types and the assemblage is known as Bundelkhand Granitoid Complex (BGC). The BGC contains a wide variety of plutonic and hypabyssal rocks

Table 36.1 Generalised Stratigraphic Succession of Uttar Pradesh (after GSI, 2012)

Sr. Gr.	Group	Formation	Member/lithology
	Newer alluvium	Channel alluvium/ colluvium	Grey micaceous fine to coarse-grained sand, silt and clay
		Terrace alluvium	Cyclic sequence of grey micaceous sand, silt and clay
		Fan alluvium/ Bhat alluvium/ Ramnagar alluvium	Light grey to khaki, silt-clay and cross-bedded fine sand with gravel and pebbles
	Older alluvium	Varanasi alluvium	Polycyclic sequence of oxidised khaki to brown silt-clay with kankar and brown to grey fine to medium micaceous sand gravel (divisible into silt-clay sandy and rudaceous facies)
		Banda alluvium	Reddish to deep brown quartzo-feldspathic sand with gravel lenses, silt and clay
		Laterite	Gravel and pebble of laterite and bauxite with deep cherry red to brownish-black lithomarge clay
Vindhyan	Kaimur	Scarp ≡ Mangesar/ Bhouri/ Dudauni sandstone	Red, pink, compact, blocky sandstone, khaki and greenish-grey, micaceous sandstone and siltstone
		Bijaigarh	Grey micaceous siltstone, red and yellow ochre shale and siltstone, black carbonaceous shale and ferruginous sandstone
		Markundi	Light greyish white, medium to fine-grained silicified sandstone and arkosic sandstone
		Susnai breccia ≡ Koh	Autoclastic breccia with an angular fragment of porcellainitic shale interbedded in gritty to sandy matrix
		Ghurma	Black carbonaceous, micaceous, yellow-brown and light grey porcellainitic shale and thin bands of siderites ironstone
		Ghaghar ≡ Sasaram	Coarse to medium-grained pinkish sandstone
	Semri	Rohtasgarh ≡ Tirohan/ breccia	Alaur member: Thinly bedded limestone with argillaceous and cherty intercalations Rudauli member: Finely laminated black porcellainitic shale with cherty limestone nodule Kurail member: Very fine-grained limestone, dolomitic at the base with sand and mud dykes
		Basuhari ≡ Glauconitic, Rampur Pandwafall Sandstone	Khandura member: Fine laminated flaggy greenish shale and porcellainitic shale Mungadih member: Greenish glauconitic medium-grained sandstone

(Continued)

Table 36.1 (Continued) Generalised Stratigraphic Succession of Uttar Pradesh (after GSI, 2012)

Sr. Gr.	Group	Formation	Member/lithology
		Bargawan ≡ Fawn and Salkhan limestone	Nauka Tola member: Fawn to yellow-brown cherty dolomitic limestone
		Kheinjua ≡ Koldaha Shale, Olive Shale	Olive to greenish-grey, khaki, splintery shale with calcareous interbeds and partings
		Chopan ≡ Deonar	Light grey, greenish porcellainitic shales agglomeratic beds and arkosic sandstone
		Kajrahat	Siliceous, cherty, dolomitic limestone, blocky and slabby limestone with argillite interbeds
		Arangi	leached purplish porcellainitic shale and black calcareous shales
		Phaterwar (≡Basal Conglomerate, Deolond)	Gritty to pebbly sandstone, medium-grained sandstone, siltstone and basal conglomerate
Bundelkhand Granitoids	Basic-Ultrabasic Intrusives		Dolerite, lamprophyre, olivine basalt, kersanite, gabbro
	Acid Intrusives		Aplite, pegmatite, quartzofeldspathic vein, quartz reef/vein, pyrophyllite
	Porphyries		Granite, rhyolite, diorite, dacite
	Granite		Coarse-grained; Medium-grained; and Fine-grained grey and pink granite
			Porphyritic coarse-grained and Porphyritic grey and pink granite
			Fine-grained and Medium-grained leucogranite
	Gneisses		Augen gneiss, granite gneiss, porphyritic granite gneiss, migmatite

dominated by granites of several generations, gneisses, migmatites and leucogranites. In the Sonbhadra region, gneiss is the dominant rock and the assemblage is termed as Dudhi Gneissic Complex. The Dudhi Gneissic Complex consists mainly of granite gneiss, migmatites and non-foliated, massive younger granite with enclaves of metamorphites and veins of pegmatite, aplite and quartz. This is the westward continuation of the Chhotanagpur Granite Gneiss Complex of Bihar/Jharkhand.

The Proterozoic sequence is constituted by the rocks of the Vindhyan Supergroup, Mahakoshal Group and Bijawar Group. The Bijawar Group is represented by a sequence of ferruginous quartzite, carbonate, phosphorite, sandstone and tuffaceous rocks. The Mahakoshal Group includes metasediments with interlayered metavolcanic and granitic bodies intruding it.

The Vindhyan sequence has been sub-divided into four groups namely Semri, Kaimur, Rewa and Bhander in ascending order and the sequence is resting unconformably over the Mahakoshal and Bijawar groups. The Semri Group includes carbonates, tuffs, shale and minor sandstone, often glauconitic. The Kaimur Group consists of a thick arenite-argillite sequence. The Rewa Group is represented by an alternate sequence of argillite and arenite. The Bhander Group consists of shale, greenish shale, siltstone, reddish-brown to purple-pink, spotted sandstone with shale partings and quartz arenite at the top.

In a small area in the Sonbhadra district, the rocks belonging to the Gondwana supergroup of the Permo-Carb period have been exposed, which are composed of conglomerate, sandstone, gritty at places, pink and pale green shales (GSI, 2018).

Detached outcrops of basic volcanics (dolerite to basalt), representing the Deccan trap, are found as capping over BGC, Bijawar Group and Vindhyan supergroup in the Bundelkhand region in the south of Uttar Pradesh.

36.4.1 Mineral resources

The major part of Uttar Pradesh is covered by the Gangetic alluvium in the north, whereas the southern part is covered by Peninsular terrain. The Peninsular part of the state is covered by the rocks of the Archean to Mesozoic age, which is rich in the availability of various minerals. A detailed list of the numerous mineral deposits of economic and industrial value available in the state is presented in Table 36.2.

Table 36.2 Mineral exploration based on the geological investigation carried out in the State of Uttar Pradesh (GSI, 2012; DGMUP, 2020; DSR, 2020)

Major minerals

Minerals	Details
Refractory industry	Crystalline marble (Sonbhadra region)
Cement industry	Limestone (cement grade), Shale and Gypsum (Sonbhadra region)
	Low PCE fire clays: (not suitable for any industrial purposes, due to their non-plastic nature and high porosity)
	Diaspore (Jhansi, Lalitpur)
	Andalusite
	Sillimanite (Sonbhadra region)
Glass industry	High-grade silica sand (Prayagraj region)
Aluminium industry	Bauxite (high-grade containing >45% alumina) in Chittrakut, high alumina clays, diaspore (Banda, Mirzapur and Varanasi, Chandauli region)
Fertiliser industry	Gypsum, limestone, pyrite, deposits of phosphatic rock (Lalitpur region) at Jhansi
Construction industry	Building stone (Bundelkhand Granites and Vindhyan Sandstones)
	Marble, ballast, sand, Morrum, brick earth, lime, roofing slates, low-grade limestone and dolomite (Sonbhadra and Banda region)
	Pyrophyllite (Jhansi, Lalitpur)
	Ochre (Banda/Chitrakoot)
Miscellaneous Industries	SMS grade dolomite (suitable for steel melting), calcite, arsenic, antimony: Sonbhadra region
	Medium to low-grade coal: Sonbhadra district
	Low-grade iron ore (25%–30% iron) copper, Platinum, Phosphorite, Uranium, Baryte: In Lalitpur
	Gold (placer gold): Bijnor, Gonda, Barabanki, Beharaich, Faizabad, Ambedker Nagar, Lalitpur, Sonbhadra
	Diamond: Sonbhadra, Banda
	Molybdenum: In Sonbhadra, Hamirpur
	Agate: In Banda

(Continued)

Table 36.2 (Continued)   Mineral exploration based on the geological investigation carried out in the State of Uttar Pradesh (GSI, 2012; DGMUP, 2020; DSR, 2020)

Minor minerals

Defined in Section 3 of the Mines and Minerals (Regulation and Development) Act, 1957

Locally available and have local use, not used in any major industries

Building and construction materials

Limestone, marble or marble chips, brick earth, saltpetre, building stone

Slabs and Ashlar, sized dimensional stones (sand Stone, Quartzite)

Millstone and hand chakki (sand Stone, Quartzite)

Khandas and boulders: Granite and Dolostone, Sandstone and Quartzite sized up to 25 × 25 × 25 cm

Gitti ballast: One metre or large sandstone quartzite

Morrum, sand other than ordinary sand used for the specified purpose, Kankar, Bajri, ordinary earth

Brick earth (found in all the districts, except the hilly districts): strip-mining the ground in 2–6-metre-deep pits, generally in the agricultural land

Building stone: In southern part of Uttar Pradesh known as Bundelkhand (Patia, chauka, gitti, ashlar, soling, chakki, khanda, block, patia, stonewares, soling, patal).

Category I sand (coarse and clean sands without flaky minerals)

Category II sands (Morrum, Kankar, Salt petre or Shora, Reh, Bajri, and boulder)

36.4.2   Aquifer systems

Systematic reviews of the available data of the distribution of the principal aquifer systems in the state reveal that the principal aquifer systems are lying majorly in the alluvium bed (91.28%) and the minor contribution by the banded gneissic complex (4%), quartzite (2.32%), shale (1.05%), schist (0.63%), sandstone (0.48%) and basalt (0.25%). No aquifer was retained by limestone, granite, Charnockite, Khondalite, Gneiss, Intrusive Laterite, or any Unclassified strata (CGWB, 2012).

36.5   PROPERTIES OF SOILS AND ROCKS

The borehole lithologs show various soils with their description, consistency and colour of each stratum. Considering the borehole lithologs across the various locations of the state of Uttar Pradesh, in Tables 36.3–36.6, standard penetration tests (SPT) results are given as N-values (measured SPT) for these boreholes, where N' and N'' show the SPT 'N' value corrected for overburden and SPT 'N' value corrected for dilatancy, respectively.

Tables 36.3 and 36.4 provide the borehole data up to 10 m depth in Agra and Prayagraj (formerly known as Prayagraj) districts, respectively. Soil dry density ranges between 1.61 and 1.79 g/cm^3. The soil in the Agra region is majorly lean clay (normally consolidated) in upper strata up to 6 m depth and beyond that deposits of poorly graded sand, sticky plastic clay and traces of silty clay were observed. Similarly, in the Prayagraj region, an upper layer to a shallow depth of lean silt has been identified, and in addition, a lean clay with a small deposit of a thin layer of silty clay has been noted.

Table 36.3 Typical boring log with SPT N-value of a location in Agra, Uttar Pradesh

Depth (m)	Field density (g/cm³) Wet	Dry	Field moisture content (%)	Observed SPT value	Specific gravity	Classification as per IS:1498 (1970)	N	N'	N"
1.50	1.95	1.71	14.86	13	2.65	CL	13	18	17
3.00	1.98	1.70	16.45	15	2.65	CL	15	18	17
4.50	2.02	1.72	17.76	17	2.64	CL	17	18	17
6.00	2.01	1.69	19.32	19	2.65	CL	19	18	17
7.50	2.07	1.71	21.15	20	2.65	CL-ML	20	17	16
9.00	1.81	1.61	12.61	22	2.62	SP	22	18	17
10.00	1.85	1.63	13.51	25	2.65	ML	25	20	18

Table 36.4 Typical boring log with SPT N-value of a location in Prayagraj, Uttar Pradesh

Depth (m)	Field density (g/cm³) Wet	Dry	Field moisture content (%)	Observed SPT value	Specific gravity	Classification as per IS:1498 (1970)	N	N'	N"
1.50	1.87	1.64	14.16	11	2.63	ML	11	16	16
3.00	2.00	1.69	18.24	13	2.64	CL	13	15	15
4.50	2.11	1.78	18.43	16	2.65	CL-ML	16	17	16
6.00	2.13	1.79	18.84	24	2.65	CL-ML	24	23	19
7.50	2.15	1.77	21.74	26	2.66	CL	26	22	19
9.00	2.16	1.76	22.85	30	2.66	CL	30	24	20
10.00	2.16	1.77	22.81	32	2.66	CL	32	24	20

Table 36.5 Typical bore log with SPT N-value of a location in Jhansi, Uttar Pradesh

Depth (m)	Field density (g/cm³) Wet	Dry	Field moisture content (%)	Observed SPT value	Specific gravity	Classification as per IS:1498 (1970)	N	N'	N"
1.50	1.99	1.63	21.69	5	2.66	CI	5	7	11
3.00	2.01	1.62	24.13	5	2.66	CI	5	6	11
4.50	2.01	1.67	20.23	7	2.65	CI	7	7	11
6.00	2.02	1.66	21.27	8	2.63	CI	8	8	12
7.50	1.96	1.67	17.23	9	2.65	SM	9	8	12
9.00	2.02	1.69	19.46	14	2.66	SM	14	11	13
10.50	2.03	1.62	25.4	20	2.66	CI	20	15	15
12.00	2.03	1.63	24.72	22	2.65	CI	22	16	16
13.50	2.01	1.70	17.86	24	2.66	CI	24	16	16
15.00	2.02	1.68	19.9	27	2.66	CI	27	17	16
16.50	2.03	1.65	22.81	30	2.66	CI	30	18	17
18.00	1.82	1.52	19.65	28	2.60	SP	28	17	18

Table 36.6 Typical bore log with SPT N-value of a location in Lalitpur, Uttar Pradesh

Bore/pit no.	Depth (m)	Field density (g/cm³) Wet	Dry	Field moisture content (%)	Observed SPT value	Specific gravity	Classification as per IS:1498 (1970)
1	1.50	1.88	1.79	5.66	>50	2.68	GM-GP
	3.00	2.02	2.01	0.30	>50	2.69	GM-GW
2	1.50	1.85	1.77	4.65	>50	2.67	GM-GP
	3.00	1.98	1.97	0.30	>50	2.68	GM-GW
3	1.00	2.12	2.11	0.10	>50	2.70	GM-GP
4	0.50	2.08	2.07	0.20	>50	2.69	GM-GW
5	1.50	2.09	2.02	3.50	>50	2.67	GM-GW
	3.00	2.07	2.01	0.70	>50	2.68	GM-GW
6	1.50	2.05	2.03	0.50	>50	2.70	GM-GW
	3.00	2.13	2.08	0.40	>50	2.71	GM-GW
7	1.50	1.84	1.67	0.30	>50	2.67	GM-GP
	3.00	2.02	2.01	0.10	>50	2.69	GM-GW
8	0.50	2.01	2.00	0.10	>50	2.71	GM-GW

Obtained data of the soil deposits from the various boreholes of Gonda and Kannauj districts, it is noted that the water table was noted at a depth of 6 and 8 m, respectively. The soil in these regions is mainly silty sand and poorly graded sand in the upper and lower strata. A small layer of inorganic silt with low to medium compressibility deposit at a deeper depth of 16 m below the ground level was noted in few boreholes of the Gonda district. Representative borehole data obtained from the southern part of the state are given in Tables 36.5 and 36.6, for Jhansi and Lalitpur districts. Even residing in these locations in the rocky belt of the state, a wide variation in the soil was noted. In the Lalitpur region, well-graded gravel with silt from poorly graded gravel silt was obtained at a shallow depth up to 3 m. This is observed because, this region lies in the rocky belt of the state, which is also the source of various minerals. However, medium compressibility clay with a small, deposited layer of silty sand was also found in the Jhansi district.

36.5.1 Variation of N-value with depth

The N-values for each lithologs are given in tabular form in Tables 36.3–36.6. The SPT N-values are an important governing factor in liquefaction assessment/triggering analyses, calculation of bearing capacity for various civil engineering structures, and indirect assessment of various soil properties using available empirical relationships. Classification of sites based on any one of the three parameters, i.e., undrained shear strength, SPT N-values and shear wave velocity values, to evaluate earthquake effects is done by seismic site characterisation. A typical variation of SPT N-values observed from various sites in Uttar Pradesh is given in Tables 36.3–36.6.

As the geographical area of Uttar Pradesh is fairly large, it is difficult to present the variation in spatial soil stratigraphy from a geotechnical point of view in a comprehensive form. However, a few representative pictures of the soil strata obtained from the few locations across the state covering Lucknow, Jhansi, Lalitpur and Gorakhpur are

Figure 36.1 A view of excavated soil at a location in Lucknow, Uttar Pradesh.

Figure 36.2 A view of excavated soil at a location in Jhansi, Uttar Pradesh.

Figure 36.3 A view of excavated soil at a location in Lalitpur, Uttar Pradesh.

Figure 36.4 A view of excavated soil at a location in Gorakhpur, Uttar Pradesh.

shown in Figures 36.1–36.4, respectively. In view of the wide variation in the soil property in the state, it is recommended that a detailed soil investigation should be carried out before any foundation/earth structure design is undertaken at any given site.

36.6   USE OF SOILS AND ROCKS AS CONSTRUCTION MATERIALS

The geological survey carried out for mineral exploration reveals the presence of numerous mineral deposits in the state used as construction materials. A brief account of some such minerals is given here (DGMUP, 2020):

36.6.1   Fire clays

In Bansi-Misra-Makrikhoh areas in the southwest part of the Sonbhadra district, about 3 million tonnes of low pyrometric cone equivalent (PCE) fire clays reserves are available. Its high porosity and non-plastic nature make it unsuitable for any industrial purposes. However, these clays are blended with suitable plastic clays and used for making BIS 6:1983 (R 2016) standard firebricks.

36.6.2   Sand

All the rivers in the state contain sand deposits on its bed. The coarse and clean sands (excluding flaky portion) are termed as Category 1 sand, which is preferred in construction work, like masonry, due to higher strength and lower construction cost.

36.6.3   Morrum

Morrum is available in the rivers flowing through Jhansi, Jalaun, Lalitpur, Mahoba, Hamirpur, Fatehpur, Sonbhadra and Banda districts. The weathering and disintegration of granitic rocks is another source of morrum. 'Red Morrum', which is in fact laterite soils are obtained from old weathering surfaces of elevated ground. These are used for spreading on kutcha roads.

Sandstones and granites: The Vindhyan sandstones and Bundelkhand granites are extensively quarried and used as building stones and ballast.

36.6.4   Limestone, shale and gypsum

Cement grade limestone, shale and gypsum are required as raw materials in the cement industry. About 430 million tonnes reserves of cement grade limestone are available and deposits of shale are found with or in the vicinity of the limestone deposits.

36.6.5   SMS grade dolomite

About 6.5 million tonnes reserves of SMS grade dolomite are reported by the Directorate of Geology and Mining in the state in Bari-Bagmana, Chopan, Sinduria and Karamdand areas of the Sonbhadra district. These are suitable for the steel melting shop.

36.6.6   Pyrophyllite

The geological investigation reveals the presence of about 0.5 million tonnes of pyrophyllite in the Lalitpur and Hamirpur districts. Pyrophyllite is used as a polishing agent and fillers in rubber, a carrier for insecticides and pesticides, and as an extender in paints.

36.7   OTHER GEOMATERIALS

Geomaterials derived from industrial waste, mining waste, or other sources can be utilised in geotechnical applications. Some of them are discussed below.

36.7.I Fly ash

Fly ash is a finely divided residue that is produced from the combustion of pulverised coal in electric power generating plants. Over 25 million tonnes of fly ash were produced from 18 Thermal Power Stations in the state during the year 2017–18 and about

55% of fly ash was utilised. The following are the different modes of the utilisation of fly ash (CEA, 2018):

1. Fly ash is used as a pozzolanic material by the cement industry in the manufacturing of Portland Pozzolana Cement. Moreover, building materials like bricks, tiles, blocks and so on are manufactured by using fly ash which saves the fertile topsoil.
2. Many favourable properties of fly ash make it suitable to stabilise the subgrade of road pavements and the construction of embankments. Topsoil can be saved by using fly ash for the reclamation of low-lying areas as a substitute for soil/sand.

36.7.2 Mine wastes

A large volume of mine wastes such as clay-rich tailings, waste rock, red mud is produced in mining operations in the state of Uttar Pradesh. These wastes are used in geotechnical structures in the following ways.

36.7.3 Clay-rich tailings

Clay-rich tailings are used in the manufacturing of bricks, floor tiles and cement.

36.7.4 Red mud

Bauxite red mud is solid alkaline waste generated by Bayer's process in aluminium refineries. Due to the unique physical and chemical properties of red mud, it can be used in several aspects (Patel and Pal, 2015). The applications which have been in use on a commercial scale are for making crude and fine ceramics, such as tiles, floor tiles, bricks, road construction, as a component in making OPC and special cement and cement mortar, fillers in the rubber and plastic industry and pigment in the production of paints.

36.7.5 Waste rock

Waste rock can be used as a backfill, aggregate in road construction, or feedstock for cement and concrete.

36.8 FOUNDATIONS AND GEOTECHNICAL STRUCTURES

In general, an open shallow foundation is suitable for light to medium loaded structures. In the town Amethi, the rural houses under PMGAY (Pradhan Mantri Gramin Awaas Yojana) scheme used spread footings as the foundation. However, to find the depth and type of foundation for heavily loaded structures like dams, bridges and high-rise buildings, geotechnical investigations and subsoil explorations should be carried out to determine the properties and behaviour of existing strata.

A large part of the state lies in Earthquake Damage Risk Zones IV and III. As per the earthquake history of Uttar Pradesh, mostly, the earthquake struck the western districts of the state. Special care should be taken in the design of foundations in these areas. According to the guidelines of the Uttar Pradesh State Disaster Management Plan, the foundations of buildings should be tied together well and tied firmly with the walls as well. However, the exact design method of foundation depends upon the type of structure being designed.

Geotechnical structures include earthen structures such as dams, earth retaining walls, embankments, slopes, canals and so on, and foundations of buildings, bridges and so on. These projects require a detailed site investigation for obtaining information about the subsurface conditions of the proposed construction site. The investigation should also include the assessment of risk from natural hazards such as earthquakes, floods, landslides and so on.

Many flyovers and railway over bridges have been built in Uttar Pradesh in seismically active zones, using TENAX geogrid reinforced soil system. The flexibility of the reinforced soil structure provides resistance to the structure, even in high-intensity earthquakes, while preventing structures from collapsing.

36.9   NATURAL HAZARDS

Over the years recurring natural hazards have been causing severe damage in the state in terms of loss of life and property. According to the State Disaster Management Authority, Floods, Droughts, Fires, and earthquakes are significant natural disasters in Uttar Pradesh (UPSDMA, 2020). The state is facing an estimated loss of hundreds of crores of rupees annually due to these disasters.

36.9.1   Tectonic movement

Based on the Vulnerability Atlas of India, Uttar Pradesh is divided into three seismic zones- Earthquake High Damage Risk Zone IV, Moderate Damage Risk Zone III and Low Damage Risk Zone II. The Terai belt districts of the state and entire districts of Muzaffarnagar, Saharanpur, Baghpat, Meerut, Bijnor, Ghaziabad, Amroha, Gautam Buddh Nagar, Rampur, Bulandshahr, Moradabad in western UP fall under Earthquake High Damage Risk Zone IV.

Earthquake is an unavoidable, unpredictable, infrequent phenomenon. Its parameters are its location, its destructive energy and the depth of focus below the ground level. The earthquake induces lateral forces on a structure that creates a base overturning moment that exceeds the available overturning resistance due to gravity loads where structures fail (Yim and Chopra, 1984). Adoption of earthquake resistance measures in the constructions like the use of Base Isolation Technology, Disaster Resistant Pier System and so on in areas that fall within Earthquake Damage Risk Zones III and IV can reduce the destruction caused by seismic activity.

36.9.2   Floods

Floods are among the most common natural hazards in the state affecting approximately 27 lakh hectares area almost every year. They cause extensive damage to infrastructure, the economy and devastation to human settlements. Flooding damages the existing foundations in mainly two ways: due to the force of fast-moving water and due to the seepage of water causing or widening foundation cracks, which subsequently weaken the structural integrity of any geotechnical structures. The flood creating rivers in the state are Ganga, Yamuna, Ramganga, Gomti, Sharda, Ghaghara, Rapti and Gandak. Out of the total geographical area of 240.93 lakh hectares of the state, around 73.06 lakh hectares area is at risk of flooding.

36.10   CASE STUDY

This case study on raft foundation of a residential building presents the field and laboratory test data collected at the proposed site, analysis and interpretation of the entire data, recommendations on the type of foundations to be provided for various structures, and the corresponding values of allowable soil pressure for their design. The soil investigation was conducted to determine the allowable bearing pressure for the proposed raft foundation with dimensions of 25 m × 30 m for a B + G + 16 storey residential building planned in Ghaziabad district and the details are briefly are discussed below.

The raft foundation was planned at depths of 4.5 and 5.5 m, and the soil exploration up to a depth of 50 m at ten specified borehole locations was conducted. At the proposed site, a thick clay layer was visible at many open excavations.

36.10.1   Field investigations

In view of what has been stated above, it was decided to conduct the following field tests:

1. Borings (ten numbers) to be advanced to a depth of 50.0 m or refusal, whichever is earlier, using a 150 mm diameter casing pipe for protecting the borehole walls from caving in.
2. SPT – Ten numbers to be conducted simultaneously during boring in each borehole at 1.5 m interval initially up to 9 m and at 3 m interval up to a depth of 50.0 m or refusal whichever is earlier.
3. Dynamic cone penetration tests (DCPT) – Twenty numbers, to be extended to a depth of 50.0 m or refusal, whichever is earlier.
4. Plate load tests (PLT) – Four numbers, to be conducted in test pits at depths of 4.5 to 5.5 m on 300 mm × 300 mm size square test plate.
5. Collection of undisturbed and representative soil samples from different depths during borings and plate load test pits for laboratory tests.

SPTs were conducted as per IS:2131 (1965). The observed values were corrected for overburden and dilatancy wherever applicable. In general, the refusal has been found to occur between depths of 15 and 20 m. However, low values have also been encountered due to intermediate clay layers. Due to the presence of gravel in the soil mass, SPT could not be conducted in two boreholes. The standard penetration resistance at different depths in various borehole locations has been found to vary from 22 to 63.

Dynamic cone penetration tests (18 numbers) were conducted at various locations spread across the area (distributed uniformly) under investigation to identify the presence of weak strata or soft pockets, if any, in the soil mass beneath the ground surface.

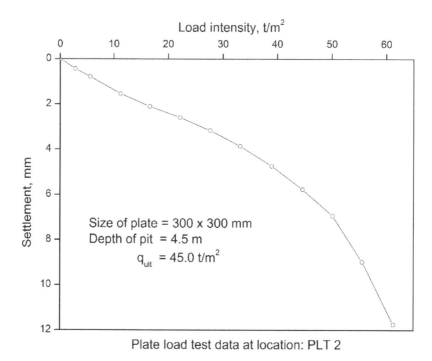

Plate load test data at location: PLT 2

Figure 36.5 The plot of load intensity versus plate settlement obtained from a plate load test in Ghaziabad, Uttar Pradesh.

The refusal has been found to occur at about 9–10 m in all the cases. The findings from dynamic cone penetration tests (IS:4968, 1976) suggest that penetration resistance is low (<10) up to 4.0 m and then increases consistently with depth.

PLT were conducted according to IS:1888 (1981) with monotonic loading at four locations on 300 mm × 300 mm size test plate at depths of 4.5, 4.5, 5.5 and 5.5 m, respectively. Figure 36.5 shows, the plot of load intensity versus plate settlement obtained at a selected test location PLT 2.

This plot shows a nearly non-linear variation of load intensity versus plate settlement for the tests at all locations. At all PLT locations, i.e., PLT 1–4, the ultimate bearing capacity of the test plate has been obtained by the double tangent method. For locations PLT 1 to 4, the ultimate bearing capacity of the plate was found to be 45.0 t/m^2 except at location PLT 4, where it was observed to be 42.5 t/m^2 only.

36.10.2   Laboratory investigations

The soil samples obtained from boreholes were used in the laboratory for various classification and identification tests on soils. Laboratory and field investigations were followed by detailed analysis and interpretation of results to arrive at a representative subsoil profile, computation for bearing capacity, settlements and allowable bearing pressure. The laboratory testing for the classification tests programme includes mechanical sieve analysis for studying the grain size distribution of soil and Atterberg

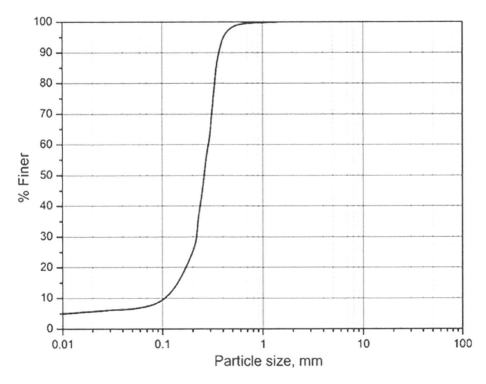

Figure 36.6 The typical grain size distribution curve for soil obtained from a borehole at 15 m depth.

Limit tests for studying the plasticity characteristics of soil as per IS:1498 (1970). A typical grain size distribution curve obtained from the sieve analysis is presented in Figure 36.6.

The result of soil classification tests states that the soil strata, in general, consist of silt/clay of low compressibility (ML/CL) up to 10.5 m depth below existing ground level, followed by poorly graded sand (SP) up to about 16.5 m depth, followed again by a clay layer of low compressibility up to a depth of 28.5 m, and further trailed by poorly graded sand (SP) layer up to the depth of exploration 50.0 m. Undisturbed soil samples, wherever applicable, were collected during borings for the Unconfined compression and consolidation tests. The unconfined compressive strength lies in the range of 10.4 to 15.4 t/m^2. The compression index varies from 0.095 to 0.135 representing the presence of highly compressible soil strata.

36.10.3   Design criteria

Foundations, in general, are designed for two criteria: (i) foundations must be safe against shear failure and (ii) foundations should not settle excessively. IS:1904 (1986) gives limiting values of the total settlement, differential settlement and angular distortion for certain types of structures including the framed buildings and the silo structures. These 12 limiting values have been specified for: (i) sand and hard clay and (ii) plastic clays. Accordingly, the permissible settlements for raft foundations of reinforced concrete structures are 100 mm for sand and hard clay and 125 mm for plastic clay. The permissible total settlement as specified by the code (IS:1904, 1986) is considered in the present situation as 100 mm.

36.10.4   Allowable soil pressure

The allowable soil pressure can be estimated in the present situation, based on the (i) projected average value of penetration resistance in the zone of influence; (ii) plate load test data (IS:6403, 1981); (iii) consolidation tests data and (iv) unconfined compressive strength (UCS) test data.

36.10.5   Shear failure consideration

Considering the depth of the foundation to be 4.5 m, a representative value of corrected N below this depth has been taken as 27 based on penetration resistance. Corresponding to the representative SPT N-value, the UCS value is taken as 32.4 t/m^2 (Bowles, 1997). The ultimate bearing capacity works out to be 104.08 t/m^2.

Based on plate load test data, the values of net safe bearing capacity have therefore been worked out at all the test locations. Considering a factor of safety of 3.0, the most conservative value of net safe bearing capacity of the foundation works out to be 14.16 t/m^2.

Based on the UCS test data, a representative value of UCS is taken as 10.4 t/m^2. By considering, raft resting on clay soil, the ultimate bearing capacity of the foundation is found to be 333.41 t/m^2 and with a consideration of a factor of safety of 3.0, the magnitude of net safe bearing capacity has been worked out to be 11.13 t/m^2.

36.10.6   Settlement criterion

It is well-known that the settlement of a footing resting on soils with the presence of significant plastic fines can be extrapolated from the settlement experienced by the test plate at the same loading intensity. For the proposed raft foundation of 25 m, a permissible settlement of foundation was given to be 100 mm, which corresponds to 1.2 mm settlement of the test plate. The corresponding values of allowable pressure from the load intensity versus settlement curves work out to be in the range of 10.0 to 16.0 t/m^2. The allowable pressure, therefore taken as the minimum of these values i.e., 10.0 t/m^2.

Due to the presence of clayey strata between 4.5 and 10.5 m depth, and again between 16.5 and 28.5 m at some locations, the settlement caused by the consolidation will be the decisive factor for deciding the permissible soil pressure based on the settlement criterion. The compression index has been found to vary from 0.095 to 0.13 representing the presence of highly compressible soil strata. The allowable bearing pressure corresponding to a permissible settlement of 100 mm (due to consolidation) has been calculated. Calculations for allowable bearing pressure have been carried out for the specified depth of foundation i.e., 4.5 and 5.5 m, and net allowable bearing pressures were found to be 9.25 and 10.68 t/m^2, respectively. A comparison of the recommended allowable bearing pressure based on shear failure and settlement criterion is summarised in Table 36.7.

Table 36.7 Recommended allowable bearing pressure based on various criteria and field and laboratory tests

Criterion	Based on	Allowable pressure (t/m^2)
Shear failure	SPT penetration resistance	34.69
	Plate load test	14.1
	UCS test	11.1
Settlement	Plate load test	10.0
	Consolidation test	9.25@4.5 m depth
		10.68@5.5 m depth

36.10.7   Liquefaction analysis

In the present soil exploration, the water table was found at 12 m depth. Till the desired depth of exploration of 50 m, a considerable amount of fine silty sand was also observed. So, it is essential to check the liquefaction potential of these sandy soils. SPT N-values available from a borehole have been used for analysing the liquefaction potential. To check for the worst condition, the water table is assumed at ground level. The average shear stress (τ_{av}) developed at any depth due to any probable earthquake is taken as $0.65\,\gamma h\,(a_{max}/g)\,r_d$, where h and γ are the depth and unit weight of soil and r_d is the reduction factor obtained from curves given by Seed and Idriss (1971). Maximum acceleration, a_{max}, is the design earthquake acceleration is considered $0.184 \times 10^{0.320M} \times D^{-0.8}$, where D is the maximum epicentral distance in km and M is the magnitude of design earthquake on the Richter scale. For the present location of the site, the values of M and D are taken as 8.0 and 200 km respectively, which give a_{max}/g ratio equal to 0.07052. It is observed that at all points, shear stress (τ_h) is greater than the design average shear stress (τ_{av}), which indicates that the present sandy strata are safe against liquefaction.

36.10.8   Recommendations

Based on the field and subsequent laboratory testing, the recommendations for the design of foundations for the proposed B + G + 16 storey building at Ghaziabad are summarised here. The soil strata, in general, consists of alternate layers of clay and poorly graded sand. The representative subsoil strata comprise silt/clay of low compressibility (ML/CL) up to 10.5 m depth below the existing ground level, followed by sand (SP) up to a depth of 16.5 m, which is again followed by a layer of clay of low compressibility up to 28.5 m depth. After this depth, poorly graded sand (SP) continues up to the maximum depth of the exploration, i.e., 50 m below the existing ground surface. Variation of dynamic cone penetration resistance with depth suggests that the penetration resistance is low (<10) up to 4.0 m depth and then increases consistently with depth. SPT data validate the observations made during dynamic cone penetration tests. The settlement criterion governs the allowable bearing pressure. For the proposed raft foundation, the recommended values of net allowable bearing pressure are 9.0 t/m^2 at 4.5 m depth and 10.0 t/m^2 at 5.5 m depth, respectively. Adequate measures should be taken in design against earthquake forces. It is recommended that suitable drainage measures must be planned and provided during and after the construction of the building complex to prevent excessive seepage of water into the clay mass. The above recommendations have been made based on limited investigations conducted at the site during field and subsequent laboratory testing. However, if during the construction, any deviation is observed regarding the soil type and the nature of the strata, advice may be sought from a competent authority.

36.11   CONCLUDING REMARKS

Since the geographical area of Uttar Pradesh is large, it is quite challenging to characterise the subsoil throughout the state with the help of limited information available from the few boreholes. Following are some generalised facts and conditions regarding subsoil of the state and related issues:

*   Uttar Pradesh is located in the north-central Gangetic plains of India, which claims more than half of the state's area, consisting of alluvial soil. The soils developed from the alluvium beds are deposited by the rivers Ganga and Yamuna and their tributaries into the vast Gangetic plains.
*   Topographically, Uttar Pradesh is divided into Terai-Bhabar, Alluvial plains and the southern plateau. The southernmost stratum of Gangetic plains in Uttar Pradesh is rendered by hard and varied topography of hills, highlands and plateaus.
*   The lowland area of the Shivalik Hills and Terai region is dispersed with marshes, thick forests and swamps rich in clay. This area is composed of fine alluvium primarily sand, clay, silt and gravel.
*   Soils in the southernmost part of the state are generally developed from Vindhyan rocks. The major soil-forming rocks are granite, schists, gneiss, shales, sandstones, phyllite, quartzite and so on.
*   The Peninsular part of the state is covered by rocks of the Archean to Mesozoic age, which is rich in the availability of various minerals. The presence of numerous mineral deposits leads to the availability of an abundant amount of construction materials including sand, morrum, fire clays, sandstones, granites, dolomites, limestone, shale and gypsum.
*   Geomaterials derived from industrial and mining waste such as fly ash, red mud and waste rocks are abundantly available in the state for utilisation in geotechnical applications.
*   The districts falling under the Terai belt and western Uttar Pradesh are Earthquake High Damage Risk Zones. In those regions, the presence of shallow water tables and compressible soil stratum increases the probability of the failure of structures.
*   Around 30% area of the state is flood-prone, which may pose a high risk to existing foundations of the structures.

ACKNOWLEDGEMENT

The authors of this chapter would like to express their sincere appreciation to all the well-wishers for encouraging, supporting and providing their valuable insights for this book chapter. They would also like to gratefully acknowledge the efforts and extend appreciation to Mr Sagar Jaiswal and Ms Ananya Srivastava (PG students, Civil Engineering Department, MMMUT Gorakhpur, Uttar Pradesh) for their assistance in formatting and compilation of data in this book chapter. The authors also extend their sincere thanks to the Ventech Engineers, Sharda Nagar, Kanpur (Uttar Pradesh) for providing the requested borehole data of the state, which had enhanced the quality of the chapter.

REFERENCES

BIS 6:1983 (R2016). Indian Standard Specifications for Moderate Heat Duty Fireclay Refractories. Bureau of Indian Standards, New Delhi.

Bowles, J.E. (1997). *Foundation Analysis and Design*, 5th edition, McGraw-Hill Companies, Inc, New York.

CEA (2018). Report in fly ash generation at coal/lignite based thermal power stations and its utilization in the country for the year 2017-18, December 2018, New Delhi.

CGWB (2012). *Aquifer systems of India*, Central Ground Water Board, Ministry of Water Resources.

DGMUP (2020). Directorate of Geology and Mining. Government of Uttar Pradesh. http://dgmup.gov.in/en/page/major-minerals-list, accessed 28 October 2020.

DSR (2020). District survey report for (Planning & Execution of) minor mineral excavation (In-situ Rock)-Hamirpur. https://hamirpur.nic.in/notice/mining/, accessed 28 November 2020.

FARMECH (2021). Mechanization and Technology Division, Department of Agriculture, Cooperation and Farmers Welfare, Ministry of Agriculture and Farmers Welfare, Government of India. https://farmech.dac.gov.in/FarmerGuide/UP/UI.htm, accessed 9 April 2020.

GSI (2012). *Geology and mineral resources of the states of India*. Part XIII: Uttar Pradesh and Uttarakhand, 2nd edition, Miscellaneous Publication No. 30.

GSI (2018).Briefing Book of Northern Region updated upto March 2018, Northern Region Lucknow.

IS:1498 (1970). Classification and Identification of Soils for Civil Engineering Purposes. Bureau of Indian Standards, New Delhi.

IS:1888 (1981). Method of Load Tests for Soils. Bureau of Indian Standards, New Delhi.

IS:1904 (1986). Structural Safety of Buildings, Shallow Foundations. Bureau of Indian Standards, New Delhi.

IS:2131 (1965). Methods for Standard Penetration Test for Soils. Bureau of Indian Standards, New Delhi.

IS:4968 (1976). Methods of Subsurface Soundings for Soils, Part-I for Dynamic Cone Penetration Tests. Bureau of Indian Standards, New Delhi.

IS:6403 (1981). Code of Practice for Determination of Bearing Capacity of Shallow Foundations. Bureau of Indian Standards, New Delhi.

NNRMS-ISRO (2012). Land use and land cover map of Uttar Pradesh. https://www.isro.gov.in/earth-observation/land-use-cover, accessed 31 July 2020.

NNRMS-ISRO (2019). Land use and land cover map of Uttar Pradesh. https://www.isro.gov.in/earth-observation/land-use-cover, accessed 4 August 2020.

Patel, S., and Pal, B.K. (2015). Current status of an industrial waste: red mud an overview. *International Journal of Latest Technology in Engineering, Management and Applied Science*, Vol. 4, No. 8, pp. 1–16.

Seed, H.B., and Idriss, I.M. (1971). Simplified procedure for evaluating soil liquefaction potential. *Journal of Soil Mechanics and Foundations Division*. ASCE, Vol. 97, No. 9, pp. 1249–1273.

UPSDMA (2020). Uttar Pradesh State Disaster Management Plan for Earthquake, Government of Uttar Pradesh. http://upsdma.up.nic.in/, accessed 17 August 2020.

Yim, C.S., and Chopra, A.K. (1984). Earthquake response of structures with partial uplift on Winkler foundation. *Earthquake Engineering and Structural Dynamics*, Vol. 12, No. 2, pp. 263–281.

Chapter 37

West Bengal

Kaushik Bandyopadhyay, Abhipriya Halder, Saptarshi Nandi,
Bappaditya Koley, and Subhajit Saraswati
Jadavpur University

CONTENTS

37.1   INTRODUCTION

West Bengal, a constituent republic state of India, is situated in the eastern part of the country. It extends from the base of Darjeeling Himalayas in the north to the Bay of Bengal in the south and from the edge of Chotanagpur highlands in the west to the border of Bangladesh and Assam in the east. The map of this state is given in Figure 37.1. There are 23 districts in all.

DOI: 10.1201/9781003177159-37

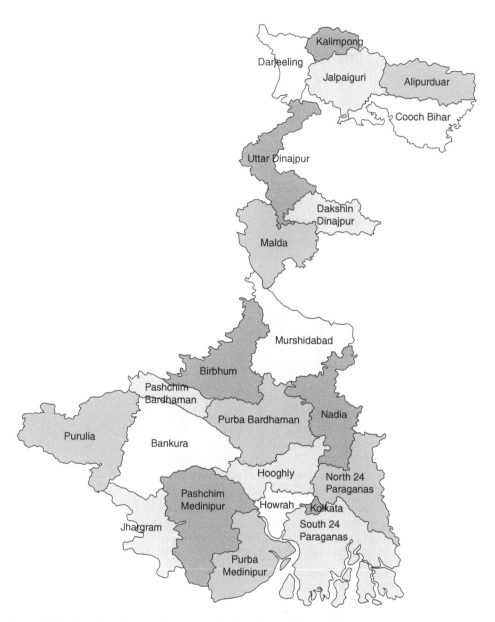

Figure 37.1 Map showing the districts of West Bengal (https://districts.ecourts.gov.in/wb, accessed on 13 January 2021).

Nature has gifted its bounty to the state of West Bengal in a variety of ways, be it natural beauty or geological and geographical divisions. The delta of the river Ganga is part of South Bengal. About 1% of its area is mountainous, lying in the far north. Moreover, 6% of the total land area is within the plateau fringe and Purulia triangle of upland along the western border. Based on these features, West Bengal may be divided into seven physical regions (Mukherjee, 1956):

- Northern mountain region
- Western plateau fringe
- Plains comprising northern and southern regions
- Terai region
- Rarh region
- Sunderbans delta
- Coastal fringe

37.1.1 Northern mountain region

The location of the northern mountain region is on the north-western part of West Bengal and it belongs to the Eastern Himalayan range. The whole of Darjeeling district except the Siliguri division and some parts of Jalpaiguri district is covered within this region. Sedimentary and metamorphic rocks are mainly present in this region. The northern part is confronted with rising mountain ranges in the Himalayas and downslope to hills on the border of the Jalpaiguri district, the hills finally roll to humid plains known as 'Dooars' ('Dooar' means 'the door to Bhutan' in Bengali).

37.1.2 Western plateau fringe

This region is situated in the western part of West Bengal. This plateau fringe is a rolling upland, with small, isolated hills standing here and there. This plateau region connecting the Rajmahal hills of Bihar and Chotanagpur is the part of the Chotanagpur plateau. This region is made of old igneous rocks mainly granite and gneiss of the Archaean era as well as coal-bearing mudstone and quartzite rocks of the Carboniferous period. Purulia district (100 m) has the highest mountain Ayodhya hill (677 m) in this region. Because of long and continuous erosion, the whole region has been transformed into an undulating peneplain interspersed by small monadnocks locally known as tila. These rocky plains descend eastward to merge with the higher slope of the alluvial plain.

37.1.3 Northern plains

Except for the northern hilly mountainous and western plateau areas, the remaining is the plain area. The North Bengal plain starts from the south of the Terai region and continues up to the left bank of the Ganges. Ganga river flows from the west to the east and divides the plain into northern and southern parts. This plain is formed mainly by the alluvium of the Ganga River and its branches. The eastern part known as 'Barind' or 'Barendrabhumi' consists of an undulating plain and is made of old alluvium which is a part of the Ganges delta. The western part on the other hand is made of new alluvium and in this section, the river Kalindi meets the Mahananda river. The part of Malda lying to the north of river Kalindi is known as Tal. This is lowland and covered with swamps and beels (small water bodies) whereas the area in the south of Kalindi is very fertile and is known as Diara. The western plain has been largely built up by the silt brought by the western tributaries of the Bhagirathi River.

37.1.4   Terai region

This region extends to the Siliguri division of the Darjeeling district, north and eastern part of Jalpaiguri and northern part of Dinajpur. The Terai zone is made of alternate layers of clay and sand, with a high-water level that creates many springs and wetlands. The entire region is made of sand, gravel and pebbles laid down by the Himalayan rivers like the Teesta, Torsa, Raidak, Jaldhaka, Sankosh and several other small rivulets.

37.1.5   Rarh region

This region intervenes between the southern Ganges delta and the western plateau region. This region is believed to be created from the soil from the Deccan plateau. This region is dominated by laterite soil.

37.1.6   Sundarbans delta

The Sundarbans delta is the largest mangrove forest in the world. 'Sundari' trees are found in abundance in this region and hence it is known as 'Sundarbans'. It consists of the Hoogly river estuary and newly created Ganga delta characterised by tidal creeks, mudflats and newly formed islands. This area has been created by deposition of silt by its numerous rivers, namely Hoogly, Matla, Jamira, Gosaba, Saptamukhi and Haribhanga river and their tributaries. The formation of the delta is an ongoing process and new bars, islands are being created along the rivers and at the river mouth.

37.1.7   Coastal fringe

The coastal plain region is on the extreme south of the state. Part of the district of Purba (East) Medinipur along the Bay of Bengal constitutes the coastal fringe. The topography in this strip of land is related to the sea. This emergent coastal plain is made up of sand and mud deposited by rivers as well as wind. Parallel to the coast are colonies of sand dunes and marshy areas.

37.2   MAJOR TYPES OF SOILS AND ROCKS

This section describes the major types of soils and rocks found in this region.

37.2.1   Geology of the Darjeeling Himalayas

The entire Himalayas were formed by the ongoing collision of peninsular India with the Asian plate. Consequently, the mountains of the Himalayas rose in a succession of thrusts to the tableland formed by the Tibetan highlands.

Overlooking the piedmont plains of the Darjeeling – Jalpaiguri districts, the thick-bedded sandstone with minor siltstone, shale and conglomerate beds of the Siwalik constitutes the southernmost Himalayan ranges. The Siwaliks are thickest to the West of Mangzing Khola, especially in the Gish, Lish, Teesta and Mahanadi and Hill cart road sections. It is very thin between Mangzing and Murti river sections, east of which the Siwalik belt attains some considerable width in Kuma hills, in Naxal Khola and Jaldhaka sections. This belt is followed to the north by a wide zone of slates, phyllites, carbonaceous beds (coal and coaly slates) and quartzites and conglomerates with some dolomite beds of various descriptions belonging to the Gondwana and the Daling groups. These are generally commonly low-grade rocks mapped previously as belonging to the Damudas of the Dalings.

Recent work has revealed the presence of both the Talchirs and the Damudas within the Gondwanas. Thin belts of the Buxas are also noticeable. The so-called Daling phyllites, slates and quartzites are further sub-divided into two subdivisions, namely the Garubathans and the Reyangs. A belt of highly sheared granite gneiss, locally associated within the Dalings and at other times within the Chunthangs, is met with and continuous with the Lingtse granite of eastern Sikkim.

The gneiss rocks, which were hitherto mapped as Darjeelings, were found to belong to two distinct groups, namely the Darjeelings and the Chunthangs (=Paro of Bhutan), the latter being described from north to eastern Sikkim (Ray and Chakraborty, 1971). Besides, the augen gneisses of the north-eastern part of the district belong to the Kangchenjunga gneissic group (Chubakha of North Sikkim). Stratigraphically and tectonically, Darjeeling and Sikkim Himalayas are mostly covered by Precambrian Metapelites of low to medium grade and Buxa Carbonate quartzite association belonging to the Daling Group of high-grade granite gneisses namely Kangchenjunga Gneiss/Darjeeling Gneiss Formation and deformed granite gneiss namely Lingtse granite Gneiss. The Palaeozoic Mesozoic rocks include Gondwana-equivalent Rishi group (Damuda group, Rangit pebble state groups) and Tethyan groups of rock namely Everest Pelite Formation, Everest Limestone Formation, Lachi Formation and Tso Lhamo Formation.

The Daling Group Formation is classified into three groups of rocks: Buxa, Rejang, and Gorubathan Formation. The Buxa formation essentially comprises an alternating sequence of thin, grey cherty mature quartzite, chert, fine-grained and finely laminated to massive grey dolostone, pyritous sericitic and variegated slates. The Reyang formation comprises thick-bedded, ortho to proto quartzite, variegated phyllite/slate with minor impersistent beds of crystalline carbonates and conformable metabasites from the type section around Reyang in the Teesta River valley in the Darjeeling district being transitional between the Gorubathan and the Buxa Subgroups. The Gorubathan formation sequence comprises inter banded chlorite sericite schist/ phyllite, quartzite, meta greywacke, pyritiferrous black slate/carbon phyllite and basic meta volcanics.

The Central Crystallines Gneissic Complex has been classified into Kangchenjunga Gneiss/Darjeeling Gneiss and Chungthang Formation. The Kangchenjunga Gneiss/Darjeeling Gneiss Formation consists of gneisses, dominantly comprising quartz, feldspar and biotite (with minor amounts of other minerals). It has been classified into three types, i.e., (i) banded/streaky gneisses/migmatites,

(ii) sillimanite granite gneisses and (iii) augen bearing biotite gneiss with/without garnet, kyanite and sillimanite. The Chungthang Formation mainly covers the main rock types of this formation which are quartzites, garnet-kyanite-staurolite bearing biotite schist, calc-silicate rock, graphitic schist and amphibolite. Lingtse granite gneisses are essentially constituted of coarse to medium-grained, foliated to strongly lineated granite mylonite. These are streaky, banded, augen gneisses or porphyroblastic gneisses and are traversed by concordant and discordant pegmatite veins (GSI, 2012).

The Damuda group formation comprises sandstone, calcareous sandstone, shale and carbonaceous shale with thin coal beds. The sandstone and siltstone at places exhibit cross-stratification and channel structures. The Rangit pebble state group is represented by diamictite, dark slate, dark grey claystone, bands and lenses of sandstone. The pebbles essentially comprise quartzite, calcareous quartzite, sandstone, limestone, stromatolitic dolomite, slaty phyllite and granite gneiss (De, 1982).

37.2.2   Bengal basin

The geological study of the Bengal basin could not clearly precisely demarcate the boundaries between the Pleistocene and the recent series of sediments in the vertical sequence. Recent sediments comprise sand, silt and clay with the latter two being dominant. The Bengal basin consists of mainly two environments, viz., flood plain or Backswamp environment and the Meander Belt environment. However, overlapping of deposits of different environmental conditions is also recognised (Ghosh and Gupta, 1968).

Basically, there are two horizons of soil layers encountered, e.g., the first horizon representing soft deposits and the second or lower horizon representing stiffer soil deposits. As for river channel deposits, basically, cohesionless materials have been encountered all throughout the depth of the channels. The existence of peat beds and decomposed timber in the first horizon suggests that the deposition took place in a swampy environment and was not exposed on the surface. The alternate sequence of sand and clay suggests a sequential change in the channels of the river system responsible for their deposition. A wide range of variations in texture and colour have been encountered for alluvial deposits. At some locations, the layers resemble more or less like fluvial deposits. The alternate layers of silt/fine sand and clay are very small in thickness of the order of a few inches. These layers cause significant changes in the engineering properties of soil, which often is ignored. These factors give rise to a problem particularly in the case of deep excavations and underground construction as discussed later.

37.3   PROPERTIES OF SOILS AND ROCKS

Schematic diagrams showing different soil layers obtained from boring log data of some of the soil exploration works are shown in Figures 37.2 to 37.5. Deviations from these layers may also be observed at some locations.

Photographs of some open pits are also presented in Figures 37.6 and 37.7.

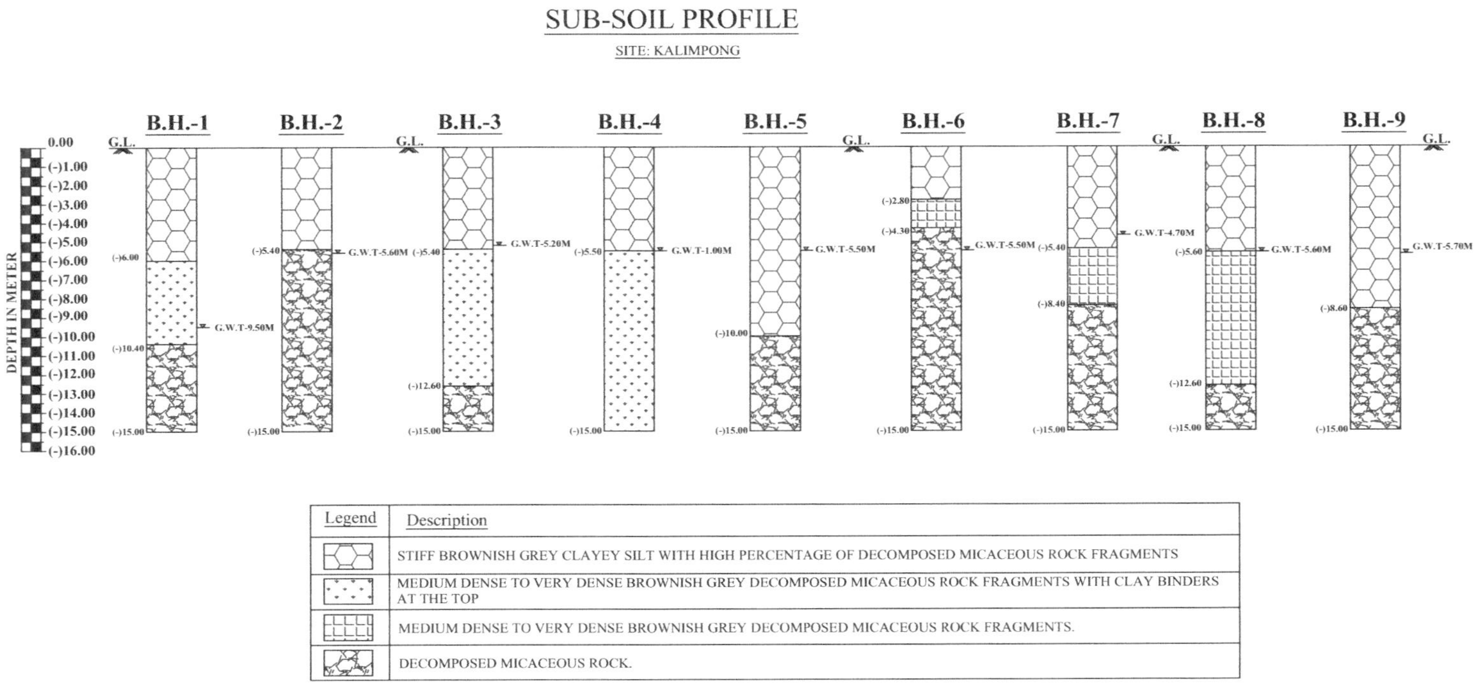

SUB-SOIL PROFILE

SITE: KALIMPONG

Figure 37.2 Typical soil profile for the project on construction of the proposed development of stadium, indoor games building at Kalimpong, West Bengal.

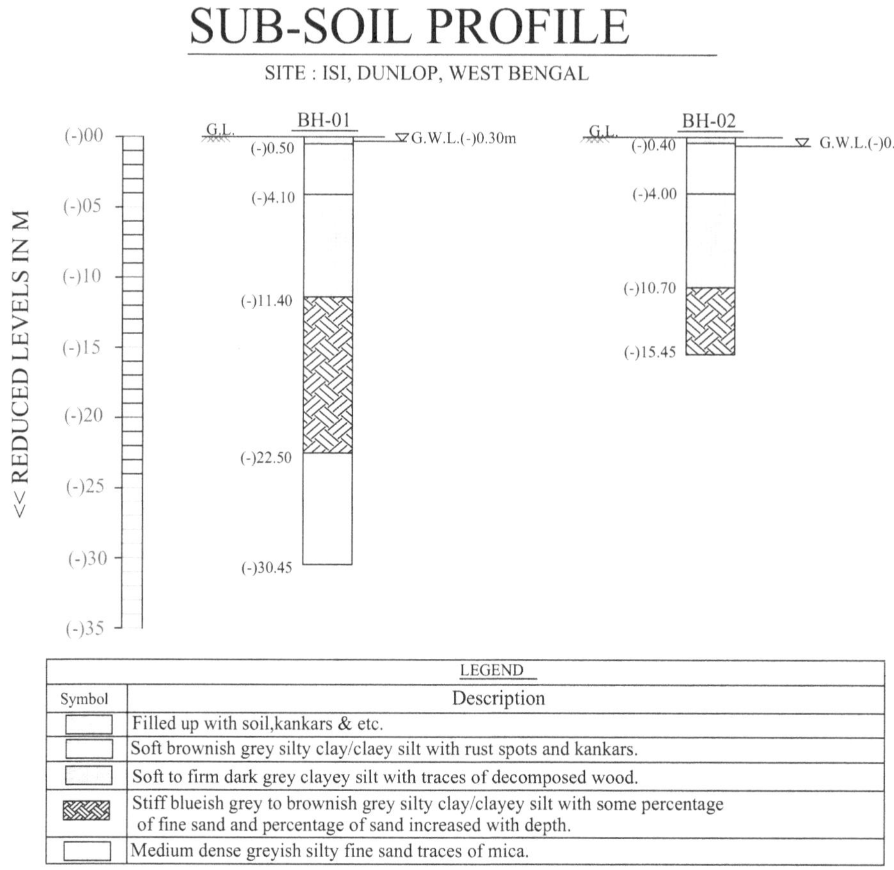
SUB-SOIL PROFILE
SITE : ISI, DUNLOP, WEST BENGAL
<< REDUCED LEVELS IN M
(-)00
(-)05
(-)10
(-)15
(-)20
(-)25
(-)30
(-)35
BH-01
G.L.
G.W.L.(-)0.30m
(-)0.50
(-)4.10
(-)11.40
(-)22.50
(-)30.45
BH-02
G.L.
G.W.L.(-)0.60m
(-)0.40
(-)4.00
(-)10.70
(-)15.45
LEGEND
Symbol
Description
Filled up with soil,kankars & etc.
Soft brownish grey silty clay/claey silt with rust spots and kankars.
Soft to firm dark grey clayey silt with traces of decomposed wood.
Stiff blueish grey to brownish grey silty clay/clayey silt with some percentage of fine sand and percentage of sand increased with depth.
Medium dense greyish silty fine sand traces of mica.

Figure 37.3 Typical soil profile for a project work of bridge at ISI, Dunlop, Kolkata.

Typical engineering geological information and geotechnical properties of soils/rocks as observed in some boreholes are summarised in Figure 37.8. Variations in the soil profile from these are also common and may be found in some locations.

Broadly, it is observed that the soil in the plain land portion can be classified under most of the predominant geological formations as (Som and Das, 2003): (i) alluvial soils, (ii) organic soils, (iii) laterite and expansive soils, (iv) marine deposits and (v) boulder deposits.

37.3.1 Alluvial soils

The large part of the Bengal basin within the Indo-Gangetic plains is covered by the sedimentary deposits, which were formed by the rivers and their tributaries. The thickness of this deposit is often greater than 100 m above the bedrock. These deposits mostly comprise layers of sand, silt and clay depending upon the position of the river

SUB-SOIL PROFILE

SITE: KRISHNAGAR

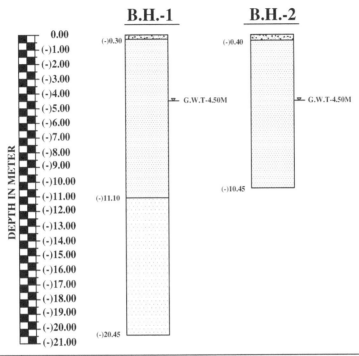

Legend	Description
	Filled up by soil and sand etc.
	Loose to medium dense light grey silty fine sand with mica
	Dense light grey silty fine to medium sand with mica.

Figure 37.4 Typical soil profile for a project work of bridge at Krishnagar, Nadia.

away from the source. Conventional shell and auger boring using the rotary mud circulation technique are found to be suitable for carrying out the exploration work. By and large, parts of South and North 24 Parganas, Hooghly, Kolkata and Howrah exhibit these types of deposits.

37.3.2 Organic soils

The organic soils contain a large quantity of organic/vegetable matter in various stages of decomposition. In organic clay, vegetable matter is sometimes intermixed with decomposed timber pieces and peat. These decomposed timber pieces are mainly buried

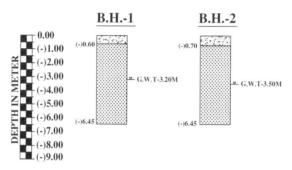

Figure 37.5 Typical soil profile for a project work of bridge at Birbhum.

Figure 37.6 View of an open pit for soil exploration work in Kalimpong.

Sundari trees of the Sunderbans (Gupta, 2009). Peat is characterised by a high liquid limit and spongy structure with low specific gravity. Consequently, they possess low shear strength and high compressibility. The top 10–12 m of the normal Calcutta deposit contains common organic clay of the Bengal basin (Som and Das, 2003). The presence of a thin layer of peat is a very common phenomenon at many locations.

Figure 37.7 View of an open pit for the construction of emergency access road for approaching undershoot/overshoot area, Main Fire Station at Kolkata Airport.

Conducting SPT tests or collecting samples in such deposits becomes very difficult. Often the reading of the SPT is nil or sometimes it shows a very small questionable value. Sometimes the sampler slips into the hole as the shear strength of the soil is very low. Under such situations, conducting a dilatometer test (DMT) or cone penetration test (CPT) can be better suited (Figure 37.9).

In the case of a normal Calcutta deposit, the undrained shear strength (c_u) is determined from either the unconfined compression tests or the unconsolidated undrained triaxial tests. When the partings of silt/fine sand are present as fine layers, these tests will give values that are significantly different from the actual field ones. Especially, it becomes significant when there is rain, or the water table is high. This is because the permeability of the layers is considerably increased due to the presence of the fine sand partings. This simultaneously causes faster consolidation of the clay strata than was expected. As a result, the drained shear strength parameters would prevail. This means that the drained cohesion value which is generally much less than the undrained one coupled with an appreciable increase in the friction angle of the silt/fine sand fraction becomes the guiding criteria for design (Bandyopadhyay, 2014a). During the design stage, if these factors are not considered, actual field earth pressures would be significantly different from that of the calculated ones.

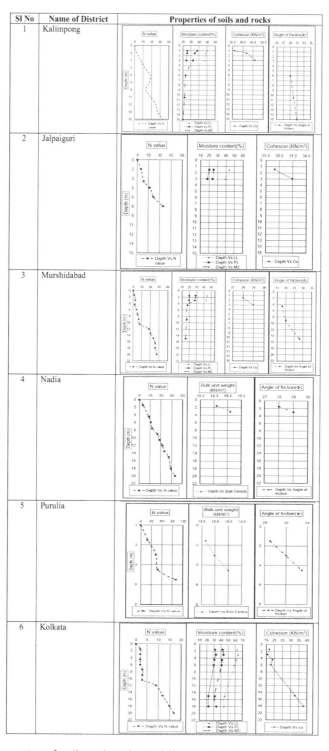

Figure 37.8 Properties of soils and rocks in different districts.

Figure 37.9 Photograph showing dilatometer test (DMT) being conducted in Kolkata.

37.3.3 Laterite and expansive soils

This soil is found mostly in the western arid part of West Bengal, viz., the eastern part of Bankura, the west-central part of Birbhum, middle Bardhaman, part of Murshidabad, West Medinipur and eastern parts of Purulia districts. Laterite soils are residual soils formed by the decomposition of rock-forming oxides of iron and aluminium. These are locally known as murram. Some of the laterite soils exhibit properties like expansive soils, which contain expansive clay mineral-like kaolinite with a high swelling potential.

37.3.4 Marine deposits

India has a long coastline along with the eastern and western parts. The east coast merges with the Bay of Bengal. A part of West Bengal that falls within this zone (e.g., parts of districts like Purba Medinipur and South 24 Parganas) comprises soils that are generally soft and often contain organic matter. Like in organic soil, conducting SPT tests or collecting samples in such deposits become very difficult and it is preferable to conduct a DMT or CPT.

In deltaic areas, this soil has typical characteristics of soft soil resembling London clay, in particular Cambridge soil. Consequently, constitutive modelling by the cam-clay model is sometimes well suited for such soils (Bandyopadhyay, 2009). Literature also reveals that normal Calcutta soil has similar values of liquid limits and plastic limits with those of Boston blue clay and Chicago clay. Values of Atterberg limits of soils of Haldia (Purba Medinipur) have close proximity to those of London clay, Norwegian quick clay and Shellhaven clay (Som and Das, 2003).

37.3.5 Boulder deposits

Generally, the boulder deposits are found in hilly terrains where the high velocity of the flowing river carries along with it large pieces of boulders. As a result, the land area lying mostly at the foot of the Darjeeling Himalayas, e.g., parts of Siliguri and

Jalpaiguri, comprises such deposits. In many instances, it has been experienced that installing driven piles or well foundations becomes risky. This is because when the pile hits any boulder, alignment gets distorted and eventually the pile has to be sacrificed. In the soil exploration report, the presence of such an obstacle sometimes gets over-looked or missed. For well foundations, the most common phenomenon is tilting of the well.

37.4   USE OF SOILS AND ROCKS AS CONSTRUCTION MATERIALS

Murram or laterite soils are mainly used for road construction. The distribution of laterites and lateritic soils is limited to parts of the western plateau fringe of West Bengal, com-prising the eastern part of Bankura, west-central part of Birbhum, middle Bardhaman, part of Murshidabad, West Medinipur and eastern parts of Purulia districts.

There are many coal-fired thermal power plants in the southern part of West Bengal to cater to the electricity needs of the state for various purposes right from industrial to domestic uses. These plants produce huge quantities of fly ash. The use of fly ash as a waste resource is encouraged by the Government as well as by private organisations. The fly ash is used in road construction, embankment construction, construction of fly ash bricks, filling material and so on.

37.5   FOUNDATIONS AND OTHER GEOTECHNICAL STRUCTURES

Foundations and other geotechnical structures are decided based on soil strength, that is, the bearing capacity and settlement of foundation soil. Here, these are explained by considering the physiographic divisions of West Bengal.

In the northern mountain region and western plateau fringe, normally shallow/ open foundations are used for different constructions as the bearing capacity of soil is normally high.

In the northern plains, normally shallow/open foundation is used for up to three-storeyed buildings but when the load from the structure (e.g., bridges) is high, normally pile or well foundation is the suitable option for the foundation.

In the Terai region, normally shallow/open foundations are used for different con-structions as the bearing capacity of soil is high. In this belt, sometimes boulder is present. In the Teesta river, the scouring effect is high; so the pile and well foundation is used for bridge and tower structures. During the construction of wells, tilting of wells is a common phenomenon in this part of the country owing to the presence of boulder deposits. Obstruction is often met during the construction of pile foundations. The soil exploration report does not always depict all such impediments. Therefore, extra caution should be exercised during the execution of such constructions.

In the Rarh region, normally shallow/open foundation is used for different con-structions as the bearing capacity of soil is normally high due to the presence of the murram type of soil. Black cotton soils or more precisely expansive soils are present in some parts of Birbhum, Bankura and Paschim Medinipur; so raft/pile foundation is also used in this region as a countermeasure.

In the southern plains, Sunderbans delta and coastal fringe, by and large, shal-low/open foundation is used for (G + 2) storeyed buildings but when the load from the

superstructure is high, e.g., tall structures and bridges, normally pile or well foundation respectively are suitable options for the foundation.

While a shallow/open foundation is to be adopted, care should be taken to consider some typical characteristics of the inherent soil stratification in the region. In
many cases, there is a top fill layer that may extend from 1 to even 3 m for filling up water bodies below the existing ground level. In such cases, the foundation depth should
be judiciously chosen. In most cases, either foundation is placed below this fill layer
when the depth of fill is small, or some ground improvement technique is adopted when
the depth is more. For residential buildings, this is accomplished either by replacing
the fill layer with some engineered filling materials like sand or by introducing timber
piles (locally known as Shalballah pile). If large-scale ground improvement is required,
for example for the construction of highways and embankments, preloading with sand
drains and prefabricated vertical drains (PVD) is the most popular method of ground
improvement. In addition to this, filling of the low-lying and swampy areas is accomplished by fly ash from nearby thermal power plants. For oil storage tanks, the most
popular ground improvement method is the use of stone columns. Unlike conventional
pile foundations which transfer the load to a deep bearing stratum, both timber piles
and stone columns work as soil reinforcements over a small depth to increase the shear
strength of the soil. When deep foundations in the form of piles are adopted, calculation of the design capacity should be done keeping in mind the local prevalent practice
for tackling difficult unforeseen problems during actual construction. There are some
varying schools of thoughts. One school opines that for bored piles, while calculating
the vertical capacity, only the skin friction values are to be considered. This is because
chances are there that due to poor construction practices, soil in the borehole may get
mixed up with concrete resulting in a degraded value of the end-bearing. But sometimes this gives rise to highly conservative designs. Another school opines that the
factor of safety for end-bearing should be higher than that of skin friction to account
for this highly conservative design.

Apart from all these, practice is also prevailing in this region, which takes into
consideration the correction recommended by Meyerhoff when a pile tip rests on a
sand layer underlain by a clay layer or in a thick sand layer below a clay layer. In this
method, equal weightage is given for both skin friction and end-bearing with the latter
reduced by a correction factor when Meyerhoff's theory is applicable.

While calculating the lateral capacity of piles in this region, it has often been
found economical, considering the plastic analysis (Bandyopadhyay, 2014b) by applying Brom's theory instead of elastic analysis as recommended by the Bureau of Indian
Standards (IS 2911 (Part 1/Sec 2), 2010).

Another important aspect is the consideration of negative skin friction. This is
particularly critical when construction takes place within the city limits. Because land
is very scarce and costly particularly in and around Kolkata, the developers sometimes
choose recently filled ponds or other water bodies for the construction of high-rise
buildings. In such situations, caution should be exercised to take into consideration
the additional load on the pile due to this negative skin friction or down-drag force.

Ideally, the field load test data should be relied upon and it should be compared
with the theoretical value. Finally, the design capacity should be decided.

Tunnels were laid in the past during the maiden metro construction of the country
through the alluvial deposit of Kolkata. During that time, the tunnel was constructed

by shield tunnelling technique under a canal (Bagjola) near Belgachhia. Later, in recent times, the tunnelling operation has been successfully completed through the alluvial deposit under the Hooghly river for the east-west metro line. In this project, the earth pressure balancing mechanism by use of a tunnel boring machine has been employed.

37.6   OTHER GEOMATERIALS

This section includes different types of wastes (fly ash, mine tailings and so on), which are utilised or can be utilised in geotechnical and other construction applications.

There are a number of thermal power plants, steel plants and cement plants in West Bengal. Huge quantities of fly ash, bottom ash, blast furnace slag and so on are produced from these plants. Fly ash can be effectively used as a valuable waste resource for various useful purposes, e.g., road construction, highway embankment construction, filling of low-lying areas, as a partial replacement of cement in the construction industry, and so on. Slag can be used for the manufacture of cement, replacement of coarse aggregates in the construction industry and so on.

Most of these power plants are coal-fired and the ash collected is broadly classified into the following categories depending upon the place of collection in the plant:

- Bottom ash – pulverised fuel ash collected from the bottom of boilers by any suitable process.
- Fly ash – pulverised fuel ash extracted from flue gases by any suitable process such as by cyclone separator or electro-static precipitator.
- Pond ash – fly ash or bottom ash or both mixed in any proportion and conveyed in the form of water slurry and deposited in ponds or lagoons.
- Mound ash – fly ash or bottom ash or both mixed in any proportion and conveyed or carried in dry form and deposited dry.
- Siliceous pulverised fuel ash – pulverised fuel ash with reactive calcium oxide less than 10%, by mass. This fly ash is normally produced from burning harder and older anthracite or bituminous coal and has pozzolanic properties. This corresponds to Class F fly ash as per ASTM C 618-19 (2019).
- Calcareous pulverised fuel ash – pulverised fuel ash with reactive calcium oxide not less than 10% by mass. This fly ash is normally produced from lignite or sub-bituminous coal and has both pozzolanic and hydraulic properties. This corresponds to Class C Fly ash as per ASTM C 618-19 (2019).

The fly ash, being a pozzolanic material, reacts with lime under normal temperature and pressure to form cementitious compounds. Fly ash containing relatively more amorphous silica is more reactive. Alternatively, the addition of a chemical activator such as sodium silicate (water glass) to a Class F ash can lead to the formation of a geopolymer.

Typical physical and chemical properties of bottom ash obtained from the Bandel thermal power plant are given in Tables 37.1 and 37.2 (Kundu, 2018), respectively.

Mine tailings: In Asansol (Paschim Bardhaman), Paschim Medinipur and Jhargram area, there are several mines. Mine tailings from these may be effectively used.

Table 37.1 Physical properties of bottom ash (Bandel Thermal Power Plant) (after Kundu, 2018)

Sl. No.	Parameters	Test results
1	Bulk unit weight (kN/m^3)	9.14
2	Moisture content (%)	1.0

Table 37.2 Chemical properties of bottom ash (Bandel Thermal Power Plant) (after Kundu, 2018)

Sl. No.	Parameters	Test results
1	Specific gravity	2.02
2	Loss on ignition (%)	4.90
3	SiO_2 (%)	82.4
4	Na_2O (%)	0.06
5	CaO (%)	1.73
6	Al_2O_3 (%)	5.37
7	Fe_2O_3 (%)	0.79
8	K_2O (%)	0.17
9	MgO (%)	0.43

37.7 NATURAL HAZARDS

This section includes details of earthquakes, landslides/erosion, floods and so on focussing on how they affect foundations and other geotechnical structures.

37.7.1 Natural hazards in hilly terrain

Landslides in addition to frequent earthquakes have remained the most common form of natural disaster in the Darjeeling Himalayas. Landslides occur due to both natural and manmade causes. Amongst the natural causes are unstable geology, slope conditions, high rainfall intensity and seismic activity. Human risk factors include massive soil erosion caused by deforestation and human activity like construction on cut slopes. Severe landslides have occurred in the Darjeeling and Kalimpong districts almost 16 times during the present century (Rao, 2009).

The young mountains of the Himalayas have risen in a sequence of E-W trending folds and thrusts over the past 20 million years in a geological process that is still ongoing. This accounts for the very high seismicity across most of the region and the entire Darjeeling hill region straddles the area by the Himalayan thrusts. In recent times, the Himalayan earthquake of the 18th of September 2011, to the north of Dr Graham's Homes (Figure 37.10), damaged it to an extent that required its withdrawal from public use. Apart from the earthquake, the chapel has also suffered a series of other ailments – mostly weather and maintenance related. These got accumulated over the years and made things worse requiring urgent efforts for their remedy. While several buildings on the campus have subsequently been repaired and put back to use, the Katherine Graham Memorial Chapel is yet to be restored (Bandyopadhyay, 2020).

Figure 37.10 Crack in the façade of Dr Graham's Chapel, Kalimpong, West Bengal (Bandyopadhyay, 2020).

37.7.2 Natural hazards within the Bengal basin

Flood is a very common phenomenon in the South Bengal region and part of regions in the bordering areas with North Bengal, viz., Malda, Murshidabad and Nadia. Every year during the monsoon season, the land areas adjoining the river Damodar and Teesta get flooded destroying many houses and cultivation.

Within the Bengal basin, it is the alluvial deposit that is of concern owing to its susceptibility to liquefaction caused by earthquakes. The recent geological activities in the region depict various small to medium range earthquakes. The broad shelf zone of the basin is demarcated in the west by the Precambrian outcrops and in the east by the Eocene Hinge Zone. The major divisions of the tectonic units of the Bengal basin are (i) shelf zone, (ii) hinge, (iii) deep basin and (iv) western shear zone from the west to the east. It is this Eocene Hinge Zone that is the most prominent tectonic feature in the Bengal basin and is called Calcutta-Mymensingh Hinge Zone. Its alignment is 'S' shaped and swings at Jagli and Contai (East Midnapore) areas (Chakraborty et al., 2004; Bandyopadhyay et al., 2015).

Literature also reveals that if there is a moderate to high-intensity earthquake, there may be a significant post-earthquake settlement of soil and consequent damage to structures in this region where an alluvial deposit is encountered (Bandyopadhyay and Bhattacharjee, 2016).

Care should be taken when designing foundations against seismicity/liquefaction. In the case of a normal Calcutta deposit when it consists primarily of clayey soil, normally these remain nonsusceptible to liquefaction, though caution should be exercised when dealing with sensitive clays, which can exhibit strain-softening behaviour similar to that of liquefied soil (Kramer, 2009). Accordingly, proper checks are to be made against any significant strength loss. Also while determining the lateral capacity of piles, shear strength of soft soil lying in the upper layers of the soil profile should be ignored following the recommendations of the BIS code on the earthquake resistant design of structures (IS 1893, 2016).

In the case of alluvial deposits, a detailed liquefaction analysis of all the layers lying within the significant depth should be carried out following the latest version of the aforesaid BIS code of practice. In case the factor of safety against liquefaction falls below the permissible value, appropriate measures should be taken either through implementing suitable ground improvement techniques for installing shallow foundations or going for deep foundations avoiding the vulnerable layers depending upon the importance of the structure and cost involvement.

Due to the occurrence of some frequent small intensity earthquakes in this region and subsequent modifications in the BIS code of practice (IS 1893, 2016), this region has been raised to higher seismic zones from the earlier lower ones. As a consequence, it is advisable that while dealing with soils of this region, due importance should be given for the proper evaluation of dynamic soil properties in addition to all other index and shear strength properties. For better performance and increased confidence, both field and laboratory tests are recommended to determine stiffness, damping, Poisson's ratio and density. Field tests should include both low-strain tests (e.g., seismic Reflection, seismic Refraction and seismic Cross-Hole) as well as high-strain tests (e.g., standard penetration tests, CPT, DMT, PMT, etc.). Laboratory tests in the low-strain category may include the Resonant Column test, Ultrasonic Pulse test and so on and in the high-strain category, it may include Cyclic Triaxial test, Cyclic Direct Simple shear and Torsional Shear test.

37.8   CASE STUDIES AND FIELD TESTS

In this section, a case study on the failure of deep excavation for a multistorey building in Kolkata with reference to the soil in this area is discussed.

A deep excavation scheme of about 9 m was planned in the southern fringes of the city for a proposed 2B + G + 8 storeyed building. About 100 numbers structural piles of diameter 600 mm and length about 20 to 22 m were constructed at the site. All along the periphery of the excavation, contiguous bored piles of 450 mm diameter and 12 to 13 m length were proposed. Inside the bored piles, a skin wall of 250 mm thickness all around the excavation was proposed. Subsequently, from the design data, it was proposed that a two-level strut (at 2.00 and 5.20 m below the existing ground level) would be installed. Accordingly, excavation up to 4.0 m was made at some portions and the skin wall was cast up to this level. At this stage, cracks and distresses appeared in some neighbouring buildings and the construction was stopped. Horizontal struts were provided at four corners of the area as shown in Figure 37.11 to arrest further distress.

Subsequently, the whole design was revised, and a new scheme was proposed for the construction. The bottom level of the basement raised from −8.80 to −7.20 m below the existing ground level. The foundation design was changed from a pile raft to a basement raft. Initially, a portion of the raft was constructed at the central portion

Figure 37.11 A view of horizontal struts placed at corners to prevent cave-in.

Figure 37.12 Inclined struts resting on centrally cast raft to support the skin wall.

of the area. Stays were supported from this portion of the raft up to the skin wall to withstand the lateral earth pressure as shown in Figure 37.12.

Subsoil profile revealed that basically soft clay extended up to the depth of 15 m from the ground level, i.e. beyond the depth of excavation with N values ranging between 3 and 4. At some locations, the excavation had been done up to 8 m without any lateral support. As Terzaghi's stability factor ($N = \gamma H / C$) for some locations became more than 4, a plastic zone began to form in the clay near the lower corners of the excavation. Under this circumstance, the simple assumption that the surface of sliding extends as the arc of a circle becomes almost invalid. In contrast, the wedge behind the cut merges with the plastic zone bounded by a surface sliding much further from the edge of the cut and much deeper into the subsoil. Consequently, the lateral earth pressure is no longer hydrostatic, rather it assumes a parabolic distribution yielding a higher value of the lateral earth pressure. As a result, the coefficient of active earth pressure is no longer determined by Coulomb's or Rankine's theory. The value as proposed by Terzaghi should have been considered. This would give the depth of the circumferential bored piles much longer than what was provided (the actual depth provided was 10–12 m whereas at least 15–16 m was needed) (Bandyopadhyay, 2014a). Since the cut was in soft clay, the possibility of bottom heave should have been checked. The factor of safety against bottom heave was calculated and was found to be less than unity. This shows that the chance of bottom heave was imminent. In fact, there had been instances of soil collapse at the bottom of excavation at some pockets. The following lessons are worth to be learnt:

- Designers should be more careful about choosing the various soil parameters.
- Only exploratory boring results are not sufficient.
- Actual site-specific data should be checked.
- Quality control to ensure proper sequence of excavation and placement of struts and constant monitoring of the work along with the alignment of the piles are of utmost importance.
- Earth pressure calculations should take into effect the choice of earth pressure coefficients regarding site-specific soil.
- During the design stage, all aspects of foundation design and its niceties should be carefully considered.
- During construction, design guidelines should be strictly adhered to.
- Any unforeseen happening should immediately be reported for corrective measures by the site engineer.
- Proper study in the light of geomorphological and sedimentological character of the area should be made. Without this knowledge, no scheme of hypothesis would stand the test of time.

37.9   GEOENVIRONMENTAL IMPACT ON SOILS AND ROCKS

In the hills, one of the major problems faced by environmental issues is the disposal of solid wastes. An effort is being made at the Government level as well as by some non-Government organisations to resolve this issue; however, the implementation of the benefits to a large extent is still a distant dream.

The population of Kolkata city has reached about 46 million. The MSW generation in the city estimates to be about 3,000 tonnes per day. Dhapa landfill (Figure 37.13) on the eastern fringes of the city is a dumpsite without any bottom lining system (Halder et al., 2019).

It is spread over an area of 25 ha and its waste height is about 30 m above the ground. It has been observed that the waste in the city comprises 51% biodegradables and 12% recyclables. The C&D waste amounts to 17% of the total waste. The annual rainfall of the city is 1,650 mm. There is a quaternary aquifer that is sandwiched between a 40 m thick clay layer on the top and the tertiary clay existing at an average depth of about 296 m.

Figure 37.13 View of Dhapa landfill (Halder et al., 2019).

The top clay layer is underlain by a sequence of fine to coarse sand horizons mixed occasionally with gravel. The aquifer often is found to be silty and micaceous. The hydraulic gradient in the area has been estimated at about 1 m/km. The landfill is being closed for the waste with the construction of a composite cover system. People living in the vicinity complain of various foul gases coming out of the dumpsite due to the burning of methane coming in contact with air and causing environmental pollution.

37.10   CONCLUDING REMARKS

Broadly speaking, the following key points are significant regarding the aspects of soils, rocks and related problems within the state of West Bengal:

*   Sedimentary and metamorphic rocks are found in the hilly regions of Darjeeling and Kalimpong with thick-bedded sandstone, minor siltstone and shale in the plains of Darjeeling–Jalpaiguri districts followed to the north by a wide zone of slates, phyllites, carbonaceous beds (coal and coaly slates) and quartzites.
*   Old igneous rock mainly granite and gneiss of the Archaean era and coal-bearing mudstone and quartzite rocks of the Carboniferous period are found in the western part of this state.
*   The Terai zone is made of alternate layers of clay, sand, gravel and pebbles.
*   The western plateau region is dominated by laterite soil.
*   Bengal basin consists of mainly two environments, viz., flood plain or Backswamp environment and the Meander Belt environment with overlapping of deposits at some places. Basically, there are two horizons of soil layers e.g., the first horizon representing soft deposits and the second or lower horizon representing stiffer soil deposits. As for river channel deposits, mainly cohesionless materials have been encountered. The existence of peat beds and decomposed timber is also noticed in the first horizon.

- The Sundarbans delta has been created by the deposition of silt by its numerous rivers.
- The coastal fringe in the extreme south of the state is made up of sand and mud deposited by rivers and wind.
- Some parts of the Bengal basin are susceptible to liquefaction. As a result, while designing foundations, due importance should be given to the proper evaluation of dynamic soil properties in addition to all other index and shear strength properties.
- By and large, pile foundations are suitable for the construction of high-rise buildings and well foundations for bridges in this region.
- For other parts, generally shallow foundations in the form of isolated/strip footings may be provided for the construction of small to medium-rise buildings.
- Choice of other types of foundations may also be made based on the judgement of the engineer-in-charge and results of field tests.

ACKNOWLEDGEMENT

The kind permission of M/S B.S. Geotech Pvt. Ltd., Hooghly to publish the different district-wise soil data and contributions from Mr Sumit Mukherjee and Mr Santosh Kr. Dey (Directors of M/S B.S. Geotech Pvt. Ltd.) are also highly acknowledged.

Valuable suggestions for improvement of this chapter from Dr A. Ghosh, Professor & Head, Department of Civil Engineering, Indian Institute of Engineering Science and Technology, Shibpur, West Bengal, India are thankfully acknowledged.

Last but not the least, the DMT test was performed with the equipment received from Studio Prof. Marchetti s.r.l., Rome, Italy and the probe was pushed by means of a TG 63 penetrometer from PAGANI, Calendasco, Italy. The kind patronages of these two companies are gratefully acknowledged.

REFERENCES

ASTM C 618-19 (2019). Standard Specification for Coal Fly Ash and Raw or Calcined Natural Pozzolan for Use in Concrete, ASTM International, West Conshohocken, PA, 2019.

Bandyopadhyay, K. (2009). *Proceedings of the National Conference on Geotechnics of Infrastructure*, Kolkata, India.

Bandyopadhyay, K. (2014a). Features of Calcutta subsoil and a failure case study. *Keynote lecture at the Workshop on Forensic Geotechnical Engg.*, Guru Nanak Dev Engg. College, Ludhiana, India.

Bandyopadhyay, K. (2014b). Deep foundations – load carrying capacities and some field experiences. *Proceedings of Indian Geotechnical Conference IGC-2014*, December 18–20, Kakinada, India.

Bandyopadhyay, K. (2020). Report on Guidance, Material Tests in Laboratory and Other Procedures for Restoration of Heritage Structures of Graham's Chapel, Kalimpong, West Bengal, West Bengal Heritage Commission, Kolkata, India.

Bandyopadhyay, K. and Bhattacharjee, S. (2016). Comparative study of subsoil profiles obtained by SDMT and SPT tests and subsequent determination of settlement of post-earthquake condition. *Journal of Japanese Geotechnical Society, Special Publication, International Workshop on Geotechnics for Resilient Infrastructure at Second Japan India Workshop in Geotechnical Engineering*, Vol. 3, No. 2, pp. 90–96.

Bandyopadhyay, S., Das, S., Kar, N.S. (2015). Discussion paper: 'Changing river courses in the western part of the Ganga-Brahmaputra Delta'. *Geomorphology*, Vol. 250, pp. 442–453.

Chakraborty, P., Pandey, A.D., Mukherjee, S. and Bhargava, A. (2004). Liquefaction assessment for micro-zonation of Kolkata city. 13th WCEE Vancouver, B.C., Canada.

De, A.K. (1982). A note on the detailed examination by traverse survey along selected sections of the Rangit valley Gondwana Belt in Sikkim. Unpublished Report, GSI (F.S. 1978–79).

Ghosh, P.K. and Gupta, S. (1968). Subsoil character of Calcutta region. *Convention in Institution of Engineers (India)*, 1968.

Gupta, S. (2009). Consulting Geotechnical Engineer. Personal communication.

Halder, A., Nandi, S., Bandyopadhyay, K. and Reddy, K.R. (2019). Final cover construction and slope stability assessment of waste dump – a case study. *Proceedings of the 1st International Conference on Sustainable Waste Management through Design*, Springer Nature Switzerland AG 2019 H. ICSWMD 2018, LNCE 21, pp. 1–9, 2019. https://doi.org/10.1007/978-3-030-02707-0_11.

GSI (2012). Geology and Mineral Resources of Sikkim. Geological Survey of India, Miscellaneous Publication, No. 30, Part XIX – Sikkim, 1–65.

https://districts.ecourts.gov.in/wb, accessed on 13 January 2021.

IS 1893 (Part 1) (2016). Indian Standard on Criteria for Earthquake Resistant Design of Structures, General Provisions and Buildings, Bureau of Indian Standards, New Delhi, India.

IS 2911 (2010). Indian Standard on Design and Construction of Pile Foundations – Code of Practice for Bored Cast In-situ Concrete Piles, Part 1/Sec 2, Bureau of Indian Standards, New Delhi, India.

Kramer, S. L. (2009). *Geotechnical Earthquake Engineering*. Third Impression, 2009, Pearson Education in South Asia.

Kundu, B. (2018). Study on Effect of Replacement of Fine Aggregate by Bottom Ash in Concrete. Master's Thesis, Department of Construction Engineering, Jadavpur University, Kolkata.

Mukherjee, S.N. (1956). A Brief Agriculture Geography of West Bengal. Directorate of Agriculture, West Bengal, Calcutta, August 19–20.

Rao, P. (2009). A Presentation to the Save the Hills Seminar on Landslides & Earthquakes in the Darjeeling Himalaya, 5 November 2009, Tindharia.

Ray, K.K. and Chakraborty, S. (1971). Studies on the Geology of Darjeeling District, West Bengal and the Daling Chu Base Metaldeposit. G.S.I. Published Report for 1970–71.

Som, N.N. and Das, S.C. (2003). *Theory and Practice of Foundation Design*. PHI Learning Pvt. Ltd., New Delhi, India.

Index

Note: **Bold** page numbers refer to illustration.